Lecture Notes in Networks and Systems

Volume 104

The series "Lecture Notes in Networks and Systems" publishes the latest developments in Networks and Systems - quickly, informally and with high quality. Original research reported in proceedings and post-proceedings represents the core of LNNS.

Volumes published in LNNS embrace all aspects and subfields of, as well as new challenges in, Networks and Systems.

The series contains proceedings and edited volumes in systems and networks, spanning the areas of Cyber-Physical Systems, Autonomous Systems, Sensor Networks, Control Systems, Energy Systems, Automotive Systems, Biological Systems, Vehicular Networking and Connected Vehicles, Aerospace Systems, Automation, Manufacturing, Smart Grids, Nonlinear Systems, Power Systems, Robotics, Social Systems, Economic Systems and other. Of particular value to both the contributors and the readership are the short publication timeframe and the world-wide distribution and exposure which enable both a wide and rapid dissemination of research output.

The series covers the theory, applications, and perspectives on the state of the art and future developments relevant to systems and networks, decision making, control, complex processes and related areas, as embedded in the fields of interdisciplinary and applied sciences, engineering, computer science, physics, economics, social, and life sciences, as well as the paradigms and methodologies behind them.

**** Indexing: The books of this series are submitted to ISI Proceedings, SCOPUS, Google Scholar and Springerlink ****

More information about this series at http://www.springer.com/series/15179

Kai-Uwe Sattler · Duy Cuong Nguyen ·
Ngoc Pi Vu · Banh Tien Long ·
Horst Puta
Editors

Advances in Engineering Research and Application

Proceedings of the International Conference on Engineering Research and Applications, ICERA 2019

Springer

Editors
Kai-Uwe Sattler
Department of Computer Science
and Automation
TU Ilmenau
Ilmenau, Thüringen, Germany

Ngoc Pi Vu
Faculty of Mechanical Engineering
Thai Nguyen University of Technology
Thai Nguyen, Vietnam

Horst Puta
Institute for Automation
and Systems Engineering
TU Ilmenau
Ilmenau, Thüringen, Germany

Duy Cuong Nguyen
Faculty of Electronic Engineering
Thai Nguyen University of Technology
Thai Nguyen, Vietnam

Banh Tien Long
Hanoi University of Science and Technology
Hanoi, Vietnam

ISSN 2367-3370 ISSN 2367-3389 (electronic)
Lecture Notes in Networks and Systems
ISBN 978-3-030-37496-9 ISBN 978-3-030-37497-6 (eBook)
https://doi.org/10.1007/978-3-030-37497-6

This Springer imprint is published by the registered company Springer Nature Switzerland AG
The registered company address is: Gewerbestrasse 11, 6330 Cham, Switzerland

Keynote Speakers

Data-Driven Discovery from a Data Management Perspective: Challenges, Solutions, and Applications

Kai-Uwe Sattler

Technische Universität Ilmenau, Germany
kus@tu-ilmenau.de

Abstract. Nowadays, data represents an important asset in many domains of science, engineering, and economy. Thus, data-driven discovery—i.e., capturing, modeling, analyzing, visualizing, and interpreting huge datasets—opens new ways to approach major challenges in our society. However, handling massive amounts of data and making sense of it require not only appropriate technology but also an interdisciplinary approach combining expertise from data management, data science, and domain experts.

In this talk, we discuss challenges and solutions of data-driven discovery, particularly in the field of engineering from a data management perspective by presenting selected research results. Furthermore, we discuss challenges and approaches related to education of students and prospective experts in this area.

Thin Shells in Science and Engineering: Theory, Finite Element Formulation, Constitutive Models and Applications

Roger A. Sauer

RWTH Aachen University, Germany
sauer@aices.rwth-aachen.de

Abstract. This presentation introduces a computational formulation for nonlinear thin shells that is suitable to model a wide range of surface and interface problems in solid and fluid mechanics. The theory and the finite element equations are fully formulated in curvilinear coordinates, thus allowing for a very general description of deforming surfaces and their use in coupled multi-field problems. The formulation is based on isogeometric finite elements, since they provide global C1 continuity across the surface. At patch boundaries within the isogeometric discretization, a constraint is used in order to maintain the surface continuity. Various constitutive models are discussed in order to describe classical engineering shells, 2D nanomaterials, and fluidic membranes and tissues. The robustness and accuracy of the formulations are demonstrated through several classical engineering benchmark examples. Further applications include carbon nanotubes, lipid bilayer membranes, and arteries.

HiPIMS Technology: Description and Applications

Pierre-Yves Jouan

Jean Rouxel Materials Institute - University of Nantes
Pierre-Yves.Jouan@univ-nantes.fr

Abstract. The improvement in the properties of thin films deposited by sputtering over the past 50 years is closely linked to technological advances. After having worked for a long time on the quality of the vacuum and having freed oneself of the residual pollution, the increase of the ionic densities as well as the control of the energy of these ions appeared as crucial parameters.

During the 1990s, the coupled induced plasma (ICP) technique using a radiofrequency polarized antenna and the use of electron cyclotron resonance (ECR) significantly increased the ionization of the sputtered metal vapor.

In 1999, V. Kouznetzov et al. have published the first work on high-power pulsed magnetron sputtering (HiPIMS) and have, in a relatively short time, attracted the attention of many researchers and industry leaders. This HiPIMS technology consists of providing the target with pulses of 10 to 1000 μs at frequencies of the order of 100 Hz to a few kHz. The instantaneous power dissipated during the pulses can reach several tens of kW, corresponding to currents of some ten amperes for a discharge voltage of 600 to 1000 V with an average power close to that used in the other techniques.

The advantages of HiPIMS over conventional sputtering are the control of a pulsed high voltage that ionizes a very high percentage of the target material without overheating, creating a dense plasma (10^{13} ions cm^{-3}). This produces high-performance dense coatings with good adhesion, extreme smoothness, and good compliance on substrates with complex geometries.

Another useful application of HiPIMS is pretreatment of the substrate surface prior to application of the coating to "clean" a surface of impurities for better adhesion. HiPIMS bombards the surface with high-energy gas ions, eliminating the natural oxide layer that can be found on most substrates to be coated. The high ionization of metallic plasma produces ideal deep etching and ion implantation for high-performance products such as automotive parts, metal cutting tools, and decorative and non-corrosive finishes.

Unfortunately, HiPIMS has a major drawback which is a lower deposition rate compared to other physical deposition techniques.

In this talk, an in-depth description of the HiPIMS technique will be presented followed by examples of applications such as functional deposits (NiO, NiN, CrN, and AlN) and surface preparation (Ni).

Contents

A Combination of Finite Impulse Response Neural Networks, ARIMA and Principal Component Analysis for Forex Market Prediction

Hà Gia Sơn(✉)

Viet-Hung University, Hanoi, Vietnam
hagiason@Gmail.com

Abstract. Real-world time series data are complex and a single forecasting model may not able to capture data patterns well. This paper presents a novel method for time series forecasting and its application in predicting foreign exchange rates. The proposed method uses a hybrid of finite impulse response (FIR) neural networks and autoregressive integrated moving average (ARIMA) model using time-variable parameters to forecast time series variables, and uses principal component analysis (PCA) to combine results of variables to get better final result. Experimental results show that the proposal obtains higher prediction performance than other methods for mid-term and long-term forecasting.

Keywords: FIR neural networks · ARIMA model · PCA · Time series prediction

1 Introduction

Time series forecasting is an important area of machine learning. It has many applications in various fields such as business, engineering, environometrics, economics, medicine, politics, social sciences. In this research, we study time series models for predicting foreign exchange rates, which plays an important role in business and policy making. Two of the most popular time series models are autoregressive integrated moving average (ARIMA) model and artificial neural networks (ANN).

ARIMA models [1, 2] are widely used because they can yield prediction results which are more accurate than other conventional econometrics model, especially for short-term forecasting. It is flexible and can be applied to various types of real-world time series data. ARIMA models are effective when working with linear data. However, their limitation is the pre-assumed linear form of the model and therefore, they can not capture nonlinear time series patterns.

ANN are also popular in time series prediction because of several advantages [3]. They have the capability of modeling nonlinear data and the models are adaptive formed based on training data. This data-driven approach does not require theoretical guidance and expert knowledge. Finite impulse response (FIR) neural networks [4] are a variation of ANN. FIR neural networks inherite advantages of hybrid neural networks

K.-U. Sattler et al. (Eds.): ICERA 2019, LNNS 104, pp. 1–10, 2020.
https://doi.org/10.1007/978-3-030-37497-6_1

and have lightweight architecture. Many methods using FIR neural networks for time series prediction problems have reported promising results [3, 5, 6].

In time series prediction, many researchers proposed hybrid approaches which combine two or more models [7–16]. The motivation of the hybrid models is that they can utilize both model's advantages to get a better model. Real-world time series data often contain both linear and nonlinear patterns. Both ARIMA and ANN models have their unique advantages in linear modeling and nonlinear modeling, respectively. Therefore, there are many hybrid approaches which choose ARIMA and ANN models for the combination and yield more general and more accurate forecasting model than traditional models. Work from Aburto [10] considers the time series as a composition of various series. They use ARIMA to model the original time series and represent the error associated to the forecast as another time series, which shall be modeled by an ANN. The hybrid forecast is an addition of an ARIMA process and an ANN model. Method in [14] applies a similar approach with ARIMA and an ANN(p, q, d) model. Khandelwal et al. [9] use Discrete Wavelet Transform (DWT) to decompose time series data into linear and nonlinear parts, then use ARIMA to model linear part and use an ANN to model nonlinear part. Quin et al. [11] proposed a hybrid model which combines ARIMA and Deep Belief Network for red tide forecasting. In this work, Particle swarm optimization is used to enhance the speed of model training. In [16], ARIMA is combined with adaptive neuro fuzzy inference systems (ANFIS) to forecast energy consumption. In [13], authors calculate the prediction results by combining ARIMA and FIR neural network using the following equation:

$$Y = \beta_0 + \beta_1 FIR + \beta_2 ARIMA \tag{1}$$

where Y is the combined prediction result, FIR is the prediction result by FIR neural networks, ARIMA is the prediction result by an ARIMA model and $\beta_0, \beta_1, \beta_2$ are parameters. Equation (1) is also a multiple linear regression model. The parameters β_1 and β_2 are determined by solving the multiple linear regression. The main contribution of [13] is that the parameters are not fixed but vary as a linear function of time.

In time series analysis, principal component analysis (PCA) has been used to get more accurate results from multivariate time series data. Suppose that we have a multivariate time series dataset and want to forecast a key variable among multiple variables of that data. Authors of [17] use PCA to identify variables related to the key variable, then FIR neural networks were applied to each variable, and finally calculate the final results by multiple linear regression with fixed parameters. PCA is a statistical procedure that does not require any statistical hypotheses. It utilizes observations of variables in some planes to analyse correlation of variables. In [17], after applying PCA to observations of variables, we can obtain vectors of variables and use these vectors to analyse the correlation between variables. Variable Xk has an effect on variable Xi if and only if the angle between two vectors is not equal to 90° and the length of vector Xi is greater than the length of vector Xk.

In this paper, we improve the hybrid method in [13] by using basic economic functions instead of liner function for time-variable parameters. Furthermore, our method uses (PCA) to get higher accuracy for foreign exchange rates prediction.

2 Our Proposal

In this section, we introduce our proposal for multivariate time series prediction. In some applications, e.g., Forex market, we have multivariate time series data and want to forecast only one key variable. For example, exchange rate data have 4 variables: Price, Open, High, Low and the key variable is Price. Our method uses ARIMA, FIR neural networks and PCA to solve this problem. Our method consists of 3 steps.

- Step 1: Apply PCA to identify variables related to the key variable, based on the method presented in [17]. We called these identified variables and the key variable as selected variables. For example, in exchange rate analysis, if the key variable is Price and related variables are Open and High then the selected variables are Price, Open and High.
- Step 2: Use a hybrid of ARIMA model and FIR neural network to predict each selected variable. This model is an extension of the model proposed in [13] and will be described in the following part.
- Step 3: After having all prediction of selected variables, we recalculate the prediction of the key variable by a multiple linear regression model as in [17].

In Step 2, for each selected variable, an ARIMA model and an FIR neural network are constructed separately and then they are combined using Eq. (1) with parameters varying as a function of time t. In [13], the functions for time-variable parameters are linear functions, which have the form

$$\beta_i = \alpha_{i0} + \alpha_{i1} t \tag{2}$$

where α_{i0}, α_{i1} (i = 0, 1, 2) are parameters for the linear function. Therefore, we get the equation for the prediction value at time t as:

$$Y_t = \alpha_{00} + \alpha_{01} t \; + \; \alpha_{10} FIR_t + \; \alpha_{11} t FIR_t \; + \; \alpha_{20} ARIMA_t + \alpha_{21} t ARIMA_t \tag{3}$$

where FIR_t and $ARIMA_t$ are value at time t predicted by FIR neural networks and ARIMA model, respectively.

Let $A_0 = \alpha_{00} + \alpha_{01} t$, $A_1 = \alpha_{10}$, $A_2 = \alpha_{20}$, $A_3 = \alpha_{11}$, $A_4 = \alpha_{21}$ and $F_1 = FIR_t$, $F_2 = ARIMA_t$, $F_3 = tFIR_t$, $F_4 = tARIMA_t$. Then, we get the following equation:

$$Y_t = A_0 + A_1 F_1 + A_2 F_2 + A_3 F_3 + A_4 F_4 \tag{4}$$

Therefore, we can calculate A_0, A_1, A_2, A_3, A_4 by solving the multiple linear regression.

In our proposal, instead of using only linear function (2) as in [13], we use linear function and the following basic economic functions:

$$\beta_i = \alpha_{i0} + \alpha_{i1} t + \alpha_{i2} t^2 \tag{5}$$

$$\beta_i = \alpha_{i0} + \alpha_{i1} \frac{1}{t} \tag{6}$$

$$\beta_i = \alpha_{i0} + \alpha_{i1} \ln t \tag{7}$$

We calculate the each parameters β_1 and β_2 in (1) by five methods: four method using these four functions and a method calculate β_i as fixed parameters. Hence, we have $5 \times 5 = 25$ functions and use multiple linear regression to get the parameters' value. Then among 25 functions, we choose one with the smallest mean absolute error (MEA).

For example, if we use function (5) for β_1 and function (7) for β_2 then from (1) we get:

$$Y_t = \beta_0 + (\alpha_0 + \alpha_1 t + \alpha_2 t^2) * FIR + (\alpha_3 + \alpha_4 \ln t) * ARIMA \tag{8}$$

Thus,

$$Y_t = \beta_0 + \alpha_0 * FIR + \alpha_1 t * FIR + \alpha_2 t^2 * FIR + \alpha_3 * ARIMA + \alpha_4 \ln t * ARIMA \tag{9}$$

Then we can use multiple linear regression to calculate and use these parameters to predict future value of time series.

From this idea we have a following combined method:

Input : Data file // ARIMA file data ,FIR, results, time t and columns dong1,dong2,dong3 would contain time function
Output: MAEFIR, MAEARIMA, MAEkethop.

Step1. Put data into columns dong1, dong2, dong3 through using the expressions t*f1, t2*f, $\ln t$ *f, $\frac{1}{t}$ *f

Step2. Combine with the autoregressive algorithm if i >2 and j>2 then do hq2b // hq2 bis the autoregressive algorithm with 2 variables
else do hqboi // hqboi is the multiple autoregressive algorithm

Step 3. Find minimum MAE in the verification domain and calculate the combined values
dulieu.kethop =beta1+beta2*fir + beta3*arima
dulieu.mekethop= |dulieu.ketqua - dulieu.kethop|

Step 4. Calculate prediction MAE and find the minimum prediction MAE

Step 5. Calculate the forecast results :
dulieu.dubao= dulieu.kethop

Algorithm 1: Combine the neural network FIR with model ARIMA through the mobile method

Verify the combined neural network FIR with ARIMA method algorithm through the mobile method applied into Forex market: The randomized taken from this website data, contain 1200 records and 33 variables satisfy the conditions of verification from 9/5/2013 **to** 29/12/2017, with the first 1000 data to estimate model; the 100 next data to verify; the 100 next data to forecast and to compare, see the following table.

Table 1. Data of verification

Items	Website
Bren Oil (last)	https://vn.investing.com/commodities/brent-oil
Oil of London (last)	https://vn.investing.com/commodities/london-gas-oil
Oil (Fuel) (last)	https://vn.investing.com/commodities/heating-oil
Copper (last)	https://vn.investing.com/commodities/copper?cid=959211
Zinc (last)	https://vn.investing.com/commodities/zinc?cid=956470
Lead (last)	https://vn.investing.com/commodities/lead?cid=959207
Calf (last)	https://vn.investing.com/commodities/live-cattle
Lean (last)	https://vn.investing.com/commodities/lean-hogs
US Soybean Oil (last)	https://vn.investing.com/commodities/us-soybean-oil
US Soybean (last)	https://vn.investing.com/commodities/us-soybeans
US Dried Soybean (last)	https://vn.investing.com/commodities/us-soybean-meal
US Wheat (lần cuối)	https://vn.investing.com/commodities/us-wheat
England Wheat (last)	https://vn.investing.com/commodities/london-wheat
Exchange rate (US Dollar/Switzerland franc) (last, oen, high, low)	https://vn.investing.com/currencies/usd-chf
Exchange rate (EU euro/Jamaica dollar) (last, oen, high, low)	https://vn.investing.com/currencies/eur-jmd
Share Value of vinaconex (last, oen, high, low)	https://vn.investing.com/equities/vinaconex.,jsc-historical-data
Share Value of casio-computer (last, oen, high, low)	https://vn.investing.com/equities/casio-computer-co.,-ltd.-historical-data
Share Value of home-depot (last, oen, high, low)	https://vn.investing.com/equities/home-depot-historical-data

Results of verification of algorithm 2-combination of the neural network FIR with the ARIMA model through the mobile method applied to the Forex market: after use of software SPSS to forecast through the level exponent method by the method of leveling, and the software Grelt to orecast through VAR, VECM, ARIMA, neural network FIR methods to forecast for the variables, we have the following table:

Table 2. MAE of methods

No	Proposed method	FIR	ARIMA	Smoothing	VAR	VECM	MIN MAE	Minimum MAE method
1	6.3251	0.7828	6.3603	7.8549	7.9790	9.3435	6.3251	FIR
2	48.1323	7.0916	48.4333	45.6030	58.1477	74.8475	7.0916	FIR
3	29.4197	209.3700	101.2480	3059.0393	1409.6355	280.1619	29.4197	*Proposed method*
4	61.4270	82.7980	71.6221	686.6207	386.2843	277.6406	61.4270	*Proposed method*
5	71.3301	98.8896	71.9886	201.0250	92.8215	125.1397	71.3301	*Proposed method*
6	27.5566	62.5520	40.6979	128.8297	109.8149	185.6742	27.5566	*Proposed method*
7	390.7892	958.5850	377.5557	2121.9788	2812.7402	1567.4402	377.5557	ARIMA
8	9.2073	0.8644	9.3076	17.8794	10.6519	7.9922	0.8644	FIR
9	0.2184	0.2338	0.2492	5.9370	0.5875	0.5802	0.2184	*Proposed method*
10	2.6827	5.9829	2.6599	159.9546	38.6461	58.7291	2.6599	ARIMA
11	2.5932	2.6317	2.7739	85.6184	17.7970	25.5020	2.5932	*Proposed method*
12	3.3716	3.8058	3.4015	95.6514	26.9042	16.8964	3.3716	*Proposed method*
13	0.6475	0.6880	0.4408	16.2273	5.8394	3.0560	0.4408	ARIMA
14	0.0201	0.0815	0.1148	1865.3615	2074.7788	37.5257	0.0201	*Proposed method*
15	0.0152	0.3324	0.0255	15395.6178	2101.7233	640.6381	0.0152	*Proposed method*
16	0.0283	0.7928	0.6351	16766.7503	2293.8222	175.6888	0.0283	*Proposed method*
17	0.0327	0.0978	0.2505	15764.3708	2022.4786	118.0061	0.0327	*Proposed method*
18	198.1083	968.1300	222.3244	9284.7913	4792.7514	2379.9855	198.1083	*Proposed method*
19	315.0106	856.5600	489.4387	7717.2310	5342.3910	2444.4606	315.0106	*Proposed method*
20	562.0226	970.3100	914.1786	8911.3757	4676.8591	2397.3941	562.0226	*Proposed method*
21	334.8794	995.2900	381.7813	8107.7276	5287.9357	2443.1229	334.8794	*Proposed method*
22	1.5062	2.0313	1.5130	3.6896	2.4500	1.5090	1.5062	*Proposed method*
23	1.3897	1.6470	1.4050	3.3818	2.4700	1.6180	1.3897	*Proposed method*
24	0.6872	1.6038	0.7186	3.7071	2.5680	1.6830	0.6872	*Proposed method*
25	1.2936	1.4782	1.3203	3.5057	2.4000	1.4120	1.2936	*Proposed method*
26	3.6111	14.7280	8.0593	436.1484	199.5818	53.3694	3.6111	*Proposed method*
27	6.5650	15.7320	10.4114	436.6267	196.4767	50.5287	6.5650	*Proposed method*
28	18.4452	15.4620	17.6449	393.8926	205.8767	58.4306	15.4620	FIR
29	14.1179	14.8760	14.2079	463.5738	199.0569	49.4306	14.1179	*Proposed method*
30	2.1339	2.7026	2.3285	4.3332	19.3482	10.9273	2.1339	*Proposed method*
31	1.3235	4.0396	1.3344	8.4581	19.0586	7.1560	1.3235	*Proposed method*
32	7.7838	3.4760	7.7376	8.8619	19.2460	7.6854	3.4760	FIR
33	9.0496	3.9374	8.9974	5.3503	18.9678	11.1809	3.9374	FIR

Analysing this table shows that the minimum MAE method is proposed one (estimation of 24/33 = 72,73%), and next method the neural network FIRR (6/33 = 18,18%), and then model ARIMA (3/33 = 9, 09%), and other models are absent. After using the analysing software Keel, applying the verification method Wilcoxon with ($\alpha = 0.05$), and the initial condition H0 is the method in the thesis and the effective methods we get the following table:

Table 3. Comparison of proposed in this paper algorithm with FIR, ARIMA, smoothing, VAR, VECM ones

TT	Comparing the proposed method with	R+	R−	Exact P-value	Confidence-interval	Exact-confidence	Hypothesys
1	FIR	440.0	121.0	0.003542	[−90.1038, −0.31505]	0.95158	Ignoring hypothesys H_0
2	ARIMA	464.0	97.0	6.51E−4	[−6.58825, −0.05995]	0.95158	Ignoring hypothesys H_0
3	SMOOTHING	545.0	16.0	3.934E−8	[−3, 965.06005, −185.87405]	0.95158	Ignoring hypothesys H_0
4	VAR	561.0	0.0	2.328E−10	[−1, 211.80245, −92.65405]	0.95158	Ignoring hypothesys H_0
5	VECM	552.0	9.0	7.684E−9	[−327.07385, −24.9933]	0.95158	Ignoring hypothesys H_0

Because the parameter Exact P-value of all cases are below than the level ($\alpha = 0.05$) we can ignore hypothesis H0, the parameter R+ of all cases is so bigger than R- then we can include "If data are the economic index from Forex market in particular and from economic fields in general, the method proposed in this paper is more effective than 5 above-mentioned methods".

3 Experiment

3.1 Dataset and Evaluation Metrics

In this research, we use the dataset of the exchange rate between British pound and US dollar which contains weekly observations from 1980 to 1993. This dataset can be downloaded at https://www.investing.com/currencies/gbp-usd-historical-data. The data have 731 data points in the time series with 4 variables Price, Open, High, Low. The last 52 data points are used as test data. This dataset has been widely used in time series forecasting researches. Evaluation metrics for this experiment are a mean squared error (MSE), MEA and mean absolute percentage error (MAPE).

The experimental results of the proposed method are compared with results of ARIMA model, ANN model, FIR model, our hybrid model without Step 3, Zhang's hybrid model [15], Khashei's hybrid model [7] and Khandelwal's hybrid model [9].

3.2 Experimental Setup

In experiment, we implemented the proposed method and other methods in [15] and [9].

In Step 1 of the proposed method, to forecast the Price variable, we use PCA and calculate the length of vectors of variables and angle between vector of Price variable and vectors of other variables. The results are presented in Table 1. According to this result, the High vector's length is 0.9994 and angle with Price vector is 60.04° which are significantly smaller than 90°. Thus, this variable has effect on the Price variable. The Open vector has similar length with the Price vector and the angle between the Low vector and the Price vector is 90°. Hence, the Open variable and the Low variable have hardly any effect on the Price variable. Consequently the Price variable and the High variable are selected variables for Step 2.

We used Gretl and Fir software to predict the Price variable and the High variable using the combination of ARIMA models and FIR neural networks with time-variable parameters as presented in Sect. 2. Table 2 shows the results of this step. The combination method has a MAE of 0.005063 for Price variable and 0.005066 for High variable, which is lower than MAE of ARIMA and FIR models indicating better prediction performance (Table 5).

Table 4. Value and angle of variables

Variable	Length	Angle between vectors of variables			
		Price	Open	High	Low
Price	1.0000	–	61.05	60.04	89.03
Open	1.0000	61.05	–	61.03	65.05
High	0.9994	60.04	61.03	–	60.05
Low	0.9993	89.03	65.05	60.05	–

Table 5. Results of step 2

Variable	ARIMA parameters	FIR parameters	MAE of ARIMA	MAE of FIR	MAE of the combination
Price	(0, 1, 0)	1 × 3 × 2 × 1	0.005358	0.005135	0.005063
High	(2, 1, 1)	1 × 2 × 3 × 1	0.005715	0.005217	0.005066

In Step 3, the prediction of the Price variable is recalculated by multiple regression model. We get three regression parameters $\hat{\beta}_0 = -0.00091810$, $\hat{\beta}_1 = 0.02429589$, $\hat{\beta}_2 = 0.97604929$, and the achieve the final MAE of 0.004951 (Table 6).

Table 6. Comparison of the performance of the proposed model with those of other forecasting models

Model	1 month		6 months		12 months	
	MAE	MSE	MAE	MSE	MAE	MSE
ARIMA	0.005016	3.68493	0.0060447	5.65747	0.0053579	4.52977
ANN	0.004218	2.76375	0.0058823	5.65507	0.0052513	4.52657
FIR	0.005011	3.44200	0.0052420	**3.37804**	0.0051350	3.23623
Zhang's hybrid model	0.004146	2.67259	0.0058823	5.65507	0.0051212	4.35907
Khashei's hybrid model	**0.003972**	**2.39915**	0.0053361	4.27822	0.0049691	3.64774
Our proposed model without Step 3	0.004660	4.06420	0.0058270	3.48990	0.0050635	**3.08996**
Our proposed model	0.004564	3.895356	**0.0052284**	**3.25959**	**0.0049509**	**3.04568**

Table 7. Comparison of the performance of the proposed model with those of Khandelwal's hybrid model [9]

Model	MAPE
ARIMA	6.46017
ANN	4.73684
FIR	4.41813
Zhang's hybrid model	4.53310
Khandelwal's hybrid model	4.32839
Our proposed model without Step 3	4.33813
Our proposed model	**4.31998**

4 Results and Analysis

Table 3 gives the forecasting results for the exchange rate dataset with ARIMA model, an ANN model, FIR model, our hybrid model without Step 3, Zhang's hybrid model [15], Khashei's hybrid model [7]. Table 4 shows the comparison of the proposed model and Khandelwal's hybrid model [9] with MAPE metrics. From these tables, we can conclude that the results of ARIMA model are worse than other models. The reason is that ARIMA model can not capture nonlinear patterns in time series data. In general, hybrid models give better results than single models. For short-term forecasting (1 month), Khashei's hybrid model significantly outperforms other models. For mid-term forecasting (6 months), FIR model gives the best performance, however our proposed method obtains comparable results. For long-term forecasting (12 months), our proposed model outperforms other models. Table 4 shows that our proposed model give the best result with MAPE metric. In comparison with our proposed model without Step 3, our proposed model is better with MAE and MAPE metrics (Table 7).

5 Conclusion

In this paper, we have introduced a novel hybrid model for time series forecasting. Our model combines ARIMA model and FIR neural network based on multiple linear regression with parameters varying as a function of time. We also use PCA to obtain more accurate prediction results. Experiments are conducted with an exchange rate dataset. The experimental results show that the proposed method gives the best results for mid-term and long-term forecasting. In future work, we will investigate other ways of combining different models to get more general and accurate hybrid model.

References

1. Damodar, N.: Basic Econometrics. The Mc-Graw Hill, London (2004)
2. Michael, K.: Evans Practical Business Forecasting. Blackwell Publishers Ltd., a Blackwell Publishing company, Bodmin, Cornwall (2002)
3. Zhang, G., Patuwo, B.E., Hu, M.Y.: Forecasting with artificial neural networks: the state of the art. Int. J. Forecast. **14**, 35–62 (1998)
4. Wan, E.A.: Temporal backpropagation for FIR neural networks. In: IJCNN International Joint Conference on Neural Networks, IEEE (1990)
5. Wan, E.A.: Finite impulse response neural networks for autoregressive time series prediction. In: Proceedings of the NATO Advanced Workshop on Time Series Prediction and Analysis, Sante Fe, NM (2003)
6. Kim, H.J.: Time series prediction using an interval arithmetic FIR network. Neural Inf. Process. Lett. Rev. **8**(3), 39–47 (2005)
7. Mehdi, K., Bijari, M.: A novel hybridization of artificial neural networks and ARIMA models for time series forecasting. Appl. Soft Comput. **11**(2), 2664–2675 (2011)
8. Faruk, D.Ö.: A hybrid neural network and ARIMA model for water quality time Series prediction. Eng. Appl. Artif. Intell. **23**(4), 586–594 (2010)
9. Khandelwal, I., Adhikari, R., Verma, G.: Time series forecasting using hybrid ARIMA and ANN models based on DWT decomposition. Procedia Comput. Sci. **48**, 173–179 (2015)
10. Aburto, L., Weber, R.: Demand Forecast in a Supermarket using a Hybrid Intelligent System. Department of Industrial Engineering, University of Chile, pp 143–151 (2012)
11. Qin, M., Li, Z., Du, Z.: Red tide time series forecasting by combining ARIMA and deep belief network. Knowl.-Based Syst. **125**, 39–52 (2017)
12. Barak, S., Sadegh, S.S.: Forecasting energy consumption using ensemble ARIMA–ANFIS hybrid algorithm. Int. J. Electr. Power Energy Syst. **82**, 92–104 (2016)
13. Thành, N.C., Sơn, H.G.: A combination of FIR neural networks and ARIMA model with time-variable parameters to improve time series forecasting performance. J. Mil. Sci. Technol. 170–179 (2017)
14. Khashei, M., Bijari, M.: Which methodology is better for combining linear and nonlinear models for time series forecasting? J. Ind. Syst. Eng. **4**(4), 265–285 (2011)
15. Zhang, G.P.: Time series forecasting using a hybrid ARIMA and neural network model. Neurocomputing **50**, 159–175 (2003)
16. Barak, S., Sadegh, S.S.: Forecasting energy consumption using ensemble ARIMA–ANFIS hybrid algorithm. Int. J. Electr. Power Energy Syst. **82**, 92–104 (2016)
17. Sơn, H.G.: A forecasting method for mutual dependant multivariate time series using FIR neural networks. J. Mil. Sci. Technol. **49** (2017)

A Magnetic Wheeled Robot for Steel Bridge Inspection

Anh Q. Pham[1,2], Hung M. La[1], Kien T. La[3], and Minh T. Nguyen[3(✉)]

[1] Duy Tan University, Da Nang, Vietnam
phamquyenanh.qb@gmail.com
[2] University of Nevada, Reno, NV 89557, USA
hla@unr.edu
[3] Thai Nguyen University of Technology, Thai Nguyen, Vietnam
lakien524@gmail.com, nguyentuanminh@tnut.edu.vn

Abstract. There are thousands of steel bridges with different structures and designs that have been built around the world. Although they have different structures and designs, they have in common that they need regular quality checks to avoid possible unfortunate accidents. Often, the inspection is being carried out manually, and the inspectors need to bring the testing equipment to climb the tall bridges to perform inspection. This job is dangerous and difficult. To support the inspectors, we present the design and construction of a robotic system that can assist them to perform steel bridge inspections to help minimize difficulties and dangers as well as increase productivity in the quality inspection. The robot can move on bridges with a square or circular steel structure. It is capable of carrying several types of sensors for navigation and mapping. The collected data is stored in an onboard computer and simultaneously sent to the ground station for processing in time. The robot can also mark suspicious locations to facilitate locating repairs. The results of laboratory tests in fact show the feasibility of robot design.

Keywords: Climbing robot · Bridge inspection · Magnetic wheeled robot

1 Introduction

A bridge is built to connect two areas of land that can be divided by ditches, rivers, streams, or other foreign objects such as roads, railways, etc. The construction of bridges has made the traffic continually contributing to promoting economic and social development. However, bridges are aging, overloaded in use, but not regularly checked and maintained, which have caused extremely tragic bridge collapses. There are hundreds of bridges in the condition, which should be checked for maintenance to ensure safe use. The need for regular steel bridge inspection and maintenance is extremely necessary. However, this work is currently done manually by inspectors through visualization or using expensive and specialized equipment for analysis and evaluation. It is dangerous and difficult to bring bulky equipment to climb high to perform the inspection. Not to mention climbing for a long time, fatigue can make the

K.-U. Sattler et al. (Eds.): ICERA 2019, LNNS 104, pp. 11–17, 2020.
https://doi.org/10.1007/978-3-030-37497-6_2

inspectors' assessment no longer accurate and stable. Therefore, the use of technology including robots to assist inspectors to perform testing is perfectly suitable to reduce the dangers and difficulties as well as increase productivity, accuracy and stability for the inspection process and look up. More and more research has applied advanced robot technology to automate the bridge structure inspecting process over the past two decades. La et al. [1–3] has developed an automated robot system that is integrated with advanced non-destructive evaluation (NDE) sensors to inspect and evaluate bridge decks. Lim et al. [4, 5] has developed a robotic system to inspect the bridge deck cracks. Besides, there have been studies including checking other bridge structures. For example, Mazumdar et al. [6] proposed a robot that uses strong permanent magnet pins to move through a steel structure to inspect the steel bridge. Different types of magnetic wheeled robots have been also reported in recent developments [7–9].

The objective of this paper is to present the construction and demonstration of a mechatronic system, which is designed to support the structural steel bridge inspection. A robot designed with four permanent magnet wheels, which can adhere to steel surfaces, controlled via the wireless connection and can move freely on flat and curved steel surfaces. At the same time, the robot can bring a few sensors to detect image cracks, structural mapping and mark fault location structures. With advanced mechanical design, the robot can carry heavy loads (about 10 lb or 4.5 kg) while climbing on both inclined and reverse surfaces. The collected data is stored in the minicomputer equipped on the robot and sent to the ground station for monitoring and processing in real time. In addition, magnetic field sensors, inertial measurement unit (IMU), encoder and range sensors are also integrated to help the robot map and move safely on steel surfaces. Magnetic force analysis has been conducted to determine the possible structures that robots can work with.

2 Overall Design

A robot design with four powerful permanent magnet wheels is proposed to be able to move flexibly on steel surfaces without consuming any energy for adhesion. The robot control system is divided into two classes: low-level and high-level controllers. The low-level controller receives commands from the high-level controller and then performs the appropriate operations such as wheel control, camera control, reading of the sensor value and transmits to the high-level processor as required. Robot system with integrated sensors is shown in Fig. 1. The high-level controller resides in an onboard computer equipped on the robot to handle complex problems like image processing, mapping and communication with ground stations. The robot is equipped with a camera to capture images of suspected locations, while helping inspectors at the ground station to observe the robot working easily. The use of 2-axis Pan-Tilt set will make the camera's vision more flexible. To orient in 3D space we use a smart 9-DOF IMU sensor and encoders to extract robot's velocity and position. Besides, a safe moving robot is an important criterion, so we use 4 IR sensors at the four corners of the robot along with intelligent algorithms to help the robot recognize it's position in advance to move safely. An 8 bit-AVR microcontroller is equipped as the control center of the low-end controller, and Intel NUC industrial computer is used as a high-level controller. High-end controller using a robot operating system (ROS) to localize, navigate and collect data. To transmit data between the robot and the ground station we use a wireless LAN connection. The moderate robot works on cylindrical and square structures but at the same time ensures safety in operation. The following sections will cover the design of robots to solve these problems.

Fig. 1. Steel climbing robotic system with integrated sensors.

Table 1. Robot parameters

Length	192.70 mm
Width	271.96 mm
Height	183.5 mm
Weight	3.6 kg
Drive	4 motorized wheels

Table 2. Motor parameters

Torque	525 g/mm (2S Li-Po)
Speed	0.12 s/60° (2S Li-Po)
Length	40.13 mm
Width	20.83 mm
Height	39.62 mm
Weight	71 g
Voltage	6–8.5 V (2S Li-Po battery)

3 Mechanical Design

A robot design with four wheels is proposed to increase the adhesion as well as to take advantage of the robot movement flexibility. Besides the wheels and motors can automatically fold according to the contact surface to move. The robot dimension is shown in Fig. 2, the parameter of the robot is displayed in the Table 1, and motor parameters are listed in the Table 2.

Each wheel is made up of a plastic frame and 60 permanent magnets (diameter 10 mm × 5 mm length) surrounded, similar design to [7]. In addition to the four motor-mounted wheels, the middle part of the robot is also equipped with four small wheels with permanent magnets (12.7 mm × 25.4 mm × 6.35 mm) to increase the ability to grip and keep the robot balanced.

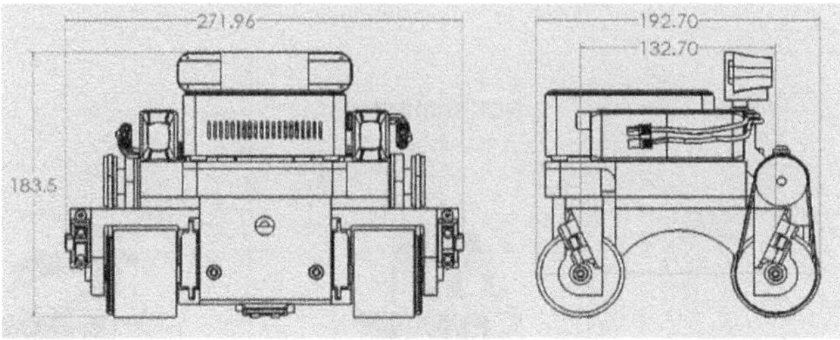

Fig. 2. The robot dimension.

A bridge may have many steel pipes of different diameters, hence the robot needs to be designed to be able to flexibly move on different sized cylindrical steel pipes. If a hard design is left, the robot cannot be moved on steel pipes of small size. Obviously, the magnet wheels are inaccessible, clinging to the steel surface. To avoid this issue, the robot design consists of three main parts: the middle body consists of passive wheels and magnets, the left and right wheel sections must be motor-mounted magnet wheels, and they are linked to the middle body through the swivel. The swivel joints allow the robot to flexibly change to cling to the surface of the steel cylinder that the existing designs [7, 9, 10] may not be able to perform.

4 Circuit and Software Design

The robot needs to be small and compact to be easy to move without being entangled. Therefore, the control system also needs to be designed to be compact, but it must meet the operating capacity of the robot. The low-level controller is designed in detail as shown in Fig. 3. The center of the low-level control circuit was using At mega 2560 at speed up to 16 MHz. Equipped with up to 54 digital pins and 16 analog pins for increased connectivity with more peripherals.

Besides, equipped with four UART communication ports, one SPI, one I2C also allows connection with many existing modules such as IMU, GPS,…, etc. An intelligent 9-axis Absolute Orientation Sensor is also integrated into the circuit to help orient the robot when operating more accurately. A MEMS accelerometer, magnetometer, and gyroscope are combined together and processed by a high-speed ARM Cortex-M0 based processor to digest all the sensor data, abstract the sensor fusion and real-time requirements away, and spit out data, so we can use in quaternions, Euler angles or vectors. Based on feedback from the encoders connected with the wheels, we designed a PID motor controller to synchronize left, right wheels speed. The gravity of

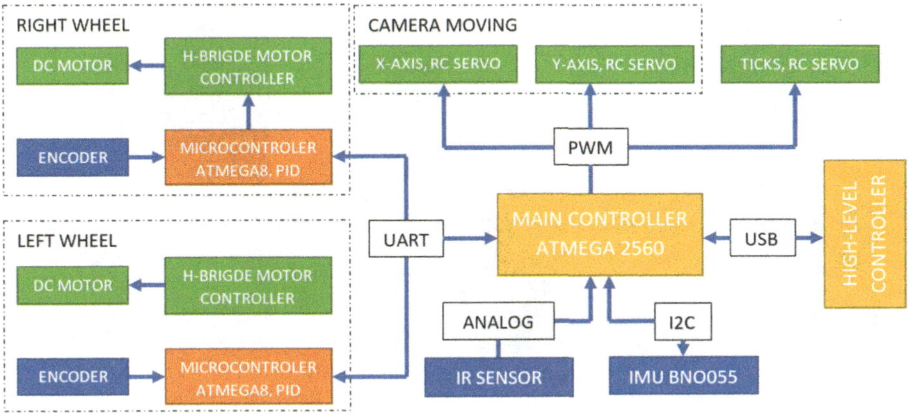

Fig. 3. The low-level control diagram.

the robot will cause the wheels spinning, hence we applied the PID position controller to keep the robot stationary. If the value from the encoder is changed when the robot is in a stop state, this means that the robot is moving under the action of gravity. At this time, the PID position controller will calculate to give a signal to control the spinning wheels in the opposite direction with a sufficient force to keep the robot not move. We used ROS to build and develop a control software system for this robot. We have created ROS topics to control robot movement, camera movement or read the value of IR sensor, IMU, encoders, 3D structure camera.

5 Experimental Result

Both indoor and outdoor experiments have been carried out to confirm the effectiveness of the robot. Indoor tests are in the laboratory environment on small steel bars that are attached to each other while outdoor testing conducted on the bridge connecting the two buildings on the University of Nevada, Reno campus. The ability to move and hold position is evaluated in both experiments. We used 3 of the 3S1P 11.1 V 2200 milliampere-hours (mAh) battery to power the robot for one hour of the test. A laptop that can connect to a wireless LAN is used as a ground station to observe robot activity and sensor's data. The robot shows the ability to adhere to flat steel surfaces, move flexibly, and its ability to hold a good position by PID position controller with many different poses as shown in Fig. 4. The maximum velocity of the robot is approximately 30 cm/s. In outdoor experiments, we also conducted experiments to demonstrate self-tuning ability to adhere to curved steel surfaces, move flexibly as well as its ability to hold a good position by PID position controller. Figure 5 shows the robot changing its body shape to cling in order to adhere to curved steel surfaces. Figure 6 shows positions of the robot moving on the bridge as well as its ability to hold the position on the curved steel surface.

Fig. 4. Robot moving onflat steel surface and transiting from one surface to the other.

Fig. 5. Robot itself changes to cling to the cylindricalsteel structure.

Fig. 6. Robot moving on a steel bridge at the University of Nevada's campus.

6 Conclusion and Future Work

This work describes the design and implementation of steel climbing robot capable of moving on many steel surfaces on different structures. The results also show that the robot moves more safely when equipped with infrared sensors and encoders, which help the robot keep a good position when needed. In addition, IMU and 3D cameras are also equipped on the robot for later localization. With its climbing capability, the proposed robot can assist the inspectors in steel bridge or steel structure inspection. In the future, we also plan to upgrade the robot so that it can carry non-destructive sensors such as eddy current sensors to inspect defects inside steel structures.

References

1. La, H.M., Lim, R.S., Basily, B.B., Gucunski, N., Yi, J., Maher, A., Romero, F.A., Parvardeh, H.: Mechatronic systems design for an autonomous robotic system for high-efficiency bridge deck inspection and evaluation. IEEE/ASME Trans. Mechatron. **18**(6), 1655–1664 (2013)
2. La, H.M., Gucunski, N., Kee, S.-H., Nguyen, L.V.: Data analysis and visualization for the bridge deck inspection and evaluation robotic system. Vis. Eng. **3**(1), 1–16 (2015)
3. La, H.M., Gucunski, N., Kee, S.H., Nguyen, L.V.: Visual and acoustic data analysis for the bridge deck inspection robotic system. In: The 31st International Symposium on Automation and Robotics in Construction and Mining (ISARC), pp. 50–57 (2014)
4. Lim, R.S., La, H.M., Shan, Z., Sheng, W.: Developing a crack inspection robot for bridge maintenance. In: IEEE International Conference on Robotics and Automation (ICRA), pp. 6288–6293 (2011)
5. Lim, R.S., La, H.M., Sheng, W.: A robotic crack inspection and mapping system for bridge deck maintenance. IEEE Trans. Autom. Sci. Eng. **11**(2), 367–378 (2014)
6. Mazumdar, A., Asada, H.H.: Mag-foot: a steel bridge inspection robot. In: IEEE/RSJ International Conference on Intelligent Robots and Systems, pp. 1691–1696 (2009)
7. La, H.M., Dinh, T.H., Pham, N.H., Ha, Q.P., Pham, A.Q.: Automated robotic monitoring and inspection of steel structures and bridges. Robotica **37**(5), 947–967 (2019)
8. Nguyen, S.T., La, H.M.: Development of a steel bridge climbing robot. In: The Proceedings of the IEEE/RSJ International Conference on Intelligent Robots and Systems (IROS), Macau, China, 3–8 November (2019)
9. Pham,N.H., La, H.M.: Design and implementation of an autonomous robot for steel bridge inspection. In: the 54th Annual Allerton Conference on Communication, Control, and Computing (Allerton), pp. 556–562 (2016)
10. Pham, N.H., La, H.M., Ha, Q.P., Dang, S.N., Vo, A.H., Dinh, Q.H.: Visual and 3D mapping for steel bridge inspection using a climbing robot. In: The 33rd International Symposium on Automation and Robotics in Construction and Mining (ISARC), Auburn, Alabama, USA, pp. 1–8 (2016)

A New Refined Forecasting Model Based on the High - Order Time-Variant Fuzzy Relationship Groups and Particle Swam Optimization

Nghiem Van Tinh(✉)

Thai Nguyen University of Technology, Thai Nguyen University, Thai Nguyen, Vietnam
nghiemvantinh@tnut.edu.vn

Abstract. In this paper, a novel refined fuzzy time series (FTS) model is presented to handle three main issues, viz., determination of effective length of intervals, handling of fuzzy logical relationships (FLRs) and determination of forecasted output values. In forecasting models, lengths of intervals always affect the results of forecasting. So, we use particle swarm optimization (PSO) technique to find the optimal length of intervals in the universe of discourse. Most of the existing forecasting models simply ignore the repeated FLRs without any proper justification or accept the number of recurrence of the FLRs without considering the appearance history of these fuzzy sets in the grouping fuzzy logical relationships process. Therefore, in this study, we consider the appearance history of the fuzzy sets on the right-hand side of the FLRs to establish the fuzzy logical relationship groups, called the time - variant fuzzy relationship groups (TV-FRGs). Furthermore, many researchers suggest that high-order FLRs improve the forecasting accuracy of the models. Therefore, we also use the high-order fuzzy relationships in order to obtain more accurate forecasting results. Proposed model is applied to forecast two numerical datasets (enrollments data of the University of Alabama, dataset of Gasonline Price in Viet Nam). The results indicate that the proposed model gets a higher average forecasting accuracy rate to forecast enrolments of the University of Alabama than the existing methods based on the high-order FTS.

Keywords: Enrollments · Gasonline · Forecasting · Fuzzy time series · Time – variant fuzzy logical relationship groups · Particle swarm optimization

1 Introduction

Forecasting is the prediction of future events based on past or present experiences. However, forecasting of these events with 100% accuracy may not be possible, but their forecasting accuracy and the speed of forecasting process can be improved. To resolve this problem, many researchers proposed several different models based on fuzzy time series concept. In this study, we also introduce a novel refined forecasting model, which is developed by combining fuzzy time series theory with PSO for

K.-U. Sattler et al. (Eds.): ICERA 2019, LNNS 104, pp. 18–34, 2020.
https://doi.org/10.1007/978-3-030-37497-6_3

handling other problems. The major objective of designing such a hybridized model is explained as follows:

For fuzzification of time series data set, determination of lengths of intervals of the historical time series data set is very important. In most of the FTS models in articles [2, 3, 11, 12, 22–24, 27, 28], the lengths of the intervals were kept the same. No precise reason is mentioned for using the fixed lengths of intervals. Huarng [4] pointed out that the different lengths of intervals in the universe of discourse can affect the forecasting result and a proper choice of the length of each interval can greatly improve the forecasting accuracy rate. So, for creating the effective lengths of intervals, the particle swarm optimization technique is used to adjust the initial length of each interval in the universe of discourse by minimizing MSE value in this paper.

After generating the intervals, time series data set is fuzzified based on the fuzzy time series theory. Each fuzzified historical data value is then used to create the FLRs. Still most of the existing fuzzy time series models ignore repeated FLRs [2, 3, 5, 10, 14–18]. These FTS models do not consider the identical FLRs during forecasting. They simply use the FRGs by discarding the repeated FLRs. The ignorance of repeated FLRs in the FRG is not properly justified. Another approach, Yu [28] argued that recurrent fuzzy logical relationships should be considered during forecasts. Yu said that the recent FLRs are more important than the previous ones. However, this scheme of establishing FLRs is not justifiable for each of forecasting time, because it does not consider the appearance history of the fuzzy sets on the right-hand side of the FLRs. To resolve these problems, we use a method [20] for establishing fuzzy relationship group in the model. In this approach, the FRGs are determined by considering the appearance history of the fuzzy sets on the right-hand side of the FLRs. That is, only the fuzzy sets on the right - hand side appearing before the fuzzy sets on the left-hand side of the FLR at forecasting time is put together to form FRG. The advantage of using such an approach is that the model can capture more persuasive information of the FLRs based on their chronological order. Furthermore, many researchers show that high-order FLRs improve the forecasting accuracy of the models [3, 5, 14–18]. Therefore, in this study, we also employ the high-order fuzzy relationships for building the forecasting model. We verify the effectiveness of the proposed model using the two following real-world data set: (1) university enrollments dataset of Alabama [2], (2) dataset of Gasonline Price in Viet Nam. The empirical study on these datasets shows that the performance of proposed model is better than those of any existing models.

The rest of this paper is organized as follows: Sect. 2 presents related works for FTS models. In Sect. 3, a brief review of the basic concepts of FTS and algorithms are introduced. In Sect. 4, an improved forecasting model based on the high – order FTS and PSO algorithm is demonstrated. Section 5 compares and evaluates the forecasting performance of the proposed model with the existing methods on the enrolment data of the University of Alabama and the Gasonline price data. Finally, conclusion and future work are discussed in Sect. 6.

2 Related Works

Fuzzy time series forecasting model is applied to deal with various problems, such as forecasting university enrollments, temperature prediction, crop productions, sales, road accidents and financial forecasting. In a conventional time series model, the recorded values of a special dynamic process are represented by crisp numerical values. But, in a fuzzy time series model, the recorded values of a special dynamic process are represented by linguistic values. Based on the fuzzy time series theory, the first forecasting model was introduced by Song and Chissom [23, 24], which was used to forecast the time series values based on linguistic values. They presented the FTS model by means of fuzzy relational equations involving max–min composition operation, and applied the model for forecasting the enrollments in the University of Alabama. Unfortunately, their method had many drawbacks such as huge computation when the fuzzy rule matrix is large and lack of persuasiveness in determining the length of intervals in the universe of discourse. Therefore, Chen [2] proposed the first-order FTS model by using simple arithmetic calculations instead of max-min composition operations [23] for better forecasting accuracy. Later, many studies provided some improvements in Chen [2] method in terms of determining of lengths of intervals, fuzzification, fuzzy logical relationships and defuzzification techniques.

To further enhance forecasting accuracy of model, many researchers proposed various FTS models. For example, Huarng [3] presented an effective approach which can properly adjust the lengths of intervals. He pointed out that the different lengths of intervals in the universe of discourse can affect the forecasting result. Researchers [17, 27] proposed computational methods of forecasting based on the high-order FLRs to overcome the drawback of fuzzy first-order forecasting models [2, 11]. Singh [22] proposed a new method in FTS forecasting. This model has the advantage that it minimizes the time of complicated computations of fuzzy relational matrices. Huarng et al. [12] exploited neural networks to construct FTS model. Their model was used to forecast stock index and obtained better forecasting result.

In recent years, many researchers have proposed various artificial intelligence based models to solve complex problems in forecasting. For example, Lee et al. [17] presented the method for forecasting the temperature and the TAIFEX based on the high – order FRGs and genetic algorithm. They also used simulated annealing techniques [16] to adjust the length of each interval in the universe of discourse for increasing the forecasting accuracy rate. By introducing genetic algorithm, Chen & Chung [5] presented the high-order FTS model to forecast the enrollments of students of University of Alabama. Particle swarm optimization technique has been successfully applied in many applications to partition the universe of discourse as can be found in [4, 7, 8, 10, 14, 15, 18–21]. Based on Chen's model [2], Kuo et al. [14] introduced a new hybrid forecasting model which combined fuzzy time series with PSO algorithm to find the proper length of each interval. Hsu et al. [18] proposed a new method for the temperature prediction and the TAIFEX forecasting, based on two-factor high-order FLRs and PSO. Huang et al. [10] proposed the hybrid forecasting model based on PSO algorithm and the refinement which aggregates the global information of fuzzy relationships with the local information of latest fuzzy fluctuation to calculate the forecasted value in the forecasting stage. In addition, Dieu et al. [20] introduced the concept of time-variant fuzzy logical relationship groups and used it in the determining of fuzzy logical relationship stage for the forecasting problems. Park et al. [21] presented a new method using PSO and two-factor high-order FTS with the aim to increase the forecasting accuracy. Chen and Bui [4] used the PSO technique to obtain not only optimal partition of intervals but also optimal weight vectors. They proposed the forecasting model based on optimal partitions of intervals and optimal weighting vectors of two-factor second-order fuzzy-trend logical relationship groups simultaneously for forecasting the TAIFEX and the NTD/USD exchange rates. Cheng et al. [8] presented a fuzzy forecasting method for forecasting the TAIFEX based on fuzzy logical relationships (FLRs), PSO techniques, the K-means clustering algorithm and similarity measures on the subscripts of fuzzy sets. Chen and Jian [7] developed a fuzzy forecasting algorithm based on PSO and similarity measures between the subscripts of fuzzy sets. Another way for determining optimal intervals is considered as the main factor affecting the performance of forecasting. That is a variety of clustering techniques which have been developed to minimize forecast errors. Among them clustering methods as rough fuzzy C-means [1], K-means [25, 26], fuzzy C-means [9], automatic clustering [6] are introduced in recent works.

3 Basic Concepts of Fuzzy Time Series and Algorithms

3.1 Basic Concepts of Fuzzy Time Series

This section briefly introduces the concepts and notations related to fuzzy time series as follows. Let $U = \{u_1, u_2, \ldots, u_n\}$ be an universal set; a fuzzy set A_i of U is defined as $A_i = \{f_A(u_1)/u_1 +, f_A(u_2)/u_2 \ldots + f_{Ai}(u_n)/u_n\}$, where f_{Ai} is a membership function of a given set A, f_{Ai}: $U \rightarrow [0, 1]$, $f_{Ai}(u_i)$ indicates the grade of membership of u_i in the fuzzy set A, $f_{Ai}(u_i) \in [0, 1]$, and $1 \le i \le n$. The definition of fuzzy time series is briefly reviewed as follows:

Definition 1: Fuzzy time series [23, 24].
Let $Y(t)(t = .., 0, 1, 2..)$, a subset of R, be the universe of discourse on which fuzzy sets $f_i(t)(i = 1, 2\ldots)$ are defined and if F(t) be a collection of $f_1(t), f_2(t), \ldots$, then F(t) is called a FTS on Y(t)(t . . ., 0, 1, 2...).

Definition 2: Fuzzy logic relationships (FLRs) [23, 24].
The relationship between F(t) and F(t − 1) can be denoted by $F(t-1) \rightarrow F(t)$. Let $A_i = F(t)$ and $A_j = F(t-1)$, the relationship between F(t) and F(t − 1) is denoted by fuzzy logical relationship $A_i \rightarrow A_j$ where A_i and A_j refer to the current state or the left - hand side and the next state or the right-hand side of fuzzy time series.

Definition 3: m - order fuzzy logical relationships [23, 24].
Let $F(t)$ be a fuzzy time series. If $F(t)$ is caused by $F(t-1), F(t-2), \ldots, F(t-m+1) F(t-m)$ then this fuzzy relationship is represented by $F(t-m), \ldots, F(t-2), F(t-1) \rightarrow F(t)$ and is called an m - order fuzzy time series.

Definition 4: Fuzzy logical relationship groups (FRGs) [2].
Fuzzy logical relationships with the same fuzzy set in the left-hand side of the fuzzy relationships can be grouped into a fuzzy logical relationship group. Suppose there are exists fuzzy logical relationships as following: $A_i \rightarrow A_{k1}$, $A_i \rightarrow A_{k2}, \ldots, A_i \rightarrow A_{km}$; these fuzzy logical relationship can be grouped into the same FLRG as: $A_i \rightarrow A_{k1}, A_{k2}, \ldots, A_{km}$.

The same fuzzy set appears more than once on the right hand side, according to Chen 2, it can be only counted one time but Yu model [28], the recurrence of fuzzy set can be admitted.

3.2 Particle Swarm Optimization Algorithm

PSO, first introduced by Eberhart and Kannedy [13] in 1995, is a random searching algorithm based on group cooperation that can solve the near optimal solution of any kind of optimization problems [14] and is inspired by simulating the social behaviour of animals, such as fish schooling, birds flocking and the swarm theory. It is particle swarm optimization that initializes each particle randomly, and then finds the optimal solution through iteration. At each step of optimization, the particles update themselves by tracking their own best position and the best particle. To get the optimal solution, the particles update their own speed and positions according to the following formulas:

$$V_{id}^{k+1} = \omega^k * V_{id}^k + C_1 * Rand() * \left(P_{best\_id} - X_{id}^k\right) + C_2 * Rand() * \left(G_{best} - X_{id}^k\right) \quad (1)$$

$$X_{id}^{k+1} = X_{id}^k + X_{id}^{k+1} \quad (2)$$

$$\omega^k = \omega_{max} - \frac{k * (\omega_{max} - \omega_{min})}{iter\_max} \quad (3)$$

where,

- X_{id}^k is the current position of a particle id in k-th iteration;
- V_{id}^k is the velocity of the particle id in k-th iteration, and is limited to $[-V_{max}, V_{max}]$, where V_{max} is a constant pre-defined by user.
- P_{best_id} is the position of the particle id that experiences the best fitness value.
- G_{best} is the best one of all personal best positions of all particles within the swarm.
- Rand() is the function can generate a random real number between 0 and 1 under normal distribution.
- C_1 and C_2 are acceleration values which represent the selfcondence coefficient and the social coefficient, respectively.
- ω is the inertia weight factor according to Eq. (3).

3.3 High – Order Time – Variant Fuzzy Relation Groups Algorithm

Suppose there are fuzzy time series F(t), t = 1, 2,…, q which are presented by fuzzy sets as follows: $A_{i1}, A_{i2}, \ldots, A_{iq}$.

Based on the Definitions 5 and 6 of the time - variant fuzzy logical relationship groups, an algorithm is proposed as follows:

Algorithm 1: The m - order TV- FRGs algorithm

Input: Historical time series dataset
Output: Time –variant fuzzy logical relationships groups (TV – FRGs)
Begin
1: Define the universe of discourse U and partition U into equal length intervals
3: Fuzzify historical data into fuzzy sets
3: Define fuzzy logical relationships
4: initialize the m-order TV – FRGs t=m ; F(1),F(2),..,F(m-1) →F(m) or $A_{j2},\ldots,A_{jm} \rightarrow A_{k1}(m)$
5: ***fort:*** *= m to q* ***do***
for *h: = m* ***down to*** *1* ***do***
  Establish all m- order FLRs $A_{j2}(t-m),\ldots,A_{jm}(t-1) \rightarrow A_{k1}(t)$
end for
6: ***for*** *v: = 1 to t-1* ***do***
for *h = 1 to v* ***do***
if *there are fuzzy logical relationship* $A_{j2},\ldots,A_{jm} \rightarrow A_{k2}(h)$ *at the same left - hand side* ***then*** *add* A_{k2} *into fuzzy logical relationship groups as follows:* $A_{j2},\ldots,A_{jm} \rightarrow A_{k1}, A_{k2}$
  end ***for***
 end ***for***
end.

---

4 A Refined Forecasting Model Based on the High–Order TV-FRGs and PSO Algorithm

In this section, a novel refined forecasting model which combined the high – order TV-FRGs and PSO algorithm is introduced. The detail of the proposed model is presented as follows:

4.1 Forecasting Model Based on the High – Order TV-FRGs

To verify the effectiveness of the proposed model, all historical enrolments data [2] from 1971s to 1992s are used to illustrate the high - order FTS forecasting process. The step-wise procedure of the proposed model is detailed as follows:

Step 1: Define the universe of discourse U
Assume Y(t) be the historical data of enrolments at year t ($1971 \leq t \leq 1992$). The university of discourse is defined as $U = [D_{min}, D_{max}]$. In order to ensure the forecasting values bounded in the universe of discourse U, we set $D_{min} = I_{min} - N_1$ and $D_{max} = I_{max} + N_2$; where I_{min}, I_{max} are the minimum and maximum data of Y(t); N_1 and N_2 two proper positive real values which denote the buffers to adjust the lower bound and the upper bound of the universe of discourse and to cover the noise of the testing data.

From the historical data, we obtain $I_{min} = 13055$ và $I_{max} = 19337$. Thus, the universe of discourse is defined as $U = [I_{min} - N_1, I_{max} + N_2] = [13000,\ 20000]$ with $N_1 = 55$ and $N_2 = 663$.

Step 2: Partition U into equal length intervals
Divide U into equal length intervals. Compared to the previous models in article [2, 14] and for convenience to demo of the forecasting example here, we firstly divide U into seven intervals, $u_1, u_2, \ldots, u_7$, respectively. The length of each interval is $L = \frac{D_{max} - D_{min}}{7} = \frac{20000-13000}{7} = 1000$. Thus, the seven intervals are defined as follows: $u_i = [13000 + (i-1)*L,\ 13000 + i*L)$, with $(1 \le i \le 7)$ gets seven intervals as:
$u_1 = [13000,\ 14000)$, $u_2 = [14000, 15000)$, ..., $u_6 = [18000, 19000)$, $u_7 = [19000, 20000)$.
Step 3: Define the fuzzy sets for each interval
Each of interval in Step 2 represents a linguistic variable of "enrolments" in 2. For seven intervals, there are seven linguistic values which are A_1 = "not many", A_2 = "not too many", A_3 = "many", A_4 = "many many", A_5 = "very many", A_6 = "too many", and A_7 = "too many many" to represent different regions in the universe of discourse on U, respectively. Each linguistic variable represents a fuzzy set A_i and its definitions is described in (4) and (5) as follows.

$$A_i = a_{i1} \backslash u_1 + a_{i2} \backslash u_2 + \ldots + a_{ij} \backslash u_j + \ldots + a_{i7} \backslash u_7 \tag{4}$$

$$a_{ij} = \begin{cases} 1 & j = i \\ 0.5 & j = i-1 \quad or \quad j = i+1 \\ 0 & otherwise \end{cases} \tag{5}$$

Here, the symbol '+' denotes the set union operator, $a_{ij} \in [0, 1] (1 \le i \le 7, 1 \le j \le 7)$, u_j is the jth interval of U. The value of a_{ij} indicates the grade of membership of u_j in the fuzzy set A_i. For simplicity, the different membership values of fuzzy set A_i are selected according to Eq. (5). According to Eqs. (4) and (5), a fuzzy set contains 7 intervals. Contrarily, an interval belongs to all fuzzy sets with different membership degrees. For example, u_1 belongs to A_1 and A_2 with membership degrees of 1 and 0.5 respectively, and other fuzzy sets with membership degree is 0.

Step 4: Fuzzy all historical enrolments data
In order to fuzzify all historical data, it's necessary to assign a corresponding linguistic value to each interval first. The simplest way is to assign the linguistic value with respect to the corresponding fuzzy set that each interval belongs to with the highest membership degree. For example, the historical enrolment of year 1972 is 13563, and it belongs to interval u_1 because 13563 is within [13000, 14000). So, we then assign the linguistic value "not many" (e.g. the fuzzy set A_1) corresponding to interval u_1 to it. Consider two time serials data Y(t) and F(t) at year t, where Y(t) is actual data and F(t) is the fuzzy set of Y(t). According to Eq. (4), the fuzzy set A_1 has the maximum membership value at the interval u_1. Therefore, the historical data time series on date

Y(1972) is fuzzified to A_1. The completed fuzzified results of the enrollments are listed in Table 1.

Table 1. The results of fuzzification according to enrollments data

Year	Actual data	Fuzzy sets	Membership degree	Linguistic value
1971	13055	A_1	[**1** 0.5 0 0 0 0 0]	"not many"
1972	13563	A_1	[**1** 0.5 0 0 0 0 0]	"not many"
–	–	–	–	–
1991	19337	A_7	[0 0 0 0 0 0.5 **1**]	"too many many"
1992	18876	A_6	[0 0 0 0 0.5 **1** 0.5]	"too many"

Step 5. Create all m – order fuzzy relationships ($m \geq 2$).
Based on Definition 3. To establish a m-order fuzzy relationship, we should find out any relationship which has the type $F(t-m), F(t-m+1), \ldots, F(t-1) \rightarrow F(t)$, where $F(t-m), F(t-m+1), \ldots, F(t-1)$ and $F(t)$ are called the current state and the next state, respectively. Then a m - order fuzzy relationship is got by replacing the corresponding linguistic values as follows: $A_{im}, A_{i(m-1)}, \ldots, A_{i2}, A_{i1} \rightarrow A_k$. For example, supposed $m = 3$, a fuzzy relationship $A_1, A_1, A_1 \rightarrow A_2$ is got as F(1971), F(1972), F(1973) $\rightarrow$ F(1974). Continue with the examples above and based on Table 1, all 3rd - order FLRs are shown in Table 2.

Table 2. The complete the 3rd - order fuzzy logical relationships

Year	No of FLR	3rd – order fuzzy relations	LinguisticvariableF(t)
1974	1	$A_1, A_1, A_1 \rightarrow A_2$	F(1971), F(1972), F(1973) $\rightarrow$ F(1974)
1975	2	$A_1, A_1, A_2 \rightarrow A_3$	F(1972), F(1973), F(1974) $\rightarrow$ F(1975)
–	–	–	–
1992	19	$A_4, A_7, A_7 \rightarrow A_6$	F(1989), F(1990), F(1991) $\rightarrow$ F(1992)
1993	20	$A_7, A_7, A_6 \rightarrow$ #	F(1990), F(1991), F(1992) $\rightarrow$ F(1993)

Step 6: Establish all m – order time – variant fuzzy relationship groups
In previous studies [2, 14] all the fuzzy logical relationships having the same fuzzy set on the left-hand side or the same current state can be grouped into a same fuzzy logical relationship group. But, according Algorithm 1, the appearance history of the fuzzy sets on the right-hand side of fuzzy logical relationships need more considertion. From this viewpoint and based on Table 2, we can obtain 20 groups of the high -order TV- FRGs which are shown in column 5 of Table 3. Where, the first 19 groups of the high – order fuzzy logical relationships are called the trained patterns (or in training phase), and the last one is called the untrained pattern (or in testing phase).

Step 7. Calculate the forecasting values for all high – order TV- FRGs
To defuzzify the fuzzified data and to obtain the forecasted values, a new defuzzification technique is developed to calculate the forecasted values for all high – order TV-FRGs in training phase. Then we use the master voting (MV) scheme for the untrained pattern proposed in work [14] for high – order TV-FRGs in testing phase.

For the training phase, we estimate all forecast values for all high – order TV-FRGs based on fuzzy sets on the right-hand side or next state within the same group. For each group in column 5 of Table 3, we divide each corresponding interval of each next state into p sub-intervals with equal length, and calculate a forecasted value for each group according to Eq. (6).

$$\text{forecasted}_{\text{output}} = \frac{1}{\text{n}}\sum\nolimits_{\text{j}=1}^{\text{n}} subm_{kj} \tag{6}$$

where, $(1 \leq j \leq \text{n})$

- n is the total number of next states or the total number of fuzzy sets on the right-hand side within the same group.
- $subm_{kj}$ is the midpoint of one of p sub-intervals (means the midpoint of ^{k}th sub-interval in which the historical data belong to this sub-interval, $1 \leq k \leq$ p) corresponding to j-th fuzzy set on the right-hand side, where the highest level of A_{kj} takes place in this interval.

For the testing phase, based on the master vote scheme, we calculate forecasted value for a group which contains the unknown linguistic value of the next state according to Eq. (7); where the symbol w_h means the highest votes predefined by the user, m is the order of the fuzzy logical relationship, the symbols M_{t1} and M_{ti} denote the midpoints of the corresponding intervals of the latest past and other past linguistic values in the current state. From Table 3, it can be shown that group 20 has the fuzzy relationship A_7, A_7, $A_6 \rightarrow \#$ as it is created by the fuzzy relationship $\text{F}(1990), \text{F}(1991), \text{F}(1992) \rightarrow \text{F}(1993)$; since the linguistic value of F(1993) is unknown within the historical data, and this unknown next state is denoted by the symbol '#'. Then, calculate the value for "#" based on the current state of group G_{20} which is determined by Eq. (7)

$$Forecated_{for\#} = \frac{(M_{t1} * w_h) + M_{t2} + \ldots + M_{ti} + \ldots + M_{tm}}{w_h + (\text{m} - 1)}; \text{ With } 1 \leq \text{i} \leq \text{m} \tag{7}$$

Based on the forecasted rules eated in (6) and (7), we complete forecasted results for all 3rd - order TV- FRGs which are listed in column 6 of Table 3.

Table 3. The complete forecasted values for all TV-FRGs

Year	At time	Fuzzy set	No group	3rd – order TV-FRGs	Forecasted value
1974	T = 4	A_2	G1	A_1, A_1, $A_1 \rightarrow A_2$	14625
1975	T = 5	A_3	G2	A_1, A_1, $A_2 \rightarrow A_3$	15375
–	–	–	–	–	–
1992	T = 22	A_6	G19	A_4, A_7, $A_7 \rightarrow A_6$	18875
1993	T = 23	N/A	G20	A_7, A_7, $A_6 \rightarrow \#$	18667

From column 6 of Tables 3 and 1, we complete forecasted results for the enrolments of University of Alabama during the period from 1971 to 1992 based on 3rd-order fuzzy time series model with seven intervals which are listed in Table 4.

Table 4. The complete forecasted outputs based on the 3rd – order fuzzy time series

Year	Actual data	Fuzzy set	Forecasted value	Year	Actual data	Fuzzy set	Forecasted value
1971	13055	A_1	N/A	1983	15497	A_3	15375
1972	13563	A_1	N/A	1984	15145	A_3	15125
–	–	–	–	–	–	–	–
1980	16919	A_4	16875	1992	18876	A_6	18875
1981	16388	A_4	16375	1993	N/A	#	18667
1982	15433	A_3	15375	**MSE**			**96447**

To evaluate the performance of the proposed model, the mean square error (MSE) is employed as an evaluation criterion to represent the forecasted accuracy. The MSE value is calculated as follows:

$$MSE = \frac{1}{n}\sum\nolimits_{i=m}^{n}(F_i - R_i)^2 \quad (8)$$

where, R_i denotes actual data at year i, F_i is forecasted value at year i, n is the number of the forecasted data, m is the order of the fuzzy logical relationships.

To verify the superiority of the proposed model without using PSO based on the third – order FTS under seven intervals, five existing high – order forecasting models (e.g., the model [27] model [12], model [14], model [10] and the model [22] are selected for comparison. A comparison of the forecasted results based on the MSE value (8) is presented in Fig. 1. In Fig. 1, the MSE value of the proposed model portrays small error rate in comparison with the model compared.

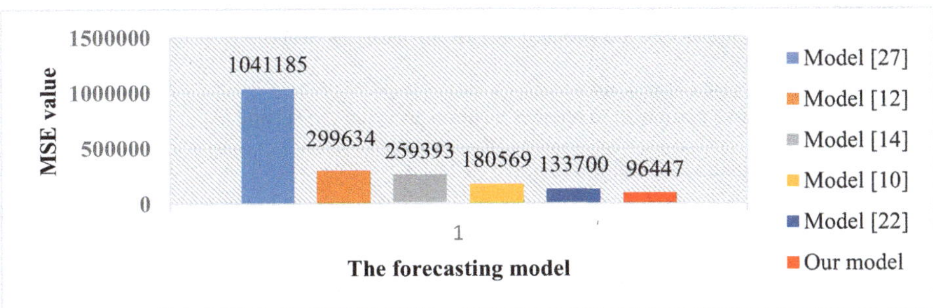

Fig. 1. A comparison of the MSE values for various models based on high – order FTS model under seven intervals

4.2 Forecasting Model Based on the High – Order TV-FRGs and PSO Algorithm

To improve forecasted accuracy of the proposed model, the effective lengths of intervals, TV- FRGs and defuzzification techniques which are three main issues presented in this paper. A novel refined hybrid method for forecasting problems is developed to adjust the length of each of the interval in the universe of discourse without increasing the number of intervals by the MSE value (8). In the proposed model, each particle exploits the intervals in the universe of discourse of historical data Y(t). Let the number of the intervals be n, the lower bound and the upper bound of U on historical data Y(t) be b_0 and b_n, respectively. Each particle is a vector consisting of n−1 elements b_i where $1 \leq i \leq n-1$ and $b_i \leq b_{i+1}$. Based on these n−1 elements, define the n intervals as $u_1 = [b_0, b_1]$, $u_2 = [b_1, b_2]$,,..., $u_i = [b_{i-1}, b_i]$,... and $u_n = [b_{n-1}, b_n]$, respectively. When a particle moves to a new position, the elements of the corresponding new vector need to be sorted to ensure that each element b_i $(1 \leq i \leq n-1)$ arranges in an ascending order. The complete steps of the proposed model are presented in Algorithm 2.

Algorithm 2: The HV-FRGs-PSO algorithm

1. **Input**: historical time series data
2. **Output**: The forecasting results and the MSE value (Gbest = min(pbest))

Initialize:
- ✓ Define the universe of discourse U
- ✓ Generate random population of N particles including (number of order, number of interval, positions X_{id} and velocities V_{id} of all P_n particles)
- ✓ Initialize the value of the weight factor w;

3. **while** the stop condition (maximum iterations or minimum MSE criteria) is not satisfied **do**

3.1. **for** particle id, $(1 \leq i \leq P_n)$ **do**
- ✓ Define linguistic terms based on the current position of particle id
- ✓ Fuzzify all historical data by Step 4 in Subsection 4.1
- ✓ Create all m – order fuzzy logical relationships by Step 5 in Subsection 4.1
- ✓ Construct all m – order time -variant FRGs by Step 6 in Subsection 4.1
- ✓ Calculate and defuzzification forecasting outputs by Step 7 in Subsection 4.1
- ✓ Compute the MSE values for particle id based on Eq. (8)
- ✓ Update the personal best position (pbest) of particle id according to Eq. (8).

end for

3.2. Update the global best position(Gbest) of all particles according to the MSE values mentioned above.

4. **for** particle id, $(1 \leq i \leq p_n)$ **do**
- ✓ move particle id to another position according to (1) and (2)

end for
- ✓ update ω according to Eq. (3)

end while

End.

5 Experimental Results

5.1 Input Data

In this paper, the proposed model is utilized to forecast the enrollments of University of Alabama with the whole historical data [2], the period from 1971 to 1992 which is shown in Fig. 2(a) and handles other forecasting problems, such as the empirical data for the Gasonline price in Viet Nam covering the period from January 4, 2018 to July 2, 2019 which is shown in Fig. 2(b). We have implemented the experiments using visual studio 2017 environment with C# programming language on an Intel Core i7 PC with 8 GB RAM. The essential parameters of the proposed model for forecasting enrolments and Gasonline price are listed in Table 5.

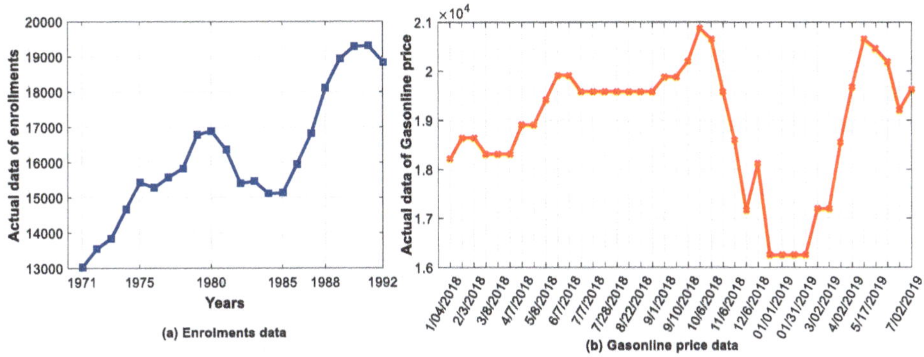

Fig. 2. The curves represent the actual data of enrolments and gasoline price

Table 5. Parameters used for forecasting enrolments [2] and Gasonline price

Description for the parameters	Values of enrolments	Values of gasonline price
Number of particles	50	30
Maximum number of iterations	400	200
The value of inertial weigh ω be linearly decreased	1.4 to 0.4	1.4 to 0.4
The coefficient $C_1 = C_2$	2	2
The velocity be limited to	[−100, 100]	[−100, 100]
The position be limited to	[13000, 20000]	[6200, 7600]

5.2 Experimental Results for Forecasting Enrolments

To verify the forecasting effectiveness of the proposed model under different number of intervals and different high - order FRGs, five FTS models named as model [29], model [30], CC06b [5], HPSO [14] and AFPSO [10] are examined and compared. The forecasted accuracy of the proposed method is estimated using the MSE (8) the parameters are expressed in column 2 of Table 5. The proposed model is executed 10 runs, and the best result of runs is taken to be the final result. A comparison of the forecasting accuracy with various orders and different number of intervals among the models [5, 10, 14, 29, 30] and the proposed model are listed in Table 6.

Table 6. A comparison of the forecasted results of the proposed model with the existing models based on different high – order fuzzy time series.

Years	Actual data	Model [30]	Model [29]	CC06b [5]	HPSO [14]	AFPSO [10]	Our model
1971	13055	N/A	N/A	N/A	N/A	N/A	N/A
1972	13563	N/A	N/A	N/A	N/A	N/A	N/A
1973	13867	13813	13845	N/A	N/A	N/A	N/A
1978	15861	15895	15895	N/A	N/A	N/A	15923.2
–	–	–	–	–	–	–	–
1991	19337	19428	19346	19334	19337	19335	19316
1992	18876	19046	18822	18910	18882	18882	18853
MSE		**6825**	**1121**	**1101**	**234**	**173**	**16.33**

From Table 6, it is obvious that the proposed model has a MSE value of **16.33** which is the lowest among all forecasting models compared. The main difference among all the compared models is the fuzzy logical relationship group algorithms used to forecast. Five forecasting models in model [29], CC06b [5], model [30], HPSO [14] and the AFPSO [14] used the time – invariant fuzzy relation groups algorithm to defuzzify forecasting output, while this study has proposed a method that benefits from the time - variant fuzzy relationship groups algorithm. As shown in Table 6, three models; the HPSO, AFPSO and our model all use the PSO algorithm, but our proposed model gets smaller MSE values in forecasting. To be clearly visualized, the following Fig. 3 shows the performance of the enrolments forecasting. The convergence of the best objective values (MSE) and computational time for PSO based the 7th-order FTS with number of interval of 14 are also depicted in Fig. 3.

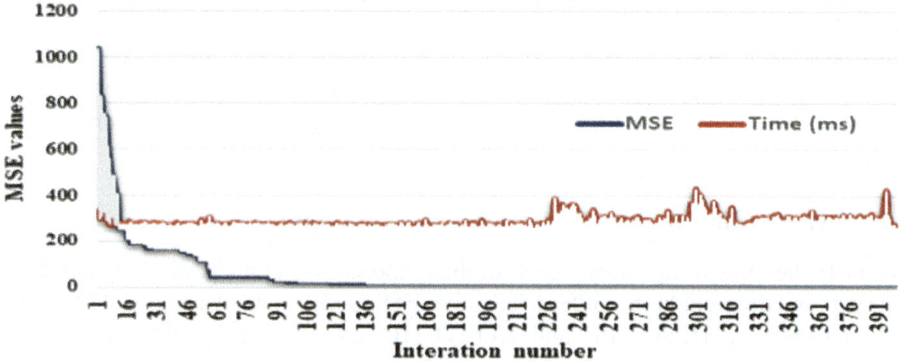

Fig. 3. Convergence Speed and computational time of the proposed model using PSO algorithm

In addition, we also perform 10 more runs to be compared with various high-order forecasting models under seven intervals such as C02 model [3], CC06b model [5], HPSO model [14] and AFPSO model [10] The detail of comparison is shown in Table 7. The trend in forecasting enrollments based on the high - order FTS under various orders by MSE value can be visualized in Fig. 4.

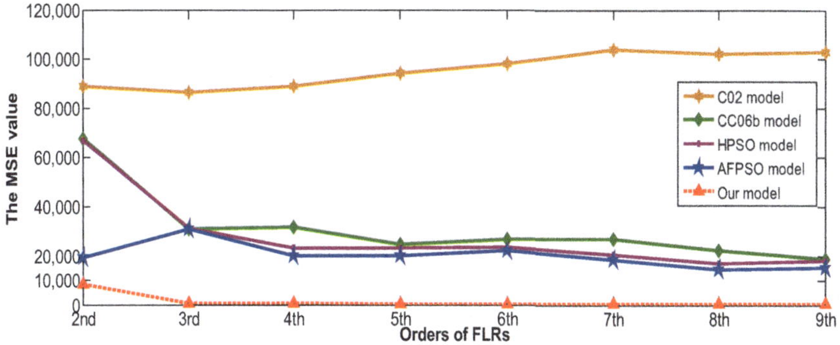

Fig. 4. A comparison of the MSE values for 7 intervals with various high-order FRGs.

Table 7. A comparison of the MSE value between our model and its counterpart under different number of orders and the number of interval is 7.

Models	Number of orders								Average
	2	3	4	5	6	7	8	9	
C02	89093	86694	89376	94539	98215	104056	102179	102789	95868
CC06b	67834	31123	32009	24984	26980	26969	22387	18734	31373
HPSO	67123	31644	23271	23534	23671	20651	17106	17971	28121
AFPSO	19594	31189	20155	20366	22276	18482	14778	15251	20261
Our model	**8712**	**1008**	**736**	**664**	**667**	**627**	**383**	**566**	**1670.4**

During the simulation, the number of intervals is kept (the number of intervals = 7) with different high – order FTS for the existing models and our model. A comparison of MSE value is listed in Table 7. From Table 7, it can be seen that the accuracy of the proposed model is improved significantly. Particularly, our model gets the lowest MSE value of **383** with 8th-order fuzzy relations and the average MSE value of the proposed model is **1670.4**, which is the smallest among six forecasting models compared.

5.3 Experimental Results for Forecasting Gasonline Price

In this subsection, we apply the proposed method for forecasting the gasonline price in Viet Nam which was taken from the site https://vnexpress.net/kinh-doanh or https://www.gso.gov.vn/Default.aspx?tabid=217 and presented in Fig. 2(b). The forecasted accuracy of the proposed model is estimated using the MSE (8). To verify the superiority of the proposed model under various order of FRGs, During the simulation with parameters expressed in column 3 of Table 6, the number of intervals is kept fixed (number of intervals = 14). The forecasting performance of the proposed model is shown in Table 8.

Table 8. The forecasting results of the proposed model for gas online in Viet Nam based on different orders FTS

Date	Actual data	2nd-order	3rd-order	4th-order	5th-order	6th-order	7th-order
1/4/2018	18240	–	–	–	–	–	–
1/19/2018	18670	–	–	–	–	–	–
2/3/2018	18670	18688.3	–	–	–	–	–
2/21/2018	18340	18322.6	18326	–	–	–	
–	–	–	–	–	–	–	–
6/17/2019	19230	19260.3	19188.8	19176.6	19205.3	19209.5	19243.2
7/02/2019	19650	19639.3	19634.8	19709.8	19673.6	19640.7	19650.4
7/03/2019	N/A	19672	19666.3	19826.9	19977.4	19971.5	19793.8
MSE		**12444.03**	**6428.6**	**3763**	**1084.9**	**630.93**	**103.55**

As shown in Table 8, the proposed model gets the smallest forecasting error rate by the MSE value of 103.55 based on 7th- order for forecasting Gasonline in Viet Nam. Also from Table 8, it can be seen that the forecasting trend for the next day is gradually increasing.

6 Conclusion

In this paper, a new refined FTS forecasting model which combines the high – order FTS and PSO algorithm is presented. To reconcile the drawbacks of the conventional high-order FTS model which use the fuzzy relation groups, the time - variant high - order FRGs is used in this paper. By adopting the time - variant high - order FRGs which helps a more effective use of the historical data, it has been proved to be more suitable for practical use. The paper also proposes a new defuzzification method to calculate the forecasted output values, which has been the main contribution issue to propose the forecasting model in this paper. In addition, The PSO techniques are used to get the optimal partition of the interval in the universe of discourse. Among the heuristic optimization technique, the comparative results show the PSO method was generally found to perform better than other algorithms in terms of success rate and solution quality. By combining the TV-FRGs with PSO algorithm, the performance of the proposed forecasting model can be improved significantly. From the empirical study for forecasting enrolments and Gasonline price, the experimental results show that the proposed model outperforms its counterparts based on high – order FTS with various orders and different interval lengths. The detail of comparison was presented in

Tables 6, 7, and 8 and Figs. 1 and 4. These results are very promising for the future work on the development of fuzzy time series in real-world forecasting applications.

References

1. Bose, M., Mali, K.: A novel data partitioning and rule selection technique for modelling high – order fuzzy time series. Appl. Soft Comput. (2017). https://doi.org/10.1016/j.asoc.2017.11.011
2. Chen, S.M.: Forecasting enrolments based on fuzzy time series. Fuzzy Set Syst. **81**, 311–319 (1996)
3. Chen, S.M.: Forecasting enrolments based on high-order fuzzy time series. Cybern. Syst. **33** (1), 1–16 (2002)
4. Chen, S.M., Phuong, B.D.H.: Fuzzy time series forecasting based on optimal partitions of intervals and optimal weighting vectors. Knowl.-Based Syst. **118**, 204–216 (2017)
5. Chen, S.M., Chung, N.Y.: Forecasting enrolments using high-order fuzzy time series and genetic algorithms. Int. J. Intell. Syst. **21**, 485–501 (2006)
6. Chen, S.M., Tanuwijaya, K.: Fuzzy forecasting based on high-order fuzzy logical relationships and automatic clustering techniques. Expert Syst. Appl. **38**, 15425–15437 (2011)
7. Chen, S.-M., Jian, W.-S.: Fuzzy forecasting based on two-factors second-order fuzzy-trend logical relationship groups, similarity measures and PSO techniques. Inf. Sci. **391–392**, 65–79 (2017)
8. Cheng, S.H., Chen, S.-M., Jian, W.S.: Fuzzy time series forecasting based on fuzzy logical relationships and similarity measures. Inf. Sci. **327**, 272–287 (2016)
9. Egrioglu, E., Aladag, C.H., Yolcu, U.: Fuzzy time series forecasting with a novel hybrid approach combining fuzzy c-means and neural networks. Expert Syst. Appl. **40**(3), 854–857 (2013)
10. Huang, Y.L., et al.: A hybrid forecasting model for enrollments based on aggregated fuzzy time series and particle swarm optimization. Expert Syst. Appl. **7**(38), 8014–8023 (2011)
11. Huarng, K.: Effective lengths of intervals to improve forecasting in fuzzy time series. Fuzzy Sets Syst. **123**(3), 387–394 (2001)
12. Hwang, J.R., Chen, S.M., Lee, C.H.: Handling forecasting problems using fuzzy time series. Fuzzy Sets Syst. **100**, 217–228 (1998)
13. Kennedy, J., Eberhart, R.: Particle swarm optimization. In: Proceedings of the 1995 IEEE International Conference on Neural Networks, 4, Perth, Australia, pp. 1942–1948 (1995)
14. Kuo, I.-H., et al.: An improved method for forecasting enrolments based on fuzzy time series and particle swarm optimization. Expert Syst. Appl. **36**, 6108–6117 (2009)
15. Kuo, I.-H., et al.: Forecasting TAIFEX based on fuzzy time series and particle swarm optimization. Expert Syst. Appl. **2**(37), 1494–1502 (2010)
16. Lee, L.-W., Wang, L.-H., Chen, S.-M.: Temperature prediction and TAIFEX forecasting based on high order fuzzy logical relationship and genetic simulated annealing techniques. Expert Syst. Appl. **34**, 328–336 (2008)
17. Lee, L.W., et al.: Handling forecasting problems based on two-factors high-order fuzzy time series. IEEE Trans. Fuzzy Syst. **14**, 468–477 (2006)
18. Hsu, L.-Y., et al.: Temperature prediction and TAIFEX forecasting based on fuzzy relationships and MTPSO techniques. Expert Syst. Appl. **37**, 2756–2770 (2010)

19. Dieu, N.C., Van Tinh, N.: Fuzzy time series forecasting based on time-depending fuzzy relationship groups and particle swarm optimization. In: Proceedings of the 9th National conference on Fundamental and Applied Information Technology Research (FAIR'9), pp. 125–133 (2016)
20. Van Tinh, N., Dieu, N.C.: An improved method for stock market forecasting combining high-order time variant fuzzy logical relationship groups and particle swam optimization. In: Proceedings of the International Conference, ICTA, pp. 153–166 (2016)
21. Park, J.I., Lee, D.J., Song, C.K., Chun, M.G.: TAIFEX and KOSPI 200 forecasting based on two-factors high-order fuzzy time series and particle swarm optimization. Expert Syst. Appl. **37**, 959–967 (2010)
22. Singh, S.R.: A simple method of forecasting based on fuzzy time series. Appl. Math. Comput. **186**, 330–339 (2007)
23. Song, Q., Chissom, B.S.: Forecasting enrolments with fuzzy time series - Part I. Fuzzy Sets Syst. **54**(1), 1–9 (1993)
24. Song, Q., Chissom, B.S.: Fuzzy time series and its models. Fuzzy Sets Syst. **54**(3), 269–277 (1993)
25. Tian, Z.H., Wang, P., He, T.Y.: Fuzzy time series based on K-means and particle swarm optimization algorithm. In: Man-Machine-Environment System Engineering. Lecture Note in Electrical Engineering, vol. 406, pp. 181–189. Springer (2016)
26. Tinh, N.V., Dieu, N.C.: A novel forecasting model based on combining time-variant fuzzy logical relationship groups and K-means clustering technique. In: Proceedings of the 9th National Conference on Fundamental and Applied Information Technology Research (FAIR10) (2017). https://doi.org/10.15625/vap.2017.0002
27. Yu, H.K.: A refined fuzzy time-series model for forecasting. Phys. A Stat. Mech. Appl. **346** (3-4), 657–681 (2005)
28. Yu, H.K.: Weighted fuzzy time series models for TAIFEX forecasting. Phys. A Stat. Mech. Appl. **349**(3–4), 609–624 (2005)
29. Wang, H., Guo, J., Feng, H., Jin, H.: An improved forecasting model of fuzzy time series. Appl. Mech. Mater. **678**, 64–69 (2014). [3rd International Conference on Mechatronics and Control Engineering (ICMCE 2014)]
30. Xu, W.H., Guo, J.C., Feng, H., Jin, H.L.: A fuzzy time series forecasting model based on data differences. 1 ICT IN EDUCATION. Frontiers in Computer Education, pp. 15–18 (2014). 2nd International Conference on Frontiers in Computer Education (ICFCE 2014), Wuhan, China (2014)

A Novel Determination of Boundaries of Cable Forces for Cable-Driven Parallel Robots with Frequency Constraint by Using Differential Evolution Algorithm

Sy Nguyen-Van[1,2(✉)] and Kwan-Woong Gwak[2]

[1] Thai Nguyen University of Technology, 3/2, Tich Luong, Thai Nguyen, Vietnam
vansytnut92@gmail.com, vansy@tnut.edu.vn
[2] Sejong University, Seoul 05006, Korea

Abstract. The vibration problem of cable-driven parallel robots has not been well studied because most of the existing methods use the spring element to model cables. In addition, to improve dynamic performance or to avoid resonance for a cable-driven parallel robot moving along a certain trajectory, its structure should be under the frequency constraint. However, the energy consumption is not minimized due to using the fixed values of boundaries of cable force in tension distributions. To solve these two problems, this study sought to use a finite element method for vibration analysis of cable-driven parallel robots. To validate the feasibility of the finite element program, vibration analysis is designed for a spatial cable-driven parallel robot and its results are compared with those of the commercial software, SAP2000. Additionally, optimization of values of boundaries of cable forces is proposed to improve dynamic performance and to save energy consumption by using differential evolution (DE). The total sum of cable forces which are related to energy consumption is the objective function. From our obtained results, it shows that the reduced percentage of the maximum cable tension between the optimized and fixed boundary cases is 10.5%.

Keywords: Cable-driven parallel robots · Finite element method · Vibration analysis · Tension distribution algorithm · Boundaries of cable forces · Frequency constraint · Differential evolution

1 Introduction

One of the considerable parallel robots is the cable-driven parallel robot (CDPRs). Over the past decade, this system has been noticed by researchers. In this type, rigid links are replaced by flexible cables. The end-effector is controlled by changing the cables' lengths. Due to using such flexible cables, the dynamic capability of CDPRs is more advanced than that of rigid parallel robots [1]. As a result, the main advantages of CDPRs are large-workspace, inexpensive cost, lightweight, flexible reconfiguration, and transportation [2].

K. U. Sattler et al. (Eds.): ICERA 2019, LNNS 104, pp. 35–46, 2020.
https://doi.org/10.1007/978-3-030-37497-6_4

However, the vibration of CDPRs is well known as one of the most critical problems because cables are highly capable of vibration in both axial and transversal directions due to the inevitably flexible characteristics of cables [1]. Thus, in this paper, the finite element method (FEM) which was reported in [3] is first applied to study the vibration of CDPRs. The relationship between the natural frequency and the values of boundaries used tension distribution is also investigated.

The vibration or stiffness is very critical for CDPRs because it affects the dynamic performance and position control accuracy. One of the possible methods to improve stiffness or dynamic performance is to apply the frequency constraint for CDPRs moving along a certain trajectory. This method can not only enhance its dynamic performance but also avoid resonance in some cases because the natural frequency is a good parameter to evaluate the stiffness and dynamic performance of structures.

For CDPRs, the natural frequency and stiffness of CDPRs depend on the applied cable tensions. The applied cable tension also depends on the boundaries of cable forces used in the algorithm of tension distribution, T_{min} and T_{max}. The higher the values of the boundaries of cable forces are, the bigger the values of the tension applied to cable elements [4, 5] and the bigger the values of the natural frequency of CDPRs are. However, the values of T_{min} and T_{max} in most of the algorithm of tension distribution are fixed [6–10] and they are mostly chosen as random values.

It should be noted that for CDPRs moving along a trajectory under frequency constraint, by increasing the fixed values of boundaries of cable forces, the energy consumption is not minimized [4]. Indeed, at each pose along a trajectory, the natural frequency of CDPRs is different, and by using such fixed values of boundaries of cable forces, the applied forces (the energy consumption) at some poses are bigger than those that are needed to make the natural frequency bigger than the frequency constraint.

Thus, from the previous discussion, to improve dynamic performance and to save energy consumption, this paper presents a novel determination of the boundaries of cable forces for a spatial cable-driven parallel robot moving along the trajectory under frequency constraint. The total sum of applied tensions to CDPRs related to energy consumption is considered as the objective function. The natural frequency of CDPR's structure is calculated by the finite element program. Then, Differential Evolution (DE) algorithm is used to get the optimal values of the boundaries of cable forces.

2 Dynamics of Cable-Driven Parallel Robots

This section presents dynamic equations of a general cable-driven parallel robot, as shown in Fig. 1.

The cable robot is driven by n cables. Each cable is attached to the base and the end effector at A_i and B_i, respectively. The ith cable and its length are denoted by the vector $\mathbf{l}_i$ and its norm. S_i stands for the unit vector of the cable i^{th}. According to the global frame O_O, the locations of the point A_i and the centroid of the end-effector (the point O_1) are denoted by vector $\mathbf{a}_i$ and $\mathbf{p}$, respectively. The reference frame O_1 is located at the centroid of the end-effector and the location of the point B_i is the vector $\mathbf{b}_i$ [10]. Then, the first kinematic equation for CDPRs is given as follows:

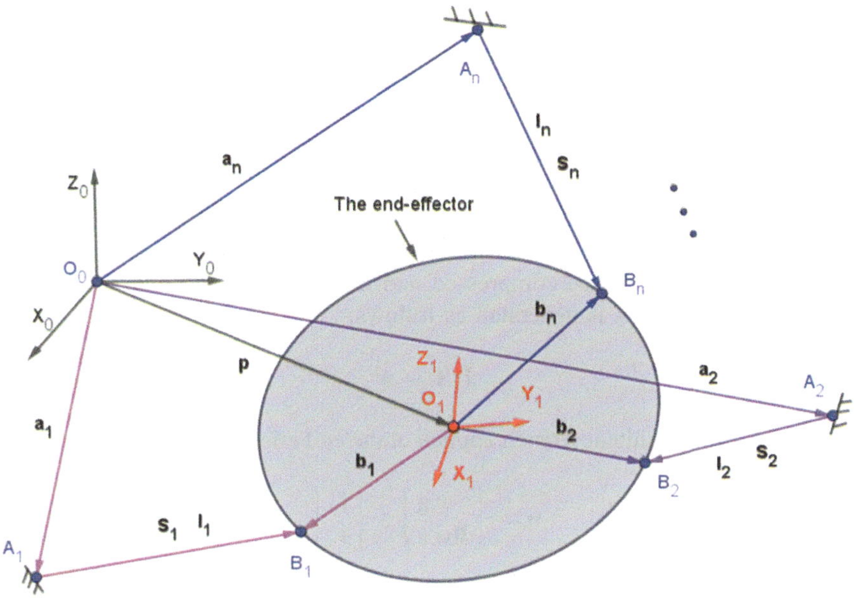

Fig. 1. A general scheme of cable robots [1]

$$\mathbf{p} = \mathbf{a}_{\mathrm{i}} + \mathbf{l}_{\mathrm{i}} - \mathbf{b}_{\mathrm{i}} \quad (i = 1, 2, \ldots, n) \tag{1}$$

According to (1), the other kinematic equations are given as follows:

$$\mathbf{l}_{\mathrm{i}}^2 = [\mathbf{p} - \mathbf{a}_{\mathrm{i}} + \mathbf{b}_{\mathrm{i}}]^{\mathrm{T}}[\mathbf{p} - \mathbf{a}_{\mathrm{i}} + \mathbf{b}_{\mathrm{i}}] \tag{2}$$

$$\dot{\mathbf{L}} = \mathbf{J}\mathbf{t} \tag{3}$$

$$\dot{\mathbf{L}} = [\,\dot{l}_1 \quad \dot{l}_2 \quad \ldots \quad \dot{l}_n\,]^T \tag{4}$$

$$\mathbf{J} = \begin{bmatrix} \mathbf{S}_1 & \mathbf{S}_2 & \ldots & \mathbf{S}_{\mathrm{n}} \\ \mathbf{b}_1 \times \mathbf{S}_1 & \mathbf{b}_2 \times \mathbf{S}_2 & \ldots & \mathbf{b}_{\mathrm{n}} \times \mathbf{S}_{\mathrm{n}} \end{bmatrix}^T \tag{5}$$

$$\mathbf{t} = \begin{bmatrix} \dot{\mathbf{p}}^T \\ \omega^T \end{bmatrix} \tag{6}$$

Where, $\dot{\mathbf{p}}$ and $\dot{\omega}$ denote for the linear velocity vector of the centroid and the angular velocity vector of the end-effector, respectively; $\mathbf{J}$ stands for the *n x 6* Jacobian matrix of CDPRs. Hence, the general dynamic equation of CDPRs is given as follows [1]:

$$\begin{bmatrix} m\mathbf{I}_{3\times3} & \mathbf{0}_{3\times3} \\ \mathbf{0}_{3\times3} & \mathbf{I}_P \end{bmatrix} \begin{bmatrix} \ddot{\mathbf{p}} \\ \dot{\boldsymbol{\omega}} \end{bmatrix} + \begin{bmatrix} \mathbf{0}_{3\times1} \\ \boldsymbol{\omega} \times \mathbf{I}_P \times \boldsymbol{\omega} \end{bmatrix} + \begin{bmatrix} -m\mathbf{g} \\ \mathbf{0}_{3\times1} \end{bmatrix} + \begin{bmatrix} \mathbf{f}_e \\ \boldsymbol{\tau}_e \end{bmatrix} = \mathbf{J}^T\mathbf{T} \tag{7}$$

Where, m denotes the mass of the end-effector; $\mathbf{I}_P$ denotes the inertia tensor of the end-effector about its centroid according to the global frame O_O; $\mathbf{I}_{3\times3}$ denotes the 3×3 identity matrix; $\mathbf{g}$ denotes the gravity acceleration vector; $\mathbf{f}_e$ and $\boldsymbol{\tau}_e$ denote the applied force vector and the applied moment vector of the end-effector, respectively; $\mathbf{T} = [T_1 \quad T_2 \quad \ldots \quad T_n]$ denotes the vector of cable forces. It should be noted that in CDPRs, each cable cannot be compressed and it is only pretensioned. Consequently, the vector of cable forces is generated as follows:

$$\mathbf{J}^T\mathbf{T} = \mathbf{w} \tag{8}$$

Where, $\mathbf{w}$ is the resultant wrench applied at the end effector and is given as follows:

$$\mathbf{w} = \begin{bmatrix} -m\mathbf{g} \\ \mathbf{0}_{3\times1} \end{bmatrix} + \begin{bmatrix} \mathbf{f}_e \\ \boldsymbol{\tau}_e \end{bmatrix} \tag{9}$$

To solve the Eq. (8), the tension distribution algorithm is needed. For more detailed information, the readers are encouraged to read the comprehensive book [6]. However, for the real-time capability of tension distributions in controlling of CDPRs, closed-form method is one of the possible methods which is briefly summarized as follows:

$$\mathbf{T} = \mathbf{T}_m - \mathbf{J}^{+T}\left(\mathbf{w} + \mathbf{J}^T\mathbf{T}_m\right) \tag{10}$$

$$\mathbf{T}_m = \frac{\mathbf{T}_{\min} + \mathbf{T}_{\max}}{2} \tag{11}$$

Where, $\mathbf{J}^{+T} = \mathbf{J}\left(\mathbf{J}^T\mathbf{J}\right)^{-1}$ is the Moore-Penrose generalized matrix.

3 Vibration Analysis

This section first presents a cable formulation to model CDPRs and then some numerical examples in terms of vibration analysis for a spatial CDPR are performed.

3.1 Cable Formulation

One of the most common formulations used for modeling of cables is an equivalent truss element. This formulation is firstly introduced by Ernst (1965), for one cable element, its equivalent modulus of elasticity is given as follows:

$$E_{eq} = \frac{E}{1 + \frac{(wL)^2 AE}{12T^3}} \tag{12}$$

Where, E_{eq} and E are the equivalent modulus of elasticity and Young's modulus of cables, respectively; L is the length of the cable, A is the cross-sectional area of the cable; w is the weight per unit length and T is the applied cable tension.

3.2 Numerical Examples

For vibration analysis, a spatial cable-driven parallel robot is studied in this section, as shown in Fig. 2.

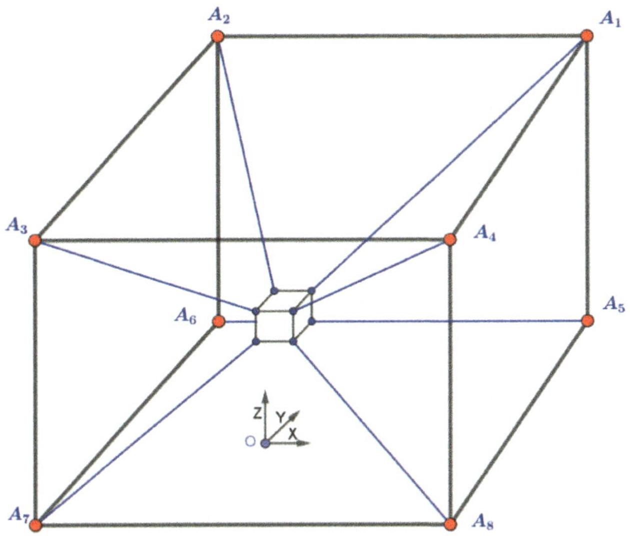

Fig. 2. A spatial cable-driven parallel robot.

This CDPR is driven by 8 cables. The ith drum is attached at point A_i. Data for the free vibration problem of the spatial CDPR are summarized in Table 1. Table 2 provides the natural frequencies of the first 10 modes of the spatial CDPR obtained by FEM algorithm and SAP2000. It shows that its natural frequencies are raised by

Table 1. Data for the free vibration problem of a spatial cable-driven parallel robot [4].

Parameters (unit)	Value	Parameters (unit)	Value
Mass of the end effector *(kg)*	25	The modulus of elasticity, $E(N/m^2)$	2.01×10^{10}
Dimensions of robot's frame *(m)*	$4 \times 4 \times 4$	The weight per unit length, γ *(N/m)*	0.251
The cross-sectional areas $A(m^2)$	50.265×10^{-6}	T_{min} *(N)* in Eq. (11)	[200, 240, 280, 320]
Position of the end-effector *x, y, z (m)*	[0; 0; 2]	T_{max} *(N)* in Eq. (11)	500

Table 2. Comparison of the fundamental natural frequency for the spatial cable-driven parallel robot.

T_{min}	200 N		240 N		280 N		320 N	
Mode	Natural frequency (Hz)		Natural frequency (Hz)		Natural frequency (Hz)		Natural frequency (Hz)	
	SAP2000	FEM	SAP2000	FEM	SAP2000	FEM	SAP2000	FEM
1	15.467	15.628	16.046	16.145	16.510	16.646	16.976	17.132
2	15.467	15.628	16.047	16.145	16.512	16.646	16.976	17.132
3	15.531	15.628	16.049	16.145	16.514	16.646	16.983	17.132
4	15.531	15.635	16.051	16.153	16.516	16.655	16.987	17.142
5	15.566	15.635	16.081	16.153	16.559	16.655	16.988	17.142
6	15.567	15.635	16.082	16.153	16.560	16.655	16.990	17.142
7	15.569	15.635	16.084	16.153	16.582	16.655	16.999	17.142
8	15.570	15.635	16.086	16.153	16.582	16.655	17.000	17.142
9	18.113	18.205	18.475	18.650	18.916	19.085	19.329	19.509
10	18.115	18.205	18.475	18.650	18.919	19.085	19.329	19.509
Error (%)		1.04%		0.62%		0.83%		0.92%

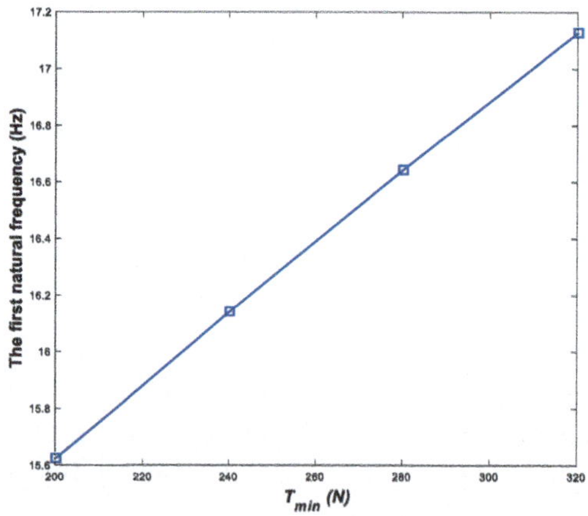

Fig. 3. The dependency of the 1st natural frequency on T_{min}.

increasing the values of T_{min}. The values of T_{min} rise from 200 N to 320 N, the 1st natural frequency rises from 15.628 Hz to 17.132 Hz, respectively, as shown in Fig. 3. In addition, it is apparent that the results gotten by FEM program are almost identical to those of SAP2000.

Figure 4 illustrates the 1st natural frequency of CDPRs along the circular trajectory with two different values of T_{min} and T_{max}. In the case of $T_{min} = 350$ N and $T_{max} = 750$ N, the stiffness of the CDPR at each pose along the trajectory is raised because the

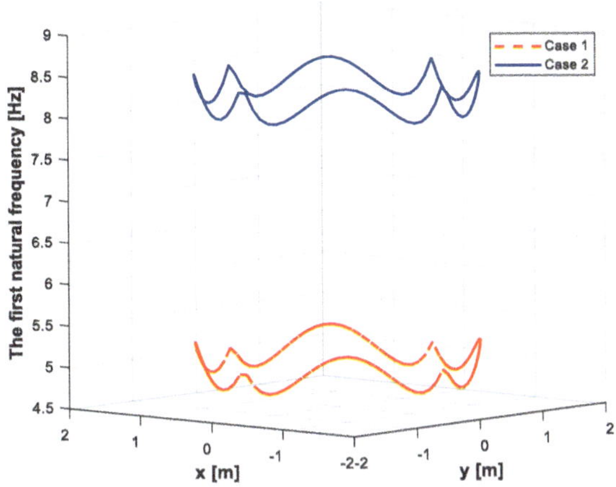

Fig. 4. The first natural frequency of CDPRs along the circle trajectory in Case 1 with T_{min} = *200* N and T_{max} = *500* N and in Case 2 with T_{min} = *400* N and T_{max} = *1000* N.

1st natural frequency is always higher than natural frequencies in the case of T_{min} = *100* N and T_{max} = *500* N.

4 Improving Dynamic Performance and Saving Energy Consumption of Cable-Driven Parallel Robots

4.1 Problem Statement

In case of CDPRs, to avoid the resonance when the external frequencies (ω_{limit}) of known external forces meet the natural frequencies of the CDPR's structures or to raise the stiffness, we can change the values of the boundaries (T_{min} and T_{max}) to make the natural frequency of CDPRs bigger than ω_{limit}. In addition, to save the energy consumption, the values of boundaries (T_{min} and T_{max}) of cable forces can be adaptively changed at each pose of the CDPRs. Hence, we choose the design variable: $x = [T_{min}, T_{max}]$. Then the objective function related to energy consumption and the constraint related to frequency constraint is given as follows:

$$\begin{aligned} &\text{Minimize}: f = \sum_{i=1}^{n} T_i \\ &\text{Subject to}: g_1(x) = \frac{\omega_1}{\omega_{\text{limit}}} - 1 \geq 0 \end{aligned} \tag{13}$$

Where, T_i is the ith cable force and n is the number of the cables; ω_1 and ω_{limit} are the first natural frequency and the external frequency, respectively.

Obviously, the Eq. (13) is a constrained function. So, to transform this constrained optimization function into a new unconstrained function, the technique in [11] is applied and the above equation can be rewritten as follows:

$$\text{Minimize} : f = \sum_{i=1}^{n} T_i + \alpha[\max(0, g_1)]^2 \tag{14}$$

$$g_1 = \begin{cases} 0 & \textit{if} \;\; \omega_1 \geq \omega_{\text{limit}} \\ g_1 & \textit{otherwise} \end{cases} \tag{15}$$

Where, α is the penalty constant and in this paper, this parameter is defined as 10^6. The flowchart of optimization is shown in Fig. 5.

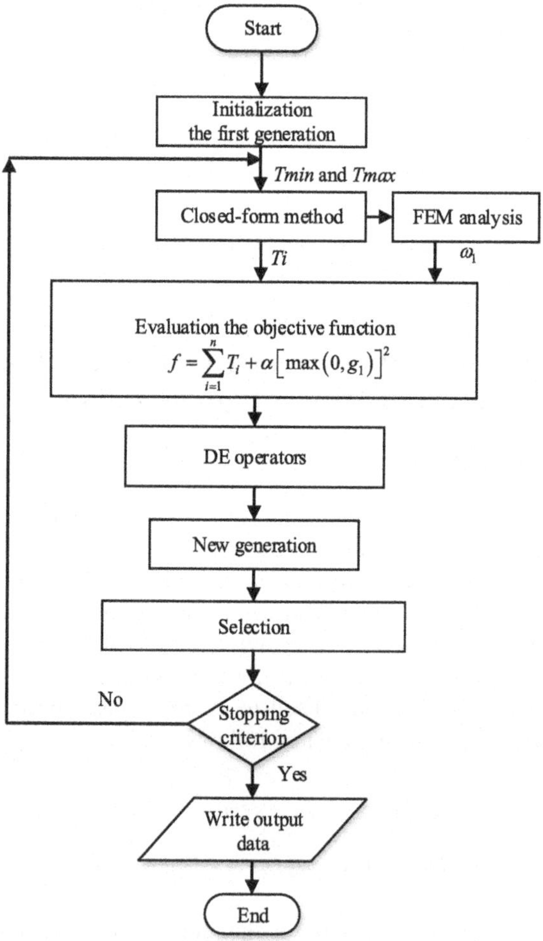

Fig. 5. Flowchart of the determination of boundaries of cable forces.

4.2 Numerical Example

This section presents optimization of boundaries of cable forces used in closed-form method for a spatial CDPR which moves along a circular trajectory. The circular trajectory for the spatial CDPR is given as follows:

$$x = r\cos\left(\frac{2\pi t}{T}\right); y = r\sin\left(\frac{2\pi t}{T}\right) \tag{16}$$

Where, r is the radius of trajectory, r = *1.5* m; t is traveling time, t = *[0, 10]*. The end-effector starts at the pose x_0 = *[1.5, 0]*. The step length is 0.01 s.

The objective is to minimize the energy consumption with frequency constraint by using Differential Evolution (DE). For DE, numbers of generations (GEN), populations (NP) and crossover-constant (CR) used in DE are 100, 50 and 0.5, respectively. The data for optimization is shown in Table 3.

Table 3. Data for optimization of boundaries of cable forces for the spatial CDPR.

Parameters (unit)	Value
Bounds for T_{min} *(N)* in Eq. (11)	[200, 400]
Bounds for T_{max} *(N)* in Eq. (11)	[500, 1000]
The frequency constraint *(Hz)*	6.5

The optimal results of the boundaries' cable forces are also compared with fixed boundaries' values, T_{min} = *350* N and T_{max} = *680* N, as shown in Fig. 6.

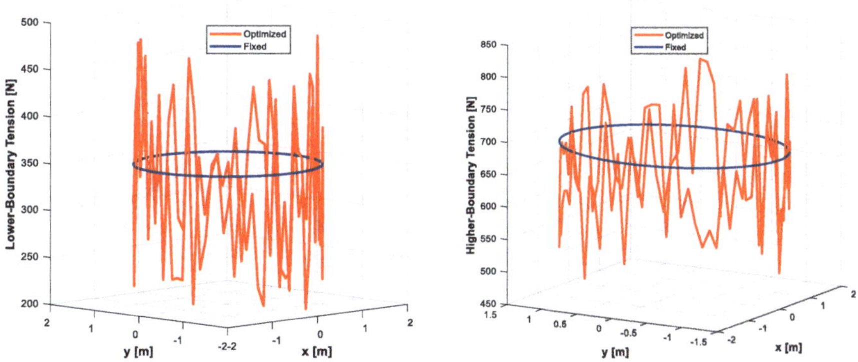

Fig. 6. The fixed and optimized boundaries along a circular trajectory.

Figure 7 shows that the 1st natural frequency obtained by optimized values of boundaries of cable forces do not violate the frequency constraints (6.5 Hz). It is apparent that as shown in Fig. 8 the cable tensions generated by the optimized values

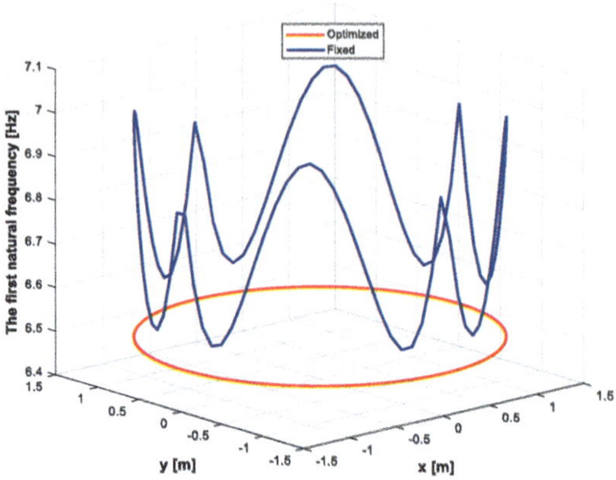

Fig. 7. The natural frequency obtained by fixed boundaries and optimized boundaries along the circular trajectory.

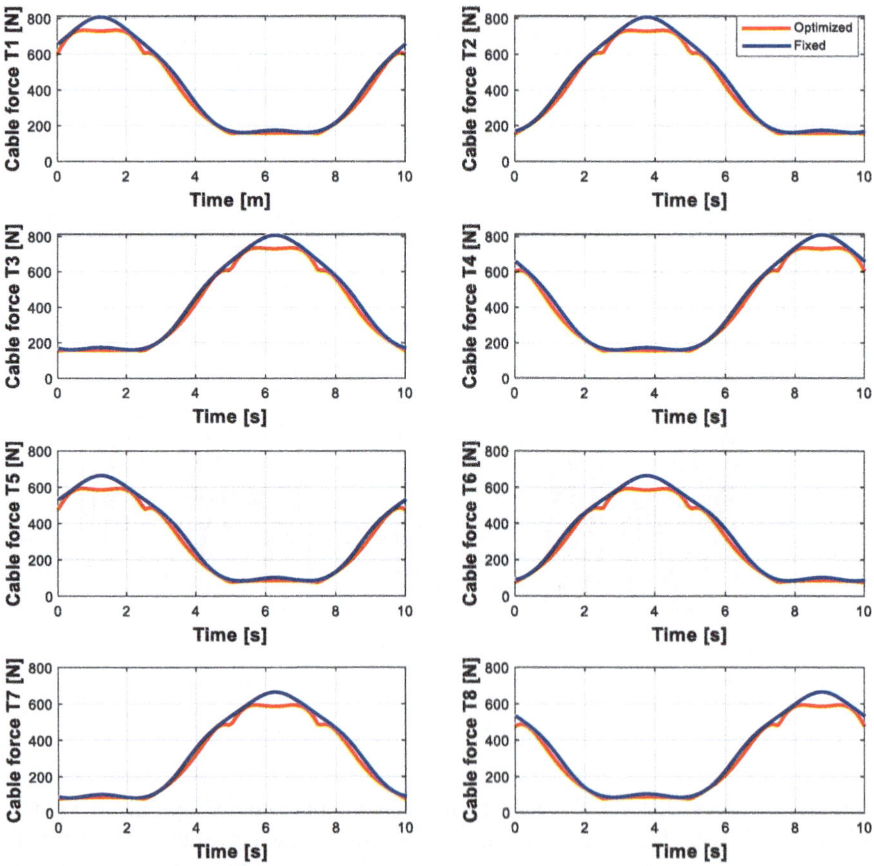

Fig. 8. The variation of cable forces according to the fixed and optimized boundaries tension with a circular trajectory for a spatial cable-driven robot.

of boundaries are always smaller than the cable tension calculated by the fixed values of boundaries.

In addition, Table 4 shows the reduced percentages of maximum tension between the optimized and fixed boundary cases. It shows that the maximum reduced percentage is 10.5%. So, the maximum tension of cables is directly related to energy consumption and the size of actuators. Hence, the optimized boundary method not only improves the dynamic performance of CDPRs but also saves the required energy and price of CDPRs.

Table 4. Comparison of maximum tension according to the fixed and optimized boundaries tension with a circular trajectory for a spatial cable-driven robot.

Cable	1st	2nd	3rd	4th	5th	6th	7th	8th	Average
Maximum cable force with optimized-boundaries (N)	740.0	739.9	739.9	740.0	599.8	599.8	599.8	599.8	669.9
Maximum cable force with fixed-boundaries (N)	812.6	812.8	812.8	812.6	669.9	670.1	670.1	669.9	741.3
Absolute difference (N)	72.6	72.9	72.9	72.6	70.1	70.3	70.3	70.1	71.5
Relative difference (%)	8.9%	9.0%	9.0%	8.9%	10.5%	10.5%	10.5%	10.5%	9.7%

5 Conclusion

In this paper, the vibration problem of a spatial cable robot is studied using a finite element program. Our results show the good accuracy of this method compared to those of SAP2000. The relationship between the natural frequency and the values of boundaries of cable forces used tension distribution were also investigated. It shows that the bigger the value of boundaries of cable forces is, the higher the value of the natural frequency of CDPRs is. Then, the optimization of boundaries of cable forces used in tension distribution algorithm with the frequency constraints by using Differential Evolution is also proposed. The applied tensions gotten by optimal boundaries are always smaller than those obtained by fixed boundaries at every pose along the trajectory. As a result, the energy and the size of actuators are also saved and the dynamic performance of CDPRs is also considered.

References

1. Diao, X., Ma, O.: Vibration analysis of cable-driven parallel manipulators. Multibody Syst. Dyn. **21**(4), 347–360 (2009)
2. Barnett, E., Gosselin, C.: Large-scale 3D printing with a cable-suspended robot. Addit. Manuf. **7**, 27–44 (2015)
3. Ernst, H.J.: Der E-modul von seilen unter berucksichtigung des durchhanges. Der Bauingenieur. **40**(2), 52–55 (1965)

4. Yuan, H., Courteille, E., Deblaise, D.: Static and dynamic stiffness analyses of cable-driven parallel robots with non-negligible cable mass and elasticity. Mech. Mach. Theory **85**, 64–81 (2015)
5. Heo, J.-M., et al.: Workspace and stability analysis of a 6-DOF cable-driven parallel robot using frequency-based variable constraints. J. Mech. Sci. Technol. **32**(3), 1345–1356 (2018)
6. Pott, A.: Cable-Driven Parallel Robots: Theory and Application. Springer, Cham (2018)
7. Dagalakis, N.G., Albus, J.S., Wang, B.-L., Unger, J., Lee, J.D.: Stiffness study of a parallel link robot crane for shipbuilding applications. J. Offshore Mech. Arct. Eng. **111**(3), 183–193 (1989)
8. Nguyen, D.Q., Gouttefarde, M.: Study of reconfigurable suspended cable-driven parallel robots for airplane maintenance. In: 2014 IEEE/RSJ International Conference on Intelligent Robots and Systems, pp. 1682–1689 (2014)
9. Izard, J.-B., et al.: Large-scale 3D printing with cable-driven parallel robots. Constr. Robot. **1**(1–4), 69–76 (2017)
10. Khosravi, M.A., Taghirad, H.D.: Robust PID control of fully-constrained cable driven parallel robots. Mechatronics **24**(2), 87–97 (2014)
11. Simon, D.: Evolutionary Optimization Algorithms. John Wiley & Sons, New York (2013)

A Prediction Method of Dynamic Cutting Forces and Machine-Tool Vibrations When Milling by Using Ball-End Mill Cutter

Nhu-Tung Nguyen[1(✉)], Yung-Chou Kao[2], Hoang Tien Dung[1], and Do Duc Trung[1]

[1] Hanoi University of Industry, Hanoi, Vietnam
tungnn@haui.edu.vn

[2] Advanced Institute of Manufacturing with Hi-Tech Innovations, National Chung Cheng University, Minxiong, Taiwan, R.O.C.

Abstract. In this paper, the cutting forces and machine-tool vibrations in ball-end mill process were modeled by using mathematical functions. All derivations of cutting forces are directly based on the tangential, radial, and axial cutting force components. The cutting force and vibration models can be formulated by a function of many parameters such as cutting force coefficients, cutter geometry, cutting conditions, and so on. The proposed cutting force model has been successfully verified by both simulation and experiment with very promising results. Besides, the machine-tool vibrations are also predicted for ball-end mill process.

Keywords: Ball-end mill · Cutting force · Cutting force prediction · Machine-tool vibration

Nomenclature

D	The diameter of cutter [mm]
R_0	The diameter of cutter [mm]
N_f	The number of flutes on the cutter
β	The helix angle on the cutter [deg]
ϕ_P	The cutter pitch angle [deg]
a	The full axial depth of cut [mm]
dz	The differential axial depth of cut [mm]
$\Psi(\phi_j(z))$	The lag angle at an axial depth of cut z [deg]
Ψ_0	The lag angle at maximum axial depth of cut $z = a$ [deg]
ϕ_j	The instantaneous immersion angle of flute number j, $(j = 1 \sim N_f)$ [deg]
$\phi_j(z)$	The instantaneous immersion angle of flute number j in z cutting depth, $(j = 1 \sim N_f)$ [deg]
$h_j(\phi_j(z))$	The instantaneous chip thickness at immersion angle ϕ_j [mm]
db	The chip width [mm]
$dS(\phi_j(z))$	The edge length of the cutting segment [mm]

K. U. Sattler et al. (Eds.): ICERA 2019, LNNS 104, pp. 47–54, 2020.
https://doi.org/10.1007/978-3-030-37497-6_5

$r(\phi_j(z))$	The radius of a circle on xy plane at an arbitrary point (P) on cutting edge [mm]
f_t	The feed per tooth [mm/tooth]
$\kappa(z)$	The axial immersion angle at z axial depth of cut
K_{tc}, K_{rc}, K_{ac}	The shearing force coefficient [N/mm²]
K_{te}, K_{re}, K_{ae}	The edge force coefficient [N/mm]
$dF_{i,j}(\phi, z)$	The differential cutting force [N]
$F_f(\phi), F_n(\phi), F_a(\phi)$	The cutting force in feed, normal, and axial direction [N]
t	The rotation time [Sec]
τ	The tool passing period [Sec]
$w_t(\phi_j)$	The dynamic displacement of current flute [mm]
$w_{(t-\tau)}(\phi_j)$	The dynamic displacement of previous flute [mm]
$x_t, y_t, x_{(t-\tau)}, y_{(t-\tau)}$	The machine tool vibrations at time t and $(t - \tau)$ [mm]
$[m_x, m_y]$	The mass matrix of machine-tool dynamic structure [kg]
$[c_x, c_y]$	The damping constant matrix of machine-tool dynamic structure
$[\omega_x, \omega_y]$	The natural frequency matrix of machine tool dynamic structure [Hz]
$[k_x, k_y]$	The stiffness matrix of machine-tool dynamic structure [N/m].

1 Introduction

There are two typical methods for the calibration of cutting force coefficients. The first method is the orthogonal to oblique cutting transformation method and the second method is the direct calibration method. In the first method, the cutting force coefficients were determined by using the shear angle, friction angle, and shear yield stress resulted from orthogonal cutting tests. By using this approach, in flat-end mill, the cutting force coefficients were determined using the oblique cutting model, the orthogonal model, and the cutting data [1, 2]. Besides, some researchers developed models to calculate cutting force coefficients that could be applied for ball-end milling using the data from orthogonal cutting tests [3, 4].

In the second method, the cutting force coefficients were estimated directly from milling tests for the specific cutter part combination. The cutting force coefficients were determined basing on the instantaneous uncut chip thickness, the cutting edge length, and the spindle speed [5]. By considering the instantaneous uncut chip thickness and the cutter rake angle simultaneously, Bayoumi et al. [6] successfully determined the cutting force coefficients. Larue and Anselmetti [7] determined the cutting force coefficients by using the measurement of cutter deflection. The cutting force coefficients were estimated by using the relationship between the instantaneous uncut chip thickness and the instantaneous cutting force [8]. Subrahmanyam et al. [9] used the corresponding maximum chip area and maximum measured force to estimate the cutting force coefficients. Besides, the cutting force coefficients were determined by using the measured average cutting force [10]. This model could be applied in ball-end mill processes [11].

In this study, the cutting force coefficients were determined through the linear force model of the ball-end mill. The main contributions of this study lie in three aspects: (1) Completing the method to determine the cutting force coefficients in ball-end mill, (2) modeling and verifying the cutting forces in ball-end mill processes, and (3) modeling and predicting the machine-tool vibrations in ball-end mill processes.

2 Mathematics of Dynamic Cutting Forces and Machine-Tool Vibrations

2.1 The Mathematics Cutting Force Model

In ball-end mill, the immersion is measured clockwise from the normal axis. Assuming that the bottom end of flute number one is designated as the reference immersion angle (ϕ_1) and the bottom end point of the remaining flute number j is at an angle (ϕ_j), then ϕ_j can be expressed as in Eq. (1), [11].

$$\phi_j = \phi_1 - (j - 1)\phi_P, j = 1 \sim N_f \tag{1}$$

where ϕ_P is the cutter pitch angle that is the lag angle from the flute number j to the flute number j + 1.

$$\phi_P = \frac{2\pi}{N_f} \tag{2}$$

When considering the cutter's helix angle, the radial lag angle Ψ at each axial depth of cut z can be expressed in Eq. (3).

$$\Psi(\phi_j(z)) = \frac{2\tan\beta}{D} z = \frac{\tan\beta}{R_0} z \tag{3}$$

The maximum radial lag angle Ψ_0 is determined at maximum depth of cut a, and can be determined by Eq. (4)

$$\Psi_0 = \frac{\tan\beta}{R_0} a \tag{4}$$

For flute number j, at an axial depth of cut z, the immersion angle is $\phi_j(z)$. It can be expressed by Eq. (5).

$$\phi_j(z) = \phi_j - \Psi(\phi_j(z)) = \phi_j - \frac{\tan\beta}{R_0} z \tag{5}$$

If zero nose radius of the cutter is assumed, the tangential, radial, and axial forces acting on a differential flute element can be expressed as in Eq. (6).

$$\begin{cases} dF_{t,j}(\phi_j(z)) = K_{tc} * h_j(\phi_j(z)) * db + K_{te} * dS(\phi_j(z)) \\ dF_{r,j}(\phi_j(z)) = K_{rc} * h_j(\phi_j(z)) * db + K_{re} * dS(\phi_j(z)) \\ dF_{a,j}(\phi_j(z)) = K_{ac} * h_j(\phi_j(z)) * db + K_{ae} * dS(\phi_j(z)) \end{cases} \tag{6}$$

where the instantaneous chip thickness $h_j(\phi_j(z))$ is determined by Eq. (7).

$$h_j(\phi_j(z)) = f_t \sin(\phi_j(z)) \sin(\kappa(z)) \tag{7}$$

So, in the determination of the total cutting force, the differential cutting forces are integrated analytically along the in-cut portion of the flute j, Eq. (8).

$$F_q(\phi_j) = \int_{z_1(\phi_j)}^{z_2(\phi_j)} dF_q(\phi_j(z)), \quad q = f, n, a \tag{8}$$

Considering the case that has more than one tooth executing the cutting processes simultaneously, the total cutting forces on the feed, normal, and axial direction can be determined by Eq. (9).

$$F_f(\phi) = \sum_{j=1}^{N_f} F_{f,j}(\phi_j), \quad F_n(\phi) = \sum_{j=1}^{N_f} F_{n,j}(\phi_j), \quad F_a(\phi) = \sum_{j=1}^{N_f} F_{a,j}(\phi_j) \tag{9}$$

2.2 The Machine-Tool Vibration Model

The vibrations in x and y directions were calculated by Eq. (10).

$$\begin{cases} [m_x]\{\ddot{x}(t)\} + [c_x]\{\dot{x}(t)\} + [k_x]\{x\} = \{F_x(t)\} \\ [m_y]\{\ddot{y}(t)\} + [c_y]\{\dot{y}(t)\} + [k_y]\{y\} = \{F_y(t)\} \end{cases} \tag{10}$$

The machine tool dynamic chip thickness was defined as in Fig. 1 and can be calculated by Eq. (11)

$$h_d(\phi_j(z)) = w_t(\phi_j) - w_{(t-\tau)}(\phi_j) \tag{11}$$

Where $w_t(\phi_j)$ and $w_{(t-\tau)}(\phi_j)$ are the dynamic displacement at rotation angle $\phi_j(z)$ of current flute and previous flute, respectively, and can be calculated by Eq. (12)

$$\begin{cases} w_t(\phi_j(z)) = x_t \sin(\phi_j(z)) + y_t \cos(\phi_j(z)) \\ w_{(t-\tau)}(\phi_j(z)) = x_{(t-\tau)} \sin(\phi_j(z)) + y_{(t-\tau)} \cos(\phi_j(z)) \end{cases} \tag{12}$$

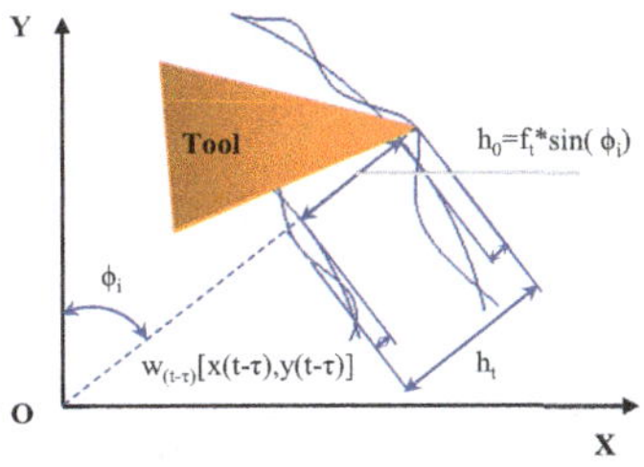

Fig. 1. The dynamic chip thickness

3 Experimental Method

3.1 Workpiece, Tool, and CNC Machine

In experimental researches, the cutter and workpiece were chosen as follows. Cutter: a HSS-Co ball-end mill with number of flutes $N_f = 2$, helix angle $\beta = 30^0$, rake angle $\alpha_r = 5^0$, and the diameter was 8 mm. The workpiece material was grey cast iron FC25. The compositions of grey cast iron FC25 are listed in Table 1 and the properties of the FC25 were the following: hardness 197–269 HB, Young's modulus = 109.5 GPa.

Table 1. Chemical compositions of grey cast iron FC25

Composite (%)						
	C	Mn	Si	P	S	Fe
Min	2.44	0.39	1.83	0.15	–	
Max	3.02	0.52	2.03	0.30	0.15	Balance

The experiments were performed at the Tongtai TWV-720A CNC machine.

3.2 Setup for Determination of Machine-Tool Dynamic Structure

In order to determine the machine-tool dynamic structure, an integrated device system that consisted of the acceleration sensor (ENDEVCO-25B-10668), hammer (KISTLER-9722A2000), signal processing box (NI 9234), and CUTPRO™ software was used. The detail setting of the measurement experiment is illustrated in Fig. 2.

a. Tool b. Acceleration sensor c. Force sensor
d. Signal processing box e. PC and software

Fig. 2. Setting of frequency response function (FRF) measurement

3.3 Setup and Measurement of Cutting Forces

A dynamometer (type XYZ FORCE SENSOR, model 624-120-5KN), processing system, and a PC were used to measure cutting forces. The detail is illustrated in Fig. 3.

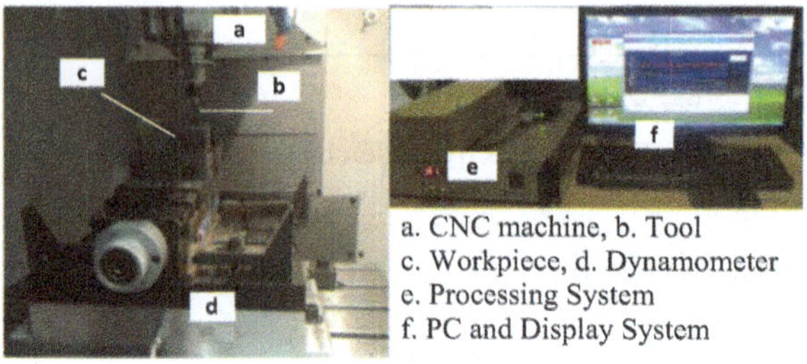

Fig. 3. Setup measurement of cutting force setting

At each depth of cut, full immersion (slotting) experiments were repeated in every feed per flute. The spindle speed was held constant at each experiment.

4 Experimental Results and Discussions

4.1 Determination of Machine Tool Dynamic Structure

Determination of the machine tool dynamic structure is very important in simulation of dynamic cutting forces and other machining characteristics. By using the measured results of the frequency response function (FRF) of the machine tool dynamic system, the machine tool dynamic structure was analyzed by CUTPRO^{TM} software. The parameters of the machine-tool dynamic structure were analyzed and listed in Table 2.

Table 2. Machine tool dynamic structure parameters

Direction	Mode No	Natural frequency [Hz]	Damping ratio [%]	Modal stiffness [N/m]	Mass [kg]
X	1	842.9752	0.898	7.8919E+08	28.1316
	2	1591.7397	5.067	4.8211E+07	0.4820
	3	2790.6410	3.940	1.2730E+07	0.0414
Y	1	991.8001	2.641	1.8614E+08	4.7933
	2	1514.4826	6.180	4.0318E+07	0.4453
	3	2785.8612	6.635	1.3818E+07	0.0451

4.2 Verification of the Dynamic Cutting Force Model

Using the calculated cutting force coefficients in Table 3 [11], the cutting forces were predicted and compared with the measured results as shown in Fig. 4.

Table 3. Prediction and comparison of cutting force coefficients

Shearing force coefficient [N/mm²]			Edge force coefficient [N/mm]		
K_{tc}	K_{rc}	K_{ac}	K_{te}	K_{re}	K_{ae}
3304.12	2148.90	725.07	67.58	5.25	−7.16

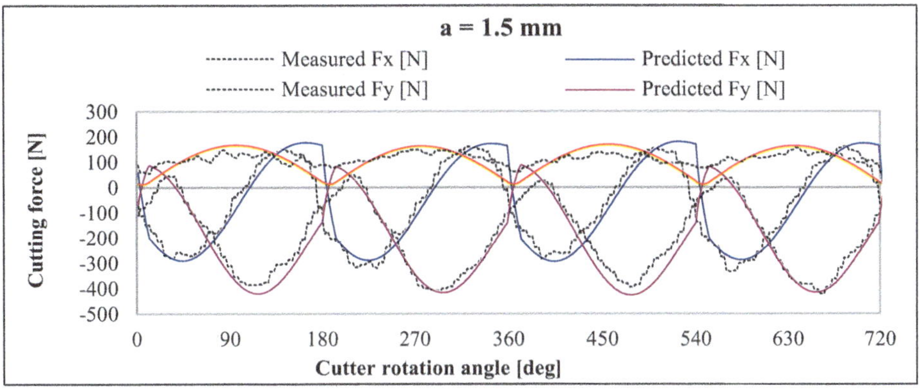

Fig. 4. Comparison between measured and predicted forces

The research results showed that those are the small deference about the shape and the amplitude of measured forces and predicted forces. The predicted results of research model quite are close to the experimental results. Therefore, the predicted results from research model agree satisfactorily with experimental results.

The reasons for the above differences were mostly originated from the noise, the deflection, the inconstancy of cutting depth, the temperature, the friction, and so on.

4.3 Prediction of Machine-Tool Vibrations

The predicted results of machine-tool vibrations were described in Fig. 5.

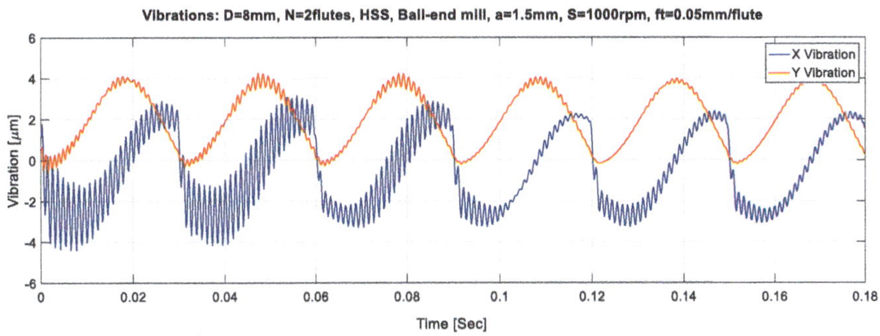

Fig. 5. The prediction of machine-tool vibrations

In order to predict the machine-tool vibrations, many input parameters were used such as tool geometry, tool material, the workpiece properties, and the cutting conditions (depth of cut, spindle speed, and so on). The cutting force coefficients and the parameters of the machine-tool dynamic structure are also used to predict the cutting forces and machine-tool vibrations.

5 Conclusions

The conclusions of this study can be drawn as follows.

By mathematical derivation, the cutting force and vibration models can be formulated by a function of many parameters such as cutting force coefficients, cutter geometry, cutting conditions, and so on.

The cutting force models have been successfully verified by both simulation and experiment with very promising results. Those proposed models could be applied to predict the cutting forces and vibrations when milling by using the ball-end mill cutter.

References

1. Altintas, Y.: Manufacturing Automation: Metal Cutting Mechanics, Machine Tool Vibrations, and CNC Design, 2nd edn. Cambridge University Press, New York (2012). ISBN 978-1-00148-0
2. Sonawane, H.A., Joshi, S.S.: Analytical modeling of chip geometry and cutting forces in helical ball end milling of superalloy Inconel 718. CIRP J. Manuf. Sci. Technol. **3**, 204–217 (2010)
3. Lee, P., Altintas, Y.: Prediction of ball-end milling forces from orthogonal cutting data. Int. J. Mach. Tools Manuf. **36**, 1059–1072 (1996)
4. Budak, E., Ozturk, E., Tunc, L.T.: Modeling and simulation of 5-axis milling processes. CIRP Ann. Manuf. Technol. **58**(1), 347–350 (2009)
5. Cheng, P.J., Tsay, J.T., Lin, S.C.: A study on instantaneous cutting force coefficients in face milling. Int. J. Mach. Tools Manuf. **37**, 1393–1408 (1997)
6. Bayoumi, A.E., Yucesan, G., Kendall, L.A.: An analytic mechanistic cutting force model for milling operations: a theory and methodology. Trans. ASME J. Eng. Ind. **116**, 324–330 (1994)
7. Larue, A., Anselmetti, B.: Deviation of a machined surface in flank milling. Int. J. Mach. Tools Manuf. **43**, 129–138 (2003)
8. Ko, J.H., Cho, D.W.: Determination of cutting-condition-independent coefficients and runout parameters in ball-end milling. Int. J. Adv. Manuf. Technol. **26**, 1211–1221 (2005)
9. Subrahmanyam, K.V.R., San, W.Y., Soon, H.G., Sheng, H.: Cutting force prediction for ball nose milling of inclined surface. Int. J. Adv. Manuf. Technol. **48**, 23–32 (2010)
10. Kao, Y.C., Nguyen, N.T., Chen, M.S., Su, S.T.: A prediction method of cutting force coefficients with helix angle of flat-end cutter and its application in a virtual three-axis milling simulation system. Int. J. Adv. Manuf. Technol. **77**(9–12), 1793–1809 (2015)
11. Kao, Y.C., Nguyen, N.T., Chen, M.S., Huang, S.C.: A combination method of the theory and experiment in determination of cutting force coefficients in ball-end mill processes. J. Comput. Des. Eng. **2**(4), 233–247 (2015)

A Solution to Adjust Kinetic of Industrial Robots Based on Alternative Trajectories

Thang Nguyen Huu, Khanh Duong Quoc, Thuy Le Thi Thu(✉), and Long Pham Thanh

Thai Nguyen University of Technology, Thai Nguyen, Vietnam
hanthuyngoc@tnut.edu.vn

Abstract. Whether new or used, industrial robots still have kinetic errors for many reasons such as quality of manufacture, assembly, wear, elastic deformation … and these errors are also changed by time. Feedback control only aims to make the generalized variable close to the theoretical calculation from the kinetic model so it is impossible to solve many of the above reasons, for example errors due to elastic deformation or wear owing to the out-of control loop. It is necessary to calibrate the robot by measuring the end manipulator error. Robots are mechatronic products so it is feasible to interfere with software to adjust hardware. This paper introduces an algorithm to build an alternate trajectory of the desired trajectory for the purpose of controlling the robot to follow this trajectory, which will move closest to the desired trajectory. This approach has the advantage of not interfering with the robot's hardware and control system structure, it also allows the quality of initial robot manufacturing and assembly to be moderate to keep prices reasonable. When kinetic errors increase over time, we only need to rebuild the trajectory instead. The calculations illustrated here show that this is a promising process when applied because of its effectiveness.

Keywords: Robot · Kinetic error · Alternative trajectory · Mechatronics · Manufacturing

1 Introduction

Industrial robot is a servo system, it maintains reverse contact between output and input within each link. That means the feedback signal of the link is taken right above that link, not from the final link. It is for this reason that some of the parameters to be monitored are still outside the control loop, such as the free movement of the center of rotation, tolerance of the DH dimensions (DH dimensions are the nominal dimensions, without tolerance), the elastic deformation of the link … consequently, even if the feedback work as designed, the final link still does not completely move as the calculated trajectory.

Some recent studies have tried to model the manipulator errors and built algorithms to correct the error. Yu and Xi [1] presented a self-correction method based on measuring the center of 04 spherical correction targets placed around the test system. Kinetic correction model is built based on Modified Denavit – Hartenberg (MDH)

K.-U. Sattler et al. (Eds.): ICERA 2019, LNNS 104, pp. 55–65, 2020.
https://doi.org/10.1007/978-3-030-37497-6_6

model. The MDH method and Hayati convention are also used by [2] to correct kinetics and enhance the accuracy of parallel 3-PRRU manipulators through the influence of errors of universal joints. Zhang et al. [3] used genetic algorithms to build mathematical equations to compensate kinetic errors for 3-order cylindrical robot coordinates. Ma et al. [4] has modeled and corrected the kinetic errors of high order passive joints. The process was performed using Chebyshev higher polynomial to represent individual errors and the laser tracker system was used to get the measured data. Laser tracker system is widely used in studies of kinetic correction of series and parallel robots [5, 6].

In addition, some studies use non-model methods to adjust robot errors because of the simplicity of the method. The authors have conducted error correction on the robot arm using interpolation techniques. Liu et al. [7] proposed a trajectory planning technique to minimize the overall error of end-effector of industrial robots. Kinetic and kinematic models were built using screw theory and Kane equations, and based on that, the final phase error was modeled by considering the effect of interpolation algorithms and flexibility of all joints. The PSO algorithm is applied to find the smallest error according to kinetic and kinematic constraints. In [8], Ying Bai et al. used Interval Type-2 Fuzzy Error Interpolation method (IT2FEI) to compensate for robot calibration accuracy in 3D workspaces. Meanwhile, Borrmann and Wollnack [9] used a laser tracker to measure the position and direction of the linear axis. B-Spline interpolation is used to model the external axis, allowing predicting the sliding axis posture. Bai et al. [10, 11] presented an online dimming interpolation method to correct the accuracy of Stewart Platform robots.

On the basis of kinetic adjustment, in this paper we introduce a method to bring the correction of the end-effector error into the control loop. This process basically includes steps to measure errors at predetermined points, builds mathematical models of errors in regression form and alternative trajectories so that when controlled according to these trajectories, the robot will move closest to the desired trajectories.

1.1 Measure Errors at Predetermined Points

The position and direction of the end-effectors when testing with this device is a general error of elastic fabrication, assembly, wear or deformation …

To accurately measure the position and direction of the end-effectors, the measurement diagram described in Figs. 1 and 2 is preferred.

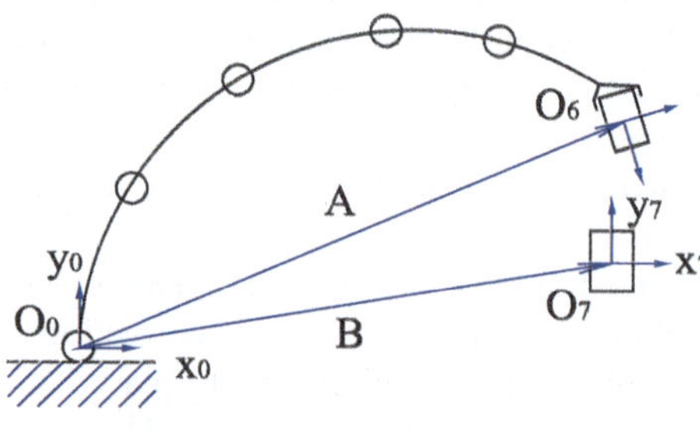

Fig. 1. Relationship of robot, sample and proximity sensor system

On Fig. 1, the sample block attached to the O_6 reference frame and clamped by the robot arm is a highly accurate cube. The position and direction of this block is described by the matrix A in the O_0 reference frame. The components in A are provided by the measuring sensors. The proximity sensor system consists of 6 units arranged on three adjacent faces of the sample block and described as Fig. 2.

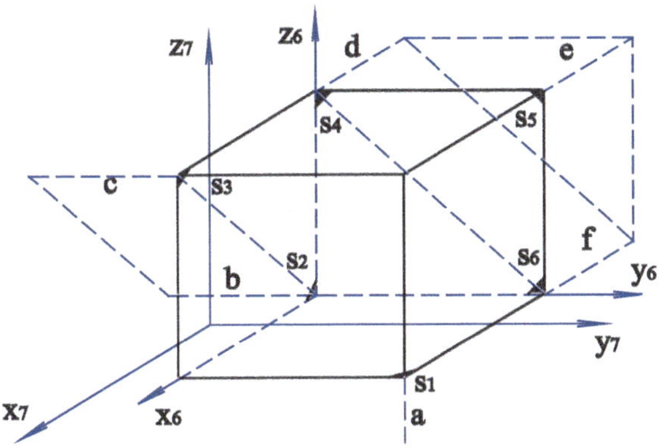

Fig. 2. Proximity sensor system

The barrier system for measuring is planes x_7O_{z7}, x_7O_{y7}, y_7Oz_7. They can be teleported to any location in the robot's working space thanks to an accurate numerically controlled drive system by ball screw parallel to the x_7, y_7, z_7 axes.

The coordinates of the center of the table are calculated according to two factors:

- Coordinates of the sliding table in three axes $x_7y_7z_7$;
- Measurement results from 6 post-processing proximity sensors.

When operating the measuring system, the proximity sensors s_1 - s_6 provide the distance values, denoted by (a, b, c … f) respectively. In particular, note that the sensor s1 measures the distance from the junction to the plane x_7O_{y7}, the sensors s_2, s_3 measure the distance from the junction to the plane x_7O_{z7}, the three sensors s_4, s_5, and s_6 measure the distance from the junction to the plane y_7Oz_7.

After the robot takes the sample block to the desired point, the sensor system is then defined by the sliding table as vector B on demand, the coordinates (a, b … .f) are recorded immediately compared to O_7 reference. The calculation of vector A from the measurement results of 6 sensors will perform as follows:

The coordinates of s_4, s_5, s_6 are used to calculate 2 vectors $\overrightarrow{s_4s_5}$, $\overrightarrow{s_5s_6}$, with $\overrightarrow{n_1} = \overrightarrow{s_4s_5} \wedge \overrightarrow{s_5s_6}$ are normal vectors. Equation P1 of this plane is written based on $\overrightarrow{n_1}$ and passing through s_4:

$$P_1 = f(\overrightarrow{\mathrm{n}_1}, \mathrm{s}_4) \quad (1)$$

$\overrightarrow{u_2} = \overrightarrow{s_2s_3}$ is direction vector of the plane x_6O_{z6}. $\overrightarrow{n_1}$ is another direction vector of the plane, so normal vector $\overrightarrow{n_2}$ of x_6O_{z6} is:

$$\overrightarrow{n_2} = \overrightarrow{u_2} \wedge \overrightarrow{n_1} \quad (2)$$

Equation P_2 of x_6O_{z6}:

$$P_2 = f(\overrightarrow{n_2}, s_3) \quad (3)$$

x_6Oy_6 has $\overrightarrow{n_1}$ and $\overrightarrow{n_2}$ as its direction vectors, the normal vector $\overrightarrow{n_3}$ is:

$$\overrightarrow{n_3} = \overrightarrow{n_1} \wedge \overrightarrow{n_2} \quad (4)$$

Equation of x_6Oy_6:

$$P_3 = f(\overrightarrow{n_3}, s_1) \quad (5)$$

The position of a sample block or the position of the arm center is determined by the intersection of three planes:

$$P = (P_1) \cap (P_2) \cap (P_3) \quad (6)$$

This vertex of the sample block, size of the cube and the way to fit the sample block in the arm will deduce the center of the arm is the managed point.

The direction of the forearm axis is determined by three vectors:

$$\overrightarrow{u} = (\overrightarrow{n_1}, \overrightarrow{n_2}, \overrightarrow{n_3}) \quad (7)$$

Comparing the values received from (6, 7) of the motor A with the vector B introduced into the control of the previous slide to get the difference between them. It is possible to determine the error that exists at each point in the robot workspace when correcting the robot error.

2 Build the Error Model

The kinetic error is proved to be interpolation due to its continuity. Because the cost of direct measurement of errors is quite large, it is only measured for sampling purposes, interpolation is to have error data at any point in the working area at a smaller cost.

Suppose the error in a survey point p_i consists of 6 components as follows:

$$\delta p_i = (\delta_x^i, \delta_y^i, \delta_z^i, \delta_{\theta x}^i, \delta_{\theta y}^i, \delta_{\theta z}^i) \quad (8)$$

Consider the triangular prism survey space as shown in Fig. 3. The error measured at the prisms of each prism gets all 6 components as (8).

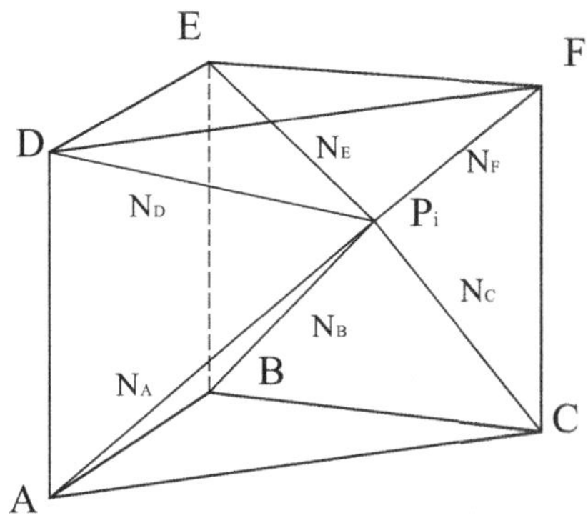

Fig. 3. Effect of sample points to the survey points

At any p_i in the triangular prism:

$$\begin{cases} \delta_x^{(i)} = N_A.\delta_x^{(1)} + N_B.\delta_x^{(2)} + \ldots + N_F.\delta_x^{(6)} \\ \ldots \\ \delta_{\theta z}^{(i)} = N_A.\delta_{\theta z}^{(1)} + N_B.\delta_{\theta z}^{(2)} + \ldots + N_F.\delta_{\theta z}^{(6)} \end{cases} \tag{9}$$

Stop factor derived from (9):

$$\begin{bmatrix} N_A \\ \ldots \\ N_F \end{bmatrix}_{(i)} = \begin{bmatrix} \delta_x^{(1)} & \ldots & \delta_x^{(6)} \\ \ldots & \ldots & \ldots \\ \delta_{\theta z}^{(1)} & \ldots & \delta_{\theta z}^{(6)} \end{bmatrix}^{-1} \cdot \begin{bmatrix} \delta_x^{(i)} \\ \ldots \\ \delta_{\theta z}^{(i)} \end{bmatrix} \tag{10}$$

In which, the right side of the square matrix is the error value of the A-F vertices of the prism. The column matrix $\left[\delta_x^{(i)} \ldots \delta_{\theta_z}^{(i)}\right]^T$ is also the error results at p_i provided by sensors. After n survey points, the results is a set of n stop value as (11):

$$\begin{bmatrix} N_A \\ \ldots \\ N_F \end{bmatrix}_{(1)} ; \ldots ; \begin{bmatrix} N_A \\ \ldots \\ N_F \end{bmatrix}_{(i)} ; \ldots ; \begin{bmatrix} N_A \\ \ldots \\ N_F \end{bmatrix}_{(n)} \tag{11}$$

Thus, experimental shape function at the kth top is determined by regression as follows:

$$(\mathrm{N}_k^{(1)}, \ldots, \mathrm{N}_k^{(i)}, \ldots, \mathrm{N}_k^{(n)}) \Rightarrow \mathrm{f}_k(x, y, z) \tag{12}$$

3 Build Alternative Trajectories

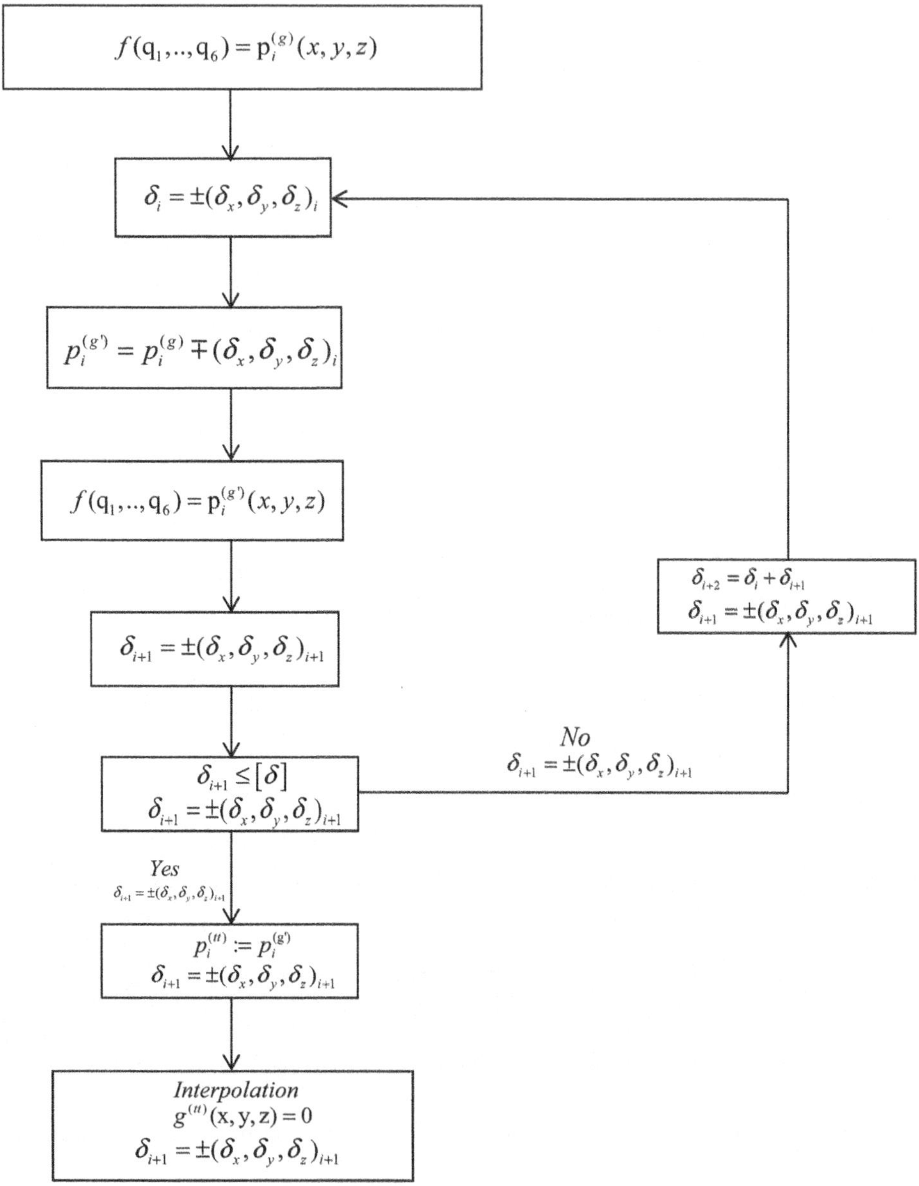

Fig. 4. Alternative trajectory diagram

Given that $g(x,y,z) = 0$ is the desired trajectory of the end-effector and $f(q_1,\ldots,q_6) = 0$ is the kinetic equation of the robot. When controlling the robot as this trajectory, under the influence of the general errors, the actual trajectory received is g'(x, y, z) = 0. Under a detailed perspective, a point $p_i^{(g)}(x, y, z)$ has been turned into a point $p_i^{(g')}(x, y, z)$, this is presented in (12) of II. It is necessary to point out a point $p_i^{(tt)}(x, y, z)$ that is when the robot is driven there, the existing errors will make it approach $p_i^{(g)}(x, y, z)$, that means the job will be completed if the locus of alternative points $p_i^{(tt)}(x, y, z)$ $\quad i = 1 \div m$ with m the key point. Completing the trajectory is then done by tertiary polynomial interpolation of these key points.

The process to find $p_i^{(tt)}(x, y, z)$ and $g^{(tt)}(x, y, z) = 0$, is described in Fig. 4.

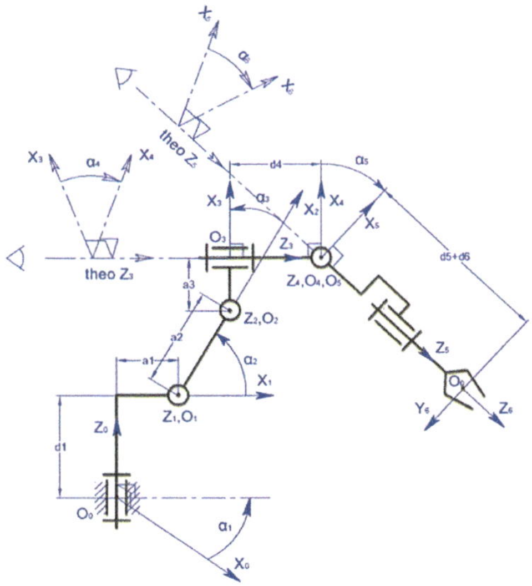

Joint	R_z	T_z	T_x	R_x
1	(α_1)	d_1	a_1	90^0
2	(α_2)	0	a_2	0
3	(α_3)	0	a_3	90^0
4	(α_4)	d_4	0	-90^0
5	(α_5)	0	0	90^0
6	(α_6)	d_5+d_6	0	0

Fig. 5. Survey joints and DH dimensions

Case Study

Basic dimensions (Fig. 5):

$$d_1 = 335,\ a_1 = 75,\ a_2 = 270,\ a_3 = 90,\ d_4 = 295,\ d_5 + d_6 = 80\,\text{mm}$$

Considering the 10-point trajectory as shown in Table 1, the actual measurements shown in Sect. 1 show that the error radius results in the range of 0.5343–0.6565 (mm);

Run the program described in Fig. 4 to get the required alternative points as shown in Table 2.

Table 1. Trajectory errors at key points

No.	Desired points			Actual points			Error			Error radius
	X	Y	Z	X2	Y2	Z2	Δx	Δy	Δz	
1	300	175	300	299.8885	174.9349	300.5185	−0.1115	−0.0651	0.5185	0.5343
2	200	275	300	199.9372	274.9140	300.5261	−0.0628	−0.0860	0.5261	0.5368
3	280	255	220	279.7621	254.7834	220.5723	−0.2379	−0.2166	0.5723	0.6565
4	260	235	220	259.8248	234.8417	220.5850	−0.1752	−0.1583	0.5850	0.6309
5	280	215	230	279.8184	214.8606	230.5742	−0.1816	−0.1394	0.5742	0.6181
6	240	255	240	239.8586	254.8499	240.5664	−0.1414	−0.1501	0.5664	0.6027
7	260	255	250	259.8315	254.8348	250.5485	−0.1685	−0.1652	0.5485	0.5972
8	280	255	250	279.7918	254.8104	250.5388	−0.2082	−0.1896	0.5388	0.6079
9	280	255	270	279.8091	254.8262	270.5175	−0.1909	−0.1738	0.5175	0.5783
10	260	235	270	259.8744	234.8865	270.5404	−0.1256	−0.1135	0.5404	0.5663

Table 2. Alternative points

No.	Alternative points			Actual points			Error			Error radius
	X1	Y1	Z1	X2	Y2	Z2	Δx	Δy	Δz	
1	300.1122	175.0656	299.4812	299.9993	174.9996	300.0003	−0.0007	−0.0004	0.0003	0.0009
2	200.0632	275.0867	299.4736	199.9995	274.9993	300.0003	−0.0005	−0.0007	0.0003	0.0009
3	280.2393	255.2178	219.4272	279.9987	254.9988	220.0005	−0.0013	−0.0012	0.0005	0.0019
4	260.1764	235.1593	219.4145	259.9988	234.9989	220.0005	−0.0012	−0.0011	0.0005	0.0017
5	280.1828	215.1404	229.4254	279.9988	214.9991	230.0004	−0.0012	−0.0009	0.0004	0.0016
6	240.1423	255.1511	239.4332	239.9990	254.9990	240.0004	−0.0010	−0.0010	0.0004	0.0015
7	260.1696	255.1662	249.4511	260.1685	255.1652	249.4515	0.1685	0.1652	−0.5485	0.0015
8	280.2093	255.1906	249.4608	279.9988	254.9990	250.0004	−0.0012	−0.0010	0.0004	0.0016
9	280.1919	255.1747	269.4822	279.9990	254.9991	270.0003	−0.0010	−0.0009	0.0003	0.0014
10	260.1265	235.1143	269.4593	259.9992	234.9992	270.0003	−0.0008	−0.0008	0.0003	0.0012

Conduct interpolation for 60 intermediate points based on the shape function presented in Sect. 2. Suppose the desired point P(x, y, z); actual control point P1(x_1, y_1, z_1) and (Δx, Δy, Δz) are errors between these 2 points. Relationship between P và P1 are:

$$\begin{cases} x_1 = x_0 + \Delta x \\ y_1 = y_0 + \Delta y \\ z_1 = z_0 + \Delta z \end{cases} \tag{13}$$

Use Minitab software to develop general rules for errors with the following results:

Δx = −0.697511 + 0.00356157 * + 0.00214686 Y − 0.00272697 Z − 5.66489e-006 XY + 7.85584e-006 YZ + 3.93517e-006 ZX − 1.81811e-008XYZ.

Δy = −0.610068 + 0.00177641 X + 0.00335958 Y − 0.00228751 Z − 4.7738e-006 XY + 2.57149e-006 YZ + 6.31545e-006 ZX − 1.36507e-008XYZ.

Δz = 0.111263 − 0.00264425 X − 0.00272709 Y − 0.00391453 Z + 7.14788e-006 XY + 1.39318e-005 YZ + 1.36715e-005 ZX − 3.3172e-008XYZ.

Determining the generalized variables according to the alternative data and the end-effector trajectory when controlling according to the alternate orbit data gives the results as shown in Fig. 6.

The solid lines show that the initial trajectory is calculated according to the desired destination, the dashed lines represent the control according to the corrected data. The result of the variation is shown in Fig. 7.

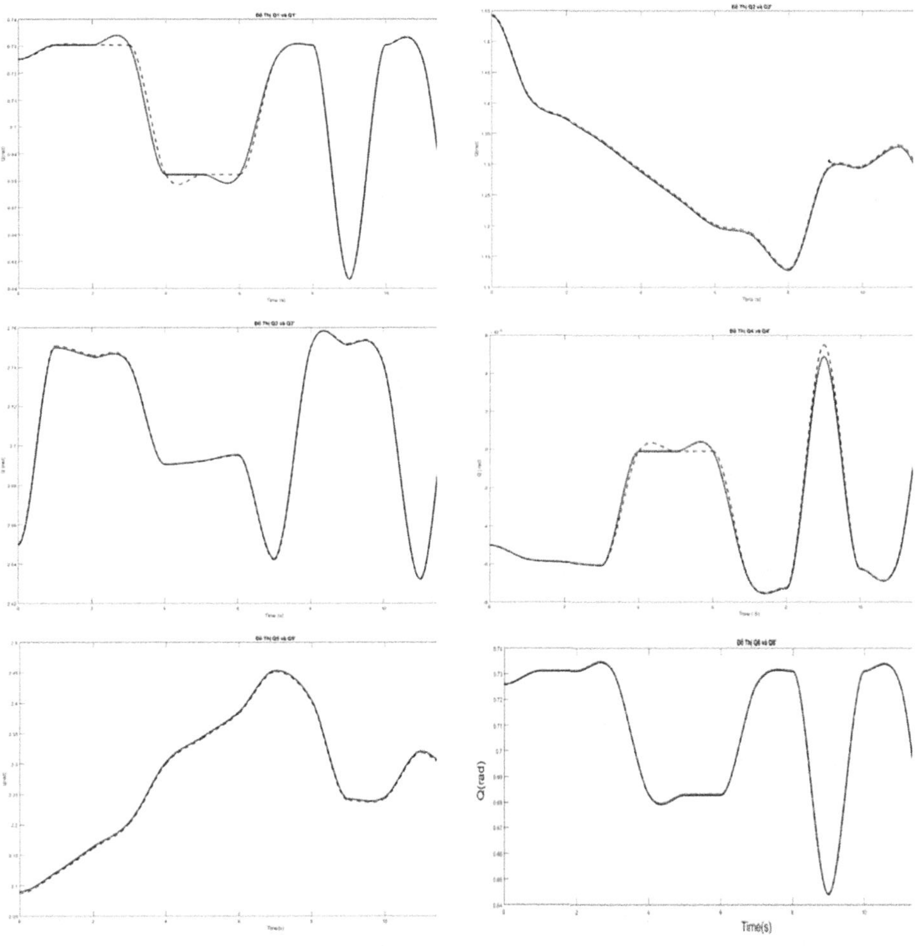

Fig. 6. Comparison of initial and alternative trajectories

The accuracy effect is clearly shown in Fig. 7, all interventions are only on software. The application for different types of robots is feasible, and it is necessary to adjust to certain cycles to keep the device always ensuring the necessary accuracy.

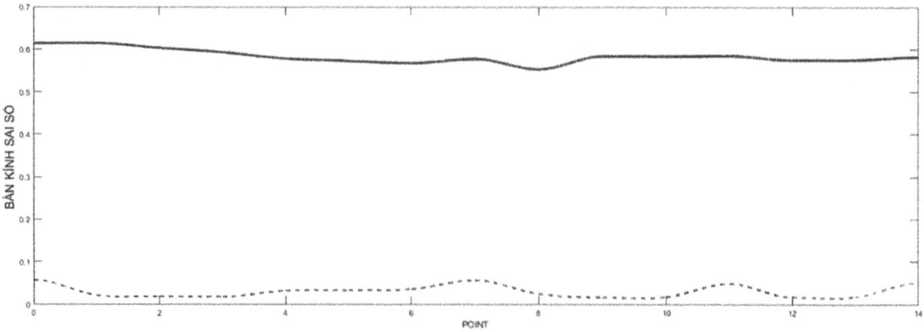

Fig. 7. Error radius before (continuous line) and after (hidden line) using alternative trajectories

4 Conclusion

Intervention in software to keep accuracy is the right direction of mechatronic systems because it allows to keep long-term accuracy with not too high quality of starting structure. This not only reduces the cost of fabrication but also extends the service life of the old device. Different from the closed-loop control in each link, in this paper, it controls the closed loop to the end-effectors, i.e. putting errors such as wear, deformation, tolerance of DH dimension into the control loops. This is the difference with the closed control signals of each link only, because the accuracy of the generalized variable and its derivatives are the only interest.

In addition to the direct test trajectory as presented in this paper, the error also holds the same for other trajectories tested on the same error model in the example. The result of using an alternative trajectory calculated from the alternative endpoint shown in this paper is one of the positive directions and it is easy to apply to practical production due to quick operation, no interference in equipment, only interfere with the preparation of control data.

Acknowledgment. This research was supported by the Thai Nguyen University of Technology (TNUT) of Vietnam (T2019-B07 project).

References

1. Yu, C., Xi, J.: Simultaneous and on-line calibration of a robot-based inspecting system. Robot. Comput. Integr. Manuf. **49**, 349–360 (2018)
2. Kong, L., Chen, G., Zhang, Z., Wang, H.: Kinematic calibration and investigation of the influence of universal joint errors on accuracy improvement for a 3-DOF parallel manipulator. Robot. Comput. Integr. Manuf. **49**, 388–397 (2018)
3. Zhang, L., Yan, X., Zhang, Q.: Design and analysis of 3-DOF cylindrical-coordinate-based manipulator. Robot. Comput. Integr. Manuf. **52**, 35–45 (2018)
4. Ma, L., Bazzoli, P., Sammons, P.M., Landers, R.G., Bristow, D.A.: Modeling and calibration of high-order joint-dependent kinematic errors for industrial robots. Robot. Comput. Integr. Manuf. **50**, 153–167 (2018)

5. Kamali, K., Joubair, A., Bonev, I.A., Bigras, P.: Elasto-geometrical calibration of an industrial robot under multidirectional external loads using a laser tracker. In: Proceedings - IEEE International Conference on Robotics and Automation (2016)
6. Sun, T., Zhai, Y., Song, Y., Zhang, J.: Kinematic calibration of a 3-DoF rotational parallel manipulator using laser tracker. Robot. Comput. Integr. Manuf. **41**, 78–91 (2016)
7. Liu, Z., Xu, J., Cheng, Q., Zhao, Y., Pei, Y., Yang, C.: Trajectory planning with minimum synthesis error for industrial robots using screw theory. Int. J. Precis. Eng. Manuf. **2**, 183–193 (2018)
8. Bai, Y., Wang, D.: On the comparison of an interval Type-2 Fuzzy interpolation system and other interpolation methods used in industrial modeless robotic calibrations. In: 2016 IEEE International Conference on Computational Intelligence and Virtual Environments for Measurement Systems and Applications, CIVEMSA 2016 - Proceedings (2016)
9. Borrmann, C., Wollnack, J.: Calibration of external linear robot axes using spline interpolation. In: Proceedings of 2014 International Conference on Model, Identification and Control, ICMIC 2014, pp. 111–116 (2015)
10. Bai, Y., Zhuang, H.: On the comparison of model-based and modeless robotic calibration based on the fuzzy interpolation technique, pp. 892–897 (2005)
11. Bai, Y., Zhuang, H.: On the comparison of bilinear, cubic spline, and fuzzy interpolation techniques for robotic position measurements. IEEE Trans. Instrum. Measur. **54**(6), 2281–2288 (2005)

Calculation of Optimum Gear Ratios of Mechanical Driven Systems Using Worm-Helical Gearbox and Chain Drive

Le Hong Ky[1], Tran Thi Hong[2], Nguyen Van Cuong[3], Luu Anh Tung[4], Nguyen Thanh Tu[4], Hoang Thi Tham[4], and Le Xuan Hung[4(✉)]

[1] Vinh Long University of Technology Education, Vinh Long 890000, Vietnam
[2] Nguyen Tat Thanh University, Ho Chi Minh 700000, Vietnam
[3] University of Transport and Communications, Ha Noi 100000, Vietnam
[4] Thai Nguyen University of Technology, Thai Nguyen 23000, Vietnam
lexuanhung@tnut.edu.vn

Abstract. A study on the calculation of the optimum gear ratios of mechanical driven systems using a worm-helical gearbox and a chain drive was presented. To find the optimum gear ratios, an optimization problem in which the height of the system was selected as the objective function was conducted. Also, the influences of the input parameters including the total gear ratio of the system, the calculated coefficient of the worm diameter coefficient, the coefficient of wheel face width and the allowable contact stress of helical gear unit, and the output torque were examined. To weigh the effect of these parameters on the optimum gear ratios, a simulation experiment was performed by computer programming. Additionally, equations to calculate the optimum gear ratios were introduced. Using these equations, the optimum gear ratios can be found easily and accurately because the equations are all in the explicit form.

Keywords: Optimum gearbox design · Gear ratio · Optimum gear ratio · Worm-helical gearbox

1 Introduction

In practice mechanical design, calculation of optimum gear ratios is a very important task. The reason of that is the dimension, the mass, and also the cost of a gearbox or a mechanical system are influenced by the gear ratios. Therefore, there have been many studies for finding the optimum gear ratios.

Until now, there were several methods which have been used for determining the optimum gear ratios. They are the graph method [1, 2], the practical method [3] and the model method [3–15]. Besides, the gear ratios have been found for different types of gearboxes such as helical gearboxes, bevel gearboxes, worm gearboxes etc.

Regardings to helical gearboxes, the optimum gear ratios were calculated for two-step gearboxes [3–8], three-step gearboxes [9–14] and four-step gearboxes [9, 13–18]. Also, the optimum ratios have been found for mechanical driven systems which use a gearbox and a V-belt drive [19–22] or a chain drive [23, 24].

K.-U. Sattler et al. (Eds.): ICERA 2019, LNNS 104, pp. 66–75, 2020.
https://doi.org/10.1007/978-3-030-37497-6_7

For optimum design of worm gearboxes, the optimum gear ratios have been calculated for two-step worm gearboxes [3, 25, 26] and for worm-helical gearboxes [26, 27].

In practice, the optimum gear ratios are usually determined with an objective function. The objectives can be the minimum mass of gears [10, 13–15, 17], the minimum cross section area of the gearbox [9, 12] or the minimum gearbox length [7, 11].

In this paper, a study on the determination of the optimum gear ratios of mechanical driven systems using a worm-helical gearbox and a chain drive was introduced. The aim of the study is to find the minimum height of the system. Also, the effects of the input parameters including the total gearbox ratio, the coefficient for calculating the worm diameter coefficient, the coefficient of wheel face width, the allowable contact stress of helical gear unit, and the output torque were inspected. To weigh them, a simulation experiment was designed and performed by computer programming. Moreover, models for calculating the optimum gear ratios were found. Using these models, the optimum gear ratios were found easily and accurately because the equations are all in the explicit form.

2 Optimization Problem

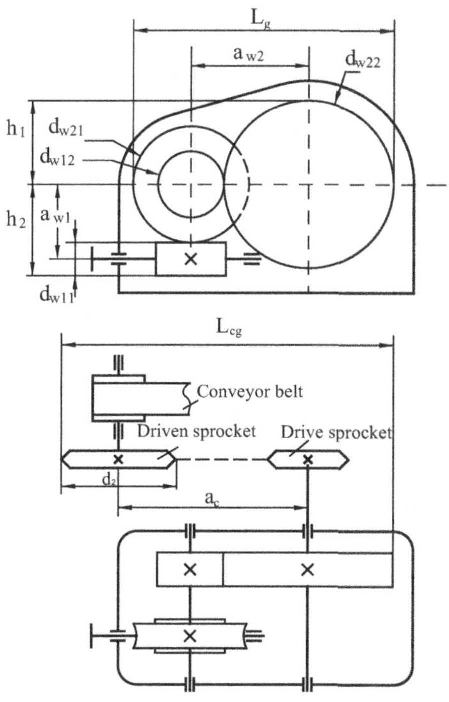

Fig. 1. Calculation schema

The height of a mechanical driven system which consists of a worm-helical gearbox and a chain drive can be determined as (Fig. 1):

$$h = h_1 + h_2 \tag{1}$$

Where, h_1 and h_2 are determined by the following equations (see Fig. 1):

$$h_1 = \max(d_{w21}/2; d_{w22}/2) \tag{2}$$

$$h_2 = \max(d_{w11}/2 + a_{w1}; d_{w22}/2) \tag{3}$$

In which, d_{w21} and d_{w22} are the pitch diameters of the driven gear of the worm unit and the helical gear unit which can be found by the following equations [29]:

$$d_{w21} = m \cdot z_2 = 2 \cdot a_{w1} \cdot z_2/(z_2 + q) \tag{4}$$

$$d_{w22} = 2 \cdot a_{w2} \cdot u_2/(u_2 + 1) \tag{5}$$

In the above equations, a_{w1} and a_{w2} are the center distance of the first and the second step; q is the coefficient of the worm pitch diameter which is determined by [29]:

$$q = k_q \cdot z_2 \tag{6}$$

In which, k_q is the coefficient for calculation q; $k_q = 0.25\ldots0.3$ [29].

From above analysis, the optimization problem is defined as:

$$\text{minimize}\, h = h_1 + h_2 \tag{7}$$

With the following constraints

$$\begin{gathered} 1 \le u_c \le 6 \\ 8 \le u_1 \le 80 \\ 1 \le u_2 \le 9 \end{gathered} \tag{8}$$

To solve the above optimization problem, it is required to calculate the center distances and the pitch diameters of the worm gear unit and the helical gear unit.

2.1 Calculating Center Distance and Pitch Diameters of Worm Drive

The center distance of the worm gear unit (mm) can be determined by [28]:

$$a_{w1} = K_a \cdot \left(K_{HV} \cdot K_{H\beta} \cdot T_{21}/[\sigma_{H1}]^2\right)^{1/3} \tag{9}$$

Where, K_a is a coefficient; $K_a = 610$ [28]; K_{HV} is the internal dynamic coefficient; $K_{HV} = 1 \div 1.3$ [28]; $K_{H\beta}$ is the load concentration coefficient; $K_{H\beta} = 1 \div 1.3$ [28]; T_{21} is the torque on the wheel (Nmm). In this case, T_{21} is determined by:

$$T_{21} = T_{out}/\left(u_2 \cdot u_c \cdot \eta_{br} \cdot \eta_c \cdot \eta_o^2\right) \quad (10)$$

In which, T_{out} is the output torque of the system (Nmm); η_{hg} is the transmission efficiency of the helical gear unit ($\eta_{hg} = 0.96\ldots0.98$ [29]); η_c is the transmission efficiency of the chain drive ($\eta_c = 0.95\ldots0.97$ [29]); η_b is the transmission efficiency of a rolling bearing pair ($\eta_b = 0.99\ldots0.995$ [29]). Choosing $\eta_{hg} = 0.97$, $\eta_c = 0.96$, and $\eta_b = 0.992$ substituting them into (10) gives:

$$T_{21} = 1.0913 \cdot T_{out}/(u_2 \cdot u_c) \quad (11)$$

Choosing $K_{hV} = 1.2$, $K_{h\beta} = 1.2$ and substituting them together with K_a and T_{21} into (9) gets:

$$a_{w1} = 709.1945 \cdot \left\{T_{21}/\left(u_2 \cdot u_c \cdot [\sigma_{H1}]^2\right)\right\}^{1/3} \quad (12)$$

Wherein, $[\sigma_{H1}]$ is the allowable contact stress of the first stage (N/mm^2); $[\sigma_{H1}]$ depends on the wheel material. With the wheel material is tinless bronze or soft grey iron, $[\sigma_{H1}]$ can be calculated by the following regression equation which was found (with the determination coefficient $R^2 = 0.9906$) from the data in [25]:

$$[\sigma_{H1}] = 5.0515 \cdot v_{sl}^2 - 49.742 \cdot v_{sl} + 189.9 \quad (13)$$

In which, v_{sl} is the slip velocity calculated by [29]:

$$v_{sl} = 0.0088 \cdot \left(P_1 \cdot u \cdot n_1^2\right)^{1/3} \quad (14)$$

If the wheel material is tin bronze, $[\sigma_{H1}]$ is calculated by [29]:

$$[\sigma_{H1}] = K_{HL} \cdot v_{sl} \cdot [\sigma_{H0}] \quad (15)$$

Where, $[\sigma_{H0}]$ is the allowable contact stress if the stress change cycle is 10^7:

$$[\sigma_{H0}] = (0.7\ldots0.9) \cdot \sigma_t \quad (16)$$

In which, σ_t is the tensile stress (N/mm^2); $\sigma_t = 260$ if $v_{sl} = 5 \div 8$); $\sigma_t = 230$ if $v_{sl} = 8\ldots12$ and); $\sigma_t = 285$ if $v_{sl} = 12 \div 25$ [29].

K_{HL} is the service life ratio which is determined as [29]:

$$K_{HL} = \left(10^7/N_{HE}\right)^{1/8} \quad (17)$$

Where, N_{HE} is equivalent loading cycle number for the wheel teeth:

$$N_{HE} = 60 \cdot n_2 \cdot t_\Sigma \tag{18}$$

With t is the service lifetime (h); n_2 is the rotational speed of the wheel (rpm).

2.2 Calculating Center Distance and Pitch Diameters of Helical Gear Drive

For the helical gear drive, the center distance a_{w2} is calculated by [29]:

$$a_{w2} = K_a \cdot (u_2 + 1) \cdot \left\{ T_{12} \cdot k_{H\beta} / \left([\sigma_H]^2 \cdot u_2 \cdot \psi_{ba2} \right) \right\}^{1/3} \tag{19}$$

Where, k_a = 43 is the material coefficient [29]; $k_{H\beta}$ is the coefficient of load concentration; $k_{H\beta} = 1.1$ [29]; T_{12} is the pinion torque which can be found by:

$$T_{12} = T_{out} / \left(\eta_{hg} \cdot \eta_c \cdot \eta_b^2 \cdot u_2 \cdot u_c \right) \tag{20}$$

Substituting $\eta_{hg} = 0.97$, $\eta_c = 0.96$ and $\eta_b = 0.992$ (as in Sect. 2.1) into Eq. (20) gives

$$T_{12} = 1.0913 \cdot T_{out} / (u_2 \cdot u_c) \tag{21}$$

Replacing k_a = 43 and $k_{H\beta} = 1.1$ and Eqs. (21) into (19) gives:

$$a_{w1} \approx 45,6998 \cdot (u_2 + 1) \cdot \left\{ T_{out} / \left(u_2^2 \cdot u_c \cdot [\sigma_H]^2 \cdot \psi_{ba2} \right) \right\}^{1/3} \tag{22}$$

2.3 Experimental Work

In this study, a simulation experiment was designed and conducted for exploring the influences of the input parameters on the optimum partial gear ratios. The experiment was designed with 2-level full factorial and 5 input parameters (Table 1). Consequently, a number of $2^5 = 32$ text runs were carried out. Aslo, to perform the experiment, based on Eqs. (7) and (8), a computer program was constructed. The different levels of the input parameters and the output results (the optimum gear ratios of the chain drive u_c and the worm drive u_1) were described in Table 2.

Table 1. Input parameters

Factor	Code	Unit	Low	High
Total transmission ratio of the system	u_t	–	50	250
Coefficient for calculation q	K_q	–	0.25	0.3
Coefficient of wheel face width of step 2	x_{ba2}	–	0.35	0.4
Allowable contact stress of step 2	AS_2	MPa	350	420

3 Results and Discussion

3.1 Calculating Center Distance and Pitch Diameters of Worm Drive

Table 2. Experimental plans and output response

Std order	Run order	Center Pt	Blocks	u_g	X_{ba1}	X_{ba2}	AS	T_{out}	u_c	u_1
2	1	1	1	250	0.25	0.35	350	100000	1.00	27.78
16	2	1	1	250	0.3	0.4	420	100000	1.00	27.78
21	3	1	1	50	0.25	0.4	350	1000000	1.00	8.00
23	4	1	1	50	0.3	0.4	350	1000000	1.00	8.00
19	5	1	1	50	0.3	0.35	350	1000000	1.00	8.00
17	6	1	1	50	0.25	0.35	350	1000000	1.00	8.00
…										
9	31	1	1	50	0.25	0.35	420	100000	1.00	8.00
20	32	1	1	250	0.3	0.35	350	1000000	1.00	27.78

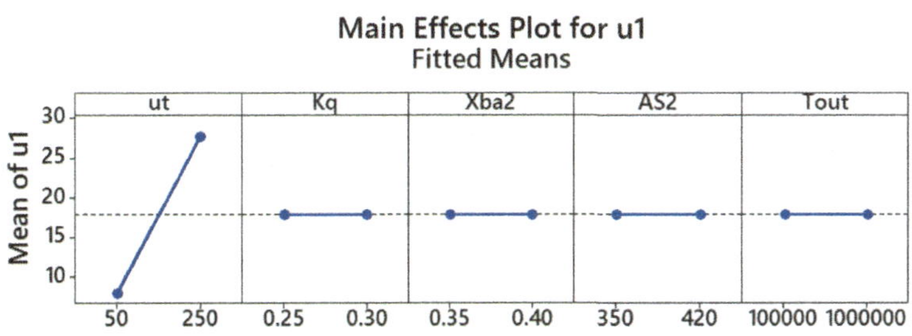

Fig. 2. Main effects plot for optimum gear ratio of the worm drive

The graph of the main effects for the optimum gear ratio of the worm drive u_1 is shown in Fig. 2. It is found from the figure that u_1 depends strongly on the total system gear ratio u_1. The special point here is that u_1 does not depend on the coefficient of calculation of the worm diameter coefficient k_q, the coefficient of wheel face width of the helical gear drive ψ_{ba2}, the allowable contact stress AS_2, and the output torque T_{out}.

Figure 3 presents the Pareto chart of the standardized effects. For the minimum gear ratio, the parameters are significant at the 0.05 level. It is reported from the chart that only the bar representing u_t(factor A) crosses the reference line. Therefore, it is the significant factor for u_1.

Figure 4 describes the estimated effects and coefficients for u_1. From the figure it was noted that only u_t has a significant effect on a response (with P-values lower than 0.05). Therefore, the optimum gear ratio of the worm drive is determined as:

$$u_1 = 3.056 - 0.009889 \cdot u_t \tag{23}$$

It is found from Fig. 4 that all values of adj-R2 and pred-R2 are 100%. It proves that Eq. (23) is greatly suitable for program data. Therefore, it can be used to determine the optimum gear ratio of the worm drive u_1.

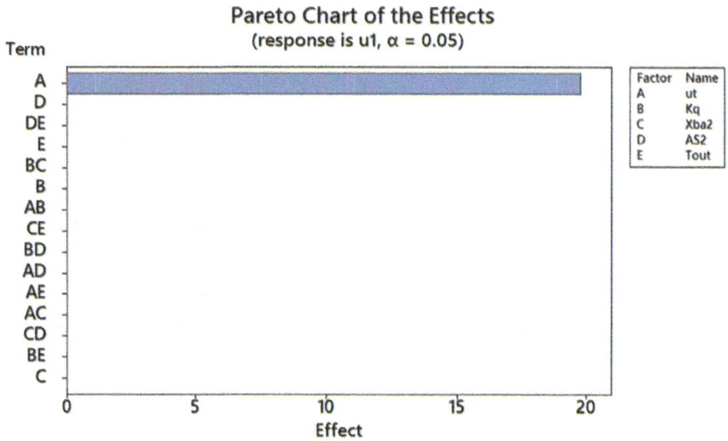

Fig. 3. Pareto chart for u_1

3.2 Determining Optimum Gear Ratio of Chain Drive

From Table 2 it is noted that the optimum gear ratios of the chain drive u_c is constant. In addition, it takes the smallest value within the allowed range ($u_c = 1$). This suggests that with the given objective function, the chain drive is only meant to transmit torque to some axis distance, but not to achieve the speed reduction target.

$$u_c = 1 \tag{24}$$

3.3 Determining Optimum Gear Ratio of Helical Gear Drive

After having (Eqs. (23) and (24)), the optimum gear ratio of the helical gear drive is easily calculated by the following formula:

$$u_1 = u_t/(u_2 \cdot u_c) \tag{25}$$

Coded Coefficients

Term	Effect	Coef	SE Coef	T-Value	P-Value	VIF
Constant		17.89	0.00	*	*	
ut	19.778	9.889	0.000	*	*	1.00
Kq	-0.000000	-0.000000	0.000000	*	*	1.00
Xba2	0.000000	0.000000	0.000000	*	*	1.00
AS2	0.000000	0.000000	0.000000	*	*	1.00
Tout	-0.000000	-0.000000	0.000000	*	*	1.00

Model Summary

S	R-sq	R-sq(adj)	R-sq(pred)
0	100.00%	100.00%	100.00%

Fig. 4. Estimated effects and coefficients for u_1

4 Conclusions

In this study, an optimum problem to calculate the optimum gear ratio of mechanical driven systems using a worm-helical gearbox and a chain drive has been solved. First, the height of the mechanical system was chosen as the target function. Next, 5 input parameters and 2-level full factorial method were selected to design and implement a simulation experiment. In addition, the effects of the input parameters including the total system gear ratio, the coefficient of calculating the worm diameter coefficient, the wheel face width coefficient, the helical gear allowable contact stress, and the output torque have been taken into account. Finally, regression formulas for determining optimum gear ratios were given. The use of these formulas is simple because they all appear in the form of an explicit function.

Acknowledgements. The work described in this paper was supported by Thai Nguyen University of Technology for a scientific project.

References

1. Kudreavtev, V.N., Gierzaves, I.A., Glukharev, E.G.: Design and Calculus of Gearboxes. Mashinostroenie Publishing, Sankt Petersburg (1971)

2. Chat, T.: Some problems of kinematics calculation of transmission mechanics system. In: Proceedings of the National Conference on Engineering Mechanics, vol. 2, Hanoi, Vietnamese, pp. 7–12 (1993)
3. Milou, G., Dobre, F., Visa, H.: Vitila: optimal design of two step gear units, regarding the main parameters. VDI Berichte no. 1230, p. 227 (1996)
4. Ngoc Pi, V.: A method for optimal calculation of total transmission ratio of two step helical gearboxes. In: Proceedings of the National conference on Engineering Mechanics, Ha Noi, p. 12 (2001)
5. Cam, N.T.H., Pi, V.N., Tuan, N.K., Hung, L.X., Thao, T.T.P.: Determining optimal partial transmission ratios of mechanical driven systems using a V-belt drive and a helical reducer with second-step double gear-sets. In: Fujita, H., et al. (eds.) ICERA 2018, LNNS 63, pp. 261–269 (2019). https://doi.org/10.1007/978-3-030-04792-4_35
6. Petrovski, A.N., Sapiro, B.A., Saphonova, N.K.: About optimal problem for multi-step gearboxes. Vestnik Mashinostroenie **10**, 13 (1987)
7. Pi, V.N.: A new study on optimal calculation of partial transmission ratios of two-step helical gearboxes. In: 2nd WSEAS International Conference on Computer Engineering and Applications (CEA'08), Acapulco, Mexico, 25–27 January, pp. 162–165 (2008)
8. Nguyen, K.T., Vu, N.P., Nguyen, T.H.C., Tran, T.P.T., Ho, K.T., Le, X.H., Hoang, T.T.: Determining optimal gear ratios of a two-stage helical reducer for getting minimal acreage of cross section. In: MATEC Web of Conferences 213, 01008 (2018). https://doi.org/10.1051/matecconf/201821301008
9. Pi, V.N., Tuan, N.K.: Optimum determination of partial transmission ratios of three-step helical gearboxes for getting minimum cross section. J. Environ. Sci. Eng. A **5**, 570–573 (2016)
10. Pi, V.N., Binh, N.D., Dac, V.Q., Quang, P.: The optimal calculation of total transmission ratio of three-step helical gearboxes for minimum mass of gears. J. Sci. Technol. **6**, 91 (2006). Engineering Universities, Vietnamese
11. Pi, V.N.: A new study on optimal calculation of partial transmission ratio of three-step helical reducers. In: the 3rd IASME/ WSEAS International Conference on Continuum Mechanics, Cambridge, UK, p. 23 (2008)
12. Pi, V.N.: A new study on optimal calculation of partial transmission ratios of three-step helical reducers for getting minimal cross section dimension. In: the 2nd WSEAS International Conference on Computer Engineering and Applications (CEA 2008), Acapulco, Mexico 25–27 January, pp. 290–293 (2008)
13. Pi, V.N.: Optimal determination of partial transmission ratios of three-step helical gearboxes with first and third-step double gear-sets for minimal mass of gear. In: American Conference on Applied Mathematics (MATH 2008), Harvard, Massachusetts, USA 24–26 March, pp. 385–388 (2008)
14. Romhild, I., Linke, H.: Gezielte Auslegung Von Zahnradgetrieben mit minimaler Masse auf der Basis neuer Berechnungsverfahren. Konstruktion **44**, 229–236 (1992)
15. Pi, V.N.: Optimal determination of partial transmission ratios for four-step helical gearboxes with first and third step double gear-sets for minimal mass of gears. In: Applied Computing Conference (ACC 2008), Istanbul, Turkey, 27–30 May (2008)
16. Pi, V.N.: Optimal calculation of partial transmission ratios for four-step helical gearboxes with first and third-step double gear-sets for minimal gearbox length. In: American Conference on Applied Mathematics (MATH 2008), Harvard, Massachusetts, USA 24–26 March, pp. 29–32 (2008)

17. Hung, L.X., Pi, V.N., Van Du, N.: Optimal calculation of partial transmission ratios of four-step helical gearboxes with second and fourth-step double gear-sets for minimal mass of gears. In: the International Symposium on Mechanical Engineering (ISME Ho Chi Minh city, Vietnam, September, pp. 21–23, September 2009
18. Pi, V.N.: Optimal determination of partial transmission ratios for four-step helical gearboxes with first and third step double gear-sets for minimal mass of gears. In: Applied Computing Conference (ACC 2008), Istanbul, Turkey, 27–30 May, pp. 53–57 (2008)
19. Pi, V.N., Thao, T.T.P., Thao, L.T.P.: A new study on optimum determination of partial transmission ratios of mechanical driven systems using a V-belt and two-step helical gearbox. Vietnam Mech. Eng. J. **10**, 123 (2015)
20. Pi, V.N., Cam, N.T.H., Tuan, N.K.: Optimum calculation of partial transmission ratios of mechanical driven systems using a V-belt and two-step bevel helical gearbox. J. Environ. Sci. Eng. A **5**, 566 (2016)
21. Tuan, N.K., Thao, T.T.P., Cam, N.T.H., Hung, L.X., Pi, V.N.: Optimum calculation of partial transmission ratios of mechanical driven systems using a V-belt and a three-step bevel helical gearbox. In: Fujita, H., et al. (eds.) ICERA, LNNS 63, pp. 469–476 (2018, 2019). https://doi.org/10.1007/978-3-030-04792-4_61
22. Pi, V.N., Tuan, N.K., Hung, L.X., Cam, N.T.H., Thao, T.T.P.: Determining optimum partial transmission ratios of mechanical driven systems using a V-belt drive and a three-stage helical reducer. In: Advances in Material Sciences and Engineering. Springer, pp. 81–88 (2019)
23. Cam, N.T.H., Pi, V.N., Tuan, N.K., Hung, L.X., Thao, T.T.P.: A study on determination of optimum partial transmission ratios of mechanical driven systems using a chain drive and a three-step helical reducer. In: Fujita, H., et al. (eds.) ICERA 2018, LNNS 63, pp. 91–99 (2019). https://doi.org/10.1007/978-3-030-04792-4_14
24. Pi, V.N., Thao, T.T.P., Tuan, D.A.: Optimum determination of partial transmission ratios of mechanical driven systems using a chain drive and two-step helical gearbox. J. Environ. Sci. Eng. B **6**, 80 (2017)
25. Chernavsky, S.A., et al.: Design of mechanical transmissions: manual for high technical schools. Mashinostroenie, Moscow, p. 560 (1984)
26. Pi, V.N., Dac, V.Q.: Calculation of total transmission ratio of two step worm reducers for the best reasonable gearbox housing structure. J. Sci. Technol. - Thai Nguyen University, Vietnamese **1**(41), 65–69 (2007)
27. Pi, V.N., Dac, V.Q.: Optimal calculation of partial transmission ratios of worm-helical gear reducers for minimal gearbox length. J. Sci. Technol. - Technical Universities, Vietnamese **61**, 73–77 (2007)
28. Karl-Heinrich Grote, K., Antonsson, E.: Springer Handbook of Mechanical Engineering. Springer, Berlin (2008)
29. Chat, T., Van Uyen, L.: Design and Calculation of Mechanical Transmissions Systems, vol. 1. Educational Republishing House, Hanoi (2007)

A Study on Determination of Optimum Gear Ratios of a Two-Stage Worm Gearbox

Luu Anh Tung[1], Tran Thi Hong[2], Nguyen Van Cuong[3], Le Hong Ky[4], Nguyen Thanh Tu[1], Le Xuan Hung[1], and Ngoc Pi Vu[1(✉)]

[1] Thai Nguyen University of Technology, Thai Nguyen, Vietnam
vungocpi@tnut.edu.vn

[2] Nguyen Tat Thanh University, Ho Chi Minh City, Vietnam

[3] University of Transport and Communications, Hanoi, Vietnam

[4] Vinh Long University of Technology Education, Vinh Long, Vietnam

Abstract. This paper presents a study on the determination of the optimum gear ratios of a two-stage worm gearbox. To find the optimum gear ratios, an optimization problem was performed. Also, the reasonable gearbox structure was chosen as the objective function of the optimization problem. In addition, the effects of the input factors including the total gearbox ratio, the electric motor speed, and the output torque were investigated. Besides, a simulation experiment was designed and performed by computer programs to evaluate the influences of these input factors on the optimum gear ratios. Moreover, some models to calculate the optimum gear ratios of a two-stage worm gearbox were suggested. As these equations are explicit, the determination of the optimum gear ratios is accurate and uncomplicated.

Keywords: Gear ratio · Optimum gear ratio · Optimum gearbox design · Two-stage worm gearbox

1 Introduction

In the design of a gearbox or a mechanical driven system, finding optimum gear ratios is one of the most important tasks. This is because the mass, the dimension, as well as the cost of the gearbox or the system depend strongly on their gear ratios. Consequently, many studies have been carried out to find the optimum gear ratios.

Until now, the optimum gear ratios of a gearbox have been determined by three basic methods including the graph method [1, 2], the practical method [3] and the model method [3–16]. Also, the gear ratios of different gearbox types such as helical gearboxes, bevel gearboxes or worm gearboxes have been found.

For helical gearboxes, the optimum gear ratios have been calculated for two-stage gearboxes [3–8], three-stage gearboxes [9–14] and four-stage gearboxes [15–18]. Also, the optimum ratios were determined not only for separate gearboxes but also for mechanical driven systems using a gearbox and a V-belt drive [19–22] or a chain drive [23, 24].

K.-U. Sattler et al. (Eds.): ICERA 2019, LNNS 104, pp. 76–84, 2020.
https://doi.org/10.1007/978-3-030-37497-6_8

Regarding worm gearboxes, the optimum gear ratios were determined for two-stage worm gearboxes [3, 25, 26] and worm-helical gearboxes [25, 27].

In previous studies, the optimum gear ratios have been found for different objectives including the minimum gearbox cross section area [9, 12], the minimum mass of gears [10, 13, 14, 16, 18], the minimum gearbox length [7, 11] etc.

In this paper, an optimization study on the calculation of the optimum gear ratios of a two-stage worm gearbox was presented. The objective of the problem is to obtain the reasonable gearbox housing structure. Besides, the influences of the input factors including the total gearbox ratio, the electric motor speed, and the output torque were investigated. Moreover, to evaluate the effects of the input factors on the optimum ratios, a simulation experiment was designed and performed by computer programming. Additionally, equations to calculate the optimum ratios were proposed. As the equations are explicit, the optimum ratios can be found accurately in a simple way.

2 Optimization Problem

For a two-stage worm gearbox, the housing structure of the gearbox is considered reasonable when the following expression is satisfied [25] (see Fig. 1):

$$a_{w2} = 2.a_{w1} \tag{1}$$

Where, a_{w1} and a_{w2} are the center distances of the first and the second stages of the gearbox.

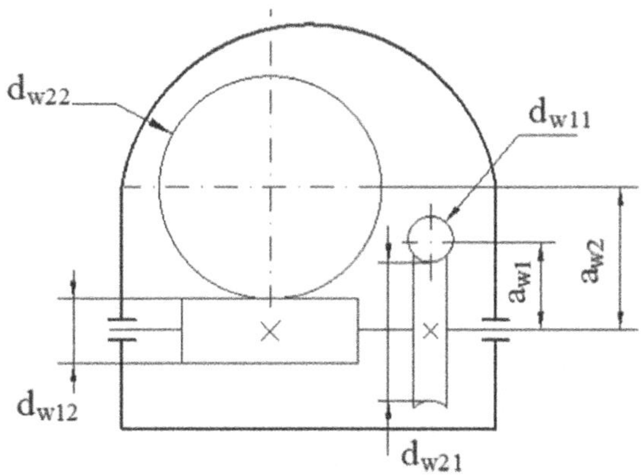

Fig. 1. Calculation schema

Therefore, the optimization problem is defined as:

Finding the optimum gear ratios of the gearbox to get: $a_{w2} = 2.a_{w1}$

With the following constraints:

$$8 \le u_1 \le 80$$
$$8 \le u_2 \le 80 \quad (2)$$

Thus, for solving the optimization problem it is essential to calculate the center distances of the first and the second stages of the gearbox.

2.1 Determining the Center Distance the First Step a_{w1}

The center distance of the first stage (mm) can be determined as [28]:

$$a_{w1} = K_a.\left(K_{HV}.K_{H\beta}.T_{21}/[\sigma_{H1}]^2\right)^{1/3} \quad (3)$$

Where, K_a is a coefficient; $K_a = 610$ [4]; K_{HV} is the internal dynamic coefficient; $K_{HV} = 1 \div 1.3$ [28]; $K_{H\beta}$ is the load concentration coefficient; $K_{H\beta} = 1 \div 1.3$ [28]; T_{21} is the torque on the wheel (Nmm); In this case, T_{21} is determined by:

$$T_{21} = T_{out}/(u_2.\eta_w.\eta_b^2) \quad (4)$$

In which, T_{out} is the output torque (Nmm); η_w is the transmission efficiency of a worm unit ($\eta_w = 0.7 \div 0.9$ [29]); η_b is the transmission efficiency of a rolling bearing pair ($\eta_b = 0.99 \div 0.995$ [29]). Choosing $\eta_w = 0.8$, $\eta_b = 0.992$ and substituting them into (4) gives:

$$T_{21} = 1.27024.T_{out}/u_2 \quad (5)$$

Choosing $K_{HV} = 1.2$, $K_{H\beta} = 1.2$ and substituting them with $K_a = 610$ and (4) into (3) gets:

$$a_{w1} = 746.01.\left\{T_{21}/\left(u_2.[\sigma_{H1}]^2\right)\right\}^{1/3} \quad (6)$$

Wherein, $[\sigma_{H1}]$ is the allowable contact stress of the first stage (N/mm^2); $[\sigma_{H1}]$ depends on the wheel material. When the wheel material is tinless bronze or soft grey iron, $[\sigma_{H1}]$ can be calculated by the following regression equation which was found (with the determination coefficient $R^2 = 0.9906$) from the data in [29]:

$$[\sigma_{H1}] = 5.0515.v_{sl}^2 - 49.742.v_{sl} + 189.9 \quad (7)$$

In which, v_{sl} is the slip velocity calculated by [29]:

$$v_{sl} = 0.0088.\left(P_1.u.n_1^2\right)^{1/3} \quad (8)$$

If the wheel material is bronze, $[\sigma_{H1}]$ is determined by [29]:

$$[\sigma_{H1}] = K_{KL}.v_{sl}.[\sigma_{H0}] \tag{9}$$

Where, $[\sigma_{H0}]$ is the allowable contact stress if the stress change cycle is 10^7:

$$[\sigma_{H0}] = (0.7 \div 0.9).\sigma_t \tag{10}$$

In which, σ_t is the tensile stress (N/mm^2); $\sigma_t = 260$ if $v_{sl} = 5 \div 8$); $\sigma_t = 230$ if $v_{sl} = 8 \div 12$ and); $\sigma_t = 285$ if $v_{sl} = 12 \div 25$ [29].

$[\sigma_{H1}]$ is the service life ratio determined by [29]:

$$K_{HL} = \left(10^7/N_{HE}\right)^{1/8} \tag{11}$$

Where, N_{HE} is the equivalent loading cycle number for the wheel teeth:

$$N_{HE} = 60.n_2.t_\Sigma \tag{12}$$

With t_Σ as the service lifetime of the gearing (h); n_2 is the wheel rotational speed (rpm).

2.2 Determining the Center Distance the Second Stage a_{w2}

For the second stage of the gearbox, calculating in the same way as for the first stage (Sect. 2.1), the center distance a_{w2} (mm) is calculated by [28]:

$$a_{w2} = K_a.\left(K_{HV}.K_{H\beta}.T_{22}/[\sigma_{H2}]^2\right)^{1/3} \tag{13}$$

Where, T_{22} is the torque on the wheel (Nmm) which can be found by:

$$T_{22} = T_{out}/\eta_b \tag{14}$$

Choosing $\eta_b = 0.992$ (as in Sect. 2.1) and substituting them into (14) gives:

$$T_{22} = 1.008.T_{out} \tag{15}$$

Choosing $K_a = 610$, $K_{HV} = 1.2$, $K_{H\beta} = 1.2$ (as in Sect. 2.1) and substituting them with T_{22} into (13) gets:

$$a_{w2} = 690/67.\left\{T_{22}/\left(u_2.[\sigma_{H2}]^2\right)\right\}^{1/3} \tag{16}$$

In which, $[\sigma_{H2}]$ is the allowable contact stress of the second stage (N/mm^2); $[\sigma_{H2}]$ can be found as in Sect. 2.1.

2.3 Experimental Work

Table 1. Input parameters

Factor	Code	Unit	Low	High
Total gearbox ratio	u_g	–	300	800
Electric motor speed	n_m	rpm	750	2910
Ratio a_{w2}/a_{w1}	Ra	–	2	2.5
Output torque	T_{out}	Nmm	10^5	10^7

To discover the influence of the input factors on the optimum gear ratios, a simulation experiment was conducted. The experiment was constructed with a 2-level full factorial design. In the experiment, 4 input factors including the total gearbox ratio u_g, the electric motor speed n_m, the ratio between the center distances of the second and the first stages R_a and the output torque T_{out} were selected for the investigation (Table 1). The experiment was therefore implemented with $2^4 = 16$ number of tests. Also, a computer program was created based on Eqs. (1) and (2) to perform the experiment. Table 2 shows the experimental plans and the optimum gear ratio of the first stage (the results of the output response).

3 Results and Discussion

Figure 2 describes the graph of the main effects for u_1 used to evaluate the influence of the input factors on u_1. It is found from the graph that it increases considerably with the growth of the total gearbox ratio u_g. Also, the optimum ratio is not affected by the electric motor speed n_m, the ratio R_a and the output torque T_{out}.

Table 2. Experimental plans and output response

StdOrder	RunOrder	CenterPt	Blocks	u_g	n_m	R_a	T_{out}	u_1
6	1	1	1	800	750	2.5	100000	10
14	2	1	1	800	750	2.5	10000000	10
3	3	1	1	300	2910	2	100000	8
10	4	1	1	800	750	2	10000000	10
15	5	1	1	300	2910	2.5	10000000	8
4	6	1	1	800	2910	2	100000	10
…	…	…	…	…	…	…	…	…
12	15	1	1	800	2910	2	10000000	10
2	16	1	1	800	750	2	100000	10

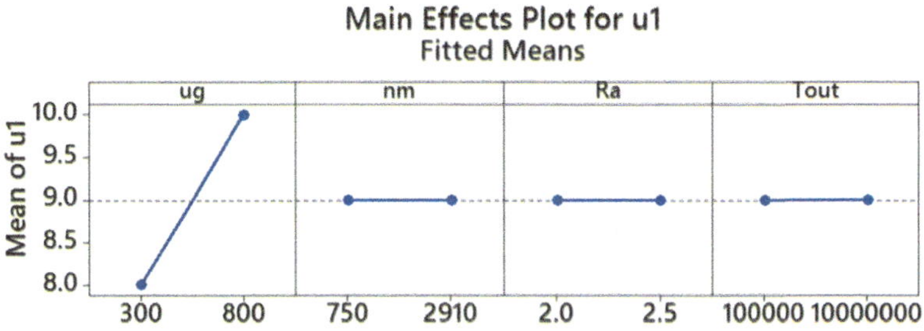

Fig. 2. Main effects plot for optimum gear ratio of the first stage

The Pareto chart of the standardized effects is described in Fig. 3. It can be observed from the chart that only the bar representing the total gearbox ratio u_g (factor A) crosses the reference line (in this case its value is 0). That means u_g is the significant factor for the optimum ratio.

Figure 4 describes the estimated effects and coefficients for u_1. It is noticed from the figure that only the constant and the total gearbox ratio u_g are influential and have values other than 0 (see column Coef. in Fig. 4). Therefore, the optimum gear ratio of the first stage is determined as:

$$u_1 = 6.8 - 0.004u_g \tag{17}$$

Equation (17) fits the data of the experiment very well because all the values of adj-R^2 and pred-R^2 are 100% (Fig. 4). Hence, this equation can be used to calculate the optimum gear ratio of the first stage u_1. After having u_1, the gear ratio of the second stage is determined by:

$$u_2 = u_g/u_1 \tag{18}$$

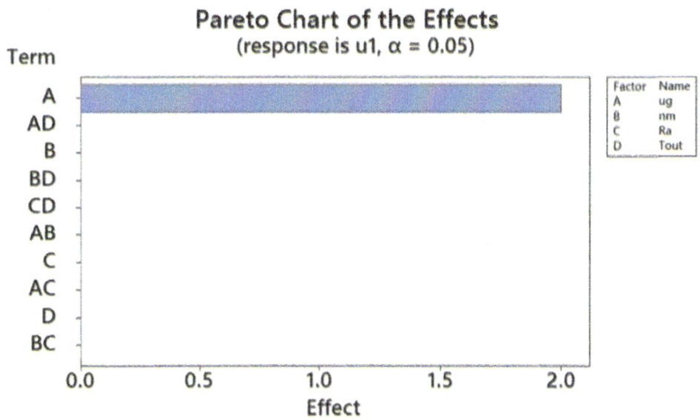

Fig. 3. Pareto chart of the standardized effects

Coded Coefficients

Term	Effect	Coef	SE Coef	T-Value	P-Value	VIF
Constant		9.000	0.000	*	*	
ug	2.000	1.000	0.000	*	*	1.00
nm	0.000000	0.000000	0.000000	*	*	1.00
Ra	-0.000000	-0.000000	0.000000	*	*	1.00
Tout	0.000000	0.000000	0.000000	*	*	1.00

Model Summary

S	R-sq	R-sq(adj)	R-sq(pred)
0	100.00%	100.00%	100.00%

Fig. 4. Estimated effects and coefficients for u_1

4 Conclusions

A study on the calculation of the optimum gear ratios of a two-stage worm gearbox to get the reasonable gearbox housing structure was conducted. In this study, the influences of the input factors including the total gearbox ratio, the electrical motor speed, the ratio between the center distances of the second and the first stages and the output torque were explored. In addition, some equations to determine the optimum gear ratios were proposed. As these equations are explicit, the optimum gear ratios can be calculated accurately and time savingly.

Acknowledgements. The work described in this paper was supported by Thai Nguyen University of Technology for a scientific project.

References

1. Kudreavtev, V.N., Gierzaves, I.A., Glukharev, E.G.: Design and Calculus of Gearboxes. Mashinostroenie Publishing, Sankt Petersburg (1971)
2. Chat, T.: Some problems of kinematics calculation of transmission mechanics system. In: Proceedings of the National Conference on Engineering Mechanics, Hanoi, Vietnamese, vol. 2, pp. 7–12 (1993)
3. Milou, G., Dobre, G., Visa, F., Vitila, H.: Optimal design of two step gear units, regarding the main parameters. VDI Berichte **1230**, 227 (1996)
4. Pi, V.N.: A method for optimal calculation of total transmission ratio of two step helical gearboxes. In: Proceedings of the National Conference on Engineering Mechanics, Hanoi, p. 12 (2001)
5. Cam, N.T.H., Pi, V.N., Tuan, N.K., Hung, L.X., Thao, T.T.P.: Determining optimal partial transmission ratios of mechanical driven systems using a V-belt drive and a helical reducer with second-step double gear-sets. In: Fujita, H., et al. (eds.) ICERA 2018. LNNS, vol. 63, pp. 261–269 (2019). https://doi.org/10.1007/978-3-030-04792-4_35
6. Petrovski, A.N., Sapiro, B.A., Saphonova, N.K.: About optimal problem for multi-step gearboxes. Vestnik Mashinostroenie **10**, 13 (1987)

7. Pi, V.N.: A new study on optimal calculation of partial transmission ratios of two-step helical gearboxes. In: 2nd WSEAS International Conference on Computer Engineering and Applications, CEA 2008, Acapulco, Mexico, 25–27 January, pp. 162–165 (2008)
8. Nguyen, K.T., Vu, N.P., Nguyen, T.H.C., Tran, T.P.T., Ho, K.T., Le, X.H., Hoang, T.T.: Determining optimal gear ratios of a two-stage helical reducer for getting minimal acreage of cross section. In: MATEC Web of Conferences, vol. 213, p. 01008 (2018)
9. Pi, V.N., Tuan, N.K.: Optimum determination of partial transmission ratios of Three-step helical gearboxes for getting minimum cross section. J. Environ. Sci. Eng. A **5**, 570–573 (2016)
10. Pi, V.N., Binh, N.D., Dac, V.Q., The, P.Q.: Optimal calculation of total transmission ratio of three-step helical gearboxes for minimum mass of gears. J. Sci. Technol. 6 Eng. Univ., 91 (2006)
11. Pi, V.N.: A new study on optimal calculation of partial transmission ratio of three-step helical reducers. In: The 3rd IASME/WSEAS International Conference on Continuum Mechanics, Cambridge, UK, p. 23 (2008)
12. Pi, V.N.: A new study on optimal calculation of partial transmission ratios of three-step helical reducers for getting minimal cross section dimension. In: The 2nd WSEAS International Conference on Computer Engineering and Applications, CEA 2008, Acapulco, Mexico, 25–27 January, pp. 290–293 (2008)
13. Pi, V.N.: Optimal determination of partial transmission ratios of three-step helical gearboxes with first and third-step double gear-sets for minimal mass of gear. In: American Conference on Applied Mathematics, MATH 2008, Harvard, Massachusetts, USA, 24–26 March, pp. 385–388 (2008)
14. Romhild, I., Linke, H.: Gezielte Auslegung Von Zahnradgetrieben mit minimaler Masse auf der Basis neuer Berechnungsverfahren. Konstruktion **44**, 229–236 (1992)
15. Pi, V.N.: Optimal determination of partial transmission ratios for four-step helical gearboxes with first and third step double gear-sets for minimal mass of gears. In: Applied Computing Conference (ACC 2008), Istanbul, Turkey, 27–30 May (2008)
16. Pi, V.N.: Optimal calculation of partial transmission ratios for four-step helical gearboxes with first and third-step double gear-sets for minimal gearbox length. In: American Conference on Applied Mathematics, MATH 2008, Harvard, Massachusetts, USA, 24–26 March, pp. 29–32 (2008)
17. Hung, L.X., Pi, V.N., Du, N.V.: Optimal calculation of partial transmission ratios of four-step helical gearboxes with second and fourth-step double gear-sets for minimal mass of gears. In: The international symposium on Mechanical Engineering (ISME Ho Chi Minh city, Vietnam, September, pp. 21–23 (2009)
18. Pi, V.N.: Optimal determination of partial transmission ratios for four-step helical gearboxes with first and third step double gear-sets for minimal mass of gears. In: Applied Computing Conference, ACC 2008, Istanbul, Turkey, 27–30 May, pp. 53–57 (2008)
19. Pi, V.N., Thao, T.T.P., Thao, L.T.P.: A new study on optimum determination of partial transmission ratios of mechanical driven systems using a V-belt and two-step helical gearbox. Vietnam Mech. Eng. J. **10**, 123 (2015)
20. Pi, V.N., Cam, N.T.H., Tuan, N.K.: Optimum calculation of partial transmission ratios of mechanical driven systems using a V-belt and two-step bevel helical gearbox. J. Env. Sci. Eng. A **5**, 566 (2016)
21. Tuan, N.K., Thao, T.T.P., Cam, N.T.H., Hung, L.X., Pi, V.N.: Optimum calculation of partial transmission ratios of mechanical driven systems using a V-belt and a three-step bevel helical gearbox. In: Fujita, H., et al. (eds.) ICERA 2018. LNNS, vol. 63, pp. 469–476 (2019). https://doi.org/10.1007/978-3-030-04792-4_61

22. Pi, V.N., Tuan, N.K., Hung, L.X., Cam, N.T.H., Thao, T.T.P.: Determining optimum partial transmission ratios of mechanical driven systems using a V-Belt drive and a three-stage helical reducer. In: Advances in Material Sciences and Engineering, pp. 81–88. Springer (2019)
23. Cam, N.T.H., Pi, V.N., Tuan, N.K., Hung, L.X., Thao, T.T.P.: A study on determination of optimum partial transmission ratios of mechanical driven systems using a chain drive and a three-step helical reducer. In: Fujita, H., et al. (eds.) ICERA 2018. LNNS, vol. 63, pp. 91–99 (2019). https://doi.org/10.1007/978-3-030-04792-4_14
24. Pi, V.N., Thao, T.T.P., Tuan, D.A.: Optimum determination of partial transmission ratios of mechanical driven systems using a chain drive and two-step helical gearbox. J. Environ. Sci. Eng. B **6**, 80 (2017)
25. Chernavsky, S.A., et al.: Design of mechanical transmissions: manual for high technical schools, Moscow, Mashinostroenie, p. 560 (1984)
26. Pi, V.N., Dac, V.Q.: Calculation of total transmission ratio of two step worm reducers for the best reasonable gearbox housing structure. J. Sci. Technol. Thai Nguyen Univ. **1**(41), 65–69 (2007)
27. Pi, V.N., Dac, V.Q.: Optimal calculation of partial transmission ratios of worm-helical gear reducers for minimal gearbox length. J. Sci. Technol. Tech. Univ. **61**, 73–77 (2007)
28. Grote, K.-H., Antonsson, E.K.: Springer Handbook of Mechanical Engineering. Springer, Heidelberg (2008)
29. Chat, T., Van Uyen, L.: Design and Calculation of Mechanical Transmissions Systems, vol. 1. Educational Republishing House, Hanoi (2007)

A Study on Determining Optimum Gear Ratios of Mechanical Driven Systems Using Two-Step Helical Gearbox with First Step Double Gear Sets and Chain Drive

Nguyen Khac Tuan[1], Tran Thi Hong[2], Nguyen Van Cuong[3], Le Hong Ky[4], Nguyen Thanh Tu[1], Luu Anh Tung[1], Le Xuan Hung[1], and Ngoc Pi Vu[1(✉)]

[1] Thai Nguyen University of Technology, Thai Nguyen 23000, Vietnam
vungocpi@tnut.edu.vn
[2] Nguyen Tat Thanh University, Ho Chi Minh 700000, Vietnam
[3] University of Transport and Communications, Ha Noi 100000, Vietnam
[4] Vinh Long University of Technology Education, Vinh Long 890000, Vietnam

Abstract. In this paper, a study on calculating optimum gear ratios of mechanical driven systems using a chain drive and a two-step helical gearbox with the first step double gear sets is introduced. To determine the optimum ratios, an optimization problem was conducted with the selection of system cross section area as the objective function. Also, several input parameters consisting of the total system ratio, the wheel face width coefficients of the both helical gear sets, the allowable contact stress and the output torque were investigated. Furthermore, a simulation experiment was run to study the impacts of these parameters on the optimum ratios. The results show that the optimum ratios can be obtained with high accuracy by using recommended models.

Keywords: Gear ratio · Optimum gear ratio · Optimum gearbox design · Helical gearbox

1 Introduction

Up to now, numerous studies have been conducted to determine the optimum gear ratios of a gearbox or a mechanical driven system. Several approaches have been employed to attain the optimum gear ratios for different gearbox types and gearboxes with different steps. The most commonly used methods are the graph method [1, 2], the practical method [3] and the model method [3–15]. In addition, the gear ratios have been found for different types of gearboxes such as helical gearboxes, bevel gearboxes and worm gearboxes.

Concerning helical gearboxes, the gear ratios have been determined for two step gearboxes [3–7], three step gearboxes [8–15] and four step gearboxes [12–19]. Regarding bevel gearboxes, they have been calculated with two-step [1, 3, 20] and three-step bevel helical gearboxes [21]. Besides, great attention has also been drawn

K.-U. Sattler et al. (Eds.): ICERA 2019, LNNS 104, pp. 85–93, 2020.
https://doi.org/10.1007/978-3-030-37497-6_9

from researchers to the calculation of the optimum gear ratios of worm gearboxes, such as two step worm gearboxes [3, 22, 23] and worm helical gearboxes [22, 24].

In efforts to determine the optimum gear ratios, many studies have been carried out not only for mechanical driven systems with a separate gearbox, but also for the systems using a gearbox and a V-belt drive [25–27] or a chain drive [28].

In this paper, an optimization study on calculating the optimum gear ratios of mechanical systems including a two-step helical gearbox with first step double gear sets and a chain drive is presented. In the study, the minimum system cross section area was selected as the objective function of the optimization problem. In addition, the influences of input parameters consisting of the total system ratio, the wheel face width coefficients of both gear steps, the allowable contact stress and the output torque were investigated. Additionally, to weigh the effects of these parameters on the optimum ratios, a simulation experiment was conducted. Moreover, some equations to find the optimum gear ratios were proposed.

2 Optimization Problem

The cross section area of the mechanical driven system using a two step helical gearbox with the first step double gear sets and a chain drive is determined by (Fig. 1):

$$A = L \cdot h \tag{1}$$

In which,

$$L = \max\left(L_g, L_c\right) \tag{2}$$

$$h = \max(d_{w21}; d_{w22}; d_2) \tag{3}$$

In Eq. (2), and are determined by (see Fig. 1):

$$L_g = d_{w11}/2 + a_{w1} + a_{w2} + d_{w22}/2 \tag{4}$$

$$L_c = d_{w22}/2 + a_c + d_2/2 \tag{5}$$

Where, a_{w1}, a_{w2} are the center distances of the first and the second steps; d_{w11}, d_{w22} are the pitch diameters (mm) of the first and the second steps, respectively; a_c, d_c are the center distance and the pitch diameter of the driven sprocket. The diameters d_{w11} and d_{w22} are found by [29]:

$$d_{w11} = 2 \cdot a_{w1}/(u_1 + 1) \tag{6}$$

$$d_{w22} = 2 \cdot a_{w2} \cdot u_2/(u_2 + 1) \tag{7}$$

In which, u_1 and u_2 are the gear ratios of the first and the second steps. It is noted that [29]:

$$u_1 \cdot u_2 = u_g \tag{8}$$

Therefore, the optimization problem is expressed by

$$\text{minimize}\, A = L \cdot h \tag{9}$$

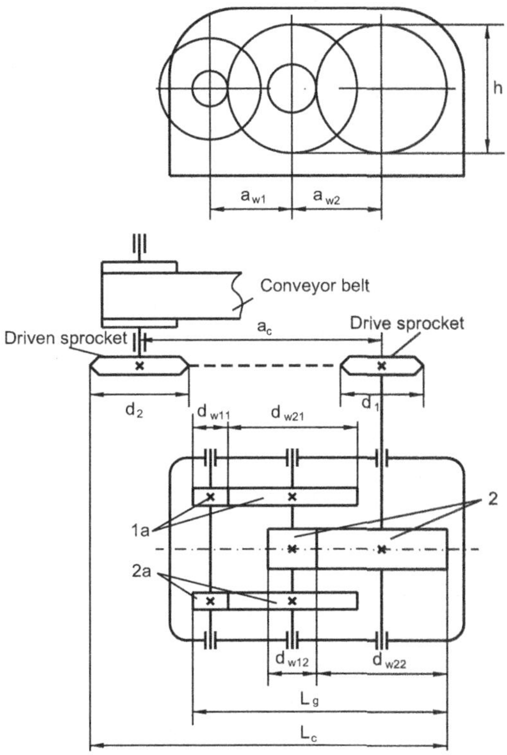

Fig. 1. Calculation schema

With the following constraints

$$\begin{aligned} 1 \le u_1 \le 9 \\ 1 \le u_2 \le 9 \\ 1 \le u_c \le 6 \end{aligned} \tag{10}$$

It is realized from the above analysis that to solve the optimization problem, the calculation of a_{w1}, a_{w2}, d_{w11}, d_{w21}, d_{w22}, a_c and d_c is required. These parameters are calculated in the following sub-sections.

2.1 Determining the Center Distance and the Pinion Pitch Diameter of the First Step

For the first step, the center distance a_{w1} can be determined by [29]:

$$a_{\mathrm{w1}} = K_a \cdot (u_1 + 1) \cdot \sqrt[3]{\frac{T_{11} \cdot k_{H\beta}}{[\sigma_H]^2 \cdot u_1 \cdot \psi_{ba1}}} \tag{11}$$

In which, k_a is the coefficient of material; $k_a = 43$ because the gear material is steel [29]; $K_{H\beta}$ is the coefficient of contact load ratio; $k_{H\beta} = 1.02 \div 1.28$ [29] and it is chosen as $k_{H\beta} = 1.1$; $[\sigma_H]$ is the allowable contact stress (MPa); if the gear material is steel, $[\sigma_H] = 350\ldots420$ (MPa); is the coefficient of the wheel face width; in this case $\psi_{ba1} = 0.4\ldots0.5$ [29]. T_{11} is the torque of the pinion (Nmm); for this gearbox, T_{11} is found by:

$$T_{11} = T_{out}/\left(2 \cdot u_g \cdot u_c \cdot \eta_{hg}^2 \cdot \eta_c \cdot \eta_{be}^3\right) \tag{12}$$

Where, η_{hg} is the efficiency of helical gear set ($\eta_{hg} = 0.96\ldots0.98$ [29]); is the efficiency of the chain drive ($\eta_c = 0.95\ldots0.97$ [29]); η_{be} is the efficiency of a rolling bearing pair ($\eta_{be} = 0.99\ldots0.995$ [29]). Choosing $\eta_{hg} = 0.97$, $\eta_{be} = 0.992$, $\eta_c = 0.96$ and substituting them into (10) gives:

$$T_{11} = 0.567 \cdot T_{out}/\left(u_g \cdot u_c\right) \tag{13}$$

. Substituting $k_{H\beta} = 1.1$ and (13) into Eq. (12) the output can be:

$$a_{\mathrm{w1}} = 36.739 \cdot (u_1 + 1) \cdot \sqrt[3]{\frac{T_{out}}{[\sigma_{H1}]^2 \cdot u_1 \cdot u_g \cdot u_c \cdot \psi_{ba1}}} \tag{14}$$

After getting a_{w1}, the pinion pitch diameter of the first step is calculated by [29]:

$$d_{\mathrm{w11}} = 2 \cdot a_{w1}/(u_1 + 1) \tag{15}$$

2.2 Calculating the Center Distance and the Gear Pitch Diameter of the Second Step

The center distance of the second step a_{w2} is found by [11]:

$$a_{\mathrm{w2}} = K_a \cdot (u_2 + 1) \cdot \sqrt[3]{\frac{T_{12} \cdot k_{H\beta}}{[\sigma_H]^2 \cdot u_2 \cdot \psi_{ba2}}} \tag{16}$$

In which, T_{12} is the pinion torque (Nmm) which can be calculated as:

$$T_{12} = T_{out}/\left(u_2 \cdot u_c \cdot \eta_{hg} \cdot \eta_c \cdot \eta_{be}^2\right) \tag{17}$$

Choosing $\eta_{hg} = 0.97$, $\eta_{be} = 0.992$ and $\eta_c = 0.96$ (as in Sect. 2.1) and substituting them into Eq. (17) gets:

$$T_{12} = 1.0913 \cdot T_{out}/(u_2 \cdot u_c) \tag{18}$$

Then, taking $k_a = 43, k_{H\beta} = 1.1$ (as in Sect. 2.1) and substituting them and (18) into (16) gets:

$$a_{w2} = 45.6998 \cdot (u_2 + 1) \cdot \sqrt[3]{\frac{T_{out}}{[\sigma_H]^2 \cdot u_2^2 \cdot u_c \cdot \psi_{ba2}}} \tag{19}$$

After having a_{w2}, the gear pitch diameter of the second step then is found by [29]:

$$d_{w22} = 2 \cdot a_{w2} \cdot u_2/(u_2 + 1) \tag{20}$$

2.3 Determining the Driven Sprocket Diameter and the Center Distance of the Chain Drive

The pitch diameter of the driven sprocket d_2 is determined as [29]:

$$d_2 = d_1 \cdot u_c \tag{21}$$

In which, d_1 is the pitch diameter of the drive sprocket which is calculated by [29]:

$$d_1 = p/\sin(\pi/z_1) \tag{22}$$

Where, z_1 is the number of teeth of the drive sprocket; z_1 can be found by [28]:

$$z_1 = 32.4 - 2.4 \cdot u_c \tag{23}$$

p is the chain pitch (mm); *p* is determined from the design power capacity P which can be determined by [29]:

$$P = P_1 \cdot k \cdot k_z \cdot k_n \tag{24}$$

Wherein, P_1 is the power rating (kW):

$$P_1 = T_1 \cdot n_1/\left(9.55 \cdot 10^6\right) \tag{25}$$

In which, n_1 is the rotational speed of the drive sprocket (rpm):

$$n_1 = n_m/u_g \quad (26)$$

$$T_1 = T_{out} \cdot \eta_c \cdot \eta_b \quad (27)$$

Where, is the efficiency of the chain drive ($\eta_c = 0.95 \div 0.97$ [29]); is the efficiency of a pair of bearings ($\eta_b = 0.99 \div 0.995$ [29]); T_1 and T_{out} are the drive and the output torques, respectively (Nmm).

k, k_z and k_n are the coefficients determined as [29]:

$$k = k_d \cdot k_p \cdot k_c \cdot k_{adj} \cdot k_{\text{lub}} \cdot k_{con} \quad (28)$$

$$k_z = 25/z_1 \quad (29)$$

$$k_n = n_{01}/n_1 \quad (30)$$

Wherein, n_{01} is the tabulated rotational speed of the drive sprocket; k_d, k_c, k_p, k_{adj}, k_{lub} and k_{con} are the coefficients of the shock factor, the center distance, the drive position, the possibility of center distance adjusting, the lubrication and the operating conditions, respectively.

2.4 Experimental Work

To examine the effects of the input factors on the optimum gear ratios, a simulation experiment was planned and conducted. In the experiment, 5 input parameters (Table 1) and a 2-level full factorial design were selected. Consequently, a number of $2^5 = 32$ tests were planned for the experiment. Based on Eqs. (9) and (10), a computer program was built to conduct the experiment. The input factor levels and the output results (the optimum gear ratios of the first step and the chain drive are presented in Table 2.

Table 1. Input parameters

Total system ratio	u_t	–	10	50
Coefficient of wheel face width of step 1	X_{ba1}	–	0.4	0.5
Coefficient of wheel face width of step 2	X_{ba2}	–	0.35	0.4
Allowable contact stress	AS	MPa	350	420
Output torque	T_{out}	Nmm	10^5	10^7

3 Results and Discussion

From the results of the experiments (Table 2) it is reported that the optimum ratios of the first step and the chain drive u_c are unchanging. In other words, they are not contingent upon the input factors. In addition, u_1 gets the maximum value of gear ratios for a helical gear unit ($u_1 = 9$). This is because when it takes the possible maximum value of a helical gear unit ($u_1 = u_{max} = 9$) it can promote the advantages of the double gear sets. Interestingly, the optimum gear ratios of the chain drive is $u_c = 1$, which means with the mentioned objective function, the chain drive only transfers the torque to a shaft at a long distance without reducing the speed.

$$u_1 = 9 \tag{31}$$

$$u_c = 1 \tag{32}$$

Table 2. Experimental plans and output response.

StdOrder	RunOrder	CenterPt	Blocks	u_g	X_{ba1}	X_{ba2}	AS	T_{out}	u_1	u_c
25	1	1	1	10	0.4	0.35	420	10000	9	1
17	2	1	1	10	0.4	0.35	350	10000	9	1
29	3	1	1	10	0.4	0.4	420	10000	9	1
6	4	1	1	50	0.4	0.4	350	100	9	1
15	5	1	1	10	0.5	0.4	420	100	9	1
19	6	1	1	10	0.5	0.35	350	10000	9	1
...										
2	31	1	1	50	0.4	0.35	350	100	9	1
28	32	1	1	50	0.5	0.35	420	10000	9	1

Equations (31) and (32) are used to find the optimum gear ratios of the first step u_1 and the chain drive u_c. After getting u_1 and u_c, the gear ratio of the second step u_2 can be determined by:

$$u_2 = u_t/(u_1 \cdot u_c) \tag{33}$$

Because $u_1 = 9$ and $u_c = 1$, Eq. (33) can be rewritten as:

$$u_2 = u_t/9 \tag{34}$$

4 Conclusions

A study on the calculating the optimum gear ratios of mechanical driven systems using a two-step helical gearbox with the first step double gear sets and a chain drive was conducted. From the results of the study, the following conclusions are given:

- The minimum system cross section area of a mechanical driven system using a two-step helical gearbox with first step double gear sets and a chain drive can be found by calculating the optimum gear ratios.

- Models to determine the optimum gear ratios of the systems were proposed to get the minimum system cross section area.
- Using proposed models, the optimum gear ratios can be found simply and accurately as they are explicit.

Acknowledgements. The work described in this paper was supported by Thai Nguyen University of Technology for a scientific project.

References

1. Kudreavtev, V.N., Gierzaves, I.A., Glukharev, E.G.: Design and Calculus of Gearboxes. Mashinostroenie Publishing, Sankt Petersburg (1971)
2. Chat, T: Some problems of kinematics calculation of transmission mechanics system. In: Proceedings of the National Conference on Engineering Mechanics, Hanoi, Vietnamese, vol. 2, pp. 7–12 (1993)
3. Milou, G., Dobre, G., Visa, F., Vitila, H.: Optimal design of two step gear units, regarding the main parameters. VDI Berichte No. 1230, p. 227 (1996)
4. Pi, V.N.: A method for optimal calculation of total transmission ratio of two step helical gearboxes. In: Proceedings of the National Conference on Engineering Mechanics, Ha Noi, p. 12 (2001)
5. Cam, N.T.H., Pi, V.N., Tuan, N.K., Hung, L.X., Thao, T.T.P.: Determining optimal partial transmission ratios of mechanical driven systems using a V-belt drive and a helical reducer with second-step double gear-sets. In: Fujita, H. et al. (eds.) ICERA 2018, LNNS, vol. 63, pp. 261–269 (2019). https://doi.org/10.1007/978-3-030-04792-4_35
6. Petrovski, A.N., Sapiro, B.A., Saphonova, N.K.: About optimal problem for multi-step gearboxes. Vestnik Mashinostroenie, Russian, no. 10, p. 13 (1987)
7. Pi, V.N.: A new study on optimal calculation of partial transmission ratios of two-step helical gearboxes. In: 2nd WSEAS International Conference on Computer Engineering and Applications (CEA 2008), Acapulco, Mexico, 25–27 January 2018, pp. 162–165 (2008)
8. Pi, V.N., Tuan, N.K.: Optimum determination of partial transmission ratios of three-step helical gearboxes for getting minimum cross section. J. Environ. Sci. Eng. A **5**, 570–573 (2016)
9. Pi, V.N., Binh, N.D., Dac, V.Q.: Phan Quang The, optimal calculation of total transmission ratio of three-step helical gearboxes for minimum mass of gears. J. Sci. Technol. 6 Eng. Univ. 91 (2006)
10. Pi, V.N.: A new study on optimal calculation of partial transmission ratio of three-step helical reducers. In: The 3rd IASME/WSEAS International Conference on Continuum Mechanics, Cambridge, UK, p. 23 (2008)
11. Pi, V.N.: A new study on optimal calculation of partial transmission ratios of three-step helical reducers for getting minimal cross section dimension. In: The 2nd WSEAS International Conference on Computer Engineering and Applications (CEA 2008), Acapulco, Mexico, 25–27 January 2008, pp. 290–293 (2008)
12. Pi, V.N.: Optimal determination of partial transmission ratios of three-step helical gearboxes with first and third-step double gear-sets for minimal mass of gear. In: American Conference on Applied Mathematics (MATH 2008), Harvard, Massachusetts, USA, 24–26 March 2008, pp. 385–388 (2008)

13. Romhild, I., Linke, H.: Gezielte Auslegung Von Zahnradgetrieben mit minimaler Masse auf der Basis neuer Berechnungsverfahren. Konstruktion **44**, 229–236 (1992)
14. Pi, V.N.: A new study on the optimal prediction of partial transmission ratios of three-step helical gearboxes with second-step double gear-sets. WSEAS Trans. Appl. Theor. Mech. **2**(11) (2007)
15. Buiga, O.: 3 stage helical speed reducer partial gear ratios optimal determination using genetic algorithms. Proc. Rom. Acad. Ser. A **20**(2), 156–163 (2019)
16. Pi, V.N.: Optimal determination of partial transmission ratios for four-step helical gearboxes with firt and third step double gear-sets for minimal mass of gears. In: Applied Computing Conference (ACC 2008), Istanbul, Turkey, 27–30 May 2008 (2008)
17. Pi, V.N.: Optimal calculation of partial transmission ratios for four-step helical gearboxes with first and third-step double gear-sets for minimal gearbox length. In: American Conference on Applied Mathematics (MATH 2008), Harvard, Massachusetts, USA, 24–26 March 2008, pp. 29–32 (2008)
18. Hung, L.X., Pi, V.N., Du, N.V.: Optimal calculation of partial transmission ratios of four-step helical gearboxes with second and fourth-step double gear-sets for minimal mass of gears. In: The International Symposium on Mechanical Engineering, ISME, Ho Chi Minh city, Vietnam, September 2009, pp. 21–23 (2009)
19. Pi, V.N.: Optimal calculation of partial transmission ratios of four-step helical gearboxes for getting minimal gearbox length. Int. J. Mech. Aerosp. Ind. Mechatron. Manuf. Eng. 22–25 (2008)
20. Pi, V.N.: A new and effective method for optimal calculation of total transmission ratio of two step bevel - helical gearboxes. In: International Colloquium on Mechanics of Solids, Fluids, Structures and Interaction, Nha Trang, Vietnam, pp. 716–719 (2000)
21. Pi, V.N.: A study on optimal calculation of partial transmission ratios of three-step bevel helical gearboxes. In: International Workshop on Advanced Computing and Applications (ACOMP 2008), 12–14 March 2008, pp. 277–286 (2008)
22. Chernavsky, S.A., et al.: Design of Mechanical Transmissions: Manual for High Technical Schools, p. 560. Mashinostroenie, Moscow (1984)
23. Pi, V.N., Dac, V.Q.: Calculation of total transmission ratio of two step worm reducers for the best reasonable gearbox housing structure (in Vietnamese). J. Sci. Technol. Thai Nguyen Univ. **1**(41), 65–69 (2007)
24. Pi, V.N., Dac, V.Q.: Optimal calculation of partial transmission ratios of worm-helical gear reducers for minimal gearbox length. J Sci. Technol. Tech. Univ. (61), 73–77 (2007)
25. Pi, V.N., Cam, N.T.H., Tuan, N.K.: Optimum calculation of partial transmission ratios of mechanical driven systems using a V-belt and two-step bevel helical gearbox. J. Environ. Sci. Eng. A **5**, 566 (2016)
26. Pi, V.N., Thao, T.T.P., Thao, L.T.P.: A new study on optimum determination of partial transmission ratios of mechanical driven systems using a V-belt and two-step helical gearbox. Vietnam Mech. Eng. J. (10), 123 (2015)
27. Pi, V.N., Tuan, N.K., Hung, L.X., Cam, N.T.H., Thao, T.T.P.: Determining optimum partial transmission ratios of mechanical driven systems using a V-Belt drive and a three-stage helical reducer. In: Advances in Material Sciences and Engineering, pp. 81–88. Springer (2019)
28. Pi, V.N., Thao, T.T.P., Tuan, D.A.: Optimum determination of partial transmission ratios of mechanical driven systems using a chain drive and two-step helical gearbox. J. Environ. Sci. Eng. B **6**, 80 (2017)
29. Chat, T., Van Uyen, L.: Design and Calculus of Mechanical Transmissions (in Vietnamese). Educational Republishing House, Hanoi (1998)

A Study on Electroless Copper Plating on Poly (Methyl Methacrylate) Through Organic Covalent Grafting

Ly Viet Anh(✉) and Ngoc Pi Vu

Thai Nguyen University of Technology, Thai Nguyen, Vietnam
Vietanh.ly4@gmail.com, Vietanh.tq3@tnut.edu.vn

Abstract. Electroless copper plating on Poly(methyl methacrylate) (PMMA) is a technology that is increasingly popular and widely used in practice. Poly (methyl methacrylate) (PMMA), also known as by the trade names, Plexiglas®, is a transparent thermoplastic often used in sheet form as a lightweight or shatter-resistant alternative to glass. Especially with the transparent plastic equipment requiring high aesthetic such as equipment in the automotive industry, toilet equipment, … coating a Copper layer on the material surface simply provides the device a high aesthetic ability, which prevents dust and also easy to be cleaned. However, most of the copper plating processes on plastics use Cr6+, which is toxic to the environment. This article presents a new approach, Copper plating on PMMA through organic covalent bonding, which is both safe and environmentally responsible.

Keywords: Electroless plating · Covalent bonding · Palladium catalyst · Chromic Acid · EDX · SEM

1 Introduction

Nowadays, transparent plastic (represented by PMMA) is increasingly widely used in the realities of everyday life because of available advantages such as light, high processing capacity, good bearing capacity, high corrosion resistance [1]. Therefore, this material has been used in many devices in practice, for example, common items in the kitchen or modern devices in the field of Electronic Microelectronics (MEMS) [1, 2]. In practice, people often use Copper (Cu) as a perfect coating on plastic substrate. Plated products according to this method are not only ensure the advantages of transparent plastic, but also are high luxurious and aesthetic [2]. However, in the treatment stage of Copper chemical plating processes on PMMA, Chromic Acid is used. It is a toxin causing cancer and endangering to human [2]. Therefore, this method is restricted to use at the European Union to minimize the harming to the environment and people [3]. To solve the above problem, this paper introduces a new method in which Copper is chemically plated on PMMA using Amine organic chemicals instead of chromium-containing chemicals in the invasion process to grip the substrate with Palladi catalyst [4, 5]. In the study, a new surface treatment process method using Hydrogen-peroxide acid solution is proposed [4, 5]. This approach not only helps protect the environment

K.-U. Sattler et al. (Eds.): ICERA 2019, LNNS 104, pp. 94–99, 2020.
https://doi.org/10.1007/978-3-030-37497-6_10

and people but also minimize the using of precious metals as catalysts (in this article, Palladi). The essence of this new method is that Hydrogen-peroxide acid, organic amines and PdC12 catalyst are used to create the complex link between PMMA and Copper plating, instead of using Chromium to create tiny holes on the PMMA plastic surface. Finally, the characteristic mechanical properties of the Copper plating layer will be evaluated by SEM, EDX and adhesive test tape standards [6–8] (Diagram 1).

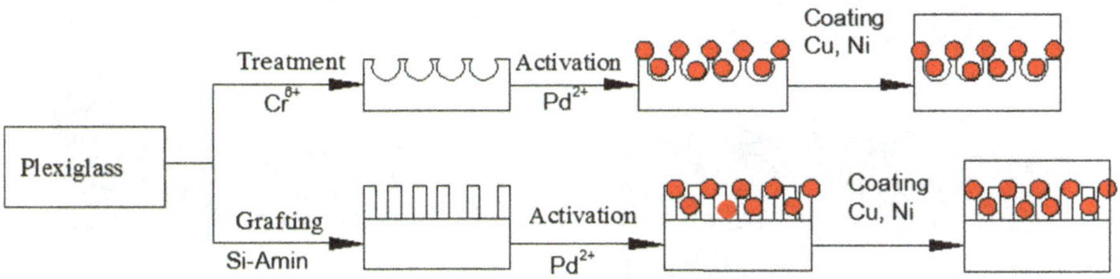

Diagram 1. The traditional Copper chemical plating processes on PMMA and our new method

2 Methodology

2.1 Sample Preparation PMMA

The PMMA substrate is cut into pieces of 40 × 40 (mm^2) which are all cleaned in the ultrasonic machine. Then, the PMMA sheets are washed twice in deionized water and alcohol 70°. Finally, they are dried by the dryer for 1–3 min.

2.2 Surface Treatment Process with a Solution of Hydro-Peroxide Acid

The solution used for pretreatment of PMMA surface includes 37% HCl solution, 50% H_2O_2 old oxide, and deionized water with a involume ratio of 1: 1: 5. The temperature needed during the process is about 80 °C in the period of 8–12 min.

2.3 Rafting of Silylated Amine Palladium Complexes

The grafting solution consists of 88% $PdCl_2$, mixed with 100 ml of HCl solution, then heated to 65 °C so that $PdCl_2$ can be transformed into H_2PdCl_4. Afterwards, a mixture of 1.12 ml of 3-propyl ethylene-diamine and 49 ml of deionized water is created to give 50 ml of solution, which is then mixed with 100 ml of H_2PdCl_4 solution above. Next, 25 ml of NH_4OH is added to the prepared solution to form a complete transplant solution, the reactions in the solution will form Silane-Amin-Palladi complex. The transplantation process on the PMMA sheets will take about 25–30 min, then the sheets are washed with deionized water and dryer.

2.4 Electroless Copper (Cu) Deposition

The solution used for Electroless deposition process Copper (Cu) onto PMMA includes $CuSO_4.5H_2O$ 10 g/l; NaOH 10 g/l; HCHO (37–41%) 10 ml/l; K-Na 50 g/l. It is noted that copper plating process is carried out at normal temperature, about 25-30 °C, the plating time is from 7–15 min. After electroless copper deposition, the PMMA sheet is washed in an ultrasonic bath with deionized water and dried by dryer (Fig. 1).

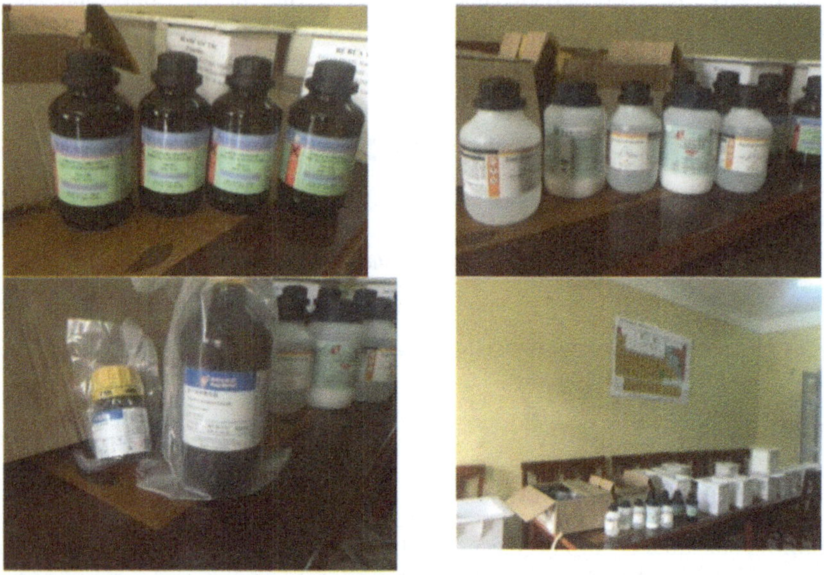

Fig. 1. Chemicals used in Experimental process of copper plating (Cu) on PMMA (Plexiglass®)

3 Results and Discussion

3.1 Assessment of Microstructured Plating Layer Through SEM Technology

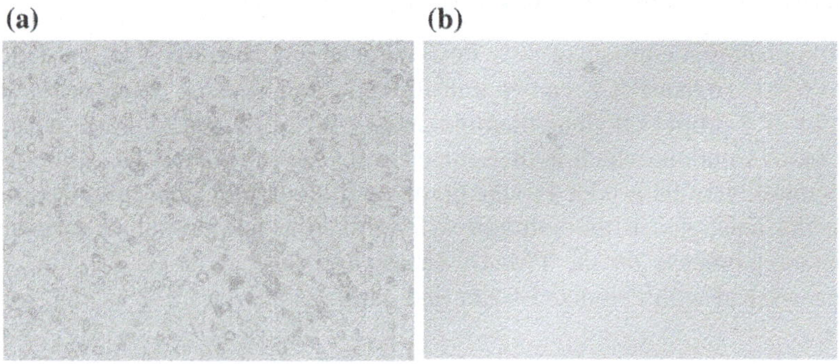

Fig. 2. Image of PMMA surface analyzed by SEM: (a) before plating, (b) after plating

The results of the experiment are recorded by the TESCAN-BRUGER SEM with the maximum magnification of 10,000 times. Figure 2 shows the microscopic analysis of the surface of the ABS substrate before and after plating. It was observed that the plated PMMA surface before plating is more rugged while the surface after plating is less membrane, held and more uniform. Moreover, Copper plating shows shiny, beautiful images with high aesthetics.

SEM magnification images with a magnification of 1500 times (Fig. 3) show that the Cu coating of the experimental process adheres well to ABS plastic substrate. After 7 min of plating time, the thickness of the coating layer shown on the SEM machine of TESCAN is about 12 μm. In addition, the coating is evenly distributed on the substrate surface even at the corner. This indicates the superiority of this method of plating.

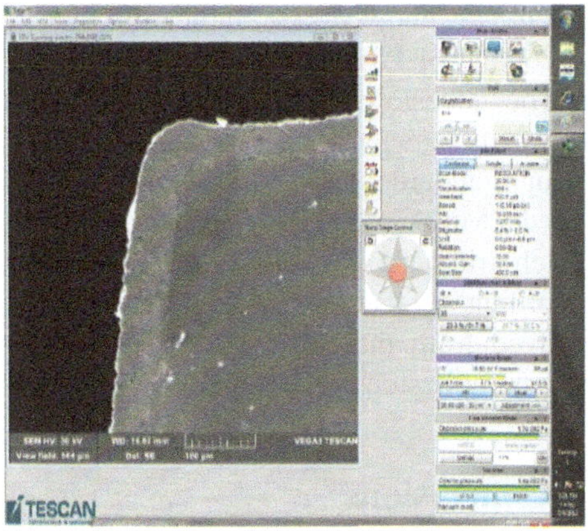

Fig. 3. Image of SEM analysis of copper plating thickness on PMMA background

3.2 Assessment of Chemical Composition Coating Through EDX Technology

EDX technology is an analytical technique used for the elemental analysis or chemical characterization of a sample. It relies on an interaction of some source of X-ray excitation with a sample. Its characterization capabilities are based on the fundamental principle that each element has a unique atomic structure allowing a unique set of peaks on its electromagnetic emission spectrum [6].

Figure 4 describes an image analysis EDX of the Copper plating layer. From the figure, the contents of elements are: Cu is 72.2%; C is 6.9%; Oxygen is 3%; iron (Fe) is 10.5%; Boron is 5.2% and Al is 2.2%. The remaining metal impurities are mixed in Cu plating.

The chemical composition of the plating layer with high copper content accounts for nearly 3/4. In addition, the O and C in covalent bonds between the metal and the PMMA substrate are the total of Cu, O and C accounting for 82.1% of the plating content. It is interesting that the maximum Pd

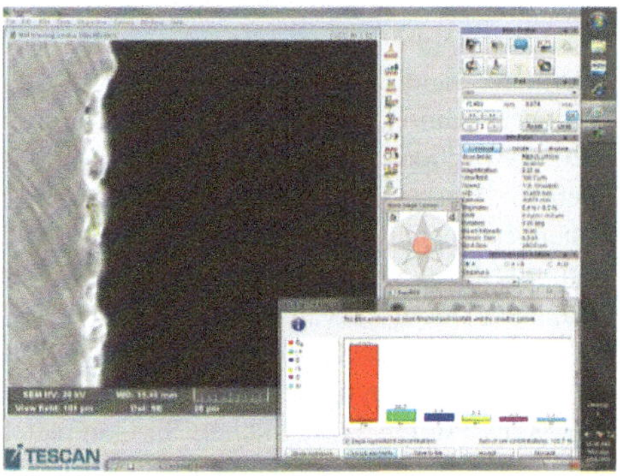

Fig. 4. Image of EDX analysis of copper plating thickness on PMMA background

content was about 0.13%, indicating that Pd was completely catalyst. This means that the content of Pd is very little and completely covered (coated) by Cu coating.

The adhesion of Cu coating on PMMA sheet is the most important factor to evaluate this method. The adhesion of the coating is checked by a tape of ASTMD - US testing standards. Using one diamond knife to cut several straight lines on the Cu-coated PMMA sheet. These lines are separated each other 1 mm, creating one of the scratch gridmade 4×4 (mm^2). This test tape is applied to the cut PMMA plastic sheet. According to the experiment, after peeling off the adhesive tape, no squares are removed from the sheet. Clearly, according to the test on the Copper coating, the required adhesion on the PMMA plate was satisfied.

4 Conclusions

The article introduces an experimental research on an advanced chemical plating process. Cu plated on the surface of insulating materials is the representative of PMMA material (a kind of plastic glass). By designing and conducting experiments and applying advanced testing techniques, the correctness of the theory of new metal plating on PMMA materials has been proved. Especially, this method does not use toxic chemicals Cr^{6+} in the process of pretreatment surface. Therefore, this method not only ensures environmental friendliness, saving precious materials as catalysts, but also leads to good properties of plating.

Acknowledgement. The work described in this paper was supported by Thai Nguyen University of Technology for a scientific project.

References

1. Nisar, A., Afzulpurkar, N., Mahaisavariya, B., Tuantranont, A.: MEMS-based micropumps in drug delivery and biomedical applications. Sens. Actuators B Chem. **130**(2), 917–942 (2008)
2. Rakhshan, V.: Marginal integrity of provisional resin restoration materials: a review of the literature. Saudi J. Dent. Res. **6**(1), 33–40 (2015)
3. European Union: European parliament and the council. Official J. Eur. Union **2005/90/EC** (EC) (907/2006), OJ L 168, 21.6.2006, 50 (2006)
4. Zeb, G., Duong, X.T., Vu, N.P., Phan, Q.T., Nguyen, D.T., Ly, V.A., Salimy, S., Le, X.T.: Chemical metallization of KMPR photoresist polymer in aqueous solutions. Appl. Surface Sci. **407**, 518–525 (2017)
5. Vu, N.P., Duong, X.T., Ly, V.A., Nguyen, D.C., Tran, M.D., Phan, Q.T., Balazinski, M., Son, L.T., Zeb, G., Le, X.T.: Electroless nickel plating onto Plexiglas® through simple covalent grafting of vinylpyridine seed layer. Mater. Des. **144**, 151–158 (2018)
6. Le, X.T., Poirier, J.-S., Michel, S.: Completely aqueous route for metallization of structural polymeric materials in micro-electro-mechanical systems. J. Electrochem. Soc., under review
7. Zeb, G., Zafar, M., Palacin, S., Le, X.T.: An application of diazonium-induced anchoring process in the fabrication of micro-electro-mechanical systems. Adv. Mater. Technol. **2**(12), 1700159 (2017)

8. Le, X.T., Viel, P., Sorin, A., Jegou, P., Palacin, S.: Electrochemical behaviour of polyacrylic acid coated gold electrodes: an application to remove heavy metal ions from wastewater. Electrochim. Acta **54**(25), 6089–6093 (2009)
9. https://vi.wikipedia.org/wiki/K%C3%ADnh_hi%E1%BB%83n_vi_l%E1%BB%B1c_nguy%C3%AAn_t%E1%BB%AD

A Vision-Based Method of Reverse Engineering for 2D CNC Machining

Huu-Cuong Nguyen[1(✉)] and Phuoc-Loc Nguyen[2]

[1] Can Tho University, Can Tho, Vietnam
nhcuong@ctu.edu.vn
[2] Kien Giang Vocational College, Rach Gia, Vietnam

Abstract. An automated vision-based method for profile reconstruction is presented in this paper in order to generate NC-code script file for 2D CNC machining from sample object. A profile reconstruction system capturing images of sample objects was developed and located on the flat table. Based on the sample image, the boundary in the image is detected by several image processing algorithms. After that the profile data is acquired and converted into NC-code commands. Experimental results show that the proposed method is effective in reverse engineering and the developed profile-reconstruction system has acquired high accuracy and flexibility.

Keywords: Reverse engineering · Profile reconstruction · CNC machining · NC-code generation · Vision-based measurement

1 Introduction

Reverse engineering and shape reconstruction play an important role in product design and manufacturing process. In mechanical engineering, reverse engineering has evolved from collecting the object data then redesign a new produce product that similar to the original object [1]. Collecting measurement data on the shape of the sample object is an important work and takes most time in reconstruction process. In most cases control of geometry in industry is carried out using devices that apply coordinate measuring technique. A wide range of methods can be rated to this technique on account of the essence of the measurement and collection of the coordinates describing a position of the individual measuring points [2, 3]. The current researches in the field of machine vision have achieved more applications in measurement data acquisition. For example, a laser-vision-based scanning system was developed and implemented to detect 3D circles and measure their radius [4]. An automated based-vision quality inspection system was proposed to measure the size and position of holes on a wheel-disc [5]. A stereo vision was developed and tested to identify common uncertainty sources [6].

In this paper, a reverse engineering method is proposed for CNC (Computer Numerical Control) machining applications such as cutting, welding, engraving… A vision-based system is developed to implement this method. The task of the system is

K.-U. Sattler et al. (Eds.): ICERA 2019, LNNS 104, pp. 100–106, 2020.
https://doi.org/10.1007/978-3-030-37497-6_11

to acquire the 2D (two-dimensional) profile data of an existing sample object, then translate the collected data into NC-code (or G-code) which is used to control the CNC machine.

2 Proposed Method

The proposed method of profile reconstruction for 2D CNC machining is illustrated as in Fig. 1. The sample object is captured by the camera, then its image is converted into grayscale image. Otsu's thresholding [7] is applied to binarize the grayscale image. Another image processing algorithms is used for contour detection [5]. In order to reduce the number of coordinate points on the detected contour, as well as reduce the NC-code to be reconstructed, a contour simplification method [8] is employed.

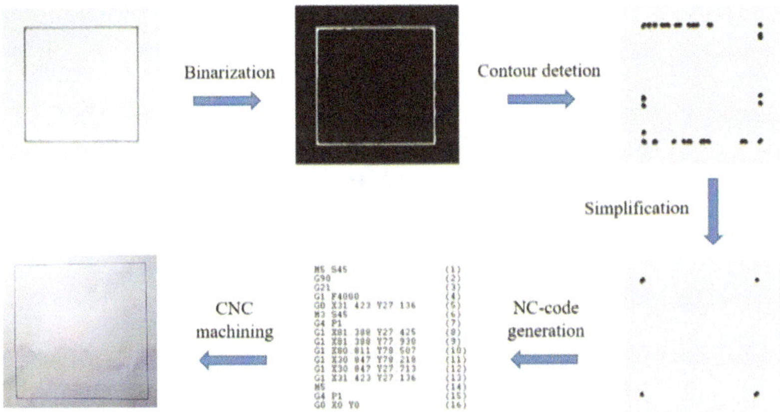

Fig. 1. Profile-reconstruction method.

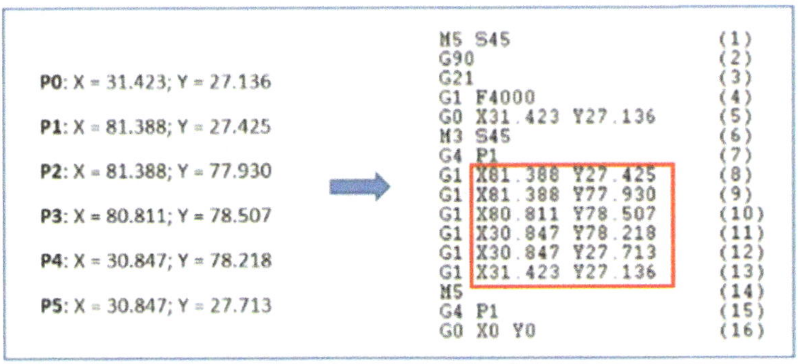

Fig. 2. NC-code generation procedure.

Based on the simplified contour data, an NC-code script file is generated by converting the coordinate points in contour data into coordinate commands in NC-code. Specifically, adjacent coordinates in the contour data is listed sequentially in the NC-code script file. A simple example for NC-code generation is described as in Fig. 2. In which, on the left is the group of points with adjacent coordinates in the contour data and on the right is the NC-code which is created in script file.

3 Profile-Reconstruction System

The proposed profile-reconstruction system was developed as shown in Fig. 3. The system includes a camera and a flat table. The camera is Logitech C270 that allows the acquisition of a 1280 × 720 pixel color image. The sample object is fixed on the flat table.

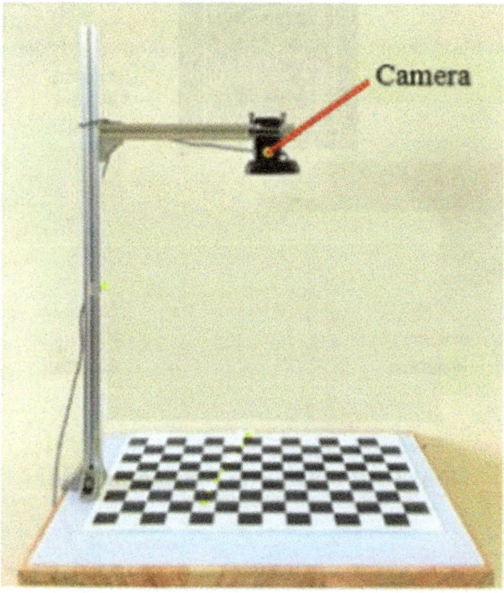

Fig. 3. Profile-data collection system.

A direct camera calibration method [9] was employed by using a chessboard as a calib-object. This calibration step aims to rectify the distortion of the captured image and estimate the relationship between the real word and image coordinates. In the Fig. 4, the left is a captured image and the right is a rectified image which is the result of the calibration.

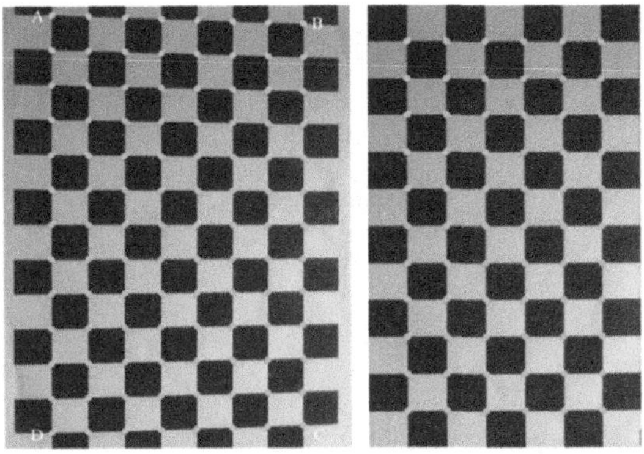

Fig. 4. Profile-data collection system.

Based on the real known dimensions of the cells of the chessboard and the pixel distances between the corners of the cells on the chessboard's image, the relationship between the real and image coordinates can be calculated using the polynomial functions

$$h = p_0 v^2 + p_1 v + p_2 \quad (1)$$

$$w = q_0 u^2 + q_1 u + q_2 \quad (2)$$

where h and w are the real height and width of the chessboard's cells, u and v are the vertical and horizontal distances, relatively.

Table 1. Calibration data.

Number of cells	Real width [mm]	Real height [mm]	Image width [pixel]	Image height [pixel]
1	29	29	101	103
2	58	58	200	202
3	87	87	302	305
4	116	116	402	405
5	145	145	502	506

A simple polynomial fitting method is applied to estimate the coefficients of the polynomial functions based in the calibration data which is listed in Table 1. The estimated coefficients are: p0 = 0.0000; p1 = 0.2886; p2 = −0.5690 and q0 = 0.0000; q1 = 0.2888; q2 = 0.0529.

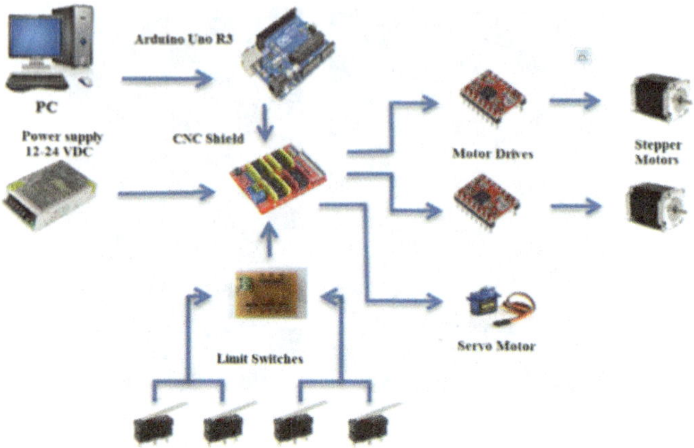

Fig. 5. Drawing system diagram

Fig. 6. Drawing system

After reconstructed, all of the real coordinates of the profile are converted into NC-code script file. A CNC drawing machine was built to implement this NC-code file and verify the accuracy of the profile-reverse system. The diagram of the CNC drawing machine is shown as Fig. 5. An Arduino Uno R3 is employed as a controller which handles two stepper motors to move a pen on a flat surface. The pen is raised and lowered by a servo motor. All of motors are driven by a CNC Shield. Figure 6 illustrates a real complete drawing machine.

4 Experimental Results

Experiments were implemented to evaluate the accuracy of the proposed profile-reconstruction system. In which, a square with a 10 cm edge and a circle with a radius of 10 cm used as samples. Reconstructed profiles were drawn which are measured in the horizontal and vertical dimensions corresponding to the X and Y directions by a caliper as shown in Fig. 7. Results of experiments are described in Table 2 shows that the proposed system achieves the accuracy of over/more than 98%.

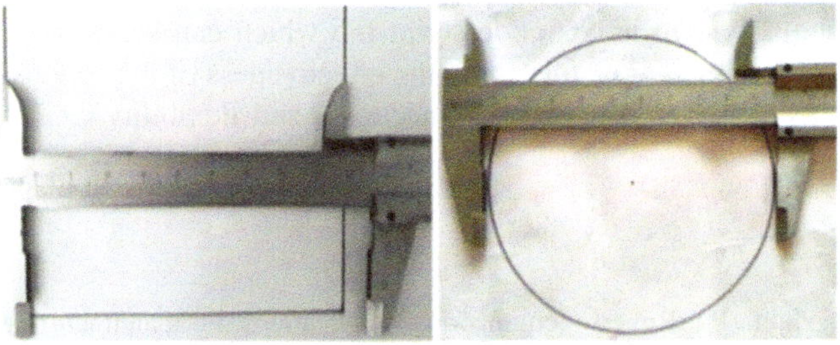

Fig. 7. Evaluate the reconstructed profile

Table 2. Experimental results.

No.	Horizontal dimension - X [cm]	Vertical dimension - Y [cm]	Horizontal error [%]	Vertical error [%]
Square				
1	9.90	10.14	1.0	1.4
2	10.12	10.02	1.2	0.2
Circle				
1	10.16	10.12	1.6	1.2
2	10.01	10.14	0.1	1.4

Some experiments were also implemented with other samples which have different shapes to assess the flexibility of the algorithm and the system. A software was built to interface with users which have a GUI as shown in Fig. 8.

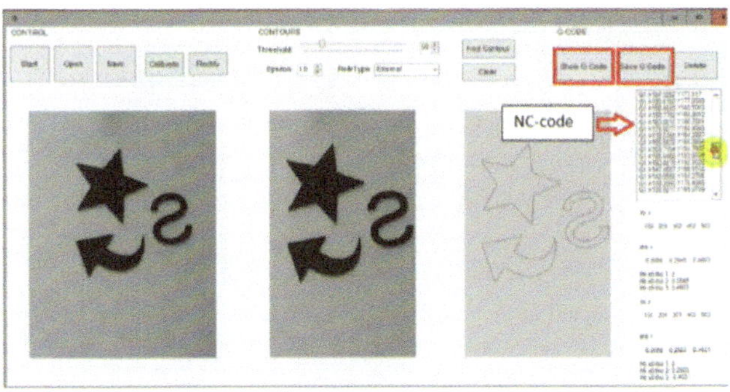

Fig. 8. GUI of profile-reconstruction software

5 Conclusion

In this study, a vision-based method of profile reconstruction for 2D CNC machining is proposed and implemented. A profile-reconstruction system is developed which consists of a sample capture mechanism, a CNC drawing machine, and a software. The output of the built system is NC-code script file which can be used for several CNC machines. The experimental results show the effectiveness of the proposed method and the developed system has achieved high accuracy and flexibility.

References

1. Rathore, N., Jain, P.: Reverse engineering applications in manufacturing industries: an overview, chapter 45. In: DAAAM International Scientific Book. DAAAM International, Vienna (2014)
2. Wozniak, A., Dobosz, M.: Metrological feasibilities of CMM touch trigger probes. Part I: 3D theoretical model of probe pretravel. Measurement **34**(4), 273–286 (2003)
3. Arman, H., Sadieh, K., Mawadah, A., Fatema, M., Abdullah, A.: Dimension effect on the ultrasonic pulse velocity. Int. J. Geogr. Geol. **6**(2), 18–25 (2017)
4. Lee, B., Nguyen, C.: Three-dimensional circle detection and radius measurement. Optik – Int. J. Light Electron Opt. **126**(24), 5412–5419 (2015)
5. Nguyen, C., Nguyen, L., Lee, B.: A vision-based wheel disc inspection system. In: Hamido, F., Nguyen, C., Vu, P., Banh, L., Hermann, P. (eds.) Advances in Engineering Research and Application ICERA 2018. LNNS, vol. 63, pp. 109–115. Springer, Vietnam (2018)
6. Quinonez, J., Sergiyenko, O., Fuentes, W., Rivas, M., Balbuena, D., Rascon, R., Mercoreli, P.: Improve a 3D distance measurement accuracy in stereo vision systems using optimization method's approach. Opto-Electron. Rev. **25**(1), 24–32 (2017)
7. Otsu, N.: A threshold selection method from gray-level histograms. IEEE Trans. Syst. Man Cybern. **9**(1), 62–66 (1979)
8. Gu, C., Kunt, M.: Contour simplification and motion compensated coding. Sig. Process. Image Commun. **7**(4–6), 279–296 (1995)
9. Nguyen, C., Lee, B.: 3D model reconstruction system development based-on laser-vision technology. Int. J. Autom. Technol. **10**(5), 813–820 (2016)

An Analysis of Lung Sound from Electronic Stethoscope with Spectrogram

Trinh Quang Duc[1], Nguyen Van Son[2], Nguyen Hoai Giang[2], Dao Huy Du[3(✉)], and Ha Ngoc Thu[1]

[1] Hanoi University of Science and Technology, Hanoi, Vietnam
duc.trinhquang@hust.edu.vn, thumedicalbvps@gmail.com
[2] Hanoi Open University, Hanoi, Vietnam
{sonnv,giangnh}@hou.edu.vn
[3] Thai Nguyen University, Thainguyen, Vietnam
daohuydu@tnut.edu.vn

Abstract. This study presents a pilot investigation for lung sounds with 3 data collected from a published library recorded by electronic stethoscope. By using the tool of spectrogram with the Gaussian window which the variance number is tuned, the sound data were analyzed. The feature of the lung sound which the frequencies ranged from 0 to 800 Hz, contain 2 part, the high frequencies and low frequencies. In normal breath, the high and low frequencies are alternated and differed 300 Hz. In the disease related to bronchi, crackles or some high frequency bulbs will be occurred. The sound pattern can be analyzed easier with the 3D surface presentation rather than the 2D cross-sectional image.

Keywords: Bioacoustic signal · Spectrogram analysis · Lung sound · Sound pattern

1 Introduction

Electronic stethoscope, nowadays, become a useful tool for medical diagnosis in tele-medicine [1]. By the measurement, the sound signal converted to the electrical signal then digitized and stored in storage device. The data can transfer to the medical doctor via internet and diagnosed with the listening. The features of the sound was used to train medical doctor specifically. In another approach, the sound also possibly analyzed with a tool of spectrogram to find the signal of diseases. There are number of authors applied the signal processing algorithms to analyze the signal such as wavelet decomposition of cardiac auscultation [2], auto-aggressive model to extract the parameters of lung sound acoustic wave in time-domain [3], Empirical Mode Decomposition to detect the crackle signal in lung sound [4], quantile vector to recognize the lung sound feature based on frequency-domain analysis [5], and time-frequency domain analysis to ex- tract the peak frequencies in spectrogram of the signal [6].

Almost of the previous studies focused on the algorithm to extract of the features of the specific sound but not the pattern of the frequencies distribution in the measured signal. The categories of the specific sound may have the features which possibly recognize by the doctor through if the visualization of the signal is detail enough.

K. U. Sattler et al. (Eds.): ICERA 2019, LNNS 104, pp. 107–113, 2020.
https://doi.org/10.1007/978-3-030-37497-6_12

The time-domain and frequency-domain analysis has drawback since the body sound produced from the bio-logical dynamic system resulting the movement of the frequencies. To treat the problem, the pattern of frequencies distribution can provide the information of the features in both time and frequency domain. In this report, the spectrogram of some recorded data were used to analyze the pattern to indicate the features of the lung sound in both normal and abnormal sounds.

2 Methods

To analyze the pattern of the frequencies distribution of the lung sound, this report used the data downloaded from the library published in website of Thinklabs Medical LLC company [7]. The data were downloaded and used as the materials of this study included: the normal breath sound [8], crackles and wheezes of a bronchiectasis patient with Cystic Fibrosis [9], and crackles of bronchiectasis [10]. Because the data is formed as mp3 file, therefore, the downloaded file should be read to extract the measured values to be a float number array.

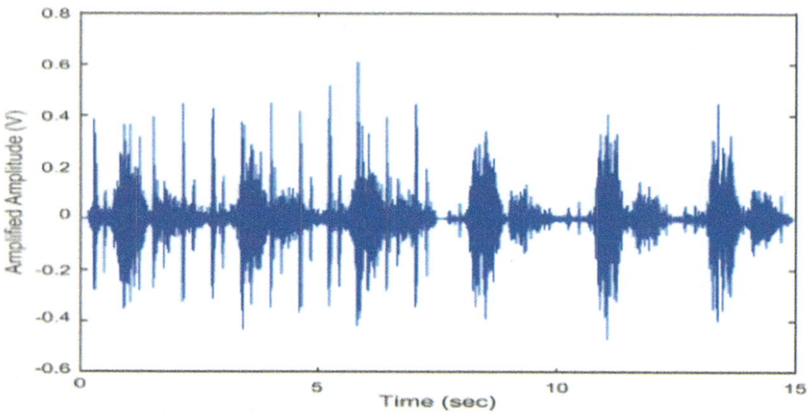

Fig. 1. The sound breath data from a normal lung [8]

Mp3 file format usually recorded with 2 channels in which left and right channel is different data to build the stereo effect. Occasionally, in some case of which when the sound sources are single, the both channels are recorded with the same measured values. In this case, the situation of the mp3 format file also recorded with the same values in both channels, thus, the data in 1 channel is considered to use instead of both. For the analysis purpose, the data in left channels were used.

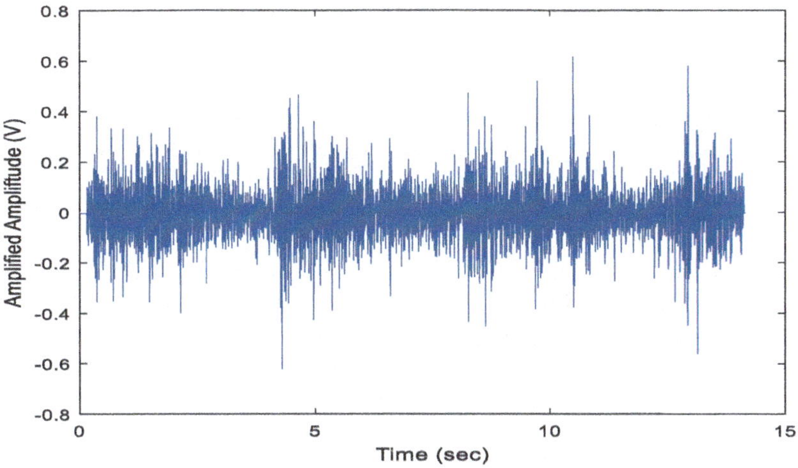

Fig. 2. The sound breath data from a bronchiectasis patient with Cystic Fibrosis [9]

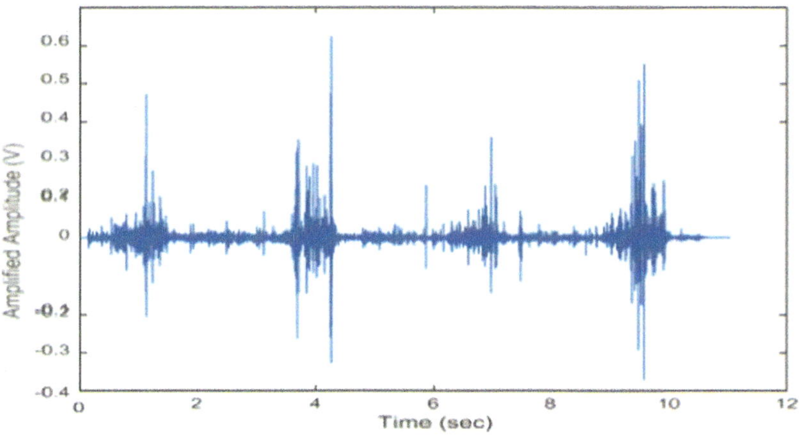

Fig. 3. The sound breath data from a bronchiectasis patient [10]

The Figs. 1, 2, and 3, present the graphs of measured data from the downloaded files. The data were used to applied the spectrogram plot the frequencies distribution of the lung sound to analyzed the feature of each sound type. The graphs show that the DC component in the data were eliminated by the filters. The length of the sound also recorded in 10 to 15 s with sample rate of 44.1 kHz. Therefore, the spectrogram should be arrange in the same range of the time length.

The conventional spectrogram is 2D image built with short-time Fourier transform to observe frequency variation in a range of time recorded. Typically, the transform use convolution with a window function in short-time length to enhance the contrast of center frequencies. In this report, the Gaussian window is used with experiment in condition change of variance value. To arrange the experiment the downloaded data are analyzed with a software written in Matlab.

3 Results and Discussion

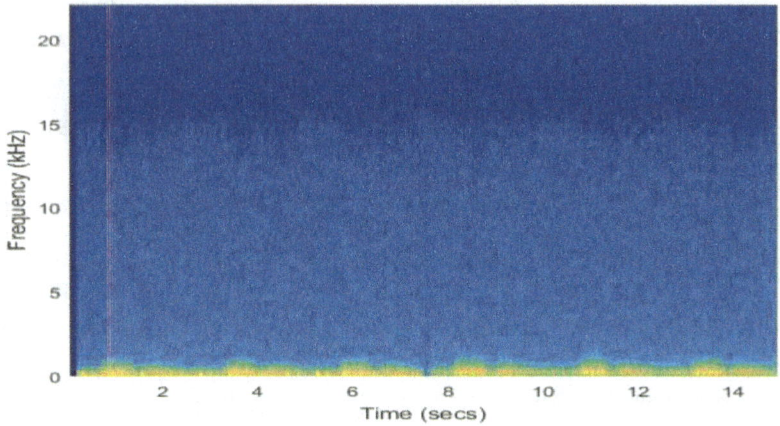

Fig. 4. The spectrogram of the normal breath sound [8] with the variance value of 8.

Figure 4 shows the spectrogram of the normal breath sound with the variance number is set as 8. The frequencies distribution ranged from 0 to 800 Hz, some higher frequencies reach to 1 kHz and occurred in time of 1 s, 3.5 s, 6 s, 8.5 s, 11 s, and 13 s. At the time of 7.5 s, the signal seems disappeared, it might be a signal to indicate the finish of a period time in the breath. In the noises range from 0 to 15 kHz.

With the change of variance number to 2, the noise level is increasing. As the Fig. 5 shown, the noise level ranged from 0 to 15 kHz in Fig. 4 seems similar to the noise level ranged from 0 to 25 kHz. This means, the variance value of the Gaussian window control the elimination of the noise level in frequency range. Therefore, the variance value should be tuned until the noise level at the highest frequency appeared at the highest frequency in the interested frequency range.

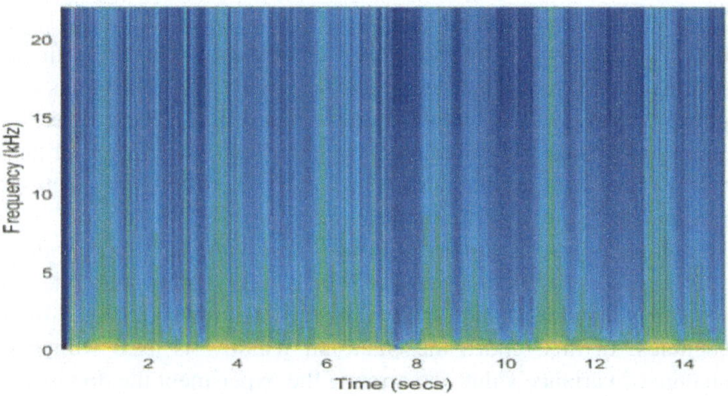

Fig. 5. The spectrogram of the normal breath sound [8] with variance value of 2.

The interested frequency range for the breath sound signal typically observed from 0 to 8 kHz. Therefore the noise level should be observed in this range for the investigation of lung sound. In series of numerical experiment, the noise level has not changed with the variance value of 4. To enlarge the signal area in contrast with the noise area, the view span is reduced to the range of 0 to 4200 Hz which located in the interested spectrum. The distribution in shorten range is shown in Fig. 6.

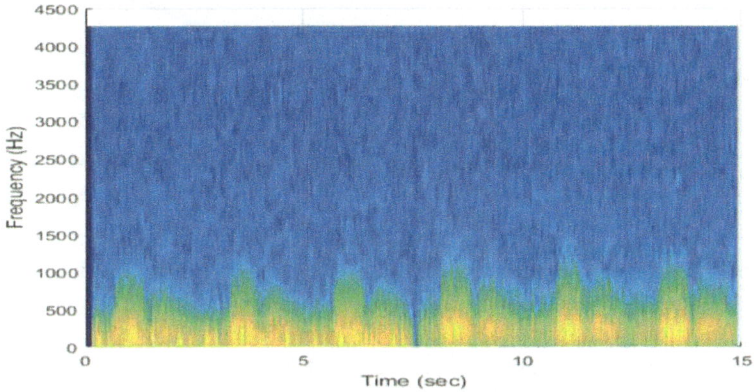

Fig. 6. The spectrogram of the normal breath sound [8] with the variance value of 4.

The 2D cross-sectional image seems hard to find the feature of the sound. To clarify the difference from the frequencies distribution of the sound pattern, the data in spectrogram are presented as 3D surface. Figure 7 shows the spectrogram in 3D surface pattern (Fig. 8).

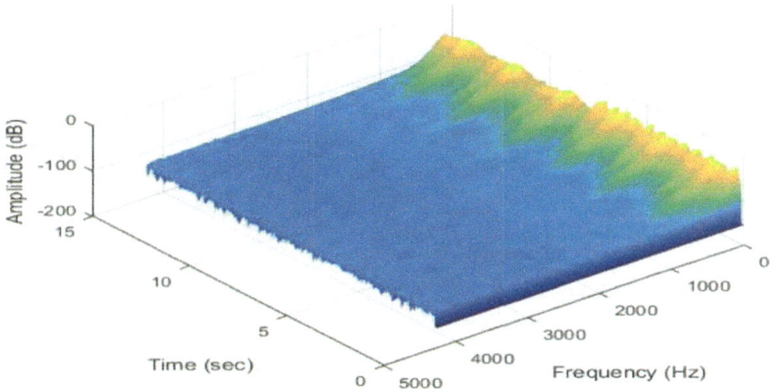

Fig. 7. The 3D surface of the spectrogram of the normal breath sound [8] with the variance value of 4.

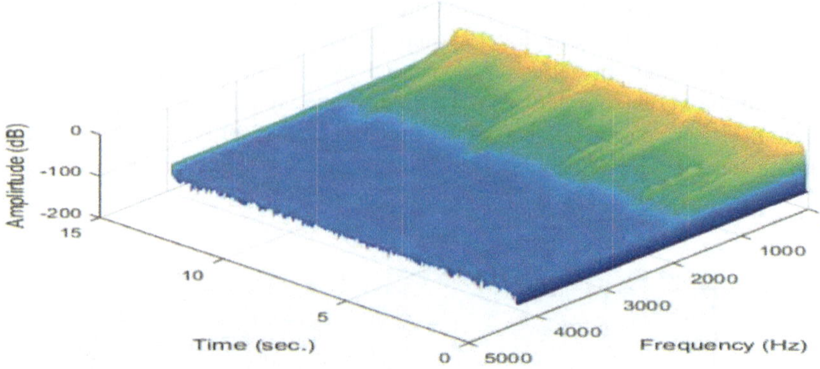

Fig. 8. The 3D surface of the spectrogram of bronchiectasis patient with Cystic Fibrosis [9] with the variance value of 4.

The difference if the bronchiectasis patient with Cystic Fibrosis is determined very clear. The frequency in the sound pattern is higher compared with the normal sound, reaches to 1500 Hz with crackles frequency especially increasing while the frequency in the normal breath sound just reaches to around 500 Hz to 800 Hz. In this bronchiectasis sound, the idle period also disappeared in 8th second. Instead, the background sound also presented and their frequency around 400 Hz. The period of the high frequency part usually occurred after 3 s. Both figure also has the pattern of the low frequency part are similar, which means the feature of the lung sounds are come from the high frequency part.

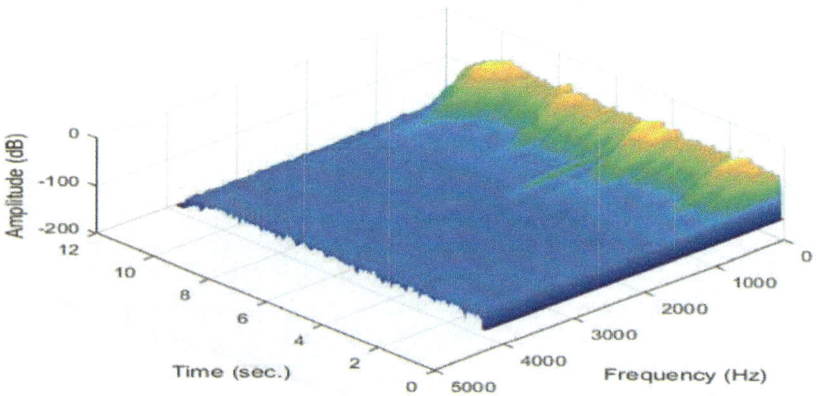

Fig. 9. The 3D surface of the spectrogram of bronchiectasis patient [10] with the variance value of 4.

Figure 9 shows the pattern of frequency distribution from the bronchiectasis patient [10]. The distribution seems similar to the normal breath. There are no crackles could be find, however, between the breath cycle, there is a high frequency reaches to

1500 Hz was occurred. The bulbs of higher frequency part is separated not clearly in comparison with the normal breath.

4 Conclusion

In this study, 3 data from the patients were used to investigate the feature of the lung sound through the tool of spectrogram with 3D surface presentation. The feature of lung sound can be identified that has 2 part of frequency distribution, the lower frequency and the higher frequency in which the parts contain the separated bulbs might be characterized from the breath rhythm. In the abnormal breath, if a patient in condition of seriously disease, in this case is bronchiectasis with cystic fibrosis, the breath sound contain crackles which their frequency reach to 1500 Hz to 2000 Hz. In the case of sickness such as the bronchiectasis patient, the crackles are very hard to find, however, the feature of the sound contain some high frequency bulb reaching to 1500 Hz with the separation of the high frequency bulb is blurred. This can be explained from the disorder of the breath when the bronchi in condition of inflammation.

Through this pilot study, the algorithm of identification the feature signal with clinical condition will be developed in the near future.

Acknowledgment. This work is financially supported by the program named as B2017-MHN.01 of Vietnam Ministry of Education and Training.

References

1. Gururajan, R., Tsai, H.-S., Chen, H.: Using digital stethoscopes in remote patient assessment via wireless networks: the user's perspective. Int. J. Adv. Netw. Appl. **03**(03), 1140–1146 (2011)
2. Gede, I.D., Wisana, H.: Design electronic stethoscope for cardiac auscultation analyzed using wavelet decomposition. Int. J. Comput. Netw. Commun. Secur. **1**(7), 310–315 (2013)
3. Alsmadi, S.S., Kahya, Y.P.: Online classification of lung sounds using DSP. In: Proceedings of the 2nd Joint Engineering in Medicine and Biology, pp. 1771–1772, IEEE Xplore Press, 23–26 October 2002
4. Lozano, M., Fiz, J.A., Jané, R.: Estimation of instantaneous frequency from empirical mode decomposition on respiratory sounds analysis. In: Proceedings of the 35th Annual International Conference of the IEEE Engineering in Medicine and Biology Society, Osaka, pp. 981–984. IEEE Xplore Press, 3–7 July 2013
5. Mayorga, P., Druzgalski, C., Gonzalez, O.H., Lopez, H.S.: Modified classification of normal lung sounds applying quantile vectors. In: Proceedings of the Annual International Conference of the IEEE Engineering in Medicine and Biology Society, San Diego, CA, pp. 4262–4265. IEEE Xplore Press, 28 August–1 September 2012
6. Rizal, A., Suryani, V.: Lung sound recognition using spectrogram and adaptive resonance theory 2 neural network (ART2). Surabaya, Indonesia (2008)
7. Lung Sound Library, Thinklabs Medical LLC
8. Normal breath sound, Lung Sound Library, Thinklabs Medical LLC
9. Alda Marques, Crackles and Wheezes—Bronchiectasis in a Patient with Cystic Fibrosis, Lung Sound Library, Thinklabs Medical LLC
10. Alda Marques, Crackles—Bronchiectasis, Lung Sound Library

An Experimental Investigation of the Cutting Forces Coefficients in Flat-End Mill Processes

Nhu-Tung Nguyen[1(✉)], Anh-Tuan Do[2], and Gia Thinh Bui[3]

[1] Hanoi University of Industry, Hanoi, Vietnam
tungnn@haui.edu.vn
[2] Hung Yen University of Technology and Education, Hải Dương, Vietnam
[3] Hai Phong University, Hai Phong, Vietnam

Abstract. In this paper, a linear force model was applied to investigate the cutting force coefficients in flat-end milling process. Six components of cutting force coefficients were determined by experimental results. This method can be applied to calculate the cutting force coefficients of Al6061-T6 for a series of cutters. The significance of each cutting force coefficient was evaluated. The shear force coefficients are the most important coefficients in calculation of cutting forces. The effect of cutter diameter on each cutting force coefficient was investigated.

Keywords: Cutting force · Coefficients · Flat-end mill

1 Introduction

Metal-cutting mechanics can be analyzed by orthogonal and oblique models that have been researched in many studies such as [1] and [2]. The procedure of cutting force modeling is generally realized by developing the experiential chip-force relationship through the cutting force coefficients. So, effective methods for calibrating the cutting coefficients are the important keys to cutting force's modeling. In the traditional mechanistic approach, edge force and shear force coefficients are calibrated for different pairs of the cutter and workpiece [2]. There were two methods for the calibration of cutting force coefficients. The first method is the orthogonal to oblique cutting transformation method and the second method is the direct calibration method.

In the first method, the shear angle, friction angle and shear yield stress resulted from orthogonal cutting test were used to estimate the cutting force coefficients [2, 3]. The cutting force coefficients were calculated from the oblique cutting model with and without end cutting edge effect [4]. In the second method, the cutting force coefficients were determined directly from milling tests. Cheng et al. determined the instantaneous cutting force coefficients depending on the instantaneous uncut chip thickness, the cutting edge length, and the spindle speed [5]. Larue et al. used the measurement of cutter deflection to calculate the cutting force coefficients [6]. Azeem et al. estimated the cutting force coefficients by considering the instantaneous cutting force [7]. The measured average cutting forces were used to calculate the cutting force coefficients [8–10].

In this study, the cutting force coefficients were determined through the linear force model of the cylindrical end mill. The shear and edge force coefficients were calculated

K.-U. Sattler et al. (Eds.): ICERA 2019, LNNS 104, pp. 114–120, 2020.
https://doi.org/10.1007/978-3-030-37497-6_13

from the experimental data of cutting forces. The main contributions of this study lie in three aspects: (1) evaluating the significance of each cutting force coefficient and (2) The effect of cutter diameter on each cutting force coefficient was investigated.

2 Calculation of Cutting Force Coefficients

The cutting force coefficients (K_{tc}, K_{rc}, K_{ac}, K_{te}, K_{re}, K_{ae}) are determined by Eq. (1), [10].

$$\begin{cases} K_{tc} = \frac{C_1\bar{F}_{fc} - C_2\bar{F}_{nc}}{C_1^2 + C_2^2} & K_{te} = \frac{C_3\bar{F}_{fe} - C_4\bar{F}_{ne}}{C_3^2 + C_4^2} \\ K_{rc} = \frac{C_2\bar{F}_{fc} + C_1\bar{F}_{nc}}{C_1^2 + C_2^2} & K_{re} = \frac{C_4\bar{F}_{fe} + C_3\bar{F}_{ne}}{C_3^2 + C_4^2} \\ K_{ac} = -\frac{\bar{F}_{ac}}{C_4} & K_{ae} = \frac{\bar{F}_{ae}}{C_5} \end{cases} \tag{1}$$

Where C_1, C_2, C_3, C_4, C_5 are calculated by Eq. (2)

$$\begin{cases} C_1 = \frac{N_f}{4\pi}\left(\begin{array}{c} -\int_{\phi_{st}}^{\phi_{st}+\Psi_a}\left[\int_0^{\frac{D}{2\tan\beta}(\phi-\phi_{st})}(\sin 2\phi_j(z))dz\right]d\phi - \int_{\phi_{st}+\Psi_a}^{\phi_{ex}}\left[\int_0^a(\sin 2\phi_j(z))dz\right]d\phi \\ -\int_{\phi_{ex}}^{\phi_{ex}+\Psi_a}\left[\int_{\frac{D}{2\tan\beta}(\phi-\phi_{ex})}^{a}(\sin 2\phi_j(z))dz\right]d\phi \end{array} \right) \\ C_2 = \frac{N_f}{4\pi}\left(\begin{array}{c} -\int_{\phi_{st}}^{\phi_{st}+\Psi_a}\left[\int_0^{\frac{D}{2\tan\beta}(\phi-\phi_{st})}(1-\cos 2\phi_j(z))dz\right]d\phi - \int_{\phi_{st}+\Psi_a}^{\phi_{ex}}\left[\int_0^a(1-\cos 2\phi_j(z))dz\right]d\phi \\ -\int_{\phi_{ex}}^{\phi_{ex}+\Psi_a}\left[\int_{\frac{D}{2\tan\beta}(\phi-\phi_{ex})}^{a}(1-\cos 2\phi_j(z))dz\right]d\phi \end{array} \right) \\ C_3 = \frac{N_f}{2\pi}\left(\begin{array}{c} -\int_{\phi_{st}}^{\phi_{st}+\Psi_a}\left[\int_0^{\frac{D}{2\tan\beta}(\phi-\phi_{st})}(\cos \phi_j(z))dz\right]d\phi - \int_{\phi_{st}+\Psi_a}^{\phi_{ex}}\left[\int_0^a(\cos \phi_j(z))dz\right]d\phi \\ -\int_{\phi_{ex}}^{\phi_{ex}+\Psi_a}\left[\int_{\frac{D}{2\tan\beta}(\phi-\phi_{ex})}^{a}(\cos \phi_j(z))dz\right]d\phi \end{array} \right) \\ C_4 = \frac{N_f}{2\pi}\left(\begin{array}{c} -\int_{\phi_{st}}^{\phi_{st}+\Psi_a}\left[\int_0^{\frac{D}{2\tan\beta}(\phi-\phi_{st})}(\sin \phi_j(z))dz\right]d\phi - \int_{\phi_{st}+\Psi_a}^{\phi_{ex}}\left[\int_0^a(\sin \phi_j(z))dz\right]d\phi \\ -\int_{\phi_{ex}}^{\phi_{ex}+\Psi_a}\left[\int_{\frac{D}{2\tan\beta}(\phi-\phi_{ex})}^{a}(\sin \phi_j(z))dz\right]d\phi \end{array} \right) \\ C_5 = \frac{N_f}{2\pi}\left(\begin{array}{c} \int_{\phi_{st}}^{\phi_{st}+\Psi_a}\left[\int_0^{\frac{D}{2\tan\beta}(\phi-\phi_{st})}dz\right]d\phi + \int_{\phi_{st}+\Psi_a}^{\phi_{ex}}\left[\int_0^a dz\right]d\phi \\ +\int_{\phi_{ex}}^{\phi_{ex}+\Psi_a}\left[\int_{\frac{D}{2\tan\beta}(\phi-\phi_{ex})}^{a}dz\right]d\phi \end{array} \right) \end{cases} \tag{2}$$

Where

D is the diameter of cutter [mm], N_f is the number of flutes on the cutter, β is the helix angle on the cutter [deg], Ψ_a is the lag angle at maximum axial depth of cut $z = a$ [deg], ϕ_{st} is the cutter entry angle [deg], ϕ_{ex} is the cutter exit angle [deg], a is the full axial depth of cut [mm], dz is the differential axial depth of cut [mm], ϕ is the instantaneous immersion angle of flute number j [deg].

The components of the linear force ($\bar{F}_{fc}, \bar{F}_{fe}, \bar{F}_{nc}, \bar{F}_{ne}, \bar{F}_{ac},$ and $\bar{F}_{ae}$) can be calculated by a linear regression of the measured cutting force data. The average cutting forces can be expressed by Eq. (3).

$$\begin{cases} \bar{F}_f = \bar{F}_{fc} f_t + \bar{F}_{fe} \\ \bar{F}_n = \bar{F}_{nc} f_t + \bar{F}_{ne} \\ \bar{F}_a = \bar{F}_{ac} f_t + \bar{F}_{ae} \end{cases} \quad (3)$$

3 Experimental Method

3.1 Experimental Procedure

The research procedure was performed by a scheme as shown in Fig. 1. Firstly, the cutting depths and spindle speeds were selected as the stable cutting condition through the cutting tests and the evaluation of the cutting force's stable. Secondly, the cutting tests were performed with variations of federate to determine the cutting force coefficients. Thirdly, the cutting force coefficients were calculated from the experimental data. Finally, the applications of the force model were proposed and performed.

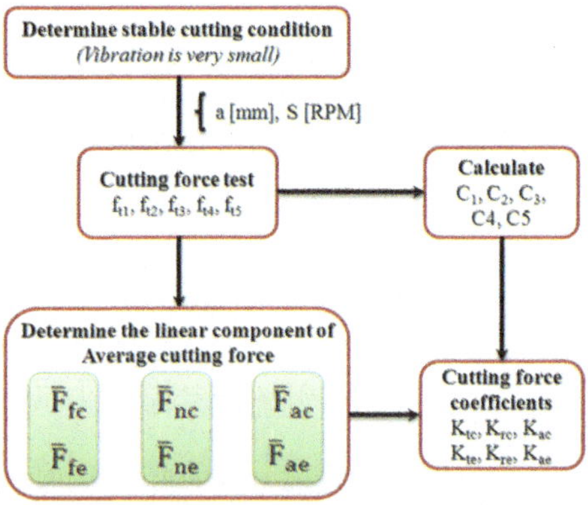

Fig. 1. Approach to identify cutting force coefficients

3.2 Experimental Conditions

In order to determine the cutting force coefficients, a series of end milling experiments were performed. The workpiece material was Al6061-T6 and its compositions are listed in Table 1.

Table 1. Chemical compositions of Al6061-T6

Composite (%)								
Al	Cr	Cu	Fe	Mg	Mn	Si	Ti	Zn
98	≤ 0.3	≤ 0.4	≤ 0.7	≤ 1.2	≤ 0.15	≤ 0.8	≤ 0.15	≤ 0.25

The properties of the Al6061-T6 were: Hardness 95 HB, Young's modulus = 68.9 GPa, Poisson's ratio = 0.33, tensile strength = 310 MPa. Cutter: The HSS (High speed steel) cylindrical end mill with the helix angle $\beta = 30°$, rake angle $\alpha_r = 5°$. The number of flutes and the cutter diameter are listed in Table 2.

Table 2. The cutting conditions of experiments

No.	D [mm]	N_f	B [deg]	a [mm]	S [RPM]	f_t [mm/flute]	Cutting type
1	8	2	30	0.2	1000	0.02; 0.04; 0.06; 0.08; 0.10	Half down
2	10	2	30	0.2	1000	0.02; 0.04; 0.06; 0.08; 0.10	Half down
3	12	2	30	0.2	1000	0.02; 0.04; 0.06; 0.08; 0.10	Half down
4	14	2	30	0.2	1000	0.02; 0.04; 0.06; 0.08; 0.10	Half down
5	16	2	30	0.2	1000	0.02; 0.04; 0.06; 0.08; 0.10	Half down
6	8	4	30	0.3	1000	0.02; 0.04; 0.06; 0.08; 0.10	Half down
7	10	4	30	0.3	1000	0.02; 0.04; 0.06; 0.08; 0.10	Half down
8	12	4	30	0.3	1000	0.02; 0.04; 0.06; 0.08; 0.10	Half down
9	14	4	30	0.3	1000	0.02; 0.04; 0.06; 0.08; 0.10	Half down
10	16	4	30	0.3	1000	0.02; 0.04; 0.06; 0.08; 0.10	Half down

The experiments were performed at a Three-axis Vertical Milling Center (DECKEL MAHO – DMC70 V hi-dyn). A Dynamometer (Type: XYZ FORCE SENSOR, Model: 624-120-5KN), processing system, and a PC were used to measure cutting forces. The detail is illustrated in Fig. 2.

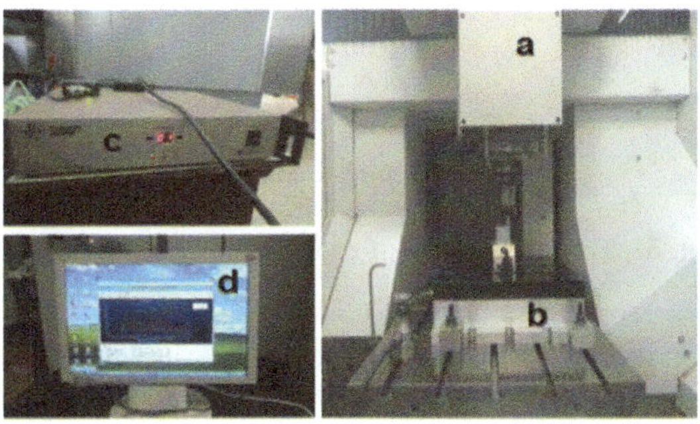

Fig. 2. Setup measurement of cutting force a. CNC machine b. Dynamometer c. Processing system d. PC and Display System

At each depth of cut, half immersion experiments were repeated in every feedrate. The spindle speed was held constant at each experiment. The experiments were performed with the different parameters, as shown in Table 2.

4 Results and Discussions

4.1 Identification of Cutting Force Coefficients

The measured cutting force data were stored to calculate the average cutting forces. The relationship of average cutting forces and the feed rate is a linear function. Considering Eqs. (1) to (3) and using the characteristic of the best line, the cutting force coefficients were calculated and listed in Table 3.

Table 3. The cutting force coefficients

D [mm]	K_{tc} [N/m^2]		K_{rc} [N/m^2]		K_{ac} [N/m^2]		K_{te} [N/m]		K_{re} [N/m]		K_{ae} [N/m]	
	N = 2	N = 4	N = 2	N = 4	N = 2	N = 4	N = 2	N = 4	N = 2	N = 4	N = 2	N = 4
8	692.3	692.5	218.4	198.1	331.9	326.8	57.7	56.9	89.4	89.8	0.22	0.25
10	696.2	698.8	216.0	195.2	333.3	329.4	58.3	57.7	89.1	89.3	0.23	0.25
12	698.7	703.1	214.6	193.3	334.2	331.1	58.6	58.2	88.9	89.0	0.24	0.26
14	700.5	706.1	213.4	191.9	334.9	332.4	58.9	58.6	88.7	88.7	0.24	0.26
16	701.9	708.4	212.5	190.9	335.4	333.4	59.1	58.9	88.6	88.5	0.25	0.26

4.2 The Significance of Each Cutting Force Coefficient

The researched results in Fig. 3 show that both two types of cutter (two flutes and four flutes), the most important cutting force coefficients are shearing force coefficients.

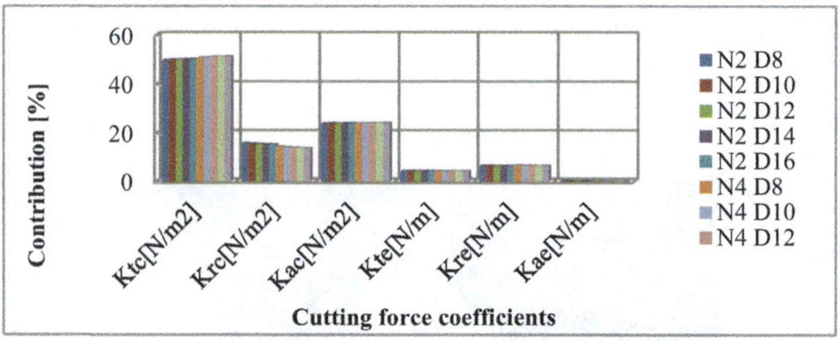

Fig. 3. The significance of each cutting force coefficients

The contribution of shearing force coefficients is larger than 80%. Including the contribution of tangential shearing force coefficient is about 50%, it is about 15% for the radial shearing force coefficient, and 22% for the axial shearing force coefficient. The contribution of each edge force coefficient is smaller than 7%.

4.3 The Effect of Diameter on the Cutting Force Coefficients

The cutting force coefficients depend on the diameter of the cutter. With shearing force coefficients, when the cutter diameter increases, the tangential shearing force coefficient and axial shearing force coefficient increase but the radial cutting shearing force coefficient decreases. These results have the same trends with the different numbers of flutes as shown in Figs. 4 and 5.

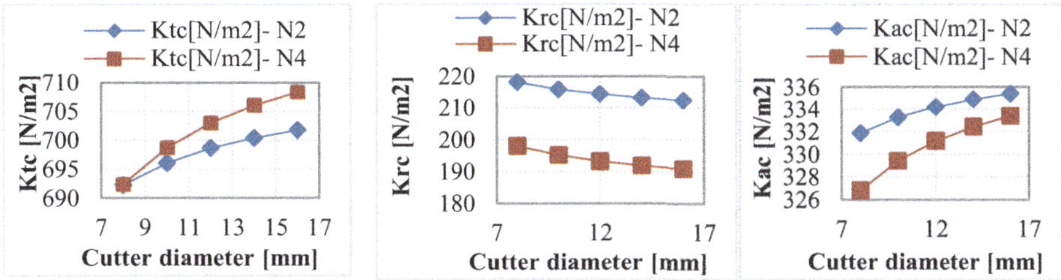

Fig. 4. Shearing force coefficient versus cutter diameter

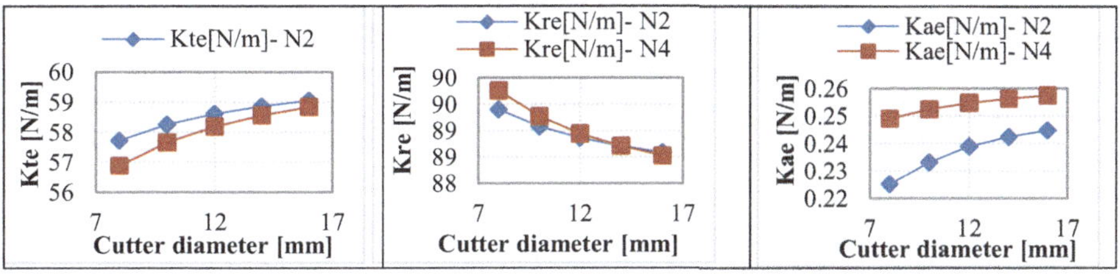

Fig. 5. Edge force coefficient versus cutter diameter

With edge force coefficients, when the cutter diameter increases, the tangential edge force coefficient and axial edge force coefficient also increase, and the radial cutting edge force coefficient also decreases.

5 Conclusions

The conclusions of this study can be drawn as follows.

For flat-end mill, the shearing force coefficients are the most important coefficients in calculation of cutting forces.

The cutting force coefficients depend on the cutter diameter. The tangential and axial force coefficients increase with the increase of diameter. Conversely, the radial force coefficients decrease with the increasing of diameter.

References

1. Merchant, M.: Mechanics of the metal cutting process. I. orthogonal cutting and a type 2 chip. J. Appl. Phys. **16**, 267 (1945). https://doi.org/10.1063/1.1707586
2. Altintas, Y.: Manufacturing Automation: Metal Cutting Mechanics, Machine Tool Vibrations, and CNC Design, 2nd edn. Cambridge University Press, Cambridge (2012). ISBN 978-1-00148-0
3. Budak, E., Altintas, Y., Armarego, E.J.A.: Prediction of milling force coefficients from orthogonal cutting data. ASME J. Manuf. Sci. Eng. **118**, 216–224 (1996)
4. Li, X.P., Li, H.Z.: Theoretical modelling of cutting forces in helical end milling with cutter runout. Int. J. Mech. Sci. **46**, 1399–1414 (2004)
5. Cheng, P.J., Tsay, J.T., Lin, S.C.: A study on instantaneous cutting force coefficients in face milling. Int. J. Mach. Tools Manuf **37**, 1393–1408 (1997)
6. Larue, A., Anselmetti, B.: Deviation of a machined surface in flank milling. Int. J. Mach. Tools Manuf **43**, 129–138 (2003)
7. Azeem, A., Feng, H.Y., Wang, L.H.: Simplified and efficient calibration of a mechanistic cutting force model for ball-end milling. Int. J. Mach. Tools Manuf **44**, 291–298 (2004)
8. Altintas, Y.: Modeling approaches and software for predicting the performance of milling operations at MAL - UBC. Int. J. Mach. Sci. Technol. **4**(3), 445–478 (2000)
9. Kao, Y.-C., Nguyen, N.-T., Chen, M.-S., Huang, S.-C.: A combination method of the theory and experiment in determination of cutting force coefficients in ball-end mill processes. J. Comput. Des. Eng. **2**(4), 233–247 (2015)
10. Kao, Y.C., Nguyen, N.T., Chen, M.S., Su, S.T.: A prediction method of cutting force coefficients with helix angle of flat-end cutter and its application in a virtual three-axis milling simulation system. Int. J. Adv. Manuf. Technol. **77**(9–12), 1793–1809 (2015)

A Study on Optimization of Manufacturing Time in External Cylindrical Grinding

Luu Anh Tung[1], Tran Thi Hong[2], Nguyen Van Cuong[3], Le Hong Ky[4], T. Muthuramalingam[5], Nguyen Huu Phan[6], Le Xuan Hung[1], and Ngoc Pi Vu[1(✉)]

[1] Thai Nguyen University of Technology, Thai Nguyen 23000, Vietnam
vungocpi@tnut.edu.vn
[2] Nguyen Tat Thanh University, Ho Chi Minh 700000, Vietnam
[3] University of Transport and Communications, Ha Noi 100000, Vietnam
[4] Vinh Long University of Technology Education, Vinh Long 890000, Vietnam
[5] SRM Institute of Science and Technology, Kattankulathur 603203, India
[6] Hanoi University of Industry, Hanoi 100000, Vietnam

Abstract. This paper introduces an optimization study on the determination of the optimum replaced grinding wheel diameter for getting minimum manufacturing time in external grinding stainless steel. In the study, several input grinding parameters including the initial grinding wheel diameter, the total depth of dressing cut, the wheel life and the radial grinding wheel wear per dress. Also, to evaluate the effects of the input parameters on the optimum replaced wheel diameter, a simulation experiment was planned and conducted. Lastly, a regression model to calculate the optimum replaced wheel diameter was suggested.

Keywords: Grinding · External grinding · Manufacturing time optimization · Replaced grinding wheel diameter

1 Introduction

Grinding is one of the most popular finished machining methods. It ensures the surface quality and the final accuracy of machining parts [1, 2]. For that reason, optimization researches to improve the quality and the performance of the grinding process have been concerned by many researchers.

Until now, various studies have been done on optimization of grinding process [3–13]. The grinding process have been learned in different types of grinding such as surface grinding [5, 7], external grinding [9, 16], and internal grinding [13, 14]. Also, the effects of coolant conditions on the grinding quality were considered in internal grinding [14] and in external grinding [15, 16]. Besides, the influences of dressing process on the grinding performance were evaluated [9, 11, 12]. The topography of the surface of different vitrified abrasive wheels with different dressing parameters was also studied in [17]. The effects of grinding process parameters on the material removal rate and the surface roughness were explored in [18].

K.-U. Sattler et al. (Eds.): ICERA 2019, LNNS 104, pp. 121–129, 2020.
https://doi.org/10.1007/978-3-030-37497-6_14

In an effort to reduce the grinding cost, several studies [19–28] have been carried out. They have focused on the determination of the optimum replaced grinding wheel diameter to get the minimum manufacturing cost. As a reason, models for calculation of the optimum replaced whell diameter were proposed for external grinding [19, 20], for internal grinding [21–25] or for surface grinding [26–28]. Nevertheless, until now, there has been no research to find the optimum replaced wheel diameter for getting the minimum manufacturing time.

This paper presents an optimization research for calculating the optimum replaced wheel diameter to minimize the manufacturing time in external grinding stainless steel. In this research, the influences of input grinding parameters including the initial grinding wheel diameter (D_O), the total depth of dressing cut (a_{ed}), the radial grinding wheel wear per dress (W_{pd}) and the wheel life (T_w) on the optimum replaced wheel diameter ($D_{e,op}$) were examined by performing a simulation experiment. Also, a regression equation to find the optimum replaced diameter was suggested.

2 Methodology

2.1 Manufacturing Time Analysis

In external grinding, the manufacturing time t_s including the auxiliary time is calculated by the following equation:

$$t_s = t_c + t_{sp} + t_{lu} + t_{d,p} + t_{cw,p} \tag{1}$$

In which, $t_c, t_{sp}, t_{lu}, t_{d,p}$ and $t_{cw,p}$ are the grinding time (h), the spark-out time (h), the time for loading and unloading the workpiece (h), the dressing time per part (h), and the grinding wheel replacing time per part, respectively.

$t_c, t_{d,p}$ and $t_{cw,p}$ can be calculated as the following equations:

$$t_c = l_w a_{e,t}/(v_{fa} \cdot f_r) \tag{2}$$

$$t_{d,p} = t_d/n_{p,d} \tag{3}$$

$$t_{cw,p} = t_{cw}/n_{p,w} \tag{4}$$

In Eq. (2), l_w(mm), $a_{e,t}$(mm), v_{fa}(mm/min), and f_r(mm/double stroke) are the part length, the total grinding depth, the axial speed, and the radial wheel feed, respectively.

v_{fa} can be found by the following regression equation [29]:

$$v_{fa} = 0.2557\, d_w^{0.0749}\, n_w^{1.1093}\, b_s^{0.9841} \tag{5}$$

And f_r is determined by [30]:

$$f_r = f_{r,tab}.c_1.c_2.c_3.c_4 \tag{6}$$

In the above equations, d_w is the workpiece diameter, n_w is the workpiece speed, and b_s is the width of grinding wheel. For grinding stainless steel, n_w is determined by [29]:

$$n_w = 9.9985.2\ d_w^{-0.966} \tag{7}$$

In Eq. (6), $f_{r,tab}$ is the tabled radial wheel feed (mm/double stoke); c_1 is the coefficient which depends on work material and tolerance grade tg; c_2 is the coefficient which is influenced by the grinding peripheral speed v_s(m/second) and the grinding wheel diameter d_s; c_3 is the coefficient considered for measurement; c_4 is the coefficient determined by the ratio of the workpiece length to its diameter.

The tabled radial wheel feed $f_{r,tab}$ and the coefficients c_1, c_2 can be determined by the following regression equations [29].

$$f_{r,tab} = 150.3605\ d_w^{-0.9331}\ v_{fa}^{-0.4922}\left(2a_{e,t}\right)^{0.2194} \tag{8}$$

$$c_1 = 0.0334\ tg^{1.6718} \tag{9}$$

$$c_2 = 0.0032\ v_s^{0.7174}\ d_s^{0.4984} \tag{10}$$

In Eqs. (3) and (4), t_d and t_{cw} are the dressing time and the replacing grinding wheel time (h). $n_{p,d}$ is the number of parts per dress and $n_{p,w}$ is the total number of parts ground by a grinding wheel which are determined by:

$$n_{p,d} = T_w/t_c \tag{11}$$

$$n_{p,w} = \left(d_{s,0} - d_{s,e}\right).n_{p,d}/\left\lfloor 2\left(W_{pd} + a_{ed,ges}\right)\right\rfloor \tag{12}$$

In the above equations, T_w is the wheel life (h), $d_{s,o}$ is the initial diameter of grinding wheel (mm), $d_{s,e}$ is the replaced grinding wheel diameter (mm), W_{pd} is the radial grinding wheel wear per dress (mm/dress), and $a_{ed,ges}$ is the total depth of dressing cut (mm).

2.2 Optimization Problem

From the above analyses, when the replaced grinding wheel diameter $d_{s,e}$ has big values, the grinding time will be small because f_r will increase (see Eqs. (2), (6) and (10)). However, with the big values of the replaced grinding wheel diameter, the total number of parts ground by a grinding wheel (Eq. (12)) is reduced. That leads to the decrease of grinding wheel replacing time per part (formula 4), and as a result, the manufacturing time will increase. From these observations, it is obvious that there will be an optimum value of the replaced grinding wheel diameter with which the manufacturing time is the smallest.

Thus, the optimization problem to find the optimum replaced grinding wheel diameter for getting the minimum manufacturing time can be expressed as

$$Minimize\ t_s = \mathrm{f}\left(d_{s,e}\right) \tag{13}$$

subject to

$$\mathrm{d}_{se,min} \leq d_{s,e} \leq d_{se,max} \tag{14}$$

2.3 Experimental Work

To learn the influences of the input parameters on the optimum replaced wheel diameter, a simulation experiment was designed. The input factors of the experiment were the initial grinding wheel diameter D_0, the grinding wheel width W_{gw}, the total depth of dressing cut a_{ed}, the radial grinding wheel wear per dress W_{pd} and the wheel life T_w (Table 1). Also, the experiment was performed with $2^4 = 16$ test runs because a full factorial design of 4 factors was selected. To perform the experiment, a computer program was created. The experimental plan and the output response (the optimum exchanged diameter D_{op}) are listed in Table 2.

Table 1. Input parameters

Factor	Code	Unit	Low	High
Initial grinding wheel diameter	D_0	mm	250	500
Grinding wheel width	W_{gw}	mm	20	60
Radial grinding wheel wear per dress	W_{pd}	mm	0.01	0.03
Total depth of dressing cut	a_{ed}	mm	0.1	0.3
Wheel life	T_w	min.	10	30

Table 2. Screening experimental works and the result values of dop

StdOrder	RunOrder	CenterPt	Blocks	D_0	W_{gw}	W_{pd}	a_{ed}	T_w	d_{op}
30	1	1	1	500	20	0.03	0.3	30	484.58
1	2	1	1	250	20	0.01	0.1	10	239.84
9	3	1	1	250	20	0.01	0.3	10	233.49
26	4	1	1	500	20	0.01	0.3	30	485.04
8	5	1	1	500	60	0.03	0.1	10	484.21
16	6	1	1	500	60	0.03	0.3	10	475.38
…									
14	31	1	1	500	20	0.03	0.3	10	475.38
32	32	1	1	500	60	0.03	0.3	30	484.58

3 Results and Discussion

The main effects of each input factors on the optimum diameter D_{op} were demonstrated in Fig. 1. It is interpreted that the initial grinding wheel diameter D_O is the largest influential factor on the optimum diameter $D_{op,}$ whereas the grinding wheel width W_{gw} and the radial grinding wheel wear per dress W_{pd} are the non-influential factors. Besides, the total depth of dressing cut a_{ed} and the wheel life T_w inconsiderably affect the optimum diameter D_{op}.

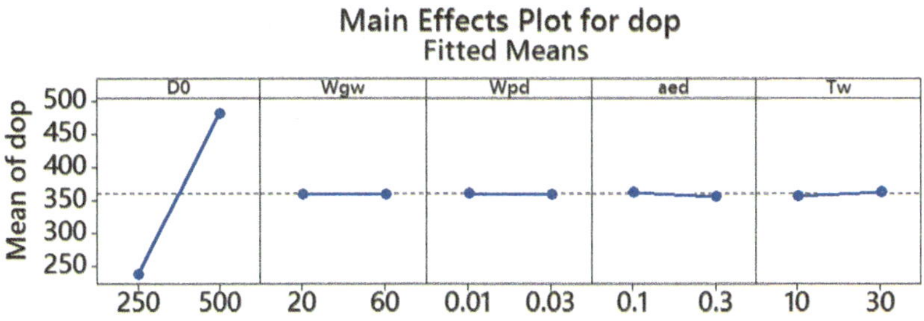

Fig. 1. Main effects plot on d_{op}

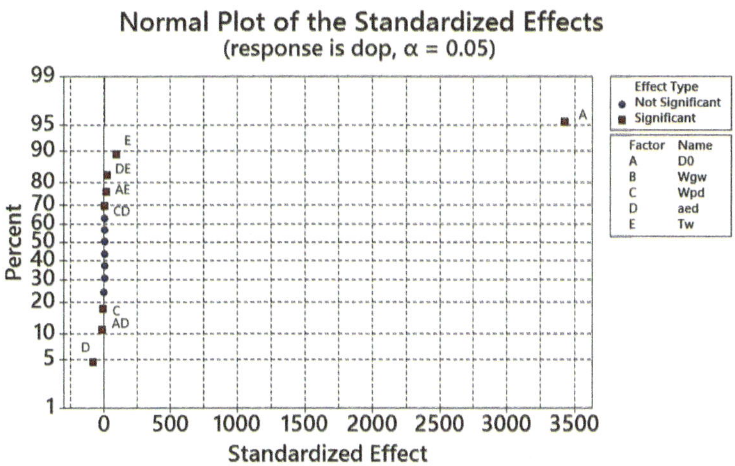

Fig. 2. Normal plot for the optimum diameter d_{op}

The normal plot of the standardized effects is displayed in Fig. 2. It is observed from the figure that the majority of the factors are distributed relatively closely to the reference line. Those factors such as the wheel life (Factor E), the total depth of dressing cut (Factor D), the radial grinding wheel wear per dress (Factor C) and the interactions DE (a_{ed} * T_w), AE (D_O * T_w), CD (W_{pd} * a_{ed}), and AD (D_O * a_{ed}) moderately influence the optimum diameter D_{op}. On the contrary, the initial grinding wheel diameter (Factor A) is located the farthest distance from the reference line, which

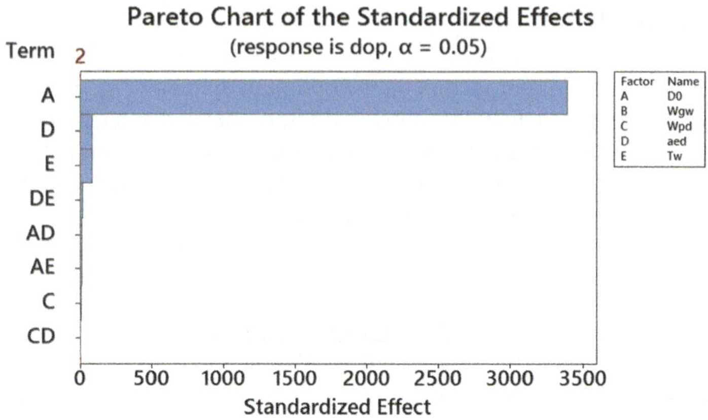

Fig. 3. Pareto chart of the standardized effects on d_{op}

indicates that the optimum diameter D_{op} is significantly affected by the parameter. Otherwise, the initial grinding wheel diameter and other factors including the wheel life (E), the radial grinding wheel wear per dress (C), the interactions DE (a_{ed} * T_w), AE (D_0 * T_w) and CD (W_{pd} * a_{ed}) have positive effects on the optimum diameter D_{op}. If their values grow, the diameter rises. Meanwhile, the total depth of dressing cut a_{ed} and the interaction AD negatively affect the optimum diameter D_{op}. It declines if their values expand.

The Pareto chart of the standardized effects on the optimum diameter d_{op} is exposed in Fig. 3. It is revealed that the initial grinding wheel diameter, the total depth of dressing cut and the wheel life significantly impact d_{op} as they are positioned on the right side of the reference line. In addition, appearing on the same side, the other factors such as the total depth of dressing cut and the interactions a_{ed} * T_w, D_0 * a_{ed} and D_0 * T_w have less significant influence on the optimum diameter d_{op}.

Table 3. Estimated effects and coefficients for d_{op}

Coded coefficients						
Term	Effect	Coef	SE Coef	T-Value	P-Value	VIF
Constant		361.430	0.036	10006.50	0.000	
DO	245.120	122.560	0.036	3393.18	0.000	1.00
Wpd	−0.6700	−0.3350	0.0361	−9.27	0.000	1.00
aed	−6.2550	−3.1275	0.0361	−86.59	0.000	1.00
Tw	6.2450	3.1225	0.0361	86.45	0.000	1.00
DO * aed	−1.1750	−0.5875	0.0361	−16.27	0.000	1.00
DO * Tw	1.1700	0.5850	0.0361	16.20	0.000	1.00
Wpd * aed	0.1750	0.0875	0.0361	2.42	0.024	1.00
aed * Tw	1.3750	0.6875	0.0361	19.03	0.000	1.00

Model summary			
S	R-sq	R-sq(adj)	R-sq(pred)
0.204323	100.00%	100.00%	100.00%

The estimated effects and coefficients of the regression function for d_{op} are demonstrated in Table 3 with the removal of the insignificant factors. It is reported from the figure that all the displayed factors and interactions receive statistical values lower than 0.05, which indicates that the initial grinding wheel D_0, the wheel life T_w, the radial grinding wheel wear per dress W_{pd}, the total depth of dressing cut a_{ed}, and the interactions including $D_0 * a_{ed}$, $D_0 * T_w$, $D_0 * T_w$, $W_{pd} * a_{ed}$, $a_{ed} * T_w$ significantly influence the response $D_{op.}$ Hence, the regression equation to obtain the optimum diameter d_{op} can be demonstrated as:

$$d_{op} = -2,485 + 0,9805.D_0 - 51.W_{pd} - 29,15.a_{ed} - 0,0008.T_w - 0,047.D_0.a_{ed} + 0,000468.D_0.T_w + 87,5.W_{pd}.a_{ed} + 0,6875.a_{ed}.T_w \quad (15)$$

It is significant that the model in Eq. (15) is the best fit for the data as evidenced by the summary results showing that the correlation coefficients equal 1 (R = 1, adj-R = 1 and pred-R = 1) (Table 3). Consequently, the model can be applied to calculate the optimum replaced grinding wheel diameter in the external cylindrical grinding process.

4 Conclusions

A study on the determination of the optimum replaced wheel diameter when external grinding stainless steel was conducted. In this study, the manufacturing time when external grinding was analyzed. Also, the effects of the input process parameters on the optimum replaced diameter were inspected scientifically by designing and conducting a simulation experiment. Moreover, a regression equation to find the optimum replaced wheel diameter was proposed. The optimum replaced diameter when external grinding stainless steel can be found quickly and easily because the proposed formula is in the form of an explicit function.

Acknowledgements. The work described in this paper was supported by Thai Nguyen University of Technology for a scientific project.

References

1. Rowe, W.B.: Principle of Modern Grinding Technology. William Andrew (2009)
2. Malkin, S., Guo, C.: Grinding Technology: Theory and Applications of Machining with Abrasives. Industrial Press (2008)
3. Chatterjee, S., Rudrapati, R., Kumarpal, P., Nandi, G.: Experiments, analysis and parametric optimization of cylindrical traverse cut grinding of aluminium bronze. Mater. Today: Proc. **5** (2), 5272–5280 (2018)
4. Vu, N.-P., Nguyen, Q.-T., Tran, T.-H., Le, H.-K., Nguyen, A.-T., Luu, A.-T., Nguyen, V.-T., Le, X.-H.: Optimization of grinding parameters for minimum grinding time when grinding tablet punches by CBN wheel on CNC milling machine. Appl. Sci. **9**(5), 957 (2019)

5. Tung, L.A., Pi, V.N., Ha, D.T.T., Hung, L.X., Banh, T.L.: A study on optimization of surface roughness in surface grinding 9CrSi tool steel by using Taguchi method. In: Fujita, H., et al. (eds.): ICERA 2018. LNNS, vol. 63, pp. 100–108 (2019). https://doi.org/10.1007/978-3-030-04792-4_15
6. Gupta, R., Shishodia, K.S., Sekhon, G.S.: Optimization of grinding parameters using enumeration method. J. Mater. Process. Technol. **112**, 63–67 (2001)
7. Wen, X.M., Tay, A.A.O., Nee, A.Y.C.: Micro-computer-based optimization of the surface grinding process. J. Mater. Process. Technol. **29**(1–3), 75–90 (1992)
8. Mekala, K., Chandradas, J., Chandrasekaran, K., Kannan, T.T.M., Ramesh, E., Babu, R.N.: Optimization of cylindrical grinding parameters of austenitic stainless steel rods (AISI 316) by Taguchi method. Int. J. Mech. Eng. Rob. Res. **3**(2), 208 (2014)
9. Tu, H.X., Thao, L.P., Hong, T.T., Nga, N.T.T., Trung, D.D., Gong, J., Pi, V.N.: Influence of dressing parameters on surface roughness of workpiece for grinding hardened 9XC tool steel. In: IOP Conference Series: Materials Science and Engineering, vol. 542, conference 1 (2018). https://doi.org/10.1088/1757-899X/542/1/012008
10. Vidal, G., Ortega, N., Bravo, H., Dubar, M., González, H.: An analysis of electroplated CBN grinding wheel wear and conditioning during creep feed grinding of aeronautical alloys. Metals **8**, 1–24 (2018)
11. Daneshia, A., Jandaghia, N., Tawakoli, T.: Effect of dressing on internal cylindrical grinding. Procedia CIRP **14**, 37–41 (2014)
12. Le, X.H., Tran, T.H., Luu, A.T., Nguyen, T.T.N., Vu, N.P.: Optimum dressing parameters for maximum material removal rate when internal cylindrical grinding using Taguchi method. Int. J. Mech. Eng. Technol. **9**, 123–129 (2018)
13. Gupta, R., Shishodia, K.S., Sekhon, G.S.: Optimization of grinding process parameters using enumeration method. J. Mater. Process. Technol. **112**(1), 63–67 (2001)
14. Hung, L.X., Lien, V.T., Pi, V.N., Long, B.T.: A study on coolant parameters in internal grinding of 9CrSi steel. Mater. Sci. Forum **950**, 24–31 (2019)
15. Tu, H.X., Jun, G., Hien, B.T., Hung, L.X., Tung, L.A., Pi, V.N.: Determining optimum parameters of cutting fluid in external grinding of 9CrSi steel using Taguchi technique. SSRG Int. J. Mech. Eng. **5**(6), 1–5 (2018). https://doi.org/10.14445/23488360/IJME-V5I6P101
16. Tu, H.X., Pi, V.N., Jun, G.: A study on determination of optimum parameters for lubrication in external cylindrical grinding base on Taguchi method. Key Eng. Mater. **796**, 97–102 (2019)
17. Palmer, J., Ghadbeigi, H., Novovic, D., Curtis, D.: An experimental study of the effects of dressing parameters on the topography of grinding wheels during roller dressing. J. Manuf. Process. **31**, 348–355 (2018)
18. Yadav, H.S., Shrivastava, R.K.: Effect of process parameters on surface roughness and Mrr in cylindrical grinding using response surface method. Int. J. Eng. Res. Technol. **3**(3) (2014)
19. Pi, V.N., The, P.Q., Khiem, V.H., Huong, N.N.: Cost optimization of external cylindrical grinding. Appl. Mech. Mater. **312**, 982–989 (2013)
20. Tu, H.X., Jun, G., Hung, L.X., Tung, L.A., Pi, V.N.: Calculation of optimum exchanged grinding wheel diameter when external grinding tool steel 9CrSi. Int. J. Mech. Eng. Robot. Res. **8**(1), 59–64 (2019)
21. Hung, L.X., Pi, V.N., Tung, L.A., Tu, H.X., Jun, G., Long, B.: Determination of optimal exchanged grinding wheel diameter when internally grinding alloy tool steel 9CrSi. In: IOP Conference Series: Materials Science and Engineering, vol. 417, 012–026 (2018)
22. Pi, V.N., Hung, L.X., Tung, L.N., Long, B.T.: Cost optimization of internal grinding. J. Mater. Sci. Eng. B **6**, 291–296 (2016)

23. Hung, L.X., Ky, L.H., Hong, T.T., Dung, H.T., Lien, V.T., Tung, L.A., Long, B.T., Pi, V.N.: A study on cost optimization of internal cylindrical grinding. Int. J. Mech. Eng. Technol. (IJMET) **10**(1), 414–423 (2019)
24. Tran, T.-H., Le, X.-H., Nguyen, Q.-T., Le, H.-K., Hoang, T.-D., Luu, A.-T., Banh, T.-L., Vu, N.-P.: Optimization of exchanged grinding wheel diameter for minimum grinding cost in internal grinding. Appl. Sci. **9**(7), 1363 (2019)
25. Le, X.H., Vu, N.P., Luu, A.T., Tu, H.X., Jun, G., Banh, T.L.: Determination of optimum exchanged grinding wheel diameter when internal grinding alloy tool steel 9CrSi. In: IOP Conference Series: Materials Science and Engineering, vol. 417, conference 1 (2018). https://doi.org/10.1088/1757-899X/417/1/012026
26. Pi, V.N., Tung, L.A., Hung, L.X., Van Ngoc, N.: Experimental determination of optimum exchanged diameter in surface grinding process. J. Environ. Sci. Eng. A **6**, 85–89 (2017)
27. Hoang, T.D., Tran, T.H., Van Cuong, N., Le, H.K., Nga, N.T.T.: An optimization study on surface grinding stainless steel. Int. J. Eng. Technol. **7**(4), 6621–6625 (2018). https://doi.org/10.14419/ijet.v7i4.29442
28. Tran, T.-H., Luu, A.-T., Nguyen, Q.-T., Le, H.-K., Nguyen, A.-T., Hoang, T.-D., Le, X.-H., Banh, T.-L., Vu, N.-P.: Optimization of exchanged grinding wheel diameter for surface grinding tool steel based on the cost analysis. Metals **9**(4), 448 (2019). https://doi.org/10.3390/met9040448
29. Pi, V.N., Khiem, V.H., Cuong, P.T.: Building formulas for calculation of cutting regime for external cylindrical grinding. Vietnam Mech. Eng. J. (12), 18–23 (2012). (in Vietnamese)
30. Kozuro, L.M., Panov, A.A., Remizovski, E.I., Tristosepdov, P.S.: Handbook of Grinding. Publish Housing of High-Education, Minsk (1981)

Applying Matrix Factorization for Predicting Click Through Rate on Advertizing in Apps on Mobile Devices

Quang H. Nguyen and Tuan-Dung Cao(✉)

Hanoi University of Science and Technology, Hanoi, Vietnam
{quangnh, dungct}@soict.hust.edu.vn

Abstract. The number of smart phone and smartphone users has increased rapidly, boosting the growth of mobile advertising. Predicting the percentage of users who click on an advertisement when displayed on an application (CTR: Click Through Rate) plays an important role in choosing an advertising push strategy for each application. However, because the field of mobile advertising is still new, the study of solutions to these problems is only for Web advertising. This paper presents a new predictive method based on Matrix Factorization (MF) algorithm. The paper also proposes a method of integrating the reliability of data into MF algorithms. The method was tested on actual data provided by Amobi advertising company with 1.7 million records including 300 advertisements and 400 mobile device applications. The AUC (Area Under The Curve) value obtained on the test set is 0.9037 with MF and 0.9411 when integrating reliability into the MF algorithm. Thus, the results show that the predictive quality is significantly improved when integrating data reliability and evaluation into MF algorithms.

Keywords: Click-Through Rate · Smartphone application · Matrix Factorization · In-app mobile advertising

1 Introduction

The number of people using smart phones and smart devices has increased exponentially since last five years and around 1.56 billion smartphones were sold worldwide in 2018 [10]. Users now tend to use smart devices for not only for communication but for other objectives through applications installed on their devices. The competition among developers makes application prices reduce. Many applications are premium, application developers benefit from advertising and IAP (In-App Purchase) for additional features. Therefore, advertising on mobile devices is growing strongly. Advertising revenues delivered on mobile devices totaled \$69.9 billion in full year (FY) 2018, a 39.7% increase from the prior FY 2017 revenues of \$50.1 billion [11].

Typically, a mobile advertising system includes ad network that acts as an intermediary between advertisers and developers (publisher). Advertisers provide ads and register to run ads with ad networks according to each campaign. Advertisers accept to pay according to user actions (such as clicking ads, installing applications…). And the

K.-U. Sattler et al. (Eds.): ICERA 2019, LNNS 104, pp. 130–137, 2020.
https://doi.org/10.1007/978-3-030-37497-6_15

developer registers with the ad network to enable the display of ads on his apps. The ad network deducts a portion of the proceeds from the advertisers to pay for the developers.

The most popular type of advertising today is CPC (cost-per-click), where advertisers pay based on the number of times the users click on the ad. The second advertising type is CPM (cost-per-impression) where advertisers pay based on the number of ad impressions. Among the ad impressions on the application, only a few times the user clicks on the ad.

This ratio is called the CTR (Click Through Rate) and CTR prediction is significant for both advertisers and ad networks. Advertisers rely on the predicted CTR to select applications or users in their ad campaigns to obtain the efficiency. Advertisers rely on ad CTR for distribution to the appropriate applications. Advertisers rely on CTR to appropriately distribute ads to applications.

There have been many studies on CTR prediction due to its importance. However, most of these works involved solving advertising prediction problems for websites. Advertising on mobile devices is quite new and has its own characteristics different from those on the Web. Therefore, we have proposed a method of predicting CTR for mobile advertising using Matrix Factorization (MF) method.

2 Related Works

The research problem is, given in a list of advertisements, how the advertising network selects and displays an advertisement for users on a Web page or on an application running on a mobile device. So far, there have been many studies concerning the finding out the relevant advertisement to Web sites. These studies mainly use two main sources of information: Web content and user behavior in the past. A great advantage of advertising on the Web site is the ability to access and analyze the content of websites displaying ads. The selection of ads based on the content of such Web sites is known as contextual advertising.

A model for contextual advertising is SOCA (Sentiment-oriented contextual advertising), was proposed by Fan and Chang [1]. This model first analyzes the content of the web page, identifies topics at sentence and paragraph levels, thereby figure out the most relevant advertisement in the set of available ads. However, the analysis task requires a lot of manual tweaking. Moreover, the content analysis in mobile device applications is almost impossible.

The second approach uses statistical information about the history of advertisement impression and users click in the past. Tagami et al. [2] uses the information about ads clicked by users among the ads shown to build a CTR prediction model. Richard et al. [3] proposed a method to integrate hierarchical information of advertisement such as an ad in a campaign and a campaign belonging to an advertiser…, to improve prediction. Yang et al. [4] introduced an advertisement selection model based on user behaviors that views clicked URLs as their attribute.

Recently, Matrix Factorization (MF) method [5] has been applied a lot in solving the problem of building a recommendation system. This method has been successfully applied in many different areas such as scientific article suggestions [6] or predicting student achievements when attending online courses [7]. For web-based advertising, Bhargav et al. [8] applied this MF method to build a suggestion system for displaying ads for a group of people. In addition, for many systems, the user-product matrix is not negative, and then the application of Non-negative Matrix Factorization (NMF) has achieved good results [9].

3 Modeling Problem

3.1 Matrix Factorization

The recommender system is a system that predicts a user's "appreciation" or "preference" for a product. In other words, the system helps suggest a product for a user to satisfy a certain goal. An effective method to build a system is to suggest matrix analysis [5].

The basic idea of matrix analysis is to represent both the user and the product as a k dimensional vector. Specifically, the vector $q_i \in R^k$ represents the product i and the vector $p_u \in R^k$ represents the user u. The value of the scalar product between the two vectors $\hat{r}_{ui} = q_i^T p_u$ will characterize the relationship between the user u and the product i, which can also be viewed as the user's level of appreciation for the product. Thus, we only need to define the vectors qi and p_u corresponding to product i and user u respectively. Once the vectors q_i and p_u are available, we then can predict the user rating for the product by calculating $\hat{r}_{ui}$.

To find q_i and p_u matching existing data, we can use singular value decomposition (SVD). However, the complicated problem is that the user-product matrix is a sparse matrix with lots of unknown values. To overcome this problem, Sarwar et al. [12] has already filled in the missing cells by itself, thereby turning the matrix into a thick matrix. However, this method makes the number of calculations needed to increase (due to the increased number of data), and the data added by us can be changed the data completely.

Therefore, another approach is to find q_i and p_u relying only on known values. Then the goal of the method is to learn q_i and p_u so that the difference between $\hat{r}_{ui}$ and r_{ui} is the smallest. This process is calculated by Eq. 1.

$$\min_{p*,q*} \sum_{(u,i)\in K} \left(r_{ui} - q_i^T p_u\right)^2 + \lambda\left(\|q_i\|^2 + \|p_u\|^2\right) \tag{1}$$

in which K is the set of pairs (u, i) that we know the value r_{ui} in the training set.

To avoid over-fitting (this is a model phenomenon that accurately represents known evaluations but is less effective at predicting unknowns), L2 regularization techniques are applied through constants λ (to the second part in Eq. 1) to make sure the coefficients are not too large. Tests with different λ values will be performed to find the optimal λ value.

The Matrix Factorization (MF) algorithm is described as follows:

Input:
+ The matrix r which size $m \times n$ in which we know the value r_{ui} ($(u,i) \in K$)
+ K is the dimension of the vector q_i and p_u
+ λ is the regularization coefficient
+ γ is learning rate

Initialization:
Randomly initialize p_{ut} and q_{it} within [0, 1], u = 1..m, i = 1..n, t = 1..K

Repeat

for each (u, i) ∈ K **do**

calculated $e_{ui} = r_{ui} - q_i^T p_u$

$$q_i = q_i + \gamma(e_{ui} p_u - \lambda q_i)$$

$$p_u = p_u + \gamma(e_{ui} q_i - \lambda p_u)$$

end for

$$cost = \sum_{(u,i) \in K} (r_{ui} - q_i^T p_u)^2 + \lambda(\|q_i\|^2 + \|p_u\|^2)$$

Until cost converging

3.2 Combine Reliability Assessment

CTR represents the probability that users click on the ad. The data show a lot of ad pairs - the application which has a number of Impressions less than 100. Therefore, with a small number of Impressions, the calculation of CTR is inaccurate, the calculated value no longer represents the relationship between promo reports and applications. Moreover, from the observation that the data set is provided, the CTR value is usually less than 10%, so pairs with a large CTR value are less accurate. From the above comment, we propose a method of assessing the reliability, which is in the range [0, 1]. This value includes two c_{imp} (Eq. 2) and c_{val} (Eq. 3) components.

$$c_{imp} = \begin{cases} 1 & impression > \alpha \\ \frac{impression}{\alpha} & otherwise \end{cases} \quad (2)$$

c_{imp} represent the reliability of the sample based on the information on the impression number: if the impression is greater than a threshold α then $c_{imp} = 1$ (completely reliable), vice versa, if the impression number is smaller α then the reliability will be < 1.

$$c_{val} \begin{cases} 1 & CTR > \beta \\ 1 - \frac{|CTR - \beta|}{|1 - \beta|} & otherwise \end{cases} \quad (3)$$

c_{val} characteristic for reliability based on CTR. If the CTR value is less than one threshold β, the sample is considered reliable, whereas reliability will be inversely proportional to CTR.

The reliability value c_{ui} will be the mean of c_{imp} and c_{val}. This value is calculated as $c_{ui} = \sqrt{c_{ui} \cdot c_{val}}$.

After some experimentation we found α = 100 and β = 0.7 gave the best results for the data set provided. Equation 4 represents the objective function when combining reliability.

$$\cos t = \sum_{(u,i)\in K} \left(c_{ui}r_{ui} - q_i^T p_u\right)^2 + \lambda\left(\|q_i\|^2 + \|p_u\|^2\right) \tag{4}$$

When using stochastic gradient descent algorithm, the vectors q_i and p_u are updated as following: $q_i = q_i + \gamma(e_{ui}p_u - \lambda.q_i)$ and $p_u = p_u + \gamma(e_{ui}q_i - \lambda.p_u)$.

4 Experiment

4.1 Test Script

The data used for testing are provided by Amobi advertising company, compiled from impression logs and clicked daily. Data were recorded in nearly two months, from October 1, 2014 until November 21, 2014. Each data component is digital information view as well as the number of clicks in each day of each pair of applications and ads. This dataset has nearly 300 ads and more than 400 applications with more than 1.7 million records.

First, we filter out the pairs of ads - the app has a greater number of impressions than 500 (the larger the impression number, the better the click/impression ratio shows exactly the probability that the user clicks on the ad) and takes out 1/3 of these pairs are used as test cases. All the remaining data is a training set. Training set is used to find p_u and q_i vectors in two cases: using reliability evaluation component, and not using this component.

4.2 Evaluation Values

After each test run obtained the set of q_i and p_u vectors, we apply on the test data and evaluate the results based on the scales: RMSE (Root-Mean-Square Error), FRED (25), and AUC (Area Under the Curve).

RMSE (Eq. 5) is the square root of the mean square error, which is a measure of the difference between the predicted value by its model and its true value. The smaller the RMSE value is, the better the model is evaluated.

$$RMSE = \sqrt{MSE(\hat{\theta})} = \sqrt{E(\hat{\theta} - \theta)^2} \tag{5}$$

A measure of the accuracy of prediction is FRED (25) - the ratio of the predicted sample to within 25% of the actual value, the better the FRED (25) is predicted.

The final CTR prediction aims to rank the compatibility between ads and applications, from which to select the ads for the application optimally, so our AUC scale is used. AUC is the area under the Receiver Operating Characteristic (ROC), which gives us the probability for a classifier to rank a randomly positive sample higher than a randomly selected negative sample. In other words, the AUC gives us information about the rating quality of the prediction. With random guesses, the approximate AUC is equal to 0.5.

4.3 Results and Discussions

To evaluate the effect of k values, we tested the values of k as 10, 15, and 20 when fixed $\lambda = 0$. The results described in Table 1 show that when the number of k increases, the predicted quality of MF in both methods decreases, the optimal value for $k = 10$.

Table 1. Experiment Summary.

Methods	RMSE	FRED (25)	AUC
MF, λ = 0, k = 10	0.0394	0.4164	0.8998
MF which a reliable rating, λ = 0, k = 10	0.0257	0.5019	0.9316
MF, λ = 0, k = 15	0.0386	0.4107	0.8861
MF which a reliable rating, λ = 0, k = 15	0.0385	0.4674	0.9118
MF, λ = 0, k = 20	0.0691	0.4042	0.8836
MF which a reliable rating, λ = 0, k = 20	0.0626	0.4337	0.8847
MF, λ = 0.003, k = 10	0.0182	0.3975	***0.9037***
MF which a reliable rating, λ = 0.003, k = 10	0.0158	0.5023	***0.9411***
MF, λ = 0.01, k = 10	0.0175	0.3336	0.9092
MF which a reliable rating, λ = 0.01, k = 10	0.0250	0.1728	0.8564
MF, λ = 0.03, k = 10	0.0292	0.0777	0.8434
MF which a reliable rating, λ = 0.03, k = 10	0.0389	0.0000	0.7199
MF, λ = 0.01, k = 10	0.0389	0.0000	0.6990
MF which a reliable rating, λ = 0.01, k = 10	0.0390	0.0000	0.6402
Application based	0.0246	0.2438	0.7770
Advertized based	0.0186	0.3171	0.8660

Displayed equations are centered and set on a separate line.

To consider the impact of regularization coefficients, we tested two methods when $k = 10$ and λ received values of 0, 0.003, 0.01, 0.03, 0.1. From the above results, we have a few comments:

- The predicted results when applying the MF algorithm are quite good, with the coefficient λ not too large, both methods give results AUC > 0.8.
- The two methods are affected by the coefficient λ: initially, when the λ increases, the predicted quality increases, after a certain increase, the quality decreases sharply. For each method, there are different optimization coefficients λ.
- There is a significant improvement when integrating reliability assessments: all scales show better results than conventional MF.

4.4 Compare with Traditional Methods

To assess the method proposed in the paper, we also perform training and testing on different algorithms and analyze the effectiveness of each algorithm based on the measurement measures. Two methods for comparison are common collaborative filtering [13, 14], based on app-based applications and adv based advertising. The similarity measure applied here is cosine measurement. The results of these tests are described in Table 1. From the results in Table 1, the conventional collaborative filtration method results in relatively good RMSE, but FRED (25) is much lower. The number of applications is much less than the number of ads, so the number of vectors to compare the similarities in adv-based will be more than in the app-based method. Because there is a lot of information to compare and find similarities, the adv-based method is for better predictive quality. In addition, the MF method incorporates reliability that outperforms other methods on all scales of evaluation. Even if it has not been improved by combining reliability, the MF method also offers better results than two common and application-based collaborative filtering methods.

5 Conclusions and Perspectives

Predicting CTR plays an important role in mobile advertising. The value affects the revenue of the developer and the advertiser, to the amount that the advertiser has to pay, affecting the user experience.

In this study, we propose a prediction method based on matrix separation. The advertising matrix - the application is split into two advertising-specific matrices which predicted unknown CTR values. In addition, we propose integrating the evaluation of reliability for input data to improve predictive efficiency. Experimental results show that this method based on matrix separation predicts quite good results. The proposed method has also been installed as a program, run daily by Amobi Company. The anticipated information of the program used by the company is used to decide the schedule for running ads for the next day.

From here, our next research direction is as follows. First, it is necessary to integrate more information about advertising and applications such as user groups, application types, etc. From that information, even if there are no new advertisements or applications, there is no feedback. From the user, we can also predict a certain part. Next, we want to develop a more complex model method, taking into account the time effect.

References

1. Fan, T.-K., Chang, C.-H.: Sentiment-oriented contextual advertising. Knowl. Inf. Syst. **23** (3), 321–344 (2010)
2. Tagami, Y., Ono, S., Yamamoto, K., Tsukamoto, K., Tajima, A.: CTR prediction for contextual advertising: learning-to-rank approach. In: Proceedings of the Seventh International Workshop on Data Mining for Online Advertising, p. 4. ACM (2013)

3. Oentaryo, R.J., Lim, E.-P., Low, J.-W., Lo, D., Finegold, M.: Predicting response in mobile advertising with hierarchical importance-aware factorization machine. In: Proceedings of the 7th ACM International Conference on Web Search and Data Mining, pp. 123–132. ACM (2014)
4. Yan, J., Liu, N., Wang, G., Zhang, W., Jiang, Y., Chen, Z.: How much can behavioral targeting help online advertising?. In: Proceedings of the 18th International Conference on World Wide Web, pp. 261–270. ACM (2009)
5. Koren, Y., Bell, R., Volinsky, C.: Matrix factorization techniques for recommender systems. Computer **8**, 30–37 (2009)
6. Wang, C., Blei, D.M.: Collaborative topic modeling for recommending scientific articles. In: Proceedings of the 17th ACM SIGKDD International Conference on Knowledge Discovery and Data Mining, pp. 448–456. ACM (2011)
7. Thai-Nghe, N., Drumond, L., Horváth, T., Nanopoulos, A., Schmidt-Thieme, L.: Matrix and tensor factorization for predicting student performance. In: CSEDU, no. 1, pp. 69–78 (2011)
8. Kanagal, B., Ahmed, A., Pandey, S., Josifovski, V., Garcia-Pueyo, L., Yuan, J.: Focused matrix factorization for audience selection in display advertising. In: 2013 IEEE 29th International Conference on Data Engineering (ICDE), pp. 386–397. IEEE (2013)
9. Lee, D.D., Seung, H.S.: Algorithms for non-negative matrix factorization. In: Advances in Neural Information Processing Systems, pp. 556–562 (2001)
10. Number of smartphones sold to end users worldwide from 2007 to 2018 (2019).https://www.statista.com/statistics/263437/global-smartphone-sales-to-end-users-since-2007/. Accessed 26 Feb 2019
11. The Interactive Advertising Bureau (IAB), IAB Internet Advertising Revenue Report Conducted by PricewaterhouseCoopers (PWC). https://www.iab.com/insights/iab-internet-advertising-revenue-report-conducted-by-pricewaterhousecoopers-pwc-2/\#year2019
12. Sarwar, B., Karypis, G., Konstan, J., Riedl, J.: Application of dimensionality reduction in recommender system-a case study. No. TR-00-043. Minnesota Univ Minneapolis Dept of Computer Science (2000)
13. Carlos, P.: Recommender Systems—User-Based and Item-Based Collaborative Filtering (2017). https://medium.com/@cfpinela/recommender-systems-user-based-and-item-based-collaborative-filtering-5d5f375a127f. Accessed 6 Nov 2017
14. Ekstrand, M.D., Riedl, J.T., Konstan, J.A.: Collaborative filtering recommender systems. Found. Trends Hum.-Comput. Inter. Arch. **4**(2), 81–173 (2011)

Approximate Response of a Non-linear Vibration Isolation System Subjected to Harmonic Excitation

Gun-Myung Lee(✉)

Research Center for Aircraft Parts Technology, Gyeongsang National University, Jinju, Gyeongnam, Korea

Abstract. A non-linear vibration isolation system which is composed of a non-linear spring and a linear damper was proposed in last research. The merits of the proposed isolation system are that the system is simple and the load carrying capacity can be controlled easily by adjusting the spring inclination angle. When the support of the isolation system is excited harmonically, the response component of the isolation system mass at the excitation frequency has been calculated approximately using three different methods: linear approximation, harmonic balance method, and higher order FRFs method. The method using higher order FRFs gives much more accurate results compared to the other methods. The error between the exact and the approximated responses does not increase monotonously with the excitation amplitude, and is within 1.53%.

Keywords: Frequency component · Harmonic balance · Higher order frequency response function · Non-linear vibration isolation system · Volterra series

1 Introduction

A vibration isolation system is used to minimize the motion transmitted to a mechanical part or system when the base or support is subjected to motion. A suspension system of a car seat is an example of a vibration isolation system.

A vibration isolation system with a non-linear spring was proposed and its characteristics were investigated in past research [1]. The proposed non-linear spring is composed of two symmetric linear springs as shown in Fig. 1. The important features of the non-linear spring are that its structure is very simple compared to other springs, for example disc springs [2], the load capacity of the spring can be adjusted easily, and the load-displacement relationship can be expressed easily, resulting in easy dynamic analysis of systems equipped with the spring.

If non-linear relationship between the spring force and the displacement can be represented by a polynomial, the equation of motion of the vibration isolation system equipped with this non-linear spring and a linear damper can be solved approximately.

In this research the equation of motion was solved approximately by three different methods: linear approximation, harmonic balance method, and higher order FRFs (Frequency Response Functions), and the results were compared with numerical analysis results.

K.-U. Sattler et al. (Eds.): ICERA 2019, LNNS 104, pp. 138–143, 2020.
https://doi.org/10.1007/978-3-030-37497-6_16

2 Description of Non-linear Equation Solving Methods

2.1 Linear Approximation

At static equilibrium of the system, the deflection of the mass is calculated. The non-linear spring is approximated by a linear spring with the stiffness, which is the slope of the tangent at the equilibrium position to the load-displacement curve. Using the transmissibility of a linear system which is the ratio of the amplitude of the mass to that of the base, the amplitude of the mass is obtained.

2.2 Harmonic Balance Method

The harmonic balance method has been used widely to study approximately the stable and unstable periodic responses of non-linear oscillator systems. In this method, the steady state response of a non-linear system subjected to a harmonic input is expressed by a sum of harmonics. In a simple case, the response is expressed as

$$y(t) = A\cos\omega \tag{1}$$

The expression for the response is inserted into a non-linear equation, and by equating coefficients of each harmonic on both sides, the amplitude of each harmonic is obtained.

2.3 Higher Order FRFs

The concept of higher order FRFs will be explained briefly. The response y(t) of a system with polynomial non-linearities to any input x(t) can be represented by the Volterra series [3].

$$\begin{aligned} y(t) &= \sum_{n=1}^{\infty} y_n(t) \\ &= \int_{-\infty}^{\infty} h_1(\tau_1)x(t-\tau_1)d\tau_1 + \int_{-\infty}^{\infty}\int_{-\infty}^{\infty} h_2(\tau_1,\tau_2)x(t-\tau_1)x(t-\tau_2)d\tau_1 d\tau_2 \\ &+ \int_{-\infty}^{\infty}\cdots\int_{-\infty}^{\infty} h_n(\tau_1,\cdots,\tau_n)x(t-\tau_1)\cdots x(t-\tau_n)d\tau_1\cdots d\tau_n + \cdots \end{aligned} \tag{2}$$

In the above equation, $h_n(\tau_1,\cdots,\tau_n)$ are higher order impulse response functions. The higher order FRFs are defined as the multidimensional Fourier transforms of these higher order impulse response functions,

$$H_n(\omega_1,\cdots,\omega_n) = \int_{-\infty}^{\infty}\cdots\int_{-\infty}^{\infty} h_n(\tau_1,\cdots,\tau_n)\prod_{i=1}^{n} e^{-j\omega_i\tau_i}d\tau_1\cdots d\tau_n \tag{3}$$

The value of $H_n(\omega_1,\cdots,\omega_n)$ is independent of the order of the arguments, $\omega_1,\cdots,\omega_n$.

When a non-linear system is subjected to sinusoidal excitation

$$x(t) = X\cos \omega t = \frac{X}{2}e^{j\omega t} + \frac{X}{2}e^{-j\omega t} \tag{4}$$

the response is expressed as follows:

$$\begin{aligned} y(t) = {} & H_1(\omega)\frac{X}{2}e^{j\omega t} + H_1(-\omega)\frac{X}{2}e^{-j\omega t} + H_2(\omega,\omega)\left(\frac{X}{2}\right)^2 e^{j2\omega t} + 2H_2(\omega,-\omega)\left(\frac{X}{2}\right)^2 \\ & + H_2(-\omega,-\omega)\left(\frac{X}{2}\right)^2 e^{-j2\omega t} + H_3(\omega,\omega,\omega)\left(\frac{X}{2}\right)^3 e^{j3\omega t} + 3H_3(\omega,\omega,-\omega)\left(\frac{X}{2}\right)^3 e^{j\omega t} \\ & + 3H_3(\omega,-\omega,-\omega)\left(\frac{X}{2}\right)^3 e^{-j\omega t} + H_3(-\omega,-\omega,-\omega)\left(\frac{X}{2}\right)^3 e^{-j3\omega t} + \cdots \end{aligned} \tag{5}$$

In this paper, a higher order FRF $H_n(\omega,\cdots,\omega,-\omega,\cdots,-\omega)$ with $n-i$ arguments of ω and i arguments of −ω is simply denoted by $H_{n,i}$. Using the notation, the above response can be expressed as:

$$y(t) = \sum\nolimits_{n=1}^{\infty} y_n(t) = \sum\nolimits_{n=1}^{\infty}\sum\nolimits_{i=0}^{n}\binom{n}{i}H_{n,i}\left(\frac{X}{2}\right)^n e^{j(n-2i)\omega t} \tag{6}$$

where $\binom{n}{i}$ is a binomial coefficient defined as $n!/i!(n-i)!$. The first order FRF is the complex ratio of the response at the excitation frequency ω to the excitation at the same frequency. By considering only the terms associated with the input $\frac{X}{2}e^{j\omega t}$ in Eq. (5) one obtains an estimate of the first order FRF for a non-linear system:

$$\begin{aligned} \hat{H}_1(\omega) &= \frac{Y(\omega)}{X(\omega)} \\ &= H_1(\omega) + \binom{3}{1}H_3(\omega,\omega,-\omega)\left(\frac{X}{2}\right)^2 + \binom{5}{2}H_5(\omega,\omega,\omega,-\omega-\omega)\left(\frac{X}{2}\right)^4 \\ &= \sum\nolimits_{m=0}^{\infty}\binom{2m+1}{m}H_{2m+1,m}\left(\frac{X}{2}\right)^{2m} \end{aligned} \tag{7}$$

The equation of motion of a single dof system with cubic stiffness subject to sinusoidal excitation becomes

$$m\ddot{y}(t) + c\dot{y}(t) + k_1 y(t) + k_2 y(t)^2 + k_3 y(t)^3 = \frac{X}{2}\left(e^{j\omega t} + e^{-j\omega t}\right) \tag{8}$$

Substituting Eq. (5) into the above equation and equating the coefficients of the $\left(\frac{X}{2}\right)e^{j\omega t}$ terms on both sides, one can obtain

$$H_1(\omega) = \frac{1}{k_1 - m\omega^2 + jc\omega} \tag{9}$$

Extending this method of harmonic probing [4], one obtains the following general expression for higher order FRFs.

$$\begin{aligned}&\binom{n}{i} H_{n,i} H_1[(n-2i)\omega]^{-1} + k_2 \sum \binom{n_1}{i_1}\binom{n_2}{i_2} H_{n_1,i_1} H_{n_2,i_2} \\ &+ k_3 \sum \binom{n_1}{i_1}\binom{n_2}{i_2}\binom{n_3}{i_3} H_{n_1,i_1} H_{n_2,i_2} H_{n_3,i_3}\end{aligned} \tag{10}$$

In the above equation the summation must be performed over all sets of n's and i's such that $n_1 + n_2 = n$ and $i_1 + i_2 = i$ in the term containing k_2, and $n_1 + n_2 + n_3 = n$ and $i_1 + i_2 + i_3 = i$ in the term containing k_3. Equation (10) shows that a higher order FRF can be expressed in terms of FRFs of lower orders. In this way all the higher order FRFs in Eq. (7) are obtained. Finally, the response at the excitation frequency, Y(ω) can be obtained using these higher order FRFs.

3 Application of the Solving Methods

3.1 Proposed Non-linear Isolation System

The proposed non-linear isolation system is composed of two symmetric linear springs, a linear damper, and a support as shown in Fig. 1. The linear springs have spring constant k, free length l_0, and are inclined by θ from the horizontal support. If the upper support of the springs moves downward by y, the length of each spring becomes

$$l = (l0\cos\theta)2 + (l0\sin\theta - y)2 \tag{11}$$

and the spring force in each spring $isk(l_0 - l)$, the resultant vertical force applying on the support becomes

$$f = 2k(l0 - l)l0\sin\theta - yl \tag{12}$$

Therefore, the two springs behave as a non-linear spring. The parameters used in this research are $k = 10,000$ N/m, $l_0 = 0.7$ m, and $\theta = 0.7752$ rad(44.42°).

When the base of the vibration isolation system is excited as shown in Fig. 1, the equation of motion of the mass is expressed as follows.

$$my = -cy - x - f(y - x) \tag{13}$$

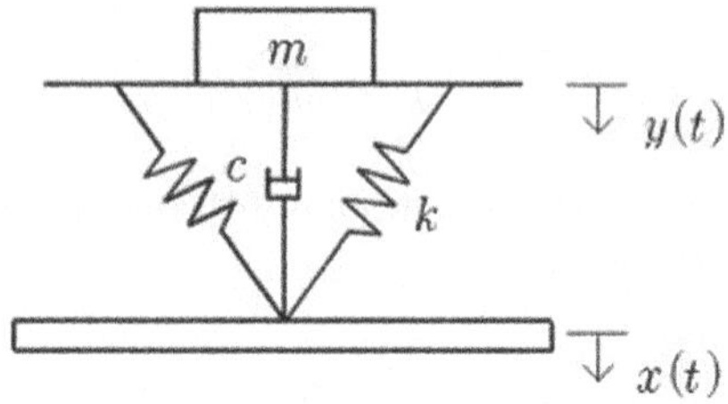

Fig. 1. Vibration isolation system composed of a non-linear spring and a linear damper.

In the above equation a dot on a letter represents differentiation with respect to time. c and $f(x)$ represent the damping constant and the spring force in Eq. (12), respectively. The excitation of the base $x(t)$ is given by $X \sin \omega t$. The displacement of the mass, y(t) is measured from the equilibrium position due to gravity. Letting z = y − x, Eq. (13) becomes

$$\begin{aligned} \mathrm{m}\ddot{z} + c\dot{z} + f(z) &= -\mathrm{m}\ddot{x} \\ &= m\omega 2X \sin \omega t \end{aligned} \tag{14}$$

Solving the above equation for z numerically and adding x, the response of the mass is obtained. Taking a Fourier Transform of y(t), its frequency component at the excitation frequency ω is obtained. The response of the system was considered for the system with the above non-linear spring and the following parameters: m = 100 kg, c = 1000 Ns/m, ω = 10 rad/s.

3.2 Approximate Solutions

The response of the mass for the sinusoidal excitation of the base was calculated by solving the equation of motion numerically. The response component at the excitation frequency ω is obtained by Fourier transform of the response. This response component is assumed to be exact. The amplitude of the response component at the excitation frequency was also calculated approximately using the methods: linear approximation, harmonic balance method, and method using higher order FRFs. The differences between exact and approximate solutions are shown in Fig. 2.

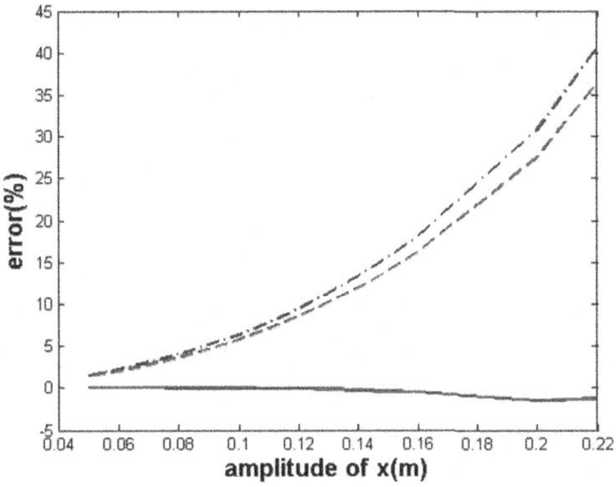

Fig. 2. Variation of the error with the excitation amplitude for linear approximation (dash-dot line), harmonic balance (dashed line), and higher order FRFs method (solid line).

4 Conclusions

A non-linear vibration isolation system which is composed of a non-linear spring and a linear damper was proposed in past research. When the base of the isolation system is excited harmonically, the response component of the mass at the excitation frequency was calculated approximately using three different methods: linear approximation, harmonic balance, and higher order FRFs method. The method using higher order FRFs give much more accurate results compared to the other methods. The error between the exact and the approximate responses does not increase monotonously with the excitation amplitude, and is within 1.53%, while the errors with the other methods are larger than 36% for large excitation amplitudes.

References

1. Lee, G.-M., Park, H.S.: Characteristics of a vibration isolation system with non-linear springs. Trans. Korean Soc. Noise Vib. Eng. **27**(7), 844–849 (2017)
2. Shin, D.H., Lee, J.Y., Oh, J.-E.: Prediction of load-displacement of the disc spring with the friction. Trans. Korean Soc. Noise Vib. Eng. **22**(4), 344–351 (2012)
3. Tomlinson, G.R., Manson, G., Lee, G.-M.: A simple criterion for establishing an upper limit to the harmonic excitation level of the duffing oscillator using the Volterra series. J. Sound Vib. **190**(5), 751–762 (1996)
4. Bedrosian, E., Rice, S.O.: The output properties of Volterra systems driven by harmonic and Gaussian inputs. Proc. IEEE **59**(12), 1688–1707 (1971)

Automatic Optical Inspection for Magnetic Particle Detection of Forging Defects

Quang-Cherng Hsu[1], Rui-Hong Ni[1], Jhan-Hong Ye[1], and Ngoc-Vu Ngo[2(✉)]

[1] National Kaohsiung University of Science and Technology, 415 Chien Kung Road, Sanmin District, Kaohsiung 80778, Taiwan, ROC
[2] Thai Nguyen University of Technology, No. 666, 3/2 Street, Thai Nguyen City, Vietnam
ngocvu@tnut.edu.vn

Abstract. The magnetic particle inspection method is to detect the surface of the ferromagnetic material and the subsurface flaw. It can be considered as a combination of two nondestructive inspection methods which are magnetic flux leakage inspection and visual inspection. The visual inspection is performed by the quality control personnel under high-power ultraviolet light. Because of the strong ultraviolet light, this method has caused damage to the eyes of quality control personnel, and the detection results are prone to errors and cause missed inspections. Therefore, this study developed an automated detection system that can be used to replace manual visual inspection. This study uses a self-made aluminum alloy semi-circular tube type cover, diffused light source, high-resolution CMOS industrial cameras and image processing algorithms. The intensity of the ultraviolet light source used in this study was derived from the illumination of the semi-circular tube type light source cover. The best UV light source parameters were available when the bilateral light source was fully open, the light source illumination angle was at 120°, the light source intensity was at level 2, and the filter was used. The average value of the RGB values and the standard deviation of defects on work-piece were analyzed. Experimental results showed that the RGB average values were 38.66, 35.99, and 26.06, respectively, and the maximum standard deviation was G standard deviation of 86.83.

Keywords: Magnetic particle inspection · Automatic optical inspection · Ultraviolet light · Defect detection

1 Introduction

Magnetic particle test (MPT) is a magnetic leakage phenomenon. It can adsorb tiny magnetic powders with fluorescent properties, and then display the enthalpy distribution. It is often used to detect the surface and subsurface enthalpy of ferromagnetic materials. The test result is still judged by the human eye, and the basis for distinguishing the defects cannot be quantified, resulting in the difference of the judgment results. Therefore, automatic optical inspection (AOI) can be used because it has a more flexible way of erection, detection equipment is relatively simple. It has become one of the methods frequently used in online detection on the industry.

K.-U. Sattler et al. (Eds.): ICERA 2019, LNNS 104, pp. 144–154, 2020.
https://doi.org/10.1007/978-3-030-37497-6_17

In a research, Ma et al. [1] detected cracks on the surface of the train axle using a magnetic particle inspection machine. This system was equipped with an industrial camera and a filter lens. It was installed with an ultraviolet light under the camera to allow the magnetic powder to display obvious yellow-green light. After the Top-hat method and the Otsu's adaptive binarization method are performed on the image processing, the result image is extracted as a feature, and the geometric feature value is calculated to obtain the size and location of the flaw. In addition, Nishimine et al. [2] developed a set of detection methods for square steel work-piece based on the magnetic particle inspection method and image detection method. The detection system was integrated to the steel work-piece manufacturing line. To implement a stable detection process, a brightness measuring device was developed. The sample was measured by an optical sensor, and the obtained measurement value was returned to the detection. The system adjusts the brightness to make the detection system more stable. In the research of Lundh [3], after using the magnetic particle inspection method for forged parts, he used the camera to capture the green cracks appearing on the surface, and then uses the image processing method to complete the noise filtering, and then identify and mark the defects. The pre-image processing sequence of the research firstly is the Top hat method, which is used as the separation of background and object, and then uses the edge search function to divide the object at the position of the image more accurately and then finds the edges and forms a contour of object. The feature extraction is performed on the contour range, and the circularity of all the feature blocks in the range, the length, the area, and the long axis angle of the main area moment are respectively taken out, and the threshold value is used as a basis for detecting the flaw. Luo et al. [4] used the crankshaft as the detection object, and used the magnetic particle inspection and CCD camera to detect the surface flaw. Since the camera can only capture images of the same plane, but the crankshaft is a cylinder with different planar surface features. Therefore, to capture the entire crankshaft image without increasing the number of cameras, this study added a motion control to the crankshaft, allowing the crankshaft to rotate, and integrated camera control to synchronize image capture. Hao et al. [5] developed an image detection system using a line scan camera for magnetic particle inspection, with the train axle as the target. Since the target member is about two meters in length and it has a diameter of about 20 cm, a large number of cameras are required to be installed due to insufficient camera field of view. Then, at the same resolution, a line scan camera was used, and at the same time, the machine was rotated to perform an image capture operation.

This study presented an automatic optical inspection system including an aluminum alloy semi-circular tube type cover, diffused light source, high-resolution CMOS industrial cameras on the existing automatic magnetic particle inspection machine. A vehicle constant velocity joint was used as a test sample, and its surface has the characteristics of metal reflection.

2 Experimental System

2.1 Semi-circular Cube Type Cover

The diffused light source used in this study was produced by the semi-circular tube type cover which was aluminum alloy material. When the light source was irradiated onto the surface of the semi-circular tube type cover, it caused the phenomenon of diffusion and reflection due to the surface characteristics of the aluminum alloy, and then reach the inside of the semi-circular tube type cover. The position of the semi-circular tube type cover in experimental system is shown in Fig. 1.

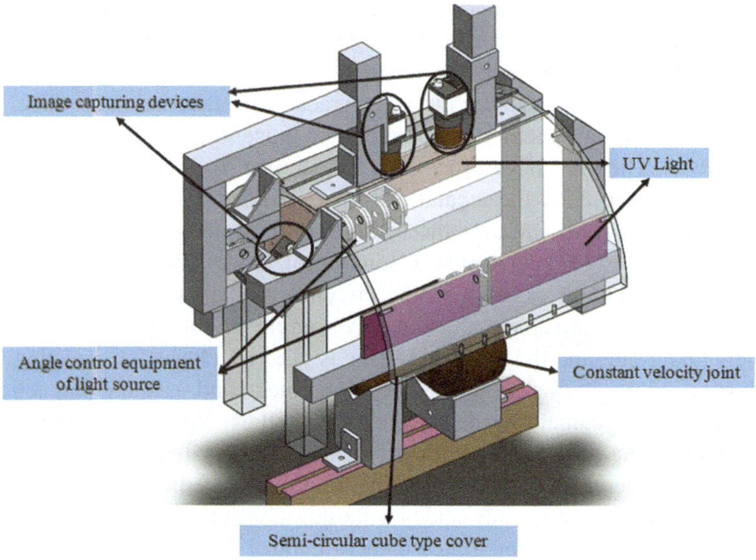

Fig. 1. Experimental system.

2.2 Diffuse Light Source

The dif- fused light source

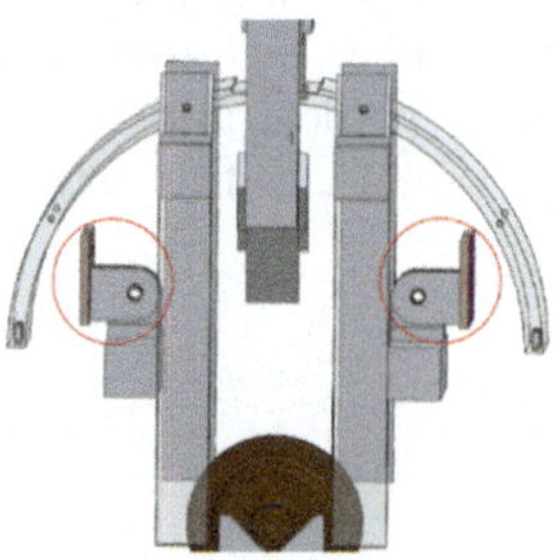

Fig. 2. Light source at 0°.

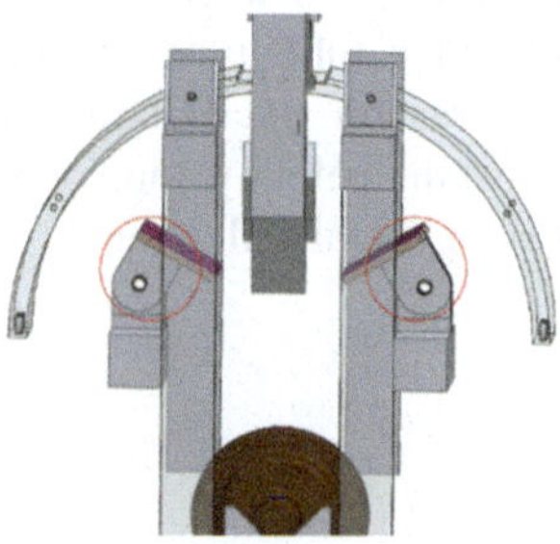

Fig. 3. Light source at 120°.

is to illuminate the surface of the work-piece so that the camera can view the surface contour and it is also necessary to display the outline of the magnetic powder accumulation caused by the sputum. The light source, which uses ultraviolet light specifically for magnetic particle inspection, increases the credibility of the test process. In this study, angle of the light source was controlled by a rotating device and the semi-circular tube type cover was irradiated at two angles that are 0° and 120°, as shown in Figs. 2 and 3, respectively.

2.3 Light Source Intensity

In this study, the intensity control of the LED light source was the knob on the power supply. The knob was controlled by the non-segment type, as shown in Fig. 4. To investigate the state of the ultraviolet light source, the illuminance must be the best conditions, and an illuminometer that specializes in measuring the ultraviolet light source was used. The illuminometer can measure two kinds of light sources separately which are ultraviolet light intensity and white light intensity, as shown in Fig. 5.

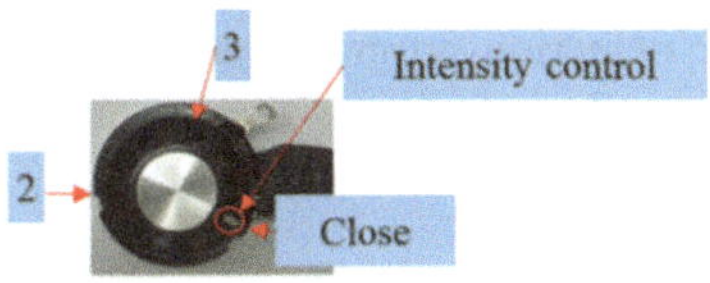

Fig. 4. Light source intensity adjustment at levels 2 and 3.

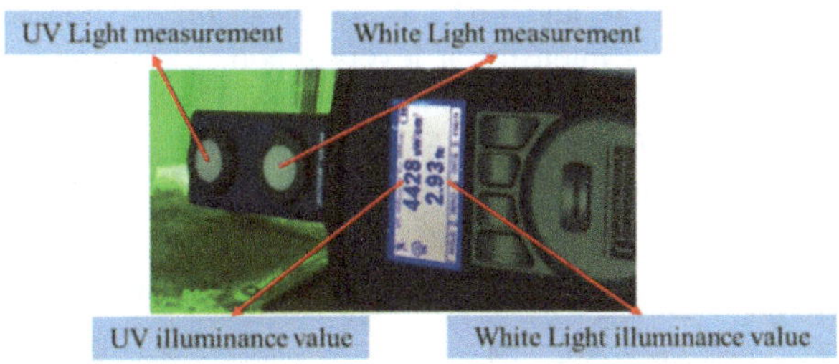

Fig. 5. Illuminometer device.

3 Contour Searching and Automatic Defect Detection

3.1 Contour Searching

The purpose of the contour searching is to find the outer contour features of objects in the image. In some image processing methods, the contour features can be directly highlighted, which is the shape of the image and the image coordinate position. The contour is a set of points representing the edge in the image. The edge detection algorithm can view the edge pixels of the contour according to the difference between the pixels, but the algorithm does not materialize the contour, but simply counts the pixels. This step is to combine the edge pixels produced by the edge detection into a contour.

3.2 Automatic Defect Detection

For the image processing and defect detection system, the defect detection was mainly based on the yellow-green light generated by the magnetic liquid accumulation, wherein the RGB Euclid distance algorithm was used to perform color separation for the light.

The defect analysis process can be described as follows: The RGB image was used to analyze the RGB distribution of the image. The average value and the standard deviation were calculated. Then, the color separation in the image processing program was returned to the image processing and detection program. The color was separated by taking the RGB average as the center and the maximum standard deviation as the radius. After obtaining the geometric feature values of the defect and non-defect contours on other original images, the distribution, the threshold value of the defect and non-defect contours can be classified. Finally, the image processing parameters were input to the fully automatic detection program for execution.

4 Results and Discussion

4.1 Ultraviolet Light Source

The illumination angle of the light source was irradiated to the semi-circular tube type cover at 0°, as shown in Fig. 6, and 120°, as shown in Fig. 7, respectively. The light source can be adjusted to the intensity at levels 2 and 3, as shown in Fig. 4. In the measurement process of the light source intensity, there was a change in the intensity testing with or without the filter, and only the ultraviolet light on the one side was turned on. The light source intensity data is shown in Table 1.

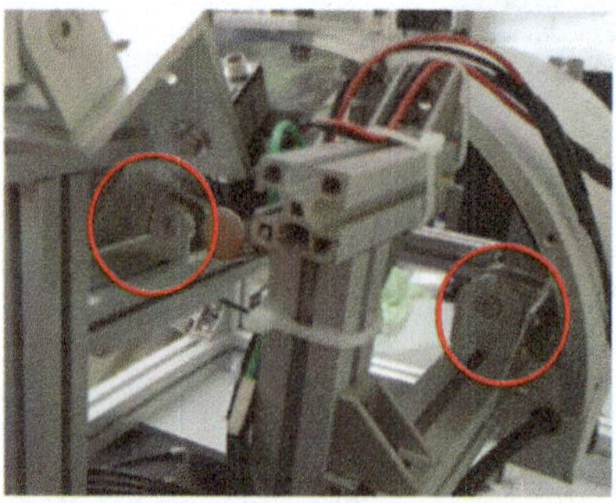

Fig. 6. Actual angle of the light source at 0°.

Fig. 7. Actual angle of the light source at 120°.

Table 1. Light source intensity data

Configuration	Ultraviolet power ($\mu W/cm^2$)	White light power (fc)
Light source at both sides, light source at 0° and light source intensity at level 3 without filter	982	0.86
Light source at one side, light source at 120° and light source intensity at level 3 without filter	5758	4.58
Light source at one side, light source at 120° and light source intensity at level 3 with filter	3074	1.74
Light source at both sides, light source at 120° and light source intensity at level 3 with filter	10580	4.79
Light source at both sides, light source at 120° and light source intensity at level 2 with filter	3196	1.5

According to the current scene detection of human eyes, the requirements for ultraviolet light and white light are as follows:

- The ultraviolet light power is usually higher than 3000 $\mu W/cm^2$.
- White light power should be less than 2.0 fc

However, when the UV lamp was illuminated, the temperature of the light plate was higher, so the influence of the light source intensity on the temperature was tested. The light source intensity was investigated at levels 2 and 3 for 15 min, and the results are shown in Table 2. When the light source intensity was level 3, the temperature reached above 70 °C. When the illumination was performed at this temperature, the life of the ultraviolet light plate would be reduced or even burned. Therefore, the best light source conditions of the study were available when the bilateral light source was fully open and light source was at 120° and the light source intensity was at level 2 with a filter.

Table 2. Temperature comparison

UV light source intensity	Temperature after 15 min (°C)
Light source intensity at level 3	74.1
Light source intensity at level 2	43.8

4.2 Image Detection System

Fig. 8. Inspection objects.

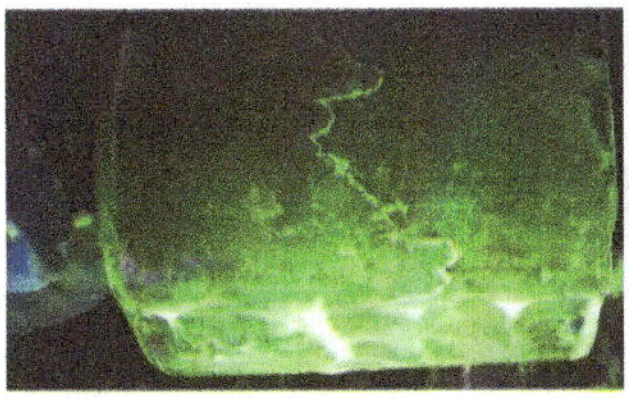

Fig. 9. Actual defect.

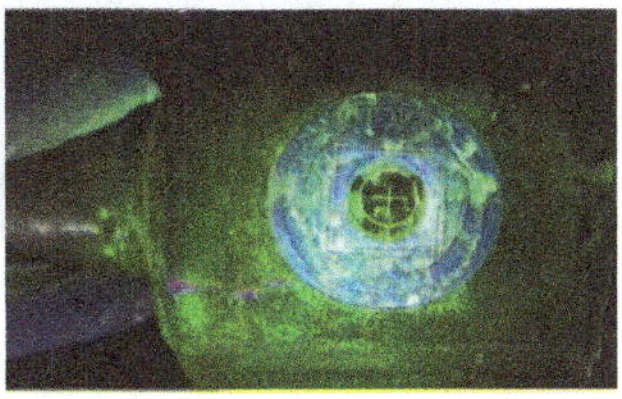

Fig. 10. Artificial defect.

Defect of the Inspection Objects. In this study, six constant velocity joints for vehicles were used, as shown in Fig. 8. For the detection, two types of surfaces were identified, one was the crack generated during molding, as shown in Fig. 9, and the other was to test the effect of magnetic particle inspection, the artificial sputum test piece is shown as Fig. 10.

Image Processing and Flaw Detection. To detect defects of the inspection objects, automatic image capture process was performed. First, according to the high-frequency signal on the image, as shown in Fig. 11(a), it was lighter than the surrounding pixels. In order to highlight the defect feature, the Top Hat operation was performed first to obtain result as in Fig. 11(b) and then the yellow-green light feature can be highlighted and the color was separated, as shown in Fig. 11(c). After getting the gray image, binarized image can be obtained, as shown in Fig. 11(d). However, there was still some noise in the binarized image, so opening operation was performed, as shown in Fig. 11 (e). Finally, the closing operation was performed to maintain the contour integrity. Figure 11(f) shows the result for contour search process. Figures 12 and 13 show the analysis process of the two defects which are defect 1 and defect 2, respectively.

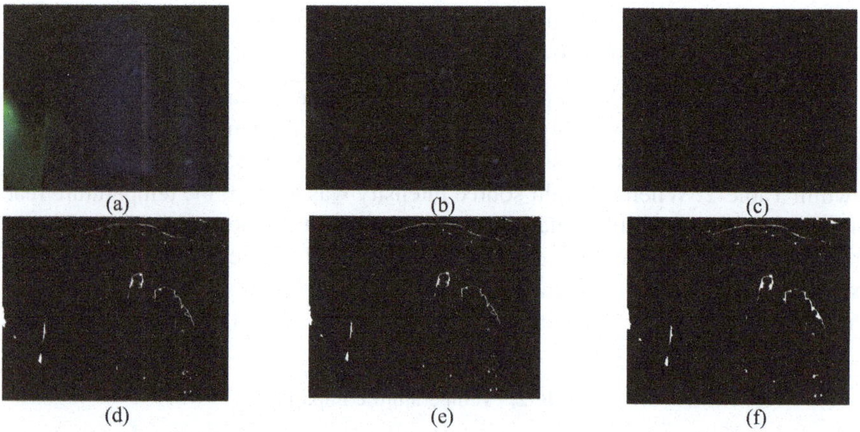

Fig. 11. Image processing to detect the type defects.

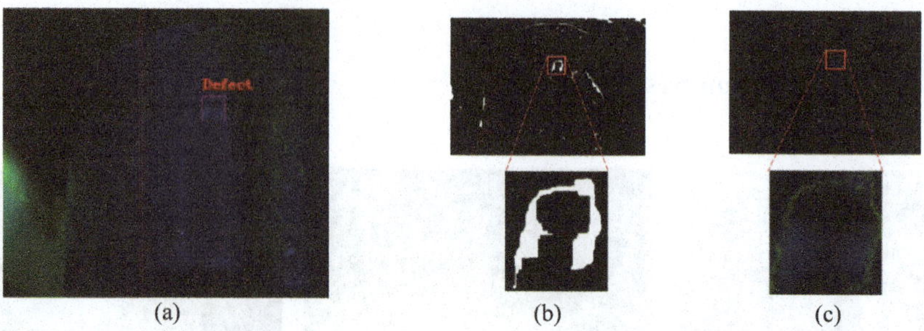

Fig. 12. Analysis process for the defect 1.

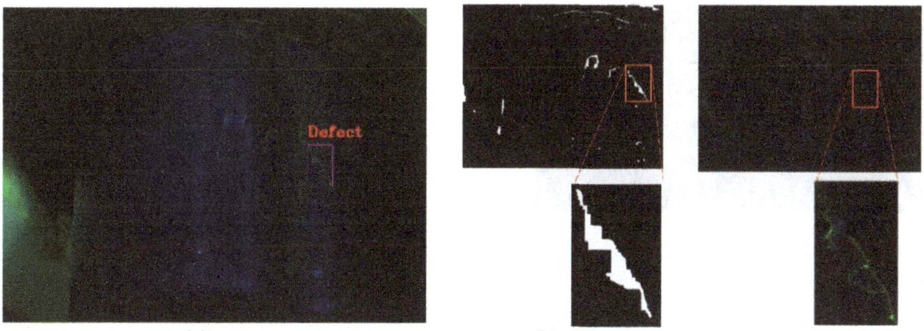

(a) (b) (c)

Fig. 13. Analysis process for the defect 2.

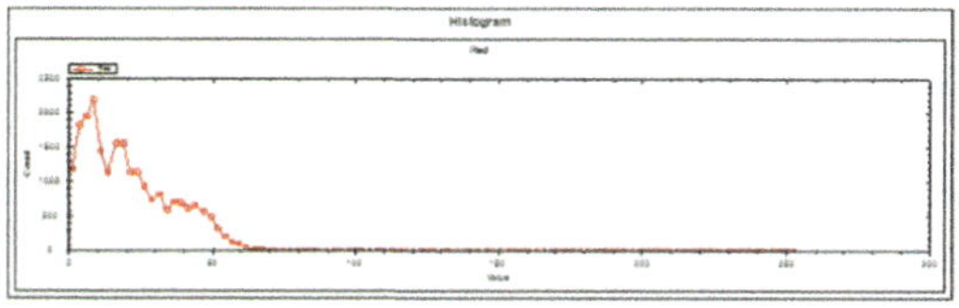

(b) Histogram of the red colour.

(a)

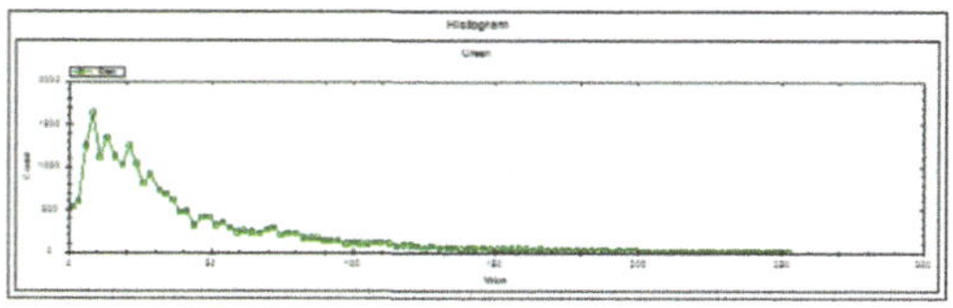

(c) Histogram of the green colour.

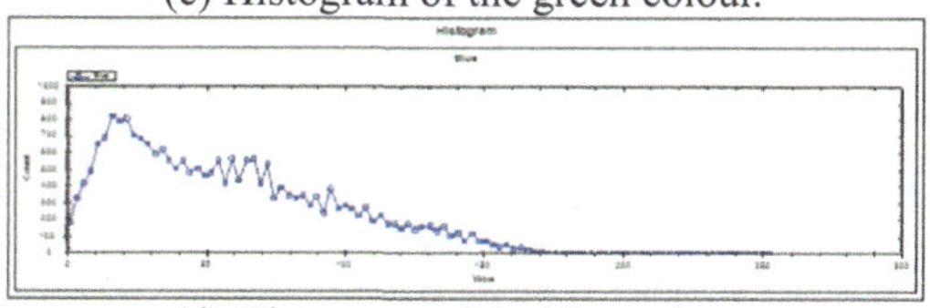

(d) Histogram of the blue colour.

Fig. 14. RGB distribution of the defect 1.

Defect Analysis. The two RGB images in Figs. 14(a) and 15(a) were respectively placed in the defect analysis program to get histograms of the red, green and blue colours, as shown in Figs. 14(b–d) and 15(b–d), respectively, distribution line map of the three colour channels from the distribution of the line graph indicated that the gray value was lower than most of the pixels. The proportion of images from defect was small, and the background pixels account for the majority. This phenomenon caused statistical distortion.

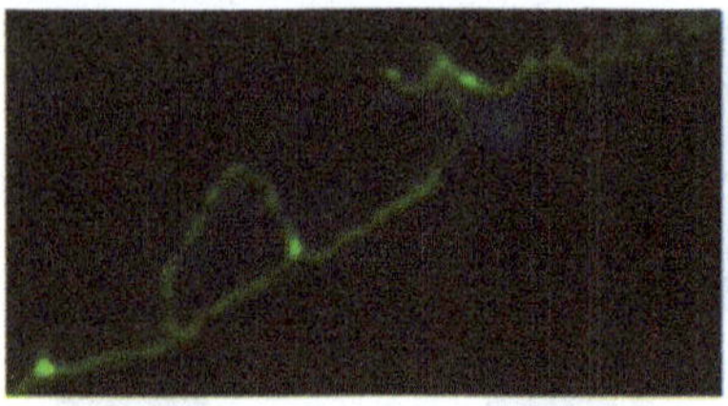

(a)

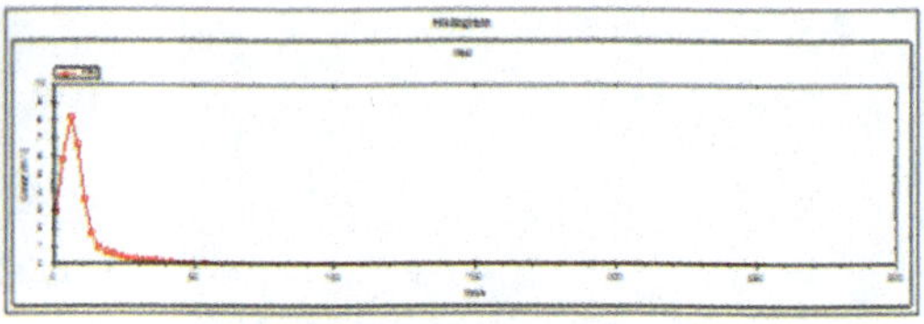

(b) Histogram of the red colour.

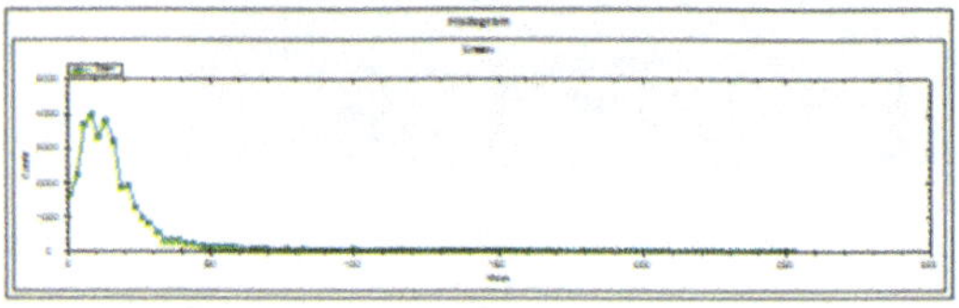

(c) Histogram of the blue colour.

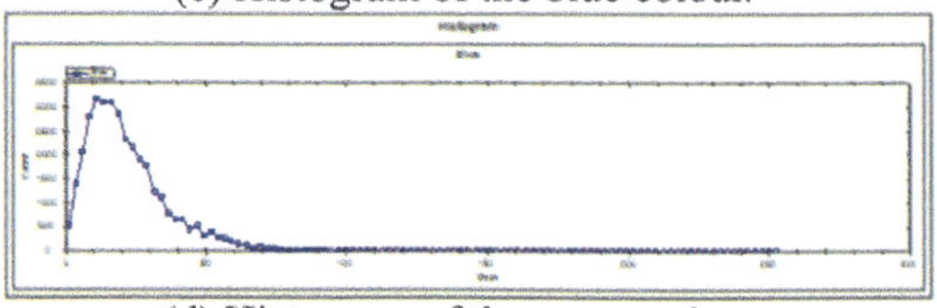

(d) Histogram of the green colour.

Fig. 15. RGB distribution of the defect 2.

To reduce the influence of the background pixel on the statistical value, according to the binarization threshold value used above, it can be estimated that the defect feature was binaryized, and it can be clearly distinguished from the background, thereby estimating that the three-channel (RGB) can use the binarization threshold to segment the background and defect. The pixels whose three-channel gray value was lower than the binarization threshold of 15 or less were excluded from the statistic process. The average value and standard deviation of the three channels are shown in Table 3.

Table 3. Defect RGB statistics

Defect name	Average value			Standard deviation		
	R	G	B	R	G	B
Defect 1	17.89	31.2	31.23	5.96	40.39	18.66
Defect 2	20.77	40.78	20.9	21	86.83	18.99
Average value	38.66	35.99	26.06			

To determine the color separation parameter according to the statistical value, since the input number of the function must be an integer, thus the integer was rounded off and the parameter was not adjusted for processing. The image after the closing operation is as shown in Fig. 16(a). As shown in Fig. 16(b), the integrity of the contour was higher than that of the previous defect image, thus the captured defect image was more complex, as shown in Fig. 16(c) and (d). Although some highlights are broken, the shape was still reliable after binarization process to maintain the integrity of the contour.

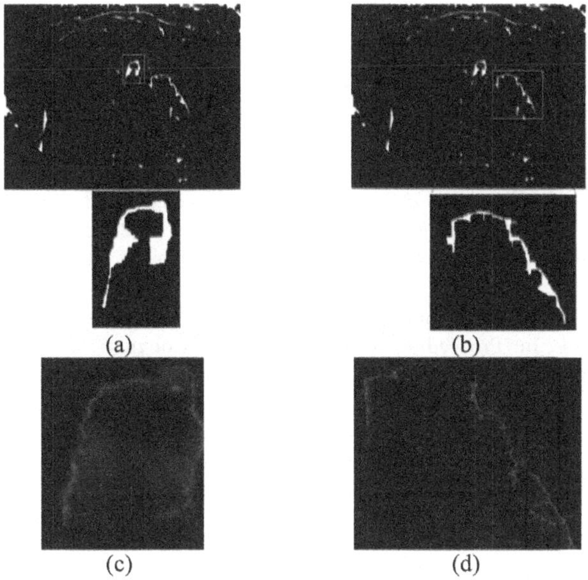

Fig. 16. Images obtained after image processing.

5 Results and Discussion

This study developed an automatic optical inspection successfully to detect defects on the surface of a vehicle constant velocity joint. Some conclusions of this work can be presented as follows:

- The ultraviolet light source was used in the illumination of the semi-circular tube type cover, and the light source state was determined by the ultraviolet light power, the white light power and the temperature of the light board at the time of detection analysis process.
- In the color separation parameter setting of image processing, the average value and standard deviation of the RGB values were analyzed by the defect analysis program, and the background pixels were removed by the binarization threshold, and the RGB average value was obtained. They were 38.66, 35.99, and 26.06, respectively, and the maximum standard deviation of the G was 86.83. Therefore, the RGB center values of the color separation of image processing were 39, 36, and 26, and the radius value was 87.
- Under the experimental conditions, the proposed system recognized two types of defects of inspection objects.

References

1. Ma, T., Sun, Z., Zhang, W., Chen, Q.: A machine vision assisted system for fluorescent magnetic particle inspection of railway wheelsets. In: AIP Conference Proceedings, vol. 1706, p. 150003. AIP Publishing (2016)
2. Nishimine, T., Tsuyama, O., Tanaka, T., Fujiwara, H.: Automatic magnetic particle testing system for square billets. In: IAS 1995, Conference Record of the 1995 IEEE Industry Applications Conference Thirtieth IAS Annual Meeting, vol. 2, pp. 1585–1590. IEEE (1995)
3. Lundh, M.: Automatic crack detection in forged metal parts. Technical report no. EX073 (2012)
4. Luo, J., Tian, Z., Yang, J.: Fluorescent magnetic particle inspection device based on digital image processing. In: Proceeding of the 11th World Congress on Intelligent Control and Automation, pp. 5677–5681. IEEE (2014)
5. Hao, H., Li, L., Deng, Y.: Vision system using linear CCD cameras in fluorescent magnetic particle inspection of axles of railway wheelsets. In: Health Monitoring and Smart Nondestructive Evaluation of Structural and Biological Systems IV, vol. 5768, pp. 442–449. International Society for Optics and Photonics (2005)

Calculating Optimum Gear Ratios of Mechanical Drive Systems Using Two-Stage Helical Gearbox with Second-Stage Double Gear Sets and Chain Drive for Minimum Gearbox Length

Nguyen Thi Hong Cam[1], Tran Thi Hong[2], Nguyen Van Cuong[3], Le Hong Ky[4], Luu Anh Tung[1], Nguyen Thanh Tu[1], Le Xuan Hung[1], and Ngoc Pi Vu[1(✉)]

[1] Thai Nguyen University of Technology, Thai Nguyen, Vietnam
vungocpi@tnut.edu.vn
[2] Nguyen Tat Thanh University, Ho Chi Minh City, Vietnam
[3] University of Transport and Communications, Hanoi, Vietnam
[4] Vinh Long University of Technology Education, Vinh Long, Vietnam

Abstract. The paper presents a study on determining optimum gear ratios of mechanical driven systems using a two-stage helical gearbox with second stage double gear sets and a chain drive. In the study, an optimization problem with the objective function as the minimum system length was solved. In addition, the effects of the input parameters including the total gearbox ratio, the wheel face width coefficient, the allowable contact stress and the output torque on the optimal transmission ratios were evaluated by conducting a simulation experiment. Moreover, several equations to find the optimum gear ratios were proposed.

Keywords: Gear ratio · Optimum gear ratio · Optimum gearbox design · Helical gearbox · Gearbox length

1 Introduction

Mechanical drive systems using gearbox are widely used in industries. For that reason, the optimal design of this system is of special interest to researchers. In the optimal drive system design, the task of determining the optimal partial gear ratios is of great importance because it helps reduce the size, the mass and, ultimately, the cost of the system.

The problem of determining the optimal partial gear ratios has been solved in different ways such as graph method [1, 2], practical method [3] and model method [3–16]. Optimal gear ratios are also defined for different gearbox types including helical gearboxes [1–19], bevel gearboxes [1, 3, 20–22], or worm gearboxes [23–26] with different gear steps such as two-step [1–8], three-step [9–15] and four-step [16–19]. The optimal gear ratios are normally determined through solving the optimal

K.-U. Sattler et al. (Eds.): ICERA 2019, LNNS 104, pp. 155–163, 2020.
https://doi.org/10.1007/978-3-030-37497-6_18

problems in which the objective functions are ordinarily the minimum gearbox length [7, 11, 25], the minimum cross section area of gearboxes [9, 12] or the minimum mass of gears [10, 13, 14, 16, 18].

Regarding mechanical drive systems using not only gearboxes, there have recently been some publications on the calculation of the optimal transmission ratios of systems including a gearbox and a V-belt drive [5, 27, 28] or a chain drive [29, 30].

This study has been done with calculating the optimal gear ratios of mechanical drive systems using a two-stage helical gearbox with second stage double gear sets and a chain drive for getting the minimum system length.

2 Methodology

The length of the mechanical drive system using Two-Stage Helical Gearbox with Second-Stage Double Gear Sets and a Chain Drive can be determined as (Fig. 1):

$$L = \max\left(L_g, L_c\right) \tag{1}$$

Where, L_g and L_c are calculated by (see Fig. 1):

$$L_g = d_{w11}/2 + a_{w1} + a_{w2} + d_{w22}/2 \tag{2}$$

$$L_c = d_2/2 + a_c + d_{w22}/2 \tag{3}$$

Hence, the optimization problem can be defined as:

$$\text{minimize } L \tag{4}$$

With the following constraints:

$$\begin{aligned} &1 \le u_1 \le 9 \\ &1 \le u_2 \le 9 \\ &1 \le u_c \le 6 \\ &u_1 \cdot u_2 \cdot u_3 \cdot u_c = u_t \end{aligned} \tag{5}$$

In which, a_{w1}, a_{w2}, d_{w11} and d_{w22} are the center distance and the pitch diameters of the first and the second stages, respectively; a_c and d_2 are the center distance and the drive sprocket diameter of the chain drive; u_1 and u_2 are the gear ratios of the first and the second stages, respectively; u_c and u_t are the gear ratios of the chain drive and the total system.

2.1 Calculating Center Distance and the Pitch Diameter of the First Step

For the first step, the center distance a_{w1} is determined by [31]:

$$a_{\mathrm{w}1} = K_a \cdot (u_1 + 1) \cdot \left[T_{11} \cdot k_{H\beta} / \left([\sigma_{H1}]^2 \cdot u_1 \cdot \psi_{ba1}\right)\right]^{1/3} \tag{6}$$

Where, $K_{H\beta}$ is the contact load ratio; $k_{H\beta} = 1.02 \div 1.28$ [31] and we can choose $k_{H\beta} = 1.1$; $[\sigma_{H1}]$ is the allowable contact stress of the first step (MPa); In practice, $[\sigma_{H1}] = 350 \ldots 420$ (MPa); K_a is the material coefficient; $K_a = 43$ [31]; ψ_{ba1} is the coefficient of wheel face width; $\psi_{ba1} = 0.3 \ldots 0.35$ [31];

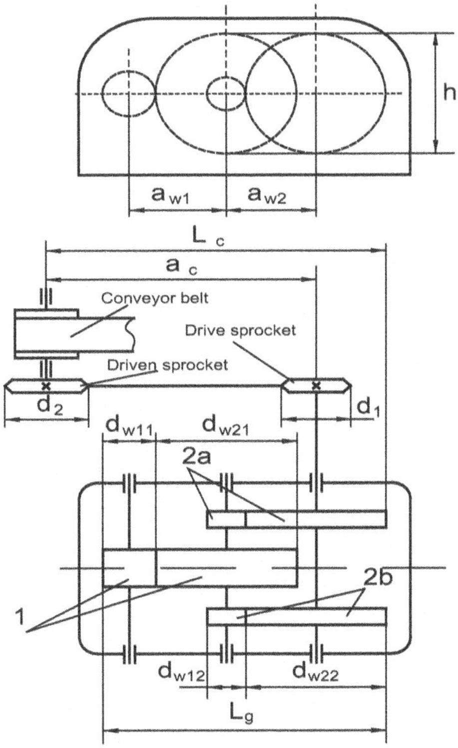

Fig. 1. Calculation schema.

T_{11} is the torque on the pinion which is determined by:

$$T_{11} = T_{out} / \left(u_g \cdot u_c \cdot \eta_{hg}^2 \cdot \eta_c \cdot \eta_b^4\right) \tag{7}$$

In which, T_{out} is the output torque (Nmm); η_{hg} is the helical gear efficiency; $\eta_{hg} = 0.96 \ldots 0.98$ [31]; η_c is the chain drive efficiency $\eta_c = 0.95 \ldots 0.97$; η_b is the efficiency of a pair of rolling bearings; $\eta_b = 0.99\ldots0.995$ [31]. Choosing $\eta_{hg} = 0.97$, $\eta_c = 0.96$, $\eta_b = 0.992$ and substituting them into Eq. (7) gives:

$$T_{11} = 1.1432 \cdot T_{out}/\left(u_g \cdot u_c\right) \tag{8}$$

Substituting $k_a =, k_{H\beta} = 1.1$ and (8) into (6) gets:

$$a_{w1} = 46.413 \cdot (u_1 + 1) \cdot \left[T_{out}/\left([\sigma_{H1}]^2 \cdot u_1 \cdot u_g \cdot u_c \cdot \psi_{ba1}\right)\right]^{1/3} \tag{9}$$

After having a_{w1}, the pitch diameter of the first step is calculated by [31]:

$$d_{w11} = 2 \cdot a_{w1}/(u_1 + 1) \tag{10}$$

2.2 Calculating Center Distance and Pitch Diameter of the Second Step

Similar to Subsect. 2.1, for the second step, the center distance a_{w2} can be found as [31]:

$$a_{w2} = K_m \cdot (u_2 + 1) \cdot \left[T_{12} \cdot k_{H\beta}/\left([\sigma_{H2}]^2 \cdot u_2 \cdot \psi_{ba2}\right)\right]^{1/3} \tag{11}$$

Where, T_{12} is the pinion torque which is calculated by:

$$T_{12} = T_{out}/\left(2 \cdot u_2 \cdot u_c \cdot \eta_{hg} \cdot \eta_c \cdot \eta_b^3\right) \tag{12}$$

Choosing $\eta_{hg} = 0.97$, $\eta_c = 0.96$, $\eta_b = 0.992$ as in Subsect. 2.1 gives:

$$T_{12} = 0.5501 \cdot T_{out}/(u_2 \cdot u_c) \tag{13}$$

Substituting (13), $K_m = 43$ and $k_{H\beta} = 1.1$ (as in Sect. 2.1) into (11) gets:

$$a_{w2} = 36.3703 \cdot (u_2 + 1) \cdot \left[T_{out}/\left([\sigma_{H2}]^2 \cdot u_2^2 \cdot u_c \cdot \psi_{ba2}\right)\right]^{1/3} \tag{14}$$

Therefore, the pitch diameter of the second step is determined by [31]:

$$d_{w22} = 2 \cdot a_{w2} \cdot u_2/(u_2 + 1) \tag{15}$$

2.3 Determining the Drive Sprocket Diameter d2

For the chain drive, the pitch diameter of the drive sprocket is calculated as [31]:

$$d_2 = d_1 \cdot u_c \tag{16}$$

In which, d_1 is the pitch diameter of the drive sprocket [31]:

$$d_1 = p / \sin(\pi/z_1) \tag{17}$$

Where, z_1 is the number of teeth of the drive sprocket; z_1 can be found by [29]:

$$z_1 = 32.4 - 2.4 \cdot u_c \tag{18}$$

p is the chain pitch (mm); p can be obtained from the design power capacity P [31]:

$$P = P_1 \cdot k \cdot k_z \cdot k_n \tag{19}$$

In which, P_1 is the power rating (kW) of the chain drive; P_1 is determined by:

$$P_1 = T_1 \cdot n_1 / \left(9.55 \cdot 10^6\right) \tag{20}$$

Wherein, n_1 is the drive sprocket revolution (rpm):

$$n_1 = n_m / u_g \tag{21}$$

$$T_1 = T_{out} / (u_c \cdot \eta_c \cdot \eta_b) \tag{22}$$

In the above equations, η_c is the chain drive efficiency ($\eta_c = 0.95 \div 0.97$ [31]); η_b is the bearing efficiency ($\eta_b = 0.99 \div 0.995$ [31]); T_1 and T_{out} are the torque on the drive sprocket and the output torque (Nmm). Choosing $\eta_c = 0.96$, $\eta_b = 0.992$ and substituting them into (22) gets:

$$T_1 = 1.0502 \cdot T_{out} / u_c \tag{23}$$

k, k_z and k_n are coefficients determined as [31]:

$$k = k_d \cdot k_p \cdot k_c \cdot k_{adj} \cdot k_{\text{lub}} \cdot k_{con} \tag{24}$$

$$k_z = 25/z_1 \tag{25}$$

$$k_n = n_{01}/n_1 \tag{26}$$

In the above equations, k_d, k_p, k_c, k_{adj}, k_{lub} and k_{con} are the coefficients denoting the influences of shock factor, the drive position, the center distance, the adjusted possibility of the center distance and the lubrication conditions, respectively; n_{01} is the tabulated number of the drive sprocket teeth.

2.4 Experimental Work

The dependence level of the optimum gear ratios on the input parameters was explored based on the implementation of a simulation experiment with a 2-level full factorial design and 6 input factors (see Table 1). Accordingly, 64 test runs were conducted. Moreover, a computer program was created based on Eqs. (4) and (5). The experimental plan and the output responses (the gear ratio of the chain drive u_c and the optimum gear ratios of the first step u_1) are described in Table 2.

3 Results and Discussion

From the experimental results (Table 2) it is interesting that the optimum ratios of the first step of the gearbox u_1 and the chain drive u_c are constant. That means they are independent of the input parameters. Additionally, u_1 always takes the maximum possible value ($u_1 = 9$) so that the size of the system is small because the torque on axis 1 is the smallest. Remarkably, the optimum gear ratios of the chain drive is $u_c = 1$. It proves that with the objective function of the problem, the chain drive has no effect on deceleration but only serves to transmit torque to the required axial distance.

$$u_1 = 9 \tag{27}$$

$$u_c = 1 \tag{28}$$

Table 1. Input parameters.

Factor	Code	Unit	Low	High
Total gearbox ratio	u_t	–	10	50
Coefficient of wheel face width of step 1	x_{ba1}	–	0.3	0.35
Coefficient of wheel face width of step 2	x_{ba2}	–	0.35	0.4
Allowable contact stress of step 1	AS_1	MPa	350	420
Allowable contact stress of step 2	AS_2	MPa	350	420
Output torque	T_{out}	Nmm	10^5	10^7

Table 2. Experimental plans and output response.

Std order	Run order	CenterPt	Blocks	u_t	Xba_1	Xba_2	AS_1	AS_2	Tout	u_c	u_1
57	1	1	1	10	0.3	0.35	420	420	10000	1.00	9.00
44	2	1	1	50	0.35	0.35	420	350	10000	1.00	9.00
39	3	1	1	10	0.35	0.4	350	350	10000	1.00	9.00
27	4	1	1	10	0.35	0.35	420	420	100	1.00	9.00
45	5	1	1	10	0.3	0.4	420	350	10000	1.00	9.00
14	6	1	1	50	0.3	0.4	420	350	100	1.00	9.00
…											
63	63	1	1	10	0.35	0.4	420	420	10000	1.00	9.00
58	64	1	1	50	0.3	0.35	420	420	10000	1.00	9.00

Equations (27) and (28) are used to determine the optimum gear ratios of the first step of the gearbox u_1 and the chain drive u_c. After having u_1 and u_c, the first step and the gear ratio of the chain drive u_c are calculated by:

$$u_2 = u_t/(u_1 \cdot u_c) \tag{29}$$

As u_1 = 9 and u_c = 1, Eq. (29) is rewritten as:

$$u_2 = u_t/9 \tag{30}$$

4 Conclusions

The study introduces an optimization problem to determine the optimal gear ratios of mechanical drive systems using a two-step helical gearbox with second step double gear sets and a chain drive to attain the minimal length of the system. To carry out the task, a simulation experiment was conducted. The results of the experiment were analyzed to evaluate the effects of the input parameters on the optimal gear ratios. In addition, the formula for determining the optimal gear ratios to reach the minimal system height was suggested.

Acknowledgements. The work described in this paper was supported by Thai Nguyen University of Technology for a scientific project.

References

1. Kudreavtev, V.N., Gierzaves, I.A., Glukharev, E.G.: Design and Calculus of Gearboxes. Mashinostroenie Publishing, Sankt Petersburg (1971). (in Russian)
2. Chat, T.: Some problems of kinematics calculation of transmission mechanics system. In: Proceedings of the National Conference on Engineering Mechanics, Hanoi, vol. 2, pp. 7–12 (1993). (in Vietnamese)
3. Milou, G., Dobre, G., Visa, F., Vitila, H.: Optimal design of two step gear units, regarding the main parameters. VDI Berichte **1230**, 227 (1996)
4. Pi, V.N.: A method for optimal calculation of total transmission ratio of two step helical gearboxes. In: Proceedings of the National conference on Engineering Mechanics, Hanoi, p. 12 (2001)
5. Cam, N.T.H., Pi, V.N., Tuan, N.K., Hung, L.X., Thao, T.T.P.: Determining optimal partial transmission ratios of mechanical driven systems using a V-belt drive and a helical reducer with second-step double gear-sets. In: Fujita, H., et al. (eds.) ICERA 2018. LNNS, vol. 63, pp. 261–269 (2019). https://doi.org/10.1007/978-3-030-04792-4_35
6. Petrovski, A.N., Sapiro, B.A., Saphonova, N.K.: About optimal problem for multi-step gearboxes. Vestnik Mashinostroenie **10**, 13 (1987). (in Russian)
7. Pi, V.N.: A new study on optimal calculation of partial transmission ratios of two-step helical gearboxes. In: 2nd WSEAS International Conference on Computer Engineering and Applications, CEA 2008, Acapulco, Mexico, 25–27 January, pp. 162–165 (2008)

8. Nguyen, K.T., Vu, N.P., Nguyen, T.H.C., Tran, T.P.T., Ho, K.T., Le, X.H., Hoang, T.T.: Determining optimal gear ratios of a two-stage helical reducer for getting minimal acreage of cross section. In: MATEC Web of Conferences, vol. 213, p. 01008 (2018). https://doi.org/10.1051/matecconf/201821301008
9. Pi, V.N., Tuan, N.K.: Optimum determination of partial transmission ratios of Three-step helical gearboxes for getting minimum cross section. J. Environ. Sci. Eng. A **5**, 570–573 (2016)
10. Pi, V.N., Binh, N.D., Dac, V.Q., The, P.Q.: Optimal calculation of total transmission ratio of three-step helical gearboxes for minimum mass of gears. J. Sci. Technol. 6 Eng. Univ., 91 (2006). (in Vietnamese)
11. Pi, V.N.: A new study on optimal calculation of partial transmission ratio of three-step helical reducers. In: The 3rd IASME/WSEAS International Conference on Continuum Mechanics, Cambridge, UK, p. 23 (2008)
12. Pi, V.N.: A new study on optimal calculation of partial transmission ratios of three-step helical reducers for getting minimal cross section dimension. In: The 2nd WSEAS International Conference on Computer Engineering and Applications, CEA 2008, Acapulco, Mexico, 25–27 January, pp. 290–293 (2008)
13. Pi, V.N.: Optimal determination of partial transmission ratios of three-step helical gearboxes with first and third-step double gear-sets for minimal mass of gear. In: American Conference on Applied Mathematics, MATH 2008, Harvard, Massachusetts, USA, 24–26 March, pp. 385–388 (2008)
14. Romhild, I., Linke, H.: Gezielte Auslegung Von Zahnradgetrieben mit minimaler Masse auf der Basis neuer Berechnungsverfahren. Konstruktion **44**, 229–236 (1992)
15. Pi, V.N.: A new study on the optimal prediction of partial transmission ratios of three-step helical gearboxes with second-step double gear-sets. WSEAS Trans. Appl. Theor. Mech. **2** (11) (2007)
16. Pi, V.N.: Optimal determination of partial transmission ratios for four-step helical gearboxes with first and third step double gear-sets for minimal mass of gears. In: Applied Computing Conference, ACC 2008, Istanbul, Turkey, 27–30 May (2008)
17. Pi, V.N.: Optimal calculation of partial transmission ratios for four-step helical gearboxes with first and third-step double gear-sets for minimal gearbox length. In: American Conference on Applied Mathematics, MATH 2008, Harvard, Massachusetts, USA, 24–26 March, pp. 29–32 (2008)
18. Hung, L.X., Pi, V.N., Du, N.V.: Optimal calculation of partial transmission ratios of four-step helical gearboxes with second and fourth-step double gear-sets for minimal mass of gears. In: The International Symposium on Mechanical Engineering, ISME, Ho Chi Minh city, Vietnam, pp. 21–23, September 2009
19. Pi, V.N.: Optimal determination of partial transmission ratios for four-step helical gearboxes with first and third step double gear-sets for minimal mass of gears. In: Applied Computing Conference, ACC 2008, Istanbul, Turkey, 27–30 May 2008, pp. 53–57 (2008)
20. Pi, V.N.: A new and effective method for optimal calculation of total transmission ratio of two step bevel - helical gearboxes. In: International colloquium on Mechanics of Solids, Fluids, Structures & Interaction, Nha Trang, Vietnam, pp. 716–719 (2000)
21. Pi, V.N., Binh, N.D., Dac, V.Q., The, P.Q.: A new and effective method for optimal splitting of total transmission ratio of three step bevel-helical gearboxes. In: The Sixth Vietnam Conference on Automation, Hanoi, pp. 175–180 (2005)
22. Tuan, N.K., Thao, T.T.P., Cam, N.T.H., Hung, L.X., Pi, V.N.: Optimum calculation of partial transmission ratios of mechanical driven systems using a V-belt and a three-step bevel helical gearbox. In: Fujita, H., et al. (eds.) ICERA 2018. LNNS, vol. 63, pp. 469–476 (2019). https://doi.org/10.1007/978-3-030-04792-4_61

23. Chernavsky, S.A., et al.: Design of Mechanical Transmissions: Manual for High Technical Schools, p. 560. Mashinostroenie, Moscow (1984)
24. Pi, V.N., Dac, V.Q.: Calculation of total transmission ratio of two step worm reducers for the best reasonable gearbox housing structure. J. Sci. Technol. Thai Nguyen Univ. **1**(41), 65–69 (2007). (in Vietnamese)
25. Pi, V.N., Dac, V.Q.: Optimal calculation of partial transmission ratios of worm-helical gear reducers for minimal gearbox length. J. Sci. Technol. Tech. Univ. **61**, 73–77 (2007)
26. Pi, V.N., Dac, V.Q.: Optimal calculation of total transmission ratio of worm-helical gear reducers. J. Sci. Technol. Thai Nguyen Univ. **4**(36), 70–73 (2005). No. 1
27. Pi, V.N., Thao, T.T.P., Thao, L.T.P.: A new study on optimum determination of partial transmission ratios of mechanical driven systems using a V-belt and two-step helical gearbox. Vietnam. Mech. Eng. J. **10**, 123 (2015)
28. Pi, V.N., Cam, N.T.H., Tuan, N.K.: Optimum calculation of partial transmission ratios of mechanical driven systems using a V-belt and two-step bevel helical gearbox. J. Environ. Sci. Eng. A **5**, 566 (2016)
29. Pi, V.N., Thao, T.T.P., Tuan, D.A.: Optimum determination of partial transmission ratios of mechanical driven systems using a chain drive and two-step helical gearbox. J. Environ. Sci. Eng. B **6**, 80 (2017)
30. Cam, N.T.H., Pi, V.N., Tuan, N.K., Hung, L.X., Thao, T.T.P.: A study on determination of optimum partial transmission ratios of mechanical driven systems using a chain drive and a three-step helical reducer. In: Fujita, H., et al. (eds.) ICERA 2018. LNNS, vol. 63, pp. 91–99 (2019). https://doi.org/10.1007/978-3-030-04792-4_14
31. Chat, T., Uyen, L.V.: Design and calculation of Mechanical Transmissions Systems, vol. 1. Educational Republishing House, Hanoi (2007)

Calculating Effects of Dressing Parameters on Surface Roughness in Surface Grinding

Do Duc Trung[1], Nguyen Hong Son[2], Tran Thi Hong[3], Nguyen Van Cuong[4], and Ngoc Pi Vu[5](✉)

[1] Faculty of Mechanical Engineering, Hanoi University of Industry, No. 298, Cau Dien Street, Bac Tu Liem District, Hanoi, Vietnam
[2] Center for Mechanical Engineering, Hanoi University of Industry, No. 298, Cau Dien Street, Bac Tu Liem District, Hanoi, Vietnam
[3] Center of Excellence for Automation and Precision Mechanical Engineering, Nguyen Tat Thanh University, Ho Chi Minh City, Vietnam
[4] Faculty of Mechanical Engineering, University of Transport and Communications, Ha Noi City, Vietnam
[5] Faculty of Mechanical Engineering, Thai Nguyen University of Technology, Thai Nguyen City, Vietnam
vungocpi@tnut.edu.vn

Abstract. In this paper, the influences of the surface grinding parameters including the dressing parameters, the grinding wheel velocity, the workpiece velocity, the depth of cut on the surface roughness are investigated. In the study, the influence of dressing parameters on surface roughness was pointed out theoretically. Also, it was found from the study results that the dressing factors have significant effects on the surface roughness. In addition, to evaluate the impacts of the dressing parameters on the surface roughness, an experiment was designed and performed. The measurement surface roughness from the experiment are in accordance with the calculated values.

Keywords: Grinding · Surface roughness · Dressing parameters

1 Introduction

In mechanical manufacturing, grinding is a popular method which is widely used in industry for precision machining. Like other machining methods, the quality of the surface finish by grinding is evaluated using many parameters. Among those, the surface roughness of the workpiece is one of the most important factors which may significantly impact on the usefulness of the workpiece [1, 2].

In grinding, the forming mechanism of surface roughness is complex and mostly dependent on many factors such as the cutting parameters, the dressing parameters, the wheel parameters, the cooling and lubrication, etc. It is claimed that dressing parameters have significant influence on the surface roughness [1–4]. There have been many studies on the influence of dressing parameters on surface roughness in grinding. These empirical models have the advantage that they require minimum efforts to develop and are used in all fields of grinding technology. However, they can only be used in a

K.-U. Sattler et al. (Eds.): ICERA 2019, LNNS 104, pp. 164–169, 2020.
https://doi.org/10.1007/978-3-030-37497-6_19

certain grinding conditions, which is the reason for the limited scope of these empirical models [5, 6].

In this paper, based on the theory of the surface grinding process, the relationship between the surface roughness and the input parameters including the dressing parameters, the grinding wheel velocity, the grinding wheel parameters, the workpiece velocity and the depth of cut is developed. This relationship is then used to calculate the influence of the dressing feed-rate and the dressing depth on the surface roughness. Also, an evaluation of the influence of the dressing parameters on the surface roughness is carried out.

2 Methodology

For surface grinding, the experimental surface roughness data tend to follow a relationship of the form [7]:

$$R_a = R_1 S_d^{0.5} a_d^{0.25} (h_m)^x \tag{1}$$

Where:
S_d and a_d are the dressing feed-rate and dressing depth, respectively; R_1 and x are the experimental coefficients; For fine grinding, $x = 0.56$ and $R_1 = \frac{0.4S_d}{b_d} + 0.2$, where b_d is the width of the single-point diamond tip [8].

h_m is the maximum of undeformed chip thickness; h_m can be calculated by the following equation [9]:

$$h_m = 2\sqrt{\frac{1}{Nr}\frac{v_W}{v_G}\sqrt{\frac{t}{d_e}}} \tag{2}$$

In which, v_w is the workpiece velocity; v_G is the grinding wheel velocity; t is the depth of cut; d_e is the equivalent wheel diameter which is determined as follows:

$$d_e = d_G \tag{3}$$

Where, d_G is the grinding wheel diameter.

r is the chip width to thickness ratio; $r = 10 - 20$ [10, 11].

N is the number of active grits per unit area which can be derived by Xu et al. [12] is as follows:

$$N = 4f\frac{1}{d_g^2}\frac{1}{\sqrt[3]{\left(\frac{4\pi}{3\epsilon}\right)^2}} \tag{4}$$

Where,
f is the fraction of grain involved in active grinding. The value of f is difficult to determine. For calculating the number of active grits per unit area, it is assumed that only one half of grain is engaged in cutting [12] and $f = 0.5$.

d_g is the equivalent spherical diameter of grain grit; d_g is given as [4, 7]:

$$d_g = 15.2/M \tag{5}$$

In which, M is the mesh number used in the grading sieve.

ϵ is the volume fraction of grain in grinding wheel. For the grinding wheels with the mesh number 80, the volume fraction $\epsilon = 12.5\% \div 37.5\%$ [13].

Substituting the Eqs. (2), (3), (4), (5) into Eq. (1) and after mathematical simplification, the value of surface roughness will be:

$$R_a = 6.8255.S_d^{0.5} a_d^{0.25} \left(\frac{0.4 S_d}{b_d} + 0.2 \right) \frac{1}{M^{0.56}} \left(\frac{1}{f.r} \right)^{0.28} \cdot \left(\frac{v_w}{v_G} \right)^{0.28} \left(\frac{4\pi}{3\epsilon} \right)^{0.187} \cdot \left(\frac{t}{d_e} \right)^{0.14} \tag{6}$$

From Eq. (6), it can be seen that the surface roughness increases if the values of the dressing feed-rate, the dressing depth, the workpiece velocity and the volume fraction increase. On the contrary, if the values of the mesh number, the fraction of particles involved in active grinding, the grinding wheel velocity, width of the single-point diamond tip and the chip width to thickness ratio grow, the surface roughness is reduced.

3 Experimental Evaluation

3.1 Experimental System

The grinding experiments were conducted on an APSG-820/2A surface grinder (Fig. 1). A 36A80LV grinding wheel with the dimensions of 180 × 13 × 31.75 was used. The component material was SKD11 steel, 60÷62HRC; and the workpiece dimensions were 50 × 10 × 10.

Fig. 1. APSG-820/2A surface grinder

The measurements were carried out with a Mitutoyo Surftest SJ-201 stylus type surface texture-measuring instrument.

3.2 Input Parameters

To evaluate the model for the surface roughness (Eq. (6)), the calculated value of the model and experimental value of the surface roughness were compared. The input values for the calculation as well as for the experiment are presented in Table 1.

Table 1. Input values for calculation and experimental of surface roughness

Parameters	Symbol	Value	Unit
Grinding wheel velocity	v_G	26	m/s
Workpiece velocity	v_w	15	m/min
Depth of cut	t	0.01	mm
Dressing feed-rate	S_d	50/60/70/80/90/100	mm/min
Dressing depth	a_d	0.005/0.01/0.015/0.02	mm
Width of the single - point diamond tip	b_d	10	mm
The mesh number used in the grading sieve	M	80	–
The fraction of grain involved in active grinding	f	0.5	–
The chip width to thickness ratio	r	10	–
The volume fraction of grain in grinding wheel	ϵ	25	%

Theoretically, the model (6) can be used to determine the surface roughness with different values of the dressing feed-rate, the dressing depth, the workpiece velocity, the volume fraction, the value of the mesh number, the fraction of particles involved in active grinding, the grinding wheel velocity, and the width of the single-point diamond tip. However, in this evaluation, only the influence of dressing feed-rate and dressing depth on surface roughness of workpiece in surface grinding was investigated.

3.3 Results and Discussion

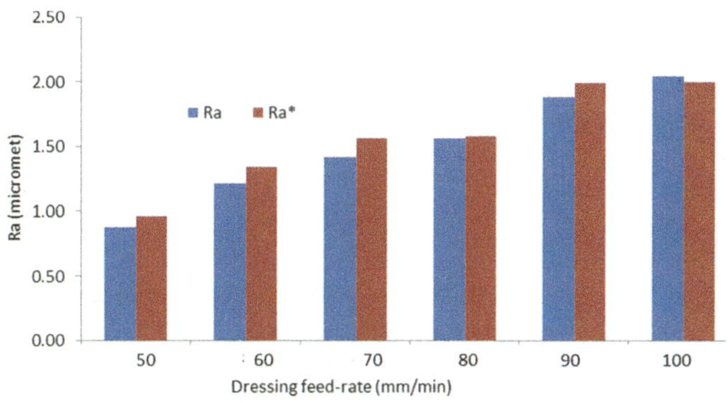

Fig. 2. *Influence of S_d on surface roughness when $a_d = 0.01(mm)$*

The influence of dressing parameters on surface roughness by calculation (R_a) and by experiment (R_a^*) are shown in Figs. 2 and 3. It can be learned from these figures that:

- The surface roughness value increases if the values of dressing feed rate and dressing depth increase.
- The measurement surface roughness from the experiment are in accordance with the calculated results.

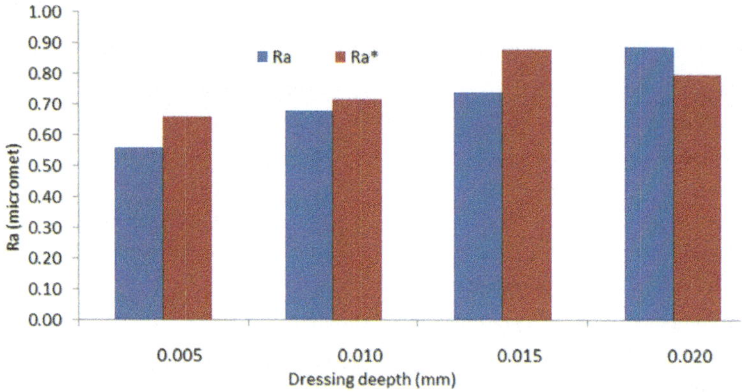

Fig. 3. *Influence of a_d on surface roughness when $S_d = 100(mm/min)$*

4 Conclusion

In the work, a study on the calculation of the surface roughness in surface grinding was introduced. The effects of the input parameters including the dressing parameters, the grinding wheel velocity, the workpiece velocity, the depth of cut on the surface roughness were explored. In addition, to evaluate the proposed model, the calculated value of the surface roughness and the experimental values were compared. It was drawn that the calculated value is very close to the experimental value. As the proposed model has been verified and proven by the experimental results, it can be used to determine the surface roughness in surface grinding.

Acknowledgements. The work described in this paper was supported by Ha Noi university of Industry and Thai Nguyen University of Technology.

References

1. Marinescu, L.D., Uhlmann, E., Brian Rowe, W.: Handbook of Machining with Grinding Wheels. CRC Press Taylor & Francis Group (2006)
2. Trung, D.D.: Study on identifying the machining parameters in centerless grinding of the 20X-carbon infiltrated steel to reduce its roundness error and surface roughness. The thesis completed at Thai Nguyen University of Technology (2016)

3. Marinescu, L.D., Hitchiner, M.: Handbook of Advances Ceramics Machining (2006). http://www.taylorandfrancis.com
4. Malkin, S.: Grinding Technology - Theory and Applications of Machining with Abrasives. Ellis Horwood – Chichester (1989)
5. Trung, D.D., Van Thien, N., Dung, H.T.: Predictive surface roughness of workpiece in surface grinding. Am. J. Mater. Res. **4**, 37–41 (2017)
6. Trung, D.D., Son, P.X.: Predictive surface roughness of workpiece in cylindrical grinding. In: The First International Conference on Material, Machines and Methods for Sustainable Development, vol. 2, pp. 831–837 (2018)
7. Malkin, S., Guo, C.: Grinding Technology - Theory and Applications of Machining with Abrasives, 2nd edn. Industrial Press, New York (2008)
8. Saad, A., Bauer, R., Warkentin, A.: Investigation of single-point dressing overlap ration and diamond-roll dressing interference angle on surface roughness in grinding. Trans. Can. Soc. Mech. Eng. **34**(2), 295–308 (2010)
9. Gopal, A.V., Rao, P.V.: A new chip-thickness model for performance assessment of silicon carbide grinding. Int. J. Adv. Manuf. Technol. **24**, 816–820 (2004)
10. Mayer, J.E., Fang, G.P.: Effect of grit depth of cut on strength of ground ceramics. Ann. CIRP **43**, 309–312 (1994)
11. Somasundaram, S., Thiagarajan, C.: Experimental evaluation of a chip thickness model based on the fracture toughness of abrasive and work material in grinding of alumina ceramics. Int. J. Mod. Eng. Res. **3**(6), 3825–3829 (2013)
12. Xu, H., Jahanmir, S., Ives, L.K.: Effect of grinding on strength of tetragonal zirconia and zirconia-toughned alumina. J. Mach. Sci. Technol. **1**, 49–66 (1997)
13. Zhang, P., Miller, M.H.: Grinding Wheel Condition Prediction and Improvement. Michigan Technological University, Houghton, MI (2002)

Calculation of Optimum Gear Ratios of Mechanical Driven Systems Using Two-Stage Helical Gearbox with First Stage Double Gear Sets and Chain Drive

Le Xuan Hung[1], Tran Thi Hong[2], Nguyen Van Cuong[3], Le Hong Ky[4], Nguyen Thanh Tu[1], Nguyen Thi Hong Cam[1], Nguyen Khac Tuan[1], and Ngoc Pi Vu[1(✉)]

[1] Thai Nguyen University of Technology, Thai Nguyen City, Vietnam
vungocpi@tnut.edu.vn
[2] Nguyen Tat Thanh University, Ho Chi Minh City, Vietnam
[3] University of Transport and Communications, Ha Noi City, Vietnam
[4] Vinh Long University of Technology Education, Vinh Long City, Vietnam

Abstract. The article reports a study on identifying the optimum gear ratios of mechanical driven systems containing a two-stage helical gearbox with the first stage double gear sets and a chain drive. To find the optimum gear ratios, the minimum system length was chosen as the objective function of the optimization problem. Moreover, the input parameters including the total system ratio, the wheel face width coefficients of the first and the second stages, the allowable contact stress and the output torque and their impacts on the optimum ratios were explored based on the design and implementation of a simulation experiment. The obtained equations showed that the optimum gear ratios can be calculated precisely and conveniently.

Keywords: Gear ratio · Optimum gear ratio · Optimum gearbox design · Helical gearbox

1 Introduction

In practice gear ratios play a significant role in mechanical engineering design. In relation to designing mechanical driven systems or gearboxes, the gear ratios heavily affect the dimension, the mass and the cost of the systems or the gearboxes. Accordingly, the calculation of optimum gear ratios of a gearbox or a mechanical driven system has attracted great attention from researchers so far.

Up to now, the optimum gear ratios have been found by several methods which include the graph method [1, 2], the practical method [3] and the model method [3–16]. Besides, the gear ratios have been identified for the optimum design of various gearbox types including helical gearboxes, bevel gearboxes and worm gearboxes. For helical gearboxes, great focus has been placed on research into this topic such as for two-stage gearboxes [3–8], three-stage gearboxes [9–15] and four-stage gearboxes [13–19]. Concerning bevel gearboxes, the optimum gear ratios have been found for two

K.-U. Sattler et al. (Eds.): ICERA 2019, LNNS 104, pp. 170–178, 2020.
https://doi.org/10.1007/978-3-030-37497-6_20

stage [1, 3, 20] and three stage bevel helical gearboxes [21]. In addition, the optimum gear ratios were determined for worm gearboxes including two-stage worm gearboxes [3, 22, 23] and worm-helical gearboxes [22, 24]. Moreover, the optimum ratios were calculated not only for gearboxes but also for mechanical driven systems with a gearbox and a V-belt drive [25–28] or a chain drive [29, 30].

This paper introduces an optimization study on the determination of the optimum gear ratios of mechanical driven systems including a two-stage helical gearbox with first stage double gear sets and a chain drive. To obtain the optimum gear ratios, an optimization problem was conducted. In the optimization problem, the minimum system length was chosen as the objective function. Also, the effects of the input factors including the total system ratio, the wheel face width coefficients of both helical gear stages, the allowable contact stress and the output torque were examined. Furthermore, to evaluate the influences of these factors on the optimum ratios, a simulation experiment was carried out. Significantly, the results have provided some models to gain the optimum gear ratios in a simple manner.

2 Optimization Problem

The length of a mechanical driven system using a two stage helical gearbox with the first stage double gear set and a chain drive can be calculated by (Fig. 1):

$$L = \max\left(L_g, L_c\right) \tag{1}$$

Where, L_g and L_c are determined by (see Fig. 2):

$$L_g = d_{w11}/2 + a_{w1} + a_{w2} + d_{w22}/2 \tag{2}$$

$$L_c = d_{w22}/2 + a_c + d_2/2 \tag{3}$$

In which, a_{w1} and a_{w2} are the center distances of the first and the second stage;d_{w11} and d_{w22} are the pitch diameters (mm) of the first and the second stage, respectively. The diameters d_{w11} and d_{w22} can be calculated as [31]:

$$d_{w11} = 2 \cdot a_{w1}/(u_1 + 1) \tag{4}$$

$$d_{w22} = 2 \cdot a_{w2} \cdot u_2/(u_2 + 1) \tag{5}$$

In the above equations, u_1 and u_2 are the gear ratios of the first and the second stages; The relation of u_1 and u_2 with the total gearbox ratio u_g is expressed by:

$$u_1 \cdot u_2 = u_g \tag{6}$$

Hence, the optimization problem can be expressed as

$$\text{Minnimize } L \tag{7}$$

With the following constraints

$$\begin{gathered} 1 \le u_1 \le 9 \\ 1 \le u_2 \le 9 \\ 1 \le u_c \le 6 \\ u_1 \cdot u_2 \cdot u_c = u_t \end{gathered} \tag{8}$$

From the above equations, to solve the optimization problem, it is needed to determine a_{w1}, a_{w2}, d_{w11}, d_{w22}, a_c and d_2. These parameters will be determined in the following sub-sections.

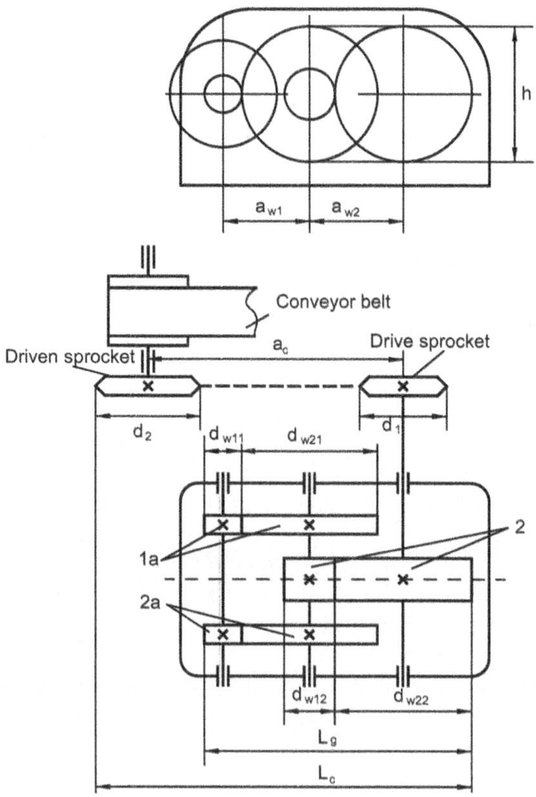

Fig. 1. Calculation schema

2.1 Determining Parameters of the First Stage Helical Gear Set

The center distance of the first stage a_{w1} is calculated by the following equation [31]:

$$a_{\text{w1}} = K_a \cdot (u_1 + 1) \cdot \left\{ T_{11} \cdot K_{H\beta} / \left([\sigma_H]^2 \cdot u_1 \cdot \psi_{ba1} \right) \right\}^{1/3} \tag{9}$$

Where, K_a is the material coefficient; If the gear material is steel $K_a = 43$ [31], $K_{H\beta}$ is the contact load ratio coefficient; $K_{H\beta} = 1.02 \div 1.28$ [31] and we can choose $K_{H\beta} = 1.1$; $[\sigma_H]$ is the allowable contact stress (MPa); In practice, if the gear material is steel, $[\sigma_H] = 350\ldots420$ (MPa); ψ_{ba} is the coefficient of the wheel face width; in this case $\psi_{ba1} = 0.4\ldots0.5$ [31]. T_{11} is the pinion torque (Nmm); For the gearbox, T_{11} is calculated by:

$$T_{11} = T_{out} / \left(2 \cdot u_g \cdot u_c \cdot \eta_{hg}^2 \cdot \eta_c \cdot \eta_{be}^3 \right) \tag{10}$$

In which, η_{hg} is the helical gear transmission efficiency ($\eta_{hg} = 0.96\ldots0.98$ [31]); η_c is the transmission efficiency of the chain drive ($\eta_c = 0.95\ldots0.97$ [26]); η_{be} is the transmission efficiency of a pair of rolling bearings ($\eta_{be} = 0.99\ldots0.995$ [31]). Choosing $\eta_{hg} = 0.97$, $\eta_{be} = 0.992$, $\eta_c = 0.96$ and substituting them into (10) gets:

$$T_{11} = 0.567 \cdot T_{out} / u_t \tag{11}$$

Substituting $K_{H\beta} = 1.1$ and (11) into (9) gives:

$$a_{\text{w1}} = 36.739 \cdot (u_1 + 1) \cdot \left\{ T_{out} / \left([\sigma_{H1}]^2 \cdot u_1 \cdot u_t \cdot \psi_{ba1} \right) \right\}^{1/3} \tag{12}$$

After having a_{w1}, the pinion pitch diameter of the first stage can be found as [31]:

$$d_{\text{w11}} = 2 \cdot a_{w1} / (u_1 + 1) \tag{13}$$

2.2 Determining Parameters of the Second Stage Helical Gear Set

As in Sect. 2.1, the center distance of the second stage a_{w2} is determined by [31]:

$$a_{\text{w2}} = K_a \cdot (u_2 + 1) \cdot \left\{ T_{12} \cdot k_{H\beta} / \left([\sigma_H]^2 \cdot u_2 \cdot \psi_{ba2} \right) \right\}^{1/3} \tag{14}$$

Where, T_{12} is the pinion torque (Nmm); T_{11} can be found by:

$$T_{12} = T_{out} / \left(u_2 \cdot u_c \cdot \eta_{hg} \cdot \eta_c \cdot \eta_{be}^2 \right) \tag{15}$$

Choosing $\eta_{hg} = 0.97$, $\eta_{be} = 0.992$ and $\eta_c = 0.96$ (as in Sect. 2.1) and substituting them into (15) gives

$$T_{12} = 1.0913 \cdot T_{out}/(u_2 \cdot u_c) \tag{16}$$

Substituting $K_a = 43$, $K_{H\beta} = 1.1$ (as in Sect. 2.1) and (16) into (14) gets:

$$a_{W2} = 45.6998 \cdot (u_2 + 1) \cdot \left\{T_{out}/\left([\sigma_H]^2 \cdot u_2^2 \cdot u_c \cdot \psi_{ba2}\right)\right\}^{1/3} \tag{17}$$

After getting a_{w2}, the gear pitch diameter of the second stage is calculated by [31]:

$$d_{w22} = 2 \cdot a_{w2} \cdot u_2/(u_2 + 1) \tag{18}$$

2.3 Determining the Driven Sprocket Diameter and the Center Distance of the Chain Drive

The pitch diameter of the driven sprocket d_2 is calculated by [31]:

$$d_2 = d_1 \cdot u_c \tag{19}$$

Where, d_1 is the pitch diameter of the drive sprocket; d_1 is found as [31]:

$$d_1 = p/\sin(\pi/z_1) \tag{20}$$

In which, z_1 is the number of teeth of the drive sprocket which is determined as [25]:

$$z_1 = 32.4 - 2.4 \cdot u_c \tag{21}$$

p is the chain pitch (mm) which can be found from the design power capacity; P is calculated by [31]:

$$P = P_1 \cdot k \cdot k_z \cdot k_n \tag{22}$$

Where, P_1 is the power rating (kW) which is calculated by:

$$P_1 = T_1 \cdot n_1/(9.55 \cdot 10^6) \tag{23}$$

Wherein, n_1 is the revolution of the drive sprocket (rpm):

$$n_1 = n_m/u_g \tag{24}$$

$$T_1 = T_{out}/(u_c \cdot \eta_c \cdot \eta_{be}) \tag{25}$$

In which, n_m is the revolution of the electric motor (rpm); η_c is the chain drive efficiency ($\eta_c = 0.95 \div 0.97$ [31]); η_{be} is the efficiency of a pair of bearings ($\eta_{be} = 0.99 \div 0.995$ [31]); T_1 and T_{out} are the torque on the drive and the output torque (Nmm).

k, k_z and k_n are coefficients determined as [31]:

$$k = k_d \cdot k_p \cdot k_c \cdot k_{adj} \cdot k_{\text{lub}} \cdot k_{con} \quad (26)$$

$$k_z = 25/z_1 \quad (27)$$

$$k_n = n_{01}/n_1 \quad (28)$$

In the above equations, k_d, k_p, k_c, k_{adj}, k_{lub} and k_{con} are the coefficients referring to the influence of the parameters including the shock factor, the drive position, the center distance of the drive, the possibility of adjusting the center distance, the lubrication and the operating conditions, respectively; n_{01} is the tabulated number of the drive sprocket teeth.

2.4 Experimental Work

Table 1. Input parameters

Factor	Code	Unit	Low	High
Total system ratio	u_t	–	10	50
Coefficient of wheel face width of stage 1	Ψ_{ba1}	–	0.4	0.5
Coefficient of wheel face width of stage 2	Ψ_{ba2}	–	0.35	0.4
Allowable contact stress	AS	MPa	350	420
Output torque	T_{out}	Nmm	10^5	10^7

A simulation experiment was implemented to explore the effects of the input parameters on the optimum gear ratios. In regard to the plan of the experiment, a 2-level full factorial design was employed. Totally, $2^5 = 32$ numbers of runs were carried out based on the number of selected parameters (Table 1). Subsequently, a computer program was established to accomplish the experiment based on Eqs. (7) and (8). The different input factor levels and the output results of the program (the optimum gear ratios of the first stage u_1 and the chain drive u_c) are demonstrated in Table 2.

3 Results and Discussion

It is realized from Table 2 that the optimum gear ratios of the first stage u_1 and the chain drive u_c are constant. That means they are not determined by any input parameters. Also, the gear ratio of the first stage u_1 has the maximum value of a helical gear set ($u_I = 9$). It can be explained that u_1 takes the maximum possible value of a helical gear set ($u_1 = u_{\max} = 9$) to promote the advantages of the double gear sets. Besides, the optimum gear ratio of the chain drive takes the lowest value ($u_c = 1$). That means, with this objective function, the chain drive merely plays a role in transferring torque to a far distance, not the speed reduction.

$$u_1 = 9 \tag{29}$$

$$u_c = 1 \tag{30}$$

Table 2. Experimental plans and output response

StdOrder	Run order	CenterPt	Blocks	u_t	Ψ_{ba1}	Ψ_{ba2}	AS (MPa)	Tout (Nm)	u_1	u_c
25	1	1	1	10	0.4	0.35	420	10000	9	1
17	2	1	1	10	0.4	0.35	350	10000	9	1
29	3	1	1	10	0.4	0.4	420	10000	9	1
6	4	1	1	50	0.4	0.4	350	100	9	1
15	5	1	1	10	0.5	0.4	420	100	9	1
19	6	1	1	10	0.5	0.35	350	10000	9	1
…										
2	31	1	1	50	0.4	0.35	350	100	9	1
28	32	1	1	50	0.5	0.35	420	10000	9	1

Equations (29) and (30) are used to determine the optimum gear ratios of the first stage u_1 and the chain drive u_c. After getting u_1 and u_c, the gear ratio of the second stage is determined by:

$$u_2 = u_t/(u_1 \cdot u_c) \tag{31}$$

As $u_1 = 9$ and $u_c = 9$, Eq. (31) can be rewritten as:

$$u_2 = u_t/9 \tag{32}$$

4 Conclusions

A study on the determination of optimum gear ratios of mechanical driven systems including a two-stage helical gearbox with the first stage double gear sets and a chain drive was carried out. Several findings can be drawn from this study:

- The minimum gearbox length of a mechanical driven system using a two-stage helical gearbox with first stage double gear sets and a chain drive can be obtained by designing with the optimum gear ratios;
- Explicit equations to calculate the optimum gear ratios of the mentioned system were found for getting the minimum system length;
- The optimum gear ratios of the gearbox can be determined accurately and simply by the proposed explicit models.

Acknowledgements. The work described in this paper was supported by Thai Nguyen University of Technology for a scientific project.

References

1. Kudreavtev, V.N., Gierzaves, I.A., Glukharev, E.G.: Design and Calculus of Gearboxes. Mashinostroenie Publishing, Sankt Petersburg (1971). (in Russian)
2. Chat, T.: Some problems of kinematics calculation of transmission mechanics system. In: Proceedings of the National Conference on Engineering Mechanics, vol. 2 (Hanoi), pp. 7–12 (1993). (in Vietnamese)
3. Milou, G., Dobre, G., Visa, F., Vitila, H.: Optimal design of two step gear units, regarding the main parameters. VDI Berichte, no. 1230, p. 227 (1996)
4. Pi, V.N.: A method for optimal calculation of total transmission ratio of two step helical gearboxes. In: Proceedings of the National Conference on Engineering Mechanics, Ha Noi, p. 12 (2001)
5. Cam, N.T.H., Pi, V.N., Tuan, N.K., Hung, L.X., Thao, T.T.P.: Determining optimal partial transmission ratios of mechanical driven systems using a V-Belt drive and a helical reducer with second-step double gear-sets. In: Fujita, H., Nguyen, D., Vu, N., Banh, T., Puta, H. (eds.) ICERA 2018. LNNS, vol. 63, pp. 261–269. Springer, Cham (2019)
6. Petrovski, A.N., Sapiro, B.A., Saphonova, N.K.: About optimal problem for multi-step gearboxes. Vestnik Mashinostroenie **10**, 13 (1987). (in Russian)
7. Pi, V.N.: A new study on optimal calculation of partial transmission ratios of two-step helical gearboxes. In: 2nd WSEAS International Conference on Computer Engineering and Applications (CEA 2008), Acapulco, Mexico, 25–27 January, pp. 162–165 (2008)
8. Tuan, N.K., Pi, V.N., Cam, N.T.H., Thao, T.T.P., Thanh, H.K., Hung, L.X., Tham, H.T.: Determining optimal gear ratios of a two-stage helical reducer for getting minimal acreage of cross section. In: MATEC Web of Conferences, vol. 213, p. 01008 (2018). https://doi.org/10.1051/matecconf/201821301008
9. Pi, V.N., Tuan, N.K.: Optimum determination of partial transmission ratios of Three-step helical gearboxes for getting minimum cross section. J. Environ. Sci. Eng. A **5**, 570–573 (2016)
10. Pi, V.N., Binh, N.D., Dac, V.Q., The, P.Q.: Optimal calculation of total transmission ratio of three-step helical gearboxes for minimum mass of gears. J. Sci. Technol. 6 Eng. Univ. **55**, 91 (2006). (in Vietnamese)
11. Pi, V.N.: A new study on optimal calculation of partial transmission ratio of three-step helical reducers. In: The 3rd IASME/WSEAS International Conference on Continuum Mechanics, Cambridge, UK, p. 23 (2008)
12. Pi, V.N.: A new study on optimal calculation of partial transmission ratios of three-step helical reducers for getting minimal cross section dimension. In: The 2nd WSEAS International Conference on Computer Engineering and Applications (CEA 2008), Acapulco, Mexico, 25–27 January, pp. 290–293 (2008)
13. Pi, V.N.: Optimal determination of partial transmission ratios of three-step helical gearboxes with first and third-step double gear-sets for minimal mass of gear. In: American Conference on Applied Mathematics (MATH 2008), Harvard, Massachusetts, USA, 24–26 March, pp. 385–388 (2008)
14. Romhild, I., Linke, H.: Gezielte Auslegung Von Zahnradgetrieben mit minimaler Masse auf der Basis neuer Berechnungsverfahren. Konstruktion **44**, 229–236 (1992)
15. Pi, V.N.: A new study on the optimal prediction of partial transmission ratios of three-step helical gearboxes with second-step double gear-sets. WSEAS Trans. App. Theor. Mech. **2** (11) 2007

16. Pi, V.N.: Optimal determination of partial transmission ratios for four-step helical gearboxes with first and third step double gear-sets for minimal mass of gears. In: Applied Computing Conference (ACC 2008), Istanbul, Turkey, 27–30 May 2008
17. Pi, V.N.: Optimal calculation of partial transmission ratios for four-step helical gearboxes with first and third-step double gear-sets for minimal gearbox length. In: American Conference on Applied Mathematics (MATH 2008), Harvard, Massachusetts, USA, 24–26 March, pp. 29–32 (2008)
18. Hung, L.X., Pi, V.N., Van Du, N.: Optimal calculation of partial transmission ratios of four-step helical gearboxes with second and fourth-step double gear-sets for minimal mass of gears. In: The International Symposium on Mechanical Engineering (ISME Ho Chi Minh City, Vietnam, pp. 21–23 (2009)
19. Pi, V.N.: Optimal determination of partial transmission ratios for four-step helical gearboxes with first and third step double gear-sets for minimal mass of gears. In: Applied Computing Conference (ACC 2008), Istanbul, Turkey, 27–30 May, pp. 53–57 (2008)
20. Pi, V.N.: A new and effective method for optimal calculation of total transmission ratio of two step bevel - helical gearboxes. In: International colloquium on Mechanics of Solids, Fluids, Structures & Interaction Nha Trang, Vietnam, pp. 716– 719 (2000)
21. Pi, V.N.: A study on optimal calculation of partial transmission ratios of three-step bevel helical gearboxes. In: International Workshop on Advanced Computing and Applications (ACOMP 2008), 12–14 March, pp. 277–286 (2008)
22. Chernavsky, S.A., et al.: Design of mechanical transmissions: Manual for high technical schools, p. 560. Mashinostroenie, Moscow (1984)
23. Pi, V.N., Dac, V.Q.: Calculation of total transmission ratio of two step worm reducers for the best reasonable gearbox housing structure. J. Sci. Technol. Thai Nguyen Univ. **1**(41), 65–69 (2007). (in Vietnamese)
24. Pi, V.N., Dac, V.Q.: Optimal calculation of partial transmission ratios of worm-helical gear reducers for minimal gearbox length. J. Sci. Technol. Tech. Univ. **61**, 73–77 (2007)
25. Pi, V.N., Cam, N.T.H., Tuan, N.K.: Optimum calculation of partial transmission ratios of mechanical driven systems using a V-belt and two-step bevel helical gearbox. J. Environ. Sci. Eng. A **5**, 566 (2016)
26. Pi, V.N., Thao, T.T.P., Thao, L.T.P.: A new study on optimum determination of partial transmission ratios of mechanical driven systems using a V-belt and two-step helical gearbox. Vietnam Mech. Eng. J. **10**, 123 (2015)
27. Tuan, N.K., Thao, T.T.P., Cam, N.T.H., Hung, L.X., Pi, V.N.: Optimum calculation of partial transmission ratios of mechanical driven systems using a V-belt and a three-step bevel helical gearbox. In: Fujita, H., et al. (eds.) ICERA 2018. LNNS, vol. 63, pp. 469–476 (2019). https://doi.org/10.1007/978-3-030-04792-4_61
28. Cam, N.T.H., Pi, V.N., Tuan, N.K., Hung, L.X., Thao, T.T.P.: Determining optimal partial transmission ratios of mechanical driven systems using a V-belt drive and a helical reducer with second-step double gear-sets. Fujita, H., et al. (eds.) ICERA 2018. LNNS, vol. 63, pp. 261–269 (2019). https://doi.org/10.1007/978-3-030-04792-4_35
29. Pi, V.N., Thao, T.T.P., Tuan, D.A.: Optimum determination of partial transmission ratios of mechanical driven systems using a chain drive and two-step helical gearbox. J. Environ. Sci. Eng. B **6**, 80 (2017)
30. Cam, N.T.H., Pi, V.N., Tuan, N.K., Hung, L.X., Thao, T.T.P.: A study on determination of optimum partial transmission ratios of mechanical driven systems using a chain drive and a three-step helical reducer. In: Fujita, H., et al. (eds.) ICERA 2018. LNNS, vol. 63, pp. 91–99 (2019). https://doi.org/10.1007/978-3-030-04792-4_14
31. Chat, T., Van Uyen, L.: Design and calculus of Mechanical Transmissions. Educational Republishing House, Hanoi (1998). (in Vietnamese)

Calculation of Optimum Gear Ratios of Two-Step Worm Gearbox

Nguyen Manh Cuong[1], Hoang Thi Tham[1], Tran Thi Hong[2], Nguyen Van Cuong[3], Le Hong Ky[4], Nguyen Thanh Tu[1], Le Xuan Hung[1], and Ngoc Pi Vu[1(✉)]

[1] Thai Nguyen University of Technology, Thai Nguyen, Vietnam
vungocpi@tnut.edu.vn
[2] Nguyen Tat Thanh University, Ho Chi Minh, Vietnam
[3] University of Transport and Communications, Ha Noi, Vietnam
[4] Vinh Long University of Technology Education, Vinh Long, Vietnam

Abstract. This paper introduces a study on the calculation of the optimum gear ratios of a two-step worm gearbox. To find the optimum gear ratios, an optimization problem in which the reasonable gearbox structure was chosen as the objective function was conducted. The influences of the input parameters including the total gearbox ratio, the electric motor speed and the output torque were explored. To evaluate the effects of these input parameters on the optimum gear ratios, a simulation experiment was designed and performed by computer programs. Moreover, equations to calculate the optimum gear ratios of a two-step worm gearbox were proposed. Using these equations, the calculation of optimum gear ratios is precise and uncomplicated.

Keywords: Transmission ratio · Gear ratio · Worm gearbox · Two-step worm gearbox · Optimum gearbox design

1 Introduction

In the optimum gearbox design, the calculation of the optimum gear ratios is considered of great importance as the gear ratios of a gearbox strongly affect its dimension, mass and cost. Consequently, the calculation of optimum gear ratios has been taken into consideration in numerous studies so far.

Until now, different methods have been applied to identify the optimum gear ratios. They include the graph method [1, 2], the practical method [3] and the model method [3–16]. Besides, the optimum gear ratios have been determined for multi-step gearboxes. For example, with the helical gearboxes, the optimum gear ratios have been calculated for two step gearboxes [3–7], for three step gearboxes [4–13] and for four step gearboxes [5, 12–16]. Also, the gear ratios have been found for several objectives including the minimum mass of gears [9, 12–14, 16] or of a gearbox [14], the minimum length of a gearbox [10], and the minimum cross section area of a gearbox [8, 11].

Recently, several studies have reported the determination of the optimum gear ratios of mechanical driven systems using a gearbox and a V-belt drive [17–20] or a chain drive [21, 22].

K.-U. Sattler et al. (Eds.): ICERA 2019, LNNS 104, pp. 179–188, 2020.
https://doi.org/10.1007/978-3-030-37497-6_21

Regarding worm gearboxes, there have not been many studies on this topic. For a two-step worm gearbox, the gear ratios can be found by a practical model [3] or calculated by the following practical equation for getting the reasonable housing structure [23]:

$$u_1 \approx u_2 \approx (u_h)^{1/2} \tag{1}$$

Moreover, for this target, the optimum gear ratio of the second step of mentioned gearbox can be chosen as $u_2 \approx 30.97$ [24].

For a worm helical gearbox, the optimum gear ratio of the helical gear set can be calculated by [23]:

$$u_2 = (0.03 \div 0.06)u_h. \tag{2}$$

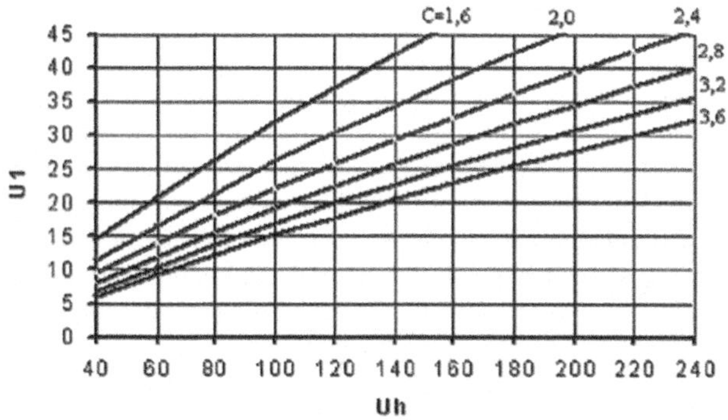

Fig. 1. Gear ratio of worm gear set versus total gear ratio [19]

Besides, to assure the oil lubrication for both steps, the gear ratio of the worm gear set is found by using the graph in Fig. 1 [2]. It can be seen from this graph that the determination of the first step gear ratio u_1 is rather complicated. Furthermore, to find u_1, it is required to choose the coefficient c. As the value of c is quite large (c = 1.6…3.6), the optimum values of u_1 cannot be reached. To ignore this defect, the optimum gear ratio of the second step u_2 with the same objective as in [2] can be calculated as [24]:

$$u_2 = 6.86 \cdot \psi_{ba2}^{1/2} \tag{3}$$

Where, $\psi_{ba2} = 0.3\ldots0.4$ is the wheel face width coefficient of the helical gear unit. Furthermore, for this gearbox type, the optimum gear ratio of the helical gear unit u_2 is found as $u_2 = [u_{2\max}] = 8\ldots10$ [25].

This paper introduces an optimization study on determining the optimum gear ratios of a two-step worm gearbox. The objective of the optimization problem is to achieve the reasonable gearbox housing structure. Also, the effects of the input parameters including the total gearbox ratio, the electric motor speed, and the output torque were taken into account. To estimate the influences of these input parameters on the optimum gear ratios, a simulation experiment was designed and accomplished by computer programming. In addition, models to determine the optimum gear ratios were suggested. As the models are explicit, the optimum gear ratios can be calculated accurately and simply.

2 Optimization Problem

For a two-step worm gearbox, the housing structure of the gearbox is considered reasonable when [20]:

$$a_{w2} = 2 \cdot a_{w1} \tag{4}$$

Where, a_{w1} and a_{w2} are the centre distances of the first and the second steps of the gearbox.

Consequently, the optimization problem is to find the optimum gear ratios of the gearbox for getting $a_{w2} = 2 \cdot a_{w1}$ (Eq. (4)).

With the following constraints

$$8 \leq u_1 \leq 80 \tag{5}$$

$$8 \leq u_2 \leq 80$$

Therefore, to solve the optimization problem it is required to determine the centre distances of both steps of the gearbox (Fig. 2).

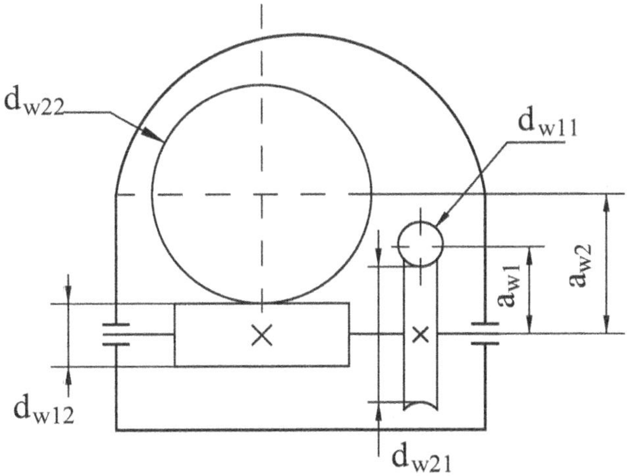

Fig. 2. Calculation schema

2.1 Determining the Center Distance of the First Step a_{w1}

The centre distance of the first step - worm gear set a_{w1}(mm) is calculated by [26]:

$$a_{w1} = K_a \cdot \left(K_{HV} \cdot K_{H\beta} \cdot T_{21}/[\sigma_{H1}]^2\right)^{1/3} \quad (6)$$

In which, K_a is a coefficient; $K_a = 610$ for involute, Archimedean, and convolute worms [26]; K_{hV} is the internal dynamic; $K_{hV} = 1\ldots1.3$ [26]; $K_{h\beta}$ is the coefficient of the load concentration; $K_{h\beta} = 1\ldots1.3$ [26]; T_{21} is the wheel torque (Nmm);

With this two-step worm gearbox, T_{21} can be found as:

$$T_{21} = T_{out}/\left(u_2 \cdot \eta_w \cdot \eta_b^2\right) \quad (7)$$

Wherein, T_{out} is the output torque (Nmm); η_w is the transmission efficiency of a worm set ($\eta_w = 0.7\ldots0.9$ [27]); η_b is the transmission efficiency of a rolling bearing pair ($\eta_b = 0.99\ldots0.995$ [27]). Choosing $\eta_w = 0.8$, $\eta_b = 0.992$ and substituting them into (7) gets:

$$T_{21} = 1.27024 \cdot T_{out}/u_2 \quad (8)$$

Similarly, taking $K_{hV} = 1.2$, $K_{h\beta} = 1.2$ and substituting them, K_a and T_{21} into (12) gives:

$$a_{w1} = 746.01 \cdot \left\{T_{21}/\left(u_2 \cdot [\sigma_{H1}]^2\right)\right\}^{1/3} \quad (9)$$

Where, $[\sigma_{H1}]$ is the allowable contact stress of the first step (N/mm^2); $[\sigma_{H1}]$ depends on the wheel material. If the wheel material is tin less bronze or soft grey iron, the following regression equation is suggested (with the determination coefficient $R^2 = 0.9906$) based on the data in [27]:

$$[\sigma_H] = 5.0515 \cdot v_{sl}^2 - 49.742 \cdot v_{sl} + 189.9 \quad (10)$$

Wherein, v_{sl} is the slip velocity which is calculated by [27]:

$$v_{sl} = 0.0088 \cdot \left(P_1 \cdot u \cdot n_1^2\right)^{1/3} \quad (11)$$

If the wheel material is tin bronze, $[\sigma_{H1}]$ can be found as [27]:

$$[\sigma_{H1}] = K_{HL} \cdot v_{sl} \cdot [\sigma_{H0}] \quad (12)$$

In which, $[\sigma_{H0}]$ is the allowable contact stress when the stress change cycle is 10^7:

$$[\sigma_{H0}] = (0.7\ldots0.9) \cdot \sigma_t \quad (13)$$

Where, σ_t is the tensile stress (N/mm^2); $\sigma_t = 260$ if $v_{sl} = 5\ldots8$; $\sigma_t = 230$ if $v_{sl} = 8\ldots12$ and $\sigma_t = 285$ if $v_{sl} = 8\ldots25$.

K_{HL} is the service life ratio which is determined by:

$$K_{HL} = \left(10^7/N_{HE}\right)^{1/8} \tag{14}$$

Wherein, N_{HE} is the equivalent loading cycle number for the wheel teeth.

$$N_{HE} = 60 \cdot n_2 \cdot t_\Sigma \tag{15}$$

With t_Σ as the service lifetime of the gearing (h), n_2 is the wheel rotational speed (rpm).

2.2 Determining the Center Distance of the Second Step a_{w2}

For the second step, the calculation is done in the same way as in Sect. 2.1, the centre distance a_{w2} (mm) can be determined by [26]:

$$a_{w2} = K_a \cdot \left(K_{HV} \cdot K_{H\beta} \cdot T_{22}/[\sigma_{H2}]^2\right)^{1/3} \tag{16}$$

In which, T_{22} is the wheel torque of the second step (Nmm):

$$T_{22} = T_{out}/\eta_b \tag{17}$$

Choosing $\eta_b = 0.992$ (as in Sect. 2.1) and substituting them into (17) gets:

$$T_{22} = 1.008 \cdot T_{out} \tag{18}$$

Choosing $K_a = 610$, $K_{hV} = 1.2$, $K_{h\beta} = 1.2$ (as in Sect. 2.1) and substituting them and T_{22} into (16) gives:

$$a_{w1} = 690/67 \cdot \left\{T_{22}/\left(u_2 \cdot [\sigma_{H2}]^2\right)\right\}^{1/3} \tag{19}$$

Where, $[\sigma_{H2}]$ is the allowable contact stress of the second step (N/mm^2); $[\sigma_{H2}]$ is determined as when calculating $[\sigma_{H1}]$ (see Sect. 2.1).

2.3 Experimental Work

As mentioned above, a simulation experiment was carried out to explore the effects of the input parameters on the optimum gear ratios. The experiment was planned with a 2-level full factorial design. In the experiment, 3 input parameters including the total gearbox ratio, the electric motor speed and the output torque were selected for the exploration (Table 1). Subsequently, the experiment was performed with $2^3 = 8$ numbers of runs. In addition, a computer program was constructed based on Eqs. (4)

and (5) to conduct the experiment. The experimental plans and the results of the output response (the optimum gear ratio of the first step u_1) are displayed in Table 2.

Table 1. Input parameters

Factor	Code	Unit	Low	High
Total gearbox ratio	u_g	–	300	800
Electric motor speed	n_m	[rpm]	750	2910
Output torque	T_{out}	[Nmm]	10^5	10^7

Table 2. Experimental plans and output response

Std order	Run order	Centre Pt	Blocks	u_g	n_m	T_{out}	u_1
7	1	1	1	300	2910	10000000	8
3	2	1	1	300	2910	100000	8
8	3	1	1	800	2910	10000000	10
5	4	1	1	300	750	10000000	37.5
6	5	1	1	800	750	10000000	10
1	6	1	1	300	750	100000	37.5
4	7	1	1	800	2910	100000	10
2	8	1	1	800	750	100000	10

3 Results and Discussion

To weigh the effects of the input parameters on the optimum gear ratio of the first step u_1, Fig. 3 prints the graph of the main effects for u_1. From the graph, u_1 decreases significantly with the increase of the total gearbox ratio u_g and the electric motor speed n_m. Also, it is not affected by the output torque T_{out}.

Figure 4 presents the Normal Plot of the standardized effects. This graph is used to describe which effects decrease or increase the response. It can be recognized from the figure that the electric motor speed n_m (factor B), the total gearbox ratio u_g (factor A) and the interaction between them (factor AB) are the most significant factors for u_1. Also, the total gearbox ratio u_g (factor A) and the electric motor speed n_m (factor B) have a negative standardized effect. If they rise, the optimum gear ratio of the first step decreases. Conversely, the interaction AB has a positive effect. When it increases, the optimum gear ratio increases.

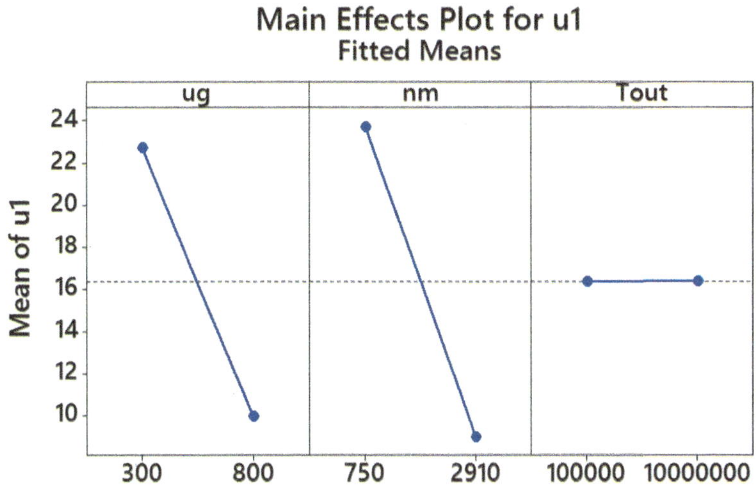

Fig. 3. Main effects plot for optimum gear ratio of the first step

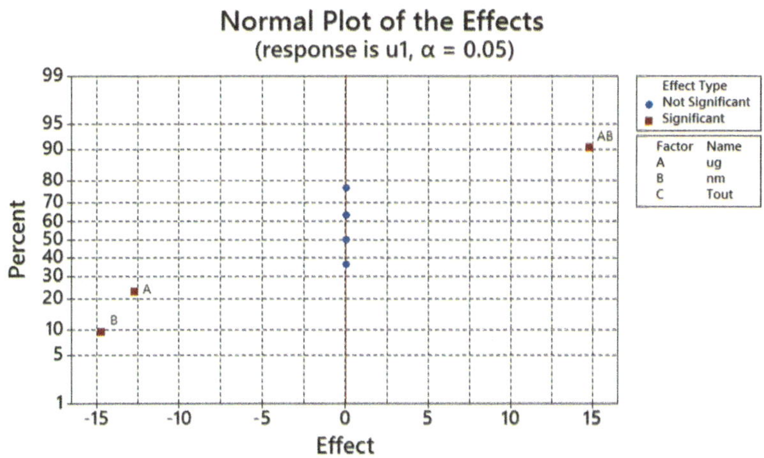

Fig. 4. Normal Plot for u_1

Figure 5 demonstrates the estimated effects and coefficients for u_1. It is noticed from the figure that the parameters which have a significant effect on a response (with P-values lower than 0.05) are the electric motor speed n_m, the total gearbox ratio u_g, and the interaction between them $u_g \cdot n_m$. Therefore, the optimum gear ratio of the first step can be calculated by the following equation:

$$u_1 = 70.39 - 0.07549 \cdot u_g - 0.02185 \cdot n_m + 0.00027 \cdot u_g \cdot n_m \quad (20)$$

Coded Coefficients

Term	Effect	Coef	SE Coef	T-Value	P-Value	VIF
Constant		16.38	0.00	*	*	
ug	-12.750	-6.375	0.000	*	*	1.00
nm	-14.750	-7.375	0.000	*	*	1.00
ug*nm	14.750	7.375	0.000	*	*	1.00

Model Summary

S	R-sq	R-sq(adj)	R-sq(pred)
0	100.00%	100.00%	100.00%

Fig. 5. Estimated effects and coefficients for u_1.

Equation (20) matches the experimental data very well as all the values of adj-R2 and pred-R2 are 100% (Fig. 5). Thus, the equation is used to determine the optimum gear ratio of the first step u_1. After having u_1, the gear ratio of the second step is calculated by the following equation:

$$u_2 = u_g/u_1 \tag{21}$$

4 Conclusions

A study on the determination of the optimum gear ratios of a two-step worm gearbox disregarding the reasonable gearbox housing structure was carried out. In the study, the effects of the input parameters including the total gearbox ratio, the electrical motor speed and the output torque were investigated. Also, equations to calculate the optimum gear ratios were proposed. Employing these equations, the optimum gear ratios are obtained simply because they are explicit.

Acknowledgements. The work described in this paper was supported by Thai Nguyen University of Technology for a scientific project.

References

1. Kudreavtev, V.N., Gierzaves, I.A., Glukharev, E.G.: Design and Calculus of Gearboxes. Mashinostroenie Publishing, Sankt Petersburg (1971). (in Russian)
2. Chat, T.: Some problems of kinematics calculation of transmission mechanics system. In: Proceedings of the National Conference on Engineering Mechanics, vol. 2, Hanoi, pp. 7–12 (1993). (in Vietnamese)
3. Milou, G., Dobre, G., Visa, F., Vitila, H.: Optimal design of two step gear units, regarding the main parameters. VDI Berichte No. 1230, p. 227 (1996)

4. Pi, V.N.: A method for optimal calculation of total transmission ratio of two step helical gearboxes. In: Proceedings of the National conference on Engineering Mechanics, Ha Noi, p. 12 (2001)
5. Cam, N.T.H., Pi, V.N., Tuan, N.K., Hung, L.X., Thao, T.T.P.: Determining optimal partial transmission ratios of mechanical driven systems using a V-belt drive and a helical reducer with second-step double gear-sets. In: Fujita, H., et al. (eds.) ICERA 2018. LNNS, vol. 63, pp. 261–269 (2019). https://doi.org/10.1007/978–3-030-04792-4_35
6. Petrovski, A.N., Sapiro, B.A., Saphonova, N.K.: About optimal problem for multi-step gearboxes. Vestnik Mashinostroenie **10**, 13 (1987). (in Russian)
7. Nguyen, K.T., Vu, N.P., Nguyen, T.H.C., Tran, T.P.T., Ho, K.T., Le, X.H., Hoang, T.T.: Determining optimal gear ratios of a two-stage helical reducer for getting minimal acreage of cross section. In: MATEC Web of Conferences 213, 01008 (2018). https://doi.org/10.1051/matecconf/201821301008
8. Pi, V.N., Tuan, N.K.: Optimum determination of partial transmission ratios of Three-step helical gearboxes for getting minimum cross section. J. Environ. Sci. Eng. A **5**, 570–573 (2016)
9. Pi, V.N., Binh, N.D., Dac, V.Q., The, P.Q.: Optimal calculation of total transmission ratio of three-step helical gearboxes for minimum mass of gears. Journal of Science and Technology of 6 Engineering Universities, p. 91 (2006). (in Vietnamese)
10. Pi, V.N.: A new study on optimal calculation of partial transmission ratio of three-step helical reducers. In: The 3rd IASME/WSEAS International Conference on Continuum Mechanics, Cambridge, UK, p. 23 (2008)
11. Pi, V.N.: A new study on optimal calculation of partial transmission ratios of three-step helical reducers for getting minimal cross section dimension. In: The 2nd WSEAS International Conference on Computer Engineering and Applications (CEA'08), Acapulco, Mexico, 25–27 January, pp. 290–293 (2008)
12. Pi, V.N.: Optimal determination of partial transmission ratios of three-step helical gearboxes with first and third-step double gear-sets for minimal mass of gear. In: American Conference on Applied Mathematics (MATH'08), Harvard, Massachusetts, USA, 24–26 March, pp. 385–388 (2008)
13. Romhild, I., Linke, H.: Gezielte Auslegung Von Zahnradgetrieben mit minimaler Masse auf der Basis neuer Berechnungsverfahren. Konstruktion **44**, 229–236 (1992)
14. Pi, V.N.: Optimal determination of partial transmission ratios for four-step helical gearboxes with first and third step double gear-sets for minimal mass of gears. In: Applied Computing Conference (ACC' 08), Istanbul, Turkey, 27–30 May (2008)
15. Pi, V.N.: Optimal calculation of partial transmission ratios for four-step helical gearboxes with first and third-step double gear-sets for minimal gearbox length. In: American Conference on Applied Mathematics (MATH'08), Harvard, Massachusetts, USA, 24–26 March, pp. 29–32 (2008)
16. Hung, L.X., Pi, V.N., Van Du, N.: Optimal calculation of partial transmission ratios of four-step helical gearboxes with second and fourth-step double gear-sets for minimal mass of gears. In: The International Symposium on Mechanical Engineering (ISME), Ho Chi Minh city, Vietnam, pp. 21–23 (2009)
17. Pi, V.N., Thao, T.T.P., Thao, L.T.P.: A new study on optimum determination of partial transmission ratios of mechanical driven systems using a V-belt and two-step helical gearbox. Vietnam Mechanical Eng. J. **10**, 123 (2015)
18. Pi, V.N., Cam, N.T.H., Tuan, N.K.: Optimum calculation of partial transmission ratios of mechanical driven systems using a V-belt and two-step bevel helical gearbox. J. Environ. Sci. Eng. A **5**, 566 (2016)

19. Tuan, N.K., Thao, T.T.P., Cam, N.T.H., Hung, L.X., Pi, V.N.: Optimum calculation of partial transmission ratios of mechanical driven systems using a V-belt and a three-step bevel helical gearbox. In: Fujita, H., et al. (eds.) ICERA 2018, LNNS, vol. 63, pp. 469–476 (2019). https://doi.org/10.1007/978-3-030-04792-4_61
20. Cam, N.T.H., Pi, V.N., Tuan, N.K., Hung, L.X., Thao, T.T.P.: Determining optimal partial transmission ratios of mechanical driven systems using a V-belt drive and a helical reducer with second-step double gear-sets. In: Fujita, H., et al. (eds.) ICERA 2018. LNNS, vol. 63, pp. 261–269 (2019). https://doi.org/10.1007/978-3-030-04792-4_35
21. Cam, N.T.H., Pi, V.N., Tuan, N.K., Hung, L.X., Thao, T.T.P.: A study on determination of optimum partial transmission ratios of mechanical driven systems using a chain drive and a three-step helical reducer. In: Fujita, H., et al. (eds.) ICERA 2018. LNNS, vol. 63, pp. 91–99 (2019). https://doi.org/10.1007/978-3-030-04792-4_14
22. Pi, V.N., Thao, T.T.P., Tuan, D.A.: Optimum determination of partial transmission ratios of mechanical driven systems using a chain drive and two-step helical gearbox. J. Environ. Sci. Eng. B **6**, 80 (2017)
23. Chernavsky, S.A., et al.: Design of mechanical transmissions: Manual for high technical schools (Moscow: Mashinostroenie) p. 560 (1984)
24. Pi, V.N., Dac, V.Q.: Calculation of total transmission ratio of two-step worm reducers for the best reasonable gearbox housing structure. J. Sci. Technol. Thai Nguyen University **1**(41), 65–69 (2007). (in Vietnamese)
25. Pi, V.N., Dac, V.Q.: Optimal calculation of partial transmission ratios of worm-helical gear reducers for minimal gearbox length. J. Sci. Technol. Tech. Univ. **61**, 73–77 (2007)
26. Grote, K.-H., Erik, K.: Antonsson: Springer Handbook of Mechanical Engineering. Springer, Berlin (2008)
27. Chat, T., Van Uyen, L.: Design and Calculation of Mechanical Transmissions Systems, vol. 1. Educational Republishing House, Hanoi (2007)

Calibration of a Laser-Plane Sensor in 5-Axis Machine Tool

Nguyen Duy Minh Phan[1(✉)], Yann Quinsat[2], and Claire Lartigue[2]

[1] The University of Danang-University of Technology and Education, Danang 550 000, Vietnam
pndminh@ute.udn.vn

[2] LURPA, ENS Paris-Saclay, Université Paris-Sud, Universite Paris-Saclay, 94235 Cachan, France

Abstract. The purpose of this paper is to propose a calibration procedure of the laser-plane sensor for On-machine Measurement (OMM) in a 5-axis machine tool. During on-machine measurement, the digitizing process of the part geometry is performed without removing the part from its set-up when the machining process is stopped. Being given the 5 degrees of freedom and the rotation of the spindle, it is possible to improve the accessibility of the sensor. The laser-plane sensor positioned in the spindle reduces the measurement time; thus allowing a fast decision-making concerning the geometrical conformity of the manufactured part and the potential machining corrections. For on-machine measurement in a 5-axis machine, it is necessary to perform a 5-axis calibration of the laser-plane sensor that is rarely addressed in literature. This calibration procedure allows to determine the orientation and position of the sensor relative to the machine frame, and to define the origin of the sensor frame.

Keywords: On-Machine Measurement · Calibration · Laser-plane sensor · 5-axis machine-tool · Scan path

1 Introduction

On-Machine Measurement (OMM) is the measurement operations for which the machine tool is the displacement system ensuring the movements of the sensor. During OMM, the measurement of the geometry of the part is performed when the machining process is stopped without removing the part. This facilitates the comparison of the machined part to its CAD model and allows a rapid decision-making concerning part geometry conformity. In a previous work, we have proposed a method to generate a laser-plane sensor scan path suitable to 5-axis scanning in a machine tool [1]. This method generates a scan path which ensures scanning quality and the control of overlap between two adjacent scan paths. The laser-plane sensor scan path is defined as the succession of positions and orientations (CE; VL; VC) (see Fig. 1). The driven point CE positions the scanning laser line in the field of view, and the couple of vectors (VL; VC) orients the sensor, with VC the director vector of the light-beam axis, and

K.-U. Sattler et al. (Eds.): ICERA 2019, LNNS 104, pp. 189–195, 2020.
https://doi.org/10.1007/978-3-030-37497-6_22

VL, the director vector of the scanningline. As the sensor takes the place of the tool in 5-axis milling machine tool, it has thus a greater accessibility due to its possible movements: 5 degrees of freedom plus the spindle rotation. The scan path expressed in the part frame as a set of coordinates (X, Y, Z, I, J, K, I*, J*, K*) is expressed in the machine-tool frame thanks to the Inverse Kinematics Transformation (IKT). In the case of a RRTTT machine tool, the IKT leads to (X, Y, Z, A, C, Wcst) in the articular space where A and C are the rotation angles of the tilt and the table of the machine, and where Wcst represents a constant value of the spindle indexation for all the sensor configurations.

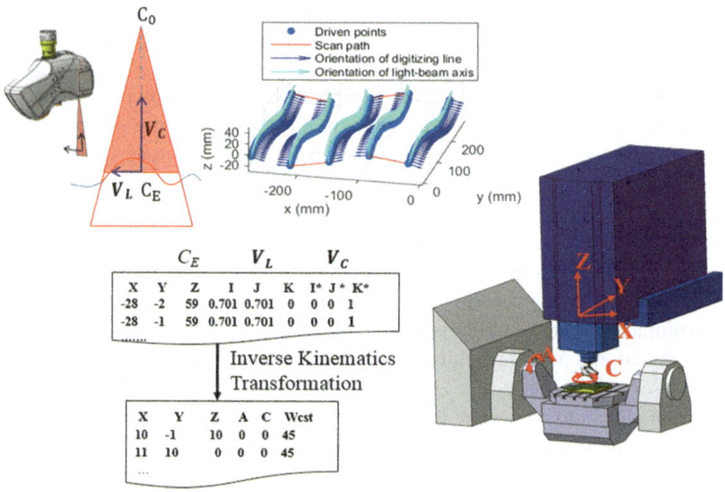

Fig. 1. Parameters defining the laser-plane sensor path for on-machine measurement.

To perform the scanning process with the proposed scan path, it is necessary to integrate and calibrate the laser-plane sensor in the machine. A 5-axis machine tool is an unconventional displacement system, whose geometric characteristics and volumetric errors are not equivalent to a CMM [2]. A 3-axis measurement with laser-plane on this machine was performed in [3]. A reference plane of 100 mm × 50 mm is scanned on 3-axis by the laser-plane sensor in the machine to evaluate the quality of the acquired data through the scanning noise and the trueness of the measurement. However, the 5-axis calibration of the sensor has not yet been addressed in the literature. The work proposed in this paper deals with a methodology for the calibration of a laser-plane sensor for measurement in a 5-axis machine tool.

2 Sensor Integration in the 5-Axis Machine

The on-machine system used consists of the Zephyr KZ25 laser-plane sensor (https://kreon3d.com/) mounted in the spindle of the machine tool Mikron UCP 710. This section explains the data flows required to operate the Kreon sensor on the 5-axis machine tool. The sensor positions and the CCD images of the laser line are analyzed by a control interface (Kreon ECU) (Fig. 2). The Kreon ECU accepts only a TTL (Transistor Transistor Logic) pulse signal for reading position signals. Polygonia software processes information from the sensor. The software allows to manage the calibration of the sensor and to visualize the measured point clouds.

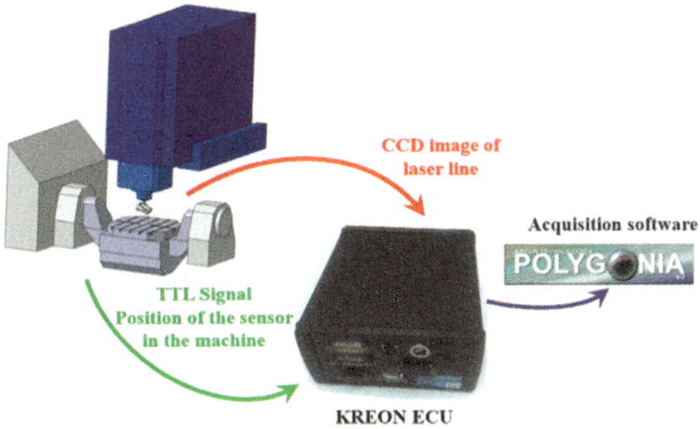

Fig. 2. Technical implementation of the Kreon laser sensor in the machine tool

For on-machine measurement on a 5-axis machine, it is necessary to carry out a 5-axis calibration of the laser-plane sensor in the machine tool.

3 Laser-Plane Sensor Calibration

The calibration procedure involves the digitization of an artifact, typically a sphere. The purpose of the calibration is to determine the orientation and position of the sensor in relation to the machine frame, and to define the origin of the sensor frame in which the digitized points will be expressed. The origin of the measurement frame is defined by the calculated center of the digitized sphere. The calibration thus makes it possible to define the transformation matrix from the sensor frame to the machine frame. For this, it is necessary to define the kinematics of the scanning system starting from the artifact to the sensor.

3.1 Kinematic of the Scanning System

The frames corresponding to the axes of the machine in Setup.exe installation file of Polygonia are defined in Fig. 3. The order of these axes is CAXYZ. The sensor calibration is performed in the Setup.exe installation file of Polygonia.

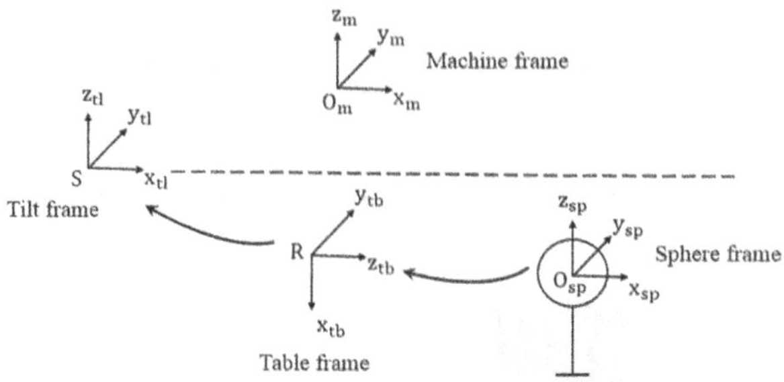

Fig. 3. Frames defined in Polygonia's *Setup.exe*

The center O_{sp} of the sphere is the origin of the measurement frame. The table frame (R; x_{tb}; y_{tb}; z_{tb}) is linked to the machine table. The table frame is obtained from the rotation around y_{sp} of the sphere frame of 90°. The transformation matrix of the sphere frame to the table frame $M_{setup}(C)$ is defined by the Eq. (1):

$$M_{setup}(C) = \begin{bmatrix} 0 & 0 & 1 & x_{O_{sp}R} \\ 0 & 1 & 0 & y_{O_{sp}R} \\ -1 & 0 & 0 & z_{O_{sp}R} \\ 0 & 0 & 0 & 1 \end{bmatrix}; M_{setup}(A) = \begin{bmatrix} 0 & 0 & -1 & x_{RS} \\ 0 & 1 & 0 & y_{RS} \\ 1 & 0 & 0 & z_{RS} \\ 0 & 0 & 0 & 1 \end{bmatrix} \tag{1}$$

where x_{RS}; y_{RS}; z_{RS} are the coordinates of the translation vector $\overrightarrow{O_{sp}R}$ between the sphere frame and the table frame.

The translation vector $\overrightarrow{O_{sp}R}$ is determined by calculating the position of the rotation center of R in the sphere frame. The tilt frame of the machine (S; x_{tl}; y_{tl}; z_{tl}) is linked to the tilt. It is obtained from the rotation around y_{tb} of the table frame of −90°. The transformation matrix of the table frame to the tilt frame is defined by $M_{setup}(A)$ in Eq. (1), where x_{RS}; y_{RS}; z_{RS} are the coordinates of the translation vector $\overrightarrow{RS}$ between the table frame and the tilt frame. The translation vector $\overrightarrow{RS}$ is identified by the fixed parameters of the machine tool. For the Mikron UCP 710: $x_{RS} = -119.995$ mm, $y_{RS} = -49.994$ mm, $z_{RS} = 0$ mm. The three transformation matrices X, Y, Z are the identity matrices (2):

$$M_{setup}(X) = M_{setup}(Y) = M_{setup}(Z) = \begin{bmatrix} 1 & 0 & 0 & 0 \\ 0 & 1 & 0 & 0 \\ 0 & 0 & 1 & 0 \\ 0 & 0 & 0 & 1 \end{bmatrix} \tag{2}$$

3.2 Sensor Calibration in the 5-Axis Machine

The sensor calibration steps in the 5-axis machine are shown in Fig. 4:

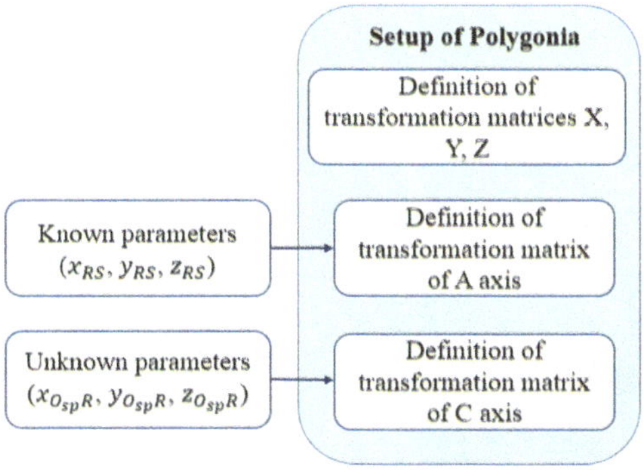

Fig. 4. Sensor calibration in the 5-axis machine

In fact, we know the values of (x_{RS}; y_{RS}; z_{RS}) which are identified by the fixed parameters of the machine tool and the three transformation matrices X, Y, Z are the identity matrices. It is therefore necessary to determine the coordinates of the rotation center in the sphere frame defined by (x_{OspR}; y_{OspR}; z_{OspR}). For this, it is necessary to perform a calibration by defining a 3-axis machine.

- Calibration on 3-axis X, Y, Z of the machine
  - Definition of the rotation matrix and translation vector for the X,Y,Z axis in Polygonia's *Setup.exe* using Eq. (2).
  - According to the Kreon procedure, the 3-axis calibration of the sensor is carried out from the digitization of a sphere with the diameter of 15.988 mm by a configuration (A = 0°, C = 0°) of the machine. Each new scan on the sphere is performed with the same machine position, but at different scanning distances d (Fig. 5). This makes it possible to change the relative position of the sphere center in the field of view of the laser plane. This 3-axis calibration result is used to determine the rotation center R of the table in the sphere frame.
- Determination of the rotation center of the table in the sphere frame.

Thanks to the result of the 3-axis calibration, it is possible to digitize the sphere with 8 positions of the table: (A = 0°, C = 0°), (A = 0°, C = 45°), (A = 0°, C = 90°), (A = 0°, C = 135°), (A = 0°, C = 180°), (A = 0°, C = 225°), (A = 0°, C = 270°), (A = 0°, C = 315°). We then obtain 8 point clouds representing the sphere for the 8 positions of the table (Fig. 6). From the point clouds, and the known radius of the sphere, we can find the 8 sphere centers for these 8 positions. We determine the circle passing through the 8 sphere centers by applying the least squares method. The coordinates of the center of the table (C_x; C_y; C_z) in the sphere frame represent the coordinates of the center of this circle. The coordinates (x_{OspR}; y_{OspR}; z_{OspR}) of the translation vector $\overrightarrow{O_{sp}R}$ in the sphere frame are then defined by (C_x; C_y; $-d_{sp-tb}$) where $-d_{sp-tb}$ is the distance from the center of the sphere to the top plane of the table. To find this distance, we scan the upper plane of the table, d_{sp-tb} is the distance along Z between the least squares plane passing through the point cloud of the upper plane of the table and the sphere center.

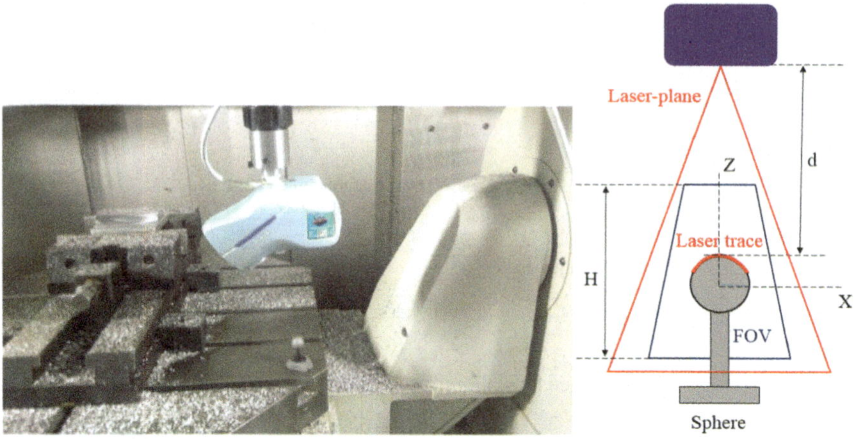

Fig. 5. Integration of laser-plane sensor in the machine tool.

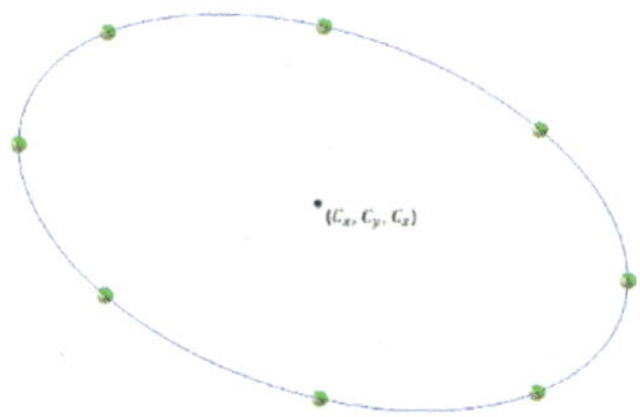

Fig. 6. Circle passing through the 8 measured centers of the sphere.

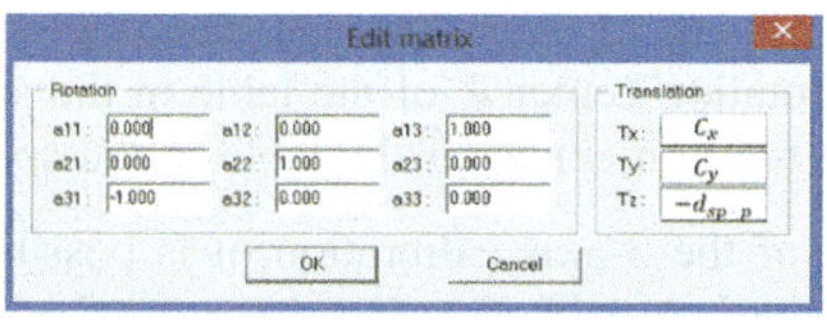

Fig. 7. Configuration of the C-axis in Polygonia's Setup.exe.

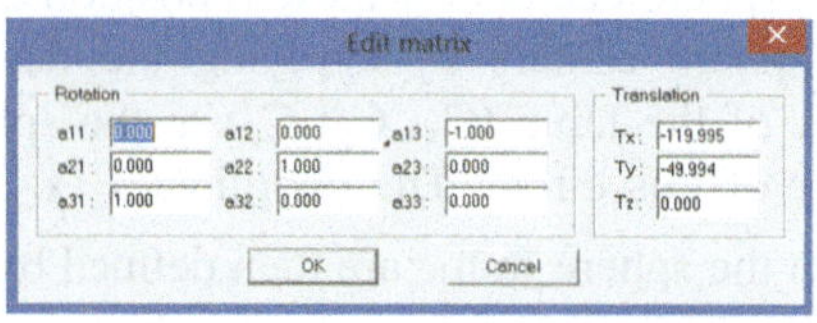

Fig. 8. Configuration of the A-axis in Polygonia's Setup.exe.

The setting parameters in Polygonia's Setup.exe for 5-axis CAXYZ is finally done by defining the rotation matrix and translation vector for the X, Y, Z, C, A axis (Figs. 7 and 8) in Polygonia's Setup.exe from the above obtained values.

4 Conclusion

This presented paper proposes the calibration procedure of laser-plane sensor in 5-axis machine tool. Thanks to this procedure, we can determine the orientation and position of the sensor in relation to the machine frame, and define the origin of the sensor frame in which the digitized points will be expressed. Future works concern the assessment of the scan path for on-machine measurement using a laser-plane in the 5-axis machine tool.

References

1. Phan, N.D.M., Quinsat, Y., Lartigue, C.: Defining scanning trajectory for on-machine inspection using a laser-plane scanner. In: Advances on Mechanics, Design Engineering and Manufacturing II, pp. 349–358. Springer (2019)
2. Quinsat, Y.: Apport de la mesure in-situ pour la maîtrise de la qualité des surfaces fabriquées. HDR thesis, ENS, Cachan (2016)
3. Dubreuil, L.: Mesure in-situ par moyensoptiques. Ph.D. thesis, Université Paris-Saclay (2017)

Classifying 3D Models Based on Transcending Local Features

Nguyen Van Tao and Nong Thi Hoa(✉)

Thai Nguyen University of Information Technology and Communication, Thai Nguyen, Vietnam
{nvtao,nthoa}@ictu.edu.vn

Abstract. Classifying 3D models is an essential task for 3D industry as it organizes objects according to categories, which helps searching 3D models perform more quickly. Many features and a suitable classifier have been used to improve classifying. However, it takes a long time for both extracting features and classifying 3D models. In this paper, a small set of transcending local features which are maximum distances from points in local regions to the center of 3D model is proposed. Then a Support Vector Machine is selected to classify 3D models based on data type of features and advantages of Support Vector Machine. The experiments are conducted on benchmark databases in Shape Retrieval Contest 2010. The results show that the approach employed to classify 3D models in an acceptable responding time of real applications is effective.

Keywords: Classifying 3D model · Extract feature · Support vector machine · 3D model · Computer vision

1 Introduction

Nowadays, 3D models have significantly increased because of the development of 3D softwares. Therefore, classifying 3D models in large databases is an important task to speed the search 3D model. Many studies of classifying 3D model have been proposed in recent years. However, short-time classification algorithms still remain a challenge because it is difficult to find a compact set of features that represents different 3D model classes. A 3D model classification algorithm usually consists of two steps, including extracting features and classifying new models. The key task of these algorithms is to find a proper set of features that shows the difference of many classes. Moreover, the smaller the number of feature is, the shorter the time for computing is. Most previous studies used many complex features to classify [1–4, 6, 7], which lead to time-consuming computing processes. In this study, a small set of transcending local features was proposed and a SVM classifier was used to decrease time as well as classify 3D models effectively. The proposed features were maximum distances from points in local regions to the center of 3D model. Then, a Support Vector Machine (SVM) [12] was selected to classify 3D models. The experiments were conducted on benchmark databases in Shape Retrieval Contest 2010 and the results show that our method has brought an effective classification in an acceptable responding time of real applications.

K.-U. Sattler et al. (Eds.): ICERA 2019, LNNS 104, pp. 196–201, 2020.
https://doi.org/10.1007/978-3-030-37497-6_23

2 Related Works

These studies can be divided into two main categories, including (i) proposing both new features and a new classifier, and (ii) selecting features from descriptors and using a available classifier.

The studies on the first class effectively improve the accuracy of classifying. Cao et al. [1] introduced a new method for classifying 3D objects by projecting a 3D object onto a spherical domain and developing a single neural network to classify the spherical projection. Features included depth variations, and contour-information viewed from different angles. Spherical projections encoded complete information of a 3D object in a single neural network capturing dependencies across different views. Shu et al. [2] proposed a new feature descriptor named Principal Thickness Images (PTI). PTI encoded the boundary surface and the voxelized constituents into three gray-scale images. They used histogram of oriented gradients to extract a feature vector for each PTI. This study extended the kernel sparse representation-based classification from 2D case to non-rigid 3D models to classify 3D models. Barra and Biasotti [4] proposed an effective 3D shape classification method that selected the most significant shape features from a big feature set by a supervised learning method. They used attributed extended Reeb graphs to encode different shape characteristics. The similarity between two graphs is addressed based on shortest paths, and the definition of kernels adapted to these descriptions. A multiple kernel learning algorithm is used to search an optimal linear combination of kernels for classification. Gao et el. [5] proposed a hypergraph analysis by constructing multiple hypergraphs for 3D objects based on their 2D views. In these hypergraphs, each vertex was an object, and each edge was a cluster of views. Therefore, an edge connected multiple vertices. They defined the weight of each edge based on the similarities between two views within a cluster. Recognition was performed based on the hypergraphs to explore the higher order relationship among objects.

The studies on the second category attract scientists', so many papers relating to the second category were published. Biao [3] used Label Propagation method based on multi-graph fusion that obtained from depth-buffer images and transformation (2D Polar-Fourier, 2D Zernike moments, and 2D Krawthcouk moments). Kassimi et al. [7] used semantic and Ontology to classify based on volume, surface area, diameter, Ferret diameter, radii, main axis, plan, equivalent spherical diameter. Foliguet et al. [6] used SVM based on Cord histograms, Extended Gaussian images, Complex Extended Gaussian Image, 3D Hough transform. Zongmin et al. [8] used SVM based on features obtaining from Kernel Principal Component Analysis. Yuesheng et al. [9] used SVM based on 2D histogram of distances and angles formed by pairs of oriented points on the surface. Siddiqi et al. [10] used medial surfaces to represent symmetries of 3D objects by a directed acyclic graph of components and a degree of invariance of parts. They classified by formulating the geometric information associated with each node along with an eigenvalue labeling of the adjacency matrix of the subgraph rooted at that node. The complex of features consumes time for computing and it is difficult to understand them.

3 New Approach

3.1 Extracting Features

First, the number of local regions of 3D models was chosen based on structure of 3D models. Then, the objects were observed to find transcending features. 20 features were recommended to use because human can recognize objects by several transcending features. Next, we divide a 3D model into local regions according to number of points of this model, each of which has the same number of points. Extracting features is done in each local regions. A feature vector of a 3D model contains values extracting from the first region to the last region. Each feature is the maximum distance from all points in a local region to the center of 3D model.

Steps for extracting features include:

Step 1: Computing the center of the 3D model. The center of a 3D model is the average position of all points.
Step 2: Formulating the transcending feature in each local region by searching the maximum distance from all points to the center of 3D model. Euclidean distance is used in computing. Values extracting from the first region to the last region form the feature vector.

3.2 Advantages of Proposed Features

The proposed features possess the three following advantages:

1. They do not depend on transformation styles of 3D models, which means 3D models can be rotated or deformed in a small level, and these features still represent many classes of objects.
2. The number of transcending features is selected based on structure of studied objects. If the number of features is small, the time for computing decreases.
3. The proposed features are easy to understand and simple to compute.

4 Experiment Results

The database of Shape Retrieval Contest 2010 (SHREC10) described as Three-dimensional nonrigid shapes in a variety of poses for non-rigid shape similarity and correspondence experiments. [11] was used in the experiments. The database contains a total of 148 objects, including 9 cats, 11 dogs, 3 wolves, 17 horses, 15 lions, 21 gorillas, 6 centaurs, 6 seahorses, 1 shark, 24 female figures, and two different male figures, containing 15 and 20 poses. 120 objects were used because these objects have the same number of points. As a result, the number of features is the same for all 3D objects.

Figure 1 shows 3D models using in experiments. Each 3D model contains 3,400 points and it can be seen that each object contains many poses. The structure of those objects is similar; thus, it is difficult to find the best feature vector of each class. 11 transcending features were applied for this data.

Fig. 1. All objects in experiments.

Based on the advantages of SVM and data type of features, a non-linear SVM with Radial Basis Function was chosen to classify. We conducted tests by using hold-out triplet data and running SVM in one-against-one mode. Each dataset is randomly divided to three subsets (training set, testing set, and validating set). The testing set and the validating set were used to test. We ran 10 times for each test. The accuracy is the percentage of patterns which is correctly classified.

4.1 Experiment 1: Classifying One Set of Human and Horse

The data is divided into two parts including 80% for training and 20% for testing. Classifying one of human and Horse includes 15 poses of David, 20 poses of Micheal, 25 poses of Victoria, and 17 poses of Horse. Table 1 presents the results of this test.

Table 1. The percentages of classifying one of human and horse

Classifying data	David-Horse	Michael-Horse	Victoria_Horse
Average	**93.33**	**92.50**	**93.75**

The data in Table 1 show that the lowest accuracy is 92.5%, which indicates that SVM working on proposed features classifies effectively.

4.2 Experiment 2: Classifying Poses of Human

The training data contain 80% patterns and the testing data include 20% patterns. The number of poses consists of 28 standing, 14 sitting, and 6 lying. Table 2 presents the results of classifying.

Table 2. The percentages of classifying poses of human

Classifying data	Standing-Other	Sitting-Other	Lying-Other
Average	86.00	95.00	96.00

The data in Table 2 show that the highest value is 96% and the lowest one is 86%. Classifying standing and other poses obtains lower values because an 11-item feature vector is difficult to represent for 25 poses of standing. Table 2 shows that SVM classifies effectively.

As can be seen in Tables 1 and 2, the accuracy of classifying is high (96%, 95%, 93.75%) if 3D models represent a typical structure of each class clearly.

4.3 Experiment 3: Time for Extracting Features

We ran the code of extracting features in C on Windows 10, CPU Intel Core i3. Features are stored in a text file. The time for extracting features includes the time of compiling and the time of running. The number of seconds is the measure of time. 3D models have 50,000 points and 3,400 points to test the change of time when computing on high-quality 3D models. We also extract 11 and 50 features to check the increase of time when increasing number of features. The time for extracting features is shown in Table 3.

Table 3. Time(s) for extracting features

Number of points	50,000	3,400
11 features	0.34486	0.2173
50 features	0.31895	0.19128

The data in Table 3 show that the number of points increases 17 times and the time increases approximately 1.6 times. The time for computing increases steadily when there is a sharp increase in the number of points of 3D models. Moreover, this time decreases when increasing the number of features. Therefore, the time for extracting features in an acceptable responding time of real applications.

5 Conclusion

In this paper, a small set of transcending local features was introduced and a SVM classifier was used to decrease the time for computing as well as classifying 3D models effectively. The proposed features were maximum distances from points in local regions to the center of 3D model. The number of local regions was selected based on structure of 3D models. The extracting features used simple math operators. We select SVM to classify 3D models based on data type of features, and the advantages of SVM. The experiments were conducted on benchmark databases in Shape Retrieval Contest 2010. The accuracy of classifying and time for extracting features show that our approach works effectively in an acceptable time.

Acknowledgment. This work has received support from the project "T2019-07-09" funded by Thai Nguyen University of Information Technology and Communication, Vietnam.

References

1. Cao, Z., Huang, Q., Karthik, R.: 3D object classification via spherical projections. In: International Conference on 3D Vision, pp. 566–574 (2017)
2. Shu, Z., Xin, S., Xu, H., Kavan, L., Wang, P., Liu, L.: 3D model classification via Principal Thickness Images. Comput. Aided Des. **78**, 199–208 (2016)
3. Leng, B., Du, C., Guo, S., Zhang, X., Xiong, Z.: A powerful 3D model classification mechanism based on fusing multi-graph. Neurocomputing **168**, 761–769 (2015)
4. Barra, V., Biasotti, S.: 3D shape retrieval and classification using multiple kernel learning on extended Reeb graphs. Vis. Comput. **30**(11), 1247–1259 (2014)
5. Gao, Y., Wang, M., Tao, D., Ji, R., Dai, Q.: 3-D object retrieval and recognition with hypergraph analysis. IEEE Trans. Image Process. **21**(9), 4290–4303 (2012)
6. Foliguet, S.P., Jordan, M., Najman, L., Cousty, J.: Artwork 3D model database indexing and classification. Pattern Recogn. **44**, 588–597 (2011)
7. Kassimi, A., Beqqali, O.E.: 3D model classification and retrieval based on semantic and ontology. Int. J. Comput. Sci. **8**(5), 108 (2011)
8. Li, Z., Wang, D., Li, B., Zhong, L.: 3D model classification using salient features for content representation. In: Sixth International Conference on Natural Computation, vol. 7, pp. 3541–3545 (2010)
9. He, Y., Tang, Y.Y.: Classification of 3D models for the 3D animation environments. In: IEEE International Conference on Systems, Man and Cybernetics, pp. 3786–3791 (2009)
10. Siddiqi, K., Zhang, J., Macrini, D., Shokoufandeh, A., Bouix, S., Dickinson, S.: Retrieving articulated 3-D models using medial surfaces. Mach. Vis. Appl. **19**, 261–275 (2008)
11. Bronstein, A.M., Bronstein, M.M., Kimmel, R.: Calculus of non-rigid surfaces for geometry and texture manipulation. IEEE Trans. Vis. Comput. Graph. **13**(5), 902–913 (2007)
12. Cortes, C., Vapnik, V.: Support vector networks. Mach. Learn. **20**(3), 273–297 (1995)

Control Parallel Robots Driven by DC Motors Using Fuzzy Sliding Mode Controller and Optimizing Parameters by Genetic Algorithm

Vu Duc Vuong[1,2(✉)], Nguyen Quang Hoang[1], and Nguyen Tien Duy[2]

[1] Hanoi University of Science and Technology, Hanoi, Vietnam
vuducvuong@tnut.edu.vn
[2] Thai Nguyen University of Technology, Thai Nguyen, Vietnam

Abstract. Parallel robots are widely used because of many acquired advantages such as rigidity, high accuracy and small dynamic link weight due to their structure as closed-loop multibody systems. However, this also leads to many difficulties in controlling the robot and requires intensive research on robot control strategies and reasonable control factors to ensure optimal control quality. This paper presents the way to control parallel robots driven by DC motors using Fuzzy Sliding Mode Controller and optimizing parameters by genetic algorithm to achieve the best optimized quality control as well as overcome the disadvantages of the basic sliding mode controller. Numerical simulations are carried out on a 3RRR parallel robot model to confirm the feasibility and effectiveness of the proposed method.

Keywords: Parallel robot · Modeling · Sliding mode controller · Fuzzy sliding mode controller · Genetic algorithm

1 Introduction

Parallel robots are commonly used in industry as well as in health due to their advantages such as rigidity, high accuracy and small dynamic link weight. In previous studies, parallel robots are often modeled in the form of a multi-object system with a loop structure using Lagrange equations, Newton-Euler equations, virtual principles, Jourdain principles or Kane equations [1–7]. Although the structure in this form offers many advantages, it leads to many difficulties in the control process, requiring in-depth research on the robot control strategy and determining reasonable control factors to ensure optimized control quality. In recent years, many control methods have been researched by scientists, including precise linearization control, gravity compensating PD control, Adaptive control, optimal control, sliding control, fuzzy control, etc. Adaptive control is a robotic control method that differs from conventional control methods in a way that the controller's parameters calibrated based on the signals in the closed loop change over time [8–10]. This control method is also suitable when the dynamic parameters of the robot are incorrect [11]. Controlled by precise linearization, the calculation controller is based on the results of the reverse dynamic problem [11].

K.-U. Sattler et al. (Eds.): ICERA 2019, LNNS 104, pp. 202–214, 2020.
https://doi.org/10.1007/978-3-030-37497-6_24

It is necessary to know how to use the dynamic parameters of the model exactly [12]. With the optimal control, for each robot, many different types of stable controls can be developed, but a set of control parameters needs to be found to ensure the optimal function and model parameters are also understood clearly [11].

Sliding control is a suitable control method for nonlinear systems such as parallel robots. The advantage of this method is its stability even when the system is noisy and the parameters are not predicted correctly. However, a major disadvantage is the phenomenon of chattering [13–16] which negatively affects the mechanical structure of the system. In addition, the controllers studied by the authors are mostly based on Differential Algebraic Equations (DAEs) that describe the motion of the robot's links. For each degree of freedom of parallel robots, an independent driven source is needed. Their influence in the motion of the system should be described directly in the dynamic equation. This paper presents the combination of the sliding controller and the fuzzy controller for modeled parallel robot control including a DC motor that drives active joints. The fuzzy component in the controller has an effect on overcoming the disadvantages while still promoting the advantages of the sliding controller. Furthermore, determining the controller parameters with complex nonlinear objects like parallel robots is difficult and does not have general guidelines. In order to optimize the control quality for the proposed controller, genetic algorithm (GA) is used to provide the optimal parameters. Numerical simulations are performed on a 3RRR parallel robot model to confirm the feasibility and effectiveness of the proposed method.

The rest of this paper is organized as follows: Sect. 2 presents a dynamical model of parallel manipulator actuated by electric motors and the transformation of the differential equations into the minimal form in space of operational coordinates. The sliding mode controller is described in Sect. 3 and Fuzzy sliding mode controller with optimized parameters by Genetic Algorithm is described in Sect. 4. Numerical simulations are shown in Sect. 5. Finally, the conclusion is given in Sect. 6.

2 Dynamical Model

2.1 Motion Equations of a Parallel Robot Driven by Electric Motors

In this section, the parallel robot having n degrees of freedom driven by n electric motors is considered as shown in Fig. 1. The system dynamic model is derived by applying Lagrange equation with multipliers and the substructure method. The Equations of motion of the systems are described by the following equations [2]:

$$\begin{aligned}(\mathbf{M}(\mathbf{q}) + \mathbf{B}\mathbf{J}_m r^2\mathbf{Z})\ddot{\mathbf{q}} + \mathbf{C}(\mathbf{q},\dot{\mathbf{q}})\dot{\mathbf{q}} + \mathbf{g}(\mathbf{q}) \\ + (\mathbf{D}+\mathbf{B}(\mathbf{D}_m+\mathbf{K}_m\mathbf{R}_a^{-1}\mathbf{K}_e)r^2\mathbf{Z})\dot{\mathbf{q}} = \mathbf{B}\mathbf{K}_m\mathbf{R}_a^{-1}r\mathbf{u} + \Phi^T(\mathbf{q})\boldsymbol{\lambda} \\ \phi(q) = 0\end{aligned} \tag{1}$$

The Eq. (1) is rewritten in compact form as

$$\mathbf{M}_s(\mathbf{q})\ddot{\mathbf{q}} + \mathbf{C}_s(\mathbf{q},\dot{\mathbf{q}})\dot{\mathbf{q}} + \mathbf{D}_s\dot{\mathbf{q}} + \mathbf{g}_s(\mathbf{q}) = \mathbf{B}_s\mathbf{u} + \mathbf{\Phi}^T(\mathbf{q})\boldsymbol{\lambda}, \tag{2}$$

$$\boldsymbol{\phi}(\boldsymbol{q}) = 0 \tag{3}$$

The details of the components in the equations are presented in the documents [17–19].

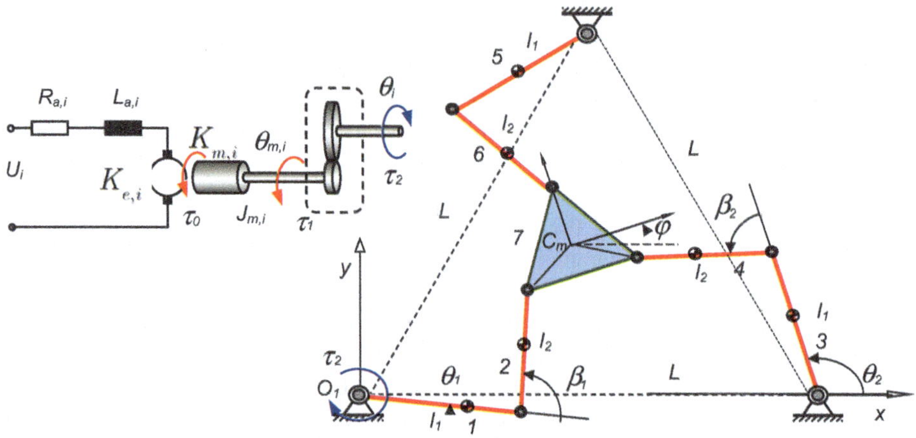

Fig. 1. Model of a 3RRR parallel planar robot with an electric motor gearbox

2.2 Equations of Motion in a Minimal Coordinate Form

The equations of motion in a redundant generalized coordinates form are convenient for deriving it as well as simulating. To design a controller for robots, the equations need to be transformed to a minimal coordinate form. The choice of the minimal coordination depends on the purpose of designing the controller in the joint space or the task space. To make transform in this section, it is assumed that Jacobi matrix $\mathbf{\Phi}(\mathbf{q}) = \partial\boldsymbol{\phi}/\partial\mathbf{q}$ has a full rank. It means that the robot does not move through singular configurations. By differentiating constraint functions (3) with respect to time (w.r.t.), one gets the velocity level constraints form as:

$$\mathbf{\Phi}_q(\mathbf{q})\dot{\mathbf{q}} = \mathbf{\Phi}_{q_i}(\mathbf{q})\dot{\mathbf{q}}_i + \mathbf{\Phi}_{q_d}(\mathbf{q})\dot{\mathbf{q}}_d = 0,\ \mathbf{q}_i = \boldsymbol{\theta},\ \mathbf{q}_d = [\boldsymbol{\beta}^T, \boldsymbol{x}^T]^T \tag{4}$$

From the above equation, we get the relationship between the independent and dependent velocity.

$$\dot{\mathbf{q}} = \begin{bmatrix} \dot{\mathbf{q}}_i \\ \dot{\mathbf{q}}_d \end{bmatrix} = \begin{bmatrix} \boldsymbol{E} \\ -\mathbf{\Phi}_{q_d}^{-1}(\mathbf{q})\mathbf{\Phi}_{q_i}(\mathbf{q}) \end{bmatrix} \dot{\mathbf{q}}_i = \boldsymbol{R}\dot{\mathbf{q}}_i \tag{5}$$

where E is the unit matrix $\boldsymbol{R} = [E, -[\boldsymbol{\Phi}_{q_d}^{-1}(\mathbf{q})\boldsymbol{\Phi}_{q_i}(\mathbf{q})]^T]^T$. By differentiating (5) w.r.t. one gets

$$\ddot{\mathbf{q}} = \boldsymbol{R}\ddot{\mathbf{q}}_i + \dot{\boldsymbol{R}}\dot{\mathbf{q}}_i \tag{6}$$

It is noted that matrix $\boldsymbol{R}$ satisfies the property $\boldsymbol{\Phi}(\mathbf{q})\boldsymbol{R} = 0$ or $\boldsymbol{R}^T\boldsymbol{\Phi}^T(\mathbf{q}) = 0$. To eliminate Lagrange multipliers in differential equations of motion, premultiplying it from link with matrix $\boldsymbol{R}^T$ ones get.

$$\boldsymbol{R}^T\mathbf{M}_s(\mathbf{q})\ddot{\mathbf{q}} + \boldsymbol{R}^T[\mathbf{C}_s(\mathbf{q},\dot{\mathbf{q}})\dot{\mathbf{q}} + \mathbf{D}_s\dot{\mathbf{q}} + \mathbf{g}_s(\mathbf{q})] = \boldsymbol{R}^T\mathbf{B}_s\mathbf{u} \tag{7}$$

By substituting $\dot{\mathbf{q}}, \ddot{\mathbf{q}}$ from (5) and (6) into (7), one gets the equations of motion in a minimal coordinate form:

$$\boldsymbol{R}^T\mathbf{M}_s(\mathbf{q})\boldsymbol{R}\ddot{\mathbf{q}}_i + \boldsymbol{R}^T\left[\mathbf{M}_s(\mathbf{q})\dot{\boldsymbol{R}} + \mathbf{C}_s(\mathbf{q},\dot{\mathbf{q}})\boldsymbol{R}\right]\dot{\mathbf{q}}_i + \boldsymbol{R}^T\mathbf{D}_s\boldsymbol{R}\dot{\mathbf{q}}_i + \boldsymbol{R}^T\mathbf{g}_s(\mathbf{q}) = \boldsymbol{R}^T\mathbf{B}_s\mathbf{u} \tag{8}$$

By defining the following matrices

$$\mathbf{M}_{q_i}(\mathbf{q}_i) = \boldsymbol{R}^T\mathbf{M}_s(\mathbf{q})\boldsymbol{R}, \mathbf{C}_{q_i}(\mathbf{q}_i,\dot{\mathbf{q}}_i) = \boldsymbol{R}^T\left[\mathbf{M}_s(\mathbf{q})\dot{\boldsymbol{R}} + \mathbf{C}_s(\mathbf{q},\dot{\mathbf{q}})\boldsymbol{R}\right],$$
$$\mathbf{D}_{q_i} = \boldsymbol{R}^T\mathbf{D}_s\boldsymbol{R}, \mathbf{g}_{q_i}(\mathbf{q}_i) = \boldsymbol{R}^T\mathbf{g}_s(\mathbf{q})$$

The differential equations of motion in the dependent coordinates form $\mathbf{q}_i$ is rewritten as follows

$$\mathbf{M}_{q_i}(\mathbf{q}_i)\ddot{\mathbf{q}}_i + \mathbf{C}_{q_i}(\mathbf{q}_i,\dot{\mathbf{q}}_i)\dot{\mathbf{q}}_i + \mathbf{D}_{q_i}\dot{\mathbf{q}}_i + \mathbf{g}_{q_i}(\mathbf{q}_i) = \boldsymbol{\tau}_{q_i} \tag{9}$$

In the Eq. (9), the following properties are still guaranteed. $\mathbf{M}_{q_i}(\mathbf{q}_i)$ is a symmetric and positive and $N_{q_i} = \dot{\mathbf{M}}_{q_i}(\mathbf{q}_i) - 2\mathbf{C}_{q_i}(\mathbf{q}_i,\dot{\mathbf{q}}_i)$ is a skew-symmetric matrix. These properties are very important for control design shown in the following section.

3 Design of a Sliding Mode Controller in a Joint Space

The purpose of the controller design is to find the law of the motor voltage so that the motion of the mobile platform tracks the given trajectory. In this section, methods of control design in the active joint space of parallel robots are presented. To design the sliding mode controller, the sliding surface is chosen as follows:

$$s = \dot{\boldsymbol{e}}_{q_i}(t) + \boldsymbol{\Lambda}\boldsymbol{e}_{q_i}(t), \ \boldsymbol{\Lambda} = diag([\lambda_1, \lambda_2, \ldots, \lambda_n]) > 0 \tag{10}$$

with the tracking error $\boldsymbol{e}_{q_i}(t) = \boldsymbol{q}_i(t) - \boldsymbol{q}_i^d(t)$ and $\dot{\boldsymbol{e}}_{q_i}(t) = \dot{\boldsymbol{q}}_i(t) - \dot{\boldsymbol{q}}_i^d(t)$.

By substituting in to Eq. (10), ones get:

$$\boldsymbol{s} = \dot{\boldsymbol{q}}_i(t) - \dot{\boldsymbol{q}}_i^d(t) + \boldsymbol{\Lambda}\boldsymbol{e}_{q_i}(t) \tag{11}$$

Defining $\dot{\boldsymbol{q}}_i^r(t) = \dot{\boldsymbol{q}}_i^d(t) - \boldsymbol{\Lambda}\boldsymbol{e}_{q_i}(t)$ ones gets

$$\boldsymbol{s} = \dot{\boldsymbol{q}}_i(t) - \dot{\boldsymbol{q}}_i^r(t),\ \dot{\boldsymbol{s}} = \ddot{\boldsymbol{q}}_i(t) - \ddot{\boldsymbol{q}}_i^r(t) \tag{12}$$

Obviously, by putting sliding surfaces $\boldsymbol{s} \to 0$ the solution $\boldsymbol{e}_{q_i}(t)$ of equation will be exponential with the negative exponents form. So, with respect to time $\boldsymbol{e}_{q_i}(t) \to 0$ and so $\boldsymbol{q}_i(t) \to \boldsymbol{q}_i^d(t)$, the mobile platform tracks the given trajectory.

To consider the stability of the system, a Lyapunov candidate function is chosen:

$$V = \frac{1}{2}\boldsymbol{s}^T\boldsymbol{M}_{q_i}(\boldsymbol{q}_i)\boldsymbol{s} \tag{13}$$

Differentiating Lyapunov function w.r.t. time, one yields:

$$\dot{V} = \boldsymbol{s}^T\boldsymbol{M}_{q_i}(\boldsymbol{q}_i)\dot{\boldsymbol{s}} + \frac{1}{2}\boldsymbol{s}^T\dot{\boldsymbol{M}}_{q_i}(\boldsymbol{q}_i)\boldsymbol{s} \tag{14}$$

From Eqs. (9), (11) and the skew-symmetric property of matrix $\boldsymbol{N}_{q_i} = \dot{\mathbf{M}}_{q_i}(\mathbf{q}_i) - 2\mathbf{C}_{q_i}(\mathbf{q}_i, \dot{\mathbf{q}}_i)$ ones obtains

$$\dot{V} = \boldsymbol{s}^T[\boldsymbol{\tau}_{q_i} - \boldsymbol{C}_{q_i}(\boldsymbol{q}_i, \dot{\boldsymbol{q}}_i)\dot{\boldsymbol{q}}_i^r - \boldsymbol{D}_{q_i}\dot{\boldsymbol{q}}_i - \boldsymbol{g}_{q_i}(\boldsymbol{q}_i) - \boldsymbol{M}_{q_i}(\boldsymbol{q}_i)\ddot{\boldsymbol{q}}_i^r] \tag{15}$$

The formula (15) suggests choosing a control law as follows:

$$\boldsymbol{\tau}_{q_i} = \boldsymbol{\tau}_{eq} + \boldsymbol{\tau}_{smc} \tag{16}$$

Where

$$\boldsymbol{\tau}_{eq} = \hat{\boldsymbol{M}}_{q_i}(\boldsymbol{q}_i)\ddot{\boldsymbol{q}}_i^r + \hat{\boldsymbol{C}}_{q_i}(\boldsymbol{q}_i, \dot{\boldsymbol{q}}_i)\dot{\boldsymbol{q}}_i^r + \hat{\boldsymbol{D}}_{q_i}\dot{\boldsymbol{q}}_i^r + \hat{\boldsymbol{g}}_{q_i}(\boldsymbol{q}_i) \tag{17}$$

$$\boldsymbol{\tau}_{smc} = -\boldsymbol{K}_{pd}\boldsymbol{s} - \boldsymbol{K}_s\mathrm{sgn}(\boldsymbol{s}) \tag{18}$$

Where $\hat{\boldsymbol{M}}_{q_i}(\boldsymbol{q}), \hat{\boldsymbol{C}}_{q_i}(\boldsymbol{q}, \dot{\boldsymbol{q}}_i), \hat{\boldsymbol{D}}_{q_i}, \hat{\boldsymbol{g}}_{q_i}(\boldsymbol{q}_i)$ are the estimators of matrix and vector $\boldsymbol{M}_{q_i}(\boldsymbol{q}), \boldsymbol{C}_{q_i}(\boldsymbol{q}, \dot{\boldsymbol{q}}_i), \boldsymbol{D}_{q_i}, \boldsymbol{g}_{q_i}(\boldsymbol{q}_i)$, $\mathrm{sgn}(\boldsymbol{s}) = [\mathrm{sgn}(s_1), \mathrm{sgn}(s_2), \ldots, \mathrm{sgn}(s_n)]^T$, matrix $\mathbf{K}_{pd}$ and $\mathbf{K}_s$ are the diagonal form and the symmetric positive definite matrices, $\mathbf{K}_{pd} = \mathbf{K}_{pd}^T > 0$, $\mathbf{K}_s = \mathbf{K}_s^T > 0$. For simplicity, these two matrices are selected to be diagonal form as

$$\boldsymbol{K}_{pd} = diag[k_{pd}^{11}, k_{pd}^{22}, \ldots, k_{pd}^{nn}],\ \boldsymbol{K}_s = diag[k_s^{11}, k_s^{22}, \ldots, k_s^{nn}].$$

Substituting control law (16) into the dynamic model of robot, one gets

$$\begin{aligned} & \boldsymbol{M}_{q_i}(\boldsymbol{q})(\ddot{\boldsymbol{q}}_i - \ddot{\boldsymbol{q}}_i^r) + \boldsymbol{C}_{q_i}(\boldsymbol{q}, \dot{\boldsymbol{q}}_i)(\dot{\boldsymbol{q}}_i - \ddot{\boldsymbol{q}}_i^r) + \boldsymbol{D}_{q_i}(\dot{\boldsymbol{q}}_i - \ddot{\boldsymbol{q}}_i^r) \\ & = -\boldsymbol{K}_{pd}\boldsymbol{s} - \boldsymbol{K}_s \mathrm{sgn}(\boldsymbol{s}) + \tilde{\boldsymbol{M}}_{q_i}(\boldsymbol{q})\ddot{\boldsymbol{q}}_i^r + \tilde{\boldsymbol{C}}_{q_i}(\boldsymbol{q}, \dot{\boldsymbol{q}}_i)\dot{\boldsymbol{q}}_i^r + \tilde{\boldsymbol{D}}_{q_i}\dot{\boldsymbol{q}}_i^r + \tilde{\boldsymbol{g}}_{q_i}(\boldsymbol{q}) \end{aligned} \tag{19}$$

or

$$\boldsymbol{M}_{q_i}(\boldsymbol{q}_i)\dot{\boldsymbol{s}} + \boldsymbol{C}_{q_i}(\boldsymbol{q}_i, \dot{\boldsymbol{q}}_i)\boldsymbol{s} + \boldsymbol{D}_{q_i}\boldsymbol{s} = -\boldsymbol{K}_{pd}\boldsymbol{s} - \boldsymbol{K}_s \mathrm{sgn}(\boldsymbol{s}) + \boldsymbol{d}_{q_i}(\boldsymbol{q}_i, \dot{\boldsymbol{q}}_i, \boldsymbol{q}_i^r, \dot{\boldsymbol{q}}_i^r) \tag{20}$$

with the following symbols:

$$\begin{aligned} & \tilde{\boldsymbol{M}}_{q_i}(\boldsymbol{q}_i) = \hat{\boldsymbol{M}}_{q_i}(\boldsymbol{q}_i) - \boldsymbol{M}_{q_i}(\boldsymbol{q}_i), \;\; \tilde{\boldsymbol{C}}_{q_i}(\boldsymbol{q}_i, \dot{\boldsymbol{q}}_i) = \hat{\boldsymbol{C}}_{q_i}(\boldsymbol{q}_i, \dot{\boldsymbol{q}}_i) - \boldsymbol{C}_{q_i}(\boldsymbol{q}_i, \dot{\boldsymbol{q}}_i), \\ & \tilde{\boldsymbol{D}}_{q_i} = \hat{\boldsymbol{D}}_{q_i} - \boldsymbol{D}_{q_i}, \;\; \tilde{\boldsymbol{g}}_{q_i}(\boldsymbol{q}_i) = \hat{\boldsymbol{g}}_{q_i}(\boldsymbol{q}_i) - \boldsymbol{g}_{q_i}(\boldsymbol{q}_i), \\ & \boldsymbol{d}_{q_i}(\boldsymbol{q}_i, \dot{\boldsymbol{q}}_i, \boldsymbol{q}_i^r, \dot{\boldsymbol{q}}_i^r) = \tilde{\boldsymbol{M}}_{q_i}(\boldsymbol{q}_i)\ddot{\boldsymbol{q}}_i^r + \tilde{\boldsymbol{C}}_{q_i}(\boldsymbol{q}_i, \dot{\boldsymbol{q}}_i)\dot{\boldsymbol{q}}_i^r + \tilde{\boldsymbol{D}}_{q_i}\dot{\boldsymbol{q}}_i^r + \tilde{\boldsymbol{g}}_{q_i}(\boldsymbol{q}_i). \end{aligned}$$

From (20) it leads to:

$$\boldsymbol{M}_{q_i}(\boldsymbol{q}_i)\dot{\boldsymbol{s}} = -\boldsymbol{C}_{q_i}(\boldsymbol{q}_i, \dot{\boldsymbol{q}}_i)\boldsymbol{s} - \boldsymbol{D}_{q_i}\boldsymbol{s} - \boldsymbol{K}_{pd}\boldsymbol{s} - \boldsymbol{K}_s \mathrm{sgn}(\boldsymbol{s}) + \boldsymbol{d}_{q_i}(\boldsymbol{q}_i, \dot{\boldsymbol{q}}_i, \boldsymbol{q}_i^r, \dot{\boldsymbol{q}}_i^r) \tag{21}$$

Substituting (21) into (14) one obtains:

$$\begin{aligned} \dot{V} & = \boldsymbol{s}^T \boldsymbol{M}_{q_i}(\boldsymbol{q}_i)\dot{\boldsymbol{s}} + \frac{1}{2}\boldsymbol{s}^T \dot{\boldsymbol{M}}_{q_i}(\boldsymbol{q}_i)\boldsymbol{s} \\ & = -\boldsymbol{s}^T[\boldsymbol{K}_{pd} + \boldsymbol{D}_{q_i}]\boldsymbol{s} - \boldsymbol{s}^T[\boldsymbol{K}_s \mathrm{sgn}(\boldsymbol{s}) - \boldsymbol{d}_{q_i}] + \frac{1}{2}\boldsymbol{s}^T\left[\dot{\boldsymbol{M}}_{q_i}(\boldsymbol{q}_i) - 2\boldsymbol{C}_{q_i}(\boldsymbol{q}_i, \dot{\boldsymbol{q}}_i)\right]\boldsymbol{s} \\ & = -\boldsymbol{s}^T[\boldsymbol{K}_{pd} + \boldsymbol{D}_{q_i}]\boldsymbol{s} - \boldsymbol{s}^T[\boldsymbol{K}_s \mathrm{sgn}(\boldsymbol{s}) - \boldsymbol{d}_{q_i}] \end{aligned} \tag{22}$$

Assuming that the system uncertainties $\boldsymbol{d}_{q_i}$ are bounded, it means $\left|\boldsymbol{d}_{q_i}\right| \leq d_0$ or $|d_i| \leq d_0$ leads to $\sum_{i=1}^{n} s_i d_i \leq \sum_{i=1}^{n} |s_i| d_{i,0}$. The Eq. (22) is rewritten as follows:

$$\begin{aligned} \dot{V} & = -\boldsymbol{s}^T(\boldsymbol{K}_{pd} + \boldsymbol{D}_{q_i})\boldsymbol{s} - \boldsymbol{s}^T \boldsymbol{K}_s \mathrm{sgn}(\boldsymbol{s}) + \boldsymbol{s}^T \boldsymbol{d}_{q_i} \\ & = -\boldsymbol{s}^T(\boldsymbol{K}_{pd} + \boldsymbol{D}_{q_i})\boldsymbol{s} - \sum_{i=1}^{n} K_s^{(ii)}|s_i| + \sum_{i=1}^{n} s_i d_i \\ & \leq -\boldsymbol{s}^T(\boldsymbol{K}_{pd} + \boldsymbol{D}_{q_i})\boldsymbol{s} - \sum_{i=1}^{n}(K_s^{(ii)} - d_{i,0})|s_i| \end{aligned} \tag{23}$$

To be sure $\dot{V} \leq 0$, the elements of the matrix $\mathbf{K}_s$ will be chosen such that $K_s^{(ii)} - d_{i,0} > 0$ or $K_s^{(ii)} > d_{i,0} > 0$. However, control law (16) has the discontinuous terms $\boldsymbol{K}_s \mathrm{sgn}(\boldsymbol{s})$ so when working, unwanted oscillations with high frequencies will appear around the sliding surface and have amplitude depending on the magnitude of the elements in the matrix. This phenomenon is called "chattering". This is the phenomenon that the trajectory slides in zig-zac path on the sliding surface to the origin.

This phenomenon may cause damage to the mechanical structure which affects the quality of control. To overcome this phenomenon, the fuzzy solution of information causing chattering in the controller is presented in the next section.

4 Design of Sliding Fuzzy Controller Parameter Optimization by Genetic Algorithm

So far, fuzzy control has confirmed its important role in modern control techniques. Fuzzy control gives considerable accuracy and capable of performance because of the simplicity in the system's structure. Wide applications of fuzzy control appear in many areas such as control and automation systems in industry and military, transportation, structural control.

The advantage of fuzzy controller is that the designer does not need to know the mathematical model of the object. The fuzzy model is described for a nonlinear input/output relationship, so it can be said that the controller is adaptable to the input values and it is effective for nonlinear objects.

There are two typical fuzzy models: Mamdani and Sugeno fuzzy models. The Mamdani fuzzy model, which has the advantages of simplicity and visualization, is easy to design and implement, but limitations on the ability to describe complex systems. The output component is the right side of the rule system, which is also represented by fuzzy sets. After performing the integration operations, it is necessary to defocus on the result fuzzy set to get the control value at the output. On the other hand, in Sugeno fuzzy model [20, 21], the output is determined by a separate function. The entire output of the rule system will be a function space. The function $f_i([x])$ is usually a polynomial according to the inputs $[x]$ but it can also be any function, depending on the description of the system output. When $f_i([x])$ is a constant, it is called the zero-order Sugeno model; this is also a special form of Mamdani model when the output fuzzy sets are singleton. Zero-order Sugeno model is quite simple in both design and installation.

In the previous section, the traditional slide controller was designed. Its drawback is to create chattering. From Eq. (18), it can be seen that τ_{smc} causes the chattering phenomenon because of $-\boldsymbol{K}_s\mathrm{sgn}(\boldsymbol{s})$. To overcome this limitation, the sliding fuzzy controller is designed to replace $\mathrm{sgn}(\boldsymbol{s})$. By fuzzilizing the sliding surface with a delta interval, based on the magnitude of the state point relative to the sliding surface, the corresponding control can be achieved (see Fig. 2). The fuzzy model is used as the zero-order Sugeno type. The sliding surface is defined as in (10).

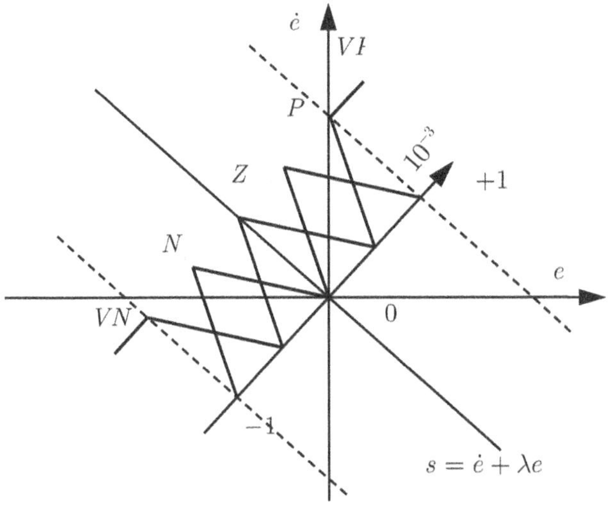

Fig. 2. Fuzzy set of variable *s*

The controller has input as surface s and the output is the control quantity u. The variable domain and fuzzy sets for the input variables are shown in Fig. 2.

In which *NB* – Negative Big, *N* – Negative, *ZE* – Zero, *P* – Positive, *PB* – Positive Big.

Output are fuzzy singleton sets defined in [−1, 1]:

$$NB = -1, N = -0.65, ZE = 0, P = 0.65, PB = 1$$

Control rule is presented in the following table:

Table 1. Rule of FSMC controller

s	*NB*	*N*	*ZE*	*P*	*PB*
u	*NB*	*N*	*ZE*	*P*	*PB*

Rule in Table 1 can be understood as follows:

$$If\ s = NB\ then\ u = NB;\ If\ s = N\ then\ u = N;\ If\ s = ZE\ then\ u = ZE;\ldots$$

Parameters Optimization by Genetic Algorithm

Genetic algorithms are the random optimized method that follows the evolution and selection of biological populations in nature [22]. Operations in Genetic algorithms include crossover, mutation and selection. Each individual is expressed simply as a chromosome consisting of multiple gene segments. Each gene segment is encoded for an optimal parameter. Then each individual is a solution of the problem with an optimal set of parameters. When an evolutionary process (repetition) is large enough, the individuals will gradually adapt to the adaptive conditions evaluated by the fitness function.

For optimization for parameters of sliding component control in the FSMC controller, these parameters include K_{pd}, λ, K_s with an objective function that was selected in standard of the absolute integral of the control deviation (IAE).

$$fitness = \sum_{k=1}^{n} |e_\theta(k)| \rightarrow \min \tag{24}$$

Where $e_\theta(k)$ is the deviation data sample at the simulation cycle k, n is the total data sample of one simulation running.

In Matlab environment, GA is a tool that we can use easily. In this study, we used function `ga(.)` with the coded gene in the real number that is type 'double'. The established values for GA include:

- `PopulationSize = 100;`
- `Generation = 3*PopulationSize;`
- `TimeLimit = 3600;`

The control scheme for parallel robot driven by DC motor is shown in Fig. 3.

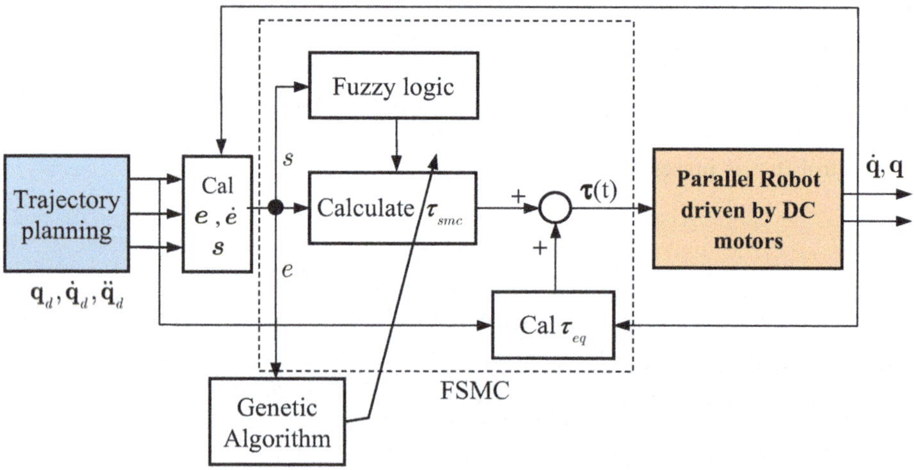

Fig. 3. Control scheme for parallel robot driven by DC motor using the FSMC controller with GA

The parameters obtained after calculation process using the gene algorithm were then used to establish the controller to perform simulation for operations of the parallel robot (3RRR) in next section.

5 Numerical Simulation

The parameters of robots are selected as follows for simulation [17, 18]:

$L_0 = 1.2$, $l_1 = 0.582$, $l_2 = 0.623$, $a = 0.32$, $b = 0.185$ [m], $m_1 = 2.072$, $m_2 = 0.750$, $m_3 = 0.978$ [kg]; $J_m = 0.0001$, $J_{C1} = 0.102723$, $J_{C3} = 0.0067192$ [kgm^2]; $d = 0.1$ [Nms/rad]; $K_m = 1.0$ [Nm/A]; $K_e = 1.0$ [Vs/rad]; $R_a = 1.00$ [Ohm]; $r = 10$.

In these simulations, the center of the platform will be moved along a circular trajectory, while its orientation is constant, $\varphi = 0[rad]$. The trajectory has a center at $(x_c; y_c) = (0.35; 0.4)$ and radius $r = 0.2$ [m].

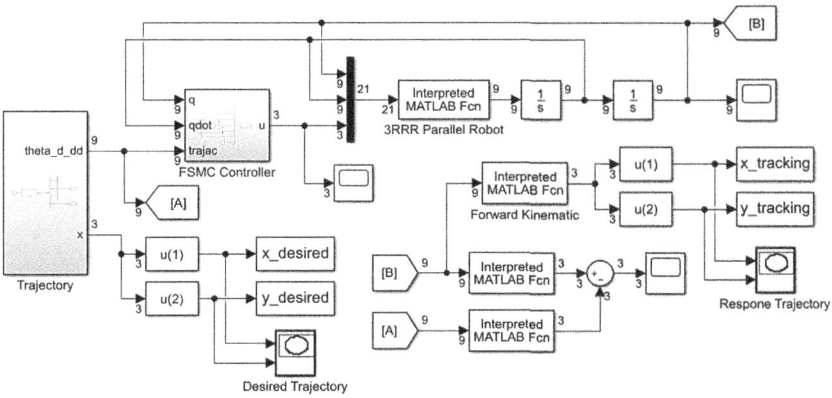

Fig. 4. Control simulation diagram of 3RRR robot with FSMC controller with GA

The simulation of controlling 3RRR parallel robot by FSMC controller on Simulink is shown in Fig. 4. On simulation diagram, "Trajectory" block is used to generate the position, velocity, and acceleration parameters of reference trajectory. The "FSMC Controller" block performs calculations based on the formulas presented above to provide control signals of parallel robot shown on the "3RRR Parallel Robot" block. Parallel robot model includes a DC motor that drives into active joints. The "Forward Kinematic" block is used to convert calculated information from joint space to task space. The information received from this block is used to draw the robot's orbit in the task space.

The simulations are performed with classic SMC controllers and FSMC controllers that suggest using fuzzy logic for overcome chattering and genetic algorithms to optimize parameters. The simulation results are given in the following figures (Fig. 9).

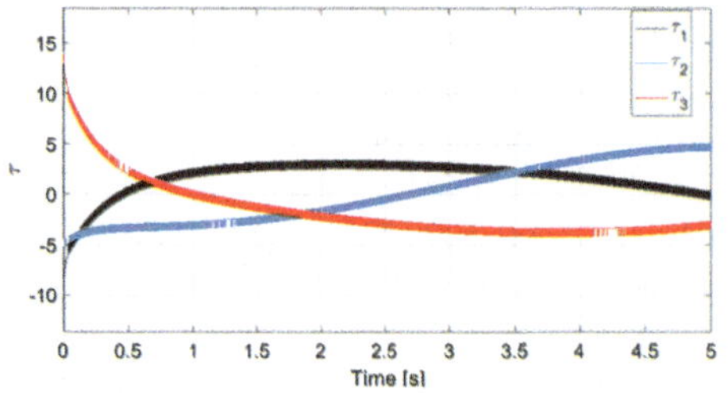

Fig. 5. Control signal with the classic sliding controller

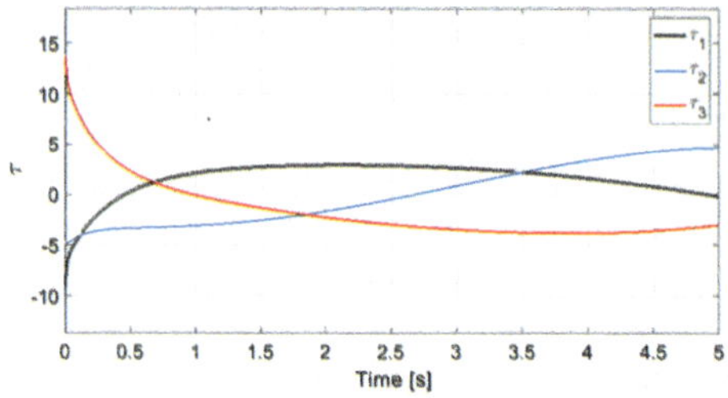

Fig. 6. Control signal with the FSMC controller

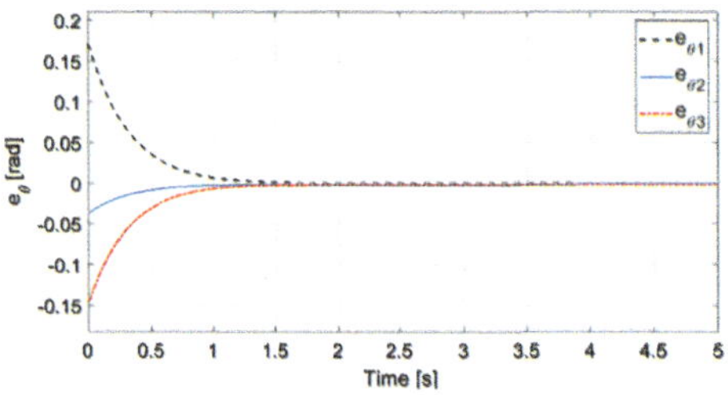

Fig. 7. Tracking errors of active joints using the FSMC controller

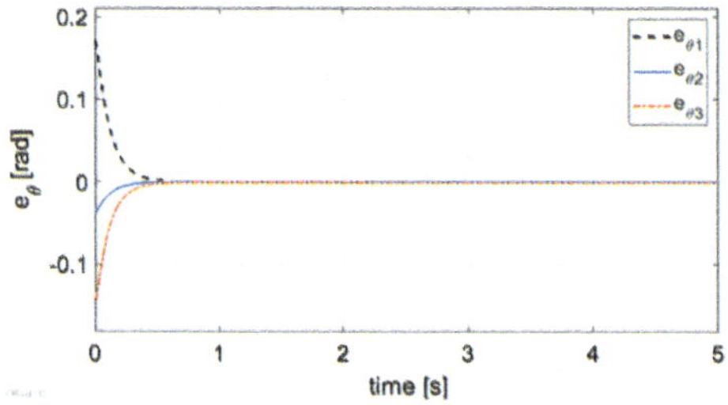

Fig. 8. Tracking errors of active joints using FSMC controller with GA

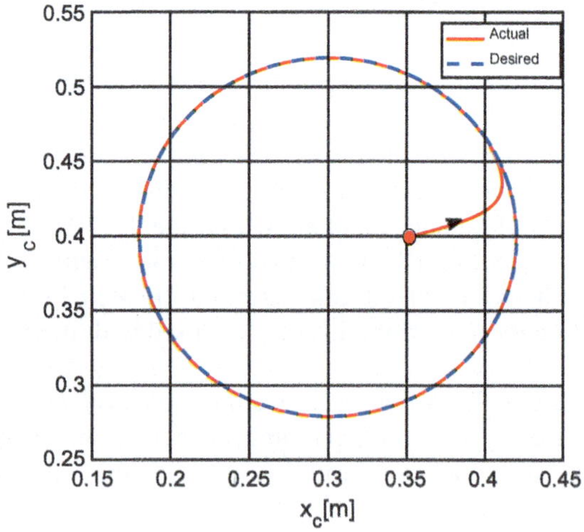

Fig. 9. The path tracked vs. the desired path with FSMC controller

The work that is fuzzy for information causing chattering in the sliding controller has been effective for the FSMC controller. The control signals are smooth curves (see Fig. 5), without shaking as in the control process using the classic sliding controller(see Fig. 4). In addition, the results in Figs. 6, 7 and 8 show that the parameters optimized by genetic algorithm give higher accuracy and the motion of the mobile platform tracks the desired trajectory after about 0.5 s.

6 Conclusion

The paper has successfully built a fuzzy sliding controller for parallel robots driven by DC motors. In particular, the fuzzy components have overcome the disadvantages of the chattering phenomenon of the sliding controller while maintaining the advantages in controlling complex nonlinear systems such as parallel robots. A dynamic model is built for control including components describing the influence of the driving source, so the simulation process is carried out closely to the actual operation of the robot. Genetic algorithms are used to provide optimal parameters for the controller. Numerical simulations are carried out on a 3RRR parallel robot model that confirms the feasibility and effectiveness of the proposed method.

References

1. De Jalon, J.G., Bayo, E.: Kinematic and Dynamic Simulation of Multibody Systems: The Real-Time Challenge. Springer, New York (2012)
2. Van Khang, N.: Dynamics of Multibody Systems. Science and Technology Publishing House, HaNoi (2007)
3. Tsai, L.-W.: Robot Analysis: the Mechanics of Serial and Parallel Manipulators. Wiley, Hoboken (1999)
4. Murray, R.M., Li, Z., Sastry, S.S., Sastry, S.S.: A Mathematical Introduction to Robotic Manipulation. CRC Press, Boca Raton (2017)
5. Angeles, J.: Fundamentals of Robotic Mechanical Systems, vol. 2. Springer, New York (2002)
6. Kane, T.R., Levinson, D.A.: Dynamics. Theory and Applications. McGraw Hill, New York (1985)
7. Shabana, A.A.: Dynamics of Multibody Systems. Cambridge University Press, Cambridge (2013)
8. Isermann, R.: Adaptive Control Systems (A Short Review). In: Isermann, R. (ed.) Digital Control Systems: Volume 2: Stochastic Control, Multivariable Control, Adaptive Control, Applications, pp. 127–140. Springer, Heidelberg (1991)
9. Le, Q.D., Kang, H., Le, T.D.: Adaptive extended computed torque control of 3 DOF planar parallel manipulators using neural network and error compensator, vol. 2, pp. 437–448 (2016)
10. Stepanenko, Y., Su, C.-Y.: Adaptive motion control of rigid-link electrically-driven robot manipulators. In: Proceedings of the 1994 IEEE International Conference on Robotics and Automation, vol. 1, pp. 630–635 (1994)
11. Siciliano, B., Khatib, O.: Springer Handbook of Robotics. Springer, Heidelberg (2008)
12. Nguyen-Tuong, D., Peters, J., Seeger, M., Schölkopf, B.: Learning inverse dynamics: a comparison. In: European Symposium on Artificial Neural Networks, no. EPFL-CONF-175477 (2008)

13. Tuan, L.A., Moon, S.-C., Lee, W.G., Lee, S.-G.: Adaptive sliding mode control of overhead cranes with varying cable length. J. Mech. Sci. Technol. **27**(3), 885–893 (2013)
14. Jouini, M., Sassi, M., Amara, N., Sellami, A.: Modeling and control for a 6-DOF platform manipulator. In: 2013 International Conference on Electrical Engineering and Software Applications, pp. 1–5 (2013)
15. Piltan, F., Sulaiman, N., Rashidi, M., Tajpaikar, Z., Ferdosali, P.: Design and implementation of sliding mode algorithm: applied to robot manipulator-a review. Int. J. Robot. Autom. **2**(5), 265–282 (2011)
16. Piltan, F., Sulaiman, N.B.: Review of sliding mode control of robotic manipulator. World Appl. Sci. J. **18**(12), 1855–1869 (2012)
17. Hoàng, N.Q., Vương, V.Đ., Quyền, N.V.: Mô hình hóa và điều khiển robot song song dẫn động bằng động cơ điện một chiều trong không gian thao tác. In: Hội nghị Khoa học toàn quốc lần thứ 2 về Cơ kỹ thuật và Tự động hóa, pp. 65–72 (2016)
18. Hoang, N.Q., Vuong, V.D., Quyen, N.V.: Modeling and model-based controller design for 3RRR planar parallel robots driven by DC motors in joint space. In: The 4th International Conference on Engineering Mechanics and Automation (ICEMA 4), vol. 4, pp. 114–123 (2016)
19. Hoang, N.Q., Vuong, V.D.: Sliding mode control for a Planar parallel robot driven by electric motors in a task space. J. Comput. Sci. Cybern. **33**(4), 325–337 (2017)
20. Bui, H.-L., Tran, D.-T., Vu, N.-L.: Optimal fuzzy control of an inverted pendulum. J. Vib. Control **18**(14), 2097–2110 (2012)
21. Ross, T.J.: Fuzzy Logic with Engineering Applications. Wiley, Hoboken (2005)
22. Kumar, M., Husian, M., Upreti, N., Gupta, D.: Genetic algorithm: review and application. Int. J. Inf. Technol. Knowl. Manag. **2**(2), 451–454 (2010)
23. Goldberg, D.E.: Genetic Algorithms in Search, Optimization and Machine Learning (1989)
24. El-Mihoub, T.A., Hopgood, A.A., Nolle, L., Battersby, A.: Hybrid genetic algorithms: a review. Eng. Lett. **11**, 124–137 (2006)

Design a Temperature Control System Using Halogen Lamp

Nam H. Nguyen[1,2], Tung X. Vu[3], Cuong K. Pham[1,2], Hai V. Bui[1,2], and Du H. Dao[3(✉)]

[1] Princeton University, Princeton, NJ 08544, USA
[2] Department of Automatic Control, School of Electrical Engineering, Hanoi University of Science and Technology, Hanoi 11615, Vietnam
nam.nguyenhoai@hust.edu.vn
[3] Thai Nguyen University of Technology, Thai Nguyen, Vietnam
daohuydu@tnut.edu.vn

Abstract. In this paper, a temperature control system based on Halogen lamp is designed for education and training purpose, which is low cost, energy saving and less time consuming for student's experimental task. The designed system consists of a box made of plastic and aluminum, a 300 W Halogen lamp, a temperature sensor, an Arduino Uno R3 based microprocessor, a triac BT137-600E by NXP Semiconductors and 220 V–50 Hz power supply. First, a transfer function from power to temperature is obtained through system identification based on a unit step response. Then, a PID controller is designed for the temperature. Finally, the temperature control system is verified through experimental tests. The results show that the temperature converges to the set-point with short settling time and small overshoot. The average time period for doing system identification and real-time control is quite small for student analyzing and performing experimental tasks. Moreover, since the cost of the designed system is cheap, it is possible to provide a system to each student for experiments. In addition, it can be also used as a testbed for verifying new control algorithms.

Keywords: Temperature control · PID · System identification

1 Introduction

For education and training purpose, it is necessary to have a real-time control system for supporting students to do experimental tasks such as system identification, controller design, and have experience of real-time control. One example of benchmark systems in industrial processes is a temperature control system, for example, temperature control in thermal power plants, heating system in steel factories and oil refine factories. In laboratories, most of temperature control systems [1, 2] were built using resistors, which consume a lot of energy and time, and have high cost. This motivates us to design a temperature control system using classical PID controller [3, 4], which is low cost, time and energy saving for education and training purpose. There are several options for running experiments of the temperature control system: (1) the PID

K.-U. Sattler et al. (Eds.): ICERA 2019, LNNS 104, pp. 215–219, 2020.
https://doi.org/10.1007/978-3-030-37497-6_25

controller is installed on the microprocessor; (2) the PID controller is carried out using Matlab/Simulink; and (3) the PID controller is run on a designed software, which is based on C/C++ programming languages. All the information about the control system such as temperature, control signal, set point and PID parameter can be seen on a display screen. Thus, it will be easier for students to tune PID parameters and collect data for writing reports. In this paper, the first option will be presented. The remaining part of the work is organized as follows. In the next section, a temperature control system is designed. In Sect. 3, the temperature system is identified. In Sect. 4, a PID controller is designed for the temperature system and some experiments are carried out. The final section gives conclusions and future works.

Fig. 1. Inside the kiln

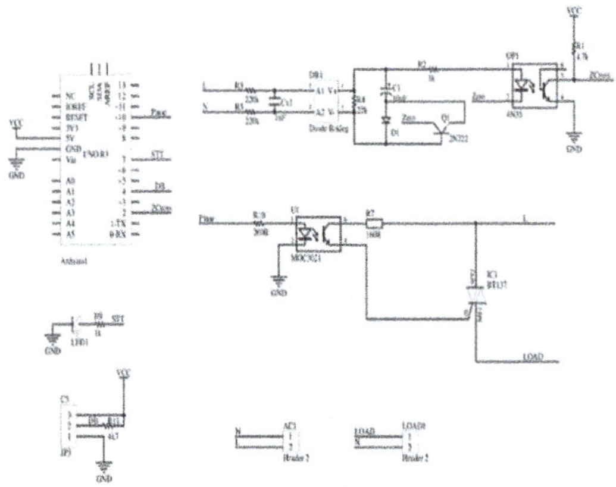

Fig. 2. Circuit diagram

2 Temperature Control System

A temperature control system is designed including a heating furnace based on Halogen lamp, a temperature sensor, a triac working as actuator, and a PID controller which can be implemented on microprocessor or computer. A kiln is designed and made of aluminum frame and plastic-aluminum wall, which has a size of 32 cm 24 cm × 24 cm (length, width and height). Figure 1 shows a picture inside the kiln, in which there are a Halogen lamp with 300 W power and a temperature sensor DS18B20. The sensor can measure a range of temperature from −55 °C to 125 °C (see [5] for more information). A circuit diagram is designed as shown in Fig. 2, which contains an Arduino UNO R3, a triac BT173 [6], an opto-isolator MOC3021 [7], an opto-coupler 4N35 [8], a diode bridge and other elements such as resistors, capacitors, transistors, headers, LED and a DC supply. The circuit mainly perform tasks such as detecting zero cross and controlling the gate of triac to adjust the current through the lamp. By changing the triac current, the temperature of the furnace will be varied correspondingly.

3 System Identification for the Furnace

In order to determine the transfer function of the furnace, a step response is obtained as shown in Fig. 3, where the type of input is step function with amplitude of 50% of the maximum power and the sampling time is $T_s = 0.64$ s. For experiments, students may use a high-order transfer function with delay to model the furnace. In this design, the furnace is modelled as a first-order system plus time delay. From the step response of the kiln, it can be seen that there is no delay at all. This can be explained that the sensor is placed quite close to the Halogen lamp. Thus, the delay can be created by moving the sensor far away from the lamp or putting a small wall to separate the sensor and the lamp. Based on the step response in Fig. 3 and using System Identification Toolbox in Matlab, the transfer function is obtained as follows.

$$P(s) = \frac{k}{T_p s + 1} \tag{1}$$

in which $k = 155.9$, $T_p = 179.2$, and $\tau = 0$. The transfer function takes power applied to the kiln as input, and temperature from the sensor as output. It will be used for PID controller design later. For training purposes, one may use other methods for system identification.

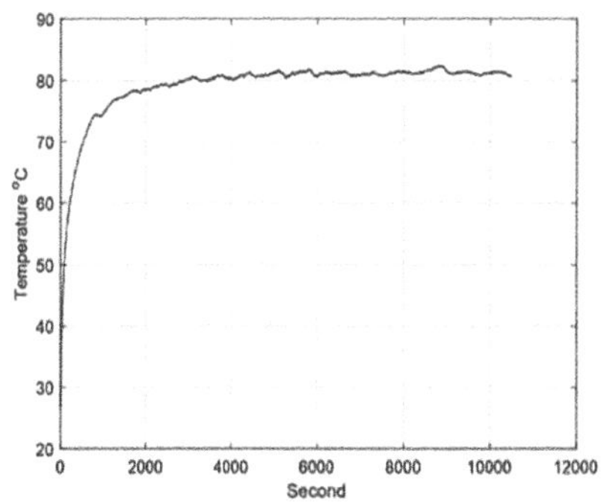

Fig. 3. Step response of the furnace

4 PID Controller Design

In this section, a PID controller is designed for the furnace. There are abundant methods [9–11] to tune PID controller. For simplicity, the PID tuner of Matlab will be used to initially determine the PID parameter. This method is used as a baseline method to compare with other methods such as genetic algorithm, experimental tuning method [12] and others. The PID controller is designed as follows

$$R(s) = K_p + \frac{K_i}{s} + K_d s \qquad (2)$$

where $K_p = 1.12$, $K_i = 0.02$ and $K_d = 0$. Then, the controller is converted into discrete-time one as follows by using trapezoidal approximation for the integral and second-order approximation for the derivative

$$u_k = u_{k-1} + a_0 e_k + a_1 e_{k-1} + a_2 e_{k-2} \qquad (3)$$

in which $a_0 = K_p + \frac{K_i T}{2} + K_d T$, $a_1 = K_p + \frac{K_i T}{2} - 2K_d/T$, $a_2 = K_d/T$, and the sampling time for the controller is T = 0.69 s. This controller will be programmed for the microprocessor. The derived digital PID controller (3) will be verified through practical tests. Figure 4 shows one of real-time control tests, where the measured temperature is on the right hand side and the control signal (power in percentage) is on the left hand side. Both the settling time and overshoot are small, they are 600 s and 1%, respectively. It means the designed temperature control system is stable as expected. This confirms that the obtained transfer function (1) is suitable and exact enough for controller design.

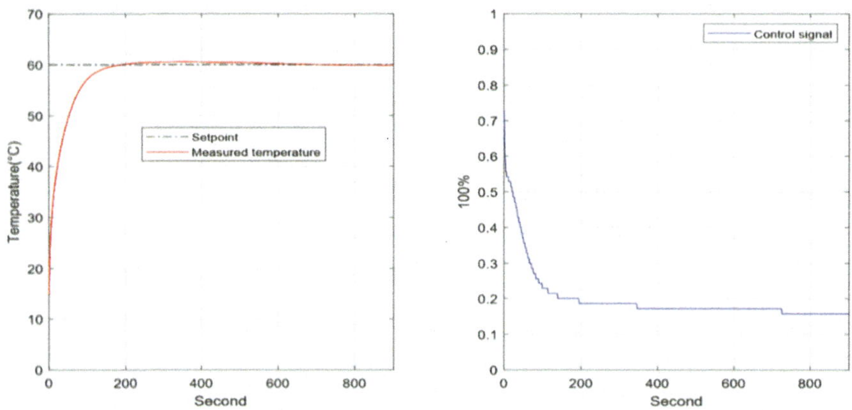

Fig. 4. Measured temperature and power under real time control test

5 Conclusions

In this work, a temperature control system for education and training purpose was designed. It consists of a microprocessor or computer based PID controller, a Halogen lamp based heating furnace, a temperature sensor and a triac for regulating the temperature. The temperature system was identified as a first-order system with zero time-delay using the System Identification Toolbox of Matlab. Then, the parameters of PID controller was obtained using the PID tuner, and the controller was converted in to digital version for programming. Some experimental tests were carried out to verify the proposed system. The results show that the proposed system works stably, can be modified to have different models, and has low cost, better time and energy saving in comparison with the resistant heating furnace [1, 2]. Thus, it is potential to apply for education and training purpose in areas of automation and control, and related fields with low cost, time and energy saving.

Acknowledgments. This research is funded by the Hanoi University of Science and Technology (HUST) under project number T2018-PC-052. This research is funded by Thai Nguyen University of Technology (TNUT) under project number T2018-B35.

References

1. Nguyen, N.H., Tran, D.T.: Neural network based model reference control for electric heating furnace with input saturation. In: 2019 First International Symposium on Instrumentation, Control, Artificial Intelligence, and Robotics (ICA-SYMP), Bangkok, Thailand, pp. 111–114 (2019). https://doi.org/10.1109/icasymp.2019.8646052
2. Nguyen, N.H., Dung, V.N., Tran, D.T.: Neural network based temperature predictor for slabs in continuous reheating furnaces. In: Regional Conference on Electrical and Electronics Engineering, Hanoi, November 2016
3. Nam, H.N., Phuoc, D.N.: Overshoot and settling time assignment with pid for first-order and second-order systems. IET Control Theory Appl. **12**(17), 2407–2416 (2018)
4. Le, H.T.T., Nguyen, N.H., Nguyen, P.D.: PID adaptive tuning with the principle of receding horizon. In: Fujita, H., Nguyen, D., Vu, N., Banh, T., Puta, H. (eds) Advances in Engineering Research and Application. ICERA 2018. Lecture Notes in Networks and Systems, vol. 63. Springer, Cham (2019)
5. Datasheets for the temperature sensor DS18B20. https://datasheets.maximintegrated.com/en/ds/DS18B20.pdf
6. Datasheets for the triac BT137-600E. https://www.mouser.com/catalog/specsheets/bt137-600e.pdf
7. Datasheets for the opto-isolator MOC3021. https://optoelectronics.liteon.com/upload/download/DS-70990019/MOC302X%20series%20201606.pdf
8. Datasheets for the optocoupler 4N35. http://pdf.datasheetcatalog.com/datasheet/fairchild/4N37.pdf
9. Firouzbahrami, M., Nobakhti, A.: Reliable computation of PID gain space for general second-order time-delay systems. Int. J. Control **90**(10), 1–13 (2017)
10. Grimholt, C., Skogestad, S.: Optimal PI and PID control of first-order plus delay processes and evaluation of the original and improved SIMC rules. J. Process Control **70**, 36–46 (2018)
11. Wang, D.J.: Synthesis of PID controllers for high-order plants with time-delay. J. Process Control **19**, 1763–1768 (2009)
12. Ziegler, J.G., Nichols, N.B.: Optimum settings for automatic controllers. Trans. ASME **64**, 759–768 (1942)

Design Axial Flux Permanent Magnet Machine for In-Wheel of Electric Vehicle

Nguyen Manh Hung[1,4], Do Manh Cuong[1], Do Nguyen Hung[1,2], and Dao Huy Du[3(✉)]

[1] Hanoi University of Science and Technology, Hanoi, Vietnam
[2] Hanoi University of Industry, Hanoi, Vietnam
[3] Thainguyen University of Technology, Thai Nguyen, Vietnam
daohuydu@tnut.edu.vn
[4] Hanoi Energy Equipment JSC, Hanoi, Vietnam

Abstract. For drive system of electric vehicle, the wheel motor is an application which requires electrical machine shape flexibility, compactness, robustness, high efficiency and high torque. Axial flux permanent magnet (AFPM) machine is a suitable solution which is directly coupled or inside the driven wheel. In this paper, the study focuses on in-wheel synchronous AFPM motors. Because of the disk shape and compactness, AFPM can be put inside the wheel. With four individual in-wheel motor, the drive ability of the car will be flexible. The motor's parameters are calculated and simulated by finite element method (FEM) on Ansys Maxwell software platform.

Keywords: Axial flux permanent magnet (AFPM) · Finite element method (FEM) · Electric vehicle (EV)

1 Introduction

The anxieties about the world petrol and gas reserves, as well as pollution and global warming issues have increased the interest in electric and hybrid vehicles. EV based on electric propulsion system, not on without internal combustion engine. All the power is based on electric power as the energy source. Some advantages of EV are no emission, more efficiency, more powerful motors, less noise, low maintenance. Recently there are many research and development works in both education and industry. Commercial EV is also available. Currently, the majority of car manufacturers offered electric cars.

Electric motor is one of important parts of EV. It has some specific requirements as follows: light weight, compact size, high power density, wide speed adjustment range. It can be DC motor or AC motor. Several main types of EV motor are DC, BLDC, induction and PM motor. Both AC and DC motor cater for the same purpose of powering up electric cars and making our planet healthier. However, the main advantage is that major electric car manufacturers prefer AC electric car motors. Today, permanent magnet (PM) machines have been widely used for EV application. This is due to having inherently more efficiency than other electric machine because of their PM excitation. Most of the completely electric vehicles have a central motor in the front, permanently connected to the drive axle. The design is similar to combustion

K.-U. Sattler et al. (Eds.): ICERA 2019, LNNS 104, pp. 220–228, 2020.
https://doi.org/10.1007/978-3-030-37497-6_26

engine cars. However, another solution has been proposed. That is smaller motors are installed directly behind the wheels called in-wheel motors. This solution is relatively lightweight because the cars do not need a power train with a gearbox and drive shafts. The total mass of the car decreases, but the wheels are heavier.

An AFPM machine have a lot of benefits such as simplified construction, compact design, high power density, high efficiency, superior torque density, possibility to add high number of poles, very low rotor losses and adjustable axial air gap. Those are reasons why producing an efficient and compact AFPM machine is ideally suited for use in vehicle. Moreover, due to its inherently short axial length and lightweight, it is suitable for an in-wheel direct drive application. In the direct drive traction system, there is no mechanical transmission system between the wheel and electric motor, therefore whole system weight is reduced [1].

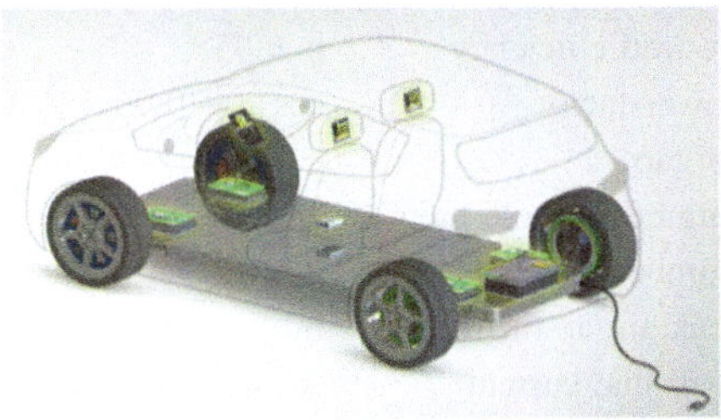

Fig. 1. A car with four in-wheel motors and inverter [2].

In Fig. 1, A car structure has four in-wheel motor of Elaphe LTD. It has no connecting rod, so weight can be reduced. With four inverters, drive ability of the car will be flexible. In-wheel motors have the speed range is from 1000–1500 rpm and a gearbox is not used.

It has been demonstrated by the previous works, in the last ten years, if the number of poles in the AFPM is large enough and the axial length is sufficiently small, its torque density is considerably larger than that obtained by traditional radial flux machines [1]. Moreover, AFPM machine can be created by stacking rotors and stators, which makes more torque.

In this paper, single-side synchronous AFPM machine is designed for EV application. Short axial length of machine will be suitable for installation in wheel. With 8 pole pairs, the motor speed is 1500 rpm, so using AFPM machine as an in-wheel drive is feasible. With direct drive system, if wheel diameter is 0.6 m, car speed can be approximately 170 km/h. According to [3], calculation shows four motors can accelerate the vehicle (EV weight is 1000 kg) from 0 km/h to 100 km/h in 10 s. The machine is designed and simulated by FEM. The analysis results will show the transient process and the magnetic field distribution of machine.

2 Design AFPM Machine

The design process diagram is shown in Fig. 2.

2.1 Choose Materials and Main Parameters

The properties of motor are also greatly dependent on the electric steel used as the core material. Electrical steels are interested in research and development. Today, Electrical steel can contribute greatly to a reduction of the magnet losses during the transfer and distribution of electrical energy to a minimum. In this research, M27 steel of Ansys is used for core of machine. On the market, electric steels of Thyssenkrupp Steel or AK Steel can be used for core. The rare earth magnets also have cheaper prices and better properties. NdFe50 magnet is selected in this study.

Some main parameters of machine are mentioned in Table 1.

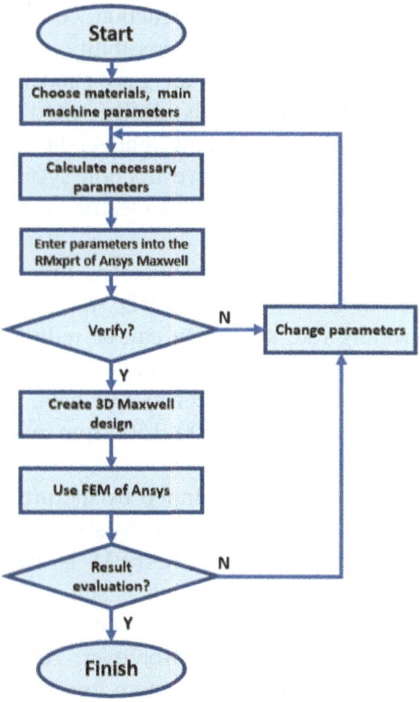

Fig. 2. The design process diagram.

Table 1. Main parameters.

Parameter	Value
Power output	Pout = 30 KW (40 HP)
Rated speed	n = 1500 rpm
Rated torque	M = 191 N.m
Desired efficiency	$\eta = 95\%$
Power factor	$\cos\varphi = 0.95$
Airgap	g = 1.5
Number of pole pairs	2p = 8
Number of slots	Z = 48
Peak airgap flux density	$B_{mg} = 0.7$
Peak line current density	$A_m = 45000$ A.m
Inner-to-outer machine radius	$k_d = {}^1/\sqrt{3}$
Number of phase	m = 3

2.2 Calculate Necessary Parameters

The formulas used to calculate the parameters are shown below [4]:

Size of machine:

$$D_o = \sqrt[3]{\frac{\varepsilon P_{out}}{\pi^2 \eta \cos\varphi \; n_s k_{w1} B_{mg} A_m k_D}} \tag{1}$$

$$D_i = k_d D_o \tag{2}$$

With $k_D = \frac{1}{8}(1+k_d)\left(1-k_d^2\right)$, $n_s = \frac{n}{60}\varepsilon = \frac{E_f}{V}$; $\boldsymbol{k_d = {}^1/\sqrt{3}}$ will give maximum electromagnet torque [5].

Choose magnet span $\alpha_M = 0.85$. Magnet thickness is [5]:

$$l_M = \frac{\mu_r B_g k_c g}{B_r - B_{mg}} \tag{3}$$

Number of stator winding turns per phase calculated on the basis of line current density [3]:

$$N = \frac{\pi D_o (1 + k_d) A_m}{4m\sqrt{2} I_a} \tag{4}$$

The number of conductors in a single coil [1]

$$N_c = \frac{a_w a_p N}{Z/m} \tag{5}$$

Stator winding is double layer and has two parallel conductors $a_w = 2$, one parallel current path $a_p = 1$. Moreover, multi-strand wire can be used for wire diameter is smaller.

The current density in the conductor can be assumed $\mathcal{J}_a = 6$ A/mm^2. The cross section area of conductor is:

$$S_a = \frac{I_a}{a_w \mathcal{J}_a} \tag{6}$$

Slot factor is k_s. The cross section of the slot is

$$S_s = \frac{2 S_a N_c}{k_s} \tag{7}$$

The minimum stator slot pitch is

$$w_{s,min} = \frac{\pi D_i}{Z} \tag{8}$$

The slot is rectangular shape. Choose slot depth d_s and calculate slot width w_s according Eq. (9). Slot width must be than $w_{s,min}$.

$$w_s = \frac{S_s}{d_s} \tag{9}$$

Rotor and Stator are made by stacking and rolling electrical steel sheets. Stacking factor of them $k_{st} = 0.95$.

Rotor length with magnetic density of rotor yoke $B_{r,y}$ [5]

$$l_r = \frac{\alpha_M B_{mg} \pi D_o}{4(2p) B_{r,y}} \tag{10}$$

Length of stator yoke with density of stator yoke $B_{s,y}$ [5]

$$l_s = \frac{\alpha_M B_{mg} \pi D_o}{4(2p) B_{s,y} k_{st}} \tag{11}$$

The below table show parameters which were calculated and suitable changed (Table 2).

Table 2. Parameters of machine.

Parameter	Value
Outer diameter	$D_o = 380$ mm
Inner diameter	$D_i = 220$ mm
Magnet thickness	$l_M = 2.8$ mm
Length of stator	$l_s = 46$ mm
Slot fill factor	$k_s = 0.6$
Depth of slot	$d_s = 30$ mm
Width of slot	$w_s = 10$ mm
Length of rotor	$l_r = 13$ mm
Conductors per single coil	$N_c = 9$
Conductors per slot	$N_s = 18$
Wire diameter	$d_a = 2.304$ mm
Number of strands	2

2.3 Simulate and Verify by RMxprt Tool of Ansys Maxwell

Machine parameters are entered into the software. Below are some simulation results.

Figure 3 shows when full load operation of machine $I_a = 95$ A,

$A_m = 43579$ A.m < 45000 A.m and $\mathcal{J}_a = 5.7$ A/mm^2 < 6 A/mm^2.

Moreover, losses of machine, $\eta = 95\%$ as expected.

In Fig. 4, mass density of material is shown. Moreover, the software calculates the total weight of the machine parts of approximately 35 kg and it is a lighter radial flux machine on the market.

	Name	Value	Units
1	Average Input Current	89325.2	mA
2	RMS Armature Current	95075.7	mA
3	Armature Thermal Load	248.449	A^2/mm^3
4	Specific Electric Loading	43579.5	A_per_meter
5	Armature Current Density	5701050	A_per_m2
6	Frictional and Windage Loss	0	mW
7	Iron-Core Loss	331844	mW
8	Armature Copper Loss	749857	mW
9	Transistor Loss	151758	mW
10	Diode Loss	27573.2	mW
11	Total Loss	1261030	mW
12	Output Power	30002800	mW
13	Input Power	31263800	mW
14	Efficiency	95.9665	%
15	Torque Angle	41.0421	deg
16	Rated Speed	1500	rpm
17	Rated Torque	191.004	NewtonMeter
18	Fundamental RMS Phase Back-EMF	120224	mV

Fig. 3. Electric parameters in full load operation.

	Name	Value	Units
1	Stator Tooth Flux Density	1.25801	tesla
2	Stator Yoke Flux Density	1.23249	tesla
3	Rotor Yoke Flux Density	1.23249	tesla
4	Magnet Flux Density	0.608023	tesla
5	Air-Gap Flux Density	0.587169	tesla
6	Stator Tooth Ampere Turns	14.7774	A.T
7	Stator Yoke Ampere Turns	7.22744	A.T
8	Rotor Yoke Ampere Turns	7.22744	A.T
9	Magnet Ampere Turns	-768.399	A.T
10	Air-Gap Ampere Turns	739.164	A.T
11	Saturation Factor	1.03955	
12	Stator Yoke Correction Factor	0.575	
13	Rotor Yoke Correction Factor	0.575	

	Name	Value	Units
1	Stator Wire Density	8933	kg_per_m3
2	Stator Core Steel Density	7650	kg_per_m3
3	Rotor Core Steel Density	7650	kg_per_m3
4	Rotor Magnet Density	7500	kg_per_m3
5	Stator Copper Weight	9.17027	kg
6	Stator Core Steel Weight	16.5827	kg
7	Rotor Core Steel Weight	7.12344	kg
8	Rotor Magnet Weight	1.34586	kg
9	Stator Net Weight	25.753	kg
10	Rotor Net Weight	8.46929	kg
11	Stator Core Steel Consumption	25.206	kg
12	Rotor Core Steel Consumption	7.12344	kg

Fig. 4. No-load magnetic variables (left) and material consumption (right).

3 Transient Analysis by FEM

With result of 2.4, a 3D model of machine is created in Fig. 5a. Because of the symmetrical machine, FEM analysis is performed on 1/16 of machine which is meshed as in Fig. 5b.

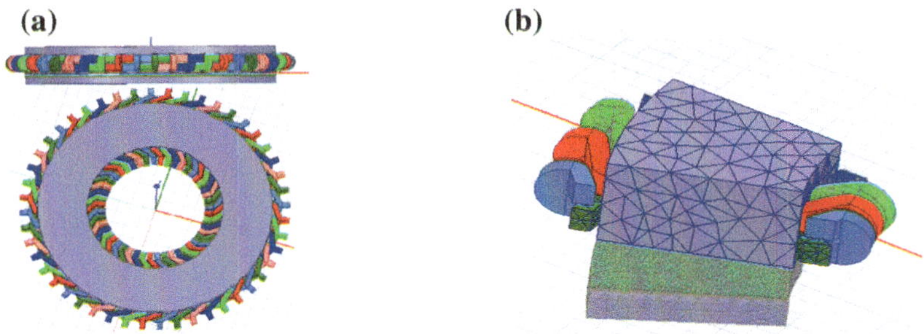

Fig. 5. 3D model of machine (a) and analyzed part of machine (b).

Some simulation results are shown as follow:

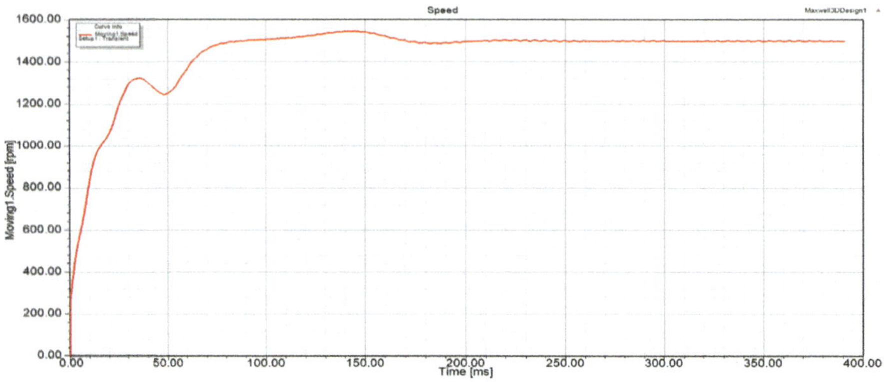

Fig. 6. Speed curve.

Figure 6 shows the speed of motor. After motor starting process, the stable speed is 1500 rpm which achieves the design speed (Fig. 7).

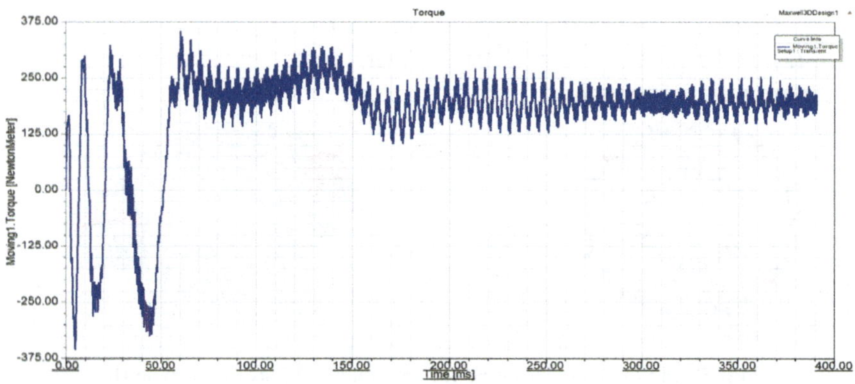

Fig. 7. Torque curve.

In transient process, max torque reaches 1.5 times of rated torque. When the motor is stable, torque oscillates around rated torque. The torque is fluctuated, because the controller with optimal control method has not been mentioned in the simulation.

In Figs. 8, stable winding current matches the results in above RMxprt.

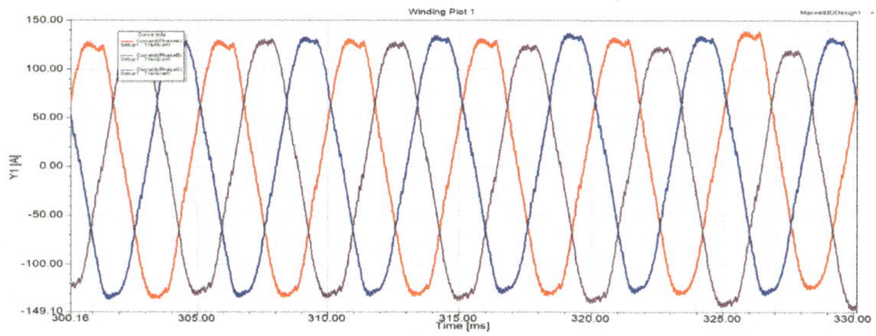

Fig. 8. Stable winding current curve.

Field results (Fig. 9).

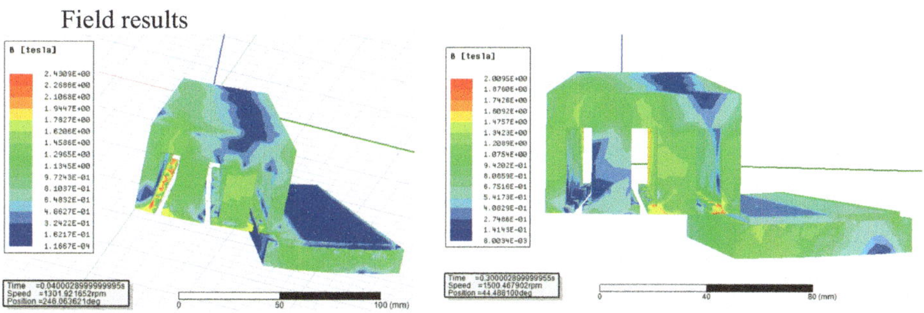

Fig. 9. Magnetic density in transient period (left) and stable period (right).

In transient period, maximum magnetic density is 2.4 T. However, it only takes place in a small part of machine and in short time. In stable period, magnetic density decreases and maximum is 2 T.

4 Conclusion

Based on simulation results, the machine has met the input expectations. It has short axial length, suitable for direct drive in-wheel application. It also achieves the rated speed and torque, and the magnetic density is suitable to B-H curve of materials as well. However, the torque is fluctuated, so it requires an optimal controller. In the next research, a controller and an electrical drive system for EV will be designed.

References

1. Hatami, H., Sharifian, M.B.B., Feyxi, M.R., Sabahi, M.: Analysis and improved design of direct drive coreless AFPM in-wheel machine for HEV application. Int. J. Tech. Phys. Prob. Eng. (2013)
2. Motors – Elaphe in-wheel electric motors. https://in-wheel.com/product-category/motors/
3. Chauhan, S.: Motor torque calculations for electric vehicle. Int. J. Sci. Technol. Res. **4**, 126–127 (2015)
4. Gieras, J.F., Wang, R.J., Kamper, M.J.: Axial Flux Permanent Magnet Brushless Machines. Kluwer Academic Publishers, Dordrecht (2005)
5. Nguyen, T.D.: Dual air-gap axial flux permanent magnet machines for flywheel energy storage systems. Nanyang Technological University (2012)

Detection and Diagnosis Gray Spots on Tea Leaves Using Computer Vision and Multi-layer Perceptron

Pham Thanh Binh[1], Tang Cam Nhung[2], and Dao Huy Du[2(✉)]

[1] Northern Mountainous Agriculture and Forestry Science Institute, Phu Tho Town, Phu Tho Province, Vietnam
[2] Thai Nguyen University of Technology, 3/2 Street, Thai Nguyen City, Vietnam
{tangcamnhung, daohuydu}@tnut.edu.vn

Abstract. This paper proposes a method to detect gray spots on tea leaves using computer vision and image processing algorithms such as collecting input images from receiving equipment, removing tea leaves from the background, delineating areas of signs of disease, creating contrast. Also, in order to establish the network training parameters, a Neural of Multi-Layer Perceptron (MLP) is going to be built with extracted identifying features on the gray spots tea leaves. When the network and training the data set have been established, they will be used in the identification process and conclusions for the subject to be tested for gray spots. The identification process after many tests has achieved results with an accuracy of 90.0%.

Keywords: Pestalozziatheae Sawada · Artificial Neural network (ANN) · Multi Layer Perceptron (MLP)

1 Introduction

In the cultivation and care of tea plants in Vietnam, only manual methods and eyed observations are used to identify diseases of tea plants through expressions on the leaves or trunks. Therefore, it is necessary to have many employees monitoring continuously, causing high cost of epidemic detection and identification when the tea growing model is increasingly expanded. In addition, the majority of tea growers have limited knowledge in tea diagnosis, which makes it difficult to control the disease. Therefore, disease detection on tea plants plays an important role, especially the detection and identification of the right disease for prevention is extremely necessary. That's why the use of automatic disease detection methods on tea plants have been highly effective.

In some studies of tea leaf identification, neural networks have been published by some scientists. In [1], the author used specific characteristics combining Gabor-HSV filter and SIFT algorithm to detect the change of tea leaf shape, then used Neural network to train the system and diseases classification. Based on this feature in [2] the

K.-U. Sattler et al. (Eds.): ICERA 2019, LNNS 104, pp. 229–237, 2020.
https://doi.org/10.1007/978-3-030-37497-6_27

color feature of the HSV system extracted from the segmentation and training process of neural networks were used with the affectivity of 80%.

Detection of disease and monitoring the status of agricultural production, embedded Java application programming language into Android to detect remote diseases of plants; communication in the system using GPRS and GMS technology [3].

This paper focuses on the use of MLP neural network combined with some image processing algorithms to detect gray spots on tea leaves. The stages of detection of gray spots for tea leaves include: input image data is included in the pretreatment to improve the quality of the predicted area with abnormal signs. The next stage is the continuous sampling process for the affected areas, then these samples are put into training in Neural network in MLP form with the futures of gray spots. Finally, subjects with abnormal signs on the leaves are taken into neural networks to determine whether or not there is a disease in these subjects. The overview diagram is given in Fig. 1 below:

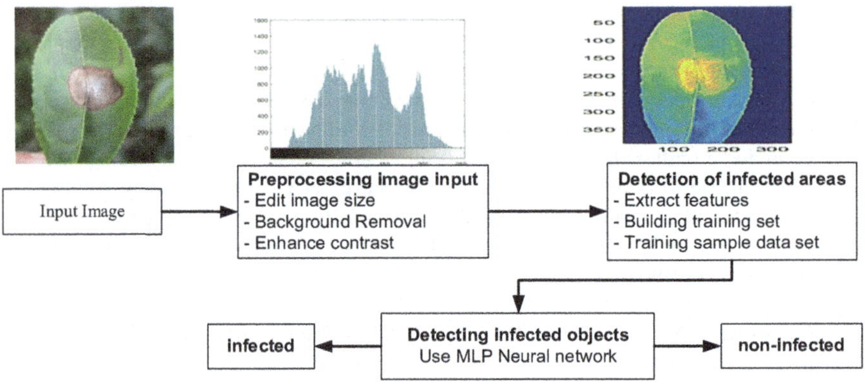

Fig. 1. Overview model of the detection and identification of gray spots on tea leaves

This paper is organized as follows: Sect. 2: Identification of gray spots on tea leaves using artificial Neural networks; Sect. 3. The proposed method for using Neural MLP network to detect gray spots on tea leaves; and Sect. 4. Implementation results.

2 Identification of Gray Spots on Tea Leaves Using Artificial Neural Networks

Gray spot disease (with scientific name: Pestalozziatheae Sawada) [4]: Gray spot disease caused by Pestalozziatheae Sawada fungal on tea leaves. The fungal appears almost of year in most tea growing areas, but the disease thrives in high humid conditions and air temperature around 25 °C–28 °C, usually from July to October and damage seriously in August – September in year.

Harmful symptoms: the lesions usually come from the leaf edge or from the middle of the leaf. The first ones are small gray brown spots, then they are larger and round, nearly round, oval, or semi-circular certain shapes, and the edges of wounds are wavy. On the lesions, there are black veins, black spots, and the surface of the disease is ash-gra. When the lesions spread to about 1/2 of the area of leaves, the leaves are shed.

Gray spot disease is considered a disease causing many losses to the quality and yield for the largest tea plant, and the spread is quite rapid if it is not promptly detected and remedied (Fig. 2).

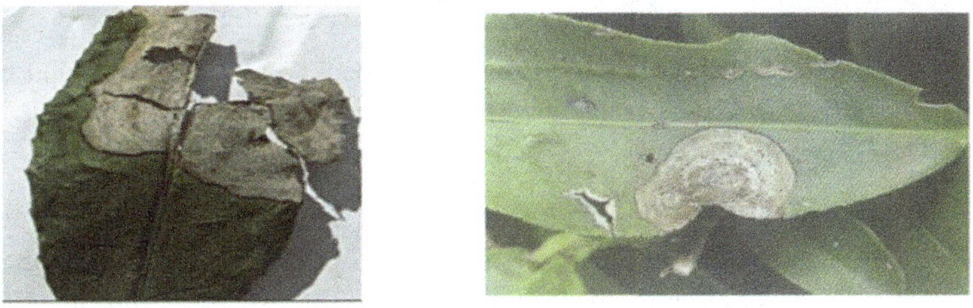

Fig. 2. Tea leaves with gray spots

3 The Proposed Method for Using Neural MLP Network to Detect Gray Spots on Tea Leaves

3.1 Pre-process Input Image Stages

These steps are used to enhance the detection of areas of unusual signs

Correction of the Input Image Size: Because the input image has different sizes, the dimensions include both height and width. These images are then formatted in a unified size of 360 × 360 pixels, which is used to accommodate image loading and processing operations so that the aspect ratio if the image is remained.

Background removal: In the system of identification of tea leaf disease areas using machine vision, the background area of the input image must be fully controlled. The dark background is usually the area used to distinguish the tea leaves from the substrate. Therefore, background pixels are always assigned low values and they are easily removed by using Heuristic algorithms to determine the threshold value μth according to the following formula [5]:

$$\mu_{th} = \frac{\mu_{MaxRep} - \mu_{medium}}{2} \quad (1)$$

where: μ_{medium} is the median of the image gray level distribution and μ_{MaxRep} is the gray level which has maximum repetition in input image.

In fact, gray levels are automatically determined for each individual image based on the Histogram chart. Figure 3 shows the image of tea leaves before and after removing the background

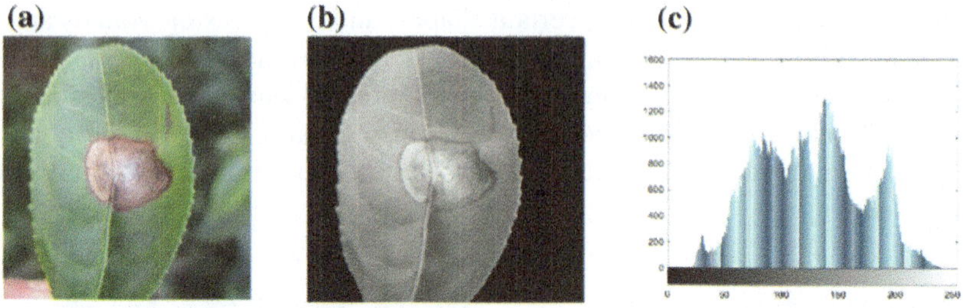

Fig. 3. (a) A sample apple image before and, (b) after background removing and binarization along with, (c) corresponding histograms of (b).

In Fig. 3, the gray level of the background is set to values less than 50 and can be visible in Fig. 3(a), but the background has been completely removed in Fig. 3(b), while in the Fig. 3(c) represents the values of μ_{medium}, μ_{MaxRep} and μ_{th} on the Histogram schema.

3.2 Segment of Infected Leaf Area

Expression of Gray Spots and Identification of Gray: According to the analysis shown on the leaves with gray spots, there are usually shapes such as: round, oval, or semi-circular, so the algorithm is applied to areas on the whole leaf of tea with the size of each area defined as 10×10 pixel. This size is determined based on signs of gray spots with the smallest areas of actual disease.

Areas around the edge of the leaf that have infected images are identified by determining the interrupted boundary.

The surface of the disease area is gray-brown: Each image segment can be represented as a two dimensional array, where every element of that array contains color information for one pixel.

In this paper, we give an algorithm to determine the disease for tea leaves, adding extra features on the surface of tea leaves and label them accordingly.

Classification using MLP Neural network is divided into 2 categories to conduct the classification test, that are:

- Categories 1: Healthy leaves and leaves with gray spots are classified.
- Categories 2: The input image data set is classified into 3 types based on the characteristics of gray spots: Type 1: healthy leaves; Type 2: leaves with abnormal signs; Type 3: gray spots leave

Method of Using MLP Neural Network to Segment Infected Areas: In this paper, we use MLP Neural network applied to pixels to determine the area of disease on tea leaves. Each pixel in the image of the tea leaves is classified into two categories - healthy or unhealthy based on the corresponding values of R (Read), G (Green), B (Blue) and H (Hue). In MLP Neural network, we use two layers and 4 inputs (corresponding to the value of R, G, B and H of pixels), 4 Neural in hidden layer and 2

Neural in output layer to classify each input pixel for one of the healthy or faulty classes. This MLP is trained to use the training set through Levenberg-Marquest algorithm [6]. The sample results of the algorithm are shown in Fig. 4, in which the infected areas are segmented compared to the non-infected and the basal areas. The

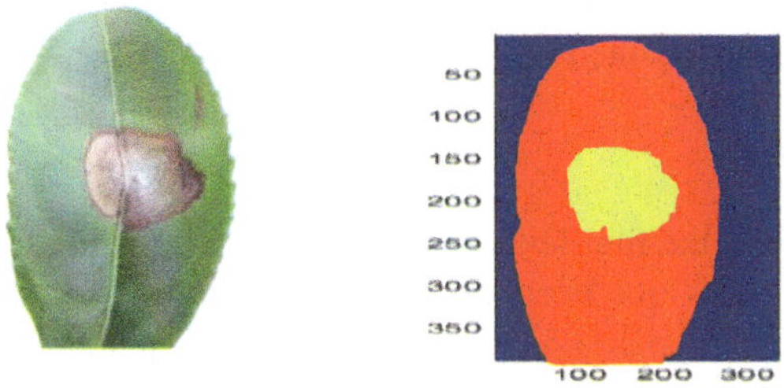

(a) (b)

Fig. 4. A sample color image after background removing (a) and result of defect segmentation (b) which are marked in yellow.

three areas are represented in three different colors, which represent the accuracy of the proposed segment algorithm.

Extract Features for Infected Areas: After identifying the faulty areas, it is necessary to select features corresponding to the signs of the disease to decide that the leaves are infected or not. We experimented on the characteristics: statistics, textural and geometric of infected areas. By using MLP Neural network algorithms and experimenting in turn for characteristics the most typical ones can be chosen. Through many practical experiments, we have selected 8 statistics, 5 textures of gray spots and 3 geometries of the infected area. Therefore, the set of image properties for tea leaves includes 16 attributes and these attributes are specified in the following sections:

- *Statistical feature:* These features are called first-order spatial statistical methods used to calculate the observed gray level value in any given region and therefore it only depends on the value of individual pixel not on the values of neighboring pixels. These features can be calculated from the Histogram chart of the pixel in the faulty area. The statistical features used in this paper are based on the characteristics of color and density including: mean and standard deviation of R (Red), G (Green), B (Blue) and H(Hue) component of defected region.

The Histogram gray schema of an image I has gray level values: $(\boldsymbol{x},\boldsymbol{y}) \in [0, \boldsymbol{K}-1]$; consists of K values, if the image has an 8bit format, K = 2K = 256. Each value of Histogram is defined as follows:

$$\boldsymbol{h}(\boldsymbol{i}) = \boldsymbol{Cardinality}\{(\boldsymbol{x},\boldsymbol{y}) | \boldsymbol{I}(\boldsymbol{x},\boldsymbol{y}) = \boldsymbol{i}\} \tag{2}$$

where: $\boldsymbol{h}(\boldsymbol{i})$: the number of pixels in image I with the value of gray intensity in the range $0 \leq i < K$

Cardinality: the number of elements in the set

$$\text{Standard deviation is defined as follows}: \sigma = \sqrt{\frac{\sum(I(x-y)-m)^2}{N}} \quad (3)$$

$$\text{Mean deviation is defined as follows}: \text{Skewness} = \frac{\sum(I(x,y)-m)^3}{N\sigma^3} \quad (4)$$

- *Textural features:* In the first-level spatial statistical method, the correlation between gray level values has not been determined, whereas second-order measures are properties of pixels in pairs. They capture spatial dependence of gray values that contribute to perception of texture so we will refer them as textural features. In this paper, we use Haralick textural features which are computed from Gray Level Co-occurrence Matrices (GLCM). [7] proposed by Haralick to extract the contrast, correlation, homogeneity, energy and entropy from GLCM as textural features.

The homogeneous gray level matrix $\boldsymbol{P_d}$ is defined as follows:

$$P_d(x) = \sum_j P_d(x,j) \text{ and } P_d(y) = \sum_i P_d(i,y) \quad (5)$$

In particular, texture features are defined as follows:

$$\text{Energy} = \sum_i \sum_j P_d^2(i,j) \quad (6)$$

$$\text{Entropy} = -\sum_i \sum_j P_d(i,j) log \sum_i \sum_j P_d(i,j) \quad (7)$$

$$\text{Contrast} = \sum_i \sum_j (i-j)^2 P_d(i,j) \quad (8)$$

$$\text{Homogeneity} = \sum_i \sum_j \frac{P_d^2(i,j)}{i+|i-j|} \quad (9)$$

$$\text{Correlation} = \sum_i \sum_j \frac{(i-\mu j)(j-\mu y)P_d(i,j)}{\sigma x, \sigma y} \quad (10)$$

Where: μx, μy are variance; σx, σy are standard deviation of $P_d(x)$.

- *Geometric features:* In this paper, we use simple shape features including: rate of infected area, infected area perimeter, average length of axis in the affected area. The rate of infected areas is the ratio of the number of pixels in the infected area compared to the whole leaf area. The perimeter of the infected area is calculated as estimated by the boundary of the affected area.

3.3 Data Set Building and Network Training

Build Data Set: Our proposed algorithm is applied on the database as follows:
The input are still images: these images are usually tea leaves taken from digital cameras with unusual signs of color and shape compared to unaffected leaves. Sample images for tea leaves include 120 that 360 × 360 pixel images. This database is classified by agricultural experts and divided into 2 categories, as following Table 1:

Table 1. Number of images for data set

Categories	Number of		
	healthy leaves	Leaves with gray spots	Leaves with abnormal signs
1	80	80	0
2	40	40	40

For each category, it will be used to extract features analyzed in the formulas (6), (7), (8), (9), (10). To statisticize the results of test for both categories, we tested on the data set several times and get the average results from the tests to give accurate results. Samples used in the experiment were randomly selected for each test. After classification, this result was compared with the classification results of experts.

Based on the set of extracted features, the Neural network in this paper consists of 16 Input layers corresponding to 16 features, 2 Output layer corresponding to the identification of leave that is infected or not. The number of hidden classes is calculated as follows [8]:

$$\text{Hidden Layer} = \frac{(Numers\ of\ Input\ layer + Numers\ of\ Onput\ layer)}{2} \tag{11}$$

The training process was developed in JAVA language and implemented on Neuroph's Framework [9].

The data set used for training Neural networks was collected including 40 healthy leaf samples, 40 leaf samples with abnormal signs and 40 samples of leaf with gray spots. The images in this template were all converted to the ***.jpg** format.

At first, with the samples put into training, each identification area was set to 10 × 10 pixels in accordance with the requirements of the system. In this paper, we use Java's imgscalr library

The next step is to apply a gray scheme to increase the contrast for the created patterns; First, extract the RGB values of each Pixel; The training template is improved and the results are put into use during the training of neural networks.

Solutions are implemented on the Java programming language and run on the Neuroph framework.

Training Network: Neural Network setup stage: To establish MLP neural network with components such as: Layers, Units, Weights and Biases, Activation functions are shown in [6] and they are the main parameter in the SIGMOID transmission function. With:

- Number of Input layers: 16
- Number of Hidden layers: 16
- Number of Output layers: 2

To create a connection from the Input layer with Hidden systems use the create Connection method function, the specified weights for each link. The process is similar to the Neural counterpart of the Hidden class with the Output layer each time when a link is formed between the Input layer and the Hidden layer. The neural network created is recorded in a file with the ***.nnet** format.

Neural network training: This is an important stage related to the optimization of parameters in the detection of tea leaf disease. In one cycle, the number of infected and non-infected tea leaf samples was included in the calculation.

- The extracted features are stored in the **dataInput** variable.
- Subsequently method addRow created new row of the training set, having the first parameter the vector with the pixel values in dataInput variable and the second value of 1.0 for infected samples and 0.0 for non-infected samples.
- A new object will be created in the **MomentumBackpropagation** class to implement back-propagation rules with the coefficient of the Moment defined. Neural network in MLP format has a set of rules set by default. The training rule set and training set is used as a parameter in the learning method of neural networks.
- The value of pixels in the segment will be stored in an array which is used as an input vector for NN network training. The infected area detected in any segment will be the output of the NN network and be equal to 1.0.

4 Implementation Results

The results are shown in the table below (Fig. 5):

Categories 1: use 40 samples to train and 40 test samples;
Categories 2: use 32 **spots** samples to train and 8 test samples;

Correctly detected	Categories 1	Average (%)	Categories 2	Average (%)
Healthy leaves	38.7 of 40	96,75%.	7.4 of 8	92.5%
Leaves with gray spots	17.9 of 20	89.5%	6.8 of 8	85.0%
Leaves with abnormal signs	0	0	7.3 of 8	90.0%

Fig. 5. The results of the test

5 Conclusion

In this paper, we have combined image processing algorithms, Neural networks with computer vision to determine gray spots on tea leaves. In the implementation process, we removed the leaf area from the background, then performed segment infected areas using the MLP neural network for each Pixel based on their R, G, B and H values. Finally, each tea leaf was assigned to a group corresponding to each expression of the general based on 16 characteristic selections including: 8 statistical features, 5 textural features and 3 geometric features and identified by MLP Neural network.

To ensure the accuracy and reality, we divided the leaves into 2 groups: the group with leaves without disease and leaves with gray spots; and the group with leaves without disease, leaves with abnormal signs and infected leaves gray spots. The results showed that after being trained in MLP, the classification achieved 89.5% of the first group and 90.0% of the second group. The results have not reached the maximum accuracy because the data collection process has not covered all the symptoms of the disease. However, this result can be completely put into practice and we need to complete the database set to cover all cases of variation of this disease so that identification process achieves higher results.

Besides, we have only identified the disease of tea leaves only on images taken directly, but in practical applications, the identification of tea leaf disease is based on different images taken according to the pictures different directions.

References

1. Jha, S., Jain, U., Kende, A., Venkatesan, M.: Disease detection in tea leaves using image processing. Int. J. Pharma Bio Sci. July 2016
2. Redd, S., Pawar, A., Rasane, S., Kadam, S.: A survey on crop disease detection and prevention using android application. IJISET – Int. J. Innov. Sci. Eng. Technol. **2**(4) (2015)
3. Shekh, S.K., Baitule, A., Narethe, M., Mallad, S.: Detection of leaf diseases and monitoring the agricultural resources using android app. Int. J. Innov. Res. Comput. Commun. Eng. **3** (2015)
4. Tiep, P.T., Giang, V.H., Hang, T.T.: Curriculum to prevent pests and diseases of tea, Ministry of Agriculture and Rural Development (2012)
5. Moallem, P., Razmjooy, N.: Optimal threshold computing in automatic image thresholding using adaptive particle swarm optimization. J. Appl. Res. Technol. **10**(5), 703–712 (2012)
6. Choras, R.S.: Image feature extraction techniques and their applications for CBIR and biometrics systems. Int. J. Biol. Biomed. Eng. **1**(1) (2007)
7. Moallem, P., Razmjooy, N., Ashourian, M.: Computer vision-based potato defect detection using neural networks and support vector machine. Int. J. Robot. Autom. **28**(2), 137–145 (2013)
8. Truong, Q., Van Vung, N., Dinh, T.Q.: Determination and recognition of disabilities on the mango peel. In: Proceedings of the Ninth National Science Conference "Basic research and application of Information Technology (FAIR'9)" (2016)
9. Perry, J.S.: Create an artificial neural network using the Neuroph Java framework, Published, 8 January 2018

Determining Optimal Gear Ratios of Mechanical Drive Systems Using Two-Stage Helical Gearbox with Second-Stage Double Gear Sets and Chain Drive for Minimal System Cross Section Area

Nguyen Thanh Tu[1], Le Hong Ky[2], Tran Thi Hong[3], Nguyen Van Cuong[4], Luu Anh Tung[1], Bui Thanh Hien[1], and Le Xuan Hung[1](✉)

[1] Thai Nguyen University of Technology, Thai Nguyen City, Vietnam
lexuanhung@tnut.edu.vn
[2] Vinh Long University of Technology Education, Vinh Long City, Vietnam
[3] Nguyen Tat Thanh University, Ho Chi Minh City, Vietnam
[4] University of Transport and Communications, Ha Noi City, Vietnam

Abstract. The present work is concerned with determining optimum gear ratios of mechanical driven systems using a two-stage helical gearbox with second stage double gear sets and a chain drive. During the research process, the construction of an optimization problem has been implemented. The minimal system cross section area was selected as the objective function. Afterwards, a simulation experiment was conducted with the aim of exploring the dependence of the optimal transmission ratios on the input factors including the total gearbox ratio, the wheel face width coefficient, the allowable contact stress and the output torque. The recommended equations can be applied to obtain the optimal gear ratios.

Keywords: Gear ratio · Optimal gear ratio · Optimal gearbox design · Helical gearbox · System cross section area

1 Introduction

Gearbox transmission systems have a significant role in industries, therefore, they are widely used in different industrial fields. However, the system size and volume have always been great concerns of manufacturers in the reduction of the system cost. Accordingly, the optimal design of this system is of particular interest to researchers. Significantly, determining the optimal partial gear ratios is greatly beneficial to the system in lowering its size, volume, and thus, its cost.

To solve the problem of determining the optimal partial gear ratios researchers have utilized different methods in their studies namely graph method [1, 2], practical method [3] and model method [3–16]. Besides, this problem has been applied to varied types of gearboxes such as helical gearboxes [1–19], bevel gearboxes [1, 3, 20–22], or worm

K.-U. Sattler et al. (Eds.): ICERA 2019, LNNS 104, pp. 238–248, 2020.
https://doi.org/10.1007/978-3-030-37497-6_28

gearboxes [23–26] with diverse gear stages including two-stage [1–8], three-stage [9–15] and four-stage [16–19]. It is reported that the problems to obtain optimal transmission ratios normally use the objective functions as the minimum gearbox length [7, 11, 25], the minimum cross-sectional area of the box number [9, 12] or minimum amount of gear [10, 13, 14, 16, 18].

Apart from the optimal transmission ratios of systems with gearboxes, considerable attention has been paid to determining the optimal transmission ratios of systems including a gearbox and a V-belt drive [5, 27–29] or a chain drive [30, 31].

This study focused on obtaining the minimal system cross section area based on calculating the optimal gear ratios of mechanical drive systems using a two-stage helical gearbox with second stage double gear sets and a chain drive.

2 Methodology

The cross section area of a mechanical drive system using two-stage helical gearbox with second-stage double gear sets and a chain drive is calculated by (Fig. 1):

$$A = L \cdot h \tag{1}$$

Where,

$$L = \max\left(L_g, L_c\right) \tag{2}$$

$$h = \max(d_{e21}; d_{w22}; d_{w23}; d_2) \tag{3}$$

In which, L_g and L_c are determined by (see Fig. 1):

$$L_g = d_{w11}/2 + a_{w1} + a_{w2} + d_{w22}/2 \tag{4}$$

$$L_c = d_2/2 + a_c + d_{w22}/2 \tag{5}$$

Thus, the optimization problem can be expressed as:

$$\text{minimize } A \tag{6}$$

With the following constraints:

$$\begin{aligned} &1 \le u_1 \le 9 \\ &1 \le u_2 \le 9 \\ &1 \le u_c \le 6 \\ &u_1 \cdot u_2 \cdot u_c = u_t \end{aligned} \tag{7}$$

Where, a_{w1}, a_{w2}, d_{w11} and d_{w22} are the center distances and the pitch diameters of the first and the second gear units, respectively; a_c and d_2 are the center distance and the drive sprocket diameter of the chain drive; u_1 and u_2 are the gear ratios of the first and the second gear units, respectively; u_c and u_t are the gear ratios of the chain drive and the system.

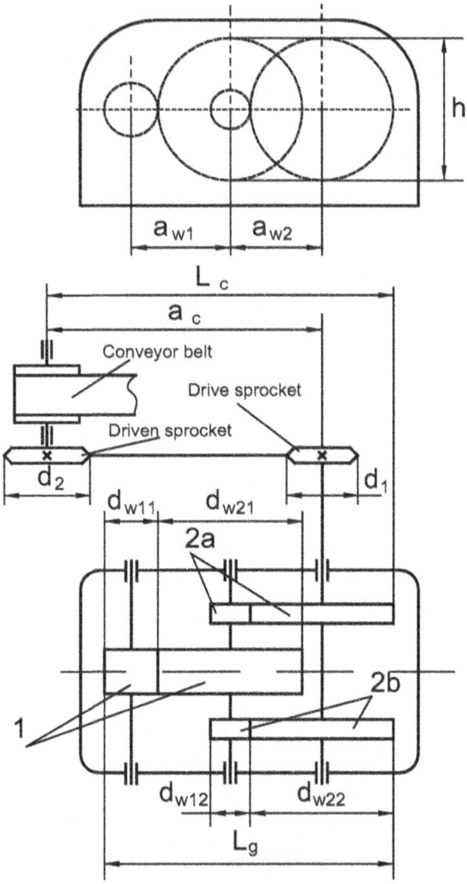

Fig. 1. Calculation schema.

2.1 Determining Center Distance and Pitch Diameter of the First Stage

The center distance a_{w1} of the first stage is found by [32]:

$$a_{w1} = K_a \cdot (u_1 + 1) \cdot \left[T_{11} \cdot k_{H\beta} / \left([\sigma_{H1}]^2 \cdot u_1 \cdot \psi_{ba1}\right)\right]^{1/3} \tag{8}$$

Where, $K_{H\beta}$ is the contact load ratio ranging from 1.02 to 1.28; [32] and the value suggested for $K_{H\beta}$ is 1.1 $k_{H\beta} = 1.1$; $[\sigma_{H1}]$ is the allowable contact stress of the first stage (MPa); In practice, $[\sigma_{H1}] = 350...420$ (MPa); K_a is the material coefficient; $K_a = 43$ [32]; ψ_{ba1} is the coefficient of the wheel face width; $\psi_{ba1} = 0.3...0.35$ [32]; T_{11} is the torque on the pinion which is determined by:

$$T_{11} = T_{out} / \left(u_g \cdot u_c \cdot \eta_{hg}^2 \cdot \eta_c \cdot \eta_{be}^4 \right) \tag{9}$$

In which, T_{out} is the output torque (Nmm); η_{hg} is the helical gear efficiency; $\eta_{hg} = 0.96...0.98$ [32]; η_c is the chain drive efficiency $\eta_c = 0.95...0.97$; η_b is the efficiency of a pair of rolling bearings; $\eta_b = 0.99...0.995$ [32]. Taking $\eta_{hg} = 0.97$, $\eta_c = 0.96$, $\eta_b = 0.992$ and substituting these values into Eq. (9) gives:

$$T_{11} = 1.1432 \cdot T_{out} / \left(u_g \cdot u_c \right) \tag{10}$$

Substituting 43 for K_a, 1.1 for $K_{H\beta}$, and Eq. (10) into (8) gets:

$$a_{w1} = 46.413 \cdot (u_1 + 1) \cdot \left[T_{out} / \left([\sigma_{H1}]^2 \cdot u_1 \cdot u_g \cdot u_c \cdot \psi_{ba1} \right) \right]^{1/3} \tag{11}$$

After receiving a_{w1}, the calculation of the pitch diameter of the first stage can be expressed by [32]:

$$d_{w11} = 2 \cdot a_{w1} / (u_1 + 1) \tag{12}$$

2.2 Calculating Center Distance and Pitch Diameter of the Second Stage

For the second stage, the center distance a_{w2} can be achieved in a similar way to Subsect. 2.1 as [32]:

$$a_{w2} = K_m \cdot (u_2 + 1) \cdot \left[T_{12} \cdot k_{H\beta} / \left([\sigma_{H2}]^2 \cdot u_2 \cdot \psi_{ba2} \right) \right]^{1/3} \tag{13}$$

Where, T_{12} is the pinion torque calculated by:

$$T_{12} = T_{out} / \left(2 \cdot u_2 \cdot u_c \cdot \eta_{hg} \cdot \eta_c \cdot \eta_b^3 \right) \tag{14}$$

With $\eta_{hg} = 0.97$, $\eta_c = 0.96$, $\eta_b = 0.992$ as in Subsect. 2.1, T13 can be given as:

$$T_{13} = 0.5501 \cdot T_{out} / (u_2 \cdot u_c) \tag{15}$$

Substituting (15), $k_m = 43$ and $k_{H\beta} = 1.1$ (as in Subsect. 2.1) into (13) gets:

$$a_{w2} = 36.3703 \cdot (u_2 + 1) \cdot \left[T_{out} / \left([\sigma_{H2}]^2 \cdot u_2^2 \cdot u_c \cdot \psi_{ba2} \right) \right]^{1/3} \tag{16}$$

Hence, the pitch diameter of the second stage is determined by [32]:

$$d_{w22} = 2 \cdot a_{w2} \cdot u_2/(u_2 + 1) \tag{17}$$

2.3 Determining the Drive Sprocket Diameter D_2

The pitch diameter of the drive sprocket is calculated by the following formula [32]:

$$d_2 = d_1 \cdot u_c \tag{18}$$

In which, d_1 is the pitch diameter of the drive sprocket [32]:

$$d_1 = p/\sin(\pi/z_1) \tag{19}$$

Where, z_1 is the number of teeth of the drive sprocket which can be calculated by [30]:

$$z_1 = 32.4 - 2.4 \cdot u_c \tag{20}$$

With p as the chain pitch (mm), p can be obtained from the design power capacity P [32]:

$$P = P_1 \cdot k \cdot k_z \cdot k_n \tag{21}$$

Wherein, P_1 is the power rating (kW) of the chain drive; P_1 is found by:

$$P_1 = T_1 \cdot n_1/(9.55 \cdot 10^6) \tag{22}$$

In which, n_1 is the drive sprocket revolution (rpm):

$$n_1 = n_m/u_g \tag{23}$$

$$T_1 = T_{out}/(u_c \cdot \eta_c \cdot \eta_b) \tag{24}$$

In these equations, η_c is the chain drive efficiency (η_c = 0.95 ÷ 0.97 [32]); η_b is the bearing efficiency (η_b = 0.99 ÷ 0.995 [32]); T_1 and T_{out} are the torques on the drive sprocket and the output torque (Nmm). Choosing $\eta_c = 0.96$, $\eta_b = 0.992$ and substituting them into (24) gives:

$$T_1 = 1.0502 \cdot T_{out}/u_c \tag{25}$$

k, k_z and k_n are the coefficients which are determined by [32]:

$$k = k_d \cdot k_p \cdot k_c \cdot k_{adj} \cdot k_{lub} \cdot k_{con} \tag{26}$$

$$k_z = 25/z_1 \tag{27}$$

$$k_n = n_{01}/n_1 \tag{28}$$

In the above equations, k_d, k_p, k_c, k_{adj}, $k_{\rm lub}$ and k_{con} are the coefficients referring to the effects of shock factor, the drive position, the center distance, the adjusted possibility of the center distance and the lubrication conditions, respectively; n_{01} is the tabulated number of the drive sprocket teeth.

2.4 Experimental Work

Table 1. Input parameters.

Factor	Code	Unit	Low	High
Total gearbox ratio	u_t	–	10	50
Coefficient of wheel face width of stage 1	x_{ba1}	–	0.3	0.35
Coefficient of wheel face width of stage 2	x_{ba2}	–	0.35	0.4
Allowable contact stress of stage 1	AS_1	MPa	350	420
Allowable contact stress of stage 2	AS_2	MPa	350	420
Output torque	T_{out}	Nmm	10^5	10^7

As mentioned above, to investigate the impacts of the input factors (Table 1) on the optimal gear ratios a simulation experiment with a 2-level full factorial design and 6 input factors was constructed. The experiment implementation was carried out with 64 test runs. Besides, based on Eqs. (4) and (5) a computer program was set up. The experimental plans and the output responses (the gear ratio of the chain drive u_c and the optimal gear ratios of the first stage u_1) are demonstrated in Table 2.

The dependence level of the optimal gear ratios on the input parameters was explored based on the implementation of a simulation experiment with a 2-level full factorial design and 6 input factors (see Table 1). Accordingly, 64 test runs were conducted. Moreover, a computer program was created based on Eqs. (6) and (7). The experimental plan and the output responses (the optimal gear ratios of the chain drive u_c and the first stage u_1) are described in Table 2.

Table 2. Experimental plans and output responses

Std order	Run order	Center Pt	Blocks	u_t	Xba_1	Xba_2	AS_1	AS_2	Tout	u_c	u_1
57	1	1	1	10	0.3	0.35	420	420	10000	1.00	5.95
44	2	1	1	50	0.35	0.35	420	350	10000	1.00	9.00
39	3	1	1	10	0.35	0.4	350	350	10000	1.00	5.35
27	4	1	1	10	0.35	0.35	420	420	100	1.00	9.00
45	5	1	1	10	0.3	0.4	420	350	10000	1.00	5.95
14	6	1	1	50	0.3	0.4	420	350	100	1.00	9.00
...											
63	63	1	1	10	0.35	0.4	420	420	10000	1.00	6.42
58	64	1	1	50	0.3	0.35	420	420	10000	1.00	9.00

3 Methodology Results and Discussion

3.1 Determining Optimum Gear Ratio of First Stage

The main effects of the input parameters on the optimum gear ratio of the first stage are exposed in Fig. 2. It is reported that the largest affected factors are u_t and T_{out} The less influential parameters include AS_1 and Xba_1 while AS_1 and AS_2 are non-influential parameters since the slope of the graph equals 0.

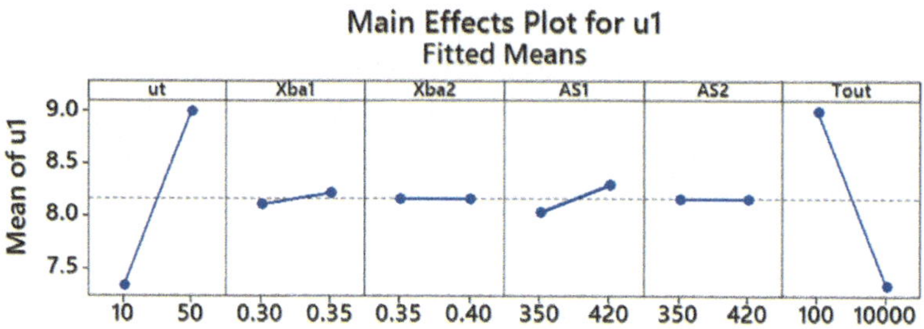

Fig. 2. Main effects plot for optimum gear ratio of u_1

The Normal Plot of the standardized effects is depicted in Fig. 3. It is interpreted that u_t (factor A), T_{out} (factor F) and the interaction AF are the most influential parameters on u_1. Meanwhile, the other factors are considerably less dominant. Besides, most of the variables including u_t, ψ_{ba2}, AS_1, and the interactions AF, BF and DF have a positive standardized effect. Accordingly, when their values grow, u_1 rises. Conversely, T_{out} and the interactions AB and AD have a negative standardized effect. u_1 expands if they decline.

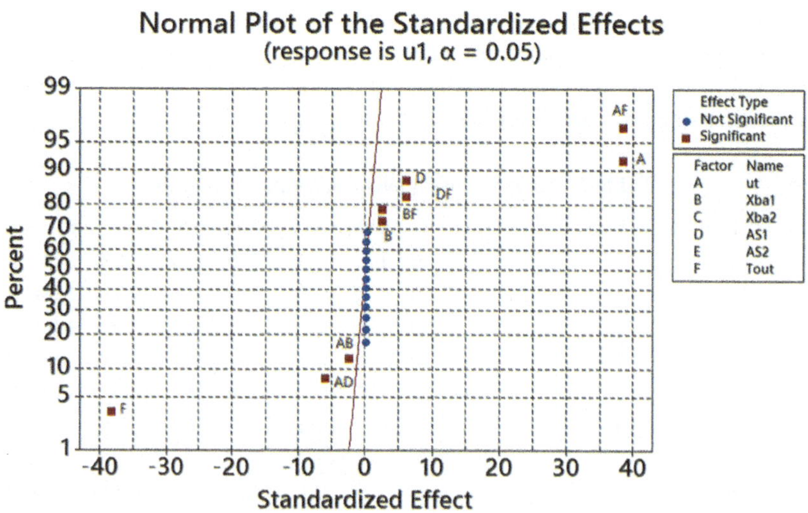

Fig. 3. Normal Plot for u_1

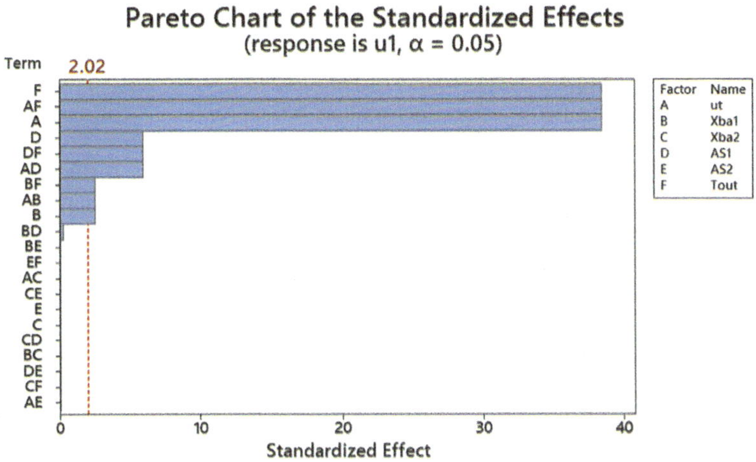

Fig. 4. Pareto chart for u_1

Figure 4 illustrates the Pareto chart of the standardized effects. It can be observed that the reference line is crossed by the bars denoting several factors including the output torque (factor F), the total gearbox ratio (factor A), the wheel face width coefficient of the bevel gear set (factor B), the helical gear set (factor D) and their interactions AF, DF, AD and AB. This occurrence means the factors are statistically significant ($\alpha = 0.05$) with the response model.

Coded Coefficients

Term	Effect	Coef	SE Coef	T-Value	P-Value	VIF
Constant		8.1669	0.0191	426.81	0.000	
ut	1.6662	0.8331	0.0191	43.54	0.000	1.00
Xba1	0.1087	0.0544	0.0191	2.84	0.006	1.00
AS1	0.2587	0.1294	0.0191	6.76	0.000	1.00
Tout	-1.6662	-0.8331	0.0191	-43.54	0.000	1.00
ut*Xba1	-0.1087	-0.0544	0.0191	-2.84	0.006	1.00
ut*AS1	-0.2587	-0.1294	0.0191	-6.76	0.000	1.00
ut*Tout	1.6662	0.8331	0.0191	43.54	0.000	1.00
Xba1*Tout	0.1087	0.0544	0.0191	2.84	0.006	1.00
AS1*Tout	0.2587	0.1294	0.0191	6.76	0.000	1.00

Model Summary

S	R-sq	R-sq(adj)	R-sq(pred)
0.153076	99.09%	98.93%	98.71%

Fig. 5. Estimated Effects and Coefficients for u_1

The estimated effects and coefficients for u_1 are demonstrated in Fig. 5. It is clarified that the effects of all the parameters on u_1 including u_t, ψ_{ba1}, T_{out} and the interactions AS_1, $u_t \cdot \psi_{ba1}$, $u_t \cdot AS_1$, $\psi_{ba1} \cdot T_{out}$, and $u_t \cdot T_{out}$ $AS_1 \cdot T_{out}$. Thus, the relation between u_1 and the significant effect factors can be expressed as follows:

$$u_1 = 5.89 + 0.1057 \cdot u_t + 3.22 \cdot \psi_{ba1} + 0.00547 \cdot AS_1 - 0.000851 \cdot T_{out} - 0.1088 \cdot u_t \cdot \psi_{ba1} - 0.000185 \cdot u_t \cdot AS_1 + 0.000008 \cdot u_t \cdot T_{out} + 0.000439 \cdot \psi_{ba1} \cdot T_{out} + 0.000001 \cdot AS_1 \cdot T_{out} \quad (29)$$

3.2 Determining Optimum Gear Ratio of Chain Drive and Second Gear Stage

From the experimental results (Table 2) it is interesting that the optimal ratios of the chain drive u_c is constant. That means it is independent of the input parameters. Additionally, the optimal gear ratios of the chain drive is $u_c = 1$. It proves that with the objective function of the problem, the chain drive has no effect on deceleration but only serves to transmit torque to the required axial distance.

$$u_c = 1 \quad (30)$$

Equations (29) and (30) are used to find the optimal gear ratios of the first stage of the gearbox u_1 and the chain drive u_c. After having u_1 and u_c, the optimal gear ratio of the second stage u_2 can be determined by the following equation:

$$u_2 = u_t/(u_1 \cdot u_c) \quad (31)$$

Because $u_c = 1$, Eq. (31) can be rewritten as:

$$u_2 = u_t/u_1 \quad (32)$$

4 Conclusions

The study reports an optimization problem with the aim of determining the optimal gear ratios of mechanical drive systems using a two-stage helical gearbox with second stage double gear sets and a chain drive to achieve the minimal cross section area of the system. A simulation experiment was performed, then the results were examined to estimate the influences of the input parameters on the optimal gear ratios. Significantly, it is recommended that the minimal system cross section area can be achieved precisely by applying the provided equations.

Acknowledgements. The work described in this paper was supported by Thai Nguyen University of Technology for a scientific project.

References

1. Kudreavtev, V.N., Gierzaves, I.A., Glukharev, E.G.: Design and calculus of gearboxes. Mashinostroenie Publishing, Sankt Petersburg (1971). (in Russian)
2. Chat, T.: Some problems of kinematics calculation of transmission mechanics system. In: Proceedings of the National Conference on Engineering Mechanics, vol. 2, Hanoi, pp. 7–12 (1993). (in Vietnamese)
3. Milou, G., Dobre, G., Visa, F., Vitila, H.: Optimal design of two step gear units, regarding the main parameters. VDI Berichte **1230**, 227 (1996)
4. Pi, V.N.: A method for optimal calculation of total transmission ratio of two step helical gearboxes. In: Proceedings of the National conference on Engineering Mechanics, Ha Noi, p. 12 (2001)
5. Cam, N.T.H., Pi, V.N., Tuan, N.K., Hung, L.X., Thao, T.T.P.: Determining optimal partial transmission ratios of mechanical driven systems using a V-belt drive and a helical reducer with second-step double gear-sets. In: Fujita, H., et al. (eds.) ICERA 2018. LNNS, vol. 63, pp. 261–269 (2019). https://doi.org/10.1007/978-3-030-04792-4_35
6. Petrovski, A.N., Sapiro, B.A., Saphonova, N.K.: About optimal problem for multi-step gearboxes. Vestnik Mashinostroenie **10**, 13 (1987). (in Russian)
7. Pi, V.N.: A new study on optimal calculation of partial transmission ratios of two-step helical gearboxes. In: 2nd WSEAS International Conference on Computer Engineering and Applications (CEA'08), Acapulco, Mexico, 25–27 January, pp. 162–165 (2008)
8. Nguyen, K.T., Vu, N.P., Nguyen, T.H.C., Tran, T.P.T., Ho, K.T., Le, X.H., Hoang, T.T.: Determining optimal gear ratios of a two-stage helical reducer for getting minimal acreage of cross section. In: MATEC Web of Conferences 213, 01008 (2018). https://doi.org/10.1051/matecconf/201821301008
9. Pi, V.N., Tuan, N.K.: Optimum determination of partial transmission ratios of three-step helical gearboxes for getting minimum cross section dimension. J. Environ. Sci. Eng. A **5**, 570–573 (2016)
10. Pi, V.N., Binh, N.D., Dac, V.Q., The, P.Q.: Optimal calculation of total transmission ratio of three-step helical gearboxes for minimum mass of gears. J. Sci. Technol. Eng. Univ. **6**, 91 (2006). (in Vietnamese)
11. Pi, V.N.: A new study on optimal calculation of partial transmission ratio of three-step helical reducers. In: The 3rd IASME/WSEAS International Conference on Continuum Mechanics, Cambridge, UK, p. 23 (2008)
12. Pi, V.N.: A new study on optimal calculation of partial transmission ratios of three-step helical reducers for getting minimal cross section dimension. In: The 2nd WSEAS International Conference on Computer Engineering and Applications (CEA'08), Acapulco, Mexico, 25–27 January, pp. 290–293 (2008)
13. Pi, V.N.: Optimal determination of partial transmission ratios of three-step helical gearboxes with first and third-step double gear-sets for minimal mass of gear. In: American Conference on Applied Mathematics (MATH'08), Harvard, Massachusetts, USA, 24–26 March, pp. 385–388 (2008)
14. Romhild, I., Linke, H.: Gezielte Auslegung Von Zahnradgetrieben mit minimaler Masse auf der Basis neuer Berechnungsverfahren. Konstruktion **44**, 229–236 (1992)
15. Pi, V.N.: Optimal determination of partial transmission ratios for four-step helical gearboxes with first and third step double gear-sets for minimal mass of gears. In: Applied Computing Conference (ACC '08), Istanbul, Turkey, 27–30 May, pp. 53-57 (2008)

16. Pi, V.N.: Optimal calculation of partial transmission ratios for four-step helical gearboxes with first and third-step double gear-sets for minimal gearbox length. In: American Conference on Applied Mathematics (MATH'08), Harvard, Massachusetts, USA, 24–26 March, pp. 29–32 (2008)
17. Hung, L.X., Pi, V.N., Van Du, N.: Optimal calculation of partial transmission ratios of four-step helical gearboxes with second and fourth-step double gear-sets for minimal mass of gears. In: The international symposium on Mechanical Engineering (ISME), Ho Chi Minh city, Vietnam, pp. 21–23, September 2009
18. Pi, V.N.: A new and effective method for optimal calculation of total transmission ratio of two step bevel - helical gearboxes. In: International colloquium on Mechanics of Solids, Fluids, Structures & Interaction Nha Trang, Vietnam, pp. 716–719 (2000)
19. Pi, V.N., Binh, N.D., Dac, V.Q., The, P.Q.: A new and effective method for optimal splitting of total transmission ratio of three step bevel-helical gearboxes. In: The Sixth Vietnam Conference on Automation, Hanoi, pp. 175–180 (2005)
20. Tuan, N.K., Thao, T.T.P., Cam, N.T.H., Hung, L.X., Pi, V.N.: Optimum calculation of partial transmission ratios of mechanical driven systems using a V-belt and a three-step bevel helical gearbox. In: Fujita, H., et al. (eds.) ICERA 2018. LNNS, vol. 63, pp. 469–476 (2019). https://doi.org/10.1007/978-3-030-04792-4_61
21. Chernavsky, S.A., et al.: Design of mechanical transmissions: Manual for high technical schools (Moscow: Mashinostroenie), p 560 (1984)
22. Pi, V.N., Dac, V.Q.: Calculation of total transmission ratio of two step worm reducers for the best reasonable gearbox housing structure. J. Sci. Technol. Thai Nguyen University **1**(41), 65–69 (2007). (in Vietnamese)
23. Pi, V.N., Dac, V.Q.: Optimal calculation of partial transmission ratios of worm-helical gear reducers for minimal gearbox length J. Sci. Technol. Tech. Univ. **61**, 73–77 (2007)
24. Pi, V.N., Dac, V.Q.: Optimal calculation of total transmission ratio of worm-helical gear reducers. J. Sci. Technol. Thai Nguyen Univ. **4**(36), No.1, 70–73 (2005)
25. Pi, V.N., Thao, T.T.P., Thao, L.T.P.: A new study on optimum determination of partial transmission ratios of mechanical driven systems using a V-belt and two-step helical gearbox. Vietnam Mech. Eng. J. **10**, 123 (2015)
26. Pi, V.N., Cam, N.T.H., Tuan, N.K.: Optimum calculation of partial transmission ratios of mechanical driven systems using a V-belt and two-step bevel helical gearbox. J. Environ. Sci. Eng. A **5**, 566 (2016)
27. Pi, V.N., Tuan, N.K., Hung, L.X., Cam, N.T.H., Thao, T.T.P.: Determining optimum partial transmission ratios of mechanical driven systems using a V-Belt drive and a three-stage helical reducer. In: Advances in Material Sciences and Engineering, pp. 81–88. Springer (2019)
28. Pi, V.N., Thao, T.T.P., Tuan, D.A.: Optimum determination of partial transmission ratios of mechanical driven systems using a chain drive and two-step helical gearbox. J. Environ. Sci. Eng. B **6**, 80 (2017)
29. Cam, N.T.H., Pi, V.N., Tuan, N.K., Hung, L.X., Thao, T.T.P.: A study on determination of optimum partial transmission ratios of mechanical driven systems using a chain drive and a three-step helical reducer. In: Fujita, H., et al. (eds.) ICERA 2018. LNNS, vol. 63, pp. 91–99 (2019). https://doi.org/10.1007/978-3-030-04792-4_14
30. Chat, T., Van Uyen, L.: Design and calculation of Mechanical Transmissions Systems, vol. 1, Hanoi, Educational Republishing House (2007)

Determining Optimum Gear Ratios of Mechanical Driven Systems Using Three Stage Bevel Helical Gearbox and Chain Drive

Tran Thi Phuong Thao[1], Tran Thi Hong[2], Nguyen Van Cuong[3], Le Hong Ky[4], Nguyen Thanh Tu[1], Le Xuan Hung[1], and Ngoc Pi Vu[1(✉)]

[1] Thai Nguyen University of Technology, Thai Nguyen, Vietnam
vungocpi@tnut.edu.vn
[2] Nguyen Tat Thanh University, Ho Chi Minh City, Vietnam
[3] University of Transport and Communications, Ha Noi, Vietnam
[4] Vinh Long University of Technology Education, Vinh Long, Vietnam

Abstract. The paper presents a study on calculating optimum gear ratios of mechanical driven systems using a three stage bevel helical gearbox and a chain drive. An optimization problem in which the system cross section area was chosen as the objective function was solved. In addition, 6 input parameters including the total system ratio, the allowable contact stress, the face width coefficients of the bevel and the helical gear sets, and the output torque were taken into account. To evaluate the impacts of these factors on the optimum gear ratios, a simulation experiment was designed and performed. From the results of the optimization problem, the effects of the input parameters on the optimum ratios were estimated and some equations to calculate the optimum gear ratios were proposed.

Keywords: Gearbox · Bevel helical gearbox · Gearbox design · Gear ratio · Optimum gear ratio

1 Introduction

In optimum mechanical driven system design, finding optimum gear ratios plays a very important role. This is because the transmission ratio is the determining factor for the volume, the size and the cost of the system.

So far, the optimal transmission ratio of mechanical driven systems has been found in different ways. They can be gained by graphs [1, 2], practical formulas [3] or mathematical models [3–16]. The optimum ratios are also determined for different gearboxes such as helical gearboxes [1–18], bevel gearboxes [1, 3, 19–21], or worm gearboxes [22–25]. Moreover, the optimum gear ratios are also identified for gearboxes with different speed levels such as two-stage gearboxes [1–8], three-stage gearboxes [9–15] and four-stage gearboxes [16–18].

Solving optimization problems with different objective functions to achieve the optimum ratios is another approach. The objective functions can be the minimum mass of gears [10, 13, 14, 16, 18], the minimum gearbox length [7, 11, 24], the minimum

K.-U. Sattler et al. (Eds.): ICERA 2019, LNNS 104, pp. 249–261, 2020.
https://doi.org/10.1007/978-3-030-37497-6_29

gearbox cross section area [9, 12] and so on. Recently, the optimum transmission ratio problems have been solved for mechanical drive systems including gearbox and a V-belt drive [5, 26, 27] or a chain drive [28, 29].

This study focuses on determining the optimum ratios of mechanical systems consisting a three stage bevel helical gearbox and a chain drive for getting minimum system cross section area based on employing a simulation experiment.

2 Optimization Problem

The cross section area of the mechanical driven system using a three stage bevel helical gearbox and a chain drive is determined by (Fig. 1):

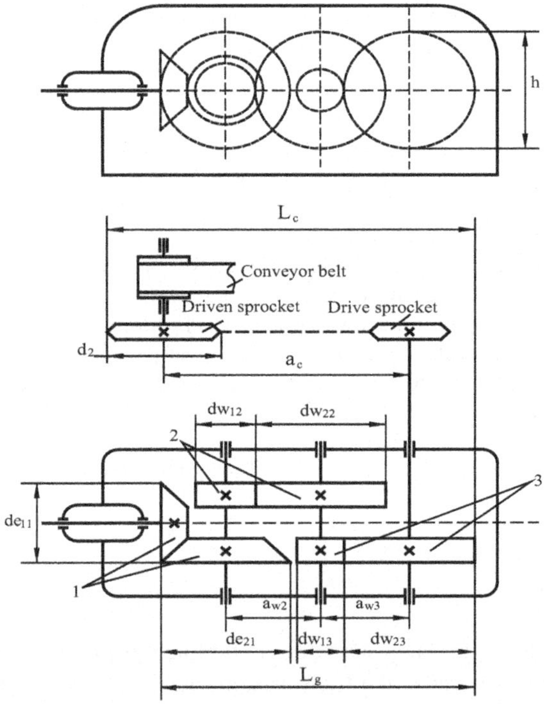

Fig. 1. Calculation schema

$$A = L \cdot h \tag{1}$$

In which,

$$L = \max\left(L_g, L_c\right) \tag{2}$$

$$h = \max(d_{e21}; d_{w22}; d_{w23}; d_2) \tag{3}$$

In which, L_g and L_c are determined by the following equations (see Fig. 1):

$$L_g = d_{e21}/2 + a_{w2} + a_{w3} + d_{w23}/2 \quad (4)$$

$$L_c = d_2/2 + a_c + d_{w23}/2 \quad (5)$$

Therefore, the optimization problem is defined as:

$$\text{minimize } A \quad (6)$$

With the following constraints:

$$\begin{aligned} 1 \le u_1 \le 6 \\ 1 \le u_2 \le 9 \\ 1 \le u_3 \le 9 \\ 1 \le u_c \le 6 \end{aligned} \quad (7)$$

In which, a_{w2}, a_{w3} are the center distance of the second and the third helical gear sets, respectively; d_{e21}, d_{w22} and d_{w23} are the pitch diameters of the first, the second and the third stage, respectively; a_c and d_2 are the center distance and the driven sprocket diameter of the chain drive.

The outer pitch diameter of the pinion of the bevel gear set can be calculated by [30]:

$$d_{e11} = 2 \cdot R_e / \sqrt{1 + u_1^2} \quad (8)$$

Where, R_e is the external cone distance and u_1 is the gear ratios of the bevel gear set.

Besides, for the third helical gear stage, d_{w23} is calculated by [30]:

$$d_{w23} = 2 \cdot a_{w3} \cdot u_3/(u_3 + 1) \quad (9)$$

With u_3 as the gear ratio of the third stage.

From Eqs. (2) and (3), it is found that to solve the optimization problem, it is recommended to determine R_e, a_{w2}, a_{w3}, a_c, d_{e21}, d_{w22}, d_{w23} and d_2.

2.1 Determining the Driven Sprocket Diameter D2

The pitch diameter of the driven sprocket is determined by [30]:

$$d_2 = d_1 \cdot u_c \quad (10)$$

Wherein, d_1 is the pitch diameter of the drive sprocket which can be found by [30]:

$$d_1 = p/\sin(\pi/z_1) \quad (11)$$

In which,

z_1 is the number of teeth of the drive sprocket which is calculated by [28]:

$$z_1 = 32.4 - 2.4 \cdot u_c \tag{12}$$

P is the chain pitch (mm); p can be obtained from the design power capacity P which can be calculated by [30]:

$$P = P_1 \cdot k \cdot k_z \cdot k_n \tag{13}$$

Where, P_1 is the power rating (kW) of the chain drive; P_1 is calculated as:

$$P_1 = T_1 \cdot n_1 / \left(9.55 \cdot 10^6\right) \tag{14}$$

In which, n_I is the revolution of the drive sprocket (rpm):

$$n_1 = n_m / u_g \tag{15}$$

$$T_1 = T_{out} / (u_c \cdot \eta_c \cdot \eta_b) \tag{16}$$

In the above equations, η_c is the efficiency of the chain drive (η_c = 0.95 ÷ 0.97 [30]); η_b is the efficiency of a pair of bearings (η_b = 0.99 ÷ 0.995 [30]); T_1 and T_{out} are the torque on the drive and the output torque (Nmm). Choosing $\eta_c = 0.96$, $\eta_b = 0.992$ and substituting them into (16) gives:

$$T_1 = 1.0502 \cdot T_{out} / u_c \tag{17}$$

k, k_z and k_n are coefficients determined as [30]:

$$k = k_d \cdot k_p \cdot k_c \cdot k_{adj} \cdot k_{\text{lub}} \cdot k_{con} \tag{18}$$

$$k_z = 25 / z_1 \tag{19}$$

$$k_n = n_{01} / n_1 \tag{20}$$

In the above equations, k_d, k_p, k_c, k_{adj}, k_{lub} and k_{con} are the coefficients indicating the effects of shock factor, the drive position, the center distance, the adjusted possibility of the center distance, the conditions of lubrication and the operating, respectively; n_{01} is the tabulated number of the drive sprocket teeth.

2.2 Determining the External Cone Distance of the Bevel Gear Set

For a straight bevel gear set, the external cone distance R_e can be calculated from the pitting resistance [30]:

$$R_e = k_R \cdot \sqrt{u_1^2 + 1} \cdot \sqrt[3]{T_{11} \cdot k_{H\beta 1} / \left[(1 - k_{be}) \cdot k_{be} \cdot u_1 \cdot [\sigma_H]^2\right]} \tag{21}$$

In which, k_R is the material coefficient; $k_R = 50$ (MPa1/3) [30]; k_{be} is the face width coefficient; $k_{be} = 0.25\ldots0.3$ [30]; $[\sigma_{H1}]$ is the allowable contact stress (MPa); $K_{H\beta1}$ is the contact load ratio for pitting resistance. From the data in [30], the following regression equationis found to determine $K_{H\beta1}$ (with the determination coefficient $R^2 = 1$):

$$K_{H\beta1} = 0.25 \cdot k^2 + 0.2 \cdot k + 1.02 \tag{22}$$

Wherein, $k = k_{be} \cdot u_1/(2 - k_{be})$.

T_{11} is the torque on the pinion of the first stage which is found by the following equation:

$$T_{11} = T_{out}/\left(u_g \cdot u_c \cdot \eta_{bg} \cdot \eta_{hg}^2 \cdot \eta_c \cdot \eta_b^4\right) \tag{23}$$

In Eq. (23), u_g is the total gearbox ratio; T_{out} is the output torque (Nmm); η_{bg} is the transmission efficiency of the bevel gear; $\eta_{hg} = 0.95\ldots0.97$ [30]; η_{hg} is the transmission efficiency of the helical gear; $\eta_{hg} = 0.96\ldots0.98$ [30]; η_c is the transmission efficiency of the chain drive $\eta_c = 0.95\ldots0.97$; η_b is the transmission efficiency of a pair of rolling bearings; $\eta_b = 0.99\ldots0.995$ [30]. Choosing $\eta_{bg} = 0.96$, $\eta_{hg} = 0.97$, $\eta_c = 0.96\eta_b = 0.992$ and substituting them into Eq. (23) gets:

$$T_{11} = 1.19 \cdot T_{out}/\left(u_g \cdot u_c\right) \tag{24}$$

Substituting $k_R = 50$ and (24) into (21) gives:

$$R_e = 52.984 \cdot \sqrt{u_1^2 + 1} \cdot \sqrt[3]{T_{out} \cdot k_{H\beta1}/\left[(1 - k_{be}) \cdot k_{be} \cdot u_1 \cdot u_g \cdot u_c \cdot [\sigma_H]^2\right]} \tag{25}$$

2.3 Determining the Center Distance and the Pitch Diameter of the Second Stage

The center distance of the second stage a_{w2} is calculated by [30]:

$$a_{w2} = k_m \cdot (u_2 + 1) \cdot \sqrt[3]{T_{12} \cdot k_{H\beta}/\left([\sigma_H]^2 \cdot u_2 \cdot \psi_{ba2}\right)} \tag{26}$$

In which, $K_{H\beta}$ is the contact load ratio for pitting resistance; $k_{H\beta} = 1.02 \div 1.28$[30] and we can choose $k_{H\beta} = 1.1$; $[\sigma_H]$ is the allowable contact stress (MPa); In practice, $[\sigma_H] = 360\ldots420$ (MPa); k_m is the material coefficient; $k_m = 43$ [30]; ψ_{ba2} is the coefficient of wheel face width; $\psi_{ba2} = 0.3\ldots0.35$ [30];

T_{12} is the torque on the pinion of the second stage; T_{12} can be determined as:

$$T_{12} = T_{out}/\left(u_2 \cdot u_3 \cdot u_c \cdot \eta_{hg}^2 \cdot \eta_c \cdot \eta_b^3\right) \tag{27}$$

Choosing $\eta_{hg} = 0.97$, $\eta_c = 0.96$, $\eta_b = 0.992$ as in Sect. 2.2 and substituting them into Eq. (27) gives:

$$T_{12} = 1.134 \cdot T_{out}/(u_2 \cdot u_3 \cdot u_c) \tag{28}$$

Substituting $k_m = 43$, $k_{H\beta} = 1.1$, $u_2 \cdot u_3 = u_g/u_1$ and (28) into (26) obtains:

$$a_{w2} = 46.2882(u_2 + 1) \cdot \sqrt[3]{\frac{T_{out} \cdot u_1}{[\sigma_H]^2 \cdot u_g \cdot u_2 \cdot u_c \cdot \Psi_{ba2}}} \tag{29}$$

After having a_{w2}, the pitch diameter of the second stage can be determined by [30]:

$$d_{w23} = 2 \cdot a_{w3} \cdot u_3/(u_3 + 1) \tag{30}$$

2.4 Determining the Center Distance and the Pitch Diameter of the Third Stage

Similar to Subsect. 2.2, the center distance of the third stage a_{w3} is determined by [30]:

$$a_{w3} = K_m \cdot (u_3 + 1) \cdot \left[T_{13} \cdot k_{H\beta}/\left([\sigma_H]^2 \cdot u_3 \cdot \psi_{ba3}\right)\right]^{1/3} \tag{31}$$

With T_{13} as the output torque which is found by:

$$T_{13} = T_{out}/\left(u_3 \cdot u_c \cdot \eta_{hg} \cdot \eta_c \cdot \eta_b^2\right) \tag{32}$$

Choosing $\eta_{hg} = 0.97$, $\eta_c = 0.96$, $\eta_b = 0.992$ as in Subsect. 2.2 gets:

$$T_{13} = 1.0912 \cdot T_{out}/(u_3 \cdot u_c) \tag{33}$$

Substituting (33), $k_m = 43$ and $k_{H\beta} = 1.1$ (as in Sect. 2.2) into (31) gets:

$$a_{w3} = 45.6983 \cdot (u_3 + 1) \cdot \left[T_{out}/\left([\sigma_H]^2 \cdot u_3^2 \cdot u_c \cdot \psi_{ba3}\right)\right]^{1/3} \tag{34}$$

In addition, the pitch diameter of the third stage then is calculated by [30]:

$$d_{w23} = 2 \cdot a_{w3} \cdot u_3/(u_3 + 1) \tag{35}$$

2.5 Experimental Work

With the aim of investigating the influences of the input parameters on the optimum gear ratios, a simulation experiment was carried out. In the test, a 2-level full factorial design accompanying 6 input parameters (Table 1) was chosen. Hence, the experiment was planned with $2^6 = 64$ number of runs. To conduct the experiment, a computer program was created based on Eqs. (6) and (7). The different levels of the input parameters and the output responses (the optimum gear ratios of the chain drive u_c and the second stage u_2) are expressed in Table 2.

3 Results and Discussions

3.1 Influences of Input Parameters on the Optimum Gear Ratio of the Chain Drive

To estimate the effects of the input factors on the optimum gear ratio of the chain drive u_c, Fig. 2 shows the graph of the main effects for u_c. From the figure it is noted that u_c depends strongly on the total system ratio u_t and the output torque T_{out}. It increases strongly when u_t increases and falls rapidly with the rise of T_{out}. Besides, the coefficient of the wheel face width of the second and the third stages ψ_{ba2}, ψ_{ba3} and the allowable contact stress AS also affect u_c but to a lesser extent than u_t and T_{out}. In addition, u_c is not affected by the coefficient of the face width of bevel gear set k_{be}.

Table 1. Input parameters

Factor	Code	Unit	Low	High
Total system ratio	u_t	–	40	70
Coefficient of the face width of bevel gear set	K_{be}	–	0.25	0.3
Coefficient of wheel face width of stage 2	X_{ba2}	–	0.3	0.35
Coefficient of wheel face width of stage 3	X_{ba3}	–	0.35	0.4
Allowable contact stress	AS	MPa	350	420
Output torque	T_{out}	Nm	10^2	10^7

Table 2. Experimental plans and output response

RunOrder	CenterPt	Blocks	u_t	K_{be}	Xba1	Xba2	AS (MPa)	Tout (Nm)	u_c	u_1	u_2
1	1	1	40	0.25	0.3	0.35	350	10000	4.00	1	4.35
2	1	1	40	0.3	0.35	0.4	350	100	4.40	1	4.11
3	1	1	70	0.25	0.3	0.4	420	10000	4.50	1	5.60
4	1	1	70	0.25	0.3	0.4	350	100	4.90	1	5.27
5	1	1	70	0.25	0.35	0.4	420	10000	4.50	1	5.87
6	1	1	40	0.25	0.35	0.4	420	10000	4.10	1	4.32
63	1	1	70	0.3	0.35	0.35	420	100	4.90	1	5.81
64	1	1	70	0.3	0.35	0.4	420	10000	4.50	1	5.87

The Normal Plot of the standardized effects is displayed in Fig. 3. It is recognized from the figure that the total system ratio u_t (factor A) and the output torque T_{out} (factor F) are the most significant factors for u_c. In addition, u_t (factor A), AS (factor E), ψ_{ba3}

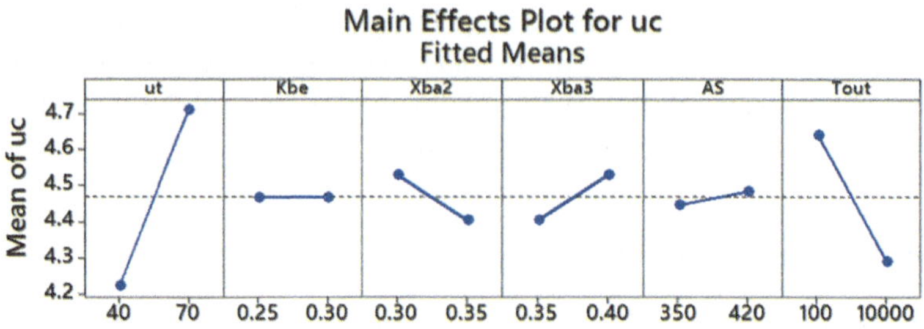

Fig. 2. Main effects plot for optimum gear ratio of the chain drive

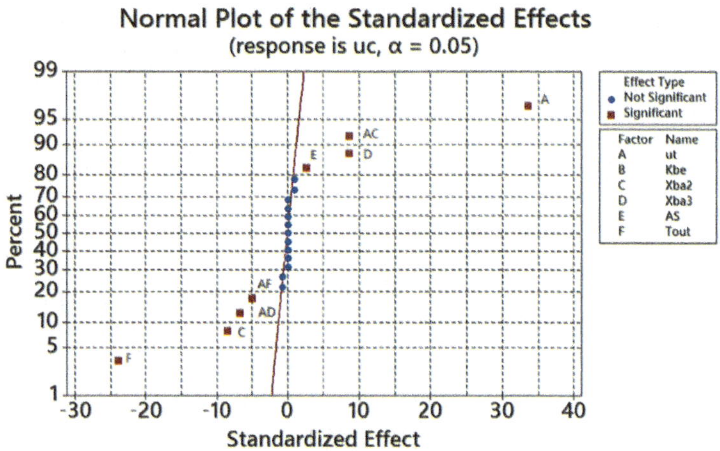

Fig. 3. Normal plot for u_c

Coded Coefficients

Term	Effect	Coef	SE Coef	T-Value	P-Value	VIF
Constant		4.46875	0.00658	678.93	0.000	
ut	0.48750	0.24375	0.00658	37.03	0.000	1.00
Xba2	-0.12500	-0.06250	0.00658	-9.50	0.000	1.00
Xba3	0.12500	0.06250	0.00658	9.50	0.000	1.00
AS	0.03750	0.01875	0.00658	2.85	0.006	1.00
Tout	-0.35000	-0.17500	0.00658	-26.59	0.000	1.00
ut*Xba2	0.12500	0.06250	0.00658	9.50	0.000	1.00
ut*Xba3	-0.10000	-0.05000	0.00658	-7.60	0.000	1.00
ut*Tout	-0.07500	-0.03750	0.00658	-5.70	0.000	1.00

Model Summary

S	R-sq	R-sq(adj)	R-sq(pred)
0.0526567	97.80%	97.48%	97.02%

Fig. 4. Estimated effects and coefficients for u_c

(factor D) and the interaction AC have a positive standardized effect, which means if they increase, u_c grows. In contrast, ψ_{ba2} (factor C), T_{out} (factor F) and the interactions AD and AF have a negative standardized effect. If they rise, u_c decreases.

Figure 4 presents the estimated effects and coefficients for u_c. It is reported that the parameters which have a significant effect on u_c include u_t, ψ_{ba2}, ψ_{ba3}, AS, T_{out} and the interactions $u_t \cdot \psi_{ba2}$, $u_t \cdot \psi_{ba3}$ and $u_t \cdot T_{out}$. Therefore, the relation between u_c and the significant effect factors can be expressed by:

$$\begin{aligned} u_c = 3.511 + 0.01463 \cdot u_t - 11.67 \cdot \psi_{ba2} + 9.83 \cdot \psi_{ba3} + 0.000536 \cdot AS \\ -0.000008 \cdot T_{out} + 0.1667 \cdot u_t \cdot \psi_{ba2} - 0.1333 \cdot u_t \cdot \psi_{ba3} - 0.000001 \cdot u_t \cdot T_{out} \end{aligned} \quad (36)$$

It is noted from Fig. 4 that the values of adj-R2 and pred-R2 are higher than 97%, which indicates that the Eq. (36) is corresponding to the experimental data (Fig. 4). Hence, this equation can be used to calculate the optimum gear ratio of the chain drive u_c.

3.2 Effect of Input Parameters on the Optimum Gear Ratio of the First Stage

It is reported from the results of the optimization problem (Table 2) that the optimal gear ratio of the bevel gear set (or the first stage of the gearbox) u_1 is 1. That means u_1 is not affected by the input parameters. In addition, with the selected objective function, the bevel gear unit does not reduce the speed but only acts to change the direction of movement to a perpendicular direction.

3.3 Effect of Input Parameters on the Optimum Gear Ratio of the Second Stage

Figure 5 illustrates the effects of six input parameters on the optimum gear ratio of the second helical gear set u_2. It is realized that the largest influence belongs to u_t, followed by ψ_{ba2}, ψ_{ba3}, T_{out} and AS. In addition, k_{be} is not affected because the slope of this graph is 0.

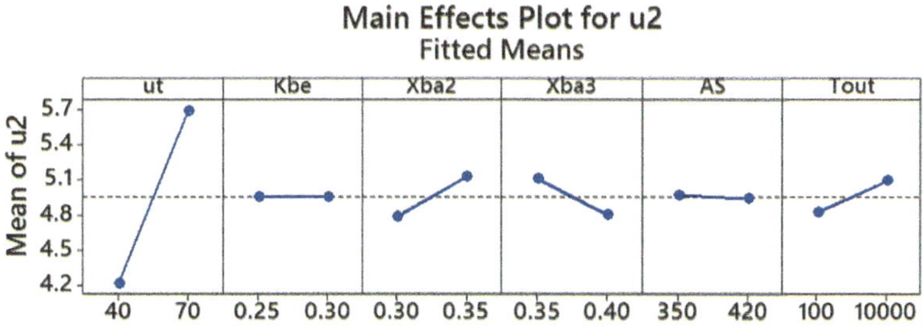

Fig. 5. Main effects plot for optimum gear ratio of u_2

Figure 6 shows the Normal Plot of the standardized effects. According to this graph, u_t is the most influential variable on u_2, followed by ψ_{ba2}, ψ_{ba3}, T_{out}, $u_t \cdot T_{out}$, $u_t \cdot \psi_{ba2}$, $u_t \cdot \psi_{ba3}$ and AS. Also, u_t (factor A), ψ_{ba2} (factor C), T_{out} (factor F) and the interactions AF and ADF have a positive standardized effect. When these factors ascend, u_2 descends. On the contrary, ψ_{ba3} (factor D), AS (factor E) and the interaction AC have a negative standardized effect. If they grow, u_2 decreases.

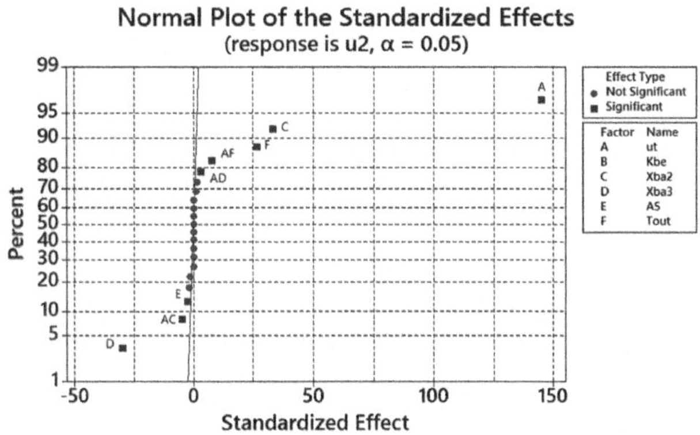

Fig. 6. Normal plot for u_2

Coded Coefficients

Term	Effect	Coef	SE Coef	T-Value	P-Value	VIF
Constant		4.95765	0.00474	1045.59	0.000	
Ut	1.47152	0.73576	0.00474	155.18	0.000	1.00
Xba2	0.33788	0.16894	0.00474	35.63	0.000	1.00
Xba3	-0.30466	-0.15233	0.00474	-32.13	0.000	1.00
AS	-0.02547	-0.01273	0.00474	-2.69	0.010	1.00
Tout	0.26726	0.13363	0.00474	28.18	0.000	1.00
ut*Xba2	-0.04965	-0.02483	0.00474	-5.24	0.000	1.00
ut*Xba3	0.03181	0.01591	0.00474	3.35	0.001	1.00
ut*Tout	0.07689	0.03845	0.00474	8.11	0.000	1.00

Model Summary

S	R-sq	R-sq(adj)	R-sq(pred)
0.0379317	99.80%	99.77%	99.73%

Fig. 7. Estimated effects and coefficients for u_2

Figure 7 illustrates the estimated effects and coefficients for u_2. The parameters found to affect significantly on u_2 are u_t, ψ_{ba2}, ψ_{ba3}, AS, T_{out} and the interactions $u_t \cdot \psi_{ba2}$, $u_t \cdot \psi_{ba3}$ and $u_t \cdot T_{out}$. Hence, the relation between u_2 and the significant effect parameters can be determined by the following equation:

$$u_2 = 2.187 + 0.05205 \cdot u_t + 10.399 \cdot \psi_{ba2} - 8.426 \cdot \psi_{ba3} - 0.000364 \cdot AS \\ -0.000001 \cdot T_{out} - 0.0662 \cdot u_t \cdot \psi_{ba2} + 0.0424 \cdot u_t \cdot \psi_{ba3} + 0.000001 \cdot u_t \cdot T_{out} \quad (37)$$

Equation (37) can be used to calculate the gear ratio of the second stage of the gearbox u_2. After having u_1, u_2 (Eq. (36)) and u_c (Eq. (37)), the gear ratio of the third stage can be reached by:

$$u_3 = u_t/(u_1 \cdot u_2 \cdot u_c) \quad (38)$$

As $u_1 = 1$, Eq. (38) can be rewritten as:

$$u_3 = u_t/(u_2 \cdot u_c) \quad (39)$$

4 Conclusions

In this paper a study on calculating the optimum gear ratios of mechanical driven systems using a three stage bevel helical gearbox and a chain drive was reported. In the problem, the minimum system cross section area was chosen as the objective function and 6 input parameters including the total system gear ratio, the face width coefficients of the bevel and the helical gear sets, the allowable contact stress, and the output torque were investigated. To evaluate the effects of these parameters on the optimum gear ratios, a simulation experiment was planned and performed. Significantly, some models to estimate the optimum gear ratios were proposed.

Acknowledgements. The work described in this paper was supported by Thai Nguyen University of Technology for a scientific project.

References

1. Kudreavtev, V.N., Gierzaves, I.A., Glukharev, E.G.: Design and calculus of gearboxes. Mashinostroenie Publishing, Sankt Petersburg (1971). (in Russian)
2. Chat, T.: Some problems of kinematics calculation of transmission mechanics system. In: Proceedings of the National Conference on Engineering Mechanics, Hanoi, vol. 2, pp. 7–12 (1993). (in Vietnamese)
3. Milou, G., Dobre, G., Visa, F., Vitila, H.: Optimal design of two step gear units, regarding the main parameters. VDI Berichte **1230**, 227 (1996)
4. Ngoc Pi, V.: A method for optimal calculation of total transmission ratio of two step helical gearboxes. In: Proceedings of the National conference on Engineering Mechanics, Ha Noi, p. 12 (2001)
5. Cam, N.T.H., Pi, V.N., Tuan, N.K., Hung, L.X., Thao, T.T.P.: Determining optimal partial transmission ratios of mechanical driven systems using a V-belt drive and a helical reducer with second-step double gear-sets. In: Fujita, H., et al. (eds.) ICERA 2018, LNNS, vol. 63, pp. 261–269 (2019). https://doi.org/10.1007/978-3-030-04792-4_35

6. Petrovski, A.N., Sapiro, B.A., Saphonova, N.K.: About optimal problem for multi-step gearboxes. Vestnik Mashinostroenie **10**, 13 (1987). (in Russian)
7. Pi, V.N.: A new study on optimal calculation of partial transmission ratios of two-step helical gearboxes. In: 2nd WSEAS Int. Conf on Computer Engineering and Applications (CEA 2008), Acapulco, Mexico, 25–27 January 2008, pp. 162–165 (2008)
8. Tuan, N.K., Pi, V.N., Cam, N.T.H., Thao, T.T.P., Thanh, H.K., Hung, L.X., Tham, H.T.: Determining optimal gear ratios of a two-stage helical reducer for getting minimal acreage of cross section. In: MATEC Web of Conferences, vol. 213, p. 01008 (2018). https://doi.org/10.1051/matecconf/201821301008
9. Pi, V.N., Tuan, N.K.: Optimum determination of partial transmission ratios of three-step helical gearboxes for getting minimum cross section. J. Environ. Sci. Eng. A **5**, 570–573 (2016)
10. Pi, V.N., Binh, N.D., Dac, V.Q.: Optimal calculation of total transmission ratio of three-step helical gearboxes for minimum mass of gears. J. Sci. Technol. Eng. Univ. **6**, 91 (2006). (in Vietnamese)
11. Pi, V.N.: A new study on optimal calculation of partial transmission ratio of three-step helical reducers. In: The 3rd IASME/WSEAS International Conference on Continuum Mechanics, Cambridge, UK, p. 23 (2008)
12. Pi, V.N.: A new study on optimal calculation of partial transmission ratios of three-step helical reducers for getting minimal cross section dimension. In: The 2nd WSEAS International Conference on Computer Engineering and Applications (CEA 2008), Acapulco, Mexico, 25–27 January, pp. 290–293 (2008)
13. Pi, V.N.: Optimal determination of partial transmission ratios of three-step helical gearboxes with first and third-step double gear-sets for minimal mass of gear. In: American Conference on Applied Mathematics (MATH 2008), Harvard, Massachusetts, USA, 24–26 March, pp. 385–388 (2008)
14. Romhild, I., Linke, H.: Gezielte Auslegung Von Zahnradgetrieben mit minimaler Masse auf der Basis neuer Berechnungsverfahren. Konstruktion **44**, 229–236 (1992)
15. Pi, V.N.: A new study on the optimal prediction of partial transmission ratios of three-step helical gearboxes with second-step double gear-sets. WSEAS Trans. Appl. Theor. Mech. **2**(11) (2007)
16. Pi, V.N.: Optimal determination of partial transmission ratios for four-step helical gearboxes with firt and third step double gear-sets for minimal mass of gears. In: Applied Computing Conference (ACC 2008), Istanbul, Turkey, 27–30 May (2008)
17. Pi, V.N.: Optimal calculation of partial transmission ratios for four-step helical gearboxes with first and third-step double gear-sets for minimal gearbox length. In: American Conference on Applied Mathematics (MATH 2008), Harvard, Massachusetts, USA, 24–26 March, pp. 29–32 (2008)
18. Hung, L.X., Pi, V.N., Du, N.V.: Optimal calculation of partial transmission ratios of four-step helical gearboxes with second and fourth-step double gear-sets for minimal mass of gears. The International Symposium on Mechanical Engineering, ISME, Ho Chi Minh City, Vietnam, pp. 21–23, September 2009
19. Pi, V.N.: A new and effective method for optimal calculation of total transmission ratio of two step bevel - helical gearboxes. In: International colloquium on Mechannics of Solids, Fluids, Structures & Interaction Nha Trang, Vietnam, pp. 716–719 (2000)
20. Pi, V.N., Binh, N.D., Dac, V.Q., The, P.Q.: A new and effective method for optimal splitting of total transmission ratio of three step bevel-helical gearboxes. In: The Sixth Vietnam Conference on Automation, Hanoi, pp. 175–180 (2005)^

21. Tuan, N.K., Thao, T.T.P., Cam, N.T.H., Hung, L.X., Pi, V.N.: Optimum calculation of partial transmission ratios of mechanical driven systems using a V-belt and a three-step bevel helical gearbox. In: Fujita, H., et al. (eds.) ICERA 2018, LNNS, vol. 63, pp. 469–476 (2019). https://doi.org/10.1007/978-3-030-04792-4_61
22. Chernavsky, S.A., et al.: Design of Mechanical Transmissions: Manual for High Technical Schools, p. 560. Mashinostroenie, Moscow (1984)
23. Pi, V.N., Dac, V.Q.: Calculation of total transmission ratio of two step worm reducers for the best reasonable gearbox housing structure. J. Sci. Technol. **1**(41), 65–69 (2007). (in Vietnamese)
24. Pi, V.N., Dac, V.Q.: Optimal calculation of partial transmission ratios of worm-helical gear reducers for minimal gearbox length. J. Sci. Technol. **61**, 73–77 (2007)
25. Pi, V.N., Dac, V.Q.: Optimal calculation of total transmission ratio of worm-helical gear reducers. J. Sci. Technol. **1**(4), 36, 70–73 (2005)
26. Pi, V.N., Thao, T.T.P., Thao, L.T.P.: A new study on optimum determination of partial transmission ratios of mechanical driven systems using a V-belt and two-step helical gearbox. Vietnam Mech. Eng. J. **10**, 123 (2015)
27. Pi, V.N., Cam, N.T.H., Tuan, N.K.: Optimum calculation of partial transmission ratios of mechanical driven systems using a V-belt and two-step bevel helical gearbox. J. Environ. Sci. Eng. A **5**, 566 (2016)
28. Pi, V.N., Thao, T.T.P., Tuan, D.A.: Optimum determination of partial transmission ratios of mechanical driven systems using a chain drive and two-step helical gearbox. J. Environ. Sci. Eng. B **6**, 80 (2017)
29. Cam, N.T.H., Pi, V.N., Tuan, N.K., Hung, L.X., Thao, T.T.P.: A study on determination of optimum partial transmis-sion ratios of mechanical driven systems using a chain drive and a three-step helical reducer. In: Fujita, H., et al. (eds.) ICERA 2018, LNNS, vol. 63, pp. 91–99 (2019). https://doi.org/10.1007/978-3-030-04792-4_14
30. Chat, T., Uyen, L.V.: Design and calculation of Mechanical Transmissions Systems, vol. 1. Educational Republishing House, Hanoi (2007)

Development of a Backlight Imaging System to Investigate Liquid Breakup in Near-Field Swirl Atomizer

Phuong X. Pham[1](✉), Nam V. T. Pham[1], Lap D. Vu[1], Kien T. Nguyen[1], Thin V. Pham[1], Vu H. Nguyen[1], Thi D. Luong[1], and Manh Q. Nguyen[2]

[1] Le Quy Don Technical University, 236 Hoang Quoc Viet, Hanoi, Vietnam
Phuongpham@lqdtu.edu.vn
[2] Weapon Institute, Hanoi, Vietnam

Abstract. A shadowgraph technique developed in this work aims to investigate the influence of swirl in air-blast atomizer such as the ones adopted in gas turbine engines. In this our initial effort, the system includes a light source, a high-speed camera, a swirl atomizer, a compressed air system, gas and liquid regulators. The light source here is a 700 W continuous LED lamb. A suitable camera is 5 to 10 kHz with a spatial resolution of 10 μm. The preliminary outcome shows that the flow rate significantly affects the spray structure and spray angle. This needs to be investigated further to understand in detailed about micro information of liquid fragments deriving from the liquid jet in atomizing zone when accounting for swirl and/or turbulent conditions and this will be done and reported in our future work.

Keywords: Swirl atomization · Backlight imaging · Spray structure · Liquid fragments

1 Introduction

Prior to combustion process in heat engines, different from gases, liquid fuels involve a number of physical processes such as atomization, droplet-droplet collision, vaporization, and mixing. These physical processes add a significant complication into the liquid fuel combustion process [1]. In the atomization process, for example, liquid fuels encompass a number of phenomena including fuel jet primary atomization [2, 3], secondary breakup [4–6] and droplet-droplet interaction [7, 8]. Primary atomization initiates the process where the bulk liquid breaks up to form droplets near the liquid surface [3, 9], followed by secondary atomization where the droplets move at a relative velocity to the surrounding environment and the aerodynamic forces cause the droplets to deform and breakup into smaller droplets and other fragments [6, 9, 10]. Correlations between internal injector information (pressure, swirl, turbulent), exit plane parameters (velocity, Weber number-We, Reynold number - Re, and Ohnesorge number - Oh), characteristics of near-field fragments and downstream parameters (drop size, velocity, and distribution) need to be investigated. Swirl is normally adopted to enhance the

K.-U. Sattler et al. (Eds.): ICERA 2019, LNNS 104, pp. 262–267, 2020.
https://doi.org/10.1007/978-3-030-37497-6_30

liquid breakup at the near-field of atomizers where primary and secondary breakup occurred but this makes the atomization process much more complicated. This technique aims to capture liquid fragments derived in the near-field and this will provide a good database for studying atomization mechanism in the future.

Break-up is generally achieved by liquid-air interfacial aerodynamic forces which can be enhanced by surrounding gas phase turbulence. One of the most important objectives of atomization studies overall is to determine the appropriate operating conditions leading to a particular desired droplet size which can ultimately optimize evaporation and mixing quality. Studies of fuel atomization have been performed using: (i) practical systems such as IC engines [11], (ii) constant volume systems or open environment systems using practical injectors [12–15], or (iii) cross flow air streams, drop towers, laboratory injectors setup also in open air environment [9]. Approach (i) and (ii) can normally help to observe only the macroscopic information of the sprays in practical systems such as spray angle and penetration. In the first two approaches, the spray is too dense and the process is too fast so that the current diagnostic capabilities are challenged. The last method of flow in a cross stream of air is a fundamental approach which is capable of quantifying microscopic parameters such as breakup length and time, breakup mechanism, and shapes and size of liquid blobs, fragments, their population and/or velocity.

Non-dimensional parameters of relevance, as mentioned briefly above, including the gas Reynolds number (Re), the liquid Ohnesorge number (Oh) and the Weber number (We), are commonly adopted to atomization studies and provided in Eqs. 1 to 3, respectively.

$$Re = \frac{\rho_g.U_g D}{\mu_g} \quad (1)$$

$$Oh = \frac{\mu_l}{\sqrt{\rho_l.\sigma_l.D}} \quad (2)$$

$$We = \frac{\rho_g.(U_g - U_l)^2 D}{\sigma_l} \quad (3)$$

where ρ_l is the liquid density; ρ_g is the gas density; U_g is the gas velocity; U_l is the liquid jet velocity; D is the liquid jet diameter; μ_g is the gas dynamic viscosity; μ_l is the liquid dynamic viscosity; and σ_l is the liquid surface tension.

Measurement techniques in non-dilute sprays remain limited with shadowgraph or backlight methods still being the most commonly used, second to Laser Doppler Anemometry/Particle Doppler Anemometry (LDA/PDA) which is limited to spherical droplets (e.g. a size range from 2–120 μm in the LDA/PDA system used in [16]). Backlight imaging has been used to observe a number of phenomena including observation of breakup morphology [17], the deformation rate of drops [18] and the qualitative evolution of ligaments [18]. The technique is also employed to observe and estimate the Rayleigh-Taylor and Kelvin-Helmholtz instability wave lengths [2, 5, 19]. To distinguish between different types of fluid elements including filaments, a number of shape quantifying parameters has been reviewed extensively in [16] in which the two most common parameters are the ratio of area over the square of perimeter of the object and the aspect ratio, respectively. Despite these advances, the characterization of fluid elements that arise from atomizing zone, especially when accounting for swirl and turbulent conditions, remains very loose and this work attempts to provide a methodical approach to this characterization, which in turn allows for a better physical understanding of the atomization process.

2 Experimental Issues

2.1 Injection and Fuel Supply System

The main components of a common fuel supply system include fuel tank, fuel pump, fuel control, filter and injector. These are not shown here but the injector's technical drawing is shown in Fig. 1. The functions of this system are to store and supply fuel to the fuel injector under a predetermined pressure and flow rate. Firstly, fuel pump supplies fuel to the fuel control unit, in which fuel flow rate is adjusted to provide required value. The fuel, under pressure, passes through filters, injector and finally exits the nozzle where atomization occurs.

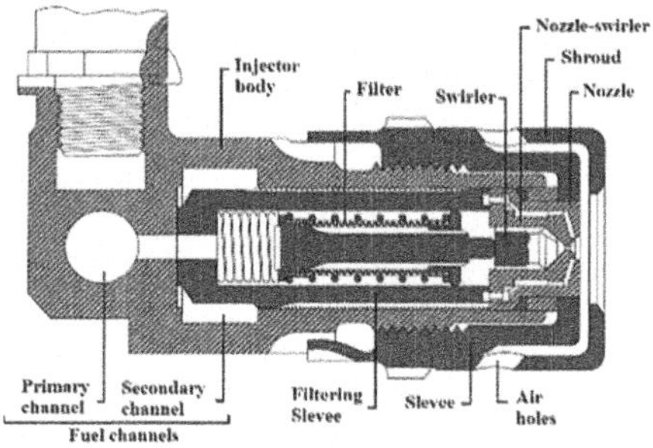

Fig. 1. Schematic of the swirl atomizer

Figure 1 shows the injector structure. For this type, the injector consists of two fuel channels (primary and secondary, respectively), swiller, nozzle and shroud. The two channels are designed so that the injector is capable to deliver a wide range of fuel flow rates. When using the primary channel, a small fuel amount can be supplied to the injector while high flow rate conditions are operated using both channels. The nozzle-swirler is located just right before the nozzle and as such it works as a nozzle for primary channel but a swirler for secondary channel. Besides, a swirler that located in the nozzle-swirler is used for the primary channel. These swirlers create a swirling motion to the fuel stream to generate a wide spray angle. It is important to note here that the nozzle-swirler and the swirler could be able to be disassembled and as such the atomization characteristics could be studied with or without the swirler. This type of injector creates a conical hollow sheet near discharge orifice. This is an airblast atomizer in which a liquid jet or sheet is exposed to air flowing at high velocity. Thus, around the face of shroud, a number of slant holes is used to supply high-velocity airstream to break up the liquid jet.

3 Camera and Back-Light Setup

Figure 2 schematically shows the experiment setup in this study. Fuel is supplied to the injector using a hydraulic system consisting of pneumatic source, fuel tank, hydraulic valves and filters (regulator). The spray structure is recorded using shadowgraph technique [16], in which cameras are used to capture the shadow of liquid fragments passing through a backlight source. The macro information of spray like spray angle will be measured using a CCD camera while micro parameters including dynamic characteristics of fuel fragments derived from near-field zone of nozzle exit require a high-speed camera (about 5000 to 20000 fps). In addition, the optical lenses are capable of adjusting magnification (from 10–20x) and are located at a distance varying between 100–300 mm from the injector centerline to avoid blurred image by fuel mist.

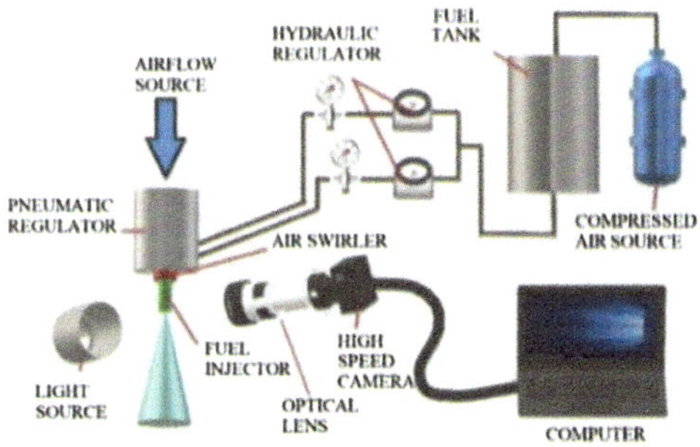

Fig. 2. Experiment setup

4 Preliminary Outcome

The main purpose of the fuel injection process is to break down fuel jet to fine droplets, which mixes with oxidants (air) to create burning mixture. This process includes the presence of a liquid core or a liquid sheet, primary breakup, secondary breakup, mixing with air and evaporation. This study shows our initial efforts to develop the system and the preliminary result reported here only limit to spray angle. It is also noted here that the airblast is not included at this stage.

Figure 3 shows an example of a spray angle. Initially, the obtained images allow to measure the spray angle under different fuel injection conditions. A Matlab code has also been developed using binary approach to measure the spray angle automatically. The reader is directed to refs [16] for further details of this technique.

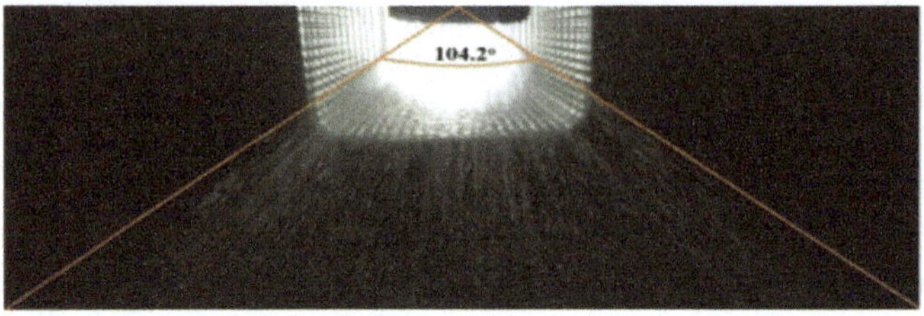

Fig. 3. Example of spray structure estimation (injection pressure = 5 bar, only primary channel used, no air-blast in this stage)

Figure 4 is the diagram of the relationship between fuel flow rate and spray angle. As the flow increases, the spray angle is enlarged significantly. This also implies that it is very important to study microstructure of this spray to understand the influence of swirl and/or turbulent conditions on the atomization characteristics. The potential output could be fuel fragment morphologies, axial and radial velocity, volume and mass flux and this is under investigation and will report in the future.

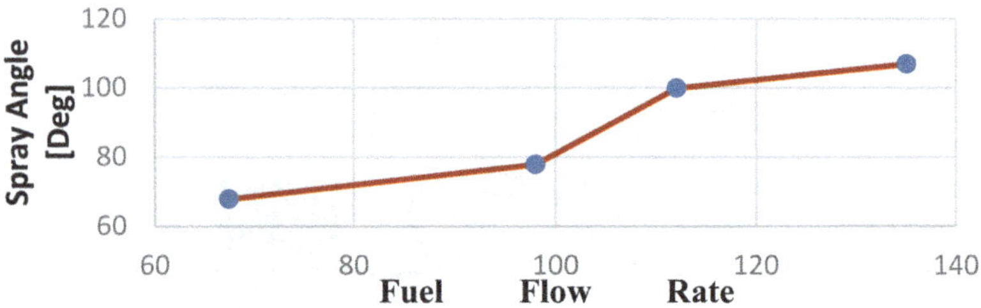

Fig. 4. Spray angle vs. fuel flow rate (primary fuel supply channel only)

5 Conclusion

A shadowgraph imaging system has been developed successfully in this study which aims to study the influence of swirl and turbulent on fuel atomization characteristics. An initial effort has also been paid to output the spray angle. The output implies that the spray angle is significantly affected by fuel flow rate and it is suggested here to study micro structure of this spray to understand the influence of swirl and/or turbulent conditions on the atomization characteristics including fuel fragment morphologies, axial and radial velocity, volume and mass flux.

Acknowledgment. This work is financially supported by Vietnam National Foundation for Science and Technology Development (NAFOSTED) under grant number 107.01-2018.310.

References

1. Ragland, K.W., Bryden, K.M.: Combustion Engineering. CRC Press, Taylor & Francis Group, Boca Raton (2011)
2. Varga, C.M., Lashera, J.C., Hopfinger, E.J.: Initial breakup of a small diameter liquid jet by a high-speed gas stream. J. Fluid Mech. **497**, 405–435 (2003)
3. Dumouchel, C.: On the experimental investigation on primary atomization of liquid streams. Exp. Fluids **45**, 371–422 (2003)
4. Hsiang, L.P., Faeth, G.M.: Near-limit drop deformation and secondary breakup. Int. J. Multiph. Flow **18**(5), 635–652 (1992)
5. Liu, A.B., Reitz, R.D.: Mechanism of air-assisted liquid atomization. Atomization Sprays **3**, 55–75 (1993)
6. Ghaemi, S., Rahimi, P., Nobes, D.S.: Assessment of parameter for distinguishing droplet shape in a spray field using image-based technique. Atomization Sprays **19**(9), 809–831 (2009)
7. Labowsky, M.: Calculation of the burning rates of interacting fuel droplets. Combust. Sci. Technol. **22**(5–6), 217–226 (1980)
8. Chiang, C., Sirignano, W.: Interacting, convecting, vaporizing fuel droplets with variable properties. Int. J. Heat Mass Transf. **36**(4), 875–886 (1993)
9. Faeth, G.M., Hsiang, L.P., Wu, P.K.: Structure and breakup properties of sprays. Int. J. Multiph. Flow **21**, 99–127 (1995)
10. Guildenbecher, D.R., Lopez-Rivera, C., Sojka, P.E.: Secondary atomization. Exp. Fluids **46**, 371–402 (2009)
11. Allen, C.A.W., Watts, K.C.: Comparative analysis of the atomization characteristics of fifteen biodiesel fuel types. Am. Soc. Agric. Eng. **43**(2), 207–211 (2000)
12. Park, S.H., Kim, H.J., Suh, H.K., Lee, C.S.: A study on the fuel injection and atomization characteristics of soybean oil methyl ester (SME). Int. J. Heat Fluid Flow **30**, 108–116 (2009)
13. Kostas, J., Honnery, D., Soria, J.: Time resolved measurements of the initial stages of fuel spray penetration. Fuel **88**, 2225–2237 (2009)
14. Kostas, J., Honnery, D., Soria, J., Kastengren, A., Liu, Z., Powell, C.F., Wang, J.: Effect of nozzle transients and compressibility on the penetration of fuel sprays. Appl. Phys. Lett. **95**, 024101 (2009)
15. Kostas, J., Honnery, D., Soria, J.: A correlation image velocimetry-based study of high-pressure fuel spray tip evolution. Exp. Fluids **51**, 667–678 (2011)
16. Pham, P.X., Kourmatzis, A., Masri, A.R.: Simultaneous volume-velocity measurements in the near-field of atomizing sprays. Meas. Sci. Technol. **28**(115203), 1–13 (2019)
17. Park, S.W., Kim, S., Lee, C.S.: Effects of mixing ratio of biodiesel on breakup mechanisms of monodispersed droplets. Energy Fuels **20**, 1709–1715 (2006)
18. Arcoumanis, C., Whitelaw, D.S., Whitelaw, J.H.: Breakup of droplets of Newtonian and non-Newtonian fluids. Atomization Sprays **6**, 245–256 (1996)
19. Hwang, S., Liu, Z., Reitz, R.D.: Breakup mechanism and drag coefficients of height-speed vaporizing liquid drops. Atomization Sprays **6**, 353–376 (1996)

Dynamic Stiffness Formulation for Vibration of FGM Stepped Annular Plates of Varying Thickness with Non-homogenous Material

Le Quang Vinh[1(✉)], Nguyen Dong Anh[2], and Nguyen Manh Cuong[3]

[1] Department of Mechanical Engineering, Viet Tri University of Industry, Viet Tri, Vietnam
vinhchc@gmail.com
[2] Institute of Mechanics, Vietnam Academy of Science and Technology, Hanoi, Vietnam
[3] School of Mechanical Engineering, Hanoi University of Science and Technology, Hanoi, Vietnam

Abstract. The article introduces a new continuous element (CE) for free vibration analysis of stepped functionally graded materials (FGMs) annular plates having non-homogenous material properties and varying thickness subjected to arbitrary boundary conditions. By applying the powerful assembly capacity of CE method, the dynamic stiffness matrix for the stepped FGMs annular plates with non-homogenous properties has been established. Continuous Element Method allows computing the natural frequencies with high accuracy for any frequency range where almost other current methods meet grand difficulties due to the number of meshing elements. Using minimum meshing for complex structures, this model accelerates the speed of computation and economies the storage capacity of computers.

Keywords: Free vibration · Stepped annular plate · Functionally graded material · Continuous Element Method

1 Introduction

Functionally graded materials (FGMs) have had many applications in various engineering fields [1–5]. Axis-symmetric structure shape kinds including conical, cylindrical shells and annular plates are very common structural elements. Therefore, a good understanding of vibration characteristics of those structures made of FGMs is necessary for designers and users. A wide range of research has been carried out on free vibration of circular and annular plates with variable thickness which mostly used the classical plate theory and a numerical solution method. Hosseini-Hashemi et al. [1] developed an exact closed-form frequency equation for free vibration analysis of circular and annular moderately thick FG plates with constant thickness based on the Mindlin's first-order shear deformation plate theory. Based on the first-order shear deformation theory, Tornabene [2] and Tahouneh et al. [3] focused on the dynamic behavior of moderately thick FG conical, cylindrical shells and annular plates. The generalized differential

K.-U. Sattler et al. (Eds.): ICERA 2019, LNNS 104, pp. 268–280, 2020.
https://doi.org/10.1007/978-3-030-37497-6_31

quadrature (GDQ) method is also used to solve a standard linear eigenvalue problem. Then, Efraim and Eisenberger [4] employed an exact element method to analyze the vibration of variable thickness annular isotropic and FGM plates based on the first order shear deformation theory (FSDT). Hosseini-Hashemi et al. [5] have presented an exact closed-form solution for free vibration of circular and annular moderately thick FG plates. Based on three-dimensional theory, dynamic analysis of multi-directional FGM annular plates is investigated by Nie and Zhong [6] using the states pace-based differential quadrature method. Dong [7] presented three-dimensional free vibration analysis of FGM annular plates using Chebyshev–Ritz method. Su et al. [8] applied the Rayleigh–Ritz method and FSDT to study the free vibrations of FM graded cylindrical, conical shells and annular plates with general boundary conditions. Xie et al. [9] presented a numerical approach based on the Haar wavelet method for the modeling and vibration analysis of FG moderately thick conical shells and annular plates.

From the review of the literature, it appears that despite a variety of methods for analytical and computational analysis of annular plate structures, it is still of need to develop a simple and efficient numerical method for vibration analysis of FGM annular plates especially for medium and high frequency ranges. In several decades, Continuous Element Method (CEM) has been developed as a new powerful tool for mathematical analysis and engineering computation. Casimir et al. [10] have succeeded in building the DSM for thick isotropic plate and shells of revolution. Recently, Thinh et al. [11] proposed new CE models for thick composite plates resting on non-homogenous foundations.

In the present work, the CEM is introduced for the vibration analysis of FGM stepped annular plates. The material properties of the shells are assumed to vary in the thickness direction according to general four-parameter power-law distributions in terms of volume fractions of the constituents and each step of the shell is made by different materials. In order to test the convergence, efficiency and accuracy of the proposed method, numerical examples are presented. The effects of various geometrical and material parameters on the natural frequencies are also discussed.

2 Theoretical Formulations

2.1 Description of the Model

Let's investigate the FGM annular plate with (x, θ, z) coordinates, see Fig. 1. x is the coordinate along the plate' generators with the origin placed at the middle of the generators, θ is the circumferential coordinate, and z is perpendicular to the plate' surfaces. R_1 and R_2 are the inner and outer radius, h and L are thickness and length. The radius coordinate $R(x)$ of a point M inside the plate is calculated as: $R(x) = R_1 + x$. The FGM shells are made of a mixture of ceramic and metal. Young's modulus E(z), density ρ(z) and Poisson's ratio μ(z) are expressed as follows

$$E(z) = (E_c - E_m)V_c + E_m,\ \mu(z) = (\mu_c - \mu_m)V_c + \mu_m,\ \rho(z) = (\rho_c - \rho_m)V_c + \rho_m \quad (1)$$

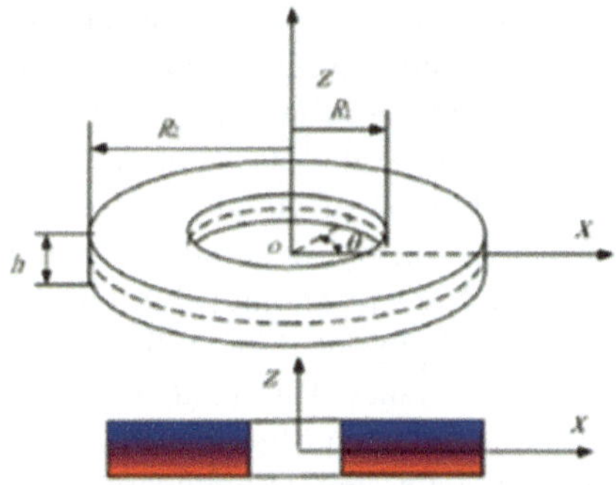

Fig. 1. Schematic sketch of structures FGM annular plates.

The subscripts c and m denote the ceramic and metallic constituents. The volume fraction V_c follows two general four-parameter power-law distributions [2, 8, 9]:

$$\begin{aligned} FGM_{I(a/b/c/p)} &: V_c = \left[1 - a\left(\frac{1}{2} + \frac{z}{h}\right) + b\left(\frac{1}{2} + \frac{z}{h}\right)^c\right]^p \\ FGM_{II(a/b/c/p)} &: V_c = \left[1 - a\left(\frac{1}{2} - \frac{z}{h}\right) + b\left(\frac{1}{2} - \frac{z}{h}\right)^c\right]^p \end{aligned} \tag{2}$$

where p is a positive real number ($0 \leq p \leq \infty$) and a, b, c dictate the material variation profile through the FG shell thickness. It is assumed that $V_c + V_m = 1$. When $p = 0$ or $p = \infty$ the FGM material becomes the homogeneous isotropic material, as:

$$\begin{aligned} p = 0 \rightarrow V_c = 1,\ V_m = 0 \rightarrow E(z) = E_c,\ \mu(z) = \mu_c,\ \rho(z) = \rho_c \\ p = \infty \rightarrow V_c = 0,\ V_m = 1 \rightarrow E(z) = E_m,\ \mu(z) = \mu_m,\ \rho(z) = \rho_m \end{aligned} \tag{3}$$

The variations of Vc versus shell thickness for different values p are illustrated in Fig. 2 where the classical volume fraction profiles, such as those reported in literature [2], are presented as special cases by setting $a = 1$ and $b = 0$. More details on the material variation profile of FGMs with different parameters p, a, b, c are available in [2].

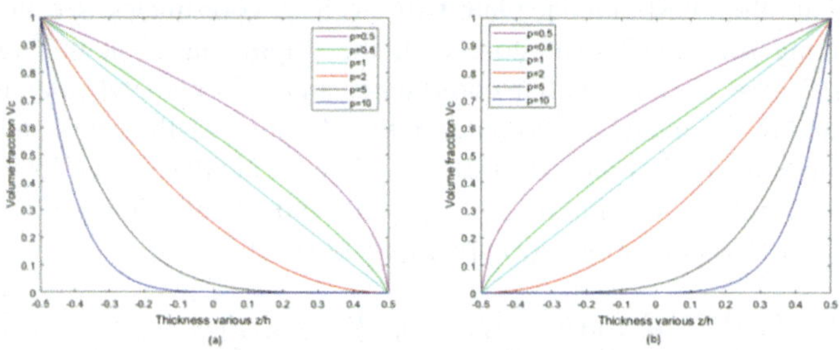

Fig. 2. Variation of the volume fraction V_c through the thickness of a shell for different values of power-law exponent p: (a) $FGM_{I(a=1/b=0/c=2/p)}$; (b) $FGM_{II(a=1/b=0/c=2/p)}$.

2.2 Kinematic Relations and Stress Resultants

The displacement components of an arbitrary point in the FG shell for the first-order shear deformation theory are expressed as given below [11]:

$$\begin{aligned} u(x,\theta,z,t) &= u_0(x,\theta,t) + z\phi_x(x,\theta,t), v(x,\theta,z,t) = v_0(x,\theta,t) + z\phi_s(x,\theta,t) \\ w(x,\theta,z,t) &= w_0(x,\theta,t) \end{aligned} \tag{4}$$

where *u, v* and *w* are the displacement components in the *x*, θ and *z* directions, respectively; u_o, v_O and w_o are the middle surface displacements of the shell in the axial, circumferential and radial directions, respectively; φ_x and φ_θ represent the transverse normal rotations of the reference surface about the θ-and x-axis. t is the time variable.

The linear strain-displacement relations in the shell space are defined as:

$$\begin{Bmatrix} \varepsilon_x \\ \varepsilon_\theta \\ \gamma_{xz} \\ \gamma_{\theta z} \\ \gamma_{x\theta} \end{Bmatrix} = \begin{Bmatrix} \frac{\partial u_0}{\partial x} \\ \frac{1}{R}\frac{\partial v_0}{\partial \theta} + \frac{u_0}{R} \\ \phi_x + \frac{\partial w_0}{\partial x} \\ \phi_\theta + \frac{1}{R}\frac{\partial w_0}{\partial \theta} \\ \frac{1}{R}\frac{\partial u_0}{\partial \theta} + \frac{\partial v_0}{\partial x} - \frac{v_0}{R} \end{Bmatrix} + z \begin{Bmatrix} \frac{\partial \phi_x}{\partial x} \\ \frac{1}{R}\frac{\partial \phi_\theta}{\partial \theta} + \frac{\phi_x}{R} \\ 0 \\ 0 \\ \frac{1}{R}\frac{\partial \phi_x}{\partial \theta} + \frac{\partial \phi_\theta}{\partial x} - \frac{\phi_\theta}{R} \end{Bmatrix} \tag{5}$$

Based on Hooke's law, the stress–strain relations of the shell are written as:

$$\begin{Bmatrix} \sigma_x \\ \sigma_\theta \\ \tau_{x\theta} \\ \tau_{xz} \\ \tau_{\theta z} \end{Bmatrix} = \begin{bmatrix} Q_{11}(z) & Q_{12}(z) & 0 & 0 & 0 \\ Q_{12}(z) & Q_{11}(z) & 0 & 0 & 0 \\ 0 & 0 & Q_{66}(z) & 0 & 0 \\ 0 & 0 & 0 & Q_{66}(z) & 0 \\ 0 & 0 & 0 & 0 & Q_{66}(z) \end{bmatrix} \begin{Bmatrix} \varepsilon_x \\ \varepsilon_\theta \\ \gamma_{x\theta} \\ \gamma_{xz} \\ \gamma_{\theta z} \end{Bmatrix} \tag{6}$$

where $Q_{ij}(z)$ are functions of thickness coordinate *z* and defined as:

$$Q_{11}(z) = \frac{E(z)}{1-\mu^2(z)},\ Q_{12}(z) = \frac{\mu(z)E(z)}{1-\mu^2(z)},\ Q_{66}(z) = \frac{E(z)}{2[1+\mu(z)]} \tag{7}$$

The stress and moment resultants are given as:

$$(N_x, N_\theta, N_{x\theta}, M_x, M_\theta, M_{x\theta}) = \int_{-h/2}^{h/2} (\sigma_x, \sigma_\theta, \tau_{x\theta}, z\sigma_x, z\sigma_\theta, z\tau_{x\theta})dz \tag{8}$$

$$(Q_x, Q_\theta) = k\int_{-h/2}^{h/2} (\tau_{xz}, \tau_{\theta z})dz \tag{9}$$

where N_x, N_θ and $N_{x\theta}$ are the in-place force resultants, M_x, M_θ and $M_{x\theta}$ are moment resultants, Q_x, Q_θ are transverse shear force resultants. The shear correction factor *k* is computed such that the strain energy due to transverse shear stresses in Eq. (9) equals the strain energy due to the true transverse stresses predicted by the three-dimensional elasticity theory [12]. For a FGM shell since k depends on shell parameters, such as material properties and boundary conditions, it is difficult to obtain the accurate value of the shear correction stresses. In this paper, k is uniformly selected by *5/6* [2]. Substituting (6) and (7) into (8) and (9) following constitutive equations are obtained:

$$\begin{Bmatrix} N_x \\ N_\theta \\ N_{x\theta} \\ M_x \\ M_\theta \\ M_{x\theta} \\ Q_x \\ Q_\theta \end{Bmatrix} = \begin{bmatrix} A_{11} & A_{12} & 0 & B_{11} & B_{12} & 0 & 0 & 0 \\ A_{12} & A_{11} & 0 & B_{12} & B_{11} & 0 & 0 & 0 \\ 0 & 0 & A_{66} & 0 & 0 & B_{66} & 0 & 0 \\ B_{11} & B_{12} & 0 & D_{11} & D_{12} & 0 & 0 & 0 \\ B_{12} & B_{11} & 0 & D_{12} & D_{11} & 0 & 0 & 0 \\ 0 & 0 & B_{66} & 0 & 0 & D_{66} & 0 & 0 \\ 0 & 0 & 0 & 0 & 0 & 0 & k.A_{66} & 0 \\ 0 & 0 & 0 & 0 & 0 & 0 & 0 & k.A_{66} \end{bmatrix} \begin{Bmatrix} \frac{\partial u_o}{\partial x} \\ \frac{\partial v_o}{R\partial\theta} + \frac{u_o}{R} \\ \frac{\partial v_o}{\partial x} + \frac{\partial u_o}{R\partial\theta} - \frac{v_o}{R} \\ \frac{\partial \phi_x}{\partial x} \\ \frac{\partial \phi_\theta}{R.\partial\theta} + \frac{\phi_x}{R} \\ \frac{\partial \phi_x}{R.\partial\theta} + \frac{\partial \phi_\theta}{\partial x} - \frac{\phi_\theta}{R} \\ \frac{\partial w_0}{\partial x} + \phi_x \\ \frac{\partial w_0}{R\partial\theta} + \phi_\theta \end{Bmatrix} \tag{10}$$

The structure materials employed in the following study are assumed to be functionally graded and linearly elastic. So, the extensional stiffness A_{ij}, the bending stiffness D_{ij}, and the extensional-bending coupling stiffness B_{ij} are respectively expressed as:

$$\begin{gathered} A_{ij} = \int_{-h/2}^{h/2} Q_{ij}(z)dz \quad B_{ij} = \int_{-h/2}^{h/2} z.Q_{ij}(z)dz \quad D_{ij} = \int_{-h/2}^{h/2} z^2 Q_{ij}(z)dz,\ i, j = 1, 2, 6 \\ A_{ij} = \int_{-h/2}^{h/2} Q_{ij}(z)dz,\ i, j = 4, 5 \end{gathered} \tag{11}$$

2.3 Equations of Motion

By means of Hamilton's principle, the equilibrium equations of motion based on FSDT can be written in terms of the force and moment resultants as [10]:

$$\begin{gathered} \frac{\partial N_x}{\partial x} + \frac{1}{R}(N_x - N_\theta) + \frac{1}{R}\frac{\partial N_{x\theta}}{\partial\theta} = I_0\ddot{u}_0 + I_1\ddot{\varphi}_x, \quad \frac{\partial Q_x}{\partial x} + \frac{1}{R}\frac{\partial Q_\theta}{\partial\theta} + \frac{1}{R}Q_x = I_0\ddot{w}_0 \\ \frac{\partial M_{x\theta}}{\partial x} + \frac{2}{R}M_{x\theta} + \frac{1}{R}\frac{\partial M_\theta}{\partial\theta} - Q_\theta = I_1\ddot{v}_0 + I_2\ddot{\varphi}_\theta, \quad \frac{\partial N_{x\theta}}{\partial x} + \frac{2}{R}N_{x\theta} + \frac{1}{R}\frac{\partial N_\theta}{\partial\theta} = I_0\ddot{v}_0 + I_1\ddot{\varphi}_\theta \\ \frac{\partial M_x}{\partial s} + \frac{1}{R}(M_x - M_\theta) + \frac{1}{R}\frac{\partial M_{x\theta}}{\partial\theta} - Q_x = I_1\ddot{u}_0 + I_2\ddot{\varphi}_x \end{gathered} \tag{12}$$

Where $[I_0, I_1, I_2] = \int_{-h/2}^{h/2} \rho(z)[1, z^1, z^2]dz$, in which $\rho(z)$ is the density of the shell per unit middle surface area. I_0, I_1 and I_2 are the mass inertias.

3 Dynamic Stiffness Matrix Formulation for FGM Annular Plate

The chosen state-vector is $y^T = \{u_O, v_O, w_O, \varphi_x, \varphi_\theta, N_x, N_{x\theta}, Q_x, M_x, M_{x\theta}\}^T$. Next, the Fourrier series expansion for state variables is written as:

$$\{u_o(x,\theta,t), w_o(x,\theta,t), \phi_\theta(x,\theta,t), N_x(x,\theta,t), Q_x(x,\theta,t), M_x(x,\theta,t)\}^T$$
$$= \sum_{m=1}^{\infty} \{u_m(x), w_m(x), \phi_{\theta m}(x), N_{x_m}(x), Q_{xm}(x), M_{x_m}(x)\}^T \cos m\theta e^{i\omega t}$$
$$\{v_0(x,\theta,t),\ \phi(x,\theta,t),\ N_{x\theta}(x,\theta,t),\ M_{x\theta}(x,\theta,t)\}^T$$
$$= \sum_{m=1}^{\infty} \{v_m(x), \phi_m(x), N_{x\theta m}(x), M_{x\theta m}(x)\}^T \sin m\theta e^{i\omega t}$$

where m is the number of circumferential waves. Substituting (13) into (10) and (11), a system of ordinary differential equations in the x-coordinate for the mth mode can be expressed in the matrix form for each circumferential mode m as [11]:

$$\frac{d\mathbf{y}_m}{dx} = \mathbf{A}_m \mathbf{y}_m \tag{13}$$

With A_m is a 10 × 10 matrix. The dynamic transfer matrix T_m is evaluated as:

$$T_m(\omega) = e^{\int_0^L A_m(\omega)dx} = \begin{bmatrix} T_{11} & T_{12} \\ T_{21} & T_{22} \end{bmatrix} \tag{14}$$

Finally, the dynamic stiffness matrix $K_{(\omega)}$ for annular plate is determined by [10]:

$$K_m(\omega) = \begin{bmatrix} T_{12}^{-1}T_{11} & -T_{12}^{-1} \\ T_{21} - T_{22}T_{12}^{-1}T_{11} & T_{22}T_{12}^{-1} \end{bmatrix} \tag{15}$$

Natural frequencies will be extracted from the harmonic responses of the structure by using the procedure detailed in [11].

4 Continuous Element for FGM Stepped Annular Plate with Non-homogenous Properties

Consider a functionally graded plate consists of 3 steps with radii r_1, r_2, ..., $r_3 = R$, and step thicknesses h_1, h_2, ..., h_n, respectively, as shown in Fig. 3. Each step can be made from different FGM materials. The plate geometry and dimensions are defined in an orthogonal cylindrical coordinate system (r, θ, z) to extract the mathematical formulations. r is the radius of the annular plate at the origin O. u, v and w are the displacement components in the x, θ and normal directions, respectively. The FGMs are composite materials, the mechanical properties of which vary gradually due to changing the volume fraction of the constituent materials usually in thickness direction.

In this study, the properties of the plate are assumed to vary through the plate thickness with a power-law distribution of the volume fractions of two materials. Unless mentioned otherwise, the top surface of the first segment ($z = h_1/2$), which assumed here the thickest part of the plate, is metal-rich whereas the bottom surface of the same segment ($z = -h_1/2$) is ceramic-rich.

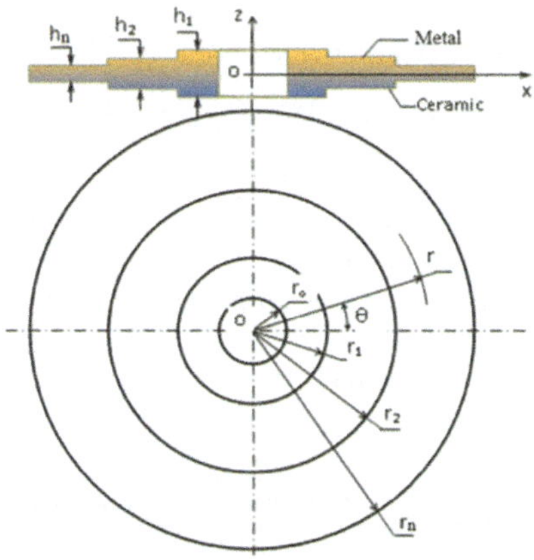

Fig. 3. Geometry of a stepped FGM annular plate.

The dynamic stiffness matrix $K_{m(\omega)}$ for the above stepped FGM annular plate will be constructed by assembling the DSM of many sections having different constant thickness and materials. First, the stepped annular plate is divided into elements. It is necessary to build separate dynamic stiffness matrices K_{seg1}, K_{seg2}, …, K_{segn} for these segments. Then, Fig. 4 describes the assembly procedure for constructing the DSM for the stepped FGM annular plate. The natural frequencies of the studied structure will be determined from this matrix by using the method detailed in [10, 11].

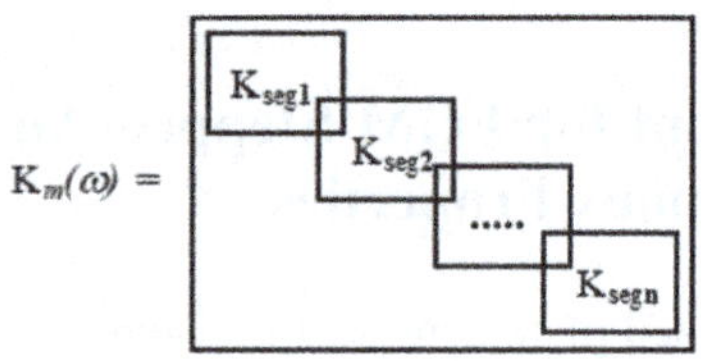

Fig. 4. Construction of the dynamic stiffness matrix for stepped FGM annular plate.

5 Numerical Results and Discussion

The present exact procedure may be applied to investigate the effects of various geometrical and material properties such as step thickness ratios, the power law index and different boundary conditions. Three configuration of functionally graded material are used with the material properties as: FGM1 (Al: E = 70 Gpa, μ = 0.3, ρ = 2707 kg/m^3; Al_2O_3: E = 380 Gpa, μ = 0.3, ρ = 3800 kg/m^3), FGM2 (Al: E = 70 Gpa, μ = 0.3, ρ = 2707 kg/m^3; Ziconia: E = 168 Gpa, μ = 0.3, ρ = 5700 kg/m^3), FGM3 (Nickel: E = 205.098 Gpa, μ = 0.31, ρ = 8900 kg/m^3; Si_3N_4: E = 322.27 Gpa, μ = 0.24, ρ = 2370 kg/m^3).

Table 1. The first eight frequencies (Hz) of $FGM2_{I(a=0/b=-0.5/c=2/p)}$ annular plates.

Mode	p = 0		p = 1		p = 5		p = 20	
	Ref. [2]	Present	Ref. [2]	Present	Ref. [2]	Present	Ref. [2]	Present
1	70.13	65.89	68.71	67.6	66.36	66.09	68.84	64.7
2	127.5	128.7	124.95	131.94	120.75	130	125.11	128.02
3	127.5	128.72	124.95	131.94	120.75	130	125.11	128.02
4	212.58	208.06	208.35	213.26	201.35	210.04	208.58	206.79
5	212.58	208.06	208.35	213.26	201.35	210.04	208.58	206.79
6	296.99	277	291.13	284.09	281.41	278	291.33	272.48
7	316.87	301.87	310.64	309.46	300.29	304.24	310.82	299.19
8	316.87	301.87	310.64	309.46	300.29	304.24	310.82	299.19

*Error(%) = |(CEM − Ref. [2])/Ref. [2] | × 100.

5.1 Validation of the Present Model

The proposed continuous element model is validated by comparison with solutions available in the literature and with results by FEM. Table 1 shows the first ten frequencies for aluminum/zirconia annular plates with different power-law exponents (i.e. $p = 0, 1, 5,$ and 20). The geometrical parameters of the FGM annular plates are $R_1 = 0.5$ m, $h = 0.1$ m, $R_2 - R_1 = 1.5$ m, $\alpha = 90^o$ having the Free-Clamped (F-C) boundary condition. The present results are compared with the research of Tornabene [2] employing the differential quadrature method base on FSDT. A good agreement is noticed through this table which validates the precision and reliability of our formulation. Small errors are found between our results and those of [2] varying from 309.46 (0.38%) to 130(7,66%). Table 2 shows the comparison of the first eight frequencies (Hz) of FGM annular plates with different thicknessess and Clamped–Clamped (C-C) boundary condition. The FG plates are composed of Aluminum and Al_2O_3. The outer radius of annular plates is $R_2 = 1$ m while the inner radius of annular plates is taken to be $R_1 = 0.2$ m.

Table 2. The first eight frequencies (Hz) of $FGM1_{I(a=1/b=0/c/p=1)}$ annular plates.

Frequencies (Hz)	h = 0.1				h = 0.2			
	Ref. [2]	Ref. [9]	Present	Error (%)	Ref. [2]	Ref. [9]	Present	Error (%)
f1	1152.56	1152.62	1127.16	2.21	1866.04	1866.19	1904.24	2.04
f2	1195.98	1196.04	1178.62	1.46	1936.20	1936.22	1981.43	2.33
f3	1379.03	1379.08	1378.01	0.08	2247.49	2247.51	2310.48	2.80
f4	1755.15	1755.19	1763.61	0.48	2840.26	2840.27	2928.84	3.12
f5	2274.94	2274.98	2284.75	0.43	3467.37	3467.37	3513.27	1.32
f6	2870.38	2870.43	2881.3	0.38	3586.08	3586.09	3707.69	3.39
f7	2891.44	2891.61	2877.82	0.48	3847.22	3847.22	4098.49	6.53
f8	2961.24	2961.41	2953.6	0.26	4147.27	4147.34	4320.1	4.17

*Error(%) = |(CEM − Ref. [9])/Ref. [9] | × 100.

Two different thicknesses $h = 0.1$ m and $h = 0.2$ m (corresponding to moderately thick plates) have been used. Meanwhile, it can be noted that the bold-faced values in Table 2 indicate the in-plane modes. The numbers in the brackets (n, m) indicate the circumferential and axial wave numbers, for which the frequencies are detected. It is seen that excellent agreement of the results is achieved. From Table 2, it is worth noting that the frequencies of the annular plates increase with the increases of thickness. The reason for this is that the thickness tends to increase the stiffness of the FGM shells and plates and thus increase the natural frequencies. As analyzed above, it can be concluded that the proposed solution is efficient and accurate in predicting nature frequencies of FG annular plates.

Table 3. Comparison study of first 9 natural frequencies (Hz) for FGM1 annular plate with one step variation

B. C's	Mode (m, n)	(0,1)	(1,1)	(2,1)	(0,2)	(3,1)	(1,2)	(4,1)	(5,1)	(2,2)
F-C	Ref. [1]	110.283	222.398	382.815	479.039	592.647	745.643	781.936	958.931	966.951
	FEM	110.330	222.480	382.950	479.210	592.840	745.830	782.170	959.220	967.190
	CEM	98.90	223.90	386.00	443.35	594.00	706.90	764.35	918.60	993.90
S-S	Mode (m, n)	(1,1)	(0,1)	(2,1)	(3,1)	(4,1)	(0,2)	(1,2)	(5,1)	(2,2)
	Ref. [1]	190.507	196.641	296.228	467.710	628.415	654.177	678.894	790.892	799.965
	FEM	190.550	196.710	296.330	467.870	628.630	654.370	679.080	791.170	800.210
	CEM	192.10	206.40	314.45	468.65	611.80	615.85	642.60	757.90	763.55

*Error(%) = |(CEM − Ref. [1])/Ref. [1] | × 100.

A comparative study for evaluation of first 9 natural frequencies (Hz) of stepped FGM1 annular plates between the present exact solution and the finite element analysis is carried out in Table 3. From Table 3 shows the natural frequencies of F–C and S–S FGM1 annular plates with one step variation. The power law index p is equal to 1. The step locations and thicknesses are selected as follows: $r_0 = 0.25$ m; $h_1 = 0.2$ m; $r_1 = 1$ m; $h_2 = 0.1$ m; $r_2 = 2$ m. Results of Table 3 reveal that very good agreement is achieved for the stepped annular FG plates. But the present results are closer to the FE results on the basis of the FSDT (Shell 281). This is due to the fact that both methods are based on the hypothesis of FSDT. It is also noticeable that the difference between the present results is very small and does not exceed 5% for the worst case. This closeness is apparent even for higher vibrating modes. It is evident that the dynamic stiffness has a small size and the CE model needs less number of elements but assures the same precision of results compared to those of FEM. Therefore, CEM becomes an interesting alternative for dealing with medium and high frequency range where FEM meets difficulties due to the large number of elements as well as the tiny element dimensions.

5.2 Influences of Shell Parameters

Figure 5 presents the effect of boundary conditions on natural frequencies of stepped $FGM1_{II(a=1/b=0/c=2/p)}$ annular plates with different power-law exponent p and circumferential wave number m. The FGM annular plates with two step variation are composed of aluminum and Al_2O_3 and geometric properties used are: $r_0 = 0.25$ m, $h_1 = 0.24$ m, $r_1 = 1$ m, $h_2 = 0.2$ m, $r_2 = 2$ m, $h_3 = 0.16$ m, $r_3 = 4$ m. It can be observed from these figures that the natural frequency with the C-C boundary condition is greater than the boundary condition of C-F and when the number of power-law exponent p increases then the natural frequency of the stepped annular plates decreases. These results show that one could easily vary the natural frequencies of the stepped FGM annular plates by varying the power-law exponent p and choosing different boundary conditions.

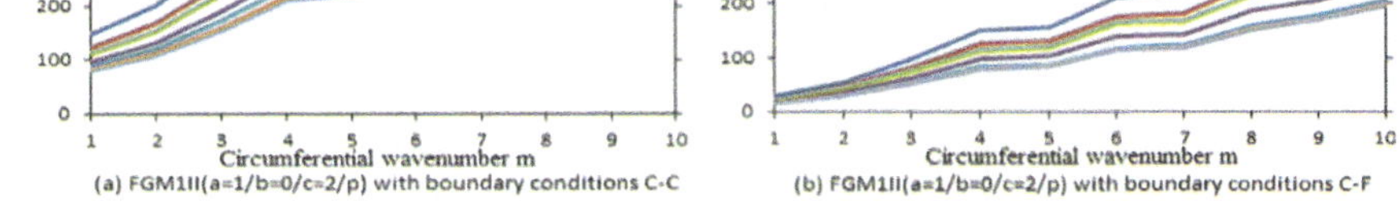

Fig. 5. Influence of boundary conditions on natural frequencies of the stepped FGM annular plate for various values of the power-law exponent p

Table 4 presents the effect on natural frequencies when the stepped of structures made of FMG material has different properties. Type A structure has the first stepped made of FGM1 material, the second stepped made of FGM2 material, the third stepped made of FGM1 material; type B has the first stepped made of FGM1 material, the second stepped made of FGM2 material, the third stepped made of FGM2 material, type C has the first stepped made of FGM1 material, the second stepped made of

Table 4. Influence of FGM configuration on natural frequencies (Hz) of the stepped FGM annular plate with boundary condition (C-C).

Frequencies (Hz)	Type FGM				
	A	B	C	D	E
f1	90.6	73.9	81.6	72.0	95.3
f2	95.9	77.1	87.1	74.5	100.1
f3	127.4	99.6	114.1	102.0	130.9
f4	178.8	136.7	155.8	149.1	180.6
f5	233.1	179.5	205.8	196.5	235.7
f6	239.8	193.2	218.3	198.4	240.5
f7	242.1	200.0	226.8	206.3	243.5
f8	287.6	226.7	263.9	239.9	291.5
f9	309.4	236.1	270.5	242.3	309.6
f10	361.0	278.0	329.1	290.1	366.6

FGM2 material, the third stepped made of FGM3 material, type D has the first stepped made of FGM1 material, the second stepped made of FGM3 material, the third stepped made of FGM2 material; type D has the first stepped made of FGM1 material, the second stepped made of FGM3 material, the third stepped made of FGM2 material. The volume fraction function has $FGM_{I(a=1/b=0.5/c=1/p=2)}$ and geometric properties used are: $r_0 = 0.25$ m, $h_1 = 0.24$ m, $r_1 = 1$ m, $h_2 = 0.2$ m, $r_2 = 2$ m, $h_3 = 0.16$ m, $r_3 = 4$ m.

The change of the first 10 natural frequencies in Table 4 is reasonable, which is explained as follows: FGM configuration of type E annular plate has the highest rigidity. Therefore, the first 10 natural frequencies are also highest and the natural frequency decreases for FGM configurations of type A, type C, type B. FGM configuration in type D has firm stiffness lowest is therefore the smallest natural frequency obtained.

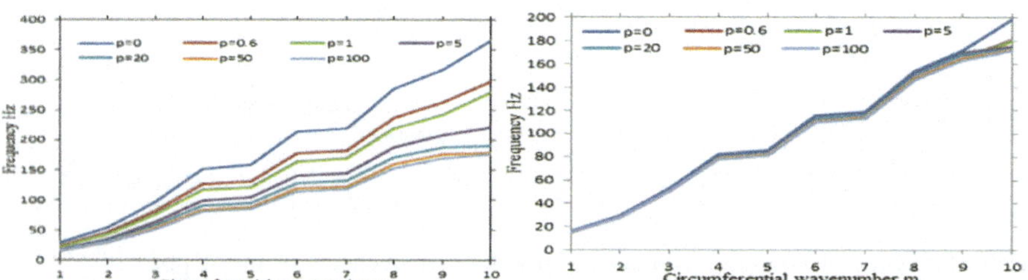

(a) FGM1I(a=1/b=0/c=2/p) with boundary conditions C-F b) FGM2I(a=1/b=0/c=2/p) with boundary conditions (C-F)

Fig. 6. Influence of FGM configuration on natural frequencies of the stepped FGM annular plate for various values of the power-law exponent p

Next, the influences of the effects of the FGM configuration are discussed. The effects of the FGM configuration are studied by solving the frequencies of two stepped FGM annular plates: Type FGM1 is composed of aluminum and Al_2O_3 and Type

FGM2 are composed of aluminum and Zirconia. The geometrical parameters of the shell are taken to be $r_0 = 0.25$ m, $h_1 = 0.24$ m, $r_1 = 1$ m, $h_2 = 0.2$ m, $r_2 = 2$ m, $h_3 = 0.16$ m, $r_3 = 4$ m. In addition, the influence of the value of p, which affects the constituent volume fraction, can be seen from Fig. 6. For the Type FGM1and Type FGM2, the frequencies corresponding to each circumferential wave number increase as the power-law exponent p increases. In fact this is expected, because a larger p means that the shell has a smaller aluminum component, and that its stiffness is thus increased. Moreover, it can be seen that the natural frequencies for a Type FGM1are higher than a Type FGM2 because Type FGM1 material has higher elastic modulus than type FGM2.

It is concluded that our formulation is accurate and can be efficiently used in any range of frequencies. Obtained results in Table 4 and Figs. 5 and 6 are new for the studied structures and they can be served for comparison with other methods.

6 Conclusions

This research has successfully constructed a new CE model for the vibration analysis of thick FGM stepped annular plates with non-homogenous material and thickness. This CE completes the existing library of Dynamic Stiffness Matrix for metal, composite and FGM structures. Various test cases confirm that Continuous Element Method allows computing the natural frequencies with high accuracy for any frequency range. Using minimum meshing for complex structures, this model accelerates the speed of computation and economies the storage capacity of computers. The effects of shell thickness, step material as well as geometry properties and boundary conditions on the dynamic behaviors of the stepped FGM annular plates with non-homogenous properties have been investigated in detailed. The presented results using CE formulations for such complex structures have been demonstrated for the first time. These results can be used as a benchmark for other methods or approaches. The developed CE model can efficiently be used for the analysis of FGM shell on elastic foundation, joined conical-cylindrical-conical shell in medium and high frequencies where almost other current methods meet grand difficulties due to the number of meshing elements.

References

1. Hosseini-Hashemi, S., Derakhshani, M., Fadaee, M.: An accurate mathematical study on the free vibration of stepped thickness circular/annular Mindlin functionally graded plates. Appl. Math. Model. **37**, 4147–4164 (2013)
2. Tornabene, F.: Free vibration analysis of functionally graded conical, cylindrical shell and annular plate structures with a four-parameter power-law distribution. Comput. Methods Appl. Mech. Eng. **198**, 2911–2935 (2009)
3. Tahouneh, V., Yas, M.H.: 3-D free vibration analysis of thick functionally graded annular sector plates on Pasternak elastic foundation via 2-D differential quadrature method. Acta Mech. **223**, 1879–1897 (2012)
4. Efraim, E., Eisenberger, M.: Exact vibration analysis of variable thickness thick annular isotropic and FGM plates. J. Sound Vib. **299**, 720–738 (2007)

5. Hosseini-Hashemi, S., Es'haghi, M., Karimi, M.: Closed-form vibration analysis of thick annular functionally graded plates with integrated piezoelectric layers. Int. J. Mech. Sci. **52**, 410–428 (2010)
6. Nie, G.J., Zhong, Z.: Dynamic analysis of multi-directional functionally graded annular plates. Appl. Math. Modell. **34**, 608–616 (2010)
7. Dong, C.Y.: Three-dimensional free vibration analysis of functionally graded annular plates using the Chebyshev-Ritz method. Mater. Des. **29**, 1518–1525 (2008)
8. Su, Z., Jin, G.Y., Shi, S.X., Ye, T.G., Jia, X.Z.: A unified solution for vibration analysis of functionally graded cylindrical, conical shells and annular plates with general boundary conditions. Int. J. Mech. Sci. **80**, 62–80 (2014)
9. Xie, X., Jin, G.Y., Ye, T.G., Liu, Z.G.: Free vibration analysis of functionally graded conical shells and annular plates using the Haar wavelet method. Appl. Acoust. **85**, 130–142 (2014)
10. Casimir, J.B., Nguyen, M.C., Tawfig, I.: Thick shells of revolution: derivation of the dynamic stiffness matrix of continuous elements and application to a tested cylinder. Comput. Struct. **85**, 1845–1857 (2007)
11. Thinh, T.I., Cuong, N.M., Ninh, D.G.: Dynamic stiffness formulation for vibration analysis of thick composite plates resting on non-homogenous foundations. Compos. Struct. **108**, 684–695 (2014)
12. Reddy, J.N.: Mechanics of Laminated Composites Plates and Shells. CRC Press, New York (2003)

Effects of Suspension Design Parameters of a Semi-trailer Truck on Vehicle Ride Comfort and Road Surface Friendliness

Le Van Quynh(✉), Bui Van Cuong, Le Xuan Long, and Do Van Quan

Faculty of Automotive and Power Machinery Engineering,
Thai Nguyen University of Technology, Thai Nguyen, Vietnam
lequynh@tnut.edu.vn

Abstract. The objective of this study is to analyze the effects of suspension design parameters of a semi-trailer truck on vehicle ride comfort as well as road surface friendliness. A half-dynamic vehicle model with 12° of freedom is established under random road excitation. The weighted root mean square (RMS) of acceleration responses of the vertical driver's seat, cab's pitch and dynamic load coefficient (DLC) is chosen as an objective function which uses Matlab/Simulink software to simulate the vehicle dynamics model and calculate the objective functions. The effects of suspension design parameters of a semi-trailer truck such as stiffness and damping coefficients on the objective functions are analyzed respectively. The results show that the effects of the stiffness and damping parameters in the suspension systems on the vehicle ride comfort as well as road surface friendliness are very obvious. The worse the vehicle ride comfort as well as road surface friendliness is, the greater the stiffness of the suspension systems is. However, the vehicle ride comfort improves with the increase of the damping of the suspension systems. Therefore, the study provides a theoretical basis for the suspension system design of semi-trailer trucks.

Keywords: Semi-trailer truck · Suspension system · Dynamic model · Ride comfort · Road friendliness

1 Introduction

The suspension system is one of the most important systems to improve vehicle ride comfort as well as reduce the negative impacts on the road surface. Therefore many scientists are interested in researching and evaluating the impact. The effect on vehicle ride comfort including the 3D dynamic model with 13-DOF (degree of freedom) is established for simulation, and the weighted r.m.s acceleration responses of the vertical driver's seat, pitch and roll angle of the cab are chosen as objective functions. The influence of different vehicle dynamics system parameters on the vertical vibration of driver's seat, pitch and roll of the cab which include the stiffness and damping of vehicle suspension, tires, cab suspension and the driver's seat suspension was analyzed by Van Quynh et al. [1]. A half rigid-elastic vibration model of the vertical dynamic response with 6 degree of freedom (DOF) was developed by Jie et al. [2] to analyze vehicle ride. The analytical study of the performance indices of articulated truck semi-trailer during

K.-U. Sattler et al. (Eds.): ICERA 2019, LNNS 104, pp. 281–289, 2020.
https://doi.org/10.1007/978-3-030-37497-6_32

three different cases to improve the driver comfort was presented by Abdelkareem et al. [3]; For the effect on road surface friendliness, Shi and Cai developed a 3D vehicle-pavement interaction model to analyse dynamic loads, and concluded that pavement response is significantly higher in rough pavements under dynamic load conditions than under static load conditions [4]. The 3D nonlinear dynamic model of a typical heavy truck was developed by Van Quynh et al. [5]. The impact factors of dynamic tire loads are used to evaluate the dynamic interaction between heavy vehicles and the study results provide both the warning limits of road surface roughness and the limits of corresponding dynamic parameters for the 5-axle heavy truck. Van Cuong et al. [6] developed a three-dimensional nonlinear dynamic model of heavy 15 DOF (degree of freedom) based on Zhou Changfeng model to evaluate influence of heavy truck operating condition on dynamic load coefficient. Long et al. [7] developed a full dynamic model of a heavy vehicle equipped with three different suspension systems to evaluate the effect of suspension characteristics on dynamic load coefficient (DLC). Van Quynh et al. [8] developed a 3D dynamic model with 14° of freedom with the dynamic models of the traditional and new air suspension systems to compare the performance of the air suspension systems for reducing the negative impacts on the road surface when the vehicle moves on different road conditions. The geometric structural parameters of the vehicle system and the nonlinear characteristics of shock absorber and leaf springs were precisely described by Yongjie et al. [9] to analyze the effects on tire dynamic load and dynamic load coefficient (DLC). The effects of driving conditions and suspension parameters on dynamic load-sharing and road-friendliness of the semi-trailer were analyzed by Yikai [10]. Van Quynh et al. [11] developed a three-dimensional vehicle-pavement coupled model with 14 of freedom to analyze the influence of semi-trailer truck operating conditions on road surface; For improving both vehicle ride comfort and road friendliness, a three-dimensional nonlinear dynamical model of a typical heavy truck with 16-DOF was established by Van Liem et al. [12] to analyze and evaluate the performance of the air suspension system of heavy trucks with semi-active fuzzy control. A half-vehicle dynamic model with three control cases including the seat controlled, the cab controlled, and the vehicle controlled was respectively established by Nguyen et al. [13] to evaluate the performance of heavy truck's semi-active isolation systems.

The rest of this paper is organized as follows: a dynamic model for a half semi-trailer truck with 12 d.o.f established under the random road excitation according to the International Standards Organization (ISO) 8608 [15] is presented in Sect. 2. In Sect. 3, Evaluation indicators such as the weighted root mean square (RMS) of acceleration responses of the vertical driver's seat, cab's pitch and dynamic load coefficient (DLC) are proposed. Section 4 presents the results, and the discussion and the conclusions are given in Sect. 5.

2 Vehicle Dynamic Model

2.1 Full Vehicle Dynamic Model

A 5 - axle semi-trailer truck with dependent leaf spring suspension systems for all axles of vehicle is selected for analyzing the influence of tire parameters of vehicles on road surface friendliness. A half semi-trailer truck dynamic model with 12° of freedom is presented by the reference [16], as shown in Fig. 1.

In Fig. 1, k_{ti} and c_{ti} are the stiffness and damping coefficients of the tires, respectively; k_i and c_i are the stiffness and damping coefficients of the passive suspension systems of vehicles, respectively; k_{c1}, k_{c2}, k_s and c_{c1}, c_{c2}, c_s are the passive suspension systems of cabs and driver seats respectively; m_i are the unsprung mass of the tractor and trailer axles, respectively; m_{ff} and m_{fr} are the sprung mass of the tractor and trailer bodies, respectively; m_c and m_s are the mass of the cab body and the driver seat; I_{ff}, I_{fr} and I_c are the mass moment of inertia of tractors, trailers and cabs; z_i, z_{ff}, z_{fr}, z_c và z_s are the vertical displacements of the axles, tractor body, trailer body, cab and driver seat, respectively; φ_c, φ_{ff} and φ_{fr} are the pitch angle displacements of the cab, tractor and trailer bodies, respectively; l_j are the distances; v is the speed of vehicle ($i=1\div5$; $j=1\div10$).

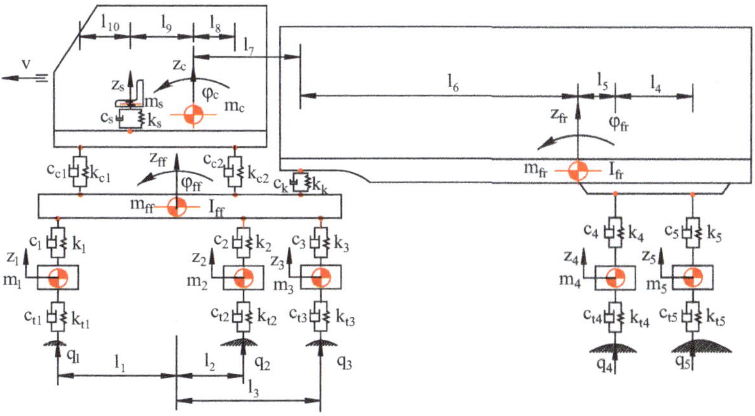

Fig. 1. Articulated truck semi-trailer dynamic model

Equations of Motion: The equations of vehicle motion can be formulated in different ways such as Lagrange's equation, Newton-Euler equation, Jourdain's principle. However, in order to facilitate the description of vehicle dynamic systems using computer simulation, a combined method of the multi-body system theory and D'Alembert's principle is chosen for this study. The multi-body system theory is used to separate the system into subsystems which are linked by the force and moment equations. D'Alembert's principle is used to set up force and moment equations to describe vehicle dynamic subsystems. The general dynamic differential equation for the articulated truck semi-trailer is given by the following matrix form:

$$M\ddot{z} + C\dot{z} + Kz = C_t\dot{q} + K_tq \quad (1)$$

in which M is the mass matrix; C is the damping matrix of the suspension system; K is the stiffness matrix of the suspension system; Ct is the damping matrix of the wheel system; Kt is the stiffness matrix of the wheel system; z is the vector of displacement; q is the vector of excitation of road surface.

2.2 Road Surface Roughness

Road surface roughness plays an important role in analyzing driver ride comfort. The random excitation of road surface roughness can be represented with a periodic modulated random process. The general form of the displacement PSD of the road surface roughness is determined by the experimental formula [6, 14]:

$$S_q(n) = S_q(n_0)\left(\frac{n}{n_0}\right)^{-\omega} \tag{2}$$

Space frequency n is the reciprocal of the wavelength λ - wave numbers in a meter. n_0 is reference space frequency defined as 0.1 m^{-1}. $S_q(n)$ is PSD of road surface under the reference space frequency $S_q(n_0)$ known as the road surface roughness coefficient and ω is the frequency index which decides the frequency configuration of PSD of road surface ($\omega = 2$).

The road surface roughness is assumed to be a zero-mean stationary Gaussian random process. It can be generated through an inverse Fourier transformation:

$$q(t) = \sum_{i=1}^{N} \sqrt{2S_q(n_i)\Delta n}\cos(2\pi n_k t + \varphi_i) \tag{3}$$

ϕ_i is random phase uniformly distributed from 0 to 2π.

In this study, typical road surface roughness is adopted according to the standard ISO 8068 [15] and computer simulation result of the typical road surface roughness ISO 8068 class B is shown in Fig. 2.

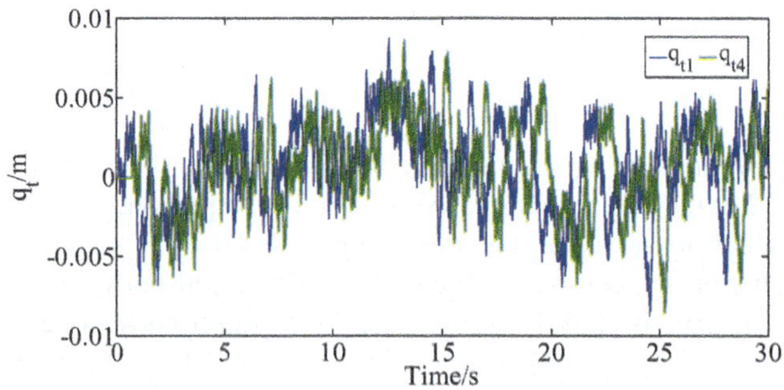

Fig. 2. Typical road surface roughness according to the standard ISO 8068 class B.

3 Evaluation Indicators

3.1 Vehicle Ride Comfort

Currently there are many methods to evaluate the vehicle ride comfort such as frequency-domain method, time-domain method, etc. This study is based on ISO 2631-1 (1997), vibration evaluation based on the basic evaluation method including measurements of the weighted root-mean-square (rms) acceleration is defined by:

$$a_w = \left[\frac{1}{T} \int_0^T a_w^2(t)dt \right]^{1/2} \quad (4)$$

where $a_w(t)$ is the weighted acceleration (translational and rotational) as a function of time, m/s^2; T is the duration of the measurements.

In this way, for indications of likely reactions to various magnitudes of overall vibration in the public transport a synthetic index-called the weighted RMS acceleration, a_w can be calculated from formula Eq. (4) and the RMS value of the vertical acceleration in vehicles would be compared with the values in Table 1.

Table 1. Comfort levels related to a_w threshold values

$a_w/(m.s^{-2})$	Comfort level	$a_w/(m.s^{-2})$	Comfort level
< 0.315	Not uncomfortable	0.8 ÷ 1.6	Uncomfortable
0.315 ÷ 0.63	A little uncomfortable	1.25 ÷ 2.5	Very uncomfortable
0.5 ÷ 1.0	Fairly uncomfortable	>2	Extremely uncomfortable

3.2 Dynamic Tire Load

Dynamic tire load would lead to stress and strain of road surface. The long-term accumulation of road surface plastic deformation causes the destruction of roads, such as cracks and rutting. In this paper, in order to analyze the effects of suspension design parameters of a semi-trailer truck on road surface friendliness, the dynamic load coefficient (DLC) is chosen as an objective function which is defined by a ratio of the root mean square of the vertical dynamic tire force over static load [4–9, 11] as follows:

$$DLC = \frac{F_{t,rms}}{F_s} \quad (5)$$

$F_{t,rms}$ and F_s are the root mean square of the vertical dynamic and the static tire force. The value of the DLC is in range of 0.05 to 0.3 under normal operating conditions. It may reach zero when the wheels move on a special smooth road or increase up to 0.4 when the tires of the axles spend a significant proportion of their time disconnecting the road surface [18].

4 Simulation and Discussion

In order to solve the general dynamic differential equation for a semi-trailer truck presented in Sect. 2.1, Matlab/Simulink software is used with a set of parameters of the articulated truck semi-trailer by the references [16, 17]. The simulation results of the acceleration responses of the vertical driver's seat, cab's pitch and the vertical dynamic tire loads acting on road surface at 3rd and 5th axles when the vehicle moves on the ISO class B road surface at v = 20 m/s with fully loaded are shown in Fig. 3.

Figure 3 could determine the values of the weighted root mean square (RMS) of acceleration responses of the vertical driver's seat, cab's pitch and dynamic load coefficients (DLC) at 3rd and 5th axles, the response values such as $a_{ws} = 0.2456$ m/s^2, $a_{wcphi} = 0.0444$ rad/s^2 and $DLC_{t3} = 0.0272$; $DLC_{t5} = 0.0438$. The values are satisfied in the allowed range (in Sect. 3) for vehicle ride comfort as well as road surface friendliness. The effects of suspension design parameters of a semi-trailer truck such as stiffness and damping coefficients on a_{ws}, a_{wcphi} and DLC_{t4} values are being presented in the following in sections.

4.1 Effect of Stiffness Coefficients

Stiffness of vehicle suspension system is responsible for reducing the vibrations transmitted from the road surface into the chassis as well as reducing the dynamic force, vibrations transmitted from the engine to the road surface. To analyze the effect of stiffness coefficients of vehicle suspension systems on a_{ws}, a_{wcphi} and DLC_{t4} values, the values of the stiffness coefficients of vehicle suspension systems k = [0.2, 0.4, 0.6, 0.8, 1.0, 1.2, 1.4, 1.6, 1.8, 2.0] xk_0 where $k_0 = [k_1, k_2, k_3, k_4, k_5]^T$ are analyzed when the vehicle moves the ISO class B road surface at v = 20 m/s and full load, where k_0 is used to designate the stiffness coefficients of vehicle suspension systems in the reference document [17]. Results of the analysis of the effect of stiffness coefficients of vehicle suspension systems on a_{ws}, a_{wcphi} and DLC_{t4} at 4th axle values are shown Fig. 4.

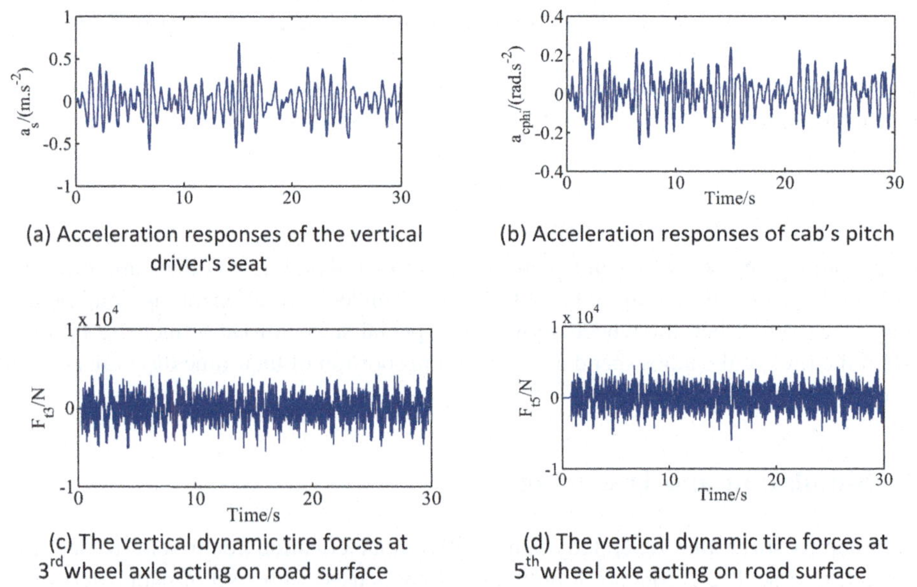

(a) Acceleration responses of the vertical driver's seat

(b) Acceleration responses of cab's pitch

(c) The vertical dynamic tire forces at 3rd wheel axle acting on road surface

(d) The vertical dynamic tire forces at 5th wheel axle acting on road surface

Fig. 3. Acceleration responses of the vertical driver's seat, cab's pitch and the vertical dynamic tire loads acting on road surface at 3rd and 5th axles.

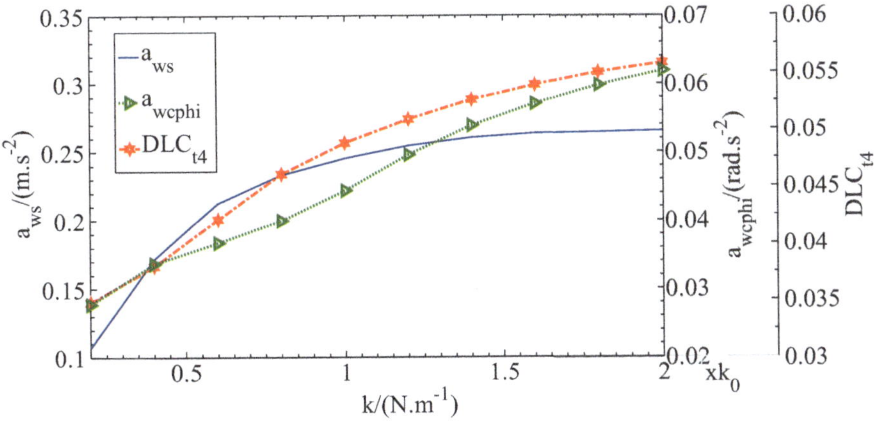

Fig. 4. Effect of stiffness coefficients of vehicle suspension systems on a_{ws}, a_{wcphi} and DLC_{t4} at 4th axle values.

From the results in Fig. 2, it is shown that the values of the stiffness coefficients of vehicle suspension systems increase, the values of a_{ws}, a_{wcphi} and DLC_{t4} at 4th axle increase, which means driver's ride comfort as well as road surface friendliness are becoming worse and worse. As the values of the stiffness coefficients of vehicle suspension systems increase from $0.8k_0$ to $2k_0$, the values of a_{ws}, a_{wcphi} and DLC_{t4} at 4th axle increase from 5.1 to 8.1%, 11.0 to 39.4% and 5.9 to 14.3% compared with the suspension stiffness values of the original vehicle.

4.2 Effect of Damping Coefficients

Damping coefficient in vehicle suspension takes responsibility to reduce vibration amplitude transmitted to the vehicle as well as to improve vehicle ride comfort and road surface friendliness. To analyze the effect of damping coefficients of vehicle suspension systems on a_{ws}, a_{wcphi} and DLC_{t4} values, the values of the damping coefficients of vehicle suspension systems $c = [0.2, 0.4, 0.6, 0.8, 1.0, 1.2, 1.4, 1.6, 1.8, 2.0]\ xc_0$ where $c_0 = [c_1, c_2, c_3, c_4, c_5]^T$is analyzed when the vehicle moves the ISO class B road surface at v = 20 m/s and full load, where c_0 is used to designate the damping coefficients of vehicle suspension systems in the reference document [17]. Results of the analysis of the effect of damping coefficients of vehicle suspension systems on a_{ws}, a_{wcphi} and DLC_{t4} at 4th axle values are shown in Fig. 5.

Figure 6 shows that the values of the damping coefficients of vehicle suspension systems increase, the values of a_{ws}, a_{wcphi} and DLC_{t4} at 4th axle decrease, which means the vehicle ride comfort as well as road surface friendliness is improved until the values are $c > 2.0c_0$. As the values are $c \leq 2.0c_0$, the values of a_{ws}, a_{wcphi} and DLC_{t4} at 4th axle increase the most quickly in the direction of the uncomfort driver and unfriendly road surface. As the values of the damping coefficients of vehicle suspension systems decrease from $0.2c_0$ to $2c_0$, the values of a_{ws}, a_{wcphi} and DLC_{t4} at 4th axle increase from 11.4 to 9.5%, 13.7 to 8.3% and 11.3 to 2.9% compared with the suspension damping values of the original vehicle.

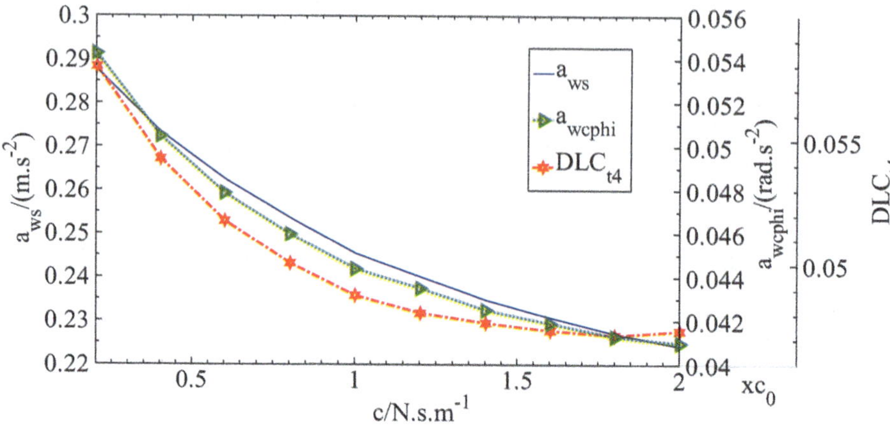

Fig. 5. Effect of damping coefficients of vehicle suspension systems on a_{ws}, a_{wcphi} and DLC_{t4} at 4th axle values

5 Conclusions

In this study, a half-vehicle dynamic model of a semi-trailer truck is established for analyzing the effects of suspension design parameters of a semi-trailer truck on vehicle ride comfort as well as road surface friendliness when the vehicle moves on the ISO class B road surface at v = 20 m/s and full load. The major conclusions that can be drawn from the analysis results are as follows:

(1) The greater the values of the stiffness coefficients of vehicle suspension systems become, the greater the values of a_{ws}, a_{wcphi} and DLC_{t4} at 4th axle are, which means both driver's ride comfort and road surface friendliness are getting worse and worse.

(2) The values of the damping coefficients of vehicle suspension systems increase whereas the values of a_{ws}, a_{wcphi} and DLC_{t4} at 4th axle decrease. That means not only the vehicle ride comfort but also the road surface friendliness improve until the values are $c > 2.0c_0$. The values decrease from $0.2c_0$ to $2c_0$, a_{ws}, a_{wcphi} and DLC_{t4} at 4th axle increase from 11.4 to 9.5%, 13.7 to 8.3% and 11.3 to 2.9% compared with the suspension damping values of the original vehicle.

(3) The study results are the theoretical basis for the researcher group to continue studying the optimal design parameters of vehicle suspension system with a view to improving road surface friendliness, vehicle ride comfort as well as reducing vehicle noise.

References

1. Van Quynh, L., Zhang, J., et al.: Influence of heavy truck dynamic parameters on ride comfort using 3D dynamic model. Dongnan Daxue Xuebao (Ziran Kexue Ban)/J. Southeast Univ. (Nat. Sci. Ed.) **43**(4), 763–770 (2013)
2. Jie, L., Wenzhu, W., Xiong, G., Zhenwei, Z.: Study on the influence of different factors on heavy truck ride comfort. SAE Technical Paper Series (2016)
3. Abdelkareem, M.A., Makrahy, M.M., et al.: An analytical study of the performance indices of articulated truck semi-trailer during three different cases to improve the driver comfort. Proc. Inst. Mech. Eng. Part K: J. Multi-Body Dyn. **232**(1), 84–102 (2017)
4. Shi, X.M., Cai, C.S.: Simulation of dynamic effects of vehicles on pavement using a 3D interaction model. J. Transp. Eng. **135**(10), 736–744 (2009)
5. Van Quynh, L., Zhang, J., Liu, X., Wang, Y.: Nonlinear dynamics model and analysis of interaction between vehicle and road surfaces for 5-axle heavy truck. J. Southeast Univ. **27**(4), 452–457 (2011)
6. Van Cuong, B., Van Quynh, L., Long, L.X.: Influence of heavy truck operating condition on dynamic load coefficient. In: Advances in Engineering Research and Application, ICERA 2018, pp. 372–379 (2018)
7. Long, L.X., Hong, T.T., Quynh, L.V., Van Cuong, B.: Performance analysis of the hydro-pneumatic suspension system of heavy truck. Int. J. Mech. Eng. Technol. (IJMET) **9**(13), 1128–1139 (2018)
8. Le, V.Q.: Comparing the performance of suspension system of semi-trailer truck with two air suspension systems. Vibroengineering PROCEDIA **14**, 220–226 (2017)
9. Yongjie, L., Shaopu, Y., et al.: Numerical and experimental investigation on stochastic dynamic load of a heavy duty vehicle. Appl. Math. Model. **34**(1), 2698–2710 (2010)
10. Yikai, C., Jie, H., Mark, K., et al.: Effect of driving conditions and suspension parameters on dynamic load-sharing of longitudinal-connected air suspensions. Sci. China Technol. Sci. **56**(3), 666–676 (2013)
11. Van Quynh, L.: Influence of semi-trailer truck operating conditions on road surface friendliness. Vibroengineering PROCEDIA **16**, 67–72 (2017)
12. Van Liem, N., Jianrun, Z., Van Quynh, L., Renqiang, J., Xin, L.: Performance analysis of air suspension system of heavy truck with semi-active fuzzy control. J. Southeast Univ. (Engl. Ed.) **33**(2), 159–165 (2017)
13. Nguyen, V.L., Van Quynh, L.: Ride comfort performance of heavy truck with three control cases of semi-active isolation systems. Vibroengineering PROCEDIA, Vol. 22, p. 93–98 (2019)
14. Dodds, C.J., Robson, J.D.: The description of road surface roughness. J. Sound Vib. **31**(2), 175–183 (1973)
15. Mechanical vibration—Road surface profiles—Reporting of measured data (1995)
16. Van Quynh, L., Hien, V.T., Cong, N.T.: Influence of tire parameters of a semi-trailer truck on road surface friendliness. Int. Res. J. Eng. Technol. (IRJET) **6**(6), 3674–3678 (2019)
17. Van Quynh, L.: Simulation and optimal design of suspension system for semi-trailer truck. Science Research Project (T2017-B31) (2018)
18. Buhari, R., Md Rohani, M., et al.: Dynamic load coefficient of tyre forces from truck axles. Appl. Mech. Mater. **405**(408), 1900 (1911)

Effects of the Die Inlet Angle and Axial Feed on Rotary Swaged Ti-6Al-4V Alloy Rods

Dinh Xuan Ta[1,2(✉)], Van Thao Le[2], and Van Canh Nguyen[3]

[1] National University of Science and Technology - MISiS, Moscow, Russia
dinhxuanta@gmail.com
[2] Le Quy Don Technical University, Hanoi, Vietnam
[3] Hanoi University of Industry, Hanoi, Vietnam

Abstract. This study addresses a comprehensive model to simulate the rotary swaging process of titanium alloy (Ti-6Al-4V) rods with a small diameter. One of the most important factors that affect the heterogeneity of products is the die inlet angle. Furthermore, the axial feed at each stroke has a strong influence on the homogeneity of the workpiece. Through numerical simulation based on 3D finite element code, Qform 9, the effect of the die inlet angle on the distributions of the strain, strain rate, and neutral plane are studied. It is can be found that the heterogeneity of strain and strain rate occurs during deformation in die angles is small. The axial feed at each stroke effects slightly on the forging force but considerably on the heterogeneity of the workpiece. The results of the simulation can be used to predict the quality of products, or as a guideline for the procedure of the die design.

Keywords: Rotary swaging · Simulation · Plastic deformation · Ti-6Al-4V

1 Introduction

Rotary swaging (RS) is known as one of the most effective processes of incremental forming, which has extensive applications due to the production of high precision and ultrafine grains [1, 2]. The process allows to reduce the cross-section of rods and tubes made of different metals, such as steel alloys, titanium, tungsten, and high-temperature superalloys. The forming of workpieces takes place in the swaging head in small steps by the radial oscillating movement of the tools [3].

In the literature, some researchers have studied RS using various approaches, such as the finite element method (FEM), and slab method. Moumi et al. [4] studied the material flow during in feed rotary swaging with a 2D model by ABAQUS package. Rong et al. [2, 5] analyzed the influence of axial feeding velocity on stress-strain field and temperature of rotary swaging process of pure magnesium using the general finite element software program MSC/Marc. Piela et al. [6] used a two-dimensional (2D) finite element model to analyze the metal flow in rotary swaging including an experimental verification. However, a comprehensive three-dimensional (3D) model for hardly deformed materials (e.g. titanium) is not yet considered. In RS, all plastic deformations of materials occur in the inlet zone.

K.-U. Sattler et al. (Eds.): ICERA 2019, LNNS 104, pp. 290–295, 2020.
https://doi.org/10.1007/978-3-030-37497-6_33

In this study, a full 3D model for hot rotary swaging is presented for investigating the effect of die inlet angle on the rotary swaging of solid rods with a small dimension. The results of this work allow us to understand the effect of the die inlet angle on the heterogeneous strain, the locations of the neutral plane and the forging forces on the tools.

2 Materials and Methods

In this study, the 3D finite element code, Qform 9, is used to simulate and investigate complicated deformation of round titanium alloy rods in hot RS process. The workpiece was formed from a cylinder with a diameter of 20 (mm) and a length of 120 (mm) to a diameter of 16 (mm). The chemical compositions of Titan alloy (Ti-6AL-4V) are shown in Table 1 and its plastic behavior was chosen in the library of the program Qform 9.

Table 1. Chemical compositions of the Ti-6Al-4V alloy (wt%).

Al	V	Zr	Fe	O	H	Si	N	C	Ti
5.3–6.8	3.5–5.3	≤ 0.3	≤ 0.6	≤ 0.2	≤ 0.15	≤ 0.1	≤ 0.05	≤ 0.1	Balance

The process parameters for RS process include spindle speed of 700 (rpm), tool displacement (stroke) of 2 (mm), cylinder rollers of 8 so the angle rollers rotate around the spindle in one impact of 45°, the time between stroke in the order of 0.029 s. These parameters were chosen based on the requirement that can be operated in the RS machine of two dies (RKM-2) and material properties.

In order to investigate the effects of die inlet angle (α) and axial feeding at each stroke (s) on material behavior, these parameters were varied as follows: α = 12°, 16°, 20°, 24° and s = 0.8; 1; 1.2 (mm). These values are also suitable for difficultly deformed materials.

The RS process was conducted at a temperature of 800 °C that can ensure the stability of the workpiece due to the high deformation rate. The shear stress at the interface between the billet and the tools is expressed as the Coulomb law of friction. The frictional factor between the dies and workpieces was determined 0.15 according to classic Coulomb model.

3 Results and Discussion

3.1 Strain Field Analysis

The strain distribution in the cross-section of workpieces, which was modeled with four types of die inlet angle and s = 1 (mm), is presented in Fig. 1. It can be seen that the process with the smallest die angle (α = 12°) shows a more heterogeneous strain distribution, where the difference between the minimum strain value and maximum strain value is of 1.09 (mm/mm), while the difference in the cases of the die angle equal to 16°, 20°, and 24° is 1.55, 1.71, and 1.75 (mm/mm), respectively. It is clearly illustrated in Fig. 1(a), the medium strain expands over the center a considerable area. When the die angle increases, the small strain area changes increase with the increase of inlet die angle. Moreover, the high values occur in the boundary of the workpiece with a thin layer (Fig. 1b, c, d).

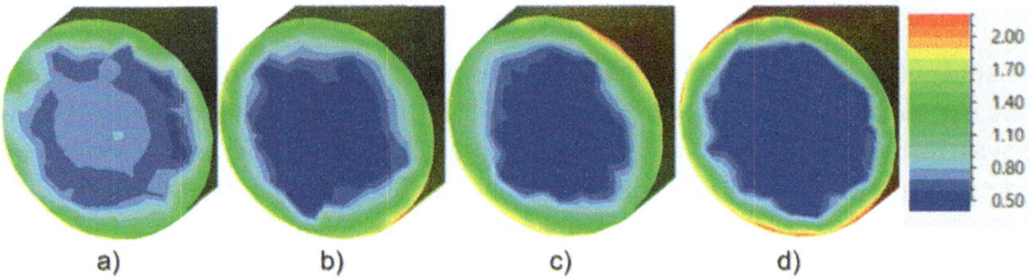

Fig. 1. Strain distribution in cross-section for the case: α = 12° (a), 16° (b), 20° (c), and 24° (d).

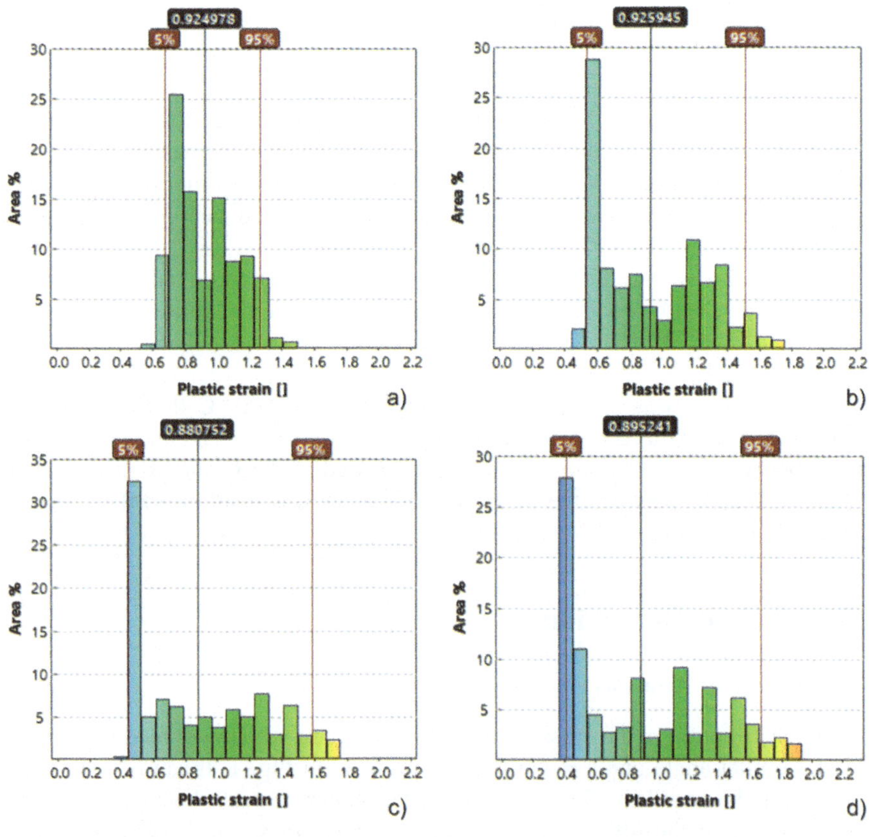

Fig. 2. Strain distribution as area percentages for α = 12° (a), 16° (b), 20° (c), and (d) 24°.

In order to more understand the strain distribution in the cross-section, a graphic of the area percentage/strain values was created, as shown in Fig. 2. The graphic analysis shows that 90% in the area has deformation degree in a range of 0.7–1.3 when the die inlet angle is equal to 12°and in a range of 0.6–1.5 for $\alpha = 16°$. With an increase of die angle, the range of the deformation degree expands in both directions. In particular, the proportion of small strain degree of a range of 0.4–0.5 increases dramatically with the increase of die angle. Examining the scheme in Fig. 2(c) it may be concluded that the total of the degree of plastic deformation less than 0.5 is bigger than 25% in area, while it is bigger 30% with the deformation degree less than 0.7 with the die angle of 24°.

As shown previously, it clearly depicts that deformations are indeed homogeneous in a small die inlet angle. However, the decrease of the die angle may be caused to increase the force on the die. Therefore, it is essential that the force must carry out carefully. Figure 3 shows the results of the simulation of the maximum forging force (F) depends on the die angle and the axial feeding of the workpiece. It can be found from Fig. 3 that the influence of the axial feeding on the maximum force is small, but the die angle affects significantly.

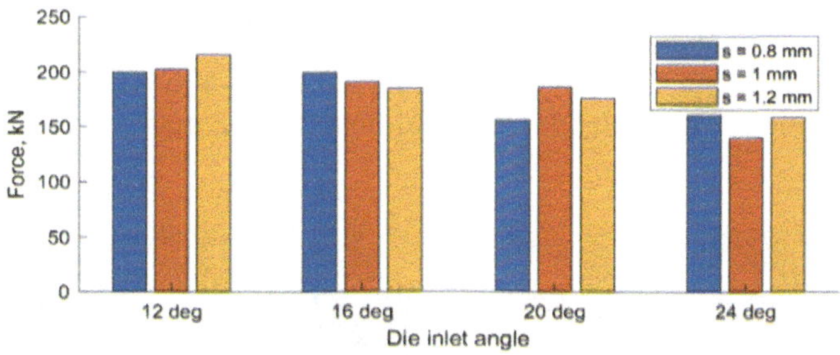

Fig. 3. Dependence of the forging force F on the die angle and the axial infeed workpiece

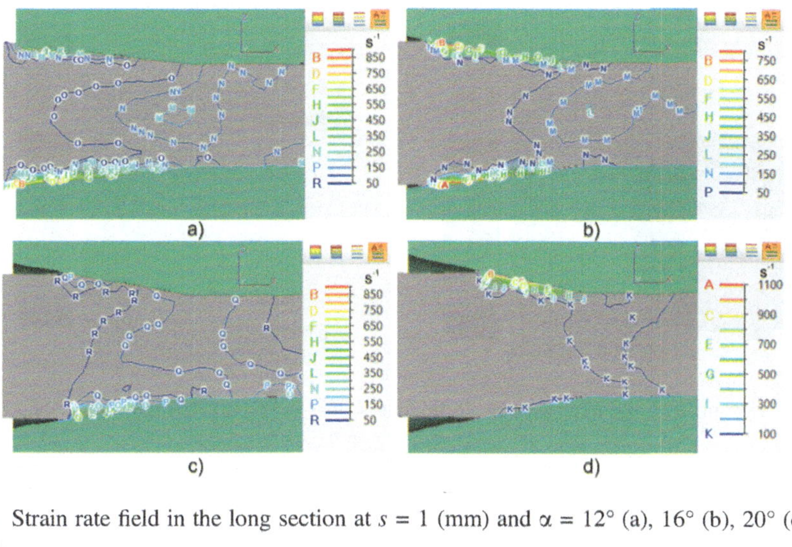

Fig. 4. Strain rate field in the long section at $s = 1$ (mm) and $\alpha = 12°$ (a), 16° (b), 20° (c), and 24° (d).

3.2 Strain Rate Analysis

In Fig. 4, the strain rate states in different die inlet angle are shown. This pictures clearly reveals that the high strain rate zone is concentrated at the contact between dies and workpiece with a value of more than 350 (s-1). In the deformation zone, a range of the strain rate between 50 and 100 (s-1) is distributed uniformly throughout the central area with $\alpha = 12°$ to 20°. It is also can be seen that a small zone of the high strain rate is set at 150–100 (s-1) for die angle of 12° and 16°. A high strain rate can lead to finer-grained microstructures, therefore a heterogeneity of strain rate also is an important parameter that affects the quality of the product.

3.3 Neutral Plane Locations

Another important parameter in the rotary swaging is the location of the neutral plane because of the stability of the workpiece. If the neutral plane lies outside the zone of deformation since the material would simply slip away from the radially approaching hammer dies without any deformation taking place [7]. Figure 5 shows the location of the neutral plane (n) in the dependence on the die angle with infeed of 1 (mm). As illustrated in Fig. 5c and d, the neutral plane is located closer to the undeformed part of the workpiece, while the positions of the neutral plane are in the zone of deformation with the angle of 12°–16°. In addition, the neutral cure has a parabolic shape toward the sizing zone in the center so that the material flow may be more stable.

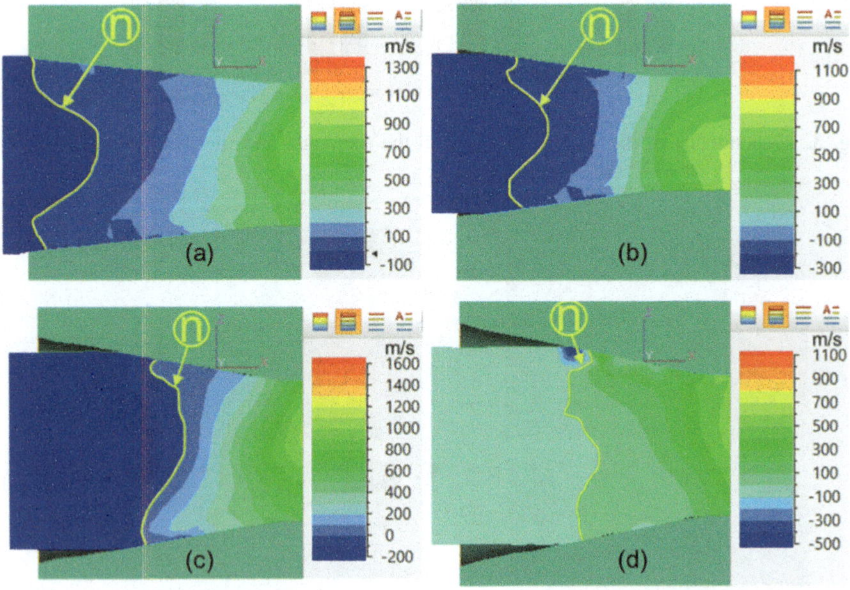

Fig. 5. Locations of the neutral plane in the cases of $\alpha = 12°$ (a), 16° (b), 20° (c), and 24° (d).

4 Conclusions

In this study, a 3D finite element code was used to the simulate of the rotary swaging process for the purpose of analyzing strain-stress field, location of the neutral plane in the workpiece and the required force on the dies. The main results of this paper can be summarized as follows:

(1) The process parameters such as die inlet angle, infeed displacement significantly affect on the stress-strain fields and the forging force, especially for materials which are difficult to deform, so that it is necessary to carry out and simulate for the process design.

(2) The inhomogeneity of deformation increases with the die inlet angle and decreasing with this tendency. Therefore, they are determined by the workpiece materials, the strength of the die structural analysis and the machine conditions.

(3) The simulation of rotary swaging is a meaningful method for the prediction of the quality of products. By modeling, the complicated process of rotary swaging with the technical parameters such as die inlet angle, the axial infeed of workpiece can be understood.

References

1. Chuvil'deev, V.N., Kopylov, V.I., Nokhrin, A.V., Tryaev, P.V.: Effect of severe plastic deformation realized by rotary swaging on the mechanical properties and corrosion resistance of near-α-titanium alloy Ti-2.5Al-2.6Zr. J. Alloy. Compd. **785**, 1233–1244 (2019)
2. Rong, L., Nie, Z., Zuo, T.: 3D finite element modeling of cogging-down rotary swaging of pure magnesium square billet-revealing the effect of high-frequency pulse stroking. Mater. Sci. Eng. A **464**(1), 28–37 (2017)
3. Vollertsen, F.: Micro Metal Forming. Springer, Heidelberg (1997)
4. Moumi, E., Ishkina, S., Kuhfuss, B., Hochrainer, T., Struss, A., Hunkel, M.: 2D-simulation of material flow during infeed rotary swaging using finite element method. Procedia Eng. **81**, 2342–2347 (2014)
5. Rong, L., Nie, Z.R., Zuo, T.Y.: FEA modeling of effect of axial feeding velocity on strain field of rotary swaging process of pure magnesium. Trans. Nonferrous Met. Soc. China **16**(5), 1015–1020 (2006)
6. Piela, A.: Analysis of the metal flow in swaging-numerical modelling and experimental verification. Int. J. Mech. Sci. **39**(2), 221–231 (1997)
7. Lahoti, G.D., Altan, T.: Analysis of the radial forging process for manufacturing rods and tubes. J. Eng. Ind. **98**(1), 265–271 (2010)

Effects of the Tube Diameter on the Mechanical Properties of Black Phosphorene Nanotubes

Van-Trang Nguyen[1(✉)], Minh-Quy Le[2], and Danh-Truong Nguyen[1,2]

[1] Faculty of Mechanical Engineering, Thai Nguyen University of Technology, No. 666, 3/2 Street, Tich Luong Ward, Thai Nguyen, Vietnam
nvtrang@tnut.edu.vn

[2] Department of Mechanics of Materials and Structures, School of Mechanical Engineering, Hanoi University of Science and Technology, No. 1, Dai Co Viet Road, Hanoi, Vietnam

Abstract. We use molecular dynamics finite element method with Stillinger-Weber potential to study the effects of the tube diameter on the mechanical properties of black phosphorene nanotubes (BPNT). Various armchair and zigzag BPNTs with a wide range of diameters from ~12 through ~34 Å are considered. The effects of the tube diameter on the mechanical properties of zigzag tubes are more significant than those of the armchair ones. When increasing the diameter in the studied range, the Young's modulus and fracture stress increase about 10% and 8.6%; and 109% and 214% for armchair and zigzag tubes, respectively; whereas, the fracture strains decrease very slightly (~3%) for the armchair tubes and increase up to 113% for the zigzag ones. When the diameter of BPNT is sufficiently large, its mechanical properties approach to those of the black phosphorene sheet.

Keywords: Atomistic simulation · Black phosphorene nanotube · Mechanical properties

1 Introduction

Recent fabrication of two-dimensional (2D) black phosphorous (black phosphorene or α-phosphorene (α-P)) [1–5] motivates the research of this new 2D material and its nanotubes. A black phosphorene nanotube (BPNT) is geometrically formed by rolling a single-layer black phosphorene along the armchair or zigzag direction as schematically illustrated in Fig. 1. Similar to black phosphorene, BPNTs are also semiconductors with tunable bandgaps [6–10], suggesting that they have potential applications in optoelectronics. The bandgaps and electronic structure of BPNTs are also affected by the mechanical strain, see e.g. [11]. On the other hand, Li et al. [8] showed that for an armchair and zigzag BPNTs of equal diameters, the bandgap of the armchair tube is higher than that of its corresponding zigzag one. They also indicated that the bandgap of BPNT increases with the tube radius and approaches to that of the α-P monolayer.

Studies of the mechanical properties of BPNTs are particularly important for their engineering application. The 2D Young's modulus of the armchair phosphorene tubes with diameters larger than 20 Å under uniaxial tension along the zigzag direction was

K.-U. Sattler et al. (Eds.): ICERA 2019, LNNS 104, pp. 296–302, 2020.
https://doi.org/10.1007/978-3-030-37497-6_34

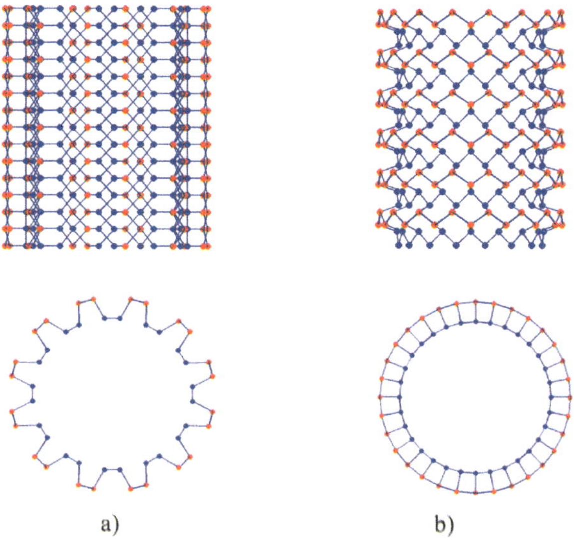

Fig. 1. Schematic illustration of atomic structures of: (a) (0, 14) armchair BPNT; and (b) (18, 0) zigzag BPNT.

estimated at ~40.6 N/m and below 30 N/m by molecular dynamics (MD) simulations with compass force field by Chen et al. [12] and density functional tight-binding study [13], respectively. These values are well lower than the value of ~86.5 N/m predicted by density functional theory (DFT) calculations [14]. Deviation between different theoretical results implies that the uniaxial tension of black phosphorene nanotubes should be further studied. Liu et al. [15] showed that missing atoms could reduce significantly the tensile performance of BPNTs.

The present work investigates through molecular dynamics finite element method (MDFEM) the effects of the tube diameters on the tensile mechanical properties of armchair and zigzag BPNTs. The Young's modulus, fracture stress and fracture strain are considered for a wide range of the tube diameter. Fracture mechanism of BPNTs is also analyzed.

2 Model and Method

Nine armchair BPNTs, namely (0, 8), (0, 10), (0, 12), (0, 14), (0, 16), (0, 18), (0, 20), (0, 22) and (0, 24); and nine zigzag ones, including (10, 0), (13, 0), (16, 0), (18, 0), (21, 0), (23, 0), (27, 0), (29, 0) and (31, 0) are considered. Their diameters fall within ~12 and ~34 Å. The length-diameter ratios L/D of these 18 BPNTs are fixed at 10. The geometric parameters of black phosphorene are taken from [16].

Stillinger-Weber potential is here used to model the P-P interatomic interactions [17]. We have successfully developed MDFEM [18] to investigate the mechanical properties of 2D hexagonal crystals, and study the mechanical properties of black phosphorene [19] and BPNTs [20, 21] with Stillinger-Weber potential. Our MDFEM code is here used to investigate the mechanical properties of BPNTs with Stillinger-Weber potential. σ and ε denote the nominal axial stress (engineering stress) and nominal axial strain (engineering strain), respectively. Young's modulus Y is determined from the first derivative of the stress-strain curve at zero strain. Here Yt and σt denote 2D Young's modulus and 2D stress, respectively. t is the tube's thickness.

3 Results and Discussion

Figure 2 show the stress-strain curves of armchair and zigzag BPNTs under uniaxial tension. At a given axial strain, the axial tensile stress of an armchair BPNT is always higher than that of the corresponding zigzag one with the same diameter. The axial tensile stress increases with an increase of the axial tensile strain, then it drops suddenly from a maximal value of stress for all tubes as indicated in Fig. 2, demonstrating a brittle fracture. Maximal stress and strain at maximal stress are fracture stress and fracture strain, respectively. Tables 1 and 2 summary results of the Young's modulus, fracture stress and fracture strain of BPNTs. Results reveal that the fracture stress of armchair (0, 24) BPNT (3.88 N/m) is about twice the value of zigzag (31, 0) one (1.8 N/m). These results are in good agreement with those from density functional-tight binding calculations [13]. In addition, the fracture strain of the armchair (0, 24) and zigzag (31, 0) tubes are 16.1% and 27%, respectively. The fracture strain of armchair BPNT and zigzag BPNT approaches to that of phosphorene sheet under tension along the zigzag direction (~16%) and along the armchair one (~28%), respectively [19, 22, 23].

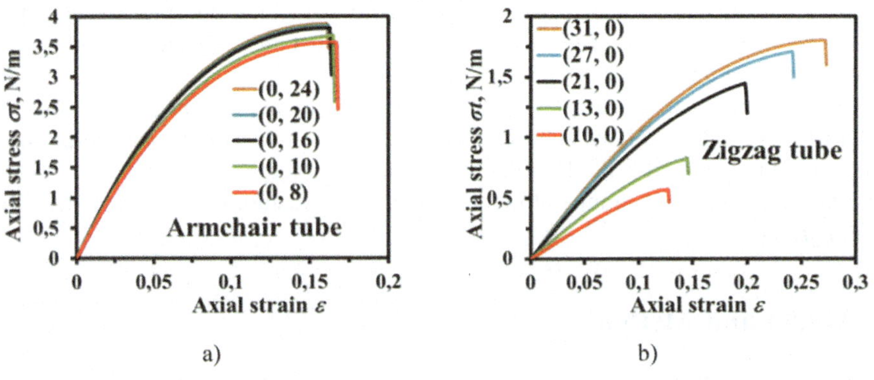

Fig. 2. Stress-strain curves of: (a) armchair; and (b) zigzag BPNTs under uniaxial tension.

Young's modulus and fracture stress of BPNTs increase with an increase of their tube diameter as indicated in Figs. 3 and 4. Using the same force field parameters (Stillinger-Weber potential), previous studies [19, 22, 23] reported that 2D Young's modulus, 2D fracture stress and fracture strain of black phosphorene are about 54–58 N/m, 4.0–4.6 N/m, and 16.0–16.2% under tension along the zigzag direction,

Table 1. Mechanical properties of armchair BPNT (tensile along the zigzag direction).

Nanotube	Young's modulus *Yt*, N/m	Fracture stress σt, N/m	Fracture strain ε, %
(0, 8)	50.36	3.573	16.6
(0, 10)	51.68	3.689	16.5
(0, 12)	52.77	3.714	16.4
(0, 14)	53.57	3.763	16.3
(0, 16)	54.21	3.805	16.2
(0, 18)	54.59	3.828	16.2
(0, 20)	54.92	3.849	16.1
(0, 22)	55.20	3.866	16.1
(0, 24)	55.42	3.880	16.1

Table 2. Mechanical properties of zigzag BPNT (tensile along the armchair direction).

Nanotube	Young's modulus *Yt*, N/m	Fracture stress σt, N/m	Fracture strain ε, %
(10, 0)	5.88	0.574	12.7
(13, 0)	7.69	0.834	14.4
(16, 0)	9.39	1.126	17.0
(18, 0)	10.08	1.249	17.5
(21, 0)	10.87	1.448	19.8
(23, 0)	11.30	1.569	21.5
(27, 0)	11.83	1.708	24.1
(29, 0)	12.08	1.775	26.0
(31, 0)	12.26	1.805	27.0

respectively; and about 12.5–12.6 N/m, 1.9–2.6 N/m, and 27.5–28.0% under tension along the armchair direction, respectively. The 2D Young's modulus and fracture stress of the (0, 24) armchair BPNT (under tension along the zigzag direction) with the largest diameter studied in the present work are estimated about 55.42 N/m (~ 106 GPa) and 3.88 N/m, respectively (see Table 1). The 2D Young's modulus and fracture stress of the (31, 0) zigzag BPNT (under tension along the armchair direction) with the largest diameter in the present study are ~ 12.3 N/m (~ 24 GPa) and ~ 1.8 N/m, respectively (see Table 2). Hence, the Young's modulus and fracture stress of BPNT approach to those of the phosphorene sheet when the tube diameter is sufficiently large.

Our predicted 2D Young's modulus of large armchair tubes (~ 55 N/m) is more reliable than that predicted by MD simulations with compass force field by Chen et al. [12] (~ 40.6 N/m) and density functional tight-binding study [13] (<30 N/m), compared to the value of ~ 86.5 N/m predicted by DFT calculations [14]. Compass force field used by Chen et al. [12] was built for phosphorus in organic compounds which is different from structure in phosphorene nanotube. So 2D Young's modulus of Chen et al. is lower than

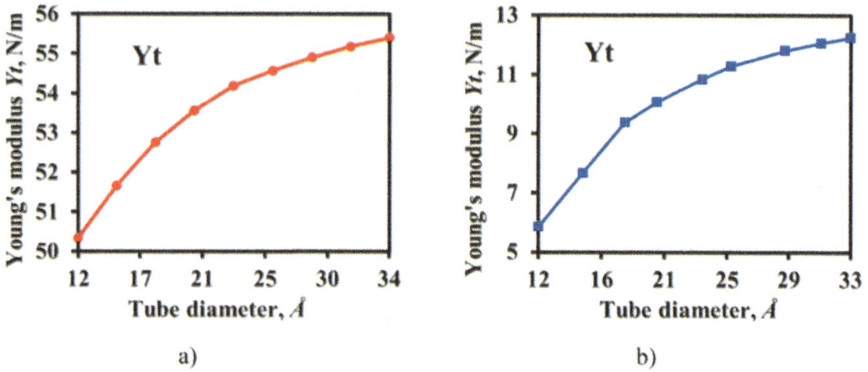

Fig. 3. Variations of the Young's modulus versus the tube diameter for: (a) armchair; and (b) zigzag BPNTs.

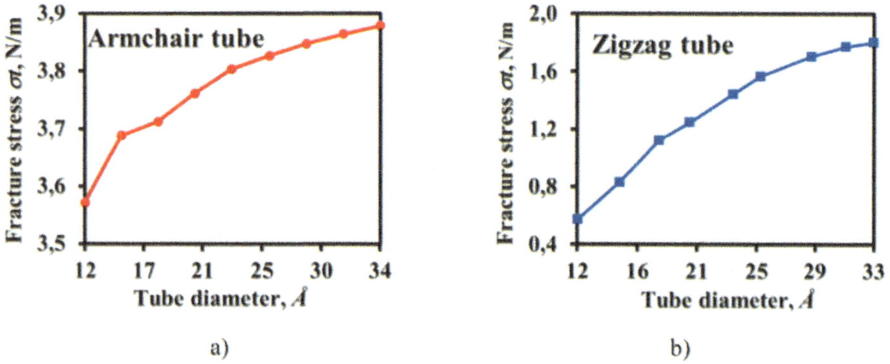

Fig. 4. Variations of the fracture stress versus the tube diameter for: (a) armchair; and (b) zigzag BPNTs.

our results. For nine zigzag BPNTs, namely (10, 0), (13, 0), (16, 0), (18, 0), (21, 0), (23, 0), (27, 0), (29, 0) and (31, 0), their 2D Young's modulus increases from 5.88 N/m to 12.26 N/m when their diameter increases from ~12.1 Å to ~34.1 Å.

These results of zigzag BPNTs agree with those from finite element analysis by Ansari et al. [24], who showed that for five zigzag BPNTs with their diameter from 14.26 to 24.24 Å, their 2D Young's modulus raises from 6.7 to 14.8 N/m. The fracture strain of the armchair BPNT decreases very slightly with an increase of its diameter as shown in Fig. 5. Sorkin et al. [13] also found that the fracture strain of armchair BPNT decreases with an increase of tube diameter in their previous density functional tight-binding calculations. The fracture strain of zigzag BPNT increases with an increase of the tube diameter.

The fracture strains of armchair BPNT and zigzag BPNT approach to that of phosphorene sheet under tension along the zigzag direction (~16%) and along the armchair one (~28%) [19, 22, 23] when the tube diameter becomes sufficiently large.

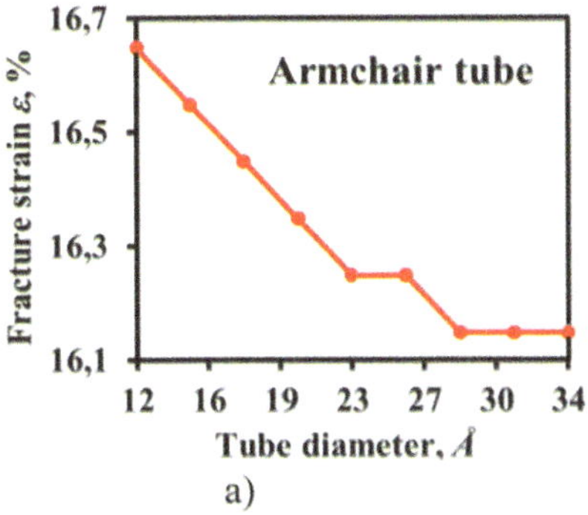

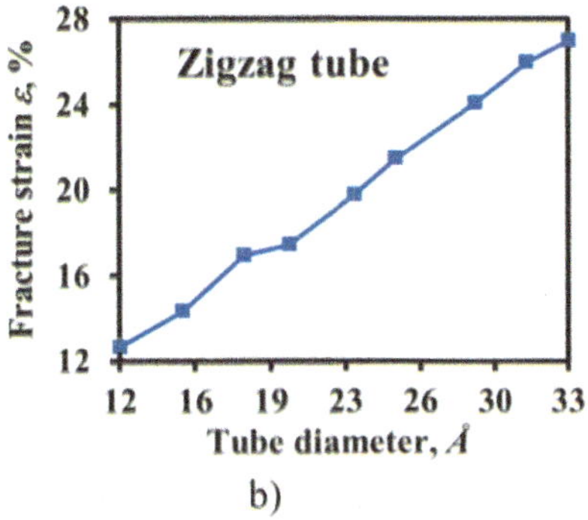

Fig. 5. Variations of the fracture strain versus the tube diameter for: (a) armchair; and (b) zigzag BPNTs.

4 Conclusions

We use molecular dynamics finite element method with Stillinger-Weber potential to study the effects of the tube diameter on the Young's modulus, fracture stress and fracture strain of 9 armchair and 9 zigzag BPNTs under uniaxial tension. Main results are summarized below:

For 2 tubes of equal diameters, the axial tensile stress of an armchair BPNT is always higher than that of the corresponding zigzag one at a given axial strain; the Young's modulus and fracture stress of an armchair BPNT are larger than those of the corresponding zigzag one.

When increasing the diameter from ~12 through ~34 Å, Young's modulus and fracture stress increase about 10% and 8.6%; and 109% and 214% for armchair and zigzag tubes, respectively. Fracture strains of armchair tubes decrease ~3%, whereas fracture strains of zigzag tubes increase up to 113% with an increase of the tube diameter in the above-mentioned range. The mechanical properties of BPNTs tend to those of black phosphorene sheet when the tube diameter becomes sufficiently large.

Acknowledgements. Van-Trang Nguyen was supported by the Thai Nguyen University of Technology, Thai Nguyen University (Vietnam) under grant number T2017-B01.

References

1. Lu, W., et al.: Plasma-assisted fabrication of monolayer phosphorene and its Raman characterization. Nano Res. **7**(6), 853–859 (2014)
2. Li, L., et al.: Black phosphorus field-effect transistors. Nat. Nanotechnol. **9**(5), 372–377 (2014)
3. Xia, F., Wang, H., Jia, Y.: Rediscovering black phosphorus as an anisotropic layered material for optoelectronics and electronics. Nat. Commun. **5**, 4458 (2014)
4. Reich, E.S.: Phosphorene excites materials scientists. Nature **506**(7486), 19 (2014)

5. Kou, L., Chen, C., Smith, S.C.: Phosphorene: fabrication, properties, and applications. J. Phys. Chem. Lett. **6**(14), 2794–2805 (2015)
6. Guo, H., et al.: Phosphorene nanoribbons, phosphorus nanotubes, and van der Waals multilayers. J. Phys. Chem. C **118**(25), 14051–14059 (2014)
7. Yu, S., et al.: A computational study of the electronic properties of one-dimensional armchair phosphorene nanotubes. J. Appl. Phys. **118**(16), 164306 (2015)
8. Li, C., et al.: Tunable bandgap and optical properties of black phosphorene nanotubes. Materials **11**(2), 304 (2018)
9. Hu, T., Hashmi, A., Hong, J.: Geometry, electronic structures and optical properties of phosphorus nanotubes. Nanotechnology **26**(41), 415702 (2015)
10. Allec, S.I., Wong, B.M.: Inconsistencies in the electronic properties of phosphorene nanotubes: new insights from large-scale DFT calculations. J. Phys. Chem. Lett. **7**(21), 4340–4345 (2016)
11. Zhang, W., et al.: Strain/stress engineering on the mechanical and electronic properties of phosphorene nanosheets and nanotubes. RSC Adv. **7**(81), 51466–51474 (2017)
12. Chen, W.-H., et al.: Mechanical property assessment of black phosphorene nanotube using molecular dynamics simulation. Comput. Mater. Sci. **133**, 35–44 (2017)
13. Sorkin, V., Zhang, Y.: Mechanical properties of phosphorene nanotubes: a density functional tight-binding study. Nanotechnology **27**(39), 395701 (2016)
14. Hao, J., et al.: Structure and elastic properties of black phosphorus nanotubes: a first-principles study. Physica Status Solidi (b), **254**(10) (2017)
15. Liu, P., et al.: Mechanical properties and fracture behaviour of defective phosphorene nanotubes under uniaxial tension. J. Phys. D Appl. Phys. **50**(48), 485303 (2017)
16. Takao, Y., Asahina, H., Morita, A.: Electronic structure of black phosphorus in tight binding approach. J. Phys. Soc. Jpn. **50**(10), 3362–3369 (1981)
17. Jiang, J.-W.: Parametrization of Stillinger-Weber potential based on valence force field model: application to single-layer MoS2 and black phosphorus. Nanotechnology **26**(31), 315706 (2015)
18. Le, M.-Q., Nguyen, D.-T.: Atomistic simulations of pristine and defective hexagonal BN and SiC sheets under uniaxial tension. Mater. Sci. Eng., A **615**, 481–488 (2014)
19. Nguyen, D.-T., et al.: Effects of various defects on the mechanical properties of black phosphorene. Superlattices Microstruct. **112**, 186–199 (2017)
20. Nguyen, V.-T., Minh-Quy, L.: Compressive buckling of black phosphorene nanotubes: an atomistic study. Mater. Res. Express (2018)
21. Nguyen, V.-T., Nguyen, D.-T., Minh-Quy, L.: Atomistic simulation of the uniaxial tension of black phosphorene nanotubes. Vietnam J. Mech. **40**, 163–169 (2018)
22. Yang, Z., Zhao, J., Wei, N.: Temperature-dependent mechanical properties of monolayer black phosphorus by molecular dynamics simulations. Appl. Phys. Lett. **107**(2), 023107 (2015)
23. Sha, Z.-D., et al.: Mechanical properties and fracture behavior of single-layer phosphorene at finite temperatures. J. Phys. D Appl. Phys. **48**(39), 395303 (2015)
24. Ansari, R., Shahnazari, A., Rouhi, S.: A density-functional-theory-based finite element model to study the mechanical properties of zigzag phosphorene nanotubes. Physica E **88**, 272–278 (2017)

Evaluation Motion Comfort of Sleeper Coach, the Type of Vehicle Is Manufactured and Assembled at Vietnam, Effect of Random Road Profile Based on Standard ISO 8608

Truong Manh Nguyen[1,2](✉), Lap Duc Vu[2], and QuangThanh Nguyen[3]

[1] National Research Institute of Mechanical Engineering, Pham Van Dong, Hanoi, Vietnam
Truong1601@gmail.com
[2] Faculty of Mechanical Engineering and Energy in Transportation, Le Quy Don Technical University, Hoang Quoc Viet, Hanoi, Vietnam
[3] Hanoi University of Industry, CauDien, Hanoi, Vietnam

Abstract. This study investigated generation of random road surface profile by Matlab software based on international standard ISO 8608 and applying this random road surface profile to evaluate motion comfort effect on passenger traveling by sleeper coach. The comfort indicator is determined by Root Mean Square (RMS) vertical acceleration indicator determined according to evaluation of ISO 2631-1 criteria. Evaluation model of sleeper coach is full car model, the model has eight degrees of freedom. The solve method is digital method with Matlab application software to appraise result. The results are shown the level of comfort at all different position of sleeper seat.

Keywords: Ride comfort · Full car model · Random road profile · Sleeper coach

1 Introduction

Research of motion comfort effect on human when travelling by vehicle is the subject for many researchers to improve the quality of vehicle. The profile of road surface is determined main source effect on the oscillation of vehicle. The road surface profile usually describe by pulse profile, step profile, sinusoidal profile… but the random profile reflect approximately real road profile [4]. The vehicle model used to evaluate are quarter model, half car model, full car model, other evaluations are used by software packages: Adam, Ansys.. [7]. The acceleration signal effect on driver and passenger is measured and evaluated according to ISO 2631-1. The random road profile is related different roughness evaluation method: IRI (International Roughness Index), DLI (Dynamic Load Index) and a_{WZ} (Whole-body Vibration Acceleration) [8]. The type of oscillation model depends on structural vehicle and components of suspension system. Depending on the requirement and scope of research, the appropriate model will be chosen. The sleeper coach is a new vehicle model which is manufactured and

K.-U. Sattler et al. (Eds.): ICERA 2019, LNNS 104, pp. 303–309, 2020.
https://doi.org/10.1007/978-3-030-37497-6_35

assembled in Vietnam. This type of vehicle has been increased quickly in recent years but the study about sleeper coach is limited. The study presents the process of generating random road profile and application to evaluate and analyze motion comfort of sleeper coach using full car model.

2 Methodology

2.1 Generation Random Road Profile

According to ISO 8608 [1] random road profile is described by power spectral density (PSD)

$$G_d(n) = G_d(n_0).\left(\frac{n}{n_0}\right)^{-w} \tag{2.1}$$

$$G_d(\Omega) = G_d(\Omega_0).\left(\frac{\Omega}{\Omega_0}\right)^{-w} \tag{2.2}$$

Where:

n, n_0: Spatial frequency and referent spatial frequency (=0,1 cycle/m);
Ω, Ω_0: Angular spatial frequency and referent angular spatial frequency ($\Omega_0 = 1$ rad/m);
w: waviness, for most of road surface profile w = 2;
$G_d(\Omega_0)$: Value of PSD at referent angular spatial frequency;
$G_d(n_0)$: Value of PSD at referent spatial frequency;

Road surface profile following ISO 8608 is divided to different road class from A to H. The A road class is the best roughness quality class.

In practical application, Power Spectral Density of vertical displacement is achieved by Fourier Transform of auto-correlation function of stochastic process.

The relation of PSD and mean square value in frequency band [3]:

$$G_d(n_i) = \frac{\psi_x^2(n_i, \Delta n)}{\Delta n} = \frac{\psi_x^2(i.\Delta n, \Delta n)}{\Delta n} \tag{2.3}$$

Where:

n_i: Generic spatial frequency value
Δn: Frequency band ($\Delta n = 1/L_m$)
i: Value change from 0 to N ($N = n_{\max}/\Delta n$).
n_{max}: Maximum sample frequency ($n_{max} = 1/B_m$)
L_m, B_m: Length of road profile and sample interval.

Some researchers [4] show that random road profile can generate by equation:

$$h_x = \sum_{i=0}^{N} A_i \cos(2\pi n_i x + \varphi_i) \tag{2.4}$$

$$A_i = \sqrt{2}\psi_x = \sqrt{2.\Delta n G_d(n_i)} \tag{2.5}$$

Substituting (2.5) to (2.4):

$$h_x = \sum_{i=0}^{N} \sqrt{2.\Delta n.G_d(i.\Delta n)} \cos(2\pi n_i x + \varphi_i) \tag{2.6}$$

Where:

φ_i: random phase angle following an uniform probabilistic distribution with in the $[0, 2\pi]$.

Substituting (2.1) in (2.6) Random road profile can describe:

$$h_x = \sum_{i=0}^{N} \sqrt{2.\Delta n.2^k.10^{-3}.\left(\frac{n_0}{i\Delta n}\right)} \cos(2\pi.i.\Delta n.x + \varphi_i) \tag{2.7}$$

Where:

x is abscissa value of road length (0, L);
k is a constant value depending from ISO 8608 road profile classification, k is integer increasing from 3 to 9 corresponding to the profile from class A to class H (Fig. 1).

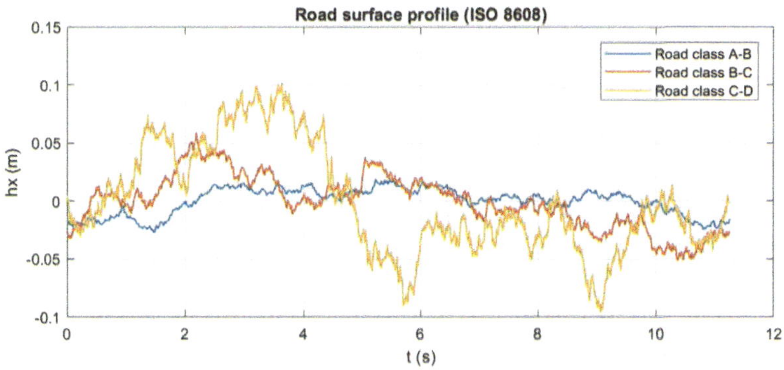

Fig. 1. Generation road profile base on ISO 8608 at V = 80 km/h.

2.2 Model of Sleeper Coach

Achieving for evaluation and analysis motion comfort of sleeper coach, model of sleeper coach is built based on analyzing structure characteristic of sleeper coach (Fig. 2).

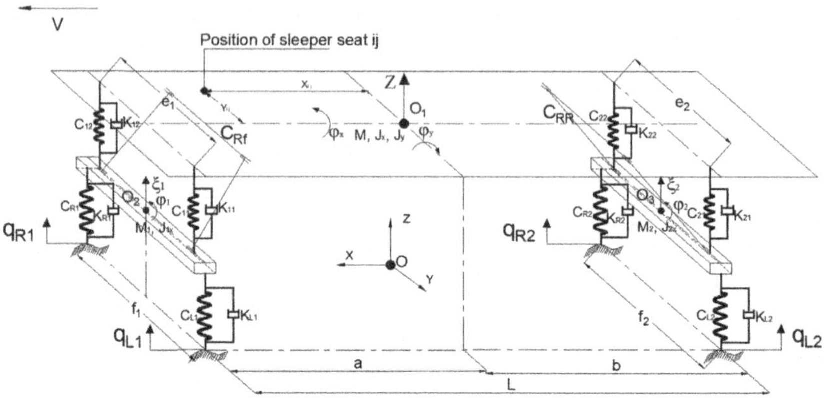

Fig. 2. Dynamic model of sleeper coach.

The oscillation model of sleeper coach is determined by lumped mass method of sprung and unsprung masses. The vehicle body is represented by sprung mass M and inertia moment J_x, J_y, while front and rear axle are represented by masses M_1, M_2 and inertia moment J_{1x}, J_{2x}. The elements of suspension connect with sprung and unsprung masses are represented by stiffness C_{11}, C_{12}, C_{21}, C_{22} and damper coefficient K_{11}, K_{12}, K_{21}, K_{22}. Model of tire is represented by stiffness C_{L1}, C_{L2}, C_{R1}, C_{R2} and damper coefficient K_{L1}, K_{L2}, K_{R1}, K_{R2}. Model of sleeper seat is replaced by model of one degree of freedom with stiffness c_{ngi} and damper coefficient k_{ngi}. The physical behavior of front and rear anti roll bar instead of stiffness C_{RR}, C_{RF}.

Applying Lagrange's equation type II, The equation system of sleeper coach are derived by eight second order differential equations.

Eights second differential equations convert and solving by state space model:

$$\begin{aligned} \dot{X} &= AX + Bu \\ Y &= CX + Du \end{aligned} \tag{2.8}$$

Where:

X is State vector (16 × 1).
$X^T = [Z, \dot{Z}, \varphi_x, \dot{\varphi}_x, \varphi_y, \dot{\varphi}_y, \xi_1, \dot{\xi}_1, \varphi_1, \dot{\varphi}_1, \xi_2, \dot{\xi}_2, \varphi_2, \dot{\varphi}_2, Z_1, \dot{Z}_1]$
Y is Output vector (16 × 1), u is Input vector (4 × 1).
A, B, C, D are State space Matrices. A (16 × 16), B (16 × 4), C (16 × 16), D (16 × 4).

The signal of vertical acceleration at each position of passenger will be taken then analyzed by Root Mean Square (RMS) indicator following by ISO 2631-1 criteria [1] (Table 1).

Table 1. Parameters of sleeper coach

No	Name of parameters	Symbol	Unit	Value
1	Sprung mass (Vehicle body)	M	kg	13403
2	Lateral Sprung mass moment of Inertia	Ix	$kg.m^2$	9664
3	Longitudinal Sprung mass moment of Inertia	Iy	$kg.m^2$	45469
4	Mass of front axle	M_1	kg	746
5	Front axle mass moment	I_1	$kg.m^2$	577
6	Mass of rear axle	M_2	kg	1355
7	Rear axle mass moment	I_2	$kg.m^2$	1049
8	Stiffness of right front sprung	C_{11}	N/m	200000
9	Stiffness of left front sprung	C_{12}	N/m	200000
10	Damper coefficient of right front sprung	K_{11}	Ns/m	20000
11	Damper coefficient of left front sprung	K_{12}	Ns/m	20000
12	Stiffness of right rear sprung	C_{21}	N/m	400000
13	Stiffness of left rear sprung	C_{22}	N/m	400000
14	Damper coefficient of right rear sprung	K_{21}	Ns/m	40000
15	Damper coefficient of left rear sprung	K_{22}	Ns/m	40000
16	Stiffness of tire (i = 1:2, j = 1:2)	C_{Lij}	N/m	1000000
17	Damper coefficient of tire (i = 1:2, j = 1:2)	K_{Lij}	Ns/m	150
18	Base length	L	m	6
19	Distance from center of front axle to center of gravity	a	m	4.098
20	Distance from center of rear axle to center of gravity	b	m	1.902
21	Distance from center of front axle to track of tire	f_1	m	1
22	Distance from center of rear axle to track of tire	f_2	m	1
23	Distance from axis of center of gravity to front sprung	e_1	m	1
24	Distance from axis of center of gravity to rear sprung	e_2	m	1
25	Stiffness of front anti roll bar	C_{RF}	N/m	30000
26	Stiffness of rear anti roll bar	C_{RR}	N/m	30000
27	Stiffness of sleeper seat (i = 1:22)	C_{ngi}	N/m	4000
28	Damper coefficient of sleeper seat (i = 1:22)	K_{ngi}	Ns/m	150

3 Result

Sleeper coach carries passenger between medium and far distance, on national road, highway then the study made B-C class, this road class is determined profile influent at Vietnam, and vehicle was operated at different vehicle speed: 60 km/h, 80 km/h, 100 km/h.

Table 2. Result of RMS value at different positions

No	Position of sleeper seat	RMS value (m/s^2)		
		V = 60 km/h	V = 80 km/h	V = 100 km/h
1	1st position of sleeper seat	0.754	1.423	1.445
2	2nd position of sleeper seat	0.202	1.119	0.958
3	3rd position of sleeper seat	0.332	0.823	0.485
4	4th position of sleeper seat	0.867	0.528	0.013
5	5th position of sleeper seat	1.403	0.232	0.459
6	6th position of sleeper seat	1.82	0.002	0.827
7	7th position of sleeper seat	0.009	1.354	1.01
8	8th position of sleeper seat	0.525	1.059	0.537
9	9th position of sleeper seat	1.061	0.763	0.065
10	10th position of sleeper seat	1.596	0.468	0.407
11	11th position of sleeper seat	2.013	0.238	0.775
12	12th position of sleeper seat	0.368	1.894	1.549
13	13th position of sleeper seat	0.183	1.589	1.062
14	14th position of sleeper seat	0.718	1.294	0.589
15	15th position of sleeper seat	1.254	0.998	0.117
16	16th position of sleeper seat	1.789	0.703	0.354
17	17th position of sleeper seat	2.206	0.473	0.723
18	18th position of sleeper seat	2.505	0.374	1.431
19	19th position of sleeper seat	2.602	0.257	1.405
20	20th position of sleeper seat	2.698	0.139	1.379
21	21st position of sleeper seat	2.794	0.022	1.353
22	22nd position of sleeper seat	2.891	0.095	1.327

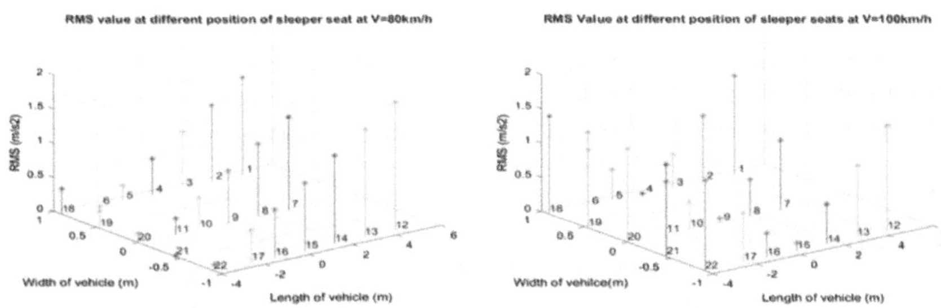

Fig. 3. RMS value at different positions of sleeper seat.

The results in Table 2 and Fig. 3 show that:

The maximum and minimum of RMS at positions of sleeper seat depend on the velocity of the vehicle.

- When the vehicle velocity is at 60 km/h, RMS value is maximum at the twenty-two sleeper seat. It is the non-comfort position on vehicle in this situation.
- The velocity is at 80 km/h and 100 km/h, RMS value is maximum at the twelth sleeper seat.

4 Conclusions

The study described and applied random road surface profile based on ISO 8608. This profile was applied on full car model to evaluate motion comfort of sleeper coach.

Three speeds 60 km/h, 80 km/h, 100 km/h were considered during crossing road surface classes B-C. In this surveys, the maximum RMS evaluation value is shown at 12th, 22nd positions of sleeper seat.

This model can be used for next advanced research as optimization parameters of suspension system of sleeper coach.

References

1. Mechanical vibration, road surface profiles, reporting of measured data, ISO std. 8608 (1995)
2. Mechanical vibration and shock_Evaluation of human exposure to whole body vibration_Part 1: General requirements, ISO, std. 2631-1 (1997)
3. Bendat, J.S., Piersol, A.G.: Random Data: Analysis and Measurement Procedures, 2nd edn. Wiley, New York (1986)
4. Agostinacchio, M., Ciampa, D., Olita, S.: The vibrations induced by surface irregularities in road pavements – a Matlab® approach. Eur. Transp. Res. Rev. **6**, 267–275 (2013)
5. Shinozuka, M.: Digital simulation of random processes and its applications. J. Sound Vibr. **25**, 111–128 (1972)
6. Yang, J.J., Zhang, Y.: An improved human biodynamic model considering the interaction between feet and ground. ResearchGate (2015)
7. Sekulic, D., Mladenovic, D.: Evaluation and analysis of vibration effects on Bus users. Horizons (2016)
8. Loprencipe, G., Zoccali, P.: Use of generated artificial road profiles in road roughness evaluation. J. Mod. Transp. **25**(1), 24–33 (2017)
9. Truong, N.M., Lap, V.D., ThanhQuang, N.: Investigating effect of changing base length to motion comfort on coach is made and assembly at Vietnam. In: Proceedings of the 1st INDO-ASEAN Conference on Innovative Approaches in Applied Sciences and Technologies, paper E.3, p. 91 (2018)

Control and Optimize Black Tea Fermentation Using Computer Vision and Optimal Control Algorithm

Pham Thanh Binh[1], Dao Huy Du[2(✉)], and Tang Cam Nhung[2]

[1] Northern Mountainous Agriculture and Forestry Science Institute, Phu Tho Town, Phu Tho Province, Vietnam
[2] Thai Nguyen University of Technology, 3/2 Street, Thai Nguyen City, Vietnam
daohuydu@tnut.edu.vn

Abstract. Black tea is a completely fermented tea. Fermentation is a very important stage in the process of producing black tea because the characteristic color of black tea depends entirely on the factors in this process, such as: humidity, temperature, and fermentation time. To control color variation during fermentation, computer vision is used to detect the color change of black tea in RGB, HSV or CIE LAB color systems. In order to get the fermented tea products of the standard, it is necessary to calculate and adjust exactly the factors affecting the quality of the fermentation process. In fact, the quality of black tea is always correlated with color. The parameters of color characteristics and sensory characteristics of black tea are used to put into the optimal control system for fermentation of black tea, ensuring that the output quality reaches the highest index. This article uses extract features of color in the CIE LAB color system combined with the predictability of RF nonlinear models to analyze the relationship between image information and quality indicators to determine control parameters for the system. Using this method will make the process of monitor and controlling black tea fermentation become simpler and both labor to monitor and the accuracy of the process achieved be higher results than using manual methods.

Keywords: Black tea fermentation · CIE lab · Random forests (RF) · Computer vision

1 Introduction

Black tea is one of the most popular beverages in the world with many good values that it brings to people. For black tea, fermentation is an extremely important stage, it determines the characteristics of tea including color, flavor, and shape of tea.

The fermented tea in the automatic line will be equipped with devices to provide temperature and humidity to create a perfect environment for fermentation. The characteristic pigments for black tea are mainly formed during fermentation. The color of black tea is a combination of many different color compounds including the following main types:

K.-U. Sattler et al. (Eds.): ICERA 2019, LNNS 104, pp. 310–319, 2020.
https://doi.org/10.1007/978-3-030-37497-6_36

- Theaflavins (TFs): TRs are compounds that produce a light yellow color and have a strong taste, TFs are directly proportional to the brightness of tea.
- Thearubigins (TRs): TRs make the color of tea become dark red, if there are too many TRs, the color of the tea is darker and affects the scent.
- Anthocyanidin: This is the main colorant for tea. This compound creates the red-pink color of the tea. The more Anthocyanidin content is, the better the beautiful black tea water color.

Black tea has marked a change in color during fermentation from turquoise to greenish-yellow, yellowish-brown and finally dark brown. The color of tea depends on the absolute content of Theaflavin (yellow) and Thearubigin (red) and depends on the content of these two color groups. In addition, in order for black tea to achieve a good quality as well as sensory color, the factors affecting fermentation include temperature, humidity and fermentation time. Color is an important feature used as a basis for optimizing the factors affecting fermentation such as temperature, humidity, fermentation time. Currently, there has been no research project to control the factors affecting color-based fermentation but only separate studies for each stage. Dong's article demonstrated that the image color characteristics and the nonlinear method can be used to evaluate changes in quality indicators (e.g. sensory quality and pigment) during the fermentation of black tea [3]. Borah [4, 5] has set up a detection system, which can be used to detect the degree of fermentation of broken black tea. In this system, images of tea leaves are collected intermittently during fermentation. Research project to build the optimal automation system for processing and preserving agricultural products of Xuan Quynh has put the optimal control technology into black tea production line to optimize automation. The process of processing and preserving tea in both OTD and CTC processing methods was presented in [7].

In general, computer vision was used to detect fermentation quality for black tea in some countries such as China, India. Currently, in Vietnam, this technology is not known and the research only focuses on tea classification and quality of fermentation for black tea. Correlation, interaction, and construction of control systems for color based on fermentation have not been studied. With the aim of overcoming deal with these shortcomings, this paper focuses on researching the color change of tea in the fermentation process using color spatial variation method and nonlinear RF regression method to provide parameters for the model to control the fermentation of black tea continuously.

2 Computer Vision and Extract Features Color

Based on machine vision technology and nonlinear RF regression algorithms, the paper focuses on research and offers optimal control methods for black tea fermentation elements based on the transformation of color space. We extracted 6 color characteristics (R, G, B, L, a*, b*) as characteristic parameters for evaluating fermentation quality, analyzing change rules, and the other relationships between image color and fermentation quality index to establish corresponding quantitative evaluation models through RF nonlinear models. The results show that the color and quality index properties differ between different stages in the fermentation process and they also have a clear correlation.

The computer vision system used includes image sensors, lights and dark rooms. The image sensor uses SONY IMX179 single lens type camera. The specifications are shown in Table 1. The dark chamber is installed at the sampling positions and the LEDs are mounted with uniform light (luminous intensity: 600Lx, 24 V voltage, 60 W power). The lighting area is a 1.5 × 1 m^2 area and the distance from the lens to the sampling area is 80 cm.

Table 1. Camera parameters

Model	ELP-USB8MP02G-SFV (5-50)	Camera parameters	MJPEG/YUV2 (YUYV)
Sensor	SONY IMX179	Lens parameter	5–50 mm varifocal lens
Lens size	1/3.2 in.	Board size/weight	41*41 mm, about 0.3 kg
Pixel size	1.4 um	Dynamic range	72.5 dB
Image area	6.18 mm× 5.85 mm	Connecting port type	USB2.0 High Speed
Max. resolution	3264 (H) 2448(V) 8 Megapixel	Free DriveProtocol	USB Video Class (UVC)
Mini illumination	0.5 lx	Sensitivity	0.65 V/lux-sec@550 nm

Sampling Method: At each position, we conducted research samples with fermentation time of 0 min, 30 min, 60 min, 90 min, 120 min and 150 min. At each of the above positions, a sample of analysis with a weight of 500 g is dried to 3–5% moisture. After that, tea is stored in glass vials, tightly closed, to give chemical and sensory analysis. Next, a camera is used to collect sample images, then each sample is put into a sample tray. The sample tray is inserted into the light chamber to receive the image. Each image has a standard of 1500 × 1000 pixels automatically, separated from the background to extract color characteristics according to TCVN-1053-86 in processing black tea. The result of the color transformation process is shown in Table 2.

Table 2. Quality standards for fermentation of tea over time

Norm	Fermentation time (minutes)							
	50	60	70	80	90	100	110	120
Color of fermented tea	Greenish brown	Blue brown	Blue brown	Blue brown	Dark red copper	Dark red copper	Copper	Copper

The color of fermented tea gradually changes over time from brownish green to dark red copper and then red copper. The colors change most clearly at 60, 90 and 110 min. From the above results, we selected the surveillance camera setpoint at positions 60', 90' and 110' on the conveyor. The location and time of selection, we conducted photos of fermented tea as a basis to study the image processing process monitoring the fermentation of black tea. The diagram of the black tea fermentation system used in the study is shown in Fig. 1.

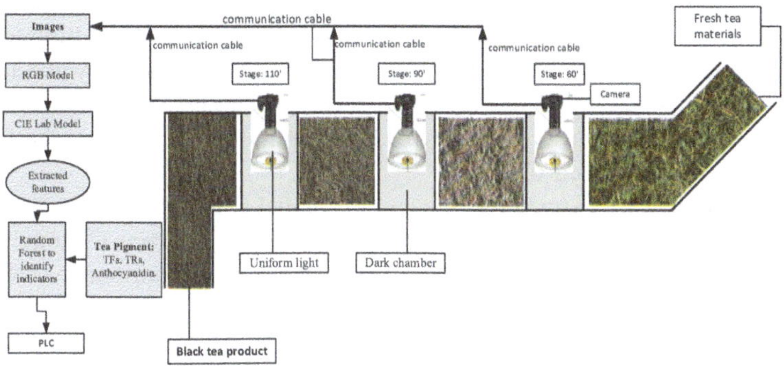

Fig. 1. Model of collecting image data in the black tea fermentation system

3 Processing and Analyzing Data

During the process of collecting images, the gloss is created due to the lack of light in tea areas that are sunken compared to the sample tray surface. Therefore, before analyzing, the Image Analysis Module will automatically exclude the background shadow of the set threshold image. The input images will be divided into: gray multi-level images, RGB images, index images (Index Image), and binary images [3]. The Histogram chart is set corresponding to the color distribution of the regions as shown in Fig. 2.

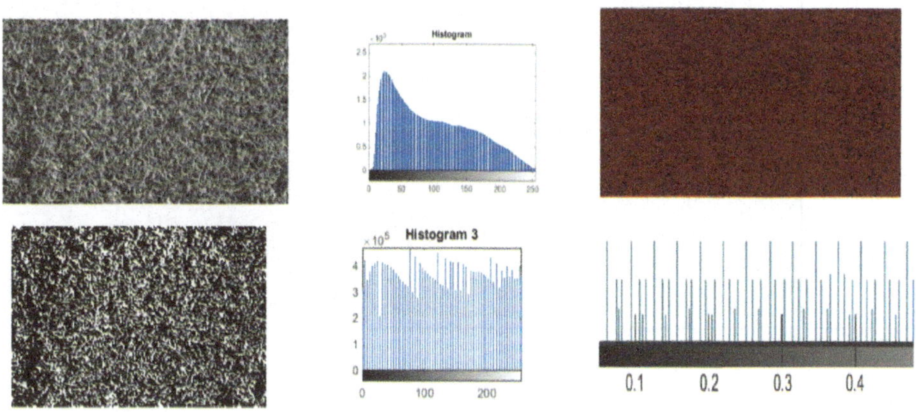

Fig. 2. Histogram chart is set corresponding to the color distribution

For a more detailed analysis of tea leaf color changing rules during fermentation, this paper analyzed the visual changes in an overview. Each selected image will be analyzed according to the image type and each of those will be represented in the Histogram chart. Through the analysis chart, we can see that the color uniformity is accurately represented when the image is represented in the index image format. During sampling, different fermentation images are arranged in chronological order and represented as an index image. We can observe a clear difference in color variation during different fermentation periods. To achieve a clearer distinction, color saturation and image brightness have been enhanced. In addition, the study converted the image from an RGB color model to an HSV color model, where S is multiplied by three values, V is doubled by value and then transferred back from the HSV model to RGB model. The result is described in Fig. 3. As shown in the results for the fermentation process, the color of the leaf is gradually changed from the brownish-green to the final dark red color, which is Red Copper.

Six color indexes are extracted using a color model corresponding to RGB models, CIE Lab: average values of R channels (Red Channel), G (Green Channel), B (Blue Channel components a (a*), component b (b*), brightness component (L*) [8]. The color change is based on the following formulas:

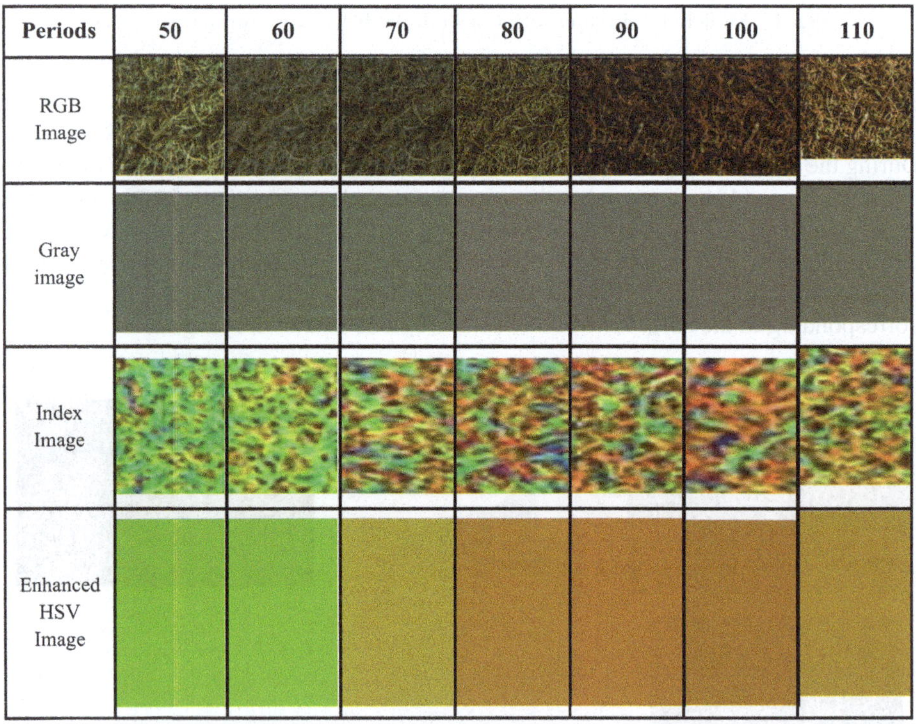

Fig. 3. Color change process in different fermentation stages

- Convert RGB color space to CIE XYZ color space:

$$\begin{pmatrix} R \\ G \\ B \end{pmatrix} = M_{RGB} \begin{pmatrix} X \\ Y \\ Z \end{pmatrix}$$

$$\text{with: } M_{RGB} = \begin{pmatrix} 3.240479 & -1.537150 & -0.498535 \\ -0.969256 & 1.875992 & 0.041556 \\ 0.055648 & -0.204043 & 1.057311 \end{pmatrix}$$

- Convert CIE XYZ color space to CIE $L^*a^*b^*$- Convert CIE XYZ color space to CIE CIE $L^*a^*b^*$ color space

$$L^* = 116.Y' - 16; a^* = 500(X' - Y'); b^* = 200(Y' - X')$$

where: $X' = f\left(\frac{X}{X_{ref}}\right); Y' = \left(\frac{Y}{Y_{ref}}\right); Z'\left(\frac{Z}{Z_{ref}}\right)$

and $f(c) = \begin{cases} c^{\frac{1}{3}} & \text{if } c > 0.008856 \\ 7.787c + \frac{116}{116} & \text{if } c \leq 0.008856 \end{cases}$

with: $X_{ref} = 0.950456; Y_{ref} = 1.000000; Z_{ref} = 1.088754.$

The value of L^* is a positive number in the range [0,100], the values of a^* and b^* range from [−127, +127].

$L^* = 116.Y' - 16$; $a^* = 500(X' - Y'); b^* = 200(Y' - X')$ in the next stage, parameters with different color characteristics will be extracted and these parameters represent the rules for changing the color of tea leaves relatively; The results are shown in Table 2. During the fermentation process, all the individual values of the color tend to decrease except for a* that tends to increase. During the fermentation period from 0–80' all color properties have been significantly changed and at the time of 110', the index a* and H respectively reached the highest and lowest mark and the color of Tea leaves at this time has the highest level of red. After this time, the red color (R, H and a*) of the fermented leaves is slightly changed, b* indicates the yellow color is still decreasing, the brown color of the fermented leaf is darker; the substance and the sensory content of black tea are reduced. According to the general analysis, the color of the leaves is relatively high compared to the TRs and TF content, the color change is completely consistent with the change of biochemical composition. The higher the TRs, the more leaves will be red, but if TRs are too high, the leaves will become darker. The higher the TFs, the brighter the leaf color. In this study, at the fermentation stage at the time of 110', the components of TFs, TRs and TBs were at their best, when the color of tea leaves became red and the organoleptic quality was good. Most of this time, these factors are completely in line with the sensory evaluation criteria of black tea.

Establish RF nonlinear model for quality indicators: With quality index ratios (TFs, TRs, TBs, and senses) of 300 samples of tea leaves used as reference values for the model, samples were divided according to the following Table 3: 150 samples were selected for training, the remaining 70 samples were used for testing samples.

Random forest algorithm (RF) is an effective learning method, which improves the accuracy of model prediction through the combination of a large number of decision trees combinations [9]. RF algorithms contain many decision tree classes and their outputs are based on the state of the output of the particular decision trees.

Based on the technique of recovering boot-strap samples, RF continuously creates training samples and prototypes. The training samples will then generate a random forest and the predicted values of the dependent variables can be generated by averaging the results of these trees. Random forests have a fast computational speed, which simplifies nonlinear functions of variables and clearly shows the interplay of variables [10]. Moreover, random forests are also unaffected by exceptions [11]. In recent years, RF algorithms have been widely used in many different industries. However, there are no documents mentioning tea leaves and predicting the image of tea leaves.

PCs (number of principle combinations) and N (number of decision trees) have a direct impact on the accuracy of the RF model. Therefore, it is necessary to optimize N and PC (within a certain range). 20 N (50–1000, with jump size equal to 50) and 9PCs (1–9, with jump size equal to 1) are chosen to optimize parameters based on RMSEC corresponding to each quality index model amount. The optimal results are illustrated in Fig. 4 like that:

- In the TFs prediction model, when PCs = 6 and N = 700, the minimum RMSEC model (0.033), RP, RMSEP, Bias, SEP, CV and RPD of the predictive set is set at 0.891, respectively. 0.058, −0,007, 0,011, 0,125 and 1,612. The result is shown in Fig. 4(A)
- The model predicts TRs, when PC = 7 and N = 100, RMSEC of the model reaches a minimum of 0.163. RP, RMSEP, Bias, SEP, CV and RPD of the prediction set are: 0890, 0.297, −0.027, 0.044, 0.081 and 1,267 respectively. The results are shown in Fig. 4(B)
- Anthocyanidin prediction model, when PC = 2 and N = 200, RMSEC of the model reaches a minimum of 0.168. RP, RMSEP, Bias, SEP, CV and RPD of the prediction set are: 0.944, 0.347, −0.068, 0.101, 0.136 and 2.636 respectively. The results are shown in Fig. 4(C)

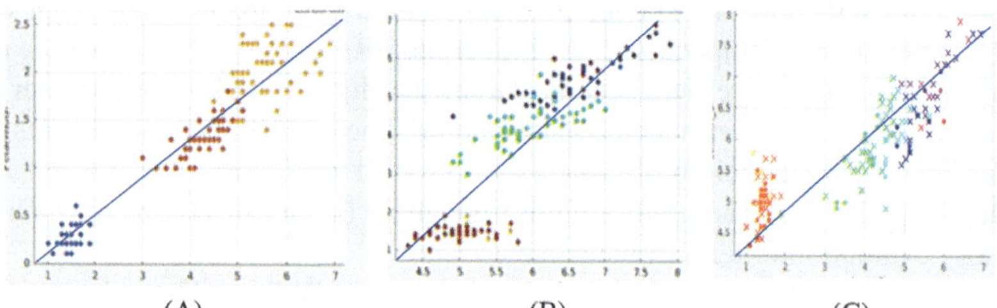

(A) (B) (C)

Fig. 4. The training results of the nonlinear RF prediction model for the color characteristics of black tea

4 Optimal Control of Parameters Based on the Color Analysis Index of the Fermentation Stages

Fermentation is the most important stage in the process of processing black tea. In this stage, it is necessary to create the most suitable temperature and humidity conditions for oxidizing enzymes to promote the necessary chemical changes. The quality, color, special taste of black tea should be determined. The most suitable temperature is about 25 °C–28 °C (≤ 30 °C). The humidity of the fermentation room is controlled at 95%–98%. In CTC black tea production technology, fermentation time is usually 40–120 min.

Structure of fermentation conveyor: The machine is mainly composed of an endless conveyor belt, made of stainless steel, with perforated holes, which is attached to two rotating increase axes. The conveyor is driven by an electric motor, can change the rotation speed, on the machine has a gear reducer to adjust the conveyor running very slowly, ensuring fermentation time. The temperature sensor is set up on the modules of the conveyor belt. According to a certain distance, we will determine the temperature of the fermented tea. Below the conveyor, the surface is a ventilation system consisting of hoppers facing upwards, the bottom attached to the pipe set, on the pipe with adjustable valves, the tube is connected to the blower and humidifying. The system is responsible for blowing wind and moisture to regulate air, cooling and ventilating tea from bottom to top. The practical black tea fermentation line is shown in Fig. 5.

Fig. 5. Overview of the line of fermenting black tea

After being loaded into the conveyor, the system will automatically determine relative humidity, temperature and fermentation time parameters (based on fermentation techniques standards). Go to the 60', 90' and 110' landmarks the system starts collecting and analyzing image data obtained from conveyors. PC indicators, N RP, RMSEP, Bias, SEP, CV and RPD are determined in turn for components: TF, TR, and Anthocyanidin. If these indicators are not up to standard, the control system will make a decision to continue to influence or stop the fermentation conditions. The algorithm diagram for the control process is shown in Fig. 6 below:

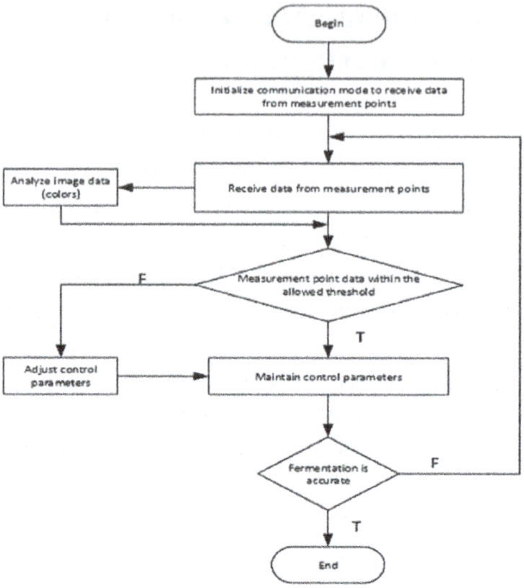

Fig. 6. Algorithm diagram controlling black tea fermentation system

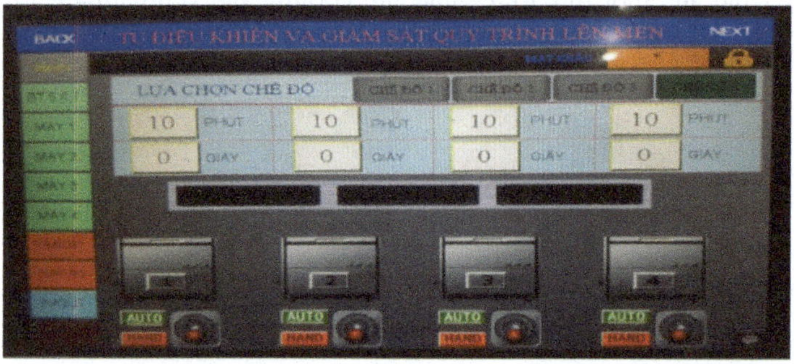

Fig. 7. The main control interface of the fermentation line

After testing the algorithm on the real model of the control program developed as shown in Fig. 7, the operating system has produced good results, meeting the standards of fermentation for black tea.

5 Conclusion

In this paper, we collected image data for each fermentation stage of black tea to select the characteristics suitable to the factors affecting the fermentation quality of tea. Through the conversion of color systems from RGB, HSV to CIE Lab combined with the RF nonlinear representation model, the standard value for important components that make up tea color black like: TF, TR and Anthocyanidin has been calculated and determined. These values will be used as input parameters for the supply chain controller: temperature, humidity and fermentation time suitable for black tea.

References

1. Thi Luu, N., Hung, L., Thành, N.T.M.: Textbook of processing black semi-finished black tea. Ministry of Agric High technology of tea production - coffee - coffee - cashew nuts Nha Trang University
2. Dong, C., Liang, G., Hu, B., Yuan, H., Jiang, Y., Zhu, H., Qi, J.: Prediction of Congou black tea fermentation quality indices from color features using non-linear regression methods. Sci. Rep. (2018)
3. Borah, S., Bhuyan, M.: Non-destructive testing of tea fermentation using image processing. Insight - Non-Destr. Testing Condition Monit. **45**, 55–58 (2003)
4. Borah, S., Bhuyan, M.: A computer based system for matching colours during the monitoring of tea fermentation. Int. J. Food Sci. Technol. **40**, 675–682 (2005)
5. Wu, X., Yang, J., Wang, S.: Tea category identification based on optimal wavelet entropy and weighted k-Nearest Neighbors algorithm. Multimed. Tools Appl. 1–15 (2016)
6. Xuan Quynh, N.: Research project to build the optimal automation system for processing and preserving agricultural products technology. Ministry of Science and Technology (2015)
7. Li, J., et al.: Variable selection in visible and near-infrared spectral analysis for noninvasive determination of soluble solids content of 'ya' pear. Food Anal. Methods **7**, 1891–1902 (2014)
8. Breiman, L.: Random Forests. Mach. Learn. **45**, 5–32 (2001)
9. Dong, L., Li, X., Xie, G.: Nonlinear methodologies for identifying seismic event and nuclear explosion using random forest, support vector machine, and Naive Bayes classification. Abstract Appl. Anal. **2014**, 1–8 (2014)
10. Pitra, Z., Bajer, L., Holeňa, M.: Comparing SVM, Gaussian process and random forest surrogate models for the CMA-ES. In: ITAT 2015: Information Technologies-Applications and Theory, pp. 186–193. CreateSpace Independent Publishing Platform, North Charleston (2015)

Experimental Study of Micro Vibration of the Ring Gear in Internal Gear Motor and Pump Under Unstable Working Condition

Trong Hoa Pham[1(✉)], Dinh Tu Nguyen[1], Thanh Cong Nguyen[1], Juergen Weber[2], and Lutz Müller[2]

[1] Faculty of Mechanical Engineering, University of Transport and Communications (UTC), No. 3 Cau Giay, Dong Da, Hanoi, Vietnam
hoagtvt100@gmail.com
[2] Institute of Fluid Power, Technical University of Dresden (TUD), Helmzholtstaße 7a, 01069 Dresden, Germany

Abstract. Analysis of micro movement of the ring gear is important for the understanding of dynamic behavior of the internal gear motor and pump. This paper presents the experimental results of micro movement of the ring gear under different working conditions by using the micro displacement sensors. An interesting phenomenon of ring gear is that the backward whirl phenomenon in case of instability has been occurred and observed. Experimental results pointed out that if the internal gear motor and pump operate under stable conditions, the ring gear center has small vibration and it whirls in forwards sense. However, when the internal gear motor and pump runs at unstable working conditions then the vibration amplitude of ring gear reaches much large value and it turns into spinning in backward sense. Due to backward whirl phenomenon, the solid contact between the outer ring gear and inner housing can occur. This is an unwanted phenomenon for internal gear motor and pump during operation.

Keywords: Micro vibration · Backward and backward whirl · Internal gear motor and pump · Micro movement

1 Introduction

Analysis of the micro movement of the rotating part in the field of rotor dynamic is very important in tribology field. For rotor dynamic field, it is well known that there are two categories of whirl motions, e.g., forward and backward whirl. Forward whirl means the rotor rotates around itself center with the direction the same as the direction of the speed meanwhile the backward whirl phenomenon means the rotor spin around itself center with the direction against to the direction of the rotating speed [1]. The forward whirl and backward whirl are quite well known in the field of rotating machinery, whereby the forward whirl mode has been observed experimentally, the backward whirl, however, only rarely. Forward whirl is a normal phenomenon. It does not cause the negative effect on performance of rotating machines. However, the backward whirl is an unwanted phenomenon. The appearance of backward whirl during operation can be lead to extreme vibration of the rotor/shaft. If the amplitude of

K.-U. Sattler et al. (Eds.): ICERA 2019, LNNS 104, pp. 320–327, 2020.
https://doi.org/10.1007/978-3-030-37497-6_37

vibration reaches up to sufficient value, it can cause the metal - to - metal contact between the relative rotating surfaces.

Lots of studies relating to forward and backward phenomenon have been conducted so far. Childs [2], in 2007, predicted the dry-friction and backward whirl between a rotor and a stator. The numerical and experimental study of the dry-friction whip and whirl was conducted by Wilkes [3] in 2010. Black [4] and Lahriri [5] theoretically and experimentally studied the forward and backward whirling orbit when rotor and stator come in contact with the dynamics of impacts and friction. The transient process from forward whirl to backward whirl of the beam and solid rotors with stiffening and softening effects were investigated by Rao in 2013 [6]. Norrick [7], in 2013, came up with a method for calculating the backward whirl amplitudes of a rotor with gyroscopic effects at each rotor speed. The contact phenomenon can then be predicted. An experimental investigation of shafts in respect to whirling was conducted for different shaft diameters by Takle [8] in 2014.

Internal gear motor and pump (IGMaP) is one of the most popular pump and motor in the field of fluid power application due to its simplicities and economy. According to Pham [9], the ring gear/housing can be considered as the journal/bearing. Some other studies relating to micro movement of internal gear motor and pump can be found in [10, 11]. This paper focusing on the analysis of micro movement of the ring gear through experiment by using the micro displacement sensor. The ring gear movement inside its housing is measured in case of instability to see whether the backward whirl has appeared or not.

2 Governing Equation

In this study, three gap sensors are used to measure the radial oil film thickness (ROFT) at three different points and then the circle equation is used to calculate the ring gear orbit and the minimum film thickness. In addition, two other sensors are used to measure the movement of the ring gear in the axial direction. Figure 1 shows the geometric relation between the housing and the ring gear (Figs. 2 and 3).

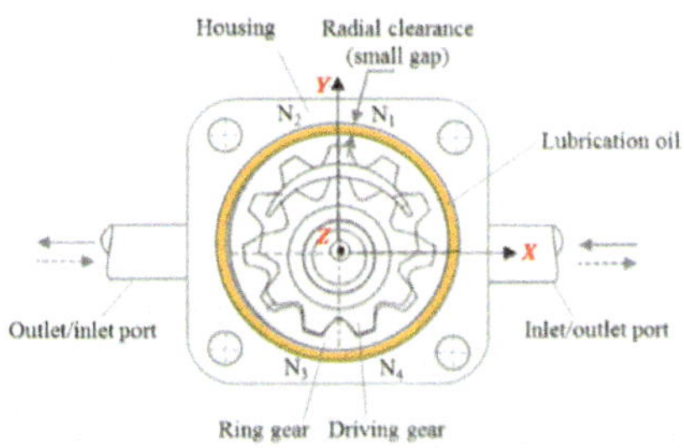

Fig. 1. Cross section of internal gear motor and pump

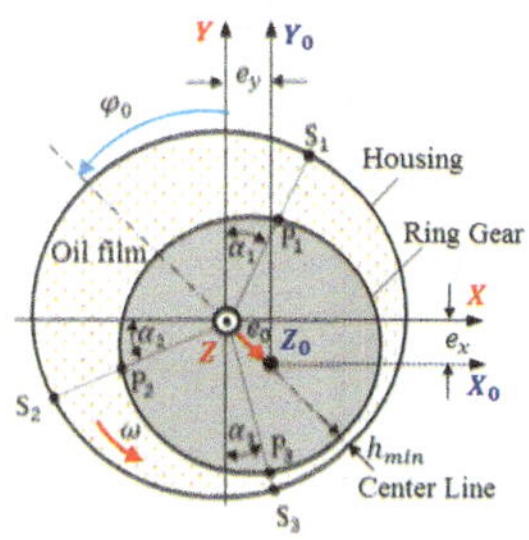

Fig. 2. Geometry of ring gear-housing

Three gap sensors are mounted on the housing in the same plane. This way, the ROFT is measured at three points and correspondingly the ring gear orbit is deduced. The assumption of inner housing is perfect circle. Based on the circle equation applied for three measured points, the ring gear center is determined as follows.

$$\begin{cases} x_m = \dfrac{(x_2^2 + y_2^2 - x_1^2 - y_1^2) - 2y_m(y_2 - y_1)}{2(x_2 - x_1)} \\ y_m = \dfrac{(x_3 - x_1)(x_2^2 + y_2^2 - x_1^2 - y_1^2) - (x_2 - x_1)(x_3^2 + y_3^2 - x_1^2 - y_1^2)}{2[(x_3 - x_1)(y_2 - y_1) - (x_2 - x_1)(y_3 - y_1)]} \end{cases} \quad (1)$$

Where, (x_m, y_m) is the position of the ring gear center. (x_1, y_2), (x_2, y_2), and (x_3, y_3) are the position of three measurement points P1, P2, P3, respectively, and r- is the radius of the ring gear. After calculation of the center position (x_m, y_m), the eccentricity will be calculated through the eccentricity in axes X and Y as follows:

$$e_0 = \sqrt{x_m^2 + y_m^2} \quad (2)$$

Minimum oil film thickness can be determined as

$$h_{min} = c - e_0 \quad (3)$$

Where, c-is the nominal radial clearance, i.e., the difference between the radius of the housing (r_h) and the radius of the ring gear (r), $c = r_h - r$.

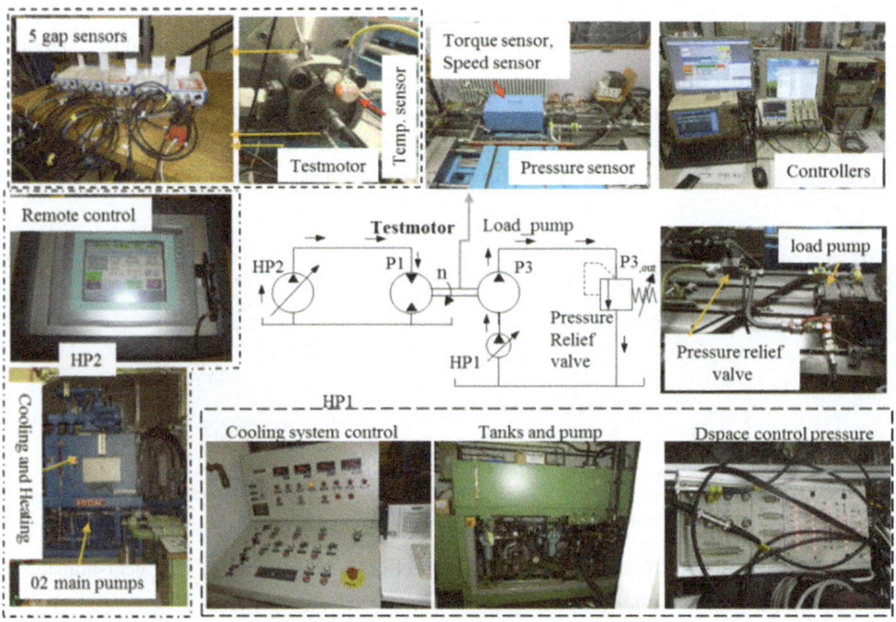

Fig. 3. Schematic of the test stand

3 Experimental Setup

A commercial internal gear motor and pump which can operate in 4-quadrant is selected for the experiment. Specifications of test IGMaP and working conditions are presented in Table 1. The test rig is designed and built in Laboratory of Institute of Fluid Power at TU Dresden Germany. The details of the test stand have been presented in Pham [10].

Table 1. Specifications of the test motor/pump

Parameter	Symbol	Value	Unit
Diameter of the ring gear	D	0.115	m
Length of the ring gear	L	0.034	m
Oil viscosity	μ (Oil HLP46)	0.042	Pa.s
Maximum speed	n	4000	rpm
Maximum pressure	Δp	250	bar
Radial clearance	c	60.10^{-6}	m
Oil working temperature	T	40° ± 1 °C	°C

4 Results and Discussion

All the setup parameters of the test ring are controlled through three PCs. All tests have been performed independently. The experimental procedure is as follows: After setting up all the parameters for test rig in personal computers, the rotating speed is speeded up to the specific values. Afterwards increasing of the external load acting on the test IGMaP is by the use of the pressure relief valve through the controller Parker PWD 00A400. The external applied load is increased until the instability of system occurred. The instability is kept for about 20 s to observe the vibration of the ring gear. After that, the external applied load is quickly reduced to ensure the safety of the whole system.

The input pressure coming into the motor and output pressure of load pump is presented in Figs. 4 and 5 respectively. The input pressure coming into test motor increases from 30 bar at 10 s up to 140 bar at 32 s before the instability happening. Meanwhile, the output pressure of the load pump increases up to 112 bar at 30.7 s. During instability the amplitude vibration of motor pressure is about 25 bar and the amplitude vibration of pump pressure is about 90 bar. This means that the amplitude of pump pressure is much larger than the amplitude of motor pressure.

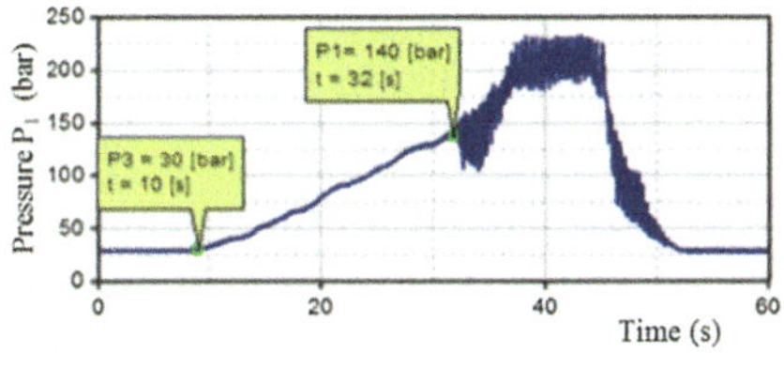

Fig. 4. Input pressure coming into motor

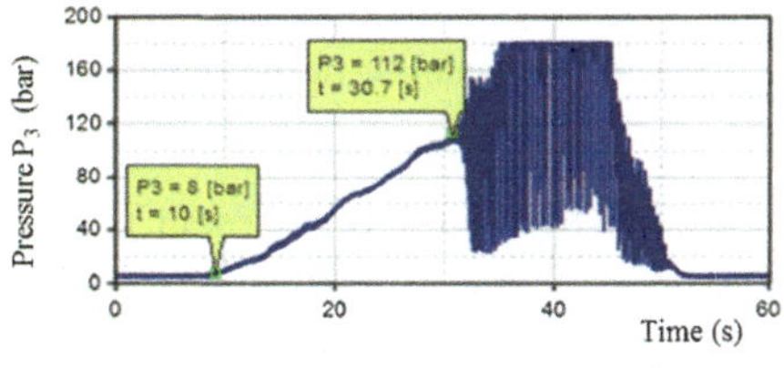

Fig. 5. Output pressure of load pump

In order to avoid the cavitation at the inlet section pipe of load pump, an oil supply system is used to ensure that the inlet pressure at section channel is kept always larger than 1 bar as seen in Fig. 6. Experimental results pointed out that the starting time of instability of supplied pressure is about 31 s. From Fig. 7 one can see that the peak to peak moment on the shaft during instability can be reached up to 130 Nm.

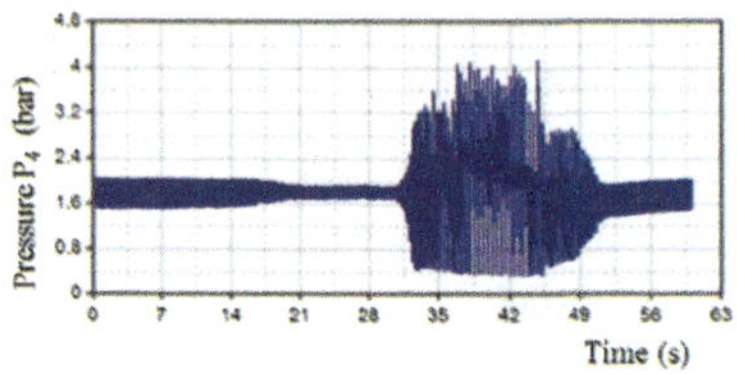

Fig. 6. Input pressure coming into load pump

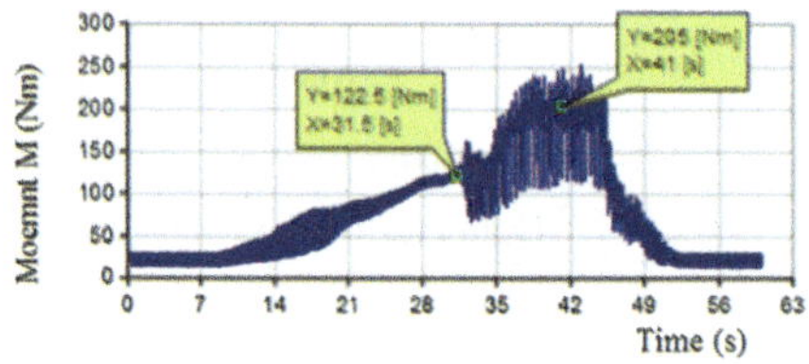

Fig. 7. Moment on the shaft

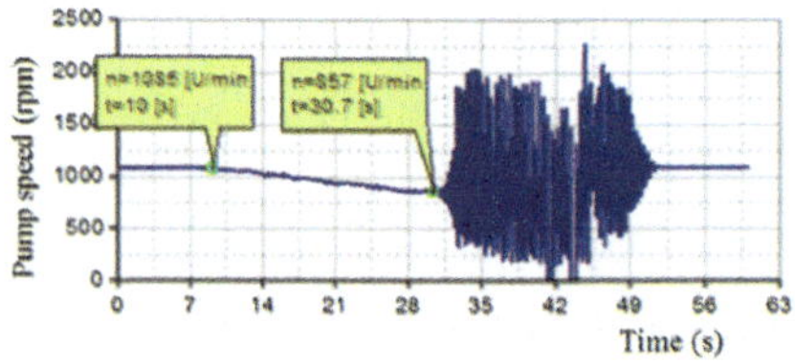

Fig. 8. Rotating speed of motor

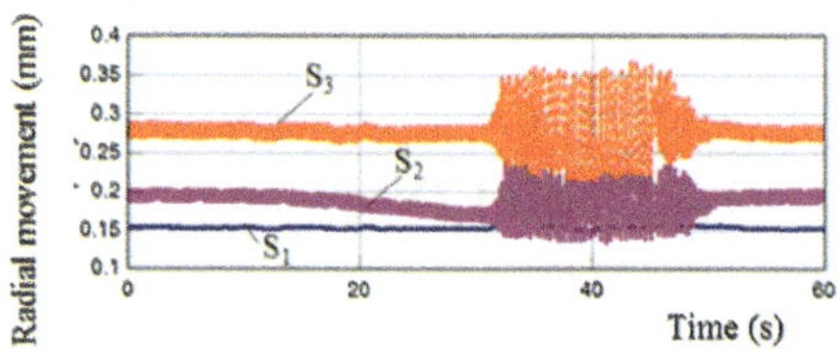

Fig. 9. Radial movement of the ring gear

The rotating speed is described in Fig. 8. One can see that the beginning time of instability of rotating speed is the same as that of pump pressure. Figure 9 depict the radial micro movement of the ring gear. It is obvious to see that the ring gear is also not stabilized under unstable working condition. Using the FFT to analyze the frequency of these parameters such as pressure, radial and axial micro movement of the ring gear, torque moment on the shaft and pressure (Figs. 10 and 11).

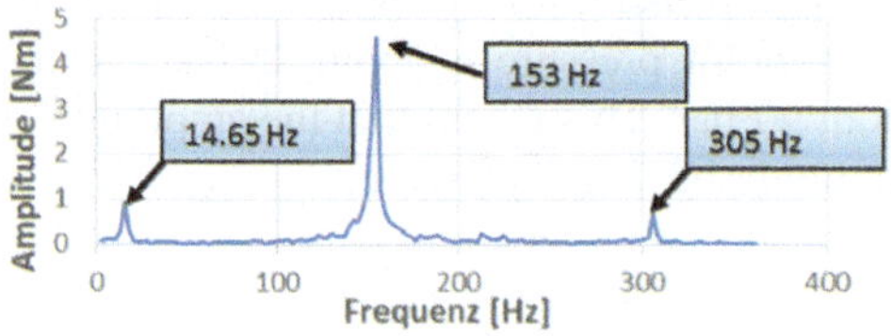

Fig. 10. Frequency of the moment

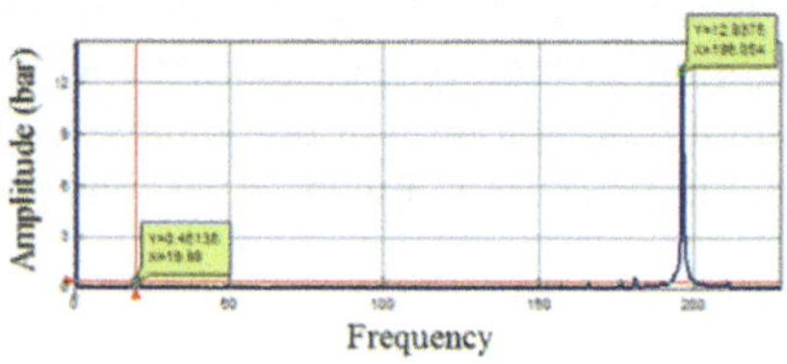

Fig. 11. Frequency of the pressure

The first frequency of moment is at a value of 14.56 Hz meanwhile the moment amplitude is about 4.6 Nm at the second frequency value of 153 Hz. For working pressure, the first frequency is equal to 19.98 Hz whilst the maximum pressure amplitude is about 12.5 bar at the second frequency value of 196 Hz. There are lots of frequency component of radial micro movement of the ring gear however both of them have the maximum amplitude at the first frequency, at a value of 14.98 Hz. Observation from these frequency characteristics of pressure, moment, radial micro movement one can see that there is no resonance frequency has been found through the FFT analysis (Fig. 12).

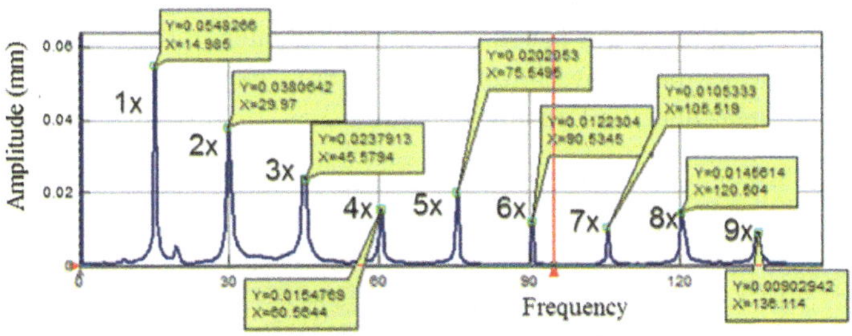

Fig. 12. Frequency characteristic of axial micro movement of the ring gear

For getting insight into the motion of the ring gear center, the Eq. (1) is used to calculate the ring gear orbit center in the software Flex Pro.9. The orbit of the ring gear center for stable and unstable working condition case of Fig. 13 is shown in Fig. 14. It is obvious that for the unstable working condition the size of ring gear orbit is much larger than that of the unstable working condition. Moreover, the ring gear center whirls in the direction the same with the direction of the rotating speed in case of stable working condition. This means that the ring gear center whirls in forward sense. However, for unstable working condition, the ring gear center spin in the clockwise direction, while the direction of the rotating speed is in the counter-clockwise direction, meaning that the ring gear rotates in backward whirl.

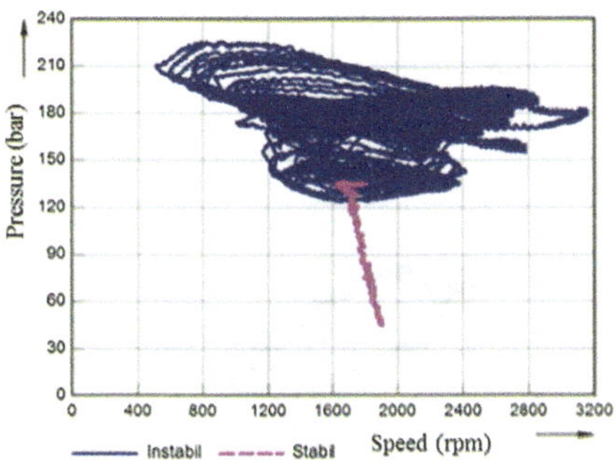

Fig. 13. Experimental case

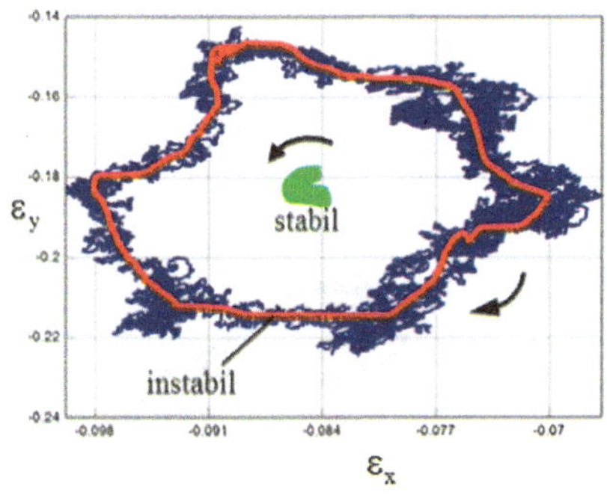

Fig. 14. Orbit of the ring gear center

The same measurement process is also performed for many other different values of the rotating speed and the operating pressure as presented in Fig. 15. The orbit of the ring gear center due to the instability phenomenon at the different levels of the operating pressure and the rotating speed is described in Fig. 16. When the motor operates stably, the ring gear orbit has a small size, however, when the instability of the motor occurs, the ring gear greatly vibrates and the orbit size becomes much larger than that of the stability. With a different pressure and rotating speed, the ring gear's orbit can reach various sizes. However, the elliptical orbit shape is almost the same. One can see from Fig. 16 that the more the speed rotates, the greater is the orbit size. Figure 16 points out that due to the great vibration of the ring gear the metal-to-metal contact between the ring gear and the housing occurs.

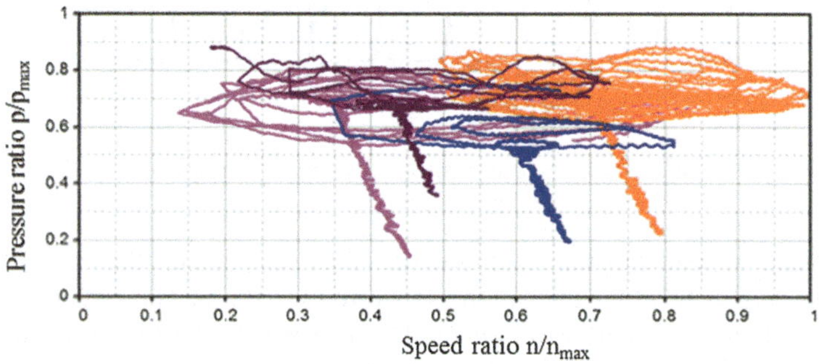

Fig. 15. Working pressure versus rotating speed

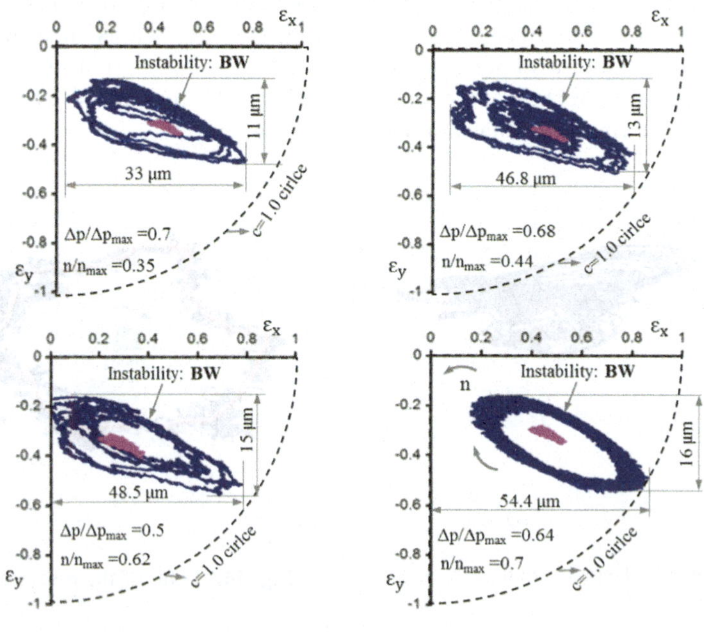

Fig. 16. The BW whirling motion of the ring gear center for different operating pressures and different rotating speeds

5 Conclusion

The micro movement of the ring gear has been analyzed in this work. Based on the experimental results presented in this paper, the following conclusions can be drawn:

1. For stable working conditions, the ring gear has small micro vibration and it spins in forward whirl. However, in case of unstable working conditions, the vibration amplitude reaches much large value and it turns into rotating in backward whirl.
2. Backward whirl is an unwanted phenomenon. The large vibration amplitude can lead to solid contact between the rotor and stator. It is, therefore, reduced the performance and the lifespan of the internal gear motor and pump.
3. Based on the obvious observation of micro movement of the ring gear through the experiment in this study one can confirm that the ring gear can whirl in backward sense under some conditions.

Acknowledgements. This research is funded by Vietnam National Foundation for Science and Technology Development (NAFOSTED) under grant number 107.03-2019.17.

References

1. Han, Y.S., Lee, C.W.: Directional Wigner distribution for order analysis in rotating reciprocating machines. Mech. Syst. Signal Process. **13**(5), 723–737 (1999)
2. Childs, D.W., Bhattacharya, A.: Prediction of dry-friction whirl and whip between a rotor and a stator. J. Vib. Acoust. **129**(3), 355–362 (2007)
3. Wilkes, J.C., Childs, D.W., Dyck, B.J., Phillips, S.G.: The numerical and experimental characteristics of multimode dry-friction whip and whirl. J. Eng. Gas Turbines Power **132**(5), 1–9 (2010)
4. Black, H.F.: Interaction of a whirling rotor with a vibrating stator across a clearance annulus. Arch. J. Mech. Eng. Sci. **10**(1), 1–12 (1968). 1959–1982
5. Lahriri, S.: on the rotor to stator contact dynamics with impacts and friction - theoretical and experimental. Dissertation (2012)
6. Rao, J.S.: Transient forward and backward whirl of beam and solid rotors with stiffening and softening effects. Adv. Vib. Eng. **12**(1), 59–74 (2013)
7. Norrick, N., Siegl, B., Alber, O.: Backward whirl during rotor-stator contact of gyroscopic rotors. In: 11th International Conference on Vibration Problems, pp. 1–10, September 2013
8. Takle, B.K.: Experimental investigation of shaft on whirling of shaft apparatus. Int. J. Sci. Eng. Technol. Res. **3**(8), 2091–2094 (2014)
9. Pham, T.H., Müller, L., Weber, J.: Dynamically loaded the ring gear in the internal gear motor/pump: mobility of solution. J. Mech. Sci. Technol. **32**(7), 3023–3035 (2018). https://doi.org/10.1007/s12206-018-0605-y
10. Pham, T.H.: Hybrid method to analysis the dynamic behavior of the ring gear for the internal gear motors and pumps. J. Mech. Sci. Technol. **33**(2), 602–612 (2019)
11. Pham, T.H., Müller, L., Weber, J.: Theoretical and experimental analysis of the effect of misaligned ring gear on performance of internal gear motors/pumps. J. Mech. Sci. Technol. **33**, 4049–4060 (2019)

Experimental Study on Effects of Process Parameters on Superplastic Deformation Ability of 7075 Aluminium Alloy Using Taguchi Method

Nguyen Manh Tien(✉), Nguyen Truong An, Tran Duc Hoan, Lai Dang Giang, and Le Trong Tan

Le Quy Don Technical University, Hanoi, Vietnam
manhtiennguyen84@gmail.com

Abstract. Superplasticity is the ability of an alloy to be permanently stretched to very high elongation within a limited range of conditions (grain size, deformation temperature and strain rate). The 7075 aluminium alloy superplastic sheet was used in this study. The AA 7075 sheets were prepared with microstructure using the thermomechanical treatment. The average grain size was 13 μm. The main purpose of this study is to investigate the influence of deformation parameters such as temperature and strain rate, on the deformation ability of 7075 aluminium alloy. The deformation ability of 7075 alloy was investigated under isothermal tensile tests. The results were analyzed to determine the optimum strain rate and temperature by using the Taguchi method and analysis of variance. The maximum superplastic elongation and the maximum stress were determined at different initial strain rates (5×10^{-4} s^{-1}, 10^{-3} s^{-1} and 1,5 $\times$ 10^{-3} s^{-1}) and temperatures (5000C, 5150C and 5300C), respectively. The optimum process parameters were obtained by analyzing the signal to noise ratio. The experimental results revealed that the effect of strain rate was more influential on the deformation ability than that of the temperature. The optimum results were achieved at the temperature of 530 °C and the strain rate of 10^{-3} s^{-1}, respectively.

Keywords: Superplastic · Deformation ability · Strain rate · Taguchi method · 7075 Al alloy

1 Introduction

Superplasticity is the ability of metals or alloys to deform plastically with a large tensile elongation without being destroyed under the impact of small stress. It depends on the strain rate as being deformed under a limited range of conditions. Superplasticity is a property of a material which can be prepared to obtain a fine stable grain size (between 5 ÷ 15 μm) by the thermomechanical treatment, when they are deformed at relatively slow strain rates normally in the range of 10^{-4} s^{-1} ÷ 10^{-2} s^{-1}, and at a temperature ≥ 0.5 T_m. Where T_m is the melting point in degrees Kelvin [1]. Barnes et al. provided a detailed report about the development of the superplastic alloys (now some 40 years

K.-U. Sattler et al. (Eds.): ICERA 2019, LNNS 104, pp. 328–334, 2020.
https://doi.org/10.1007/978-3-030-37497-6_38

old) was recently published [2]. Another factor encouraging the development of other superplastic alloys was the wish of aircraft companies to obtain a stronger alloy than AA2004 and, particularly, superplastic versions of established aircraft construction alloys such as AA7075/AA7475. Thereafter, superplasticity was demonstrated in the alloy AA8090 and it soon became clear that variants of most of the newly developed Al–Li based alloys could be rendered superplastic. Barnes [2] and Giuliano et al. [1] reported that the Aluminium and Aluminium alloy was used to manufacture complex parts in the many industries such as aerospace industries, auto industries, flying equipments, weapons, arts and etc.

Superplastic material owns the characteristic that the flow stress is highly sensitive to the strain rate. The superplastic forming process is carried out under isothermal conditions and the stress-strain rate relationship described by the following model [1]:

$$\sigma = k\varepsilon^n \dot{\varepsilon}^m \quad (1)$$

Where σ is the effective flow stress, k is a material constant, ε is the effective strain, $\dot{\varepsilon}$ is the effective strain rate, m is the strain rate sensitivity exponent, and n is the strain hardening exponent. A high value of strain rate sensitivity m is responsible for the superplastic properties of the sheet material, and thus the delay in the localized thinning.

In this work, we focused on evaluation of the ability of superplastic deformation for aluminum alloy AA7075 under different temperatures (5000C, 5150C and 5300C) and initial strain rates (5×10^{-4} s^{-1}, 10^{-3} s^{-1} and 1.5×10^{-3}). Experimental results showed the effects of process parameters on the elongation to fracture by Taguchi method. The optimum process parameters were obtained by analyzing the signal to noise ratio. The forming process regime for aluminum alloy AA7075 was dctermined based on the experiments result.

2 Experimental Procedure

2.1 Materials

The chemical composition of the analyzed AA7075 alloy is expressed in weight percent as shown in Table 1.

Table 1. Chemical composition of the analyzed AA7075 alloy (%wt)

Zn	Mg	Cu	Fe	Cr	Ti	Mn	Si	Al
5.35	2.34	1.32	0.30	0.22	0.04	0.024	0.05	Balance

The samples AA7075 with thickness of 5 mm was used for the experiments. Based on the analysis of Al-Zn-Mg-Cu aluminum alloy characteristics, the characteristics of heat treatment and deformation, it can be seen that AA7075 alloy is capable of producing a very fine-grained microstructure (~13 μm). In order to prepare a fine grained microstructure microstructure, the modified Sahraoui et al. [3] method was used to reduce the overall processing time. The thermomechanical treatment consists of different stages such as the solution treatment, the overaging, the warm rolling and the recrystallization.

2.2 Experimental Conditions and Methods

The tensile tests were carried out at temperatures of 500 °C, 515 °C and 530 °C, with different initial strain rates 5×10^{-4}, 10^{-3} and 1.5×10^{-3} s^{-1}. Nine samples numbered from 1 to 9 were used in our experiment.

Fig. 1. The tensile test equipment.

The equipment used to perform tensile tests is the Devotrans DVT FU/RDNN 50kN – CKS testing machine (Speed range from 0.001 mm/min to 500 mm/min). The machine is equipped with an electrical resistance furnace Nabertherm B180. The tensile test equipment is shown in Fig. 1.

The traditional experimental design methods require a large number of experiments, when the number of process parameters increases and hence are too complicated and difficult to use. In order to minimize the number of tests required, the Taguchi experimental design method, a powerful tool for designing high quality systems, was developed by Taguchi.

This method uses a special design of orthogonal arrays to study the entire parameter space with a small number of experiments. Results of these experiments are transformed into a signal to noise ratio to measure the deviation of quality characteristics from the desired values, a greater S/N ratio corresponds to better quality characteristics [4]. Main objectives achieved through Taguchi method are: (1) determination of optimal parameters for a process or a product; (2) estimation of the contribution of each design parameter to the quality characteristics and hence the determination of the most influential factor. In the present study, two deformation parameters were used as control factors and each parameter was designed to have three levels, denoted as 1, 2, 3 (shown in Table 2). The experimental design was according to an L_9 array based on Taguchi method (Table 3). Using Analysis of Variance (ANOVA), the effect of input parameters on deformation ability was studied.

Table 2. Input parameters and levels

Parameter	Level 1	Level 2	Level 3
Temperature, [°C]	500	515	530
Strain rate, [s^{-1}]	0.0005	0.001	0.0015

Table 3. Experimental design Taguchi L_9

Experiment no.	Temperature	Strain rate
1	1	1
2	1	2
3	1	3
4	2	1
5	2	2
6	2	3
7	3	1
8	3	2
9	3	3

3 Results and Discussions

Figure 2 shows the samples after tensile test process. Table 4 shows the elongation to fracture (δ %) obtained for different tensile testing temperatures and initial strain rates.

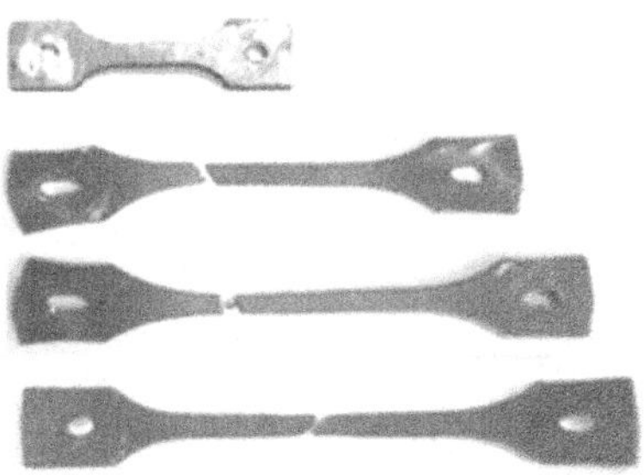

Fig. 2. The samples after tensile tests

In can be seen from Table 4 that the elongation to fracture increased with the increase in temperatures. This result could be due to softening of the material at elevated temperatures. This is in consistent with the predictions in [5] that the elongation to fracture decreases with strain rate and increases with temperature. In the Taguchi method, S/N ratio is the measurement of quality characteristics and deviation from the desired value. The signal to noise ratio was calculated using the condition: larger is better (Eq. 2).

Table 4. The elongation to fracture for tensile tests

Samples	Temperature [°C]	Strain rate [s^{-1}]	δ [%]
1	500	5×10^{-4}	150
2	500	10^{-3}	240
3	500	1.5×10^{-3}	150
4	515	5×10^{-4}	150
5	515	10^{-3}	260
6	515	1.5×10^{-3}	165
7	530	5×10^{-4}	170
8	530	10^{-3}	280
9	530	1.5×10^{-3}	180

$$\left(\frac{S}{N}\right)_k = -10\log\left(\frac{1}{n}\sum_{i=1}^{n}\frac{1}{y_k^2}\right) \tag{2}$$

Where n - number of repetition experiments (n = 1); y_k - measurement value of the elongation to fracture of the kth experiment.

By applying the equation given above, the S/N values for each experiment of L_9 (Table 4) were calculated. The S/N values are shown in Table 5.

Table 5. The S/N values for each experiment of L_9

Experiment no.	Temperature [°C]	Strain rate [s^{-1}]	S/N Ratio
1	500	5×10^{-4}	43.52
2	500	10^{-3}	47.60
3	500	1.5×10^{-3}	43.52
4	515	5×10^{-4}	43.52
5	515	10^{-3}	48.30
6	515	1.5×10^{-3}	44.35
7	530	5×10^{-4}	44.61
8	530	10^{-3}	48.94
9	530	1.5×10^{-3}	45.11

The effects of each parameter on the superplastic deformation ability are depicted in Fig. 3. The interaction between the process parameters is shown in Fig. 4. By using variance analysis method ANOVA to evaluate, the research parameters are obtained as shown in Table 6.

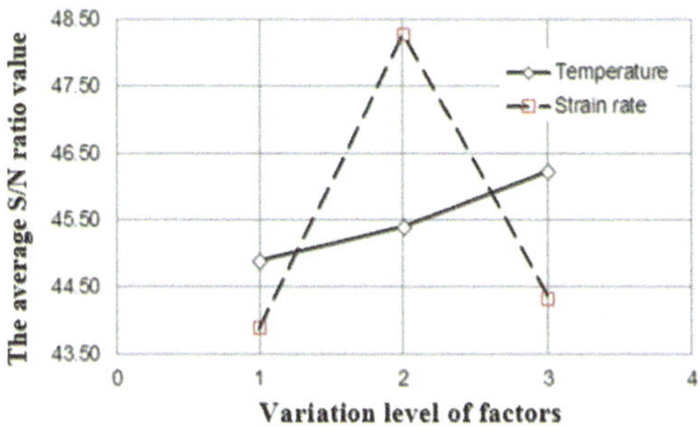

Fig. 3. The change in the average S/N noise ratio value by levels

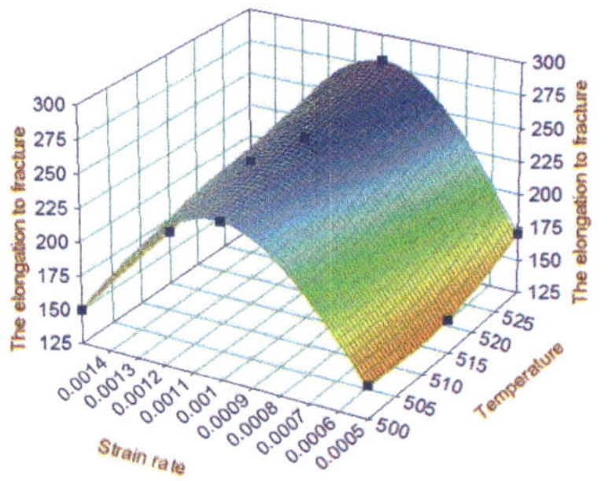

Fig. 4. Effect of process parameters on the elongation to fracture for tensile tests

From the main effect plot for S/N ratio (Fig. 3), it can be observed that the elongation to fracture increases with the increase of temperature. The largest elongations were obtained at the temperature of 530 °C. The elongations obtained lower at 500 °C and 515 °C. These lower temperatures could be insufficient for the activation of the GBS mechanism [6]. The result shown in Fig. 3 implied that the elongation to fracture increases as the strain rate increases. However, as the deformation rate is greater $10^{-3}\ s^{-1}$, the elongation is reduced. This is because the strain rate is so large that there is not enough time for the corresponding diffusion process to create favorable surfaces for sliding, so the stress increases, and the elongation to fracture decreases [7].

Table 6. Analysis of variance for S/N ratio

Level	The average S/N ratio value	
	Temperature (T)	Strain rate (ε)
1	44.88	43.88
2	45.39	48.28
3	46.22	44.33
Mean (m)	45.50	45.50
Max	46.22	48.28
Delta	1.34	4.40
Effect (%)	23.31	76.69

From the ANOVA analysis table (Table 5) and the interaction between the process parameters (Fig. 4), it can be observed that the optimum combination for the highest S/N ratio is temperature at level 3 (530 °C) and the strain rate at level 2 ($10^{-3}\ s^{-1}$). The S/N ratio of the optimal combination and the largest elongation to fracture are determined by the formula 3 and 4 [8]:

$$\left(\frac{S}{N}\right)_{opt} = m + (Max_T - m) + (Max_\varepsilon - m) = 49 \quad (3)$$

$$\delta_{\max} = 10^{\frac{(S/N)_{opt}}{20}} = 281.97(\%) \quad (4)$$

4 Conclusion

In this paper, the effects of process parameters on the elongation to fracture for tensile tests and the optimum result were determined by means of Tauguchi method and experimental data. The main conclusions from the research results of the current work can be drawn as follows:

The Taguchi's orthogonal array and analysis of variance ANOVA was used to evaluate the research parameters obtained. The results revealed that strain rate is the main parameter, which has greater influence on the superplastic deformation ability of 7075 alloy than that of the strain temperature.

Based on the S/N ratio, the elongation to fracture increases with with the increase in temperature and strain rate. However, as the deformation rate is greater 10^{-3} s^{-1}, the elongation is reduced. The optimum result for the elongation to fracture was achieved at the temperature of 530 °C and the initial strain rate of 10^{-3} s^{-1}. The objective optimization considering the elongation to fracture can be considered in future work.

References

1. Giuliano, G.: Superplastic Forming of Advanced Metallic Material. Woodhead Publishing Limited (2011)
2. Barnes, A.J.: Superplastic forming 40 years and still growing. J. Mater. Eng. Perform. **16**(4), 440–454 (2007)
3. Sahraoui, T., Hadji, M., Bacha, N., Badji, R.: Superplastic deformation behavior of 7075 aluminum alloy. J. Mater. Eng. Perform. **12**, 398–401 (2003)
4. Bagci, E., Ozcelik, B.: Analysis of temperature changes on the twist drill under different drilling conditions based on Taguchi method during dry drilling of Al 7075-T651. J. Adv. Manuf. Technol. **29**, 629–636 (2005)
5. Osakada, K.: Effects of strain rate and temperature in forming processes of metals. J. Phys. IV Colloque **07**(C3), pp. C3-XXXVII–C3-XLIV (1997)
6. Nieh, T.G., Wadsworh, J., Sherby, O.D.: Superplasticity in Metals and Ceramics. Cambridge University Press, Cambridge (2005)
7. Mahidhara, R.K.: Effect of grain size on the superplastic behavior of a 7475 alloy. J. Mater. Eng. Perform. **4**(6), 674–678 (1995)
8. Pillai, J.U., Sanghrajka, I., Shunmugavel, M., Muthuramalingam, T., Goldberg, M., Littlefair, G.: Optimisation of multiple response characteristics on end milling of Aluminium alloy using Taguchi - Grey relational approach. Measurement **124**, 291–298 (2018)

Exploiting CP in BICM-ID OFDM System

Tran Anh Thang(✉)

Thai Nguyen University of Technology, Thai Nguyen, Vietnam
trananhthang@tnut.edu.vn

Abstract. In transmission systems using OFDM (*Orthogonal Frequency Division Multiplexing*), CP (*Cyclic Prefix*) is often removed at the receiver. This is really wasteful because CP carries information, energy and time diversity. Many studies have been conducted to exploit CP to bring the necessary information for the receiver. A new technique is introduced in this paper, which uses the information and diversity of CP to improve the efficiency for the BICM-ID OFDM system. The simulation results of the system show that the best value gain of CP exploitation on the Nakagami-m channel is about 1 dB.

Keywords: BICM-ID · OFDM · Exploit CP · Nakagami-m channel

1 Introduction

OFDM transmission technology, with the advantage of saving bandwidth, possible to transmit at high speed, ability to resist the frequency selective fading and as well as narrow-band noise, is constantly being studied, expanding the scope of its applications.

The BICM-ID (Bit-Interleaved Coded Modulation with Iterative Decoding) diagram, first is proposed in the paper [1], with CM (Coded Modulation) structure increases efficiency on fading channel thanks to BI (Bit Interleved) and also on Gaussian channel thanks to ID (Iterative Decoding). The quality of the BICM-ID is comparable to the TTCM (Turbo Trellis Coded Modulation) [2]. Furthermore, the BICM-ID diagram needs only a SISO (Soft Input - Soft Output) decoder while the TTCM diagram requires two SISO decoders. Therefore, combining BICM-ID diagrams with OFDM technique will be highly effective because it advances coding structure, Interleaving, Modulation and OFDM transmission as shown in [3, 4].

In OFDM technique, CP (Cyclic Prefix) is a time guard interval. It is aimed at preventing ICI (Inter-Carrier Interference) and ISI (Inter-Symbol Interference) [5]. This interval is a copy of the end of the OFDM symbol, after that, it is inserted into the head of the symbol to form a complete OFDM symbol before being transmitted. To avoid ISI, the CP is usually longer than the channel delay and discarded at the receiver. This elimination is wasteful since CP carries information, energy and time diversity. There have been many studies to reuse or exploit information in CP for receivers such as, to estimate data to improve BER for OFDM system [6], recover energy of redundancy CP when transmitting signal via fading channel [7], exploit CP to improve the efficiency of LDPC (Low Density Parity-check Codes) decoding [8] and exploit CP for equalization in OFDM systems [9–11].

K.-U. Sattler et al. (Eds.): ICERA 2019, LNNS 104, pp. 335–341, 2020.
https://doi.org/10.1007/978-3-030-37497-6_39

This paper proposes a new technique to reuse CP for BICM-ID OFDM system. The system consists of 2 receiver branches. Branch 1 is a BICM-ID OFDM as usual, CP is reuse in the other branch. Based on iterative decoding, the soft information after decoding of the two branches will be aggregated after each loop and decided after several loops to get the information out. The simulation results on the Nakagami-*m* channel show that the quality of the proposed system is better than that of the conventional BICM-ID OFDM system.

The structure of the paper consists of 5 parts: introduction is the first; the second part introduces the CP reuse BICM-ID OFDM system diagram; the section for obtaining the CP sections is introduced in the third part 3; tissue parameters and results of the simulation are shown in the fourth part; the conclusions and the next development direction are in the last part.

2 Diagram of CP Reuse BICM-ID OFDM System

The diagram of CP reuse BICM-ID OFDM system is proposed in Fig. 1. The operation of this diagram is basically the same as the traditional BICM-ID scheme, analyzed as follows: input information **u** is encoded into **c** then is interleaved into **v**. The **v** information after passing the modulation block becomes the **s** signal set and then is taken to the OFDM transmitter. OFDM transmitter will perform IFFT transforms, synthesize into OFDM symbols and add the CP component to form the transmitted signal set on the channel.

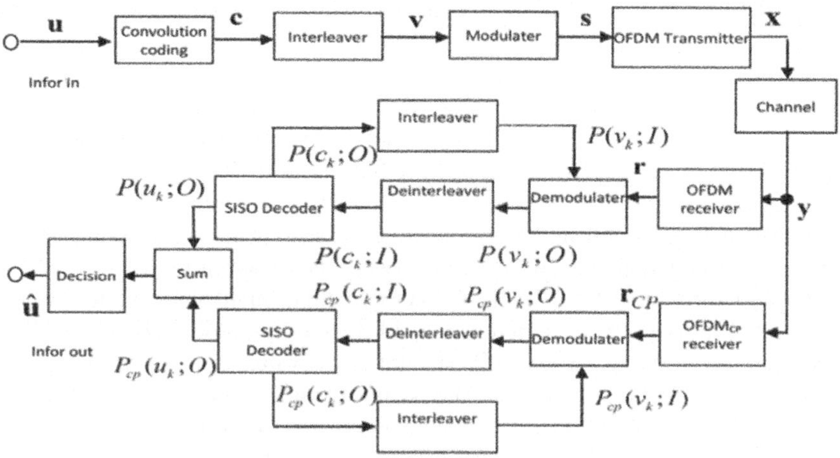

Fig. 1. Diagram of CP reuse BICM-ID OFDM system.

The received signal is:

$$\mathbf{y} = \mathbf{h} * \mathbf{x} + \mathbf{n}. \tag{1}$$

Where $\mathbf{h}$ is the Nakagami-distributed fading and $\mathbf{n}$ is AWGN with variance $\sigma^2 = \frac{N_0}{2}$ for AWGN channel $\mathbf{h} = 1$.

At the output of the channel, the received signal (OFDM symbols) is divided into two branches: Branch 1 (Normal CP) leads to OFDM receiver block to remove CP as the normal OFDM system; Branch 2 (CP reuse) leads to the OFDM_{CP} receiver block, after collecting the complete OFDM symbol, the CP segment is separated and replaced at the end of the OFDM symbol (in the body of the OFDM symbol), which is copied on the transmitter side.

The operation of branch 1 and branch 2 is parallel and independent of each other. On branch 1, OFDM symbol is implemented FFT in OFDM receiver to obtain the $\mathbf{r}$ received signals, this signal is sent to the demodulation block (often is soft demodulation block) to calculate bit metrics $P(v_k; O)$. After deinter leaver, the bit metrics become a priori probabilities $P(c_k; I)$ for the SISO decoder to calculate a posteriori bit probabilities $P(u_k; O)$ and extrinsic bit probabilities $P(c_k; O)$. The extrinsic bit probabilities, after feedback through interleaver, become a prior probability for demodulation $P(v_k; I)$ to recalculate the bit metrics for the next loop. This Iterative Decoding is introduced in [12] in detail.

Posteriori probabilities for decode bits are calculated:

$$p(v_k = b|r) \sim \sum_{s_i \in S_b^k} p(s_i|r) \sim \sum_{s_i \in S_b^k} p(r|s_i).p(s_i). \tag{2}$$

Where S_b^k is a subset of $\mathbf{s}$, including all symbols that k^{th} bit equal b, k is bit index $(k = 1, 2, 3, \ldots, M - \text{ary})$ and b is bit value $(b = 0, 1)$. $p(r|s_k)$ is probability density function of the received signal r and transmit signal s_k.

Similarly, after the OFDM_{CP} receiver, the received signal at branch 2 $\mathbf{r}_{\mathbf{CP}}$ is also iterated decoding the same as branch 1. And posteriori probability for decode bits is:

$$p(v_k = b|r_{cp}) \sim \sum_{s_i \in S_b^k} p(s_i|r_{cp}) \sim \sum_{s_i \in S_b^k} p(r_{cp}|s_i).p(s_i). \tag{3}$$

The posteriori bit probability of the two branches is taken to the sum block to add together after each loop. After several loops, a sum of the posteriori bit probabilities of the sum block is taken to the decision block to get the information out.

3 Replace CP

The received signal $\mathbf{y}$, including the CP part and the body as shown in Fig. 2a, is taken into two branches: normal CP and CP reuse. The OFDM_{CP} block (branch 2) will separate the CP segment and replace it with the corresponding part in the body (this is the copied part before transmission), then transform FFT. The length of the replacement CP segment can be taken all the CP (4/4CP), 3/4CP, 2/4CP or 1/4CP as illustrated in Fig. 2c, d, e, f, or can be taken to 1/8CP, 1/16CP. However, splitting too small CP will be difficult to implement in practice. The length of the CP segment reuse will be taken toward the end of the CP, discarding the beginning of the CP because it is often overlapped with the adjacent forward OFDM symbol (ISI).

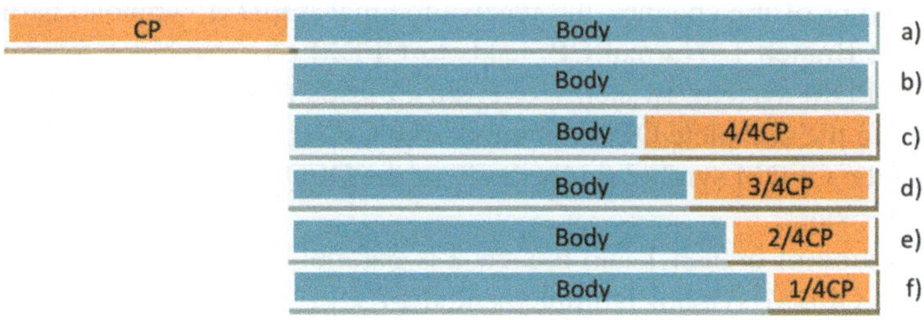

Fig. 2. Illustrate the CP sections reuse

4 Simulation Results and Discussion

4.1 Simulation Parameters for the System

The CP reuse system is simulated with MATLAB software. Simulation parameters are used convolution code [133, 171, 165] with code rate 1/3, constraint length 7, LTE block interleaving (sign: LTE-Inter) [13]. The Optimum mapping in [14] (also known as M16a) is used for modulation with 16QAM. OFDM parameter is taken as [15], where the OFDM system parameter to simulate is used an Extended CP, bandwidth 3 MHz. AWGN and Nakagami-*m* channel are used with perfect channel state information is assumed at the receiver. Overlap from symbol 1 to symbol 2 is assumed 50% of the CP. The number of iterations in the receiver system is 6.

4.2 Simulation Results

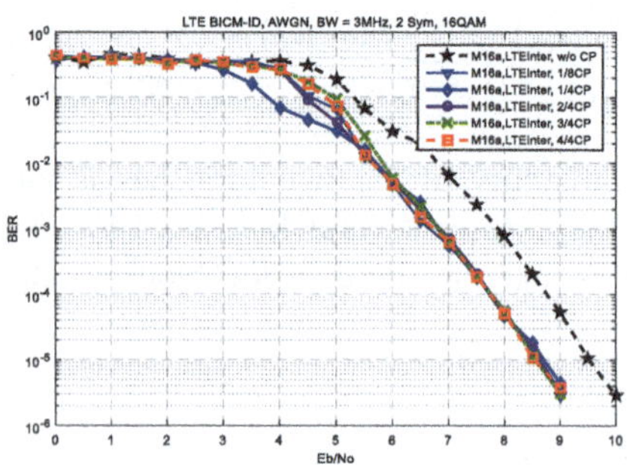

Fig. 3. Comparing the CP reuse BICM-ID OFDM system with normal BICM-ID OFDM system on AWGN channel

The results of the CP reuse system on AWGN channels is shown in Fig. 3. According to this result, the CP reuse system is compared to a normal BICM-ID OFDM system. The result of CP reuse system is better and has approximately 1 dB gain with all CP repeater lengths and the lengths of CP reuse are similar because the channel is AWGN (They do not overlap).

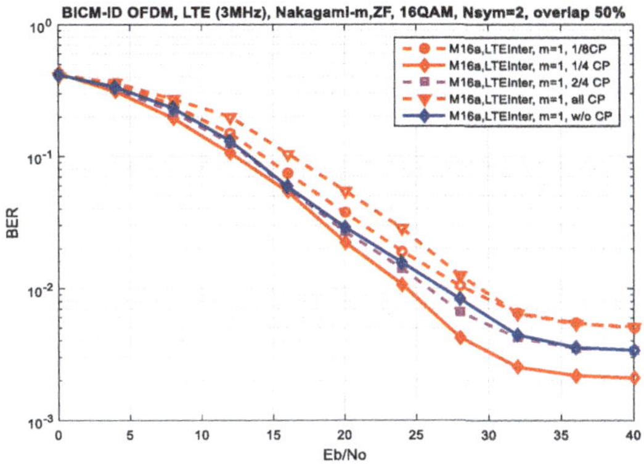

Fig. 4. Some CP lengths of the CP reuse BICM-ID OFDM system on Nakagami-m channel ($m = 1$)

The results on Nakagami–m channel is shown in Fig. 4 with $m = 1$. The results show that under the great impact of the channel (Rayleigh fading channel), both systems give poor quality. 1/4CP reuse length is the best quality, while the quality of other lengths is equal to or less than the normal systems.

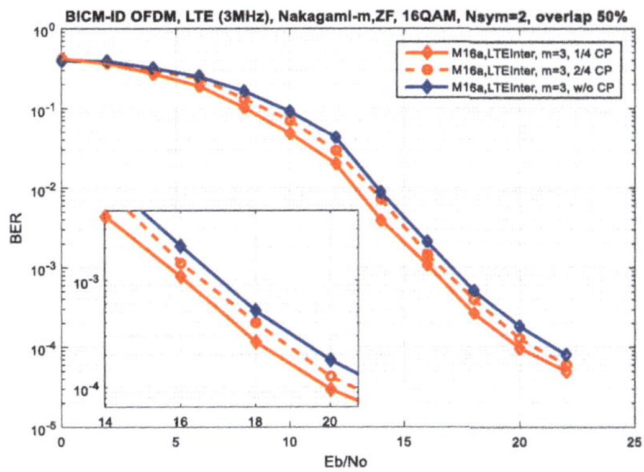

Fig. 5. Some CP lengths of the CP reuse BICM-ID OFDM system on Nakagami-m channel ($m = 3$)

With $m = 3$, the simulation results in Fig. 5 show the quality of the CP reuse system with some lengths of CP reuse and compare the normal system. 1/4CP reuse length still is the best quality and the gain achieved approximately 1 dB at BER = 10^{-3} comparing with the normal system.

When the channel is better, the effects of CP overlap are clearly shown in Fig. 6. According to the initial condition when setting up the simulation, the first symbol tail (delay) overlaps 50% on the CP of the second symbol. The length of the 3/4CP and all CP replacements have bad information causing of overlap with the previous symbol. On Nakagami channel with $m = 10$, simulation results show that two CP sections are not effective. Other CP reuse lengths (1/8CP, 1/4CP, 2/4CP), not overlapped, give better results. The best results are 1/4CP and 1/8CP. The 1/8CP reuse length has a good result because the channel is better, less errors appeared and so this length is enough to support decoding for the system. The benefit of CP reuse system using 1/4CP is about 1 dB at BER = 10^{-4} and BER = 10^{-5}.

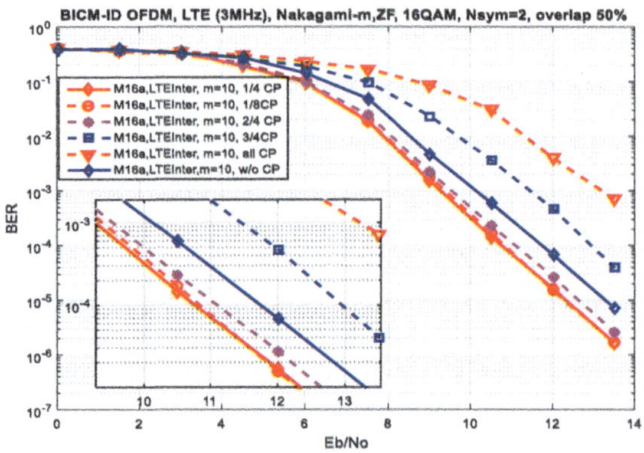

Fig. 6. Some CP lengths of the CP reuse BICM-ID OFDM system on Nakagami-m channel ($m = 10$)

5 Conclusion

In this paper, instead of removing CP at the receiver, we proposed the BICM-ID OFDM system to reuse CP. In the receivers, based on iterative decoding and a CP alternative technique, the system can exploit useful information within the CP. The simulation results of the system have shown that CP reuse sections are effective for the system when it is not overlapped with other information. Based on the simulation parameters, the paper has investigated various sections of CP reuse. The results show that 1/4CP reuse is best and achieved gain is 1 dB. However, if the length exploiting is not fit, it may cause system degradation.

References

1. Li, X., Ritcey, J.A.: Bit-interleaved coded modulation with iterative decoding. IEEE Commun. Lett. **1**(6), 169–171 (1997)
2. Li, X., Chindapol, A., Ritcey, J.A.: Bit-interleaved coded modulation with iterative decoding and 8 PSK signaling. IEEE Trans. Commun. **50**(8), 1250–1257 (2002)
3. Kiyani, N.F., Weber, J.H.: OFDM with BICM-ID and rotated MPSK constellations and signal space diversity. In: 2007 14th IEEE Symposium on Communications and Vehicular Technology in the Benelux. IEEE (2007)
4. Tran, N.H., Nguyen, H.H., Le-Ngoc, T.: Bit-interleaved coded OFDM with signal space diversity: Subcarrier grouping and rotation matrix design. IEEE Trans. Signal Process. **55**(3), 1137–1149 (2007)
5. Van Nee, R., Prasad, R.: OFDM for Wireless Multimedia Communications. Artech House Inc, London (2000)
6. Hueske, K., Götze, J.: Improving OFDM data estimation by overlapping based cyclic prefix reuse. In: 14th International OFDM Workshop (InOWo) (2009)
7. Bottomley, G.E., Wilhelmsson, L.R.: Recovering signal energy from the cyclic prefix in OFDM. IEEE Trans. Veh. Technol. **57**(5), 3205–3211 (2008)
8. Palenik, T., Farkas, P.: Exploiting cyclic prefix redundancy in OFDM to improve performance of Tanner: graph based decoding. Analog Integr. Circ. Sig. Process **69**(2–3), 143–152 (2011)
9. Yang, J., et al.: Exploiting cyclic prefix in Turbo FDE systems using factor graph. In: 2013 IEEE Wireless Communications and Networking Conference (WCNC). IEEE (2013)
10. Yang, J., et al.: A factor graph approach to exploiting cyclic prefix for equalization in OFDM systems. IEEE Trans. Commun. **61**(12), 4972–4983 (2013)
11. Xu, L., et al.: Exploiting cyclic prefix for Turbo-OFDM receiver design. IEEE Access **5**, 15762–15775 (2017)
12. Li, X., Ritcey, J.: Bit-interleaved coded modulation with iterative decoding using soft feedback. Electron. Lett. **34**(10), 942–943 (1998)
13. 3GPP: LTE; Evolved Universal Terrestrial Radio Access (E-UTRA); Multiplexing and channel coding (3GPP TS 36.212 V9.4.0 (2011-09)). 3GPP TS 36.212 (2011). www.3gpp.org/dynareport/36212.htm
14. Schreckenbach, F., et al.: Optimization of symbol mappings for bit-interleaved coded modulation with iterative decoding. IEEE Commun. Lett. **7**(12), 593–595 (2003)
15. 3GPP: LTE; Evolved Universal Terrestrial Radio Access (E-UTRA); Physical channels and modulation (3GPP TS 36.211 version 13.3.0 Release 13) 3GPP TS 36.211 (2016). www.3gpp.org/dynareport/36211.htm

Fabrication of Polymeric Micro Structures Using Improved Hot Embossing Technique

Pham Hong Phuc[1(✉)] and Dao Viet Dzung[2]

[1] School of Mechanical Engineering, Hanoi University of Science and Technology (HUST), 1-Daicoviet, Hanoi, Vietnam
phuc.phamhong@hust.edu.vn
[2] Griffith University, Gold Coast, QLD, Australia

Abstract. The paper proposes an improved hot embossing process of polymeric PMMA material using low-friction silicon molds, forcing and water-cooling subsystems integrated additionally, aiming to reduce total time of embossing cycle and extend lifetime of a master mold. The result of embossing experiments shows that the profile of PMMA micro structures (with a minimum feature size of 2 μm) using improved embossing process is sharper, higher aspect ratio and more uniformed than the counterparts, which use the traditional process without vacuum.

Keywords: Polymeric PMMA · Hot embossing · Silicon mold · DRIE process

1 Introduction

Recently, polymeric materials become an interesting study in Micro Electromechanical Systems (MEMS) due to typical advantages such as lower cost and easier process than silicon material. Another advantage of a polymeric actuator against a similar silicon one is that it needs much lower driving voltage because polymer has a lower Young's modulus, i.e. the polymeric actuator will need lower energy consumption than silicon counterpart. Therefore, polymer is recently chosen as the potential material of components in MEMS devices.

Micro fabrication of polymeric materials relies on some popular technologies, such as hot embossing, mold casting, and LIGA (German acronym for LIthographie-Galvanoformung-Abformung). Among those, hot embossing has higher accuracy and throughput than mold casting, and much simpler and lower cost than LIGA. Hot embossing technique is essentially the stamping of 3-D hard patterns into a polymer substrate by raising the temperature higher than the transition point of polymer. During the past years, this technique has been developed and used to both laboratories and industries in a variety of applications with the feature sizes are scaled down to nanometer [1].

Hot embossing of polymers is a promising alternative to replace the traditional silicon processes. Complex micromachining steps are only necessary to fabricate the master mold. From this master mold, the polymeric structures can be batch replicated by hot embossing technique. Hot embossing process has been well applied to fabricate

K.-U. Sattler et al. (Eds.): ICERA 2019, LNNS 104, pp. 342–348, 2020.
https://doi.org/10.1007/978-3-030-37497-6_40

polymer optical components [2], micro fluidic devices [3] or PMMA (polymethyl methacrylate) comb-drive actuators [4, 5].

As an example, the typical micromachining process of polymeric comb actuators utilizing hot embossing technique is illustrated in Fig. 1. In this process, steps from (*a*) to (*f*) affect strongly to an embossing time and quality of polymer products. Therefore, our research will focus on two topics: (*i*) development of improved hot embossing system and optimization of hot embossing conditions in order to reduce experimental time and raise a quality of embossed products; (*ii*) improvement of surface of silicon molds aiming to extend mold's lifetime and can be re-used.

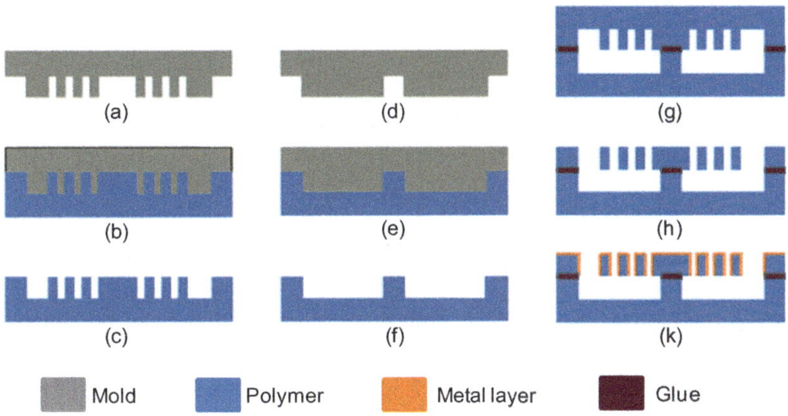

Fig. 1. Fabrication flow of polymeric comb actuators based on hot embossing technique. *(a), (d) Fabrication of master mold; (b) Hot embossing of polymeric comb actuator; (c) Actuator structure after de-molding; (e) Hot embossing of polymeric bonding pad; (f) Bonding pad after de-molding; (g) Bonding two sheets by glue; (h) Lapping, polishing and plasma etching; (k) Coating the electrodes by metal;*

2 Properties of PMMA and Applications

PMMA (polymethyl methacrylate) is a glassy polymer with an amorphous structure. It has a density of 1190 kg/m^3 and very low water absorption. The original PMMA was seen as a replacement for glass in a variety of applications and is currently used extensively in glazing applications. This material is one of the hardest polymers, rigid, glass-clear with glossy finish and good weather resistance [6].

PMMA is odorless and tasteless and physiologically safe. Therefore, it can be used for many applications such as rear lights, indicators, tachometer covers in vehicles, packaging for tablets, pills, capsules, suppositories, urine containers, sterilized equipment in medicine, pens, toys, lamp covers, switch parts, dials, control buttons, etc.

PMMA can withstand temperatures as low as −70 °C. Resistance to temperature changes is very good. The thermal stability of standard PMMA is smaller than 65 °C. When temperature is higher, PMMA is deformed more easily. It is easy for elongation and molding from 163 °C to 260 °C. This feature is a key point to select the PMMA as an attractive material for hot embossing technique, both computation and experiment [7–9]. Mechanical and physical properties of PMMA material are shown in Table 1.

Table 1. Properties of PMMA material

Property	Approximate value
Density	1190 (kg/m^3)
Poisson's ratio	0.34–0.4
Max strain	6%
Tensile strength	54–74 (MPa)
Young's modulus	2243–3243 (MPa)
Flexural strength	73–132 (MPa)
Elongation at break	2–6%
Mouth shrinkage	0.001–0.004 m/m
Coefficient of thermal expansion	5–9 × 10^{-5}/°C
Glass transition temperature	100–105 °C

3 Hot Embossing Process of PMMA Micro Structure

Hot embossing is one of the most important steps to fabricate polymeric micro structures with high aspect ratio (such as micro comb actuator, micro channel, etc.…). To prepare for embossing experiments, a hot embossing system has been assembled and set up firstly. A model of hot embossing system is shown in Fig. 2.

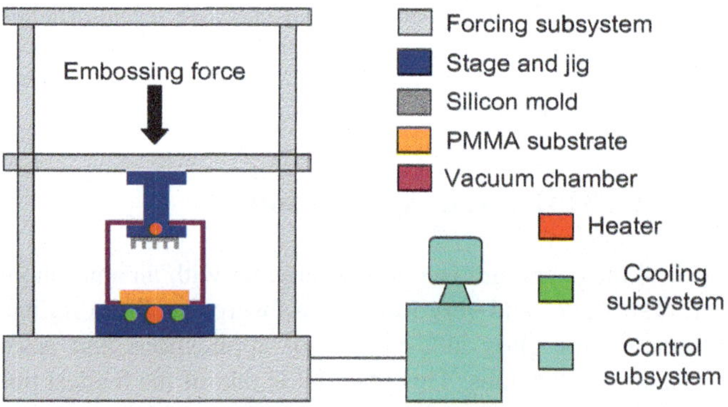

Fig. 2. The schematic view of hot embossing system

The vacuum chamber, forcing and control subsystems are AG-IS model and supplied by Shimadzu Corporation, Japan (see Fig. 2). In order to guarantee a flexibility of embossing process, the heater's location, stage and embossing jig (where the PMMA substrate and silicon mold are installed) were designed and fabricated by the authors. For reducing of a cooling time, or in other words, time of each embossing cycle, the cooling subsystem using circulation of water was designed and assembled additionally.

The suitable silicon molds with positive taper angle of sidewall are fabricated by using the single crystal silicon wafer and Deep Reactive Ion Etching (DRIE) parameters (i.e. the etching time t_e = 5 s and passivation time t_p = 4 s) as mentioned in [10].

The authors performed several embossing experiments, and summarized the improvement of hot embossing process applying for PMMA material. The entire embossing cycle and the main parameters are listed as below:

(1) Set up a silicon mold and 1 mm-thick PMMA substrate into a chamber and evacuate to a pressure under 2 kPa.
(2) Heat the silicon mold and PMMA substrate at the same time to 200 °C by heaters and maintain this temperature during molding. 100 N pre-loading of an embossing force has been installed since the beginning of the heating phase ($t_1 = 20 \div 25$ min).
(3) Insert the mold into PMMA substrate and increase the embossing force to 2000 N with the molding rate of 0.1 mm/min. Maintain the embossing force of 2000 N for 10 min (molding phase t_2).
(4) Cool the mold and PMMA to 80 °C (between solid situation and melting temperature of PMMA) by using circulation of water flow near the heaters. The cooling time t_3 is about 10 min. Keep the embossing force of 2000 N during the cooling phase.
(5) Maintain the temperature at 80 °C during de-molding phase ($t_4 \approx 30$ min). De-molding rate is 0.01 mm/min. The embossing load is decreased to zero.
(6) Vent chamber, remove the silicon mold and embossed PMMA substrate for checking.

Conditions of temperature and forcing are illustrated in Fig. 3. By using the separated heaters and water-cooling subsystem, total experiment time for improved hot embossing cycle is reduced approximately to 70 min (saving about $20 \div 30$ min in comparison with the standard hot embossing technique used in [4, 9]).

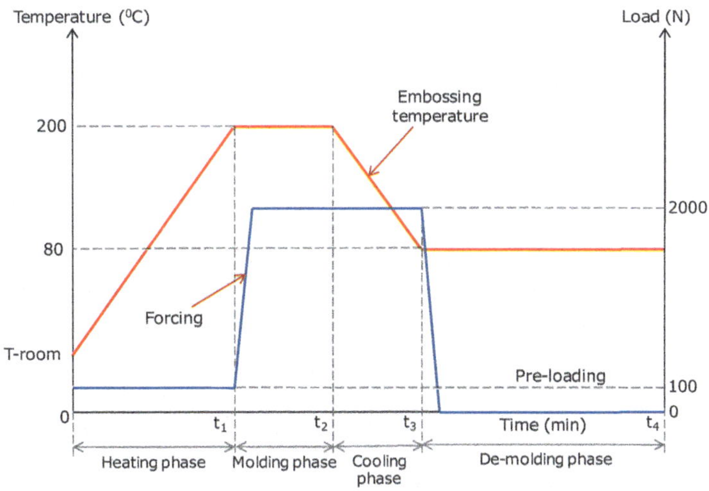

Fig. 3. Temperature, forcing and time conditions during hot embossing cycle

4 Result and Discussion

In order to prepare the low-cost master mold for hot embossing process, we need to fabricate silicon mold firstly basing on DRIE technique. Then, a post-deposition process will be applied to reduce the frictional coefficient on the surface of silicon mold. In other words, the silicon molds were plasma treated with C_4F_8 gas for 2 ÷ 3 min before hot embossing step. This post-deposition technique coats a thin layer of Teflon-like material on the surface of the mold's sidewall (as an anti-stick layer) for reducing the friction during molding and de-molding. After post-deposition step, the frictional coefficient of the silicon surface was found to be reduced approximately six times [11].

To test the performance of the fabricated silicon molds, several PMMA models of test patterns and comb actuators were hot embossed. At first, hot embossing experiments were performed with silicon molds fabricated by DRIE without post-deposition step. The result is not good and can be seen in Fig. 4. Small PMMA and silicon mold structures were broken and stuck into each other while de-molding. Next, the hot embossing experiments were continued with the silicon molds treated by post-deposition technique as discussed above. In this case, the de-molding is performed successfully because of lower friction. Fine PMMA actuators (with only 2 µm-gap of comb finger as shown in Fig. 5) were successfully embossed. Furthermore, the silicon mold was not broken after de-molding phase and can be re-used for next experiments.

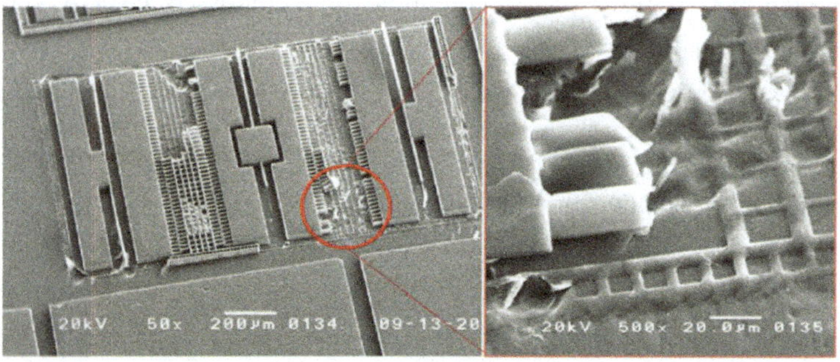

Fig. 4. The PMMA comb actuator embossed by using silicon mold without post-deposition step

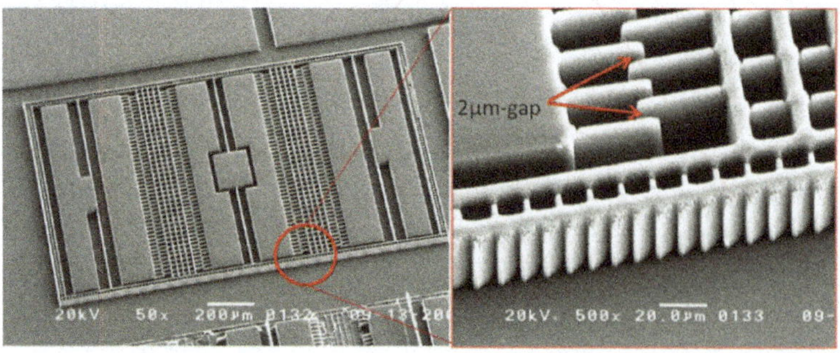

Fig. 5. The fine PMMA comb actuator embossed by using improved silicon mold

The air pressure in vacuum chamber is another important condition to obtain fine edges of the PMMA micro structure while hot embossing. At a lower pressure, the hot PMMA will be filled easier and better to the corners and micro-gaps of the silicon mold. Figure 6 shows the hot embossing results obtained in two cases: (*a*) without vacuum and (*b*) with chamber's vacuum is lower than 2 kPa. We can see that the profile of PMMA structure in Fig. 6(*b*) is sharper and more uniformed than one shown in Fig. 6(*a*). Furthermore, burr is not occurred at a bottom of structure, and evidently, height of embossed structure in Fig. 6(*b*) is higher than the counterpart in Fig. 6(*a*), i.e. higher aspect ratio can be obtained while applying vacuum in chamber is under 2 kPa.

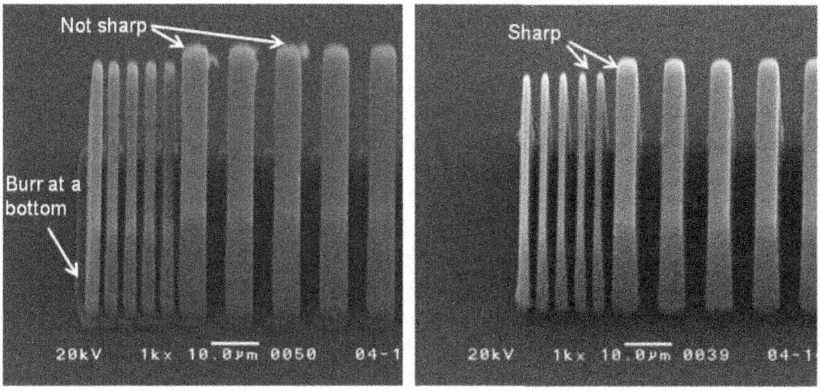

(*a*) Without vacuum (*b*) With vacuum under 2 kPa

Fig. 6. Hot embossing results of PMMA test patterns

5 Conclusion

This paper presented the fabrication process of polymeric PMMA micro structure basing on the anti-stick-coated silicon mold and improved hot embossing technique.

Improved silicon molds with positive taper angle of sidewall have been successfully achieved by DRIE process. Post-deposition technique has been done to coat a thin layer of anti-stick material on the surface of the mold's sidewall. It is a key condition to increase a lifetime of the silicon mold, aiming to lower cost of polymeric products. The edges of almost PMMA micro structures (with minimum feature size of 2 µm) embossed by new silicon molds and aided additionally by the vacuum condition of the chamber were very sharp, no burr and higher aspect ratio in opposition to the counterparts fabricated by the conventional hot embossing process.

In order to save energy and reduce the cost of MEMS devices, the embossing-based polymeric microstructures such as comb-drive actuator, micro lens, micro splitter or micro channel, etc. will be developed and used potentially in micro fluidic systems, micro optics or in micro machines.

Acknowledgment. This research is funded by Vietnam National Foundation for Science and Technology Development (NAFOSTED) under Grant number "107.01-2019.05".

References

1. Heyderman, L.J., et al.: Nanofabrication using hot embossing lithography and electroforming. Microelectron. Eng. **57–58**, 375–380 (2001)
2. Nilsson, D., Jensen, S., Menon, A.: Fabrication of silicon molds for polymer optics. J. Micromech. Microeng. **13**, S57–S61 (2003)
3. Chen, C.L., Jen, F.: Fabrication of polymer splitter by micro hot embossing technique. Tamkang J. Sci. Eng. **7**(1), 5–9 (2004)
4. Zhao, Y., Cui, T.: Fabrication of high-aspect-ratio polymer-based electrostatic comb drives using the hot embossing technique. J. Micromech. Microeng. **13**, 430–435 (2003)
5. Zhao, Y., Cui, T.: SOI wafer mold with high-aspect-ratio microstructures for hot embossing process. Microsyst. Technol. **10**, 544–546 (2004)
6. http://www.tangram.co.uk/TI-Polymer-PMMA.html
7. He, Y., Fu, J.Z., Chen, Z.C.: Research on optimization of the hot embossing process. J. Micromech. Microeng. **17**(12), 2420–2425 (2007)
8. Liu, C., Li, J.M., Liu, J.S., Wang, L.D.: Deformation behavior of solid polymer during hot embossing process. Microelectron. Eng. **87**(2), 200–207 (2010)
9. Amaya, S., Dao, D.V., Sugiyama, S.: Development of polymer electrostatic comb-drive actuator using hot embossing and ultra-precision cutting technology. J. Micro/Nanolithography, MEMS, MOEMS **8**(4), 043065 (2009)
10. Pham, P.H., Dang, L.B.: Influence of the side etching effect in DRIE on performance of electrostatic linear comb-drive actuators. Microsyst. Technol. **24**(5), 2215–2222 (2018)
11. Gao, J.X., et al.: Antistick post-passivation of high-aspect ratio silicon molds fabricated by deep-reactive ion etching. J. Microelectromech. Syst. **15**(1), 84–93 (2006)

Fault Detection for Rotating Machines in Non-stationary Operations Using Order Tracking and Cepstrum

Nguyen Phong Dien and Nguyen Trong Du(✉)

School of Mechanical Engineering, Hanoi University of Science and Technology, Hanoi, Vietnam
du.nguyentrong@hust.edu.vn

Abstract. Recently, the use of machines working under time varying load and speed is increasing rapidly in modern industry. Fault diagnosis of these technical objects plays an important role but faces great challenges, since vibration signals are mostly non-stationary due to uncertainties affected by the change of speed and load during operation. This study aims to propose and verify a signal analysis procedure, namely, a combination of order tracking technique, cepstrum analysis and a classification algorithm using support vector machine to automatically detect severe faults in rotating machines working in non-stationary conditions. The experimental example in a test rig demonstrates that the proposed method is very quick and technically simple, and can be used for monitoring the condition of a helical gearbox during speed up.

Keywords: Cepstrum · Gear vibration · Order tracking · Support vector machine

1 Introduction

Recently, the use of machines working under time-varying load and speed is increasing rapidly in the modern industry. Rotating machines operating under such non-stationary conditions have received a considerable amount of attention in the respect of vibro-acoustic diagnostics, such as marine propulsion gearbox [1], helicopter gearbox [2], diesel engines [3] and wind turbine transmission [4]. Fault diagnosis of these technical objects plays an important role but faces great challenges, since vibration signals are mostly non-stationary due to uncertainties affected by the change of speed and load during operation [5]. It is well-known that the conventional methods such as spectral analysis based on FFT are suitable for analyzing stationary vibrations, but cannot be well applied in such conditions. Complex signal analysis procedures are therefore required to extract useful diagnostic features from vibration signals. In addition, rotating machines in non-stationary operations have not been considered in the international standards for evaluation of machine vibration as ISO 10816 and ISO 7919.

Since the 90 s of the last century, a significant number of methods and techniques have been proposed for the analysis of non-stationary processes. Some of them have been used successfully for vibration diagnosis in a class of rotating equipment and

K.-U. Sattler et al. (Eds.): ICERA 2019, LNNS 104, pp. 349–356, 2020.
https://doi.org/10.1007/978-3-030-37497-6_41

machines, such as polar wavelet amplitude maps [6], principal component analysis [7] and other adaptive methods based on time-frequency analysis. However, these sophisticated signal processing tools require a complex numeric algorithm with higher computation cost that is difficult to apply in an on-line machinery diagnosis system. Because symptoms of damage in the measured signal often repeat regularly with the shaft angle of the rotating machine, order tracking or angular resampling technique [8, 9] is developed and serves as a pre-processing tool for further analysis using FFT, called order spectrum.

Cepstrum analysis was proposed in earlier phase with the first purpose for analyzing human speech. Because a cepstrum is defined as the inverse Fourier transform of the logarithm of the Fourier transform of the signal, the cepstrum analysis method has a fast algorithm and low computation cost. This method has been used in gearbox diagnostics with a number of published papers, but only in the case of stationary operation conditions. In a few recent scientific publications, complex analysis procedures using cepstrum analysis are suggested to deal with the fault diagnosis of gears operating under non-stationary conditions [10–12]. This study aims to propose and verify a signal analysis procedure, namely, a combination of order tracking technique, cepstrum analysis and a classification algorithm using support vector machine (SVM) [13] to automatically detect severe faults in rotating machines working in non-stationary conditions. The proposed procedure is verified by an experimental example in a test rig for monitoring the condition of a helical gearbox during speed up.

2 Power Cepstrum

The power cepstrum, denoted by $C_x(\tau)$, is defined as the inverse Fourier transform of the logarithmic power spectrum as follows

$$C_x(\tau) = F^{-1}[\log S_x(f)] \tag{1}$$

where $F^{-1}[]$ denotes inverse Fourier transform and $S_x(f)$ is the power spectrum of time-signal $x(t)$ and f is the frequency variable. The independent variable τ, known as quefrency, has the dimension of time.

It is known that the power cepstrum can be useful in detection of periodic structures of the frequency spectrum of a signal, and the period corresponds to the pitch occurring in the cepstral graph. This feature makes the cepstrum analysis suitable for analyzing noisy vibration signals with complex periodic structures such as gearbox signals with sideband structures. As discussed in Sect. 4, each family of sidebands of gear spectral, emphasized by the logarithmic scale, produces a peak in the cepstrum at a quefrency corresponding to the spacing between frequency lines in the sidebands. A set of such peaks can be plotted in a cepstral graph with multiple quefrencies, called rahmonics.

3 Computed Order Tracking and Order Cepstrum

As mentioned above, diagnostic symptoms of a specific faults in the vibration signal can repeat regularly in the shaft angle domain even if the angular velocity of the shaft changes. This idea leads to a signal pre-processing method called order tracking. According to this, a time-signal that is sampled at constant time increments by a measurement device, will be resampled at constant angle increments in angle-domain using a reference signal (one-pulse-per-revolution signal) and an interpolation algorithm. Order tracking becomes a common technique in vibration analysis of rotating machines as well as an aiding tool for fault diagnosis [9]. The order spectrum is obtained by taking Fourier transformation of the resampled signal. The result of the cepstrum analysis of the resampled signal using order tracking creates a cepstral graph over shaft angle, called order cepstrum. There are various ways to quantify the order spectrum to create vectors of diagnostic features for classification tasks using the standard SVM as presented in [13]. In this study, the cepstrum impulse index used as an additional feature vector:

$$f_{pulse} = \frac{C_m}{\bar{C}} \tag{2}$$

where C_m is the peak amplitude of the power cepstrum, $\bar{C}$ is defined as the mean of power cepstrum as follows:

$$\bar{C} = \frac{1}{N_c} \sum_{i=1}^{N_c} |C_i| \tag{3}$$

where C_i is the amplitude of ith quefrency and N_c the number of sampling points.

Finally, the analysis procedure is proposed as follows:

Step 1. Taking computed order tracking of the raw vibration signal using a reference signal and a standard interpolation algorithm.

Step 2. Calculation of order cepstrum of the resampled signal, the cepstrum impulse index according to Eqs. (2) and (3).

Step 3. Training SVM based on the cepstrum impulse index and solving the problem of gear condition classification

4 Experimental Results

Figure 1 shows a helical single-stage gearbox test rig for gear crack test during run up. The input shaft of the gearbox is driven by a motor which is controlled by an AC inverter to change the rotating speed from 300 rpm to 1500 rpm according to the predefined laws. The pinion and the gear have 14 and 40 teeth respectively.

The vibration measurement is taken with an accelerometer, positioned at the closest position to the test gear. An optical tachometer mounted in proximity to the input shaft is used to generate a one-pulse-per-revolution signal. The vibration signals were sampled at 10 kHz. Localized crack defect was seed in the root of the driving gear of the input shaft to simulate serious gear crack. The artificial defect was 3 mm in depth (Fig. 1c).

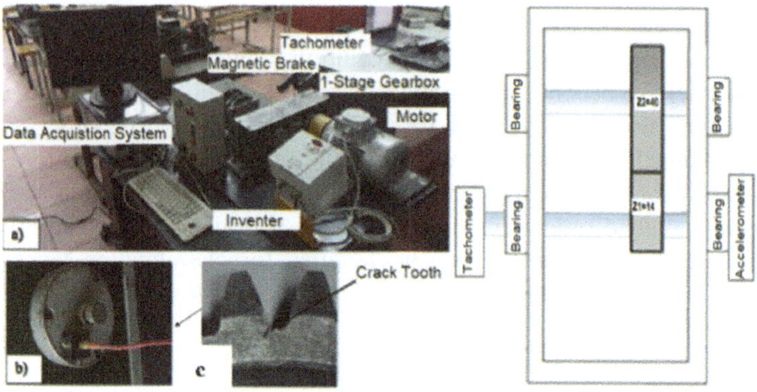

Fig. 1. Gearbox test rig and the seeded gear tooth crack

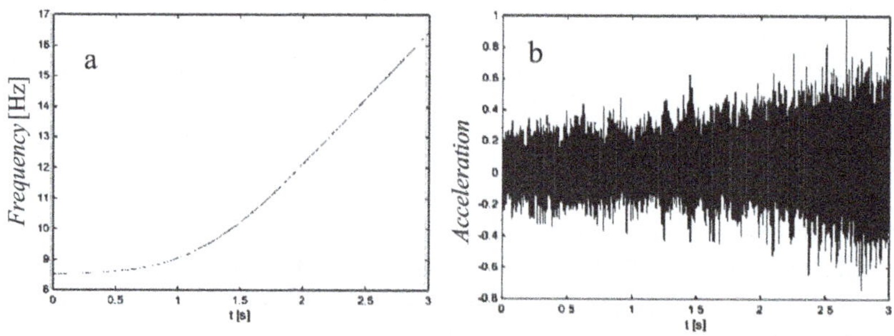

Fig. 2. Rotating frequency of the input shaft (a) and time-signal (b)

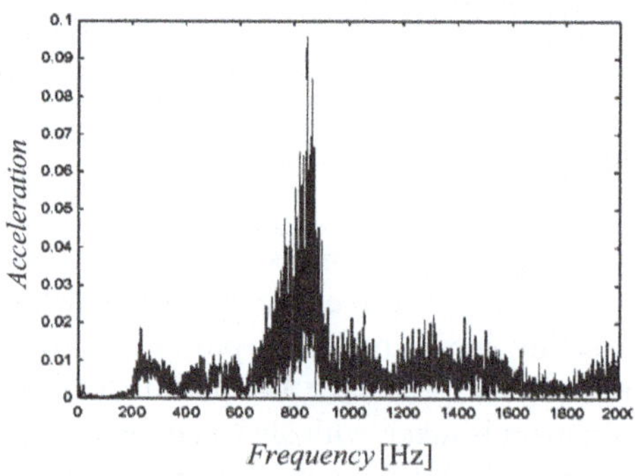

Fig. 3. Frequency spectrum of a vibration signal during speed up

The characteristic parameters of the gearbox test rig are given in Table 1. Figure 2a shows that the input shaft runs up from rotating frequency of 8.5 Hz to 16.5 Hz. The measured vibration signal with gear crack fault is displayed in Fig. 2b with strong non-stationarity in time and frequency, in which the amplitude of the vibration is increasing during speed up. It is clear that its frequency spectrum in Fig. 3 is significantly obscured by spectral smearing. The result of order tracking for the raw time-signal is plotted in shaft angle domain (Fig. 4a) and in the order spectrum (Fig. 4b). The order spectrum is dominated by the gear mesh component and its harmonics over shaft orders. However, the order spectrum is not capable of revealing the characteristic order of gear crack fault that was corrupted by the modulation effect and noise.

Table 1. Characteristic frequencies and shaft orders

Components	Parameters
Rotational frequency of the faulted gear	f_{r1}
The gear mesh frequency	$f_m = 14f_{r1}$
The characteristic frequency of gear crack	$f_{crack} = f_{r1}$
The gear mesh order	$x_m = 14$
The characteristic order of gear crack	$x_{crack} = 1$
The quefrency of the gear mesh frequency	$\hat{f}_m = 1/f_m$
The quefrency of the gear crack fault	$\hat{f}_{crack} = 1/f_{r1}$
Gear mesh component in order cepstrum at	$\hat{x}_m = 360/14 = 25.71°$
Gear crack symptom in order cepstrum at	$\hat{x}_{crack} = 360°$

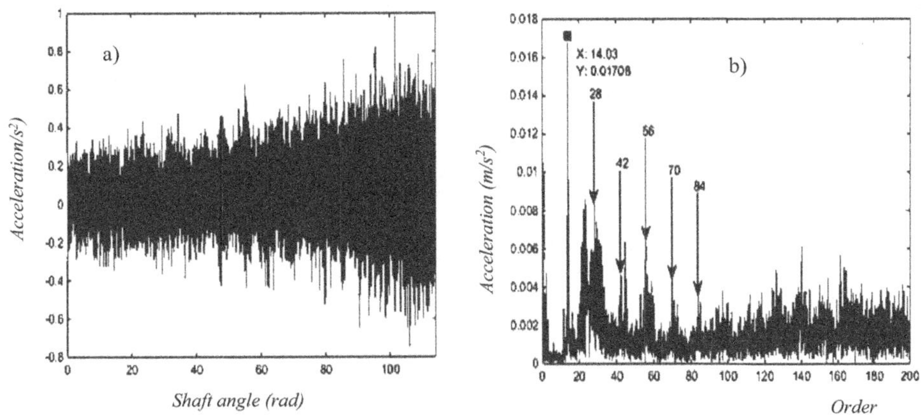

Fig. 4. The resampled signal with gear crack fault (a) and order spectrum (b)

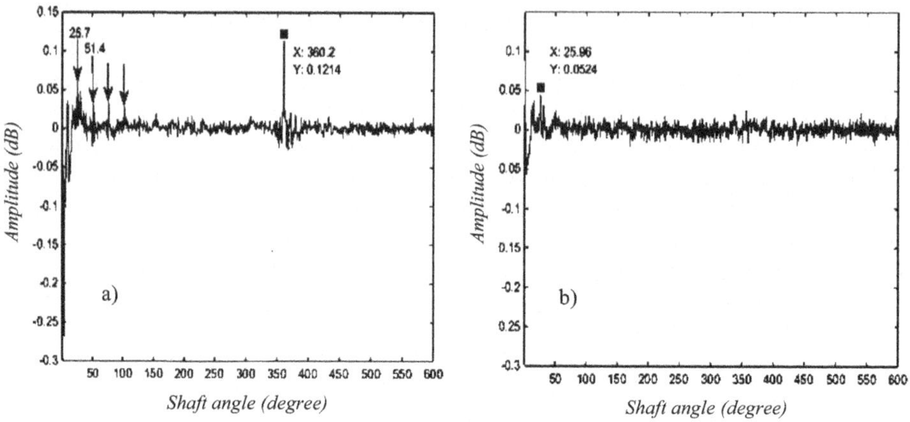

Fig. 5. Order cepstrum of the signals fault during speed up (a) with gear crack and (b) normal condition (faultless gears)

Figure 5 shows the order cepstrum for the conditions of gear crack and faultless. The characteristic order (x_{crack}) of gear crack fault, gear mesh order (x_m) and its harmonics are clearly represented in the order cepstrum. The cepstrum impulse index is calculated according to Eq. (2) for both cases and can be used as an input parameter of for SVM.

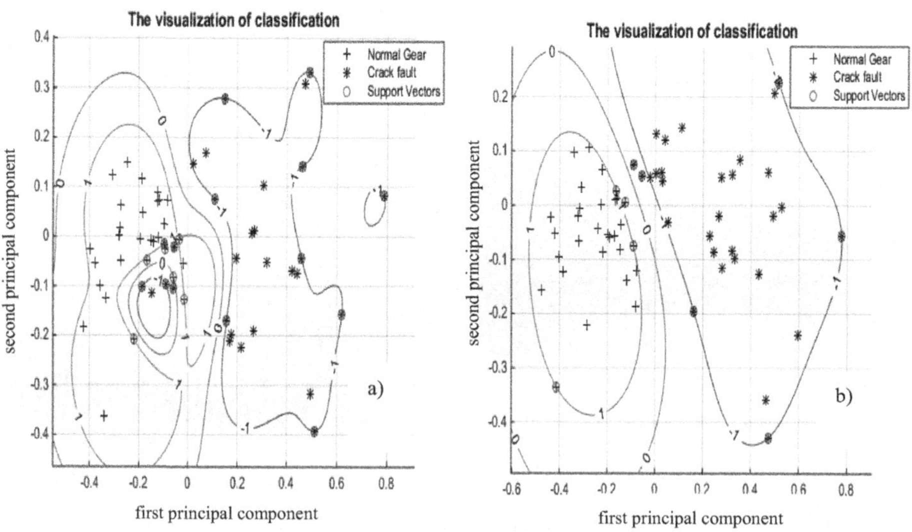

Fig. 6. Classification result using SVM with order spectrum (a) and order cepstrum (b)

To implement SVM, 30 datasets (samples) are collected at different operation conditions for each gear condition, in which 20 datasets are used to train SVM.

Figure 6a shows the classification result by SVM with common feature parameters as kurtosis, peak index, Crest factor, etc. We note that each plot in Fig. 6 includes only two principal components extracted from the feature vector to reduce dimensionality [7]. Figure 6b depicts the result by SVM using the cepstrum impulse index as an additional feature parameter. It can be seen that the proposed signal analysis procedure with order cepstrum gives better classification results.

5 Concluding Remarks

In a complex mechanical system that contains a number of vibration-excited sources, mechanical vibrations consist of different components, affected by the operating conditions (loading and speed). The conventional analysis methods such as spectral analysis have been recognized to be suitable in detecting faults at stationary operating conditions, but inefficient for the cases of time-varying speed and load.

The present study deals with a signal analysis procedure for machinery fault detection from vibration signal measured at non-stationary operation conditions. The procedure consists of some analysis tools as order tracking and cepstrum analysis.

We note that the authenticity of the resampled signal to the original time-signal, that is, the accuracy of results by computed order tracking, depends mainly on the quality and resolution of the reference signal, sampling frequency and the angular acceleration of the shaft. The experimental example in a simple test rig demonstrates that the proposed method is very quick and technically simple, and can be used for monitoring the condition of a helical gearbox during speed up.

References

1. Lia, Z., Peng, Z.: A new nonlinear blind source separation method with chaos indicators for decoupling diagnosis of hybrid failures: a marine propulsion gearbox case with a large speed variation. Chaos Solitons Fractals **89**, 27–39 (2015)
2. Elashaa, F., et al.: Application of acoustic emission in diagnostic of bearing faults within a helicopter gearbox. In: Proceedings of the Fourth International Conference on Through-life Engineering Services, vol. 38, pp. 30–36 (2015)
3. Zhao, H., et al.: A new fault diagnosis method for a diesel engine based on an optimized vibration mel frequency under multiple operation conditions. Sensors **19**, 2590 (2019). https://doi.org/10.3390/s19112590
4. Randall, R.B., Sawalhi, N., Coats, M.: Separation of gear and bearing fault signals from a wind turbine transmission under varying speed and load. In: Proceedings of the 2nd International Conference on Condition Monitoring of Machinery in Non-Stationary Operations, CMMNO 2012, pp. 3–12. Springer, Heidelberg (2012)
5. Sghaier, E.: et al.: Non-stationary operating conditions of rotating machines: assumptions and their consequences. In: Proceedings of the 6th International Conference on Condition Monitoring of Machinery in Non-Stationary Operations, CMMNO 2018, pp. 11–19 (2018)
6. Meltzer, G., Dien, N.P.: Fault diagnosis in gears operating under non-stationary rotational speed using polar wavelet amplitude maps. Mech. Syst. Signal Process. **18**(5), 985–992 (2004)

7. Zimroz, R., Bartkowiak, A.: Two simple multivariate procedures for monitoring planetary gearboxes in non-stationary operating conditions. Mech. Syst. Signal Process. **38**(1), 237–247 (2013)
8. Munck, E.D.S., Fyfe, K.R.: Computed order tracking applied to vibration analysis of rotating machinery. Mech. Syst. Signal Process. **11**(2), 187–205 (1997)
9. Singh, A.P.: Gearbox fault diagnosis under non-stationary conditions with independent angular re-sampling technique applied to vibration and sound emission signals. Appl. Acoustics **144**, 11–22 (2019)
10. Benkedjouh, T.: et al.: Gearbox fault diagnosis based on mel-frequency cepstral coefficients and support vector machine. In: CIIA 2018, IFIP Advances in Information and Communication Technology, vol. 522, pp. 220–231. Springer, Cham (2018)
11. Tian, H., et al.: Fault diagnosis of gear wearing based on order cepstrum analysis. Appl. Mech. Mater. **543–547**, 922–925 (2014)
12. Li, H., Zhang, Y., Zheng, H.: Gear fault detection and diagnosis under speed-up condition based on order cepstrum and radial basis function neural network. J. Mech. Sci. Technol. **23**, 2780–2789 (2009)
13. Dien, N.P., Du, N.T.: On a diagnostic procedure to automatically classify gear faults using the vibration signal decomposition and support vector machine. In: ICERA 2018, Lecture Notes in Networks and Systems, vol. 63, pp. 425–432. Springer, Cham (2018)

FEM Simulation for a MEMS Vibratory Tuning Fork Gyroscope

Vu Van The[1(✉)], Tran Quang Dung[1], and Do Thi Kim Lien[2]

[1] Le Quy Don Technical University, Hanoi, Vietnam
thevutb@gmail.com
[2] Viet Hung Industrial University, Hanoi, Vietnam

Abstract. This paper presents a FEM simulation for a MEMS vibratory tuning fork gyroscope with a connecting diamond-shaped frame. This structure is designed and analyzed using ANSYS software to obtain the dynamic parameters of the proposed tuning fork gyroscope. The resonant frequency of the sensing and driving mode is 11582 Hz and 11605 Hz respectively with 23 Hz mismatch frequency. The equivalent stiffness of the connecting frame in the*y*-direction is determined to be 2360 N/m and the stiffness of the flexible beams in driving and sensing are 858 and 725 N/m, respectively. The response of the proof-mass reflects the input angular velocity. The results will be used to set up in an analytical problem to study the dynamic response of the proposed tuning fork gyroscope in the future works.

Keywords: Tuning forkgyroscope · Simulation · FEM analysis

1 Introduction

Basing on the advancement of micro-electro-mechanical system (MEMS) technology, micro-machined vibratory gyroscope (MVG) has been becoming popular in science technology thanks to its advantages over traditional gyroscopes such as small size, lightweight, and low power consumption [1–4]. This device is a micro-sensor detector and determines the angular velocity or rotational angle of a body into which the MVG is integrated. The operation of the MVG is based on the "Coriolis effect" to transfer energy from the primary vibration to a secondary one [4].

A MEMS tuning fork gyroscope (TFG) includes two identical tines that vibrate in the opposite direction (anti-phase mode) [4, 5]. Thanks to this state, the performance of TFG increases significantly [5–7]. However, the traditional MEMS tuning fork structure with the direct mechanical coupling between two tines likely causes an in-phase vibratory mode [8], and also could appear some errors in the vibration of the system [9]. Therefore, TFG should be designed with the mechanical frame to guarantee the anti-phase mode in both the driving and sensing direction to considerably increase the sensitivity of the gyroscope device.

The frame connecting indirectly two tines in the TFG was designed in some shapes: a diamond-shaped frame [7, 8]; a U-shaped bar [10]; a lozenge-shaped frame [11, 12]; a structurally forced lever bar [13]; a T-shaped bar [14]; or a straight linkage bar [15]. These researches focused on simulation and modal analysis to determine the frequency

K.-U. Sattler et al. (Eds.): ICERA 2019, LNNS 104, pp. 357–365, 2020.
https://doi.org/10.1007/978-3-030-37497-6_42

of anti-phase mode and the output characteristics of the TFG model. The results showed that the connecting frame is used to reject the out-of-plane vibration and take likely the anti-phase mode. However, how the masses of the TFG respond when applying the external forces to them with the reversed phase has not been still studied yet. Besides, the response of the sensing masses in the TFG model when applying the angular velocity should be considered.

This paper presents a design for a 3D model of a TFG with a connecting diamond-shaped frame with the simple structure to create and guarantee the anti-phase mode in the first mode of the proposed model. In addition, the dynamic characteristic parameters of the proposed TFG are determined. Moreover, the mechanical response of the proof-mass when applying the angular velocity is considered in this paper. The results are obtained by using ANSYS software and used to be the input parameter for the dynamic analysis of the TFG in the continuous researches.

2 Configuration

The configuration of the proposed TFG is shown in Fig. 1. It consists of two tines defined as two single gyroscopes [16]. Every single gyroscope includes a proof-mass 1 and an outer frame 2. This outer frame is connected to the proof-mass by four elastic beams 3 and suspended on substrate thanks to four other elastic beams 4. Each of these beams is linked to the substrate (not be presented in Fig. 1) by an anchor 5 to allow the outer frame and the proof-mass to move freely in two perpendicular directions. Two

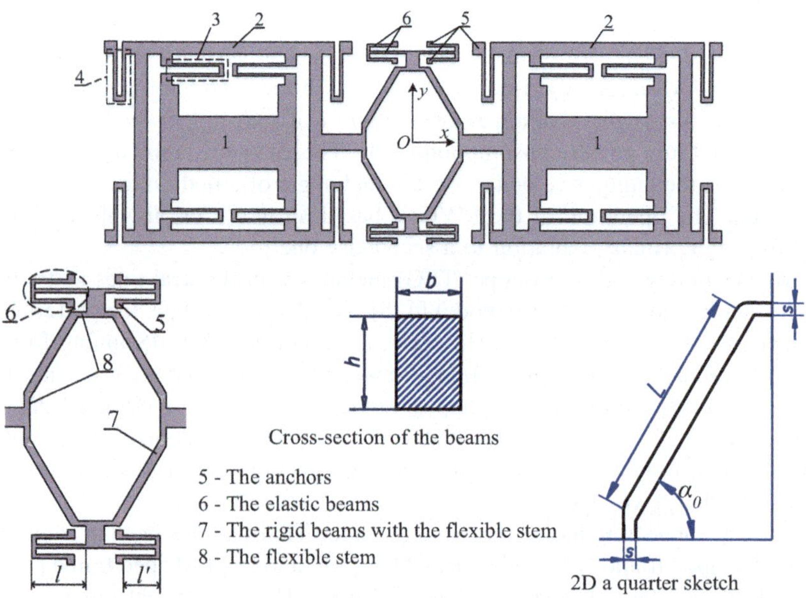

Fig. 1. D model of the proposed TFG with a diamond-shaped frame

single gyroscopes are connected through a diamond-shaped frame to create the proposed TFG. This frame has four rigid bars 7 with length L and the rectangular cross-section b × h, where h and b are the thickness and width of each bar, respectively (Fig. 2). The bars are connected to the connectors by elastic stems 8 with the width s (s < b). The configuration and dynamics analysis of this frame was shown in the previous research [17].

3 FEM Analysis

3.1 Mesh Model

The parameters of this TFG are designed using the ANSYS Workbench software shown in Table 1, and the specific material parameters of Silicon shown in Table 2.

Table 1. The parameters of the TFG

Parameter	Value
Dimensions of TFG (x × y)	5546 × 1930 μm
Thickness of structure	60 μm
Surface area of the proof-mass	894000 μm^2
Surface area of the outer frame	589160 μm^2
Length of the drive springs	500 μm
Width of the drive springs	18 μm
Length of the sensing springs	510 μm
Width of the sensing springs	16 μm
Length of the bars of the diamond frame	650 μm
Length of the flexible beam of the connecting frame	295 μm
Mass of the outer frame	0.7883×10^{-7} kg
Sensing mass	1.196×10^{-7} kg
TFG total mass	5.66×10^{-7} kg

Table 2. The properties of the Silicon material

Property	Value	Unit
Density	2330	kg/m^3
Poisson's ratio	0.28	1
Young's modulus	169×10^9	Pa
Bulk modulus	1.2879×10^5	Pa

The governing equations when carrying out the modal analysis are formed as:

$$[M]\{\ddot{q}\} + [C]\{\dot{q}\} + [K]\{q\} = \{0\} \qquad (1)$$

where [M] – masses matrix; [C] – damping matrix; [K] – stiffness matrix; and {q} – deformation vector.

In this paper, the ANSYS software is used to create, analyze and simulate the presented model in Fig. 1. The mesh model for the beams and the flexible stem is controlled by using element size of 25 μm to guarantee the elasticity of the flexible beams, whereas the rest of this TFG is defined by using maple face meshing with the element size of 400 μm to reduce element quantity and time of the calculation. The finite element type in this model is set up to be SOLID 187 (element 3D with 10 non-linear tetrahedron nudes). The element quantity for the whole TFG is 19890 with 46234 nodes. The 3D and mesh model of the proposed TFG in Fig. 2 is designed by the geometry module in ANSYS Workbench software.

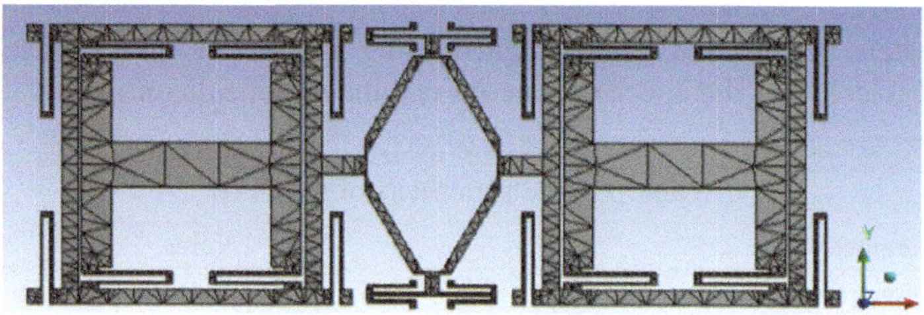

Fig. 2. Mesh model of the TFG

3.2 Equivalent Stiffness

The equivalent stiffness of the flexible beams in driving and sensing direction and the connecting frame is determined using simulation analysis in ANSYS. The unit force (1μN) is applied to the structure in the driving direction (x-direction), and the equivalent stiffness is defined from the relation between the deformation and the external force as the followed formula [18]:

$$k = \frac{F}{\delta} \tag{2}$$

where δ is the deformation of the structure after applying the external force F.

The schemas of applying an external force on the whole structures mentioned above are the same and shown in Fig. 3a. The connecting diamond-shaped frame is suppressed to make the movement for the two single gyroscopes. The deformation of this TFG in the applied force direction is shown in Fig. 3b. It is the same way to determine the equivalent stiffness of the flexible beams in driving and sensing direction.

As follow the formula (2), the equivalent stiffness of the flexible beams in driving, sensing direction and diamond connecting frame is shown in Table 3.

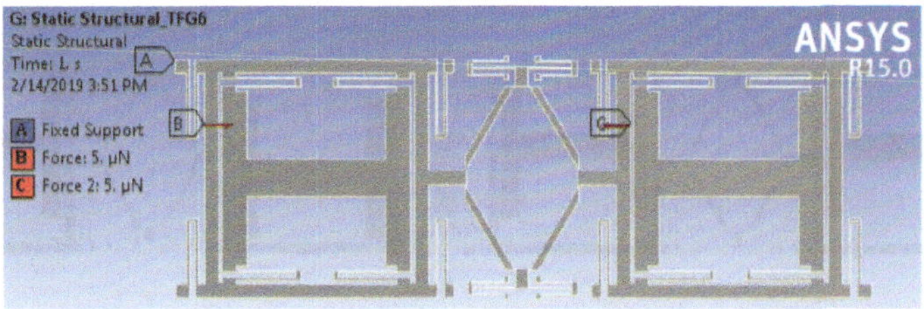

(a) Schemas of applying unit forces on the whole TFG structure

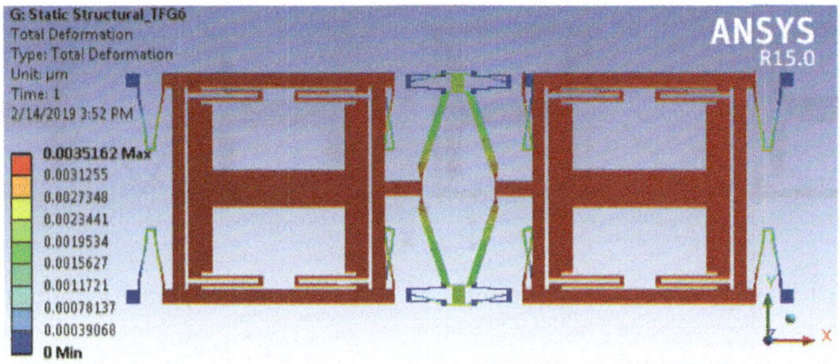

(b) Deformation of the connecting frame with applied unit forces

Fig. 3. Determining the equivalent stiffness of the proposed TFG

Table 3. The stiffness of the structures

The structure	Value (N/m)
Stiffness of the driving beams	858
Stiffness of the sensing beams	725
Stiffness of the connecting diamond-shaped frame	2360
Stiffness of the whole structure in the x-direction	1427

3.3 Modal Analysis

The two major expected modes of the proposed TFG are anti-phase driving and anti-phase sensing at 11605 Hz and 11582 Hz, respectively. The mismatch frequency determined to be the difference between the resonant frequencies of the two major modes is 23 Hz. This value allows the proposed TFG to satisfy the requirement of both the sensitivity and response time related to the operation of the gyroscope [12]. Besides, the parasitic mode appears in the fourth one at 15549 Hz and is farther than the two previous major modes (the difference is about 34%). Thus, the crosstalk effect can be reduced in the operating of this TFG [12].

The natural vibratory modes of this TFG are displayed in Fig. 4, where the two major modes (anti-phase sensing and driving mode) are shown in Fig. 4b and c, respectively. Figure 4a and 4d are the two unexpected modes in this TFG.

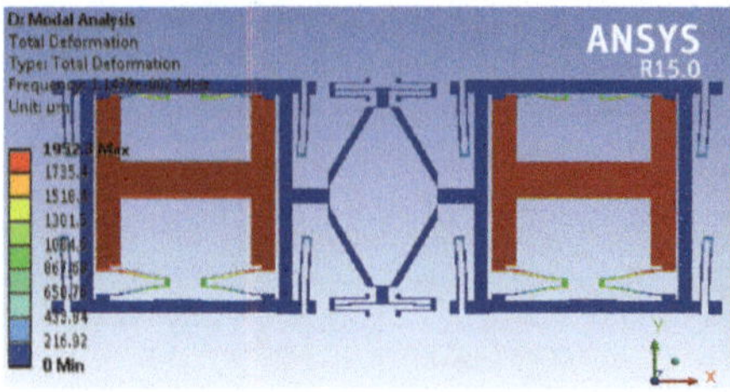

(a) The in-phase sensing mode

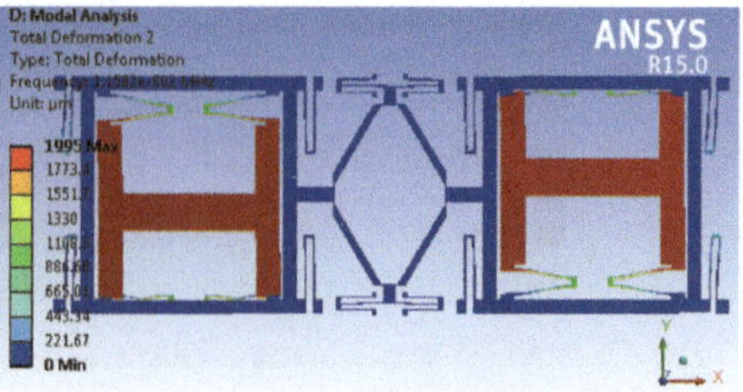

(b) The anti-phase sensing mode

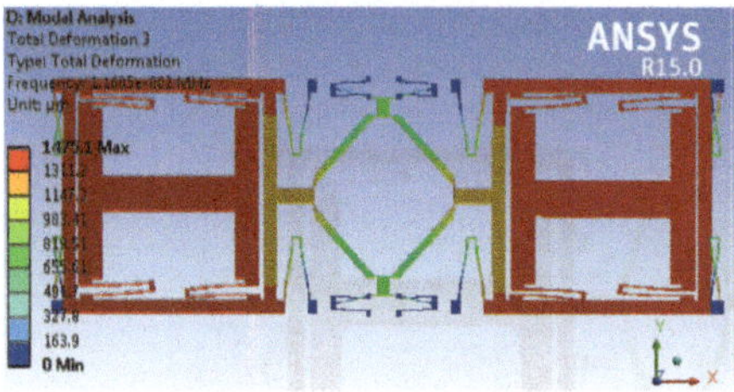

(c) The anti-phase driving mode

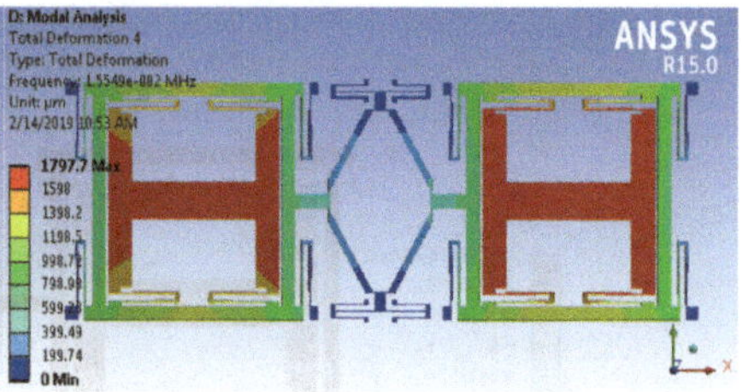

(d) One of the parasitic modes

Fig. 4. Some of the natural vibratory modes of the proposed TFG

The first four natural frequencies of this TFG are listed in Table 4.

Table 4. The first four natural frequencies of the proposed TFG

Vibration modes	Frequency (Hz)
In-phase sensing mode	11479
Anti-phase sensing mode	11582
Anti-phase driving mode	11605
Unexpected mode	15549

3.4 Harmonic Response

The harmonic response is carried out to re-determine the resonant frequency of the structure when applying the exciting force on it. The unit forces are applied to the two outer frames of the TFG in the driving direction (value of 1 μN). In order to reduce time of the calculation, the range of exciting frequency is chosen in 11.2÷12 kHz and the step of the frequency is 10 Hz. The harmonic response of the model is shown in Fig. 5.

The resonant frequency in the driving direction obtained from this analysis is 11610 Hz (Fig. 5) and agrees with the result in the modal analysis mentioned above. So, the exciting frequency of the external force applied in this TFG should be chosen following this result.

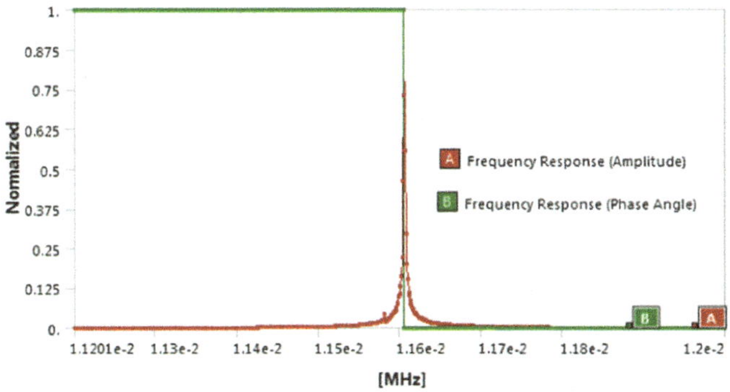

Fig. 5. Harmonic response of the proposed TFG

3.5 Transient Analysis

In this section, the dynamic response of whole TFG (include the vibration of the outer frame in driving direction and the vibration of proof-mass in sensing direction) is surveyed when applying the external forces in two sides of the TFG.

The external forces applied on this structure (F_L – in the left side and F_R – in the right side of the connecting diamond-shaped frame) are formed in harmonic function as follow:

$$F_L = F_0 \sin(2\pi ft);\ F_R = F_0 \sin(2\pi ft + \pi) \quad (3)$$

where the external frequency f is set up to be equal to the resonant frequency in the driving direction and was obtained in the previous section, F0 is the value of external force (in this case, F0 is set up to be 10 μN). The phase deviation of the two external forces is π radian to guarantee the anti-phase mode for the proposed TFG model.

The response of the structure in driving and sensing direction in case adding the angular velocity Ω (with value 25 rad/s) is shown in Fig. 6. The figures show that the motion of the two proof-masses in two tines of the connecting frame is appeared by the Coriolis effect. In other words, this motion reflects the angular velocity which the whole TFG rotates with. The relation between the input angular velocity and the maximum deformation of the proof-mass in the sense direction is shown in Fig. 7. This relation is nearly linear with R-squared 0.9368 (R-squared is defined to be a factor which shows the fit of the determined curve with the linearity) and the scale factor 0.001. The simulation result demonstrates that the proposed TFG model could be used to detect and measure the angular velocity of the rotated object which the device is attached on.

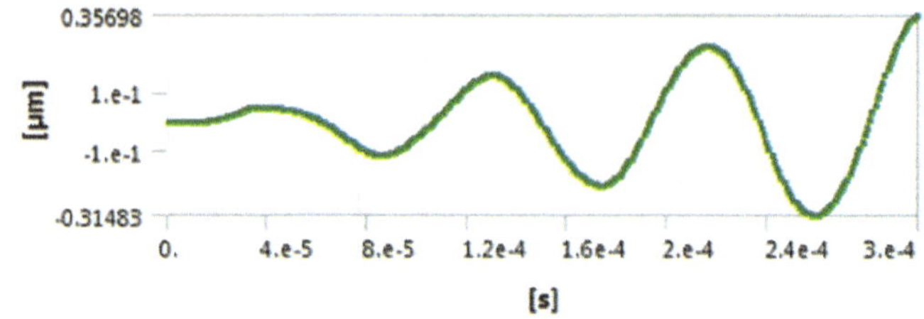

(a) Response of the outer frame in the driving direction

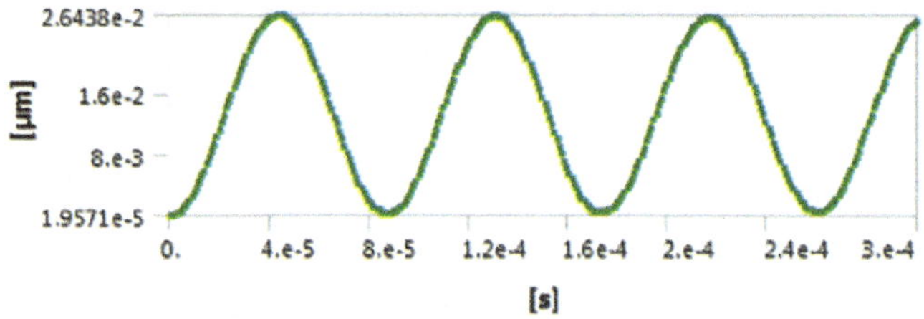

(b) Response of the proof-mass in the sensing direction

Fig. 6. Response of the TFG model with the angular velocity Ω = 25 rad/s

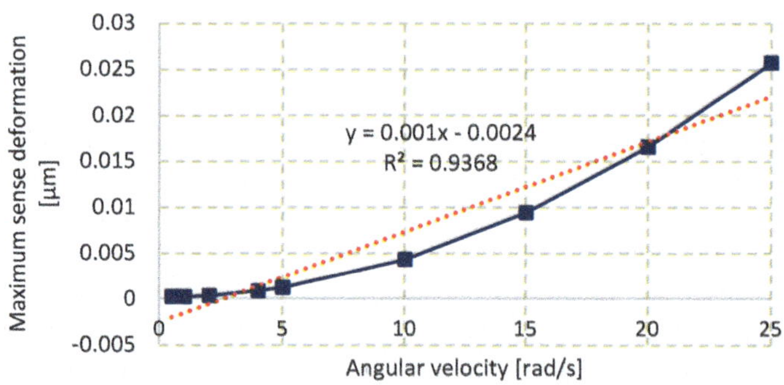

Fig. 7. Relation between input angular velocity and sense deformation of the proof-mass

4 Conclusion

This paper introduces the dynamic simulation of the MEMS vibratory gyroscope in tuning fork type with the connecting diamond-shaped frame. The geometrical parameter of the TFG is determined and established on the geometry module in ANSYS software. The FEM simulated analysis was carried out to find the equivalent stiffness in the motion directions of the proposed TFG. The resonant frequencies in driving and sensing direction are determined to be 11605 Hz and 11582 Hz, respectively. The anti-phase mode appears in both driving and sensing direction and the mismatch frequency is 23 Hz (less than 100 Hz). The different frequency between two major modes and the parasitic mode was obtained to be 34%. The relation between the input angular velocity and the maximum sense deformation is nearly linear with R-squared 0.9368 and the scale factor 0.001. The results using the FEM analysis mentioned above for the dynamic parameters of the TFG model are the basis to analyze the sensitivity of the proposed TFG in the future works.

References

1. Yazdi, N., Ayazi, F., Najafi, K.: Micromachined inertial sensors. Proc. IEEE **86**, 1640–1659 (1998)
2. Xia, D., Yu, C., Kong, L.: The development of micromachined gyroscope structure and circuitry technology. Sensors **14**, 1394–1473 (2014)
3. Shkel, A.M.: Type I and type II micromachined vibratory gyroscopes. In: Proceedings of the IEEE/ION PLANS, SanDiego, USA, pp. 586–592 (2006)
4. Acar, C., Shkel, A.M.: MEMS Vibratory Gyroscopes - Structural Approaches to Improve Robustness. Springer (2009)
5. Apostolyuk, V., Tay, F.E.: Dynamics of micromechanical coriolis vibratory gyroscopes. Sens. Lett. **2**, 1–8 (2005)
6. Choi, B., Lee, S.-Y., Kim, T., Baek, S.S.: Dynamic characteristics of vertically coupled structures and the design of a decoupled micro-gyroscope. Sensors **8**, 3706–3718 (2008)
7. Guan, Y.W., Gao, S.Q., Liu, H., Jin, L., Niu, S.: Design and vibration sensitivity analysis of a MEMS tuning fork gyroscope with an anchored diamond coupling mechanism. Sensors **16**, 468–483 (2016)
8. Guan, Y., Gao, S., Jin, L., Cao, L.: Design and vibration sensitivity of a MEMS tuning fork gyroscope with anchored coupling mechanism. Microsyst. Technol. **22**, 247–254 (2016)
9. Yoon, S.W., Lee, S., Najafi, K.: Vibration-induced errors in MEMS tuning fork gyroscopes. Sens. Actuators A: Phys. (2012). https://doi.org/10.1016/j.sna.2012.04.022
10. Guan, Y., Gao, S., Liu, H., Niu, S.: Acceleration sensitivity of tuning fork gyroscopes: theoretical model, simulation and experimental verification. Microsyst. Technol. **21**, 1313–1323 (2014)
11. Trinh, T.Q., Nguyen, L.Q., Dao, D.V., Chu, H.M., Vu, H.N.: Design and analysis of a z-axis tuning fork gyroscope with guided-mechanical coupling. Microsyst. Technol. **20**(2), 281–289 (2014)
12. Nguyen, M.N., Ha, N.S., Nguyen, L.Q., Chu, H.M., Vu, H.N.: Z-axis micromachined tuning fork gyroscope with low air damping. Micromachines. **8**(2), 42 (2017). https://doi.org/10.3390/mi8020042
13. Trusov, A.A., Schofield, A.R., Shkel, A.M.: Micromachined rate gyroscope architecture with ultra-high quality factor and improved mode ordering. Sens. Actuators A: Phys. **165**(1), 26–34 (2013)
14. Zaman, M., Sharma, A., Hao, Z., Ayazi, F.: A mode-matched silicon-yaw tuning-fork gyroscope with subdegree-per-hour allan deviation bias instability. J. Microelectromech. Syst. **17**(6), 1526–1536 (2008)
15. Huilin, S., Yongpeng, W.: Modeling and simulation for a vibratory tuning-fork MEMS gyroscope. In: Proceeding of Third International Conference on Measuring Technology and Mechatronics Automation, pp. 605–608 June 2011
16. The, V.V., Dung, T.Q., Trinh, C.D.: Dynamic analysis of a single MEMS vibratory gyroscope with decoupling connection between driving frame and sensing proof mass. Int. J. Appl. Eng. Res. **13**(17), 5554–5561 (2018)
17. The, V.V., Dung, T.Q., Hoan, V.M., Trinh, C.D.: Study of dynamic response of a diamond-shaped micro silicon frame. J. Sci. Technol. Mil. Tech. Acad. **186**, 34–42 (2017)
18. Allen, J.J.: Microelectromechanical System design. Taylor and Francis Group (2005)

Finite Element Analysis of the Lithium Diffusion in the Silicon Copper Nano-pillar

Huu-Tu Nguyen[1], Van-Trang Nguyen[2(✉)], and Minh-Quy Le[3]

[1] Faculty of Fundamental Science, Military Academy of Logistics, Hanoi, Vietnam

[2] Faculty of Mechanical Engineering, Thai Nguyen University of Technology, No. 666, 3/2 Street, Tich Luong Ward, Thai Nguyen, Vietnam
nvtrang@tnut.edu.vn

[3] Department of Mechanics of Materials and Structures, School of Mechanical Engineering, Hanoi University of Science and Technology, No. 1, Dai Co Viet Road, Hanoi, Vietnam

Abstract. When the Li+ current diffuses into the anode, its volume varies greatly, the volume of the anode can increase to 300%, resulting in a significant change in the mechanical properties of the anode. Currently, there are many structures proposed as anode for lithium-ion batteries with Si materials, for example: thin plate structure, nano-particles, nanofibers, copper nano-pillar…. We attend specially to the silicon copper nano-pillar because this structure can create a generation of batteries with fast charging capacity and large capacity. In this study, we use simulation method to find the relationship between mechanical properties (displacement, deformation, stress…) and Li+ diffusion currents during charging. That will help manufacturers to choose the optimal and reasonable structure when making batteries.

Keywords: Diffusion · Finite element analysis · Li-ion battery · Stress

1 Introduction

Studies show that there are many types of materials that can be used to make lithium battery anodes, for example: Carbon (Graphite, graphene, CNT), Silicon (Si) (Amorphous Si, Nanofibre Si, Si nano-particles) [1–4], aluminum (Al), tin (Sn), germanium (Ge) [2]…. In particular, carbon is currently the most commonly used material used as anode of lithium ion battery. However, the charge capacity of C is not high (LiC6) [5, 6], while Si is a material capable of storing a large amount of Li+ (Li×Si with x = 3.75). However, Si is currently limited in the application of battery anode because its volume varies greatly (300%) [4] when Si is filled with Li+. The large volume change makes the mechanical properties of the anode made of Si rapidly diminish after less charge and discharge cycles.

Some studies have shown different structures of anode made from Si: anode made from Si [7] nano-particles, anode is made of thin amorphous Si sheets [8, 9], the Si structure encapsulates the nanoneedle Cu-Array [10], Si copper nano-pillar [4]. The research of the authors before fully focused on studying the electrochemical ability and

K.-U. Sattler et al. (Eds.): ICERA 2019, LNNS 104, pp. 366–372, 2020.
https://doi.org/10.1007/978-3-030-37497-6_43

the ability to charge the anode. Some authors and their research groups have studied and mentioned the properties of anode made from Si such as Zhao et al. studying the mechanical behavior of anodes made from Si nano-particles, in this study, the authors studied the effect of plastic deformation field, stress, potential energy and charging rate [12]. The results of their study showed the effect of size and charging rate on the mechanical properties of Si spheres used as anodes when the small charge rate Li+ concentration is evenly distributed in the grain and low stress; or when the charging speed is high in the center of large particles. Chew et al. studied the effect of cracks and the destruction of Si thin sheets on copper substrates [8]. In their study, the authors simulated the process of charging electricity. They predicted and simulated, with the thickness of Si layer below 200 nm, the small Si membrane stress, and the crack did not occur and develop.

In this study, we use numerical simulation to determine the relationship of displacement, deformation, stress, plastic deformation with Li+ diffusion currents, and charging time.

2 Theoretical Background

When the diffused Li-ion ion flow into the anode is equal to Si, the volume of the anode increases or in other words, the Li-ion ion causes Si to be deformed. The total deformation of Si can be divided into 3 components as follows [7]: $d\varepsilon_{ij}^{tt} = d\varepsilon_{ij}^{l} + d\varepsilon_{ij}^{e} + d\varepsilon_{ij}^{p}$, here $d\varepsilon_{ij}^{l}$, $d\varepsilon_{ij}^{e}$, $d\varepsilon_{ij}^{p}$ respectively deformation components due to diffusion current, elastic deformation, plastic deformation. The elastic deformation component follows Hooke's law: $d\varepsilon_{ij}^{e} = d\left(\frac{1}{E}\left[(1+v)\sigma_{ij} - v\sigma_{kk}\delta_{ij}\right]\right)$, in which: E is elastic modulus, σ_{ij} is the stress component, v is Poisson's coefficient, δ_{ij} is the Kronecker index. The plastic deformation component is described as follows (J_2-flow rule): $d\varepsilon_{ij}^{p} = \lambda s_{ij}$, in which: $s_{ij} = \sigma_{ij} - \delta_{ij}\sigma_{kk}/3$ is the deviation stress component, and $\sigma_e = \sqrt{3s_{ij}s_{ij}/2}$ is the equivalent stress component. The deformed composition due to diffused Li-ion flow into Si is shown as follows [7]: $\varepsilon_{ij}^{l} = \delta_{ij}\ln\left[1+\Omega\overline{C}\right]/(3)$, in which: $\Omega = 2.7$ (correspondingly the volume of Si increases 270% when it is filled with Li-ion), call: $\overline{C} = C/C_{max}$ is the equivalent concentration of Li-ion ions contained in Si; $\overline{C} = 0$ when there is no Li-ion ion contained Si, $\overline{C} = 1$ when Si is full of Li-ion ions (Si_4Li_{15}). Call: $\alpha = \ln\left[1+\Omega\overline{C}\right]/\left(3\overline{C}\right)$, then we have $\varepsilon_{ij}^{l} = \delta_{ij}.\alpha.\overline{C}$ This is the thermal expansion equation if we consider $\overline{C}$ is the temperature, then α is the coefficient of expansion because of heat, so it depends on the equivalent concentration $\overline{C}$. On the other hand, the Li-ion diffusion equation enters the anode according to the first law and Frick's second law of diffusion:$j = -D\frac{\partial C}{\partial x}$ and $\partial C/\partial t = D\nabla^2 C$ [7, 8], in which: j is diffusion current, D is diffusion coefficient, t is time. According to the first law and Furier's second law of heat transfer: $q = -k\frac{\partial u}{\partial x}$ and $\frac{\partial u}{\partial t} = \frac{k}{c_p \rho}\frac{\partial^2 u}{\partial x^2}$, here: q is the heat flow, k is the heat transfer coefficient, u the calorimeter; c_p and ρ respectively the specific heat and density of the conductor. Comparing the diffusion equations of Fick with Furier's heat transfer equations, we found that the diffusion problem is equivalent to the heat transfer problem if we set $k = D$ and $c_p.\rho = 1$ then j corresponds to q and u respectively C. So to solve the Li+ diffusion problem in Si, we solve the heat transfer problem on the corresponding Si, in which we consider $T = \overline{C}$. The charging process corresponds to the heating process and vice versa the discharge process corresponds to the heat process. We use thermal and mechanical problems to simulate and study the mechanical properties of anodes [7, 8, 13] in the process of charging (which is the process of Li + ion diffusing into anode made of Si), in which $\overline{C}$ is considered temperature. Li-ion diffusion process into the structure is characterized by stable heat transfer process (trans/static), the simulation will stop when the surface of the structure reaches the maximum Li-ion concentration meaning that the surface of the structure achieves Si_4Li_{15}, corresponding to the surface temperature reaching the temperature of 1° $(\overline{C} = 1)$. Simulation results are shown through drawings and graphs.

3 Numerical Setup

Figure 2 shows representative element model used in simulation. Geometry parameters of the problem: D = 1000 nm, d = 500 nm, H1 = 1000 nm, H2 = 2000 nm, h = 150 nm.

The mechanical properties of Si are taken as follows [3, 7]:

$$E_{Si}(\overline{C}) = E_0 + \overline{C}(E_1 - E_0) \quad (1.a)$$

$$\nu_{Si}(\overline{C}) = \nu_0 + \overline{C}(\nu_1 - \nu_0) \quad (2.b)$$

$$\sigma^y_{Si}(\overline{C}) = \sigma^y_0 + \overline{C}(\sigma^y_1 - \sigma^y_0) \quad (3.c)$$

in which: $E_0 = 150\,\text{GPa}$, $\nu_0 = 0.21$, $\sigma^y_0 = 1,5\,\text{GPa}$ are respectively the Young's modulus, Poisson's ratio, and yield stress of the Si anode without Li-ion $(\overline{C} = 0)$; $E_1 = 50\,\text{GPa}$, $\nu_1 = 0.23$, $\sigma^y_1 = 0.6\,\text{GPa}$ are respectively the Young's modulus, Poisson's ratio, and yield stress of the Si anode when it is fully filled with Li+ $(\overline{C} = 1)$. Cohesive zone is used to model the imperfect interface between Si and Cu with a moderate interfacial fracture toughness $G_c = 20\ \text{J/m}^2$ [9].

Considering the Si structure encased in the copper nanotubes as shown in Fig. 1 [4], due to the uniformity of the structure, we selected a representative element consisting of a copper cylinder and a base section to conduct the simulation. The simulation results of mechanical properties of representative elements will characterize the whole structure.

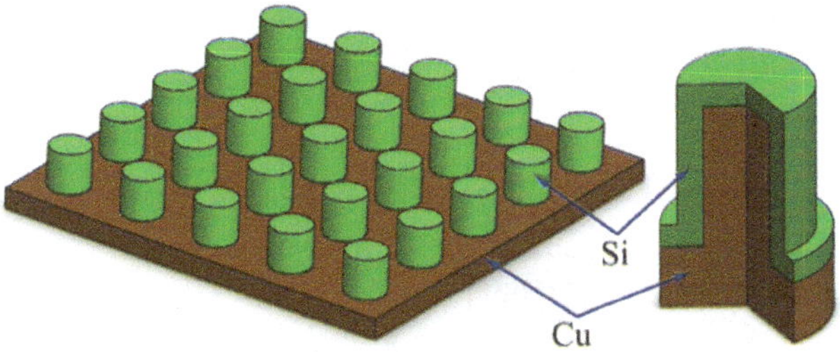

Fig. 1. Schematic illustration of silicon on nanopillar copper for anode in Li-ion batteries.

We adopt here the axisymmetric model for this representative volume as shown in Fig. 1. Hence, only one half of the model is used in FEA as shown in Fig. 2, where D = 1000 nm and h = 150 nm. Four-node bilinear axisymmetric elements are used. The copper substrate and silicon anode consist of 3222 and 11537 elements, respectively. Elasto-plastic model with bilinear strain hardening law is adopted here for copper. The mechanical properties of copper substrate are taken from: $E_{Cu} =$ 90 GPa; $\nu_{Cu} = 0.28$; $\sigma^{y}_{Cu} = 330$ MPa; $H_{Cu} = 5$ GPa [10]. The Si layer surface has a uniformly distributed heat flow, the boundary transition condition is shown in Fig. 2. The simulation will stop when the Si layer has reached the saturation concentration of Li+ ($\overline{C} = 1$) [7, 11]. Further details of numerical procedure and finite element model were presented in our previous work [13].

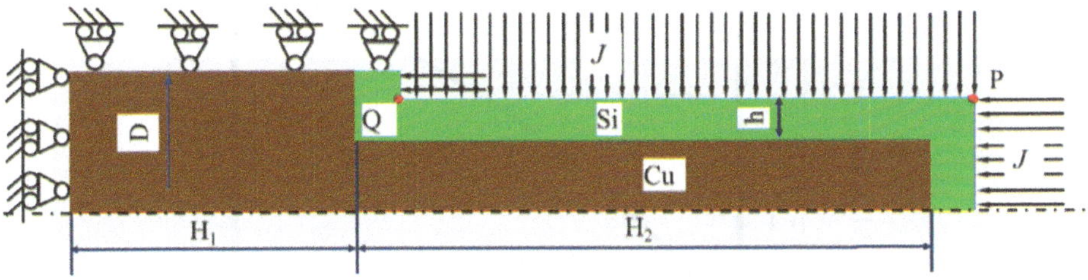

Fig. 2. A half of the representative volume used in simulation (D = 1000 nm, d = 500 nm, h = 150 nm, H_1 = 1000 nm, H_2 = 2000 nm).

4 Results and Discussion

With the value of D = 100 nm^2s^{-1}, GC = 20 J/m^2, J = 1 $nm^{-2}s^{-1}$ simulation results are shown on Figs. 3, 4, 5 and 6. Equivalent charging time is determined as τ = t/tmax. After the equivalent charging time τ = 1.08 at point Q (Fig. 2) on the structure of concentration Li+ has reached the maximum level ($\overline{C} = 1$, corresponding to Cmax = 3.75). On Fig. 3b, it shows that after charging time τ = 0.108 the entire Si layer has reached the Von Mises stress value at about 1500 MPa. In Fig. 3c, the largest plastic strain occurs at the point Q of the Si material layer. Through these figures, we see the largest concentration of Li+, the largest plastic deformation occurs at the top angle (point P) and the base angle (point Q) of the Si layer.

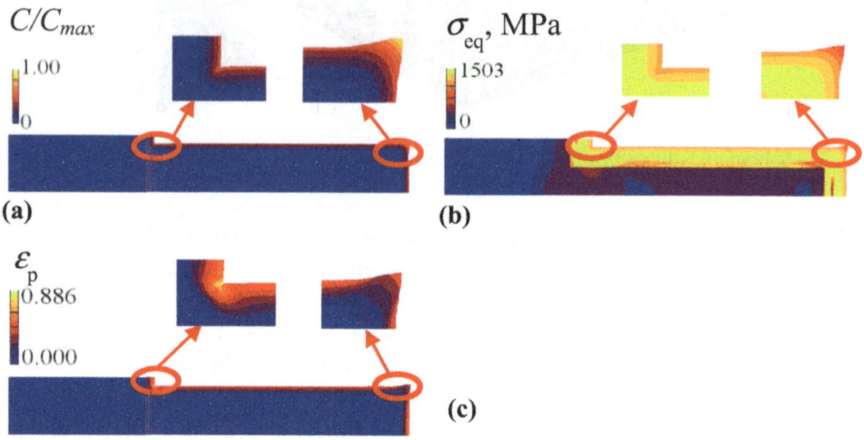

Fig. 3. Distribution of normalized concentration of Li-ion, equivalent Von Mises stress and the total equivalent plastic strain in the anode: $\tau = 0.108$, $D = 100\ \mathrm{nm^2 s^{-1}}$; $J = 1\ \mathrm{nm^{-2} s^{-1}}$, $G_c = 20\ \mathrm{J/m^2}$.

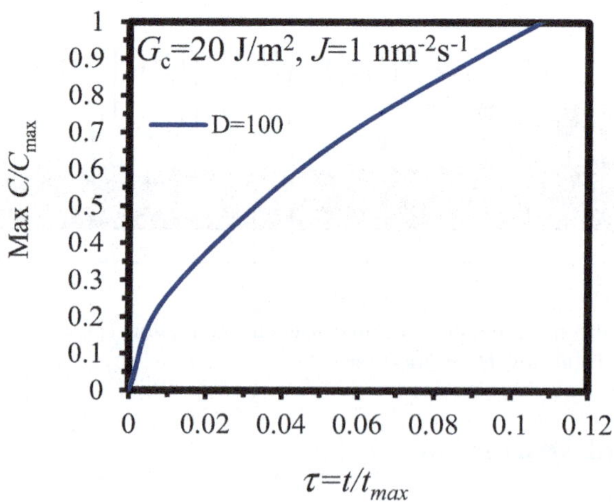

Fig. 4. Variation of the maximum concentration in the anode with $D = 100\ \mathrm{nm^2 s^{-1}}$ versus the normalized charging time.

Figures 4, 5 and 6 show the variation of the maximum concentration, maximum Von Mises stress and maximum plastic strain in the anode with $D = 100\ \mathrm{nm^2 s^{-1}}$ versus the normalized charging time.

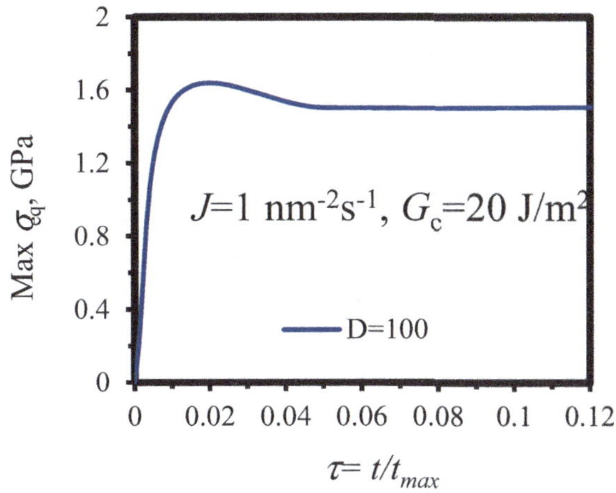

Fig. 5. Variation of the maximum Von Mises stress in the anode with D = 100 nm^2s^{-1} versus the normalized charging time.

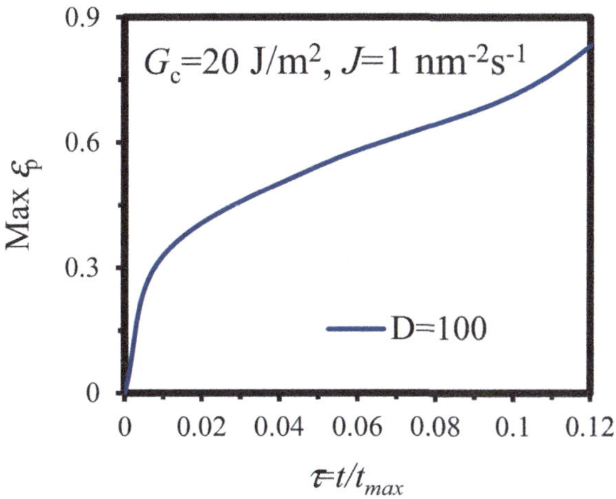

Fig. 6. Variation of the maximum plastic strain in the anode with D = 100 nm^2s^{-1} versus the normalized charging time.

5 Conclusions

In our research, we have simulated the mechanical process occurring at battery anode with Si structure encased in copper nanostructures.

The simulation results show the greatest stress, the largest plastic deformation, the largest Li+ ion concentration and the largest displacement in the structure depend on the charging time. We also determined the stress, plastic deformation, concentration of Li+ ions distributed on each material layer of the anode.

Acknowledgements. Van-Trang Nguyen was supported by the Thai Nguyen University of Technology, Thai Nguyen University (Vietnam) under grant number T2017-B01.

References

1. Obrovac, M., Christensen, L., Le, D.B., Dahn, J.: Alloy design for lithium-ion battery anodes. J. Electrochem. Soc. **154**(9), A849–A855 (2007)
2. Nitta, N., Wu, F., Lee, J.T., Yushin, G.: Li-ion battery materials: present and future. Mater. Today **18**(5), 252–264 (2015)
3. Obrovacz, M.N., Christensen, L.: Structural changes in silicon anodes during lithium insertion/extraction. Electrochem. Solid State Lett. **7**(5), A93–A96 (2004)
4. Kim, G., Jeong, S., Shin, J.H., Cho, J., Lee, H.: 3D amorphous silicon on nanopillar copper electro des as anode s for high-rate lithium-ion batteries. Ulsan National Institute of Science and Technology (UNIST), pp. 689–798
5. Noel, M., Suryanarayanan, V.: Role of carbon host lattices in Li-ion intercalation/de-intercalation processes. J. Power Sources **111**(2), 193–209 (2002)
6. Dahn, J.R., Zheng, T., Liu, Y., Xue, J.: Mechanisms for lithium insertion in carbonaceous materials. Science **270**(5236), 590 (1995)
7. Wang, H., Nadimpalli, S.P., Shenoy, V.B.: Inelastic shape changes of silicon particles and stress evolution at binder/particle interface in a composite electrode during lithiation/delithiation cycling. Extreme Mech. Lett. **9**, 430–438 (2016)
8. Chew, H.B., Hou, B., Wang, X., Xia, S.: Cracking mechanisms in lithiated silicon thin film electrodes. Int. J. Solids Struct. **51**, 4176–4187 (2014)
9. Pal, S., Damle, S.S., Kumta, P.N., Maiti, S.: Modeling of lithium segregation induced delamination of a - Si thin filmanode in Li-ion batteries. Comput. Mater. Sci. **79**, 877–887 (2013)
10. Wang, N., Hang, T., Zhang, W., Li, M.: Highly conductive Cu nanoneedle-array supported silicon film for high-performance lithium ion battery anodes. J. Electrochem. Soc. **163**(3), A380–A384 (2016)
11. Xu, R., Zhao, K.: Mechanical interactions regulated kinetics and morphology of composite electrodes in Li-onbatteries. Extreme Mech. Lett. **8**, 13–21 (2016)
12. Zhao, K., Pharr, M., Cai, S., Vlassak, J.J., Suo, Z.: Large plastic deformation in high-capacity lithium-ion batteries caused by charge and discharge. J. Am. Ceram. Soc. **94**(S1), S226–S235 (2011)
13. Nguyen-Huu, T., Le-Minh, Q.: Stress analysis of silicon-based anode in Li-ion battery. Lecture Notes in Mechanical Engineering (2017). https://doi.org/10.1007/978-981-10-7149-2_7

FPGA Implementation of Optimal PN_Sequences by Time_Multiplexing Technique

Nguyen Van Son[1], Le Chi Quynh[1], Tran Vu Kien[2], Dao Huy Du[3](✉), and Dang Khanh Hoa[4]

[1] Hanoi Open University, Hanoi, Vietnam
sonnv@hou.edu.vn, quynh.lechi@gmail.com
[2] Electric Power University, Hanoi, Vietnam
kientv@epu.edu.vn
[3] Thai Nguyen University, Thai Nguyen, Vietnam
daohuydu@tnut.edu.vn
[4] Hanoi University of Science and Technology, Hanoi, Vietnam
hoa.dangkhanh@hust.edu.vn

Abstract. In this paper, FPGA implementation of a LFSR and PSEUDO_-NOISE(PN) sequences are designed and carried out for many purposes. The motivation for that is multifold: LFSR and related PN sequences are indispensable building blocks not only for cipher sequence used in cryptography, steganography but also for BIST (build in self test system), scrambler. There have been a lot of attempts to develop specific circuit and sequences for these applications. However, the interleaving approach renders the optimal products in term of sequences length, statistical properties (distributions, autocorrelation function(ACF) and linear complexity (LC). It will be shown that the Interleaving structure can be easily expressed in time-multiplexing technique (namely Delay operation-D-transform). The equivalence between the D-transform and other well-known methods will be also highlighted in this contribution. Furthermore, the interleaving structure with outstanding LC will be given.

Keywords: Optimal PN-sequences · FPGA · Linear complexity · Trace function · D-transform

1 Introduction

Until now, representation of LFSR and PN sequences have been discussed in several studies [1–5]. The following representation methods are widely used in sequence design:

(1) The combinatorial approach, there is a one-to-one correspondence between cyclic difference sets and almost balanced binary sequences with the autocorrelation property [3, 5]. Therefore, constructing all cyclic difference sets is equivalent to finding all almost-balanced binary sequences with the desired autocorrelation property.

K.-U. Sattler et al. (Eds.): ICERA 2019, LNNS 104, pp. 373–380, 2020.
https://doi.org/10.1007/978-3-030-37497-6_44

(2) Algebraic structure method, trace function representation and analysis. The concept of trace function is widely used in representing the sequences of length $2^n - 1$. Let $n = l \times m > l$ for some positive integers l and m. The binary interleaved GMW sequence with these parameters and can be specified as

$$b_i = Tr_1^m\left(\left[Tr_m^n(\alpha^i)\right]^r\right), i = 0, 1, 2, \ldots \tag{1}$$

Where is a primitive element of $GF(2^n)$, and r is any integer relatively prime to $2^m - 1, 1 < r < 2^m - 1$.

The D-transform [1–4] of a sequence $\{b_n\}$ over $GF(p)$ (Galois filed (mod p), a primitive integer) is denoted by $D[b_n]$

$$D[b_n] = \sum_{i=0}^{m} b_i D^i, b_i \in \{GF(p)\} \tag{2}$$

Thus, the D-transform of the sequence will have the form of a polynomial in D over *GF(p)* and has been conveniently used in signal and system analysis in data transmission and CDMA. In fact, D-transform is nothing but time multiplexing, since it is exactly based on delay (D) operation. Below is a typical example for demonstration.

The D-transform of the generator sequence $\{b_n\}$ of a linear feedback shift register (LFSR) is then given b by:

$$b(D) = \frac{S(D)}{g(D)} \tag{3}$$

Where $g(D)$ of degree m is the generating polynomial of a LFSR and S(D) of degree $\leq$ m − 1 specifies the initial condition corresponding to a particular shifted version of $\{b_n\}$.

2 Methods

Equivalence between nonlinear operations in trace and D-transform. Despite the fact that Trace function and D-transform are two most popular tools in analyzing interleaved structure, there has not been any equivalence between them shown so far. In this section we will make this issue clear. It is well known [1–5] that: the interleaved structure is specified by: subsequences $\{a_n\}$ and shift sequence (interleaving order I_p^T). The subsequences $\{a_n\}$ can be determined by decomposition (decimation) of the sequence of composite length L into T subsequences. The shift sequence (or interleaving order I_p^T) determines the way the subsequences are interleaved.

Theorem 1: Let $n = l \times m > l$ for some positive integers l and m. The elements of the interleaved binary GMW or m-like sequence $\{b_n\}$ with these parameters and can be specified as:

$$b_i = tr_1^m\left(\left(tr_m^n(\alpha^i)\right)^r\right), i = 0, 1, 2, \ldots \tag{4}$$

Where α is a primitive element of $GF(2^n)$, and r is any integer relatively prime to $2^m - 1, 1 < r < 2^m - 1$.

In case r = 1 (no nonlinear operation), $\{b_n\}$ reduces to an interleaved m-sequence [3] and its elements can be expressed in trace form as:

$$b_i = \left\{tr_1^m\left(tr_m^n(\alpha^t)\right)\right\} = \left\{tr_1^m\left(tr_1^m\left(\alpha^{iT+j}\right)\right)\right\} = \left\{tr_1^m\left(\alpha^{iT} tr_1^m\left(\alpha^j\right)\right)\right\} = \left\{tr_1^m\left(\beta^{i+S_i}\right)\right\} \tag{5}$$

Where, $t = iT + j; i = 0, 1, 2, \ldots, 2^m - 1; j = 0, 1, \ldots, T - 1; \beta = \alpha^T; tr_m^n(\alpha^i) = \beta^{s_i}$; β is a primitive root over $GF(2^m)$.

Let the elements $a_i = tr_1^m\left(\beta^i\right), i = 0, 1, 2, \ldots, 2^m - 1$.

Since $\beta = \alpha^T$ is a primitive root over $GF(2^m)$, $\{a_n\}$ represents the m-sequence of length $N = 2^m - 1$ and called subsequence.

Then m-sequence $\{b_n\}$ is constructed by interleaving the cyclic shifts of $\{a_n\}$ in the manner defined by $S = S_0, S_1, \ldots, S_{T-1}$, the shift sequence (interleaving order I_p^T).

In fact, replacing $\{a_n\}$ by another m-subsequence $\{e_n\}$ over $GF(2)$ as stated in Theorem 1 above, is equivalent to proper decimation of $\{a_n\}$ by r. In trace form we have the elements

$$e_i = \left\{tr_1^m\left(\beta^{ri}\right)\right\}, i = 0, 1, 2, \ldots, 2^m - 1 \tag{6}$$

Where r is an integer with $\gcd(r, 2^m - 1) = 1$. Similarly, $\{e_n\}$ is interleaved in the same manner defined by $S = S_0, S_1, \ldots, S_{T-1}$ to form $\{C_n\}$, whose elements can be expressed as:

$$\begin{aligned} C_i &= \left\{tr_1^m\left(\beta^{r(i+S_i)}\right)\right\} = \left\{tr_1^m\left(\beta^i tr_k^n\left(\alpha^j\right)\right)^r\right\} \\ &= \left\{tr_1^m\left(tr_k^n\left(\alpha^{iT+j}\right)\right)^r\right\} \\ &= \left\{tr_1^m\left(tr_k^n\left(\alpha^t\right)\right)^r\right\} \end{aligned} \tag{7}$$

So, from an m-sequence $\{b_n\}$ we can construct a nonlinear m-like (GMW) sequence $\{C_n\}$ with higher LC (by subsequence interchanging) while maintaining the best ACF (by keeping the shift sequence unchanged).We call therefore call them as optimal PN-sequences. The two operations in D-transform and trace representation are thus equivalent. To the best of our knowledge, this fact is proved for the first in the literature. Therefore, it gains a lot of attentions when hardware implementation is considered.

Example. Let m = 3, n = 6 and let α be a primitive element of $GF(2^6)$ with primitive polynomial $b(D) = D^6 + D^5 + 1$ over $GF(2)$. Let $\{b_i\}$ denote the m-sequence generated by b(D): $\{b_i\}$ = {0 1 1 1 1 1 1 0 1 .0 1 0 1 1 0 0 1 1 .0 1 1 1 0 1 1 0 1 .0 0 1 0 0 1 1 1 0 .0 0 1 0 1 1 1 1 0 .0 1 0 1 0 0 0 1 1 .0 0 0 0 1 0 0 0 0}.

Using both methods (Trace function and D-transform) we will get the same interleaving order:

$I_p^T = \{\infty, 5, 3, 5, 6, 3, 3, 2, 5\}$, where ∞ represents Null sequence.

Nonlinear Operation Type I (subsequence interchanging) with the maximal LC value. In this paragraph we will show that not all the subsequence manipulations are equal in term of LC improvement. Our argument is again based on D-transform and DFT (discrete Fourier transform) [7].

When $G(D)$ is primitive, the LFSR sequence is an m-sequence of length $L = 2^n - 1$. Then, we have the field $GF(L = 2^n - 1)$ and there exist α, the primitive root of the field. Under this assumption, we can calculate the DFT of $\{b_n\}$ as follows:

Let $b^L = \{b_0, b_1, \ldots, b_{L-1}\}$ (it is called time domain sequence).

The DFT of b^L is called frequency domain sequence $B^L = \{B_0, B_1, \ldots, B_{L-1}\}$, where the components of B^L is given by: $B_j = \sum_{i-0}^{L-1} b_i \alpha^{ij}$ with α the primitive root in $GF(2^n - 1)$ and $j = 0, 1, \ldots, L - 1$.

The LC of $\{b_n\}$ is then calculated as follows: [7]

$$LC\{b_n\} = \mathrm{w}(BL) \tag{8}$$

With $w(.)$ denotes the Hamming weight of BL (the number of nonzero elements of BL). On the other side, the relation between the LC of spectral sequence (DFT) and the Hamming weight of the time domain sequence can also be given: $LC\{b_n\} = \mathrm{w}(b_n)$.

Since the m-like (GMW) sequences are special cases of periodic sequences, where the period $L = 2^n - 1 = T.N$, $N = 2^m - 1$, their linear complexity can be surely calculated by DFT method and is equal to the number of nonzero terms in its spectral sequence (DFT) and can be determined as [7]:

$$LC = m.l^{wr} \tag{9}$$

With $l = n/m$, $\mathrm{w}(r)$ is the Hamming weight of r(decimation).

Now, let's pay special attention to the second mapping: proper decimation by r (nonlinear operation) to prove the maximal value of m-like (GMW) sequences, which is not given in the literature. An algorithm to find the interleaved structure with largest LC: For given set $\{n, m, 1\}$, the maximal value LC_{max} of m-like sequence is reached when the subsequence $\{a_n\}$ is replaced by its reciprocal sequence.

Proof: From the second mapping it is clear that the decimation r creates the nonlinearity. From (9) we can find out the maximal value of $w(r)$. Since $\gcd(r, 2^m - 1) = 1$, $1 \le r \le 2^m - 1$, the maximal value of $r = 2^m - 2$. It is obvious that $w(2^m - 1) = m$, therefore $w(2^m - 2)_{max} = m - 1$. *e.g.* n = 10, m = 5, l = 2, $2^5-1 = 31$; w(31) = 5; $\mathrm{w}(2^5-2) = \mathrm{w}(30) = 4 \Rightarrow LC_{max} = 5.2^4 = 80$. Let see what sequence we get when decimate the subsequence $\{a_n\}$ by $r = 2^m - 2$. Let $\{e_n\}$ denotes the subsequences resulted from that decimation, then:

$$\{e_n\} = \{a_{r.n}\} \tag{10}$$

In other word, decimation is in fact sequence sampling [1, 6]: that is every r-th element of $\{a_n\}$ is selected and thus $\{e_n\}$ consist of every r-th element of $\{a_n\}$ mod N. So, we have the following relation:

For $\{a_n\} = a_0, a_1, \ldots, a_{N-3}, a_{N-2}, a_{N-1}$. Put $r = 2^m - 2 = N - 1$ we have: $\{e_n\} = a_{N-1}, a_{N-2}, a_{N-3}, \ldots, a_1, a_0$. This sequence runs in reversed order of $\{a_n\}$. That means $\{e_n\}$ is a reciprocal of $\{a_n\}$, so its characteristic polynomial is reciprocal polynomial of that of $\{a_n\}$. That completes the proof. LC of optimal PN–sequences (Table 1 and Figs 1, 2, 3, 4, 5 and 6).

Table 1. n = 16, l = 2, m = 8, $LC_{max} = 8.2^7 = 1024$.

Order	GF(2^{16})	GF(2^8)					
		101110001	100011101	110101001	100101011	101101001	100101101
1	1011010000000001	1024	16	256	64	64	256
2	1000000000101101	16	1024	64	256	256	64
3	1001110000000001	64	256	1024	16	256	64
4	1000000000111001	256	64	16	1024	64	256
5	1111100000000001	64	256	128	128	1024	16
6	1000000000111111	256	64	128	128	16	1024
7	1100101000000001	1024	16	256	64	64	256

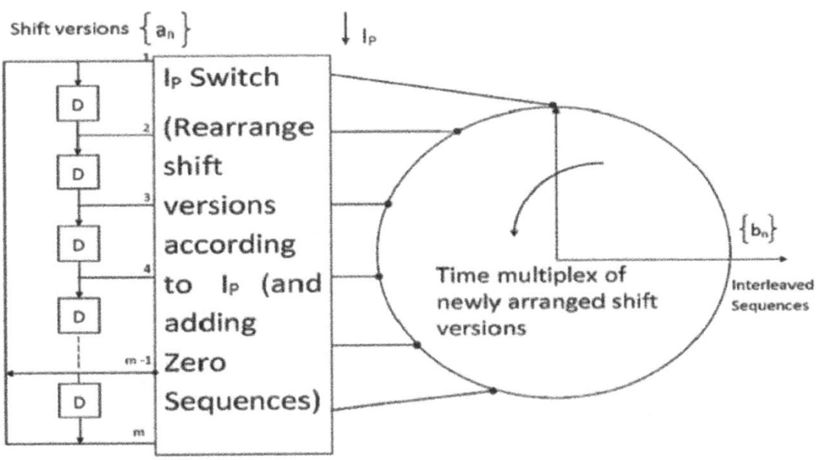

Fig. 1. Implementation of interleaved structure (for PN –generator, BUILD IN SELF TEST SYSTEM(BIST),SCRAMBLER..)

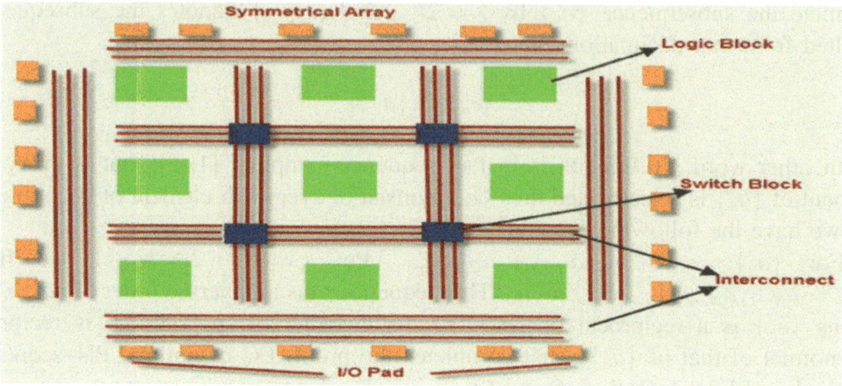

Fig. 2. General structure of the FPGA

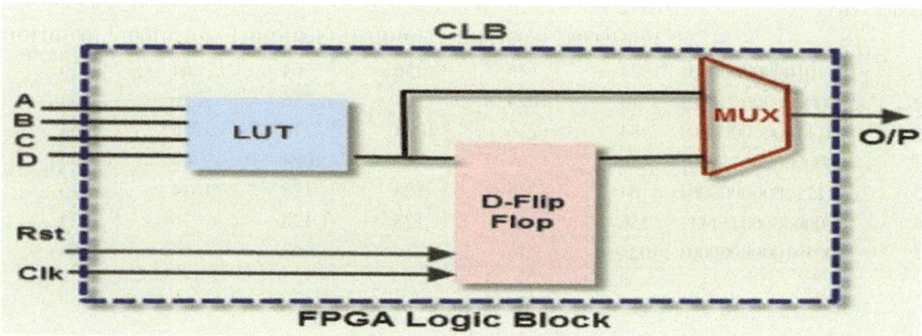

Fig. 3. FPGA logic block

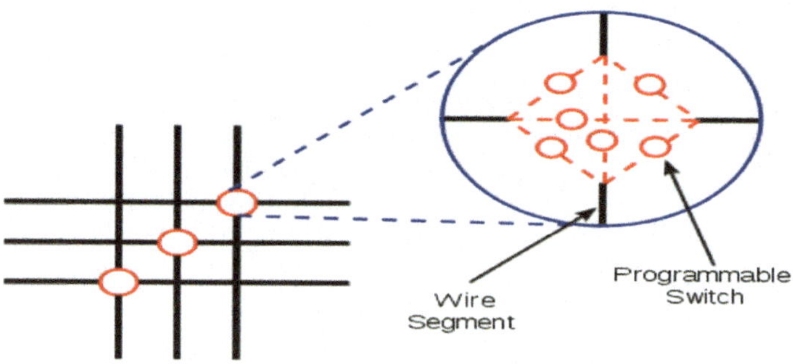

Fig. 4. Connecting block

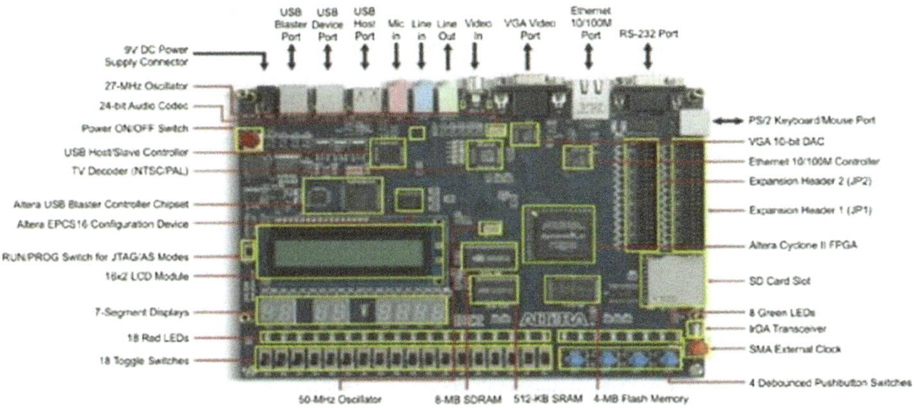

Fig. 5. FPGA PCB DE2

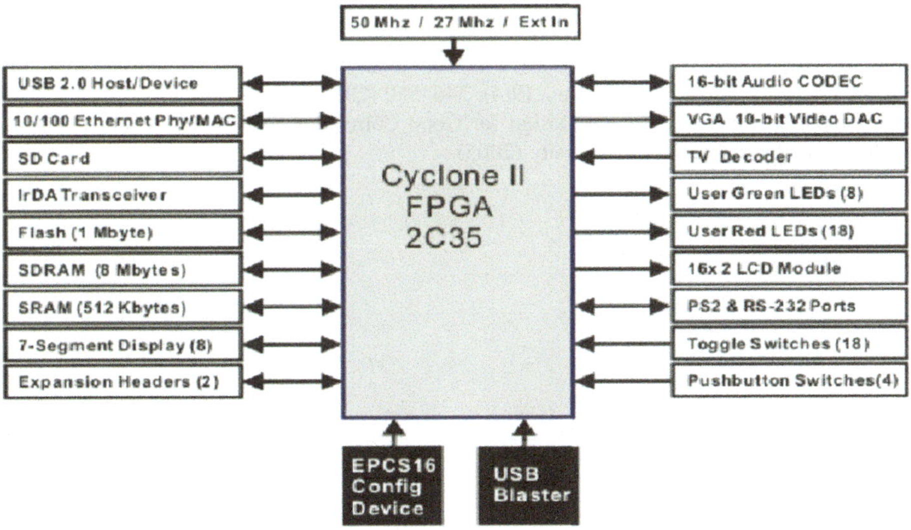

Fig. 6. Block diagram of the DE2 board

3 Conclusion

LFSR and PN-sequences have been widely used in many areas. The interleaved structure can give the rise to sequences of very large length, good statistic properties, and high LC (optimal).

These performances explained why optimal PN is widely used in compressive sensing, watermark, steganography and key stream for mobile communications. In the future we hope to discuss the Ellipic curve based PN for network security, q-ary PN sequences. NOTE-Software simulations is useful for checking the algorithm, but for practical use FPGA is the only right choice for long PN sequences! For the length of 32.767 bits, the matlab simulation software takes about 25 min on ICORE 5 PC but the FPGA version can complete it in matter of few nanoseconds! Due to limited scope of the paper, we have not been able to discuss the statistical aspect of the signal. It is needed to discuss in more detail in the future. The authors express their gratitude to the HOU for the grant.

References

1. Chi, Q.L., et al.: A hardware oriented method to generate and evaluate nonlinear interleaved sequences with desired properties. J. Inf. Eng. Appl. **6**(7), 1–12 (2016)
2. Hieu, L.M., Quynh, L.C.: Design and analysis of sequences with interleaved structure by D-Transform. IETE J. Res. **51**, 61–67 (2005)
3. He, J.: Interleaved sequences over Finite Field PhD thesis Carleton University Ottawa, Ontario (2013)
4. Stroud, C.E.: A Designer's Guide to Built-In Self-Test. Kluwer Academic Publisher (2002)
5. Zhang (Marvell), J.: EPOC scrambler EEE 802.3bn EPoC TF Meeting November, pp. 1–14 (2013)
6. Meng, R., Yan, T.: New constructions of binary interleaved sequences with low autocorrelation. Int. J. Netw. Secur. **19**(4), 546–550 (2017)
7. Solomon, S., Gong, G.: Signal Design for Good Correlation for Wireless Communication, Cryptography. Cambridge University (2005)

Handling Forecasting Problems Based on Two-Factor High-Order Fuzzy Time Series and Particle Swarm Optimization

Nghiem Van Tinh(✉) and Nguyen Tien Duy

University of Technology, Thai Nguyen, Vietnam
nghiemvantinh@tnut.edu.vn

Abstract. In real time, an observation series can depend on and be affected by many different observations. Therefore, to forecast more accurately, it is necessary to consider adding multi factors that have potential correlations in the model. This study proposes a new forecasting model based on two - factor high - order fuzzy time series (FTS) combined with particle swarm optimization (PSO) for forecasting the daily average temperature and forecasting car road accidents. Different from some previous models, in the proposed model, time - variant high-order fuzzy logical relationship groups are built to be used in the forecasting process. While, PSO is still used to find the optimal interval in the universe of discourse of each factor. Two data sets of weather temperatures and annual road traffic accidents with corresponding factors "Temperature", "Cloud density" and "Killed", "Mortally wounded", respectively are selected to demonstrate the effectiveness of the proposed model and compare it with existing models. The experimental results show that the proposed model has better forecasting accuracy than previous forecasting models based on two-factor high-order FTS.

Keywords: Fuzzy time series · Two-factor high - order FTS · Fuzzy relationship groups · PSO · Temperature

1 Introduction

Based on the fuzzy set theory [1], Song and Chissom [2, 3] proposed the FTS model for forecasting the enrollments of the University of Alabama. It has some downsides such as large amount of computation time when the fuzzy rule matrix is large and lack of persuasiveness in determining universe of discourse and the length of intervals. Chen [4] proposed an efficient first - order FTS model which consists of simple arithmetic calculations only. After that, this model became a popular FTS structure. It has been widely studied for improving accuracy of forecasting in many applications such as: university enrolments [5–16], rice production [17], dry bulk shipping index [18], forecasting pollution [19], daily average temperature [20–22], stock index [21–24], forecasting car road accidents [25–28], etc. Specifically, Lee et al., [20] presented a two-factor high-order FTS for temperature prediction and the Taiwan futures exchange (TAIFEX) forecasting. It is noted that the forecasting accuracy rates of this model mainly depend on their universe of discourse and the length of intervals. Lee, Wang,

K.-U. Sattler et al. (Eds.): ICERA 2019, LNNS 104, pp. 381–393, 2020
https://doi.org/10.1007/978-3-030-37497-6_45

and Chen [21] also developed a model for forecasting temperature and TAIFEX based on high-order FLR and genetic simulated annealing algorithm. By introducing genetic algorithm for partitioning intervals in the universe of discourse, Chen and Chung introduced two first-order [7] and high - order [8] forecasting models for forecasting the enrolments of University of Alabama. On the other hand, particle swarm optimization (PSO) technique was developed by Kennedy and Eberhart [29]. The comparative results show the PSO method was generally found to perform better than other algorithms in terms of success rate and solution quality. Also, some comparative research works for the real problems presented that the PSO-based results have better performance than those based on GA [30]. At present, the application of PSO in selecting the proper lengths of intervals in FTS forecasting model has been attracting many researchers' attention. They demonstrate that suitable selection of intervals by using PSO also increases the performance of forecasting model, as can be seen in works [9, 11, 22–24]. Motivated by these previous research works, we propose a hybrid model based on two-factor high-order FTS and PSO for increasing the forecasting accuracy. To improve the proposed model, two datasets about the daily average temperature [20] and the road traffic accident [25] are selected to examine for forecasting process. The rest of the paper is organized as follows:

In Sect. 2, the basic concepts of FTS and algorithms are briefly introduced. Section 3 presents a hybrid forecasting model which combines the two-factor high - order FTS and PSO algorithm. Section 4 makes a comparison of forecasting results of the proposed model with the existing models from two real life datasets. Conclusion and future work are discussed in Sect. 5.

2 Some Basic Concepts of FTS and Algorithms

In this section, we briefly review the basic concepts of FTS and algorithms.

2.1 Some Basic Definitions

The idea of FTS was first introduced and defined by Song and Chissom [2, 3]. Let $U = \{u_1, u_2, \ldots, u_n\}$ be an universe of discourse; a fuzzy set A of U can be defined as $A = \{\mu_A(u_1)/u_1 +, \mu_A(u_2)/u_2 \ldots + \mu_A(u_n)/u_n\}$, where μ_A: U → [0, 1] is the membership function of A, $\mu_A(u_i)$ indicates the degree of membership of ui in the fuzzy set A, $f_A(u_i)\varepsilon[0, 1]$, and $1 \le i \le n$. The basic definitions of FTS are as below:

Definition 1: Fuzzy time series [2, 3]

Let Y(t) $(t = .., 0, 1, 2..)$, a subset of R, be the universe of discourse on which fuzzy sets $f_i(t)$ $(i = 1, 2 \ldots)$ are defined and if F(t) is a collection of $f_1(t), f_2(t), \ldots$, then F(t) is called a FTS definition on Y(t) (t …, 0, 1, 2 …).

Definition 2: The one-factor m^{th}-order FTS

Let F(t) is a FTS. If F(t) is caused by $F(t-1), F(t-2)$, …, $F(t-m)$, then, the logical relationship $F(t-1), F(t-2), \ldots, F(t-m) \rightarrow F(t)$, then it is called one-factor m^{th}-order FTS

Definition 3: The two-factor m^{th}-order FTS [20]

Let $F_A(t)$ and $F_B(t)$ be two FTS. If $F_A(t)$ is caused by $(F_A(t-1), F_B(t-1))$, $(F_A(t-2), F_B(t-2)), \ldots, (F_A(t-m), F_B(t-m))$, then the fuzzy logical relationship is represented by $(F_A(t-m), F_B(t-m)), \ldots, (F_A(t-2), F_B(t-2))$, $(F_A(t-1), F_B(t-1)) \twoheadrightarrow F_A(t)$. This is called two - factor mth - order FTS model. Where, $F_A(t)$ and $F_B(t)$ are called main - factor and second - factor of FTS, respectively. $(F_A(t-m), F_B(t-m)), \ldots, (F_A(t-2), F_B(t-2))$, $(F_A(t-1), F_B(t-1))$ and $F_A(t)$ are called the left-hand side and the right-hand side of the fuzzy relationship, respectively.

Definition 4: The one – factor first – order fuzzy relationship groups [4]

The fuzzy logical relationships having the same left-hand side can be further grouped into a FRG. Assume there are exists FLRs as follows: $A_i \rightarrow A_{k1}, A_i \rightarrow A_{k2}, \ldots$, $A_i \rightarrow A_{km}$; these FLRs can be put into the same FRG as: $A_i \rightarrow A_{k1}, A_{k2}, \ldots, A_{km}$.

Definition 5: The two – factor mth – order fuzzy relationship groups [20]

Assume there are two-factor mth-order FLRs having the same left-hand side as follows:

$$(A_{im}(t-m), B_{km}(t-m)), \ldots, (A_{i1}(t-1), B_{k1}(t-1)) \twoheadrightarrow A_{j1}(t1).$$

....

$$(A_{im}(t-m), B_{km}(t-m)), \ldots, (A_{i1}(t-1), B_{k1}(t-1)) \twoheadrightarrow A_{jp}(tp).$$

These FLRs can be grouped into a two – factor mth - order FLR group as follows:

$$(A_{im}(t-m), B_{km}(t-m)), \ldots, (A_{i1}(t-1), B_{k1}(t-1)) \twoheadrightarrow A_{j1}(t1), \ldots, A_{jp}(tp).$$

2.2 Particle Swarm Optimization Algorithm (PSO)

PSO was first introduced by Eberhart and Kannedy [29]. It is a tool for dealing with optimization problems, where a set of potential solutions is represented by a swarm of particles and each particle is moved through the search space to search the optimal solution. When moving, all particles (i.e., N particles) have fitness values which are evaluated by fitness functions. The position of the best particle among all particles found so far is kept and each particle keeps its personal best position which has passed previously. At each time of moving, each jth ($1 \leq j \leq n$) element in the velocity vector $V_{id} = [v_{id,1}, v_{id,2}, \ldots, v_{id,n}]$ and each element $x_{id,j}$ in the position vector $X_{id} = [x_{id,1}, x_{id,2}, \ldots, x_{id,n}]$ of particle id are updated by the best position $P_{bestid} = [p_{id,1}, p_{id,2}, \ldots, p_{id,n}]$ encountered by the particle so far and the best position $G_{best} = \min(P^t_{best\_id})$ found by the whole population of particles according to formulas of velocity and position as follows:

$$V_{id}^{k+1} = \omega^k * V_{id}^k + C_1 * Rand() * (P_{best\_id} - x_{id}^k) + C_2 * Rand() * (G_{best} - x_{id}^k) \quad (1)$$

$$x_{id}^{k+1} = x_{id}^k + x_{id}^{k+1} \quad (2)$$

$$\omega^k = \omega_{max} - \frac{k * (\omega_{max} - \omega_{min})}{iter\_max} \quad (3)$$

The P_{best_id} for jth particle is presented as $P_{best\_id} = [p_{id,1}, p_{id,2}, \ldots, p_{id,n}]$ and calculated as:

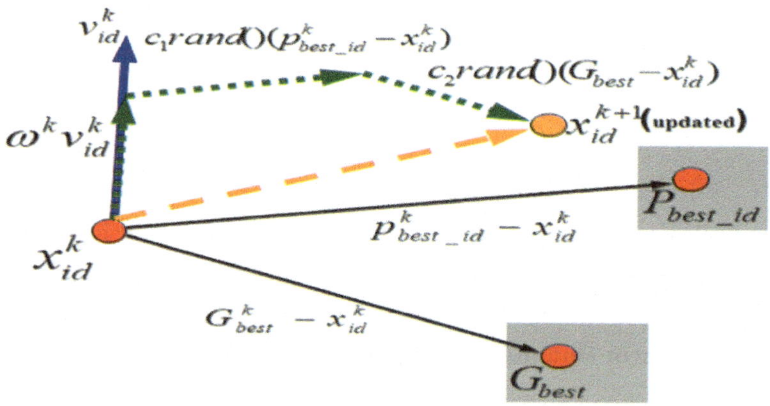

Fig. 1. Description of velocity and position updates in PSO

$$P_{best\_id}^{k+1}f(x) = \begin{cases} P_{best_{id}}^{k+1}, & if\ fitness(x_{id}^{k+1}) > P_{best_{id}}^k \\ fitness(x_{id}^{k+1}), & if\ fitness(x_{id}^{k+1}) \le P_{best_{id}}^k \end{cases} \quad (4)$$

To be clearly visualized, Fig. 1 depicts the trends of searching point of each particle by updating according to the personal best position of the particle and best positions of all particles within the swarm.

Where, X_{id}^t is the current position of a particle id at time step t; V_{id}^t is the velocity of the particle id at time step t, and is limited to $[-V_{max}, V_{max}]$, where V_{max} is a constant pre-defined by user;

- ω is the inertia weight_ENREF_3; *iter_max* is the maximum number of iterations;
- c_1 and c_2 are two learning factors which control the influence of the social and cognitive components, respectively.

3 A Hybrid Forecasting Model Combing Two – Factor High – Order FTS with PSO

In this section, a hybrid forecasting model based on PSO and two-factor high-order FTS is presented. The proposed model uses two FTSs $\boldsymbol{F_A(t)}$ and $\boldsymbol{F_B(t)}$. Where $\boldsymbol{F_A(t)}$ and $\boldsymbol{F_B(t)}$ are called the main-factor and the second-factor, respectively.

Step 1: Define the universe of discourse $(\mathbf{U_A}, \mathbf{V_B})$ and divide them into intervals

Step 2.1. Define $\mathbf{U_A}$ of the main - factor $\boldsymbol{F_A(t)}$ and $\boldsymbol{V_B}$ of the second - factor $\boldsymbol{F_B(t)}$ Let $U_A = [D_{\min A} - N_{1A}, D_{\max A} + N_{2A}]$ is universe of discourse of the main-factor and $\mathrm{V_B} = [\mathrm{D_{minB}} - \mathrm{N_{1B}}, \mathrm{D_{maxB}} + \mathrm{N_{2B}}]$ is universe of discourse of the second-factor. To define the universe of discourse, at first find the minimum value $\mathrm{D_{minA}}$, the maximum value $\mathrm{D_{maxA}}$ and the minimum value $\mathrm{D_{minB}}$, the maximum value $\mathrm{D_{maxB}}$ of the historical time series data set, respectively. In order to ensure the forecasting values bounded in the universe of discourse $\mathrm{U_A}$ and $\mathrm{U_B}$, four positive integers $\mathrm{N_{1A}}, \mathrm{N_{2A}}$ and $\mathrm{N_{1B}}, \mathrm{N_{2B}}$ are properly selected, respectively.

Initialize the PSO parameters with the predefined values and generate initial random velocity and position of all particles. The position of each particle in the PSO represents intervals in each universe of discourse $\mathrm{U_A}$ and $\mathrm{V_B}$, respectively. Particularly, each particle is an array consisting of n + m − 2 elements, where, n − 1 elements (i.e., $\{\mathrm{a_{i1}}, \mathrm{a_{i2}}, \ldots, \mathrm{a_{ik}}, \ldots, \mathrm{a_{i(n-1)}}$; with $\mathrm{a_{ik}} \geq \mathrm{a_{i(k-1)}}$, $1 \leq k \leq n-1$) represent main - factor $F_A(t)$ and m − 1 elements (i.e., $\{\mathrm{b_{i1}}, b_{i2}, \ldots, \mathrm{b_{ij}}, \ldots, \mathrm{b_{i(m-1)}}$; with $\mathrm{b_{ij}} \geq \mathrm{b_{i(j-1)}}$,

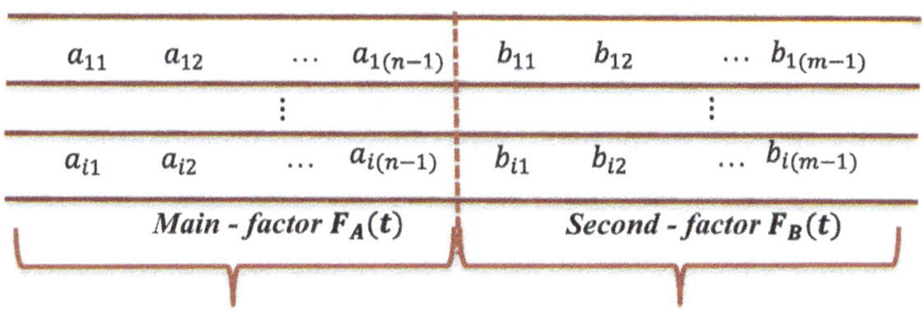

Fig. 2. The encoding format of the particles in the PSO

$1 \leq j \leq m-1$) represent the second - factor $F_B(t)$. Based on these n + m − 2 elements, define the n + m intervals as $u_1 = (D_{minA} - \mathrm{N_{1A}}, a_{i1}]$, $u_2 = (a_{i1}, a_{i2}], \ldots, u_n = (a_{i(n-1)}, D_{maxA} + \mathrm{N_{2A}}]$ and $v_1 = (\mathrm{D_{minB}} - \mathrm{N_{1B}}, b_{i1}]$, $v_2 = (b_{i1}, b_{i2}], \ldots, v_n = (b_{i(m-1)}, \mathrm{D_{maxB}} + \mathrm{N_{2B}}]$

Step 2.2: Encode the values from the position of all particles into the intervals in the PSO. Here, the two factors are encoded in a single particle as shown in Fig. 2.

Step 3: Define the linguistic terms of the two-factor FTS.

Each interval obtained in Step 2 represents a linguistic term. For n intervals of main-factor, we get n linguistic term which represented by fuzzy sets A_i and its definitions are described in (5) and (6) as follows:

$$\mathrm{A_i} = \frac{a_{i1}}{u_1} + \frac{a_{i2}}{u_2} + \ldots + \frac{a_{ij}}{u_j} + \ldots + \frac{a_{in}}{u_n} \tag{5}$$

$$a_{ij} = \left\{ \begin{array}{lc} 1 & \mathrm{j == i} \\ 0.5 \quad \text{if} & \mathrm{j == i-1 \text{ or } j == i+1} \\ 0 & \text{otherwise} \end{array} \right\} \tag{6}$$

where $a_{ij} \in [0, 1]$, $1 \leq i \leq n$, $1 \leq j \leq n$ and u_j is the jth interval of U_A. The value of a_{ij} indicates the grade of membership of u_j in the fuzzy set A_i. For simplicity, the different membership values of fuzzy set A_i are selected by according to Eq. (6).

Similarly, with m intervals of second-factor, we get m linguistic terms which represented by fuzzy sets B_i and its definitions is described in (7) and (8) as follows:

$$B_{\mathrm{i}} = \frac{b_{i1}}{v_1} + \frac{b_{i2}}{v_2} + \ldots + \frac{b_{ij}}{v_j} + \ldots + \frac{b_{i7}}{v_7} \tag{7}$$

$$b_{ij} = \left\{ \begin{array}{lc} 1 & j == i \\ 0.5 \quad if & j == i-1 \; or \; j == i+1 \\ 0 & otherwise \end{array} \right\} \tag{8}$$

where $b_{ij} \in [0, 1]$, $1 \leq i \leq m$, $1 \leq j \leq m$ and v_j is the jth interval of V_B. The value of b_{ij} indicates the grade of membership of v_j in the fuzzy set B_i.

Step 4: Fuzzify all historical data of the two - factor FTS into fuzzy sets
In order to find out an equivalent fuzzy set using Eqs. (5) and (7) in all historical data. The way to fuzzify the historical data is to find an interval to which it belongs and assign the corresponding linguistic value to it with the highest membership degree.

Step 5: Identify all two - factor high - order fuzzy logical relationships
After two FTSs $F_A(t)$ and $F_B(t)$ have been fuzzified, we can build all fuzzy relationships with different orders. The way to create all two-factor m-order fuzzy relationship is to find any relationship consisting of the type $(F_A(t-m), F_B(t-m))$, …, $(F_A(t-2), F_B(t-2))$, $(F_A(t-1), F_B(t-1))$ $\rightarrow$ $F_A(t)$, where $(F_A(t-m)$, $F_B(t-m))$, …, $(F_A(t-2), F_B(t-2))$, $(F_A(t-1), F_B(t-1))$ and $F_A(t)$ are called the current state and the next state, respectively. Then, these m - order fuzzy relationships can be replaced by the corresponding fuzzy sets.

Step 6: Establish and calculate the forecasting value for two - factor high - order FRGs

Based on Definition 5, all the fuzzy relationships having the same fuzzy sets on the left-hand side or the same current state can be put together into one fuzzy relationship group. For example: Suppose there are three – order fuzzy relationships such that: (A4, B3), (A5, B2), (A7, B2) → A8, (A4, B3), (A5, B2), (A7, B2) → A7; …. These fuzzy logical relationships can be grouped into a fuzzy relationship group (FRG) as: (A4, B3), (A5, B2), (A7, B2) → A8, A7, …

In the following, we calculate the forecasting value for all groups in the training phase and testing phase according to principles as follows:

Principle 1. Suppose there is a two - factor mth-order fuzzy logical relationship group whose type shown as follows:

$$(A_{im}, B_{im}), (A_{im-1}, B_{im-1}), \ldots, (A_{i1}, B_{i1}) \rightarrow A_{j1}, A_{j2}, \ldots, A_{jp}.$$

where the maximum membership level of A_{j1}, $A_{j2}, \ldots,$ and A_{jp} occur at intervals U_{j1}, $U_{j2}, \ldots,$ and U_{jp}, respectively. We divide each corresponding interval of fuzzy sets on the right-hand side of fuzzy relationship groups into q sub-intervals with equal length, and calculate forecasting value for each group according to Eq. (9).

$$\text{Forecasted} = \frac{1}{n}\sum\nolimits_{k=1}^{n} sub\_mid_{kj} \tag{9}$$

where, n is the total number of next states or the total number of fuzzy sets on the right-hand side within the same group;

- q is the total number of sub-intervals corresponding to kth fuzzy set on the right-hand side within the same group;
- sub_mid_{kj} is the midpoint of one of q sub-intervals corresponding to kth fuzzy set on the right-hand side where the highest level of A_{kj} occurs at this sub-interval, with $(1 \leq k \leq n; 1 \leq j \leq q)$.

Principle 2. Suppose there is a two - factor mth-order fuzzy relationship groups which has type shown as follows: (A_{im}, B_{im}), $(A_{i(m-1)}, B_{i(m-1)}), \ldots, (A_{i1}, B_{i1}) \rightarrow \#$;

where the symbol "#" denotes an unknown value, then the forecasted value of day i is calculated as follows:

$$\text{Forecasted\_for\#} = m_{i1} + \frac{\sum\nolimits_{k=2}^{m} m_{i(k-1)} - m_{ik}}{2^{k-1}} \tag{10}$$

where, $m_{i1}, m_{i2}, \ldots, m_{ik}$ is midpoints of U_{i1}, U_{i2}, …, and U_{ik} $(2 \leq k \leq m)$, respectively.

- For measurement of forecasting accuracy, the mean square error (MSE) and the mean absolute percentage error (MAPE) are employed as an evaluation criterion, defined as

$$\text{RMSE} = \sqrt{\frac{1}{n}\sum\nolimits_{i=m}^{n} (F_i - R_i)^2} \tag{11}$$

$$\text{MAPE} = \frac{1}{n}\sum\nolimits_{i=m}^{n} \left|\frac{F_i - R_i}{R_i}\right| * 100\% \tag{12}$$

Where, R_i and F_i note the actual and forecasted value of day i, respectively, n is the total number of days to be forecasted, m is the order of fuzzy logical relationship.
Step 7: Calculate the P_{best_id} based on objective values (MSE) which is calculated according to Eq. (11) for each particle and update it according to Eq. (4). And then set the G_{best} of all particles, which choose the best P_{best_id} of all particles.

Step 8: For each particle, calculate particle velocity according to Eq. (1), then update particle position according to Eq. (2).
Step 9: Update the value of the weight factor w according to Eq. (3).
Step 10: Stop if the stop criterion of the PSO, the maximum number of iteration, is satisfied. Otherwise, go to Step 2.

4 Experimental Results and Discussions

In this research, two illustrative datasets including the historical data of the daily average temperatures [20] from June 1, 1996 to September 30, 1996 in Taipei, Taiwan and the road traffic accidents dataset in Belgium [25] were selected to compare the proposed model with some other forecasting models. In the first experiment, the "temperature" was chosen as the main forecasting objective and "cloud density" as the second factor. In the second case, the researchers performed the proposed model for forecasting the road traffic accidents between 1974 and 2004, in which the "killed" was called the main factor and the "Mortally wounded" was called the second factor. The forecasting model was simulated 20 times on each dataset in each order of fuzzy relationship. Then the best result of all runs was recorded to be the final forecast result. The essential parameters of the proposed model were set for each dataset which are shown in Table 1 as follows:

Table 1. The parameters used for PSO based forecasting model

The parameters in PSO	Values of average temperature		Values of road traffic accidents	
	Average factor	*Cloud density*	*Killed factor*	Mortally wounded
The number of particles be	30		30	
Tthe max number of iterations be	150		150	
The value of inertial weight	$\omega_{max} = 0.9$ to $\omega_{min} = 0.4$		$\omega_{max} = 0.9$ to $\omega_{min} = 0.4$	
The coefficient C1 = C2 =	2		2	
The position in search range: **X** =	[23, 32]	[0, 100]	[950, 1700]	[90, 900]
The velocity in search range: **V** =	[–5, 5]	[–50, 50]	[–100, 100]	[–50, 50]

4.1 Forecasting the Daily Average Temperature

In this section, the proposed model was applied to forecast the daily average temperature [20] from June 1, 1996 to September 30, 1996 in Taipei, Taiwan. The forecasting effectiveness of the proposed model was compared with the five FTS models which are presented in papers [6, 20–22, 31]. A comparison of the forecasting accuracy of MAPE (12) between the proposed model and its counterparts under different order fuzzy relationships is listed in Tables 2, 3, 4 and 5. Like the compared models, the proposed model was set with the number of intervals for the main - factor and the second - factor

Table 2. A comparison of the average temperature forecasting accuracy of the proposed method with its counterparts in June 1996

Models		The order of the forecasting model						
		2nd	3rd	4th	5th	6th	7th	8th
Model [31]		0.28%	0.29%	0.30%	0.29%	0.29%	0.28%	0.29%
Model [20]		0.8%	0.76%	0.79%	0.76%	0.79%	0.79%	0.81%
Model [21]	$\alpha = 0.25$	0.44%	0.42%	0.42%	0.42%	0.44%	0.40%	0.40%
	$\alpha = 0.5$	0.50%	0.45%	0.42%	0.38%	0.43%	0.39%	0.46%
	$\alpha = 0.9$	0.46%	0.42%	0.44%	0.42%	0.41%	0.46%	0.39%
Model [22]		0.36%	0.34%	0.32%	0.31%	0.31%	0.28%	0.29%
Model [6]		0.25%	0.26%	0.27%	0.27%	0.27%	0.28%	0.29%
Proposed model		0.08%	0.069%	0.06%	0.071%	0.071%	0.055%	0.078%

Table 3. A comparison of the average temperature forecasting accuracy of the proposed method with its counterparts in July 1996

Models		The order of the forecasting model						
		2nd	3rd	4th	5th	6th	7th	8th
Model [31]		0.34%	0.35%	0.34%	0.34%	0.35%	0.33%	0.32%
Model [20]		0.96%	0.96%	0.98%	0.97%	1.00%	0.98%	0.99%
Model [21]	$\alpha = 0.25$	0.44%	0.42%	0.42%	0.44%	0.40%	0.40%	0.40%
	$\alpha = 0.5$	0.50%	0.42%	0.38%	0.43%	0.39%	0.46%	0.46%
	$\alpha = 0.9$	0.46%	0.44%	0.42%	0.41%	0.46%	0.39%	0.39%
Model [22]		0.34%	0.33%	0.33%	0.32%	0.32%	0.34%	0.33%
Model [6]		0.28%	0.29%	0.28%	0.29%	0.30%	0.29%	0.30%
Proposed model		0.081%	0.068%	0.059%	0.06%	0.055%	0.051%	0.062%

Table 4. A comparison of the average temperature forecasting accuracy of the proposed method with its counterparts in August 1996

Models		The order of the forecasting model						
		2nd	3rd	4th	5th	6th	7th	8th
Model [31]		0.23%	0.22%	0.22%	0.22%	0.23%	0.23%	0.22%
Model [20]		1.07%	1.06%	1.08%	1.08%	1.09%	1.07%	1.07%
Model [21]	$\alpha = 0.25$	0.44%	0.42%	0.42%	0.44%	0.40%	0.40%	0.40%
	$\alpha = 0.5$	0.50%	0.42%	0.38%	0.43%	0.39%	0.46%	0.46%
	$\alpha = 0.9$	0.46%	0.44%	0.42%	0.41%	0.46%	0.39%	0.39%
Model [22]		0.31%	0.34%	0.33%	0.33%	0.33%	0.34%	0.35%
Model [6]		0.35%	0.35%	0.35%	0.36%	0.37%	0.35%	0.36%
Proposed model		0.068%	0.067%	0.067%	0.058%	0.056%	0.056%	0.067%

Table 5. A comparison of the average temperature forecasting accuracy of the proposed method with its counterparts in September 1996

Models		The order of the forecasting model						
		2nd	3rd	4th	5th	6th	7th	8th
Model [31]		0.51%	0.49%	0.51%	0.51%	0.53%	0.50%	0.51%
Model [20]		1.01%	0.90%	0.94%	0.96%	0.95%	0.95%	0.95%
Model [21]	$\alpha = 0.25$	0.44%	0.42%	0.42%	0.44%	0.40%	0.40%	0.40%
	$\alpha = 0.5$	0.50%	0.42%	0.38%	0.43%	0.39%	0.46%	0.46%
	$\alpha = 0.9$	0.46%	0.44%	0.42%	0.41%	0.46%	0.39%	0.39%
Model [22]		0.44%	0.42%	0.42%	0.44%	0.40%	0.40%	0.40%
Model [6]		0.50%	0.42%	0.38%	0.43%	0.39%	0.46%	0.46%
Proposed model		0.46%	0.44%	0.42%	0.41%	0.46%	0.39%	0.39%

by 9 intervals and 7 intervals, respectively. From Tables 2, 3, 4 and 5, it can be seen that the proposed method gets smaller forecasting accuracy than the methods compared in the same Table. In particularly, the proposed model gives the smallest forecasting accuracy based on the 7th - order fuzzy logical relationship with the percentage error on each month, from June to September of 0.055%, 0.051%, 0.056%, 0.095%, respectively. That is, the proposed method outperforms the existing models under various high-order FLRs in dealing with the temperature prediction from June 1996 to September 1996 in Taipei, Taiwan.

Table 6. A comparison of the forecasting results between proposed model and various models based on two - factor high - order FTS.

Year	Actual data	Model [27]	Model [26]	Model [28]	Model [25]	Model [32]	Proposed model
1974	1574	—	—	—	—	—	—
1975	1460	1497	—	1506	1458	—	—
1976	1536	1497	—	1453	1467	—	—
1977	1597	1497	1497	1598	1606	1594	1596
1978	1644	1497	1497	1584	1592	1643	1643
——	——	——	——	——	——	——	
2003	1035	995	997	970	1097	1036	1033
2004	953	995	997	970	929	954	953
RMSE		**83.12**	**46.78**	**41.61**	**37.66**	**19.2**	**2.99**

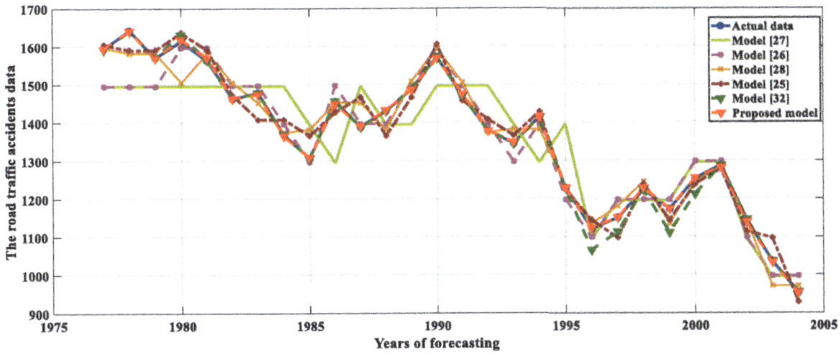

Fig. 3. The curves of the actual data and the proposed model with its counterparts

4.2 Forecasting the Road Traffic Accidents

In addition, the proposed model was also applied to forecast the road traffic accidents in Belgium [25] from 1974 to 2004 and a comparison of the forecasting results with the previous works [25–28, 32] was made. A comparison of the forecasting results using RMSE (11) is shown in Table 6. It is evident that the proposed method gets better forecasting results than the forecasting models above. More detailed comparison, for the same number of intervals for main – factor and second - factor of 13, the proposed model obtains the smallest RMSE value of 2.99 among two models [26, 32] using the 3rd - order FTS. Besides, the proposed model also has far smaller RMSE value than the models [25, 27, 28] based on first - order FTS with different number of intervals.

To be easily visualized, Fig. 3 depicts the trend of actual data compared to the trend of forecasted value between the proposed model and its counterparts. From Fig. 3, it can be seen that the curve of proposed model is closest to the actual data among five compared models. To sum up, demonstrations above show that the proposed model outperform the existing models based on two - factor high - order FTS model with different number of intervals in forecasting the road traffic accidents.

5 Conclusion

In this paper, a novel hybrid forecasting model based on two – factor high - order FTS combined with PSO is presented to forecast temperatures and road traffic accidents. The proposed method constructs two-factor high-order fuzzy logical relationship groups based on the historical data and utilizes proposed the defuzzification rule for calculating forecasting output value. In addition, PSO is also used to find the optimal intervals in the universe of discourse of each factor in this study. The experimental results show that, in many cases, the proposed method gets better forecasting performance than the existing ones. The detail of comparison was presented in Tables 2, 3, 4, 5, 6 and Fig. 3 for forecasting the daily average temperature and the road traffic accident. Even though this study shows the superior forecasting capability compared with some of previous forecasting models, we still expect that in future studies, we can focus on optimization techniques for finding the optimal order of fuzzy relationship and may use different artificial intelligence techniques in the fuzzification stage to deal with various forecasting problems.

References

1. Zadeh, L.A.: Fuzzy sets. Inf. Syst. **8**, 338–353 (1965)
2. Song, Q., Chissom, B.S.: Forecasting Enrollments with fuzzy time series – Part I. Fuzzy Sets Syst. **54**, 1–9 (1993)
3. Song, Q., Chissom, B.S.: Fuzzy time series and its models. Fuzzy Sets Syst. **54**(3), 269–277 (1993)
4. Chen, S.M.: Forecasting enrollments based on fuzzy time series. Fuzzy Sets Syst. **81**(3), 311–319 (1996)
5. Chen, S.-M., Phuong, B.D.H.: Fuzzy time series forecasting based on optimal partitions of intervals and optimal weighting vectors. Knowl. Based Syst. **118**, 204–216 (2017)
6. Chen, S.M., Tanuwijaya, K.: Fuzzy forecasting based on high-order fuzzy logical relationships and automatic clustering techniques. Expert Syst. Appl. **38**, 15425–15437 (2011)
7. Chen, S.-M., Chung, N.-Y.: Forecasting enrollments of students by using fuzzy time series and genetic algorithms. Int. J. Inf. Manag. Sci. **17**, 1–17 (2006)
8. Chen, S.-M., Chung, N.-Y.: Forecasting enrollments using high-order fuzzy time series and genetic algorithms. Int. Intell. Syst. **21**, 485–501 (2006)
9. Dieu, N.C., Tinh, N.V.: Fuzzy time series forecasting based on time-depending fuzzy relationship groups and particle swarm optimization. In: Proceedings of the 9th National Conference on Fundamental and Applied Information Technology Research (FAIR 2009). (Can Tho, 2016)
10. Huang, Y.L., et al.: A hybrid forecasting model for enrollments based on aggregated fuzzy time series and particle swarm optimization. Expert Syst. Appl. **38**, 8014–8023 (2011)
11. Kuo, I.H., et al.: An improved method for forecasting enrollments based on fuzzy time series and particle swarm optimization. Expert Syst. Appl. **36**, 6108–6117 (2009)
12. Tinh, N.V., Dieu, N.C.: Improving the forecasted accuracy of model based on fuzzy time series and k-means clustering. J. Sci. Technol. Issue Inf. Commun. Technol. **2**, 51–60 (2017)
13. Tung, H., et al.: The partitioning method based on hedge algebras for fuzzy time series forecasting. J. Sci. Technol. **54**(5), 571–583 (2016)
14. Wang, L., et al.: Determination of temporal information granules to improve forecasting in fuzzy time series. Expert Syst. Appl. **41**, 3134–3142 (2014)
15. Wei, L., et al.: Using interval information granules to improve forecasting in fuzzy time series. Int. J. Approx. Reason. **57**, 1–18 (2015)
16. Zhang, Z., Zhu, Q.: Fuzzy time series forecasting based on k-means clustering. Open J. Appl. Sci. **2**, 100–103 (2012)
17. Singh, S.R.: A simple method of forecasting based on fuzzy time series. Appl. Math. Comput. **186**, 330–339 (2007)
18. Duru, O.: A fuzzy integrated logical forecasting model for dry bulk shipping index forecasting: an improved fuzzy time series approach. Expert Syst. Appl. **37**, 5372–5380 (2010)
19. Domanska, D., Wojtylak, M.: Application of fuzzy time series models for forecasting pollution concentrations. Expert Syst. Appl. **39**, 673–7679 (2012)
20. Lee, L.W., et al.: Handling forecasting problems based on two-factors high-order fuzzy time series. IEEE Trans. Fuzzy Syst. **14**, 468–477 (2006)

21. Lee, L.-W., et al.: Temperature prediction and TAIFEX forecasting based on hight order fuzzy logical ralationship and genetic simulated annealing technique. Expert Syst. Appl. **34**, 328–336 (2008)
22. Hsu, L.-Y., et al.: Temperature prediction and TAIFEX forecasting based on fuzzy relationships and MTPSO techniques. Expert Syst. Appl. **37**, 2756–2770 (2010)
23. Kuo, I.-H., et al.: Forecasting TAIFEX based on fuzzy time series and particle swarm optimization. Expert Syst. Appl. **23**(7), 1494–1502 (2010)
24. Park, J.I., et al.: TAIFEX and KOSPI 200 forecasting based on two-factors high-order fuzzy time series and particle swarm optimization. Expert Syst. Appl. **37**, 959–967 (2010)
25. Bas, E., et al.: A modified genetic algorithm for forecasting fuzzy time series. Appl. Intell. 453–463 (2014). https://doi.org/10.1007/s10489-014-0529-x
26. Jilani, T.A., Burney, S.M.A.: Multivariate stochastic fuzzy forecasting models. Expert Syst. Appl. **35**(3), 691–700 (2008)
27. Jilani, T.A., et al.: Multivariate high - order fuzzy time series forecasting for car road accident. World Acad. Sci. Eng. Technol. **25**, 288–293 (2007)
28. Uslu, V.R., et al.: A fuzzy time series approach based on weights determined by the number of recurrences of fuzzy relations. Swarm Evol. Comput. (2013). https://doi.org/10.1016/j.swevo.2013.10.004
29. Kennedy, J., Eberhart, R.C.: Particle swarm optimization. Proceedings of IEEE International Conference on Neural Networks, vol. 4, pp. 1942–1948 (1995)
30. Panda, S., Padhy, N.P.: Comparison of particle swarm optimization and genetic algorithm for FACTS-based controller design. Appl. Soft Comput. ASOC **417**, 1–10 (2007)
31. Wang, N.-Y., Chen, S.-M.: Temperature prediction and TAIFEX forecasting based on automatic clustering techniques and two-factors high-order fuzzy time series. Expert Syst. Appl. **36**, 2143–2154 (2009)
32. Yusuf, S.M., et al.: A novel hybrid fuzzy time series approach with applications to enrollments and car road accident. Int. J. Comput. Appl. **192**(2), 37–44 (2015)

Heat Transfer Model and Critical Driving Frequency of Electrothermal V-Shaped Actuators

Dzung Tien Nguyen[1,2], Kien Trung Hoang[1], and Phuc Hong Pham[1,2](✉)

[1] Hanoi University of Science and Technology, 1-Daicoviet, Hanoi, Vietnam
phuc.phamhong@hust.edu.vn
[2] Thai Nguyen University of Technology, 666, 3-2 Street, Tichluong, Thainguyen, Vietnam

Abstract. This paper presents a finite difference heat transfer model on micro beam of electrothermal V-shaped micro actuator and a calculation method of conversion parameters in the dynamic equation. The relation between maximum displacement and a driving frequency is investigated in order to determine critical frequency ensuring the working stability of the actuator. The theoretical model is validated by finite element method (FEM) in ANSYS. The comparison of displacement results at various driving voltages between computation and simulation confirmed the high accuracy of this model. While applying the model for specific V-shaped actuator with selected dimensions, stability of the actuator can be obtained with the range of driving frequency is lower than 100 Hz. While the displacement amplitude decreases (with acceptable deviation of 5%), the critical frequencies change from 171 Hz to 207 Hz, and to be linear proportion to the square wave driving voltage with ranging from 10 V to 30 V, respectively.

Keywords: Micro electrothermal V-shaped actuator · Heat transfer · Dynamic parameters · Critical frequency

1 Introduction

During the past years, micro electromechanical system (MEMS) devices are applied in many fields thanks to their advantages such as small size, low energy consumption, easy integration and high accuracy, etc. A large number of MEMS devices need using micro actuators as the driving part so that provides power or movement. A micro electrothermal V-shaped actuator (EVA) is a kind of actuators basing on thermal expansion effect to generate a force and motion. The features of this actuator are lower driving voltage and higher output force, but working at higher temperature and lower driving frequency range are some its drawback.

There have been many studies of EVA that concentrated on some typical topics such as theoretical heat transfer model, influence of geometry dimension, stability and dynamic response. The first research topic focused on building and developing heat transfer models [1–3]. The authors built the finite difference and nonlinear thermal-mechanics models to solve a heat transfer problem in the surface-micromachining

K.-U. Sattler et al. (Eds.): ICERA 2019, LNNS 104, pp. 394–405, 2020.
https://doi.org/10.1007/978-3-030-37497-6_46

polysilicon devices, comparison between theoretical model and experiment confirmed the dependence of a heat-transfer coefficient on temperature changing (both polysilicon and medium of air). Next topic researched on influence of structural dimensions to the displacement and force of EVA [4–8]. The equations of displacement and force via average temperature of beam are established in order to investigate the effect of geometry dimension and optimize design. However, the tolerance between calculation and simulation is considerable. Relating to the topic of dynamic respond, the authors in [9] added vibration parameters into the motion equation, solved and examined the natural-vibration frequencies of the V-shaped actuator. Investigation of time and frequency respond was implemented in [10], also authors proposed the estimating equation of time constant. However, the results are approximation because the authors supposed that material coefficients (thermal conductivity, thermal expansion factor) are constant while the temperature is changing.

There were many publications related to structure and operation of EVA, but did not show/suggest clearly the influence of driving frequency, inconstant material coefficients and dynamic parameters on working quality of the system such as displacement amplitude or stability. Hence, development and continuing of these topics will be necessary in coming years.

This paper focuses on developing the finite difference equation of heat transfer model for calculating temperature distributed on V-shaped beam more exactly while considering material parameter (i.e. resistivity) depends on temperature. Then, the relation between displacement of EVA and driving frequency are determined. The influence of changing resistivity on critical frequency is also demonstrated. In addition, a method for determining conversion parameters in the dynamic equation of EVA is proposed in order to improve the accuracy of calculating result.

2 Heat Transfer and Thermal Expansion Force

2.1 Configuration of V-Shaped Actuator

The configuration of a typical EVA is shown in Fig. 1. It consists of a shuttle ① at the center, an elastic beam system ② and two fixed electrodes ③. The shuttle is suspended by symmetric beam system as a V-shaped with an initial incline angle θ in X-direction. The other end of beams is attached to fixed electrodes.

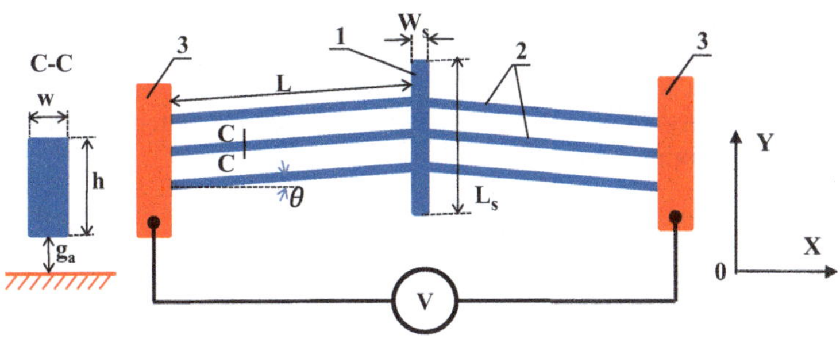

Fig. 1. Configuration of V-shaped actuator

Where: L, h and w are the length, the height and the width of a single beam, respectively; n is the number of single beam pairs; g_a is the gap between movable part and substrate; L_s, W_s are the length and the width of the shuttle, respectively.

When applying a voltage on two fixed electrodes, an electric current transmits through the thin beams and generates thermal emission. The beams will expand and push the shuttle moving in Y-direction. If voltage goes to zero, the temperature of beams decreases gradually, they will shrink and the shuttle returns to the original position.

2.2 Heat Transfer Process of V-Shaped Actuator

Heat transfer on a beam is a complex process. We assume that heat transfer on the beam is one-way and along the length of beam in X-direction. The radiation, convection and losing heat from shuttle to substrate are ignored. Considering continuous heat transfer in a pair of V-shaped beam with length of $2L$, the balancing energy equation in X-direction of a beam element i^{th} with a tiny section of Δx is as follow:

$$q_{i+1} + q_{i-1} + q_s + q_e = q_{st} \tag{1}$$

Where q_{i+1} and q_{i-1} are conductions of heat from elements (i+1) and ($i-$1) to i, respectively; q_s, q_e and q_{st} are heat energy conducted through air layer from beam to substrate, Jul heating and thermal energy stored in beam element i. They are calculated as:

$$q_{i+1} = kA\frac{T_{i+1}^{j} - T_i^j}{\Delta x};\, q_{i-1} = kA\frac{T_{i-1}^{j} - T_i^j}{\Delta x};\, q_s = k_a S\frac{T_i^j - T_0}{g_a} w\Delta x;\, q_e = J^2 \rho A \Delta x$$
$$q_{st} = C_p d\frac{T_i^{j+1} - T_i^j}{\Delta t} A\Delta x \tag{2}$$

Where k and k_a are thermal conductivity for single crystal silicon and air, respectively; T is the temperature of beam elements; T_0 is environment temperature; A is the cross-section area of beam; S is the shape factor of beam area section: $S = 0.6265\left(\frac{h}{W}\right) + 1.1188$ (determined by simulation); C_p and d are the specific heat and density of silicon; J is current density is calculated as:

$$J = \frac{V}{2\rho L} \tag{3}$$

Where $\rho = \rho_0[1 + \lambda(T - T_0)]$ is a resistivity of silicon; λ is temperature coefficient of resistivity; V is driving voltage.

From (1), (2) and (3) we have:

$$T_i^{j+1} = \frac{k}{C_p d}\frac{\Delta t}{\Delta x^2}T_{i-1}^j + \left(1 - 2\frac{k}{C_p d}\frac{\Delta t}{\Delta x^2} - \frac{k_a S}{h g_a}\frac{\Delta t}{C_p d}\right)T_i^j + \frac{k}{C_p d}\frac{\Delta t}{\Delta x^2}T_{i+1}^j + \frac{k_a S}{h g_a}\frac{\Delta t}{C_p d}T_0 + \frac{V^2}{4\rho L^2}\frac{\Delta t}{C_p d} \quad (4)$$

By setting:

$$a_1^j = a_3^j = \frac{k}{C_p d}\frac{\Delta t}{\Delta x^2};\ a_2^j = 1 - 2\frac{k}{C_p d}\frac{\Delta t}{\Delta x^2} - \frac{k_a S}{h g_a}\frac{\Delta t}{C_p d};$$

$$b^j = \frac{k_a S}{h g_a}\frac{\Delta t}{C_p d}T_0 + \frac{V^2}{4\rho L^2}\frac{\Delta t}{C_p d};\ \alpha = \frac{k}{C_p d};\ Fo = \alpha\frac{\Delta t}{\Delta x^2};\ Bi = \frac{k_a S}{h g_a}\frac{\Delta x^2}{k};$$

Finally, Eq. (4) can be written as below:

$$T_i^{j+1} = a_1^j T_{i-1}^j + a_2^j T_i^j + a_3^j T_{i+1}^j + b^j \quad (5)$$

And the coefficient: $a_2^j = 1 - 2Fo - FoBi$, the stability condition of problem is described as:

$$a_2^j \geq 0;\ \text{or}\ \Delta t \leq \frac{\Delta x^2}{(2 + Bi)\alpha} \quad (6)$$

The condition (6) enables us to estimate time increment step Δt corresponding to step length Δx.

Considering a pair of V-shaped beam with the number of length step and time step are M and N, the governing equation of distribution temperature on beam can be written in the form of vector and matrix at time stage (j + 1) as:

$$\left[T^{j+1}\right]_M = \left[A^j\right]_{M\times M}\left[T^j\right]_M + \left[B^j\right]_M (j = 1 \div N) \quad (7)$$

Where $\left[A^j\right]_{M\times M}$ is square matrix containing coefficients: a_1^j, a_2^j, a_3^j; $\left[B^j\right]_M$ is a column vector containing coefficient b^j.

The governing Eq. (7) is used to compute temperature distributing on beam at time stage (j + 1) when the temperature at previous stage j is defined.

2.3 Thermal Expansion Force

The thermal expansion force of EVA is calculated indirectly via thermal expansion of beam. The temperature of beam is determined from (7). Therefore, thermal expansion on a single beam can be expressed as:

$$\Delta L = \sum_i \alpha (T_i - T_0) \Delta x \tag{8}$$

Where α is thermal expansion coefficient.

The thermal expansion force of a single beam is computed as:

$$F_b = AE\frac{\Delta L}{L} \tag{9}$$

Where E = 169 GPa is Young's modulus of single crystal silicon; A is an area of beam's cross-section.

The total thermal force of V-shaped beams acting on shuttle in Y-direction is calculated via force scheme in Fig. 2. Thus, the total force of system including n pairs of V-beam can be expressed as:

$$F = 2nF_b \sin\theta = 2nAE\frac{\Delta L}{L}\sin\theta \tag{10}$$

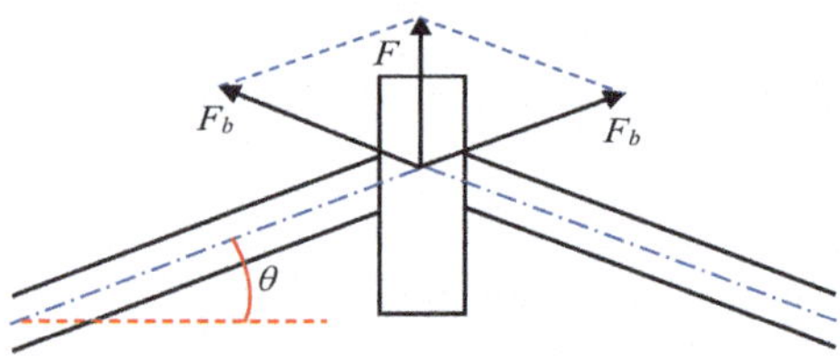

Fig. 2. Scheme of thermal expansion force acting on the shuttle

3 Conversion Dynamic Parameters

General dynamic equation of EVA converted to the displacement of shuttle is expressed as:

$$M\ddot{Y} + C\dot{Y} + KY = F \tag{11}$$

Where M, C and K are the effective mass, equivalent damping coefficient and conversion stiffness of V-shaped beams and the shuttle, respectively; F is the thermal expansion force.

3.1 Calculation of Conversion Stiffness

The conversion stiffness K of beam system in Y-direction is considered basing on principle of shuttle's equivalent displacement via an external force acts on the shuttle in Y-direction. Because the beams play a same role in system, the conversion stiffness of system is equal to the total of single beams' stiffness. A loading model determining the equivalent displacement of a single beam is shown in Fig. 3.

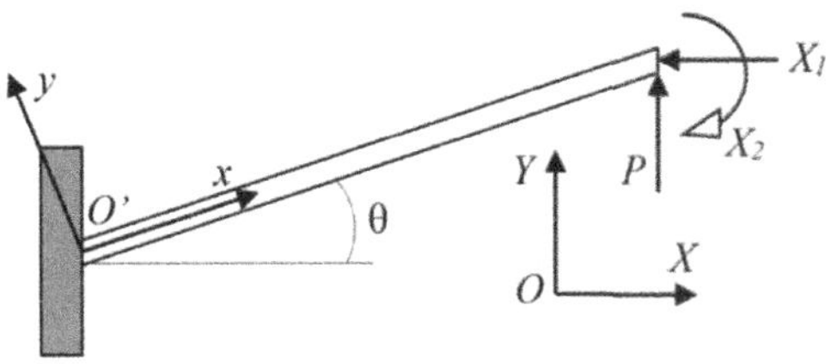

Fig. 3. Model determining equivalent displacement of a single beam

Where P is the external force acting on the shuttle per single beam; X_1 and X_2 are reaction force and moment at the end of beam. The equations calculating bending moment and axial force at position x are expressed as:

$$\begin{cases} M(x) = Px\cos\theta + X_1 x\sin\theta - X_2 \\ N(x) = P\sin\theta - X_1\cos\theta \end{cases} \tag{12}$$

X_1 and X_2 are inferred from boundary conditions:

$$\begin{cases} \varphi(L) = 0 \\ X(L) = 0 \end{cases} \tag{13}$$

Where $\varphi(L)$ and $X(L)$ are turning angle and displacement of beam at position of $x = L$. Finally, the displacement in Y-direction of the top of beam is calculated as:

$$Y = \frac{PL^3}{2E(12I\cos^2\theta + AL^2\sin^2\theta)} \tag{14}$$

The total stiffness of beam system is computed as:

$$K = \frac{nP}{Y} = \frac{2nE(12I\cos^2\theta + AL^2\sin^2\theta)}{L^3} \tag{15}$$

Equation (15) is used to determine conversion stiffness of EVA in Y-direction. It shows that K depends on beam's dimension parameters such as L, I, A and θ. Where I and A are the inertia moment and area of the beam's cross-section, respectively.

3.2 Calculation of Effective Mass

The effective mass of system is calculated basing on the rule of equivalent kinetic energy, i.e. the kinetic energy of effective mass is equal to the total kinetic energy of beam system and shuttle when actuator working.

The effective mass of a single beam is inferred from equation:

$$\frac{1}{2}m_{ef}\dot{Y}^2 = \frac{1}{2}\int_0^L \frac{m_b}{L}\dot{Y}^2(x)dx \Rightarrow m_{ef} = \int_0^L \frac{m_b}{L}\left(\frac{\dot{Y}(x)}{\dot{Y}}\right)^2 dx \tag{16}$$

where $\dot{Y}$ and $\dot{Y}(x)$ are velocities of the shuttle and cross-section of beam at position x; m_{ef} is the effective mass of a single beam; m_b is a single beam mass.

Assuming that the ratio of velocity $\dot{Y}(x)$ to $\dot{Y}$ is equal to the ratio of displacement $Y(x)$ to Y:

$$\frac{\dot{Y}(x)}{\dot{Y}} = \frac{Y(x)}{Y} \tag{17}$$

where $Y(x)$ and Y are displacement of beam's cross-section at position x and the shuttle.

From (16) and (17), the effective mass of a single beam is calculated as:

$$m_{ef} = m_b\left(\frac{1}{3} + \frac{4\cos^2\theta}{105}\right) \tag{18}$$

The total effective mass of the EVA is can be expressed:

$$M = 2n.m_{ef} + m_s = 2n.m_b.\left(\frac{1}{3} + \frac{4\cos^2\theta}{105}\right) + m_s \tag{19}$$

Where m_s is the mass of the shuttle.

3.3 Calculation of Equivalent Damping Coefficient

When EVA works, the air damping affects significantly on the movement of actuator. Here, we assume that air drag force only acts on a pairs of V-beam at front and back corresponding to the moving orientation of shuttle. Therefore, it consists of slide damping on the bottom and top of beams and shuttle, air drag on the front sides of beam and shuttle (when the shuttle moves in *Y*-direction) or the back side (when the shuttle moves in opposite direction).

The total damping force of the system is calculated as:

$$F_d = C \cdot \dot{Y} = 2n(F_{b1} + F_{b2}) + 2F_{b3} + F_{s1} + F_{s2} + F_{s3} \tag{20}$$

Where C is the equivalent damping coefficient; F_{b1} and F_{s1} are total slide damping force on the bottom of a single beam and the shuttle, they are calculated as [11]:

$$F_{b1} = \frac{\mu}{g_a}\int_0^L w\dot{Y}(x)dx \tag{21}$$

$$F_{s1} = \frac{\mu(W_s h)}{g_a}\dot{Y} \tag{22}$$

Where $\mu = 1.85 \times 10^{-5}$ Pa.s (at 20 °C) is the dynamic viscosity coefficient of air; F_{b2} and F_{s2} are slide damping force on the top of beam and shuttle, these term are insignificant and can be ignored (because the EVA works at low frequency).

F_{b3} and F_{s3} are air drag force on the area of a single beam and shuttle calculated as [11]:

$$F_{b3} = \frac{32}{3}\mu\frac{h}{2}\int_0^L \frac{\dot{Y}(x)}{L}dx \tag{23}$$

$$F_{s3} = \frac{32}{3}\mu\frac{h}{2}\dot{Y} \tag{24}$$

From Eqs. (17), (20), (21), (22), (23) and (24), the equivalent damping coefficient of system is expressed:

$$\begin{aligned} C &= \frac{F_d}{\dot{Y}} \\ &= \mu\left[2\left(n\frac{w}{g_a} + \frac{16h}{3L}\right)\frac{1}{L^3}\int_0^L x\sqrt{L^4\sin^2\theta + \cos^2\theta(3Lx - 2x^2)^2}dx + \frac{hW_s}{g_a} + \frac{16h}{3}\right] \end{aligned} \tag{25}$$

4 Result and Discussion

As an example, applying for calculation of the EVA with geometric dimension and material parameters are given in Tables 1 and 2.

Table 1. Geometric dimension parameters of EVA

$L(\mu m)$	$h(\mu m)$	$w(\mu m)$	$g_a(\mu m)$	$\theta(°)$	n	$L_s(\mu m)$	$W_s(\mu m)$
750	30	6	4	2	10	300	50

Table 2. Material properties of single crystal silicon

$d(kg/m^3)$	$E(MPa)$	ka(W/m.k)	C_p(J/kg.K)	$\rho_0(\Omega.m)$	$\lambda(1/K)$	$\alpha(1/K)$
2330	1.69×10^5	0.0257	712	400×10^{-6}	1.25×10^{-3}	2.6×10^{-6}

In order to confirm the calculation results mentioned before, the 3D model of EVA is created and simulated by ANSYS. Simulation results of temperature distribution and deformation of beam system at a voltage of 20 V are shown in Fig. 4.

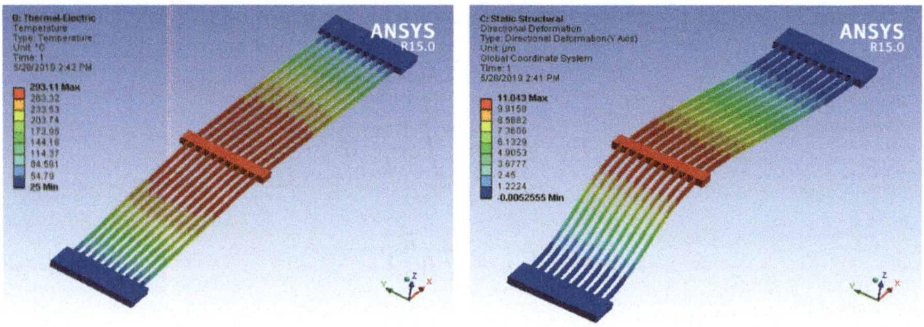

Fig. 4. Temperature and displacement of actuator while applying V = 20 V

Graph on Fig. 5 compares the maximum displacement between calculation and simulation results while applied DC voltage changing from 10 V to 35 V. The results in both cases are approximate linear and match closely (only 2 μm-deviation at 35 V, i.e. about 6.7%). The simulated displacement is larger than calculation one, can be explained by the fact that while calculation of equivalent total force F, the incline angle θ seems to be constant.

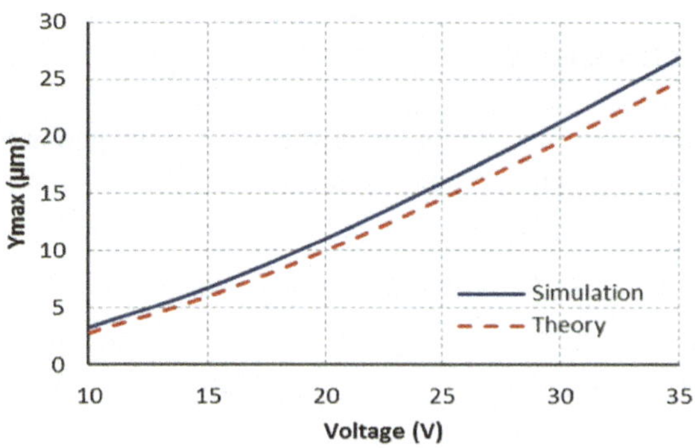

Fig. 5. Maximum displacement for both calculation and simulation

Figure 6 indicates a relation between maximum displacement of the shuttle Y_{max}, its deviation and driving frequency f at various square wave AC voltages of 10 V, 20 V and 30 V. When the frequency is lower than 100 Hz, the maximum displacement of the shuttle is nearly unchanging. Beyond this value of frequency, displacement amplitude decreases gradually, and deviation raises to 5% at various frequencies of 171 Hz, 188 Hz and 207 Hz corresponding to values of voltage are 10 V, 20 V and 30 V, respectively. These frequency values can be named as "critical" frequencies of V-shaped actuator. In other words, EVA will operate stably at the low frequency (under 100 Hz in this example) and always exists a critical frequency f_C in order to the displacement amplitude of the shuttle has an acceptable deviation from maximal value. Low frequency f_C can be explained that the working frequency of electrothermal actuator depends strongly on a thermal response, i.e. depends essentially on the time of heat transfer and thermal emission.

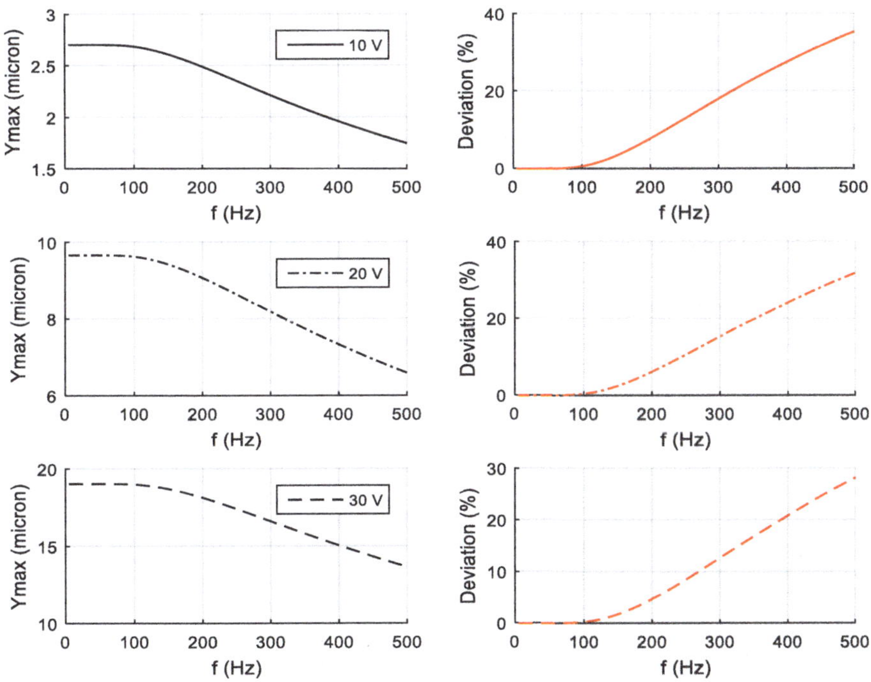

Fig. 6. Displacement and percent of deviation versus driving frequency f

Utilizing dimensions and parameters are listed in Tables 1, 2 above, linearity relation between the critical frequency f_C and amplitude of driving voltage is shown in Fig. 7. This result demonstrates clearly the influence of resistivity depending on temperature leads to the linear proportion of critical frequency values.

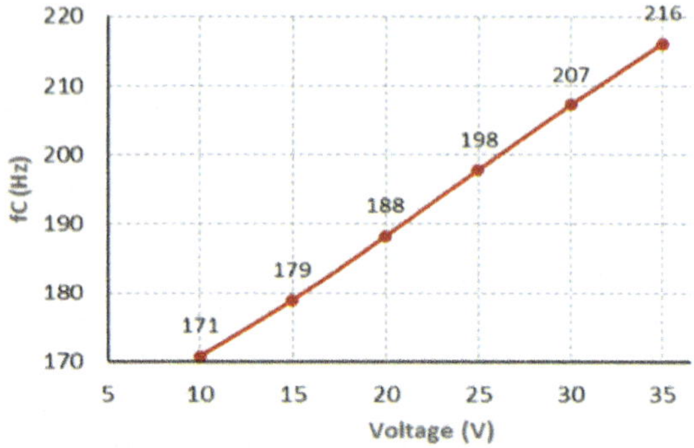

Fig. 7. Critical frequency depends on driving voltage.

5 Conclusion

This paper presented a method of using finite difference equation to solve heat transfer problem when considering the influence of resistivity depends on temperature. The equivalent dynamic parameters in difference equation of motion of V-shaped actuator are determined. Obtained results can be summarized as follow:

(i) The finite difference model of heat transfer in V-shaped beam has been developed. This model can be used to compute exactly temperature distribution in thin beam when considering resistivity depending on temperature.
(ii) The conversion parameters *K*, *M* and *C* of V-shaped actuator's dynamic equation are established via equivalent displacement principle, kinetic energy and damping force, respectively.
(iii) The relation between maximum displacement of the shuttle Y_{max} and driving frequency is determined, the critical frequencies f_C at various driving voltages is figured out in order to ensure stable operation of the V-shaped actuator.

Investigation of influence of other material parameters such as conductivity, thermal expansion and convection coefficient, etc. depending on temperature to the characteristics of V-shaped actuator will be performed in next research.

Acknowledgment. This research is funded by Vietnam National Foundation for Science and Technology Development (NAFOSTED) under Grant number "107.01-2019.05".

References

1. Lott, C.D., McLain, T.W., Harb, J.N., Howell, L.L.: Modeling the thermal behavior of a surface-micromachined linear-displacement thermomechanical microactuator. Sens. Actuators A Phys. **101**(1–2), 239–250 (2002)
2. Enikov, E.T., Kedar, S.S., Lazarov, K.V.: Analytical and experimental analysis of folded beam and V-shaped thermal microactuators. In: Proceedings of SEM X International Congress Costa Mesa, CA, USA, vol. 1 (2004)

3. Jungen, A., Pfenninger, M., Tonteling, M., Stampfer, C., Hierold, C.: Electrothermal effects at the microscale and their consequences on system design. J. Micromech. Microeng. **16**(8), 1633–1638 (2006)
4. Chiorean, R.-S.: Electro-thermo-mechanical modeling of V-beam actuators. Procedia Technol. **19**, 56–61 (2015)
5. Zhao, Y.-L., et al.: Design and characterization of a large displacement electro-thermal actuator for a new kind of safety-and-arming device. Energy Harvesting Syst. **2**(3–4), 143–148 (2015)
6. Zhang, Z., et al.: A comparison model of V- and Z-shaped electrothermal microactuators. In: 2015 IEEE International Conference on Mechatronics and Automation, ICMA 2015, pp. 1025–1030 (2015)
7. Shivhare, P., Uma, G., Umapathy, M.: Design enhancement of a chevron electrothermally actuated microgripper for improved gripping performance. Microsyst. Technol. **22**(11), 2623–2631 (2016)
8. Zhang, Z., et al.: Closed-form modelling and design analysis of V- and Z-shaped electrothermal microactuators. J. Micromech. Microeng. **27**(1) (2017)
9. Zhang, Z., Yu, Y., Zhang, X.: Vibration analysis of V-shaped beam electrothermal microactuators. In: Proceedings - 2017 2nd International Conference on Cybernetics, Robotics and Control, CRC 2017, pp. 80–84, 2018 January
10. Hickey, R., Sameoto, D., Hubbard, T., Kujath, M.: Time and frequency response of two-arm micromachined thermal actuators. J. Micromech. Microeng. **13**(1), 40–46 (2003)
11. Bao, M.H.: Analysis and Design Principles of MEMS Devices, 1st edn. Elsevier, Amsterdam (2005)

Identification of Gear Fault Signal Based on Adaptive EMD Feature Extraction and Optimal GA-LSSVM Classification Model

VietHung Nguyen[1,2](✉), TienDung Hoang[1], VanTrong Thai[1,2], QuocTuan Nguyen[1], and XuanChung Nguyen[3]

[1] Faculty of Mechanical Engineering, Hanoi University of Industry, Hanoi 10000, Vietnam

[2] College of Mechanical and Vehicle Engineering, Hunan University, Changsha 410082, China

[3] VietNam-Japan Center of Mechanical Engineering, Hanoi University of Industry, Hanoi 10000, Vietnam

Abstract. The paper proposed a new efficient diagnostic method for the various fault statuses of gear. This method identifies the gear fault exactly by an optimal classification model based on our relevant features. Firstly, the empirical mode decomposition (EMD) method is used to decompose a non-linear and non-stationary vibration signal into intrinsic mode functions (IMFs). The statistical characteristics of the first several IMFs are calculated to form a feature-matrix. Secondly, the meaningful singular values are exploited by the singular value decomposition (SVD) method, which represent as an important feature vector of the original vibration signal, named EMD-SVD. The obtained EMD-SVD feature is the necessary input to serve the classifier model. Finally, a least square support vector machine model (LSSVM) is optimally trained in the EMD-SVD feature, learning to identify the respective fault pattern with the various status of gear. In the formation of the optimal classifier model, the genetic algorithm is used to find out the best parameter pair of the LSSVM classifier model, commonly called GA-LSSVM. We experimented on gear vibration dataset, and the achieved results demonstrated that the proposed diagnostic method was effective in terms of identification accuracy in the short time.

Keywords: Gear fault · Diagnosis · Least square support vector machine (LSSVM) · Statistical characteristics · Genetic algorithm (GA)

1 Introduction

In the maintenance, gears are of more interest than most parts because they usually assume important responsibilities. These systems work especially well under the condition of healthy gears. Defects in gears normally occur on the break, or chip, which causes faults of gear. These actual fault statuses need to be identified accurately and opportunely to meet the minimal requirement for the production efficiency and to guarantee safety Therefore, an effective of diagnostic technique has been developed to reduce the maintenance cost and improve productivity.

K.-U. Sattler et al. (Eds.): ICERA 2019, LNNS 104, pp. 406–418, 2020.
https://doi.org/10.1007/978-3-030-37497-6_47

Normally, the gear fault conditions are subjectively detected by vibration signals, which are generated by an operation of gear with fault. The vibration signal is normally non-linear and non-stationary with features which are hidden deeply and cannot be discovered in fact [1]. In past decades, the time-frequency analysis methods have been formulated to probe these data and are applied to feature extraction. The methods include short-time Fourier transform (STFT) [2], wavelet transform (WT) [3], empirical mode decomposition (EMD) [4, 5], and local mean decomposition (LMD) [6, 7]. Among these methods, EMD is one of the special methods and is very well known. EMD was first proposed by Huang in 1998 [8]. It has been extensively studied and widely applied in various areas, such as process control [9], modeling [10], and voice recognition [11]. The EMD method is adequate for processing vibration signals. Moreover, EMD has become more and more attractive in the field of fault diagnosis [5, 12, 13] and it especially impresses on rotating machinery systems. EMD can decompose the vibration signal into a set of Intrinsic Mode Functions (IMFs) and a residual. The first several IMFs contain the most important information of the original signal which needs to be explored. In this work, the statistical characteristics are used to analyze the hidden features of IMFs in the time-domain and frequency-domain. These statistical characteristics is based on the root mean square (rms), skewness (s), kurtosis (k), crest factor (cf), and impulse factor (if) in the time domain, while the Shannon entropy (se) can indicate the vibration energy of the signal in the frequency domain. An adequate feature-matrix of the original vibration signal formed. However, the dimensionality of the feature-matrix can interfere with the further classification stage. We take advantage of Singular Value Decomposition (SVD) to achieve the crucial features of original signal, named EMD-SVD. These obtained singular values can exactly reflect the actual fault status of the gear. They contain the main information in the original feature-matrix, which represents as a singular feature vector of the original vibration signal. This feature vector then serves as input for the classification model, which is trained to identify the pattern statuses respectively.

Least squares support vector machine (LSSVM) is the improvement of standard SVM, which is based on statistical theory and structural risk minimization principle [1]. LSSVM [2, 3] takes the equality constraints to replace the inequality constraints, selects the least square linear system as the loss function, and transforms the solution of quadratic programming into solving linear equations. LSSVM is a powerful machine learning tool that has been employed in applications such as pattern recognition, time-series forecasting, robotics, and diagnostics [14–18]. Using SVKF for classification purposes not only requires superior input features but also requires an optimal parameter pair (γ, σ). These parameter values play an important role in the architectural design of a classifier model with high classification accuracy and stable performance. In this study, the objective is to lift restrictions, and we propose the use of the genetic algorithm (GA) to select the best parameter pair (γ, σ) for the LSSVM classifier. Based on the superiority of the algorithm, the best parameter pair (γ, σ) of the SVKF classifier model is optimally selected, named GA-LSSVM. The optimized GA-LSSVM classifier model can achieve distinguished performance in classification accuracy with a low time cost. In this model, the EMD-SVD feature provides the input for the GA-LSSVM classifier in the form of the EMD-SVD–GA-LSSVM based diagnostic technique, which is then applied to the diagnosis of various gear faults.

The remainder of this paper is organized as follows: Sect. 2 presents the EMD-SVD based adaptive feature extraction. The crucial feature set is extracted from original vibration data to serve as the classifier model. Section 3 presents the main procedures for building the optimal GA-LSSVM classifier model using in the proposed diagnostic technique. Section 4 presents two main stages: (1) adaptive feature extraction and (2) pattern classification by the optimal classifier model in architecting the new diagnostic technique. Section 4.2 presents the experimental results to demonstrate the effectiveness performance based on the diagnostic technique and discussion. Section 5 is the conclusions.

2 Extraction of Adaptive Fault Feature by EMD-SVD Method

In this section, we detail the feature extraction of the original vibration signal. Firstly, the famous self-adaptive method EMD decomposes the vibration signal into the IMFs and a residual. Adaptive statistical characteristics are then calculated and form the feature-matrix. Finally, the singular values of the feature-matrix can be obtained by the SVD method. It is a feature vector that contains the most crucial information of the original vibration signal.

2.1 Empirical Mode Decomposition Theories

The Empirical Mode Decomposition (EMD) method is an interesting method for analyzing non-stationary, non-linear data and has been applied effectively [5]. EMD analyzes the main signal based on the expansion of the basic functions. These functions are estimated by an iterative procedure called sifting. In other words, the EMD method is used to decompose a signal automatically into a set of band-limited functions, IMFs. Each IMF should satisfy two basic conditions [8]: (1) in the whole data set, the number of extreme points and the number of zero crossings must either be equal or differ by one at most; and (2) at every point, the mean value of the envelopes defined by local maxima and local minima is zero.

The EMD can separate a segment of a bearing fault signal $x(t)$ into n IMFs: $IMF_1, IMF_2, \ldots, IMF_n$, and a residue signal r. Hence, $x(t)$ can be reconstructed as a linear combination:

$$x(t) = \sum_{n=1}^{N} IMF_n + r \tag{1}$$

EMD is a sifting process and is a systematic way to extract IMFs. Figure 1 presents a flow chart of the EMD algorithm. This flow chart can be described as follows:

Given an input signal $x(t)$, $r(t) = x(t)$, $n = 0$.

Step 1: Get the local maximum and local minimum of $x(t)$

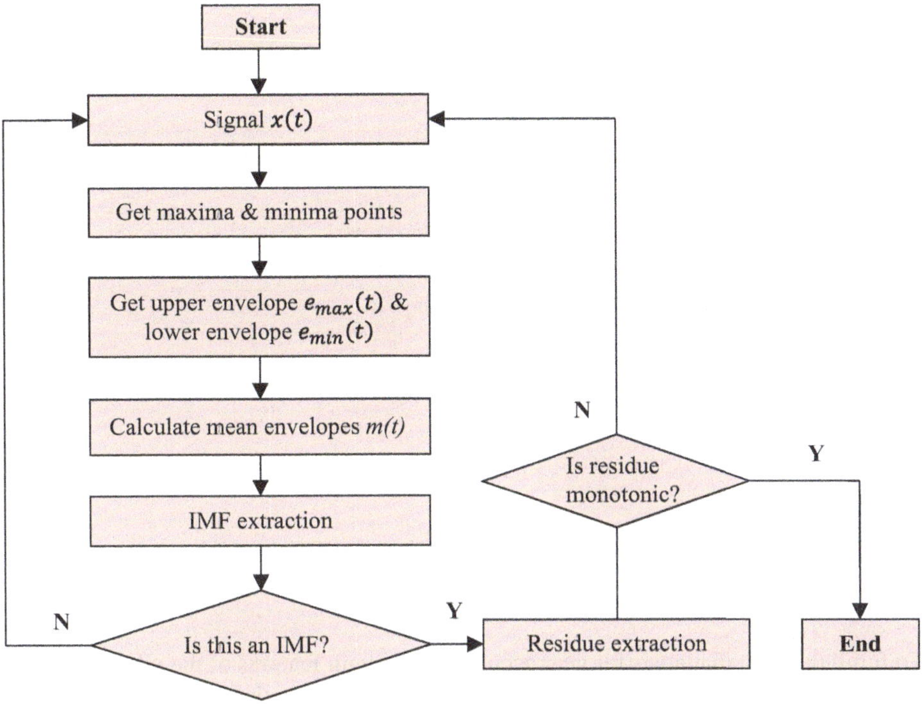

Fig. 1. The flow chart of EMD algorithm

Step 2: Get the upper envelope $e_{max}(t)$ by connecting all local maxima through cubic spline functions. Repeat the procedure for the local minima to produce the lower envelope $e_{min}(t)$.
Step 3: Calculate the mean value at every point of the envelopes:

$$m(t) = \frac{1}{2}(e_{max}(t) + e_{min}(t)) \quad (2)$$

Step 4: Calculate $h(t) = x(t) - m(t)$. If $h(t)$ satisfies the IMF condition, then $n = n + 1$, $IMF_n = h(t)$ and go to Step 5; else $x(t) = h(t)$, go to Step 1.
Step 5: Let $r(t) = r(t) - IMF_n$. If $r(t)$ is a monotonic function, end the sifting process, else $x(t) = r(t)$ and go back to Step 1.

Figure 2 illustrates an example of the EMD process, where Fig. 2(a) shows the gear vibration signal in the chipped defect and Fig. 2(b) depicts the decomposed IMFs, IMF1 to IMF7, and the residual IMF8. Each of the IMFs includes different frequency bands ranging from high to low and implies a distinct time characteristic scale.

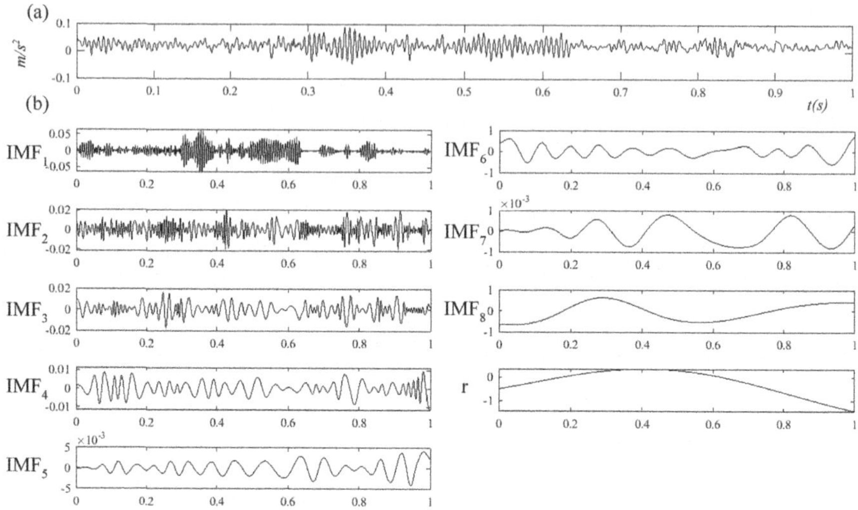

Fig. 2. A sample is analyzed by using EMD method

2.2 EMD-SVD Based Feature Extraction

To fulfill the fault diagnosis for gear accurately, the fault features of the vibration signal have a crucial role in helping the diagnosis system. The statistical characteristics are calculated from IMFs in time and frequency domains to recognize various gear defect and working conditions. They are root mean square (*rms*), skewness (*s*), kurtosis (*k*), crest factor (*cf*) and impulse factor (*if*) in time domain; Shannon entropy (*se*) in frequency domain. They are defined as follows:

$$rms = \sqrt{\frac{1}{K}\sum_{i=1}^{K} x_i^2} \tag{3}$$

$$s = \frac{K}{(K-1)(K-2)}\sum_{i=1}^{K}\left(\frac{x_i - \overline{x}}{Std}\right)^3 \tag{4}$$

Where: the *Std* parameter measures the energy in the vibration signal, $Std = \sqrt{\sum_{i=1}^{K}(x_i - \overline{x})^2/(K-1)}$.

$$k = \frac{K(K+1)}{(K-1)(K-2)(K-3)}\sum_{i=1}^{K}\left(\frac{x_i - \overline{x}}{Std}\right)^4 - \frac{3(K-1)^2}{(K-1)(K-2)} \tag{5}$$

$$cf = \frac{Peak_{max}}{Rms} \tag{6}$$

$$if = \frac{Peak_{max} - Peak_{min}}{2(Std)} \tag{7}$$

$$se = -\sum_{k=1}^{K} P_k \log P_k \tag{8}$$

Where: $P_k = |p_i|^2 / \sum_{i=1}^{K} |p_i|^2$ is the distribution of the energy probability for each spectrum component and p_i is a spectrum for $i = 1, 2, .., K$, with K the number of spectrum lines.

Suppose that the first k IMFs obtained are selected to calculate l statistical characteristics according to Eqs. (3)–(8). Variable a is constructed based on the $M^{k \times l}$ feature-matrix, which is expressed as below:

$$M = \begin{bmatrix} a_{11}a_{12}\cdots a_{1l} \\ a_{21}a_{22}\cdots a_{2l} \\ \vdots \vdots \ddots \vdots \\ a_{k1}a_{k2}\cdots a_{kl} \end{bmatrix} \tag{9}$$

Ideally, the features shown in the $M^{k \times l}$ feature-matrix are able to represent the fault information in different views. However, not all features extracted are related with the fault information. In fact, when performing the fault diagnosis, the feature-matrix needs to represent singular features to aim at diagnostic accuracy and to improve the efficiency of the process. In this work, SVD is used to exploit the singular values of the $M^{k \times l}$ feature-matrix. The singular values as well as the $f(M)$ fault feature vector are obtained:

$$f(M) = \left[f_M^1, f_M^2, \ldots, f_M^3\right] \tag{10}$$

Where: f_M^i are considered as the singular values of M. The singular values are listed in descending order automatically by the SVD function $f_M^1 \geq f_M^2 \geq \cdots \geq f_M^l$.

Based on this EMD-SVD method, the features of original vibration signal obtained are conducive features, which can give the dominant fault-related information, and the unfitting features of the feature-matrix could be ignored express the conducive features of the original vibration signal. These features serve as the input of classifier model.

3 Architecting of the GA-LSSVM Classification Model

As mentioned in the first section, this study aims at accurate diagnosis of various gear fault statuses. Based on the EMD-SVD, genetic algorithm (GA) which is used to build the proposed GA-LSSVM classifier model. This GA-LSSVM model obtained the best parameter pair obtained (γ, σ) to become the effective model and credible.

3.1 LSSVM Classification Model

Least Squares Support Vector Machine (LSSVM) was proposed by Suykens in 1999 [19] to diminish the computational burden of SVM, which applies with equality instead of inequality constraints. LSSVM has solved the following convex optimization problems with good generalization ability which try to find the optimal separating hyperplane. Suppose $\{(x_i, y_i)|i = 1, 2, \ldots, l\}$ as training set of sample number is l. The sample of $x_i, i = 1, 2, \ldots, l$ corresponds to the category of $y_i \in (-1, 1)$, the objective function and constraint condition are shown as follows:

$$\begin{cases} min(J_{LS})withJ_{LS}(w, e) = \frac{1}{2}w^T w + \frac{1}{2}\gamma \sum_{i=1}^{l} e_i^2 \\ s.t.y_i = w^T \varphi(x_i) + b + e_i,\ i = 1, 2, \ldots, l \end{cases} \tag{11}$$

Where e_i are slack variables and $\gamma \geq 0$ is a penalty factor/regularization parameter.

The values of γ will influence training result of the LSSVM model. The low value of γ means a model with high training errors and oppositely the high γ value does not permit any slack variables and consequently increases model complicity. Therefore, it is critical to find proper value for γ and it is one of LSSVM tuning parameter that should be adjusted conscientiously.

Define Lagrange function:

$$L(w, b, e, \alpha) = \frac{1}{2}w^T w + \frac{1}{2}\gamma \sum\nolimits_{i=1}^{l} e_i^2 - \sum\nolimits_{i=1}^{l} \alpha_i \left(w^T \varphi(x_i) + b + e_i - y_i\right) \tag{12}$$

Where α_i is Lagrange multiplier, which can be positive or negative. The optimum condition is as follows:

$$\frac{\partial L}{\partial w} = 0, \frac{\partial L}{\partial b} = 0, \frac{\partial L}{\partial e_i} = 0, \frac{\partial L}{\partial \alpha_i} = 0 \tag{13}$$

The matrix equation can be obtained easily as follows:

$$\begin{bmatrix} 0 & Y^T \\ Y & \Omega + {}^{I}\!/_{\gamma} \end{bmatrix} \begin{bmatrix} b \\ \alpha \end{bmatrix} = \begin{bmatrix} 0 \\ \bar{I} \end{bmatrix} \tag{14}$$

Where $Y^T = [y_1, y_2, \ldots, y_l]$; I is unit matrix $\bar{I} = [1, 2, \ldots, l]^T$; $\alpha = [\alpha_1, \alpha_2, \ldots, \alpha_l]^T$; $\Omega = y_i y_j \varphi^T(x_i)\varphi(x_j) = y_i y_j K(x_i, x_j)$; $K(x_i, x_j)$ is Kernel function of SVM.

The classifying function of LSSVM classifier finally can be obtained as follows:

$$f(x) = sgn\left(\sum_{i=1}^{l} \alpha_i y_i K(x, x_i) + b\right) \tag{15}$$

In this research, Gauss radial basic function (RBF) is utilized to map the samples into a higher dimension feature space which can make LSSVM get better performance and generalization, as shown in Eq. (16).

$$K(x,y) = exp\left(-\frac{\|x-y\|^2}{2\sigma^2}\right) \tag{16}$$

After the structure of LSSVM is determined, we need to search the optimal penalty factor γ and kernel parameter σ while the appropriate selection of two parameters will affect the learning performance of LSSVM.

In summary, the classification effect of SVKF is controlled by the parameter pair (γ, σ), where the parameter γ is mentioned in Eq. (14) and σ is a Kernel parameter mentioned in Eq. (17). These parameter pair values must be appropriately selected to obtain an effective classification accuracy, which can be done by the GA algorithm.

3.2 Optimal GA-LSSVM Classification Model

Genetic algorithm (GA), first introduced by [20], is based on the 'survival of the fittest' and the natural evolution process via reproduction, which can find the optimal solution after some iterative computations. The objective functions are often referred to as fitness functions. Three main operations in GA are selection, crossover, and mutation. The optimization process is started with random initial population of chromosomes, followed by fitness evaluation. Equation (17) is a depreciation function, which is used to evaluate the classification accuracy. The next step is selection of fittest individuals or parents for reproduction, where chromosomes with better fitness values have more potential to yield children during subsequent generation.

In this study, GA is used to select best parameter pair (γ, σ) in the LSSVM classifier model which relates to the overall ability and reliability of the classifier model. It can select the best parameter value in the space of all possible parameter values of the training process. These parameters will be used to build an optimal classifier model. This completes the training phase of the GA-LSSVM model. The flowchart of GA operation is illustrated by Fig. 3 in Sect. 4.

$$depreciation(\%) = \frac{1}{N}\sum_{i-1}^{N}\frac{Samp_{f(i)}}{Samp_{t(i)} + Samp_{f(i)}} \tag{17}$$

The excellent capability of GA combined with the generalizing capability of LSSVM to bring out synergies in the optimized GA-LSSVM model. The results in the identification accuracy and time cost demonstrated ability and reliability of GA-LSSVM which is based on the patterns of different gear faults.

4 Gear Fault Diagnosis Based on EMD-SVD-GA-LSSVM Proposed Technique

The proposed technique based on EMD-SVD-GA-LSSVM is coherently represented in this section. This technique consists of two main stages: (1) adaptive EMD-SVD feature-based feature extraction and (2) the optimal GA-LSSVM model. Firstly, we use the adaptive EMD-SVD feature extraction to extract the meaning and most responsive features of the original vibration signal, which would increase the overall reliability and classification accuracy of the classifier model. Secondly, an optimal GA-LSSVM classifier model is then used to identify the fault status with effective identification performance in which the classifier is based on the GA.

The aim of the proposed diagnostic technique EMD-SVD-GA-LSSVM is to further improve the fault diagnosis performance and make sure the diagnoses are reliable. This technique can be visualized by the flowchart presented in Fig. 3, which can be described as follows:

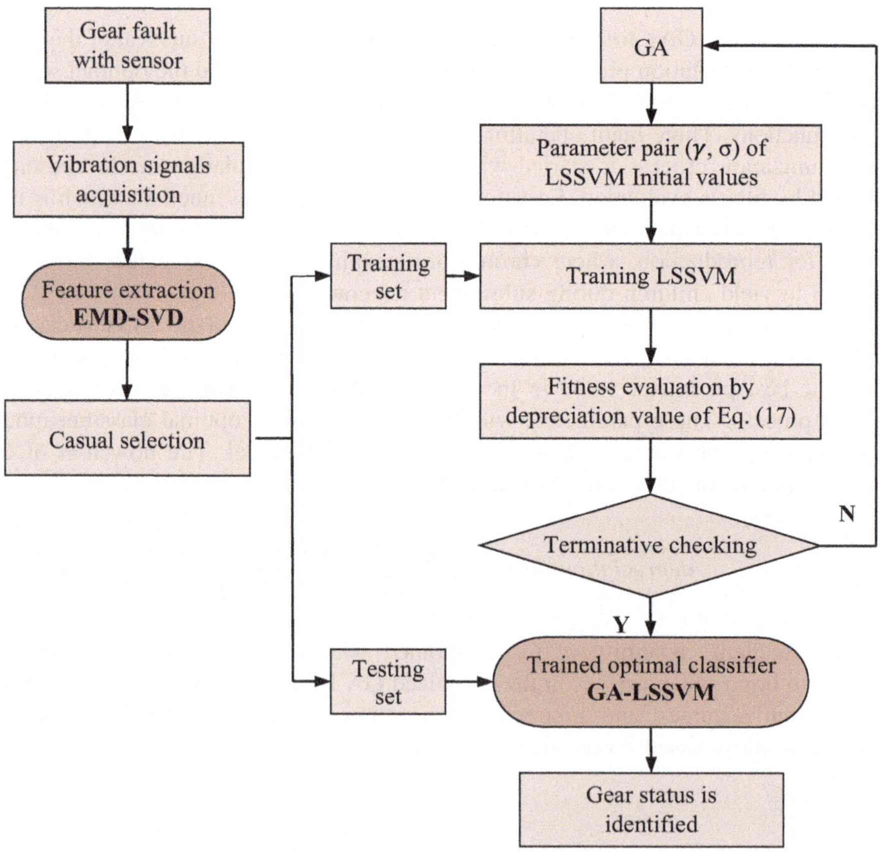

Fig. 3. The architecture of the fault diagnostic technique based on EMD-SVD and GA-LSSVM

Step 1: Data acquisition
The vibration signals of the rolling bearing components with defects are taken by using the accelerometer sensor.

Step 2: Adaptive feature extraction
The feature set is generated by the feature extraction method based on the proposed EMD-SVD. The details of this step are presented in Sect. 2.

Step 3: Fault pattern identification
The obtained feature set is divided into a training set and a testing set. The samples with the respective class labels of the training set are used to train the GA-LSSVM. The gear fault diagnostic technique based on EMD-SVD-GALSSVM demonstrated diagnostic accuracy and reliability in experiments.

Experimental Analysis
In this section, the authors report the results of an extensive performance study for evaluating the proposed EMD-SVD-GA-LSSVM diagnostic technique. This diagnosis technique is applied to identify various faults of the gear. In which, the proposed GA-LSSVM classifier model specially is stable and unbiased. To show that the classification capability of this model is superior we compared GA-LSSVM with another classifier model.

4.1 Data Acquisition

An experiment has been carried out on the small test rig shown in Fig. 4 [21] that is used for modeling different gear faults with the teeth number of the gear is 37 and module 2.5. Here, we consider three gear states that are gear with normal (NoG), with chipped tooth (CpT) and broken tooth (BkT). The moment of inertia of load is 0.03 kg. m^2. The acceleration sensor has been fixed on the bearing house with sample frequency of 1024 Hz and rotating frequency of 7 Hz. The collected vibration signal group is 60 signals in three sampled gear circumstances: NoG, CpT, and BkT, with 20 signals from each tested gear condition respectively.

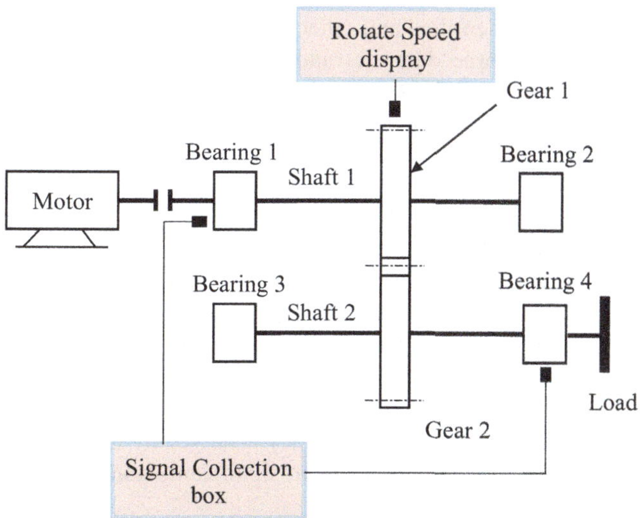

Fig. 4. Test rig

4.2 Results and Discussion

EMD-SVD based feature extraction was used to extract the feature set of the acquired vibration dataset. The adaptive parameters according to Eqs. (3)–(8) are used to calculate the statistical characteristics for the first five IMFs which are gained by the EMD method. The SVD is then used to extract the fault feature vector $\sigma(M)$. The obtained feature set is divided into the training and testing sets, which provide the input for the classifier model. The various fault statuses of the gear are respectively identified by the classifier.

Table 1. Classifier models for gear conditions

Classifier	Gear condition		
	NoG	CpT	BkT
GA-LSSVM 1	**(+1)**	(−1)	(−1)
GA-LSSVM 2	(−1)	**(+1)**	(−1)
GA-LSSVM 3	(−1)	(−1)	**(+1)**

Table 2. Detailed information of GA-LSSVMs

	EMD-SVD features				Patterns		Diagnosis task
					Training	Testing	
GA-LSSVM 1	11.23	0.29	0.10	0.02	36	24	NoG
GA-LSSVM 2	10.96	0.72	0.16	0.08	36	24	CpT fault
GA-LSSVM 3	11.62	0.81	0.13	0.03	36	24	BkT fault

In this study, the classification GA-LSSVM (i) models are designed to identify the NoG, CpT fault, BkT fault, respectively; the GA-LSSVM (i) with i = 1, 2, 3 are detailed in Table 1. It should be noted that the label of gear status is assigned to (+1) in the status that needs to be identified and other labels of gear status are assigned to (–1). The majority (60%) of sample data are randomly selected and used for training the classifier and the rest are used for testing the classifier. Table 2 show the number of training and testing samples.

To demonstrate the superior performance of the GA-LSSVM classifier model, we compare proposed classifier model with the original LSSVM model. The achieved diagnostic results are shown in Table 3. Observing these results, the GA-LSSVM classifier model obtains a higher classification accuracy result in comparison with LSSVM. The achieved EMD-SVD feature set expressed the superior quality in the optimal classifier based on the GA algorithm which is highlighted in the result table.

From the results in Table 3, based on the EMD-SVD feature set, the diagnostic accuracy of the proposed GA-LSSVM classifier reached a maximum of 100%, while the GA-LSSVM classifier achieved a maximum of 90.5%. The results objectively reveal that the percentage classification accuracy achieved by the proposed EMD-SVD and GA-LSSVM diagnosis technique is a significant result. We would like to

Table 3. Identification results and comparison

	Diagnosis accuracy (%)	Time (s)
GA-LSSVM 1	**98.64**	1.317
GA-LSSVM 2	**91.21**	1.538
GA-LSSVM 3	**100.00**	1.486
LSSVM 1	87.00	1.886
LSSVM 2	90.50	2.096
LSSVM 3	85.00	2.193

emphasize that our technique had a lower time cost than the other models, which is also clearly shown in the time column in Table 3.

5 Conclusions

The research in this paper is based on a theoretical synthesis and experiments to propose a diagnostic technique for bearing fault statuses. A new diagnostic technique integrates the adaptive feature extraction based on EMD-SVD and the ideal GA-LSSVM classifier model in experiment. Firstly, the EMD method decomposed the original vibration signals into IMFs, in which each of the first four IMFs is calculated six parameters of statistical characteristic to form the feature-matrix. The SVD method is then applied to the feature-matrix to achieve the crucial features. Finally, these features are fed into the optimal GA-LSSVM classifier model as the input for performing classification. The classification results showed the effective classification performance. In addition, the proposed methodology in this study is demonstrated in an experimental environment through application to the complex data of gear statuses to identify the actual status. The experimental results showed that the samples are identified exactly with a low time cost. Moreover, the authors are confident that, this technique can be safely applied to other fault identification fields in machinery systems with non-linear, non-stationary data.

Acknowledgement. The authors would like to express our gratitude to Faculty of Mechanical Engineering, Hanoi University of Industry and College of Mechanical and Vehicle Engineering, Hunan University for their support.

Conflicts of Interest. The authors declare that they have no conflict of interest.

References

1. Lin, J., Chen, Q.: Fault diagnosis of rolling bearings based on multifractal detrended fluctuation analysis and Mahalanobis distance criterion. Mech. Syst. Signal Process. **38**, 515–533 (2013)
2. Bartelmus, W., Zimroz, R.: Vibration condition monitoring of planetary gearbox under varying external load. Mech. Syst. Signal Process. **23**, 246–257 (2009)

3. Peng, Z.K., Chu, F.L.: Application of the wavelet transform in machine condition monitoring and fault diagnostics: a review with bibliography. Mech. Syst. Signal Process. **18**, 199–221 (2004)
4. Yan, R., Gao, R.X.: Rotary machine health diagnosis based on empirical mode decomposition. J. Vib. Acoust. **130**, 021007 (2008)
5. Lei, Y., Lin, J., He, Z., Zuo, M.J.: A review on empirical mode decomposition in fault diagnosis of rotating machinery. Mech. Syst. Signal Process. **35**, 108–126 (2013)
6. Yang, Y., Cheng, J., Zhang, K.: An ensemble local means decomposition method and its application to local rub-impact fault diagnosis of the rotor systems. Measurement **45**, 561–570 (2012)
7. Cheng, J., Zhang, K., Yang, Y.: An order tracking technique for the gear fault diagnosis using local mean decomposition method. Mech. Mach. Theory **55**, 67–76 (2012)
8. Huang, N.E., et al.: The empirical mode decomposition and the Hilbert spectrum for nonlinear and non-stationary time series analysis. Proc. R. Soc. Lond. A: Math. Phys. Eng. Sci. **454**, 903–995 (1998)
9. Luo, L., et al.: Hilbert-Huang transform, hurst and chaotic analysis based flow regime identification methods for an airlift reactor. Chem. Eng. J. **181–182**, 570–580 (2012)
10. Tang, J., Zhao, L., Yue, H., Yu, W., Chai, T.: Vibration analysis based on empirical mode decomposition and partial least square. Proc. Eng. **16**, 646–652 (2011)
11. Ambikairajah, E.: In: 2007 6th International Conference on in Information, Communications and Signal Processing, pp. 1–7 (2007)
12. Junsheng, C., Dejie, Y., Yu, Y.: The application of energy operator demodulation approach based on EMD in machinery fault diagnosis. Mech. Syst. Signal Process. **21**, 668–677 (2007)
13. Cheng, J., Yu, D., Tang, J., Yang, Y.: Application of SVM and SVD technique based on EMD to the fault diagnosis of the rotating machinery. Shock Vib. **16**, 89–98 (2009)
14. Abbasion, S., Rafsanjani, A., Farshidianfar, A., Irani, N.: Rolling element bearings multi-fault classification based on the wavelet denoising and support vector machine. Mech. Syst. Signal Process. **21**, 2933–2945 (2007)
15. Gryllias, K.C., Antoniadis, I.A.: A support vector machine approach based on physical model training for rolling element bearing fault detection in industrial environments. Eng. Appl. Artif. Intell. **25**, 326–344 (2012)
16. Lau, K.W., Wu, Q.H.: Local prediction of non-linear time series using support vector regression. Pattern Recogn. **41**, 1539–1547 (2008)
17. Long, B., Tian, S., Miao, Q. Pecht, M.: In: 2011 AUTOTESTCON, pp. 360–366. IEEE (2011)
18. Li, S., Zhou, W., Yuan, Q., Geng, S., Cai, D.: Feature extraction and recognition of ictal EEG using EMD and SVM. Comput. Biol. Med. **43**, 807–816 (2013)
19. Suykens, J.A.K., Vandewalle, J.: Least squares support vector machine classifiers. Neural Process. Lett. **9**, 293–300 (1999)
20. Holland, J.H.: Adaptation in Natural and Artificial Systems: An Introductory Analysis with Applications to Biology, Control, and Artificial Intelligence. University Michigan Press, Oxford (1975)
21. Yu, D., Yang, Y., Cheng, J.: Application of time–frequency entropy method based on Hilbert-Huang transform to gear fault diagnosis. Measurement **40**, 823–830 (2007)

Identification of Machining Conditions in the Hard Milling of Hardened SKD 61 Steel

Huu-That Nguyen[1], The-Vinh Do[2(✉)], and Nguyen-Anh-Vu Le[1]

[1] Nha Trang University, Nha Trang, Viet Nam
[2] Thai Nguyen University of Technology, Thai Nguyen, Viet Nam
thevinh8880@tnut.edu.vn

Abstract. The hard milling is machining of the hardened materials with the hardness range from 40 HRC to 70 HRC. This is an efficient solution that can be used to replace the grinding operation in the mold and die manufacturing industry dynamics and the cutting conditions must be considered and chosen properly for hard milling. This paper shows the cutting conditions for hard milling process are chosen through the stability lobe diagram of a machine tool. Firstly, the determination of the average cutting force coefficients for the tool-material pairing is carried out through slot milling experiments. Secondly, the measurement of frequency response functions for a milling machine tool is performed based on the hammer tests. Finally, the development of the stability lobe diagram is conducted by the use of the Cutpro software. The result reveals that the spindle speed and the axial depth of the cut in a stable cutting region were considered and selected for the hard milling.

Keywords: Hard milling · Stability lobe · Cutting force coefficient · Frequency response function

1 Introduction

It is known that the machine tool system is often affected by chatter vibrations during the hard machining operation. The chatter vibrations of the cutting system are generated by a self-excitation phenomenon in the production of chip thickness [1]. In hard machining, the self-excited machine tool becomes unstable and chatter vibrations grow until the cutter jumps out of the cut [2]. In general, this is one of the common phenomena which negatively affects the machine tool, the cutting tool as well as the product quality.

Therefore, the analysis of the dynamic parameters of the machine tool system is one of the very important works which need to be carried out to control the chatter vibration during the hard machining operation. Altintas and Budak [2] analyzed the structural dynamics of the machine tool and developed an analytical model for predicting the chatter vibration of the machine tool system. They noted that the chatter is a self-excited vibration which often occurs in the cutting system as machining with high depths of the cut and cutting speed. Thus, they recommended machine tool users for proper selection of depth of cuts and spindle speeds in hard machining operations. This is also consistent with the view with a study by Nguyen, et al. [3]. Furthermore, by

K.-U. Sattler et al. (Eds.): ICERA 2019, LNNS 104, pp. 419–425, 2020.
https://doi.org/10.1007/978-3-030-37497-6_48

experimental study, Thamizhmanii and his colleagues [4] showed that the causes of poor surface finish are due to chatter vibrations of a machine tool during the hard machining operation.

From the above analysis, it shows that the optimal selection of cutting process parameters based on structural dynamic analysis of machine tool can produce a good surface finish and high performance in the hard machining process.

2 Experimental Methodology

In this paper, the cutting conditions for the hard milling were chosen based on the stability lobe diagram which is generated by using the Cutpro software.

2.1 Calculation of Average Cutting Force Coefficients

Normally, the cutting force coefficients are determined for different pairs of tool and workpiece through the cutting tests [5]. In order to determine cutting force coefficients for a tool/material pair in hard milling, a small series of slot milling tests were conducted at different feed rates, but constant spindle speed and depth of cut. This study was performed on the hardened SKD 61 steel with hardness in the range of 50 HRC. With this approach, the input parameters of the average cutting coefficient tests were listed in Table 1.

Table 1. Experimental conditions for calculating the average cutting force coefficients [4, 6].

Test	Spindle speed, n (rpm)	Feed rate, f (mm/Tooth)	Axial depth of cut, a (mm)	Hardness, H (HRC)
1	850	0.050	0.3	50
2	850	0.075	0.3	50
3	850	0.100	0.3	50
4	850	0.125	0.3	50
5	850	0.150	0.3	50

The cutting tool utilized in this study were the TiAlN coated-carbide end mills from CMTec tool manufacturer and their geometry was characterized by 10 mm of diameter, 35° of helix angle, square type, with 4 flutes. Their cutting ability can reach to 60 HRC for the hardened tool steels. The tools were mounted on the machine's spindle head to carry out the various cutting tasks involved. All the experiments were performed under dry conditions at a Victor V-Center-4 Vertical machining center, which has a maximum spindle speed of 6000 rpm. The specimens were clamped on the machine table to ensure rigidity during the machining process.

The cutting force components in three directions were measured utilizing a piezoelectric three-component Dynamometer with Model 6423 (Lebow® Products Inc., Troy, MI, USA) and a System 6000 with Model 6100 scanner (Vishay Precision Group, Inc., Atlanta, GA, USA) connected to a PC employing Strain Smart force measurement software [7–9]. A digital filter which was designed based on the fast Fourier transform algorithm was integrated into the System 6000 to filter out the noise signal during the measurement. For each machining condition, 1000 data points for each of the three channels were recorded by sampling at 1000 Hz for a period of 1 s.

The average cutting force coefficients are determined from Eq. 1:

$$\begin{cases} K_{tc} = \frac{4\overline{F_{yc}}}{Na} & K_{te} = \frac{\pi \overline{F_{ye}}}{Na} \\ K_{rc} = \frac{-4\overline{F_{xc}}}{Na} & K_{re} = \frac{-\pi \overline{F_{xe}}}{Na} \\ K_{ac} = \frac{\pi \overline{F_{zc}}}{Na} & K_{ae} = \frac{2\overline{F_{ze}}}{Na} \end{cases} \quad (1)$$

Where, K_{tc}, K_{rc}, K_{ac} are cutting coefficients of the tool/work-piece material pairing in tangential, radial and axial directions (N/mm^2); K_{te}, K_{re}, K_{ae} are edge cutting coefficients of the cutting tool in tangential, radial and axial directions (N/mm); N is the number of teeth on the cutter; a is axial depth of cut; $\left(\overline{F_{xc}}, \overline{F_{yc}}, \overline{F_{zc}}\right)$ and $\left(\overline{F_{xe}}, \overline{F_{ye}}, \overline{F_{ze}}\right)$ are average cutting forces and average edge cutting forces in three directions, respectively (N).

2.2 Measurement of Frequency Response Functions

The dynamic analysis is a very important work which is utilized to identify the vibration characteristics (natural frequencies and mode shapes) of a machine tool under vibrational excitation. The objective of the machine tool dynamic analysis is to predict the maximum stable depth of cut which has not the generation of chatter vibration during the machining process. A machining tool system, which consists of the spindle, tool-holder and the cutting tool, is the most flexible component, thus, determination of dynamic behavior of this system is a necessary task for appropriate selection of cutting conditions in the hard milling process.

In practice, the determination of machine tool dynamics is conducted by tapping test. In the tap test, an impulse force hammer is used to hit the machine tool structure. This hitting will excite the structure over a certain frequency range which depend on the type of tool tip, the size of the hammer used. After the hitting, the structure will produce a response that represents the dynamics of the structure. This response can then be measured using a displacement sensor, a velocity sensor, or an accelerometer. The procedure of structural dynamic analysis of machine tool was given in Fig. 1.

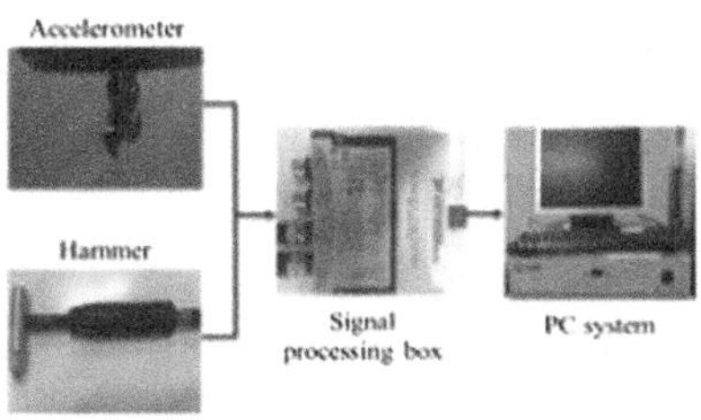

Fig. 1. Illustration of dynamic parameter measurement of machine tool

From Fig. 2, it was seen that the structural dynamics in both the X and Y directions were measured through the Kistler 9722A2000 Hammer (with a metal tip and 2.13 mV/N sensitivity) and the 8778A500 M14 accelerometer (10 mV/g sensitivity). The cutting tool was clamped on the machine's spindle head to carry out the various measurement tasks involved. The accelerometer was mounted on the relief face of the cutter by using a special wax. The output of the accelerometer and the hammer were connected to the data processing box and PC system as illustrated in Fig. 1. During tap testing, the frequency response functions (FRF) of the structure were measured in two directions by the use of the versatile transfer function measurement program in Cutpro software, which was installed in a PC system.

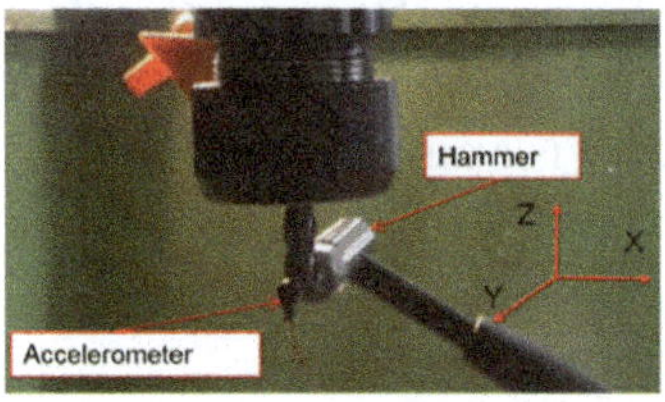

Fig. 2. Dynamic parameter measurement of the milling machine tool.

2.3 Determination of Stability Lobes by Using Cutpro Software

The window of simulation properties is displayed for the 2½ axis milling module by default when CutPro is started, as shown in Fig. 3.

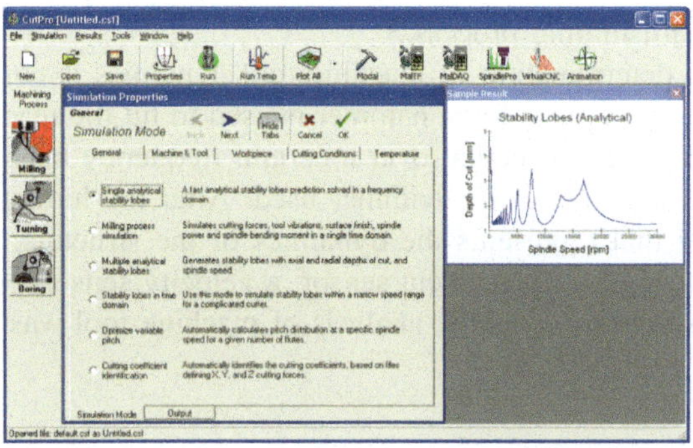

Fig. 3. Simulation window of Cutpro software in the 2½ axis milling module

The stability lobe diagram was produced from the following Eq. 2:

$$a_{lim} = -\frac{2\pi \Lambda_R}{NK_t}(1 + \kappa^2) \quad (2)$$

Where, a_{lim} is chatter-free axial depth of cut; N is the number of teeth on the cutter; Kt is cutting force coefficient; $\kappa = \Lambda_R/\Lambda_I$ is the ratio between the real part (Λ_R) and imaginary part (Λ_I) of eigenvalue of the chatter stability equation.

3 Results and Discussion

According to the acquired data of cutting forces for five different feed rates, these data files can be used in CutPro software to identify the average cutting force coefficients. After calculating, the average cutting force coefficients were given in Table 2.

Table 2. Average cutting coefficients of SKD 61 (50 HRC)

Cutting coefficients			Edge cutting coefficients		
K_{tc} (N/mm^2)	K_{rc} (N/mm^2)	K_{ac} (N/mm^2)	K_{te} (N/mm)	K_{re} (N/mm)	K_{ae} (N/mm)
757.934	−2465.626	444.493	34.428	−40.159	9.207

The calculated results of FRFs are shown in Tables 3 and 4 which may be used to determine the dynamics of the machine tool. Tables 3 and 4 show that the maximum values of the magnitude of the transfer function in X and Y directions obtained at frequencies of about 1208 Hz (Mode 3). At this frequency, the structural system may be unstable due to the lowest model stiffness.

Table 3. Generated curve-fit modal parameters for the X direction

Mode	Frequency [Hz]	Damping [%]	Residue (Re) [m/N]	Residue (Im) [m/N]	Modal stiffness [N/m]	Mass [kg]
1	707.94	4.419	−7.41E−06	−9.41E−06	2.36E+08	11.950
2	861.89	10.217	1.15E−05	−2.19E−05	1.24E+08	4.228
3	1208.52	1.902	−1.50E−05	−1.39E−04	2.71E+07	0.471
4	1364.28	0.218	−7.78E−07	−1.75E−06	2.44E+09	33.327
5	1392.88	0.405	−6.19E−07	−2.27E−06	1.92E+09	25.091
6	1798.70	8.166	2.24E−05	−3.53E−05	1.60E+08	1.255

Once the FRFs and the cutting coefficients were determined, these data were used in the Cutpro software to make the stability lobe diagram. The result of the stability lobe diagram is presented in Fig. 4.

From Fig. 4, it was seen that the chatter stability lobes create a spindle speed dependent dividing line between the unstable and stable depth of cut. These stability lobes can be used to find the best machining conditions for the measured machine tool system. From this diagram, to achieve the best surface quality and highest productivity, the cutting conditions (spindle speed and the axial depth of cut) must be considered and chosen in the stable cutting region.

Table 4. Generated curve-fit modal parameters for the Y direction

Mode	Frequency [Hz]	Damping [%]	Residue (Re) [m/N]	Residue (Im) [m/N]	Modal stiffness [N/m]	Mass [kg]
1	866.46	10.878	2.40E−05	−3.62E−05	7.55E+07	2.547
2	951.22	2.172	4.61E−06	−5.01E−06	5.95E+08	16.672
3	1207.09	2.327	1.51E−05	−7.70E−05	4.91E+07	0.856
4	1236.46	0.500	−6.28E−06	−2.63E−05	1.47E+08	2.443
5	1268.91	1.827	−2.03E−05	−2.28E−05	1.74E+08	2.745
6	1430.79	1.211	2.80E−06	−2.06E−06	2.17E+09	26.948

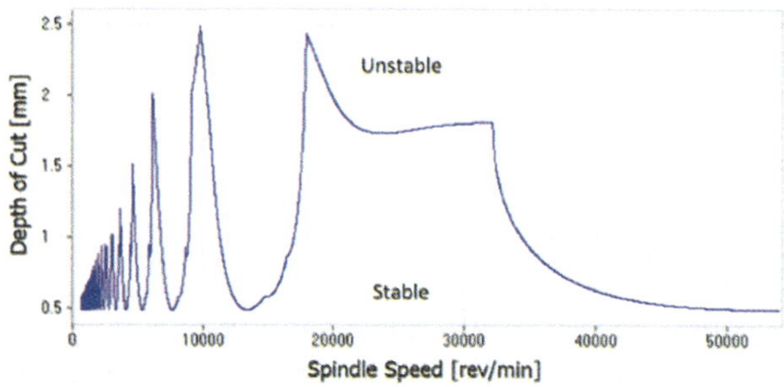

Fig. 4. Analytical stability lobe diagram of hard milling test

4 Conclusions

This study presented the determination of the machining conditions for the hard milling of SKD 61 alloy steel (50HRC) with coated carbide tool based on Cutpro software. It could be concluded as follows:

- The cutting coefficients K_{tc}, K_{rc} and K_{ac} of SKD 61 alloy steel with 50HRC were 757.934, −2465.626 and 444.493 N/mm^2, respectively. The edge cutting coefficients K_{te}, K_{re} and K_{ae} were 34.428, −40.159 and 9.207 N/mm, respectively.
- The dynamic analysis of machine tool showed that at frequency 1208 Hz in X and Y directions, the structural system may be unstable due to the lowest modal stiffness and the amplitude of vibration was maximum. This was a resonance in vibration of cutting system.
- According to this research work, the spindle speed and the axial depth of cut in a stable cutting region were selected based upon the chatter stability lobe diagram.

References

1. Altintas, Y.: Manufacturing Automation: Metal Cutting Mechanics, Machine Tool Vibrations, and CNC Design. Manufacturing Automation, pp. 56–64 (2000)

2. Altintaş, Y., Budak, E.: Analytical prediction of stability lobes in milling. CIRP Ann. **44**(1), 357–362 (1995)
3. Nguyen, N.T., Kao, Y.C., Bui, G.T., Nguyen, Q., Nguyen, Q.M., Do, T.V.: An experimental investigation of dynamic cutting forces in the stable milling processes. In: International Conference on Engineering Research and Applications, pp. 158–166. Springer, Cham (2018)
4. Thamizhmanii, S., Saparudin, S., Hasan, S.: Analyses of surface roughness by turning process using Taguchi method. J. Achiev. Mater. Manuf. Eng. **20**(1-2), 503–506 (2007)
5. Kao, Y.-C., et al.: A combination method of the theory and experiment in determination of cutting force coefficients in ball-end mill processes. J. Comput. Des. Eng. **2**(4), 233–247 (2015)
6. Nguyen, H.-T., Hsu, Q.-C.: Study on cutting forces and material removal rate in hard milling of SKD 61 alloy steel. J. Chin. Soc. Mech. Eng. **38**(1), 41–51 (2017)
7. Nguyen, H.-T., Hsu, Q.-C.: Surface roughness analysis in the hard milling of JIS SKD61 alloy steel. Appl. Sci. **6**(6), 172 (2016)
8. Do, T.-V., Hsu, Q.-C.: Optimization of minimum quantity lubricant conditions and cutting parameters in hard milling of AISI H13 steel. Appl. Sci. **6**(3), 83 (2016)
9. Vu, N.-C., Huang, S.-C., Nguyen, H.-T.: Multi-objective optimization of surface roughness and cutting forcesin hard milling using Taguchi and response surface methodology. Key Eng. Mater. **773**, 220–224 (2018)

Influence of Cutting Conditions on the Surface Roughness When Hole Turning Heat-Treated SKD11 Steel

Nguyen Hong Son[1(✉)], Hoang Xuan Thinh[1], Nhu-Tung Nguyen[2], and Do Duc Trung[2]

[1] Center for Mechanical Engineering, Hanoi University of Industry, No. 298, Cau Dien Street, Bac Tu Liem District, Hanoi, Vietnam
nguyenhongson@haui.edu.vn

[2] Faculty of Mechanical Engineering, Hanoi University of Industry, No. 298, Cau Dien Street, Bac Tu Liem District, Hanoi, Vietnam
doductrung@haui.edu.vn

Abstract. This study was performed to determine the degree of influence of the cutting parameters on the machining surface roughness when hole turning the heat-treated SKD11 steel at a hardness of 62 HRC. Three controllable factors-three levels (cutting speed, feed rate, and depth of cut) were used with the performance measurement that machining surface roughness. The results show that the feed rate has the greatest degree of impact on the surface roughness. And, the second and third factors affect on the surface roughness are cutting depth and cutting speed, respectively. The Genetic Algorithms and ANOVA method were used to determine the optimal value of the cutting speed, feed rate, depth of cut, and the optimum value of surface roughness. The optimum value of surface roughness is 0.47 μm that was obtained at cutting speed of 219.80 m/min at a feed rate of 0.05 mm/rev and a depth of cut of 0.299 mm.

Keywords: Hole turning · Surface roughness · SKD11 steel · Cutting condition · Optimization, and genetic algorithms

1 Introduction

Turning is the machining method that is used very commonly in the industry manufacturing. The amount of work that was done by the turning method is about 40% of the total machining process volume. In a factory, the number of lathe machines is usually about 25 ÷ 35% of the total tool-machine [1]. Hole turning and hole grinding of hard material are two in more surface-hole machining method. But compared to hole grinding, hole method is much more productive. The turning process is carried out by many different methods such as external surface turning, inner surface turning, flat surface turning, thread turning, … in which, inner surface turning plays a very important role. This method is often applied to machine the hole surfaces before finish machining by other machining methods such as grinding, etc. In addition, inner surface turning is also useful as a final machining method for the surfaces that require high accuracy but difficult to implement by other methods (grinding, …).

K.-U. Sattler et al. (Eds.): ICERA 2019, LNNS 104, pp. 426–435, 2020.
https://doi.org/10.1007/978-3-030-37497-6_49

Similar to other cutting methods, surface quality after hole turning is evaluated through many parameters, in which surface roughening is an important parameter. Surface roughness greatly affects on the function of machine details. So, this parameter is often selected as an evaluation indicator of machining processes [2]. Surface roughness depends on many parameters such as cutting parameter, the system stable, the workpiece hardness, etc. Cutting parameters are often heavily affect on the surface roughness, so these are often studied to improve the machining surface. The degree of influence of each parameter on the surface roughness is different [1]. The study of the effect of cutting parameters on the surface roughness will be the basis for the machining process controlling to achieve the desired roughness value. With the external surface turning method, many studies were performed to investigate the effect of cutting parameters on the surface roughness by both experimental studies [3, 4] and theoretical research [5, 6]. However, it seems that, the internal turning method has not been studied in previous researches.

Munawar et al. [7] studied experiments to investigate the influence of some parameters on surface roughness when hole turning of low-carbon steel. In that study, Taguchi method was used to investigate the influence of tool radius parameters, main angle, cutting speed, feed rate, and cutting depth on the surface roughness. The conclusions of that study are the following: tool tip radius, main tilt angle, cutting speed, feed rate have a significant effect on surface roughness, while cutting depth has a negligible effect on surface roughness. That study also showed that the optimal value of tool radius parameters, main angle, cutting speed, feed rate, and cutting depth are 0.2 mm, 9°, 175 m/min, 0.1 mm/rev and 0.6 mm, respectively. Then surface roughness has a minimum value of 2.75 μm.

Vaishnav and Sonawane [8] also applied the Taguchi method in determining the effect of the rotation speed of the workpiece, feed rate, and the flow of the oil cool on the surface roughness when hole turning the medium-carbon steel (SAE1541). Their study showed that the rotation speed of the workpiece and the feed rate had a great impact on the surface roughness, in which the feed rate had the greatest impact. The flow of coolant oil had a negligible effect on the surface roughness. This study also shows the optimal value of the rotational speed of the workpeice, the feed rate, and the flow of coolant oil were 2100 rev/min, 90 mm/min, and 20 L/min, respectively. Those cutting conditions can be ensured the surface roughness with a minimum value of 0.776 μm. An advice was given that when turning the medium-carbon steel (SAE1541), in order to improve the finished surface, the machining processes should perform with high cutting speed, and with small feed rate and small flow of coolant oil.

In this study, the hole turning experiments of SKD11 steel were conducted to examine the influence of cutting parameters on the surface roughness, then using Genetic Algorithms to determine the value of cutting parameters to machining the detail surface with the smallest roughness.

2 Experimental Method

2.1 The Experiment Setup

Workpiece and Tool

Workpiece: Experimental workpiece material is SKD11 steel. This is a type of material that is widely used in manufacturing the stamping molds, cutting tool, etc. The workpieces were heat treated to reach 62HRC hardness. The workpieces dimension is the outer diameter of 140 mm, the inner diameter of 106 mm, and the length of 25 mm as shown in Fig. 1. This product was used as the inspection bushing in mechanical. The chemical compositions of workpiece are listed in Table 1.

Fig. 1. Experimental sample

Table 1. Chemical composites of SKD 11

Composite	C	Mn	Si	Cr	Va	Mo	Ni
%	1.5	0.3	0.25	11.5	0.25	0.3	0.35

The Machine and Cutting Insert

Experiments were carried out on Doosan Lynx 220L CNC lathe at Hanoi Mechanical Industry Center, Hanoi Industry of University. The cutting insert that was used in this study is the currently popular type of cutting insert in machining the high-hardness material (after heat-treated materials). The cutting insert has the symbol DNC250 1EA19C28 0.4R of KORLOY (Korea).

Surface Roughness Measurements

The surface roughness Tester is TESA RUGOSURF 10 Roughness Gauge (Fig. 2). At each experimental, three samples were tested, each sample was measured the surface roughness three times, the value of surface roughness at each experiment was the average value of nine consecutive measurements.

Fig. 2. Setting of surface roughness measurement

2.2 Experiment Design

In this research, the cutting speed (v), feed rate (f), and depth of cut (t) were selected as control factors and their levels were expressed in the Table 2. In the experimental layout plan, the experimental plan was performed with 13 experiments and detailed as in Table 3. The value of V and t are chosen according to the recommendation of the cutting tool manufacturer. For f, according to the recommendation of the cutting tool manufacturer, the values of f are 0.05–0.3 mm/rev. However, in case of large f (f > 0.2 min/rev), the trial results show that Ra has a very high value that is not suitable for finish machining processes. Therefore, in order to determine the optimal values of V, f, and t when finish machining, the value of f is selected in accordance with the finish machining processes. Specifically, in this study, the survey area of f was chosen between 0.05–0.2 min/rev.

Table 2. Turning parameters and their levels

Machining parameters	Unit	Level 1	Level 2	Level 3
		−1	0	1
Cutting speed, v	m/min	120	170	220
Feed rate, f	mm/rev	0.05	0.125	0.2
Depth of cut, t	mm	0.05	0.175	0.3

Table 3. The experimental design and results

TT	Input parameters						Ra (μm)
	Coded factors			Actual factors			
	v	f	t	v (m/min)	f (mm/rev)	t (mm)	
1	−1	−1	−1	120	0.05	0.05	0.76
2	1	−1	−1	220	0.05	0.05	0.69
3	−1	1	−1	120	0.2	0.05	2.74

(continued)

Table 3. (*continued*)

TT	Input parameters						Ra (μm)
	Coded factors			Actual factors			
	v	f	t	v (m/min)	f (mm/rev)	t (mm)	
4	1	1	−1	220	0.2	0.05	2.87
5	−1	−1	1	120	0.05	0.3	0.6
6	1	−1	1	220	0.05	0.3	0.47
7	−1	1	1	120	0.2	0.3	2.94
8	1	1	1	220	0.2	0.3	3.07
9	0	0	0	170	0.125	0.175	1.3
10	0	0	0	170	0.125	0.175	1.28
11	0	0	0	170	0.125	0.175	1.4

3 Results and Discussions

3.1 Analysis of the Influence of Cutting Parameters on the Surface Roughness

The experimental results were investigated and listed in Table 3. Using Minitab 16 software to analyze these experimental results, the influence of cutting parameters on the surface roughness was determined and shown in Figs. 3 and 4.

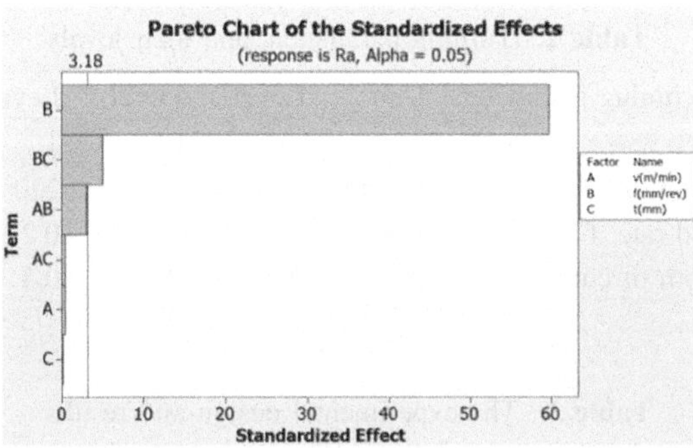

Fig. 3. Pareto graph about the effect of cutting parameters and the interaction between them on the surface roughness

The Figs. 3 and 4 show that with three cutting parameters (cutting speed, feed rate, depth of cut), only the feed rate has a significant influence on the surface roughness. The cutting speed and depth of cut as well as the interaction between parameters have little influence on the surface roughness. When increasing the value of the feed rate, the surface roughness will increase.

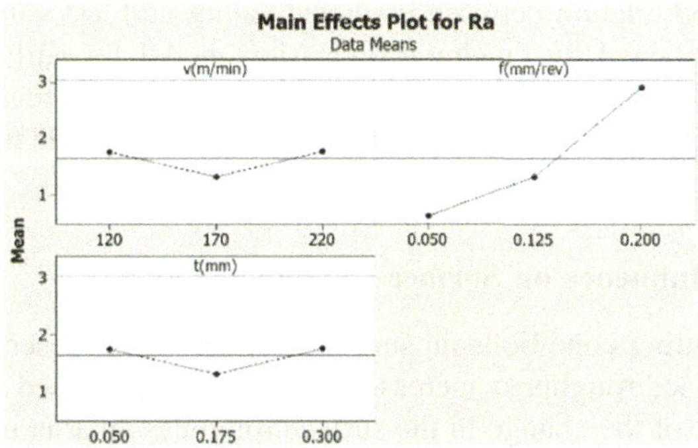

Fig. 4. The effect of cutting parameters on the surface roughness

3.2 Regression and Verification of Surface Roughness Model

Although, the parameters of cutting speed and cutting depth as well as the interaction between the parameters do not affect much on the surface roughness. However, these parameters have been removed from the regression model, because doing so will weaken the compatibility of the model [9]. On that basis, the regression function about the relationship between surface roughness and cutting parameters according to the encoding of input parameters was determined as in Eq. (1). This equation is the basis for selecting the cutting conditions to ensure the surface roughness value in each specific steel machining condition of the heat-treated SKD11 steel (Fig. 5).

$$\begin{aligned} R_a = & 0.361197 - 0.00158194 * \mathrm{v} + 10.6659 * \mathrm{f} - 1.04411 * \mathrm{t} \\ & + 0.015922 * \mathrm{v} * \mathrm{f} - 0.00136667 * \mathrm{v} * \mathrm{t} + 10.3689 * \mathrm{f} * \mathrm{t} \end{aligned} \tag{1}$$

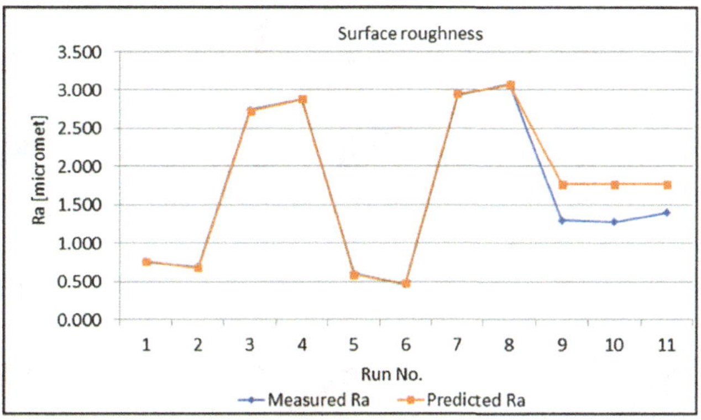

Fig. 5. Measured and predicted results of surface roughness

The verification results of surface roughness model were described in Fig. 6. As seen from this figure, the predicted results were very close to the experimental results.

There is a very good relation between predicted values and test values. The R^2 values of the equations obtained by quadratic regression model for surface roughness was found to be 95.85%. These results showed that the Quadratic regression model was shown to be successfully investigated of surface roughness in hole turning processes of heat-treated steel SKD11.

3.3 Parametric Influence on Surface Roughness

The influence of cutting conditions on surface roughness was described in Fig. 6. It is very clear that surface roughness increase with increasing of feed rate. This result is similar to the result of the change in the surface roughness that is noted in the several works such as [3, 4]. This trend can be explained that when feed rate increases, that results increase in undeformed chip thickness, and undeformed chip thickness is directly proportional to cutting forces. And then, when the cutting forces increase, the stability and damping characteristics of machine-tool system will be affected, which make more vibrations and ultimately affects the surface roughness.

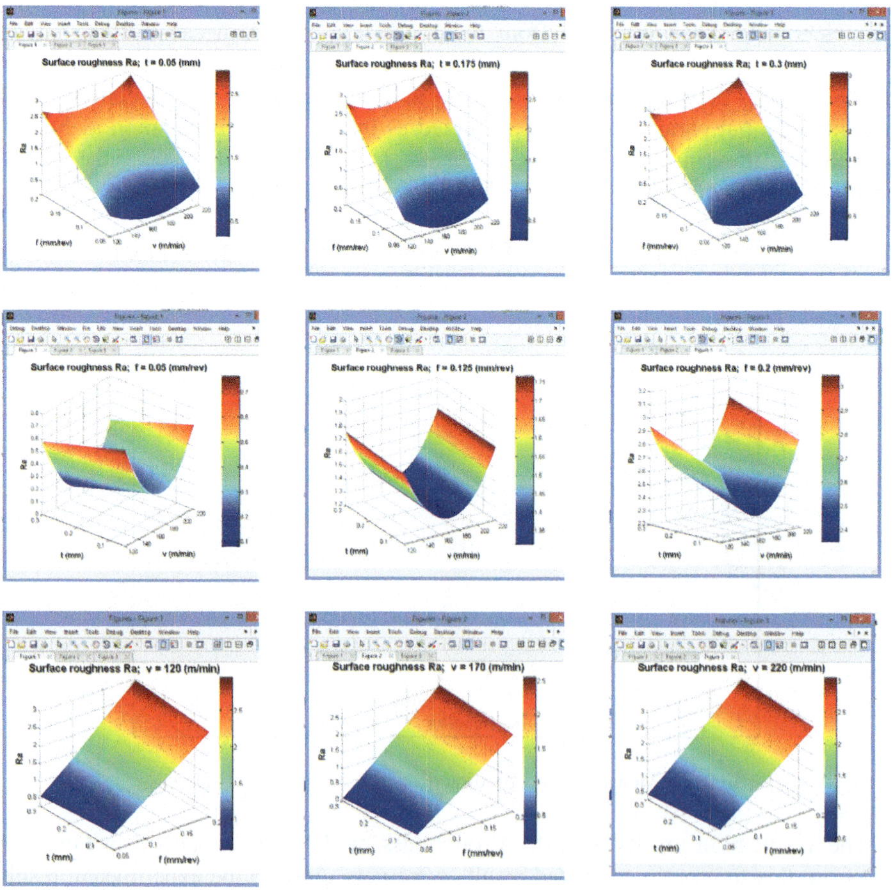

Fig. 6. Effect of cutting conditions on surface roughness

The surface roughness values exhibited decreasing tendency with increasing of axial depth of cut from 0.05 mm to about 0.3 mm. Besides, the surface roughness decreases with increasing of cutting speed from 120 m/min to about 200 m/min, and with the cutting speed increases from about 200 m/min to 220 m/min, the tendency of surface roughness is also increasing. So, in order to improve the surface roughness in the hole turning process of heat-treated SKD11 steel, the tendency of machining conditions was proposed that the feed rate decreases, the depth of cut is about 0.3 mm, and the cutting speed is about 200 m/min.

3.4 Optimization of Cutting Conditions

In this study, the Genetic algorithms were used to determine the value of the cutting parameters to ensure the smallest surface roughness. In order to solve the optimal problem, in the encoded form of the cutting mode parameters, Eq. (1) is rewritten as Eq. (2).

$$\begin{cases} R_a = function(v, f, t) \rightarrow min \\ R_a > 0 \\ 120 \leq v \leq 220 \\ 0.05 \leq f \leq 0.2 \\ 0.05 \leq t \leq 0.3 \end{cases} \tag{2}$$

Running the function of optimizing the surface roughness function in Eq. (2) with Turkkan's Excel program of evolution [10] with the basic parameters of the algorithm including population numbers, hybrid probability, and mutation probability were chosen according to [11, 12] as shown in Table 4. The results for the adaptive function were described in Fig. 7 with the optimal value of the cutting parameters in the encoded form as shown in Table 4.

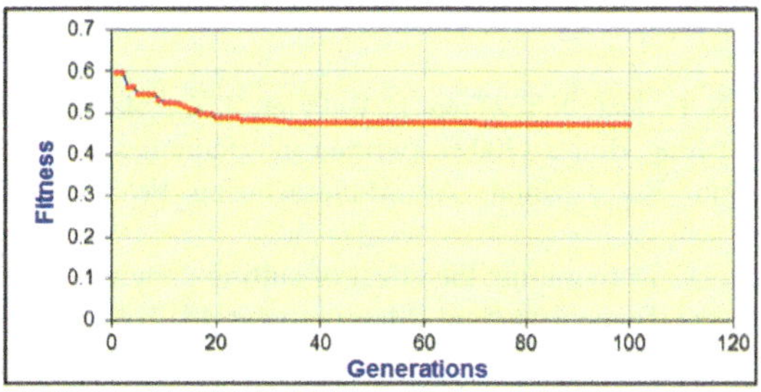

Fig. 7. Graph about the adaptive function of the surface roughness objective function

From the data in Table 4, the optimal values of cutting speed parameters, feed rate, and cutting depth are 219.80 m/min, 0.05 mm/rev, and 0.299 mm, respectively. Then the surface roughness is the smallest value, this value is 0.47 μm.

Table 4. Optimal value of parameters

Number of populations	150
MaxGeneration	100
CrossOverProb	0,25
MutationProb	0,05
v	219,80
f	0,050
t	0,299
$R_a(\mu m)$	0.47

4 Conclusions

In this study, an experimental method was performed to investigate the influence of cutting conditions on the surface roughness. Depending on the analysis of experimental results, the conclusions of this study can be drawn as follows.

- The feed rate has a great influence on the surface roughness. The cutting speed and cutting depth affects not much to the surface roughness. When increasing the value of the feed rate, the surface roughness will increase.
- The optimal value of cutting speed, feed rate, and cutting depth are 219.80 m/min, 0.05 mm/rev, and 0.299 mm, respectively. Then the surface roughness is the smallest value, this value is 0.47 μm.

Acknowledgment. The authors appreciate the generous assistance from the Mechanical Center - Hanoi University of Industry (https://www.haui.edu.vn/vn) for help in the cutting measurement experiments.

References

1. Dich, T.V., Binh, N.T., Dat, N.T., Tiep, N.V., Viet, T.X.: Manufacturing process, Science and Technics Publishing House (2003). (written in Vietnamese)
2. Gupta, H.N., Gupta, R.C., Mittal, A.: Manufacturing Process. New age international publishers (2009)
3. Asiltürk, I., Akkus, H.: Determining the effect of cutting parameters on surface roughness in hard turning using the Taguchi method. Measurement **44**, 1697–1704 (2011)
4. Sumardiyanto, D., Susilowati, S.E., Cahyo, A.: Effect of cutting parameter on surface roughness carbon steel S45C. J. Mech. Eng. Autom. **8**(1), 1–6 (2018)
5. Abburi, N.R., Dixit, U.S.: A knowledge-based system for the prediction of surface roughness in turning process. Robot. Comput. Integr. Manuf. **22**, 363–372 (2006)
6. Costes, J.-P.: A predictive surface profile model for turning based on spectral analysis. J. Mater. Process. Technol. **213**(1), 94–100 (2013)
7. Munawar, M., Chen, J.C.-S., Mufti, N.A.: Investigation of cutting parameters effect for minimization of surface roughness in internal turning. Int. J. Precis. Eng. Manuf. **12**(1), 121–127 (2011)

8. Vaishnav, M.P., Sonawane, S.A.: Analysis and optimization of boring process parameters by using Taguchi method on SAE 1541. Int. J. Eng. Sci. Inven. **3**(8), 59–63 (2014)
9. Myers, R.H., Montgomery, D.C., Anderson-Cook, C.M.: Response surface methodology: process and product optimization using designed experiments, 3rd edn. Wiley, Hoboken (2009)
10. Turkkan, N.: Floating Point Genetic Algorithm - Genetik V2.01 (2001). http://www.umoncton.ca/turk/logic.htm
11. Saravanan, R., Sachithanandam, M.: Genetic algorithm (GA) for multivariable surface grinding process optimisation using a multi – objective function model. Int. J. Adv. Manuf. Technol. **17**, 330–338 (2001)
12. Krajnik, P., Kopac, J., Sluga, A.: Design of grinding factors based on response surface methodology. J. Mater. Process. Technol. **162**, 629–636 (2005)

Influence of Lubricant Parameters on Surface Roughness of Workpiece When Grinding SKD11 Steel

Hoang Tien Dung[1], Do Duc Trung[1(✉)], Nguyen Van Thien[1], Le Hong Ky[2], and Kitikhammoune Sonpheth[1]

[1] Faculty of Mechanical Engineering, Hanoi University of Industry, No. 298, Cau Dien Street, Bac Tu Liem District, Hanoi, Vietnam
dotrung.th@gmail.com, doductrung@haui.edu.vn
[2] Vinh Long University of Technology Education, No. 73, Nguyen Hue Street, Ward 2, Vinh Long, Vietnam

Abstract. In grinding, the surface roughness of a workpiece has a significant influence on quality of the part. Surface roughness is heavily dependent upon. This article presents empirical research for determining influence of lubrication parameters on surface roughnes. The experiment is carried out with 36A60LV grinding wheel, Tectyl cool 290MC and Emulsion lubricant. Minitab 16 software is used to analyze test results to gauge the influence of lubrication parameters on surface roughnes. The regression equation that presents the relationship function of surface roughness is solved to determine the optimal value of lubrication parameters for each type of lubricant, and the values of surface roughness in two lubricant types are compared. The influence of flow and concentration of lubricant on surface roughness is quite complicated. When the flow of the solution and the solution concentration is increased, the surface roughness tends to decrease. However, if the values of these two parameters further increase, the surface roughness tends to increase. When the lubricant Tectyl cool 290MC is used, the value of surface roughness is smaller than it is when Emulsion is used. In fine grinding conditions, with Tectyl cool 290MC lubricant, if the values of the flow and concentration of lubricant are 2.62 (L/min) and 4.2 (%) respectively the surface roughness will reach the smallest value at 0.41 (μm). For Emulsion lubricant, if the values of the flow and concentration of lubricant are 2.63 (L/min) and 4.32 (%) respectively, the surface roughness will reach the smallest value at 0.49 (μm).

Keywords: Surface roughness · Grinding · SKD11 steel · Lubrication parameters · Tectyl cool 290MC · Emulsion

1 Introduction

In mechanical processing, grinding is often chosen as the final processing method for important surfaces, in which high surface precision and lustre are requested. Surface quality when grinding is assessed through many parameters, of which surface roughness is one of the most important parameters. This parameter has a great influence on

K.-U. Sattler et al. (Eds.): ICERA 2019, LNNS 104, pp. 436–447, 2020.
https://doi.org/10.1007/978-3-030-37497-6_50

the use of the surface of the machine parts. Surface roughness is often chosen as a criterion for the evaluation of the grinding process. There are many factors that affect surface roughness when grinding, among which lubrication technology is of great importance [1]. There are three popular cooling-lubrication methods used in grinding: cooling flood lubrication, minimum quantity lubrication and dry grinding. These methods have been studied and applied in grinding. However, in practical production, the method of cooling flood lubrication is still more commonly used thanks to the effects of fast heat transfer out of the cutting zone, lubrication of cutting zone, reduction of friction between surface of workpiece and grinding wheel [2]. The determination of the value of the parameters of lubrication for the workpiece surface with minor surface roughness will contribute to improving the economic and technical efficiency of the grinding process. There has been a large body of research on lubrication technology when grinding, with the main goal of such determining the influence of lubrication parameters on surface roughness of workpiece, the value of lubrication parameters, or new methods of lubricantion, or research for new solution, etc in order to process the workpiece surface with minor surface roughness.

Research has been conducted on the influence of pneumatic pressure when compressed air is blown into the grinding zone during machining SCM4 steel and SCM21 steel with CBN grinding wheel [3] as well as the influence of cold air flow pressure mixed with vegetable oil on surface roughness when grinding tool steel [4]. Other researchers also attempted to determine the optimal value of concentration, flow and pressure of lubricant when grinding 9CrSi steel with Oemeta Unimet AS 192 oil [5, 6]; the influence of workpiece velocity on surface roughness when grinding AISI1050 steel with Al2O3 grinding wheel in two cases: dry grinding and grinding with emulsion 5% [7]. Grinding process with lubricants: cooling air, emulsion 4%, and compressed air combined with mist water tested [8]. Research on grinding SUJ2 bearing steel with AL2O3 grinding wheel and CBN grinding wheel on the surface of workpiece was also conducted [9]. In this paper, the method of cooling flood lubrication with a flow of 25 L/min for four types of lubrication solutions including Emulsion, Machinery coolant, Tectyl cool 1240, Tectyl cool 1290 is used. The influence of lubricant concentration on surface roughness when grinding stainless steel 3X13 with black silicon carbide grinding wheel was experimented [10]. Empirical studies have been conducted on grinding the AISI 304 stainless steel on the surface grinder with two cooling-lubrication methods: using soluble oil and using liquefied nitrogen gas [11]. The influence of the type of lubricant on surface roughness when grinding 9XC steel with three types of cooling lubricants: Caltex Aquatex 3180 oil, AVANTIN 361I oil and JP Way oil was studied [12]. Many studies have been conducted on nozzle-formed design, position of cooling lubricant nozzle during grinding process [13] as well as the grinding the GCr15 with CBN grinding wheel in two cases: dry grinding and wet grinding [14], etc.

SKD11 steel is a type of material widely used in mechanical engineering for making dies, cutting tool, etc thanks to outstanding advantages such as high hardness, high abrasion resistance, low tempering stress, etc. This is a type of steel commonly used in grinding technology. There are many types of lubricant used in grinding technology, of which, Tectyl cool 290MC and Emulsion are the two most common, with small value of surface roughness of workpiece and low cost. However, until now, no research has shown the influence of lubrication parameters on surface roughness of workpiece, the optimal value of lubrication parameters when processing this material on surface grinder. These issue will be studied in this article to process the surface of the machine parts with minor surface roughness.

2 Materials and Methods

2.1 Location and Time

This study was conducted in the manufacturing technology room – Faculty of mechanical engineering – Ha Noi University of Industry (Vietnam) from January to March, 2019.

2.2 Experimental Component

The component material was SKD11 steel under thermal treatment at hardness of 58 ÷ 62HRC, 80 × 40 × 10 mm size. The chemical composition of major elements of SKD11 steel after thermal treatment is presented in Table 1.

Table 1. Chemical composition of SKD11 steel

Element	C	Mn		Si	Cr	V	Mo	Ni
%	1.5	0.3		0.25	11.5	0.25	0.3	0.35

2.3 Machine Tool and Grinding Wheel

The surface grinder and grinding wheel used in this study are APSG-820/2A (Fig. 1) and 36A60LV models, respectively; the size of the outer diameter x height and inner diameter of wheel are 180 × 13 × 31.75 (mm).

Fig. 1. Experimental machine-tool

2.4 Measuring Equipment

Surface roughness is measured by Tester SJ-201 (Mitutoyo - Japan), shown in Fig. 2. Each sample is measured for 3 times and the roughness value at each test is the average value of 3 consecutive measurements.

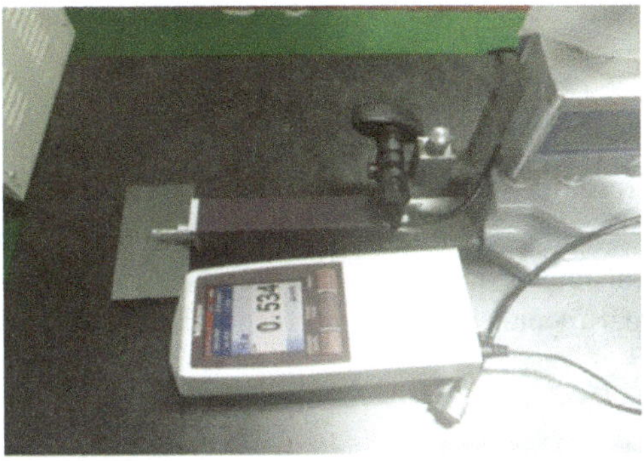

Fig. 2. SJ-201 roughness tester

2.5 Coolant

The coolant used in this research are Tectyl cool 290MC and Emulsion. These types of coolant are commonly used in cutting processing in general and processing by grinding method in particular.

2.6 Design of Experiment

The Central Composite Design (CCD) matrix is used in this research. This is the most commonly matrix for mechanical engineering optimization [15].

Table 2. Experimental design matrix

No	Flow	Concentration
1	−1	−1
2	1	−1
3	−1	1
4	1	1
5	−1.41	0
6	1.41	0
7	0	−1.41
8	0	1.41
9	0	0
10	0	0

(continued)

Table 2. (*continued*)

No	Flow	Concentration
11	0	0
12	0	0
13	0	0

For each type of lubricant, the number of experiments of the matrix includes: 2 k = 4 in original experiment, in which k = 2 identified as the number of experimental variable; 2 k = 4, representing the number of axial experiments at coding levels α and −α, where $\alpha = \sqrt{4}2^k = \sqrt{4}4 = 1.41$; the number of central points is 5 [15]. Accordingly, the experimental matrix in this research consists of 13 experiments (Table 2). The values of the experimental variables corresponding to the coding values are presented in Table 3.

Table 3. Design factors and their levels

Parameter	Symbol	Unit	Value at levels				
			−1.41	−1	0	1	1.41
Flow	F	Litre/min	1.52	1.8	2.5	3.2	3.48
Concentration	C	%	2.59	3	4	5	5.41

2.7 Grinding Condition

A grinding test is carried out according to the sequence of experimental points in Table 2 with the coding value of the parameters as shown in Table 3, for each type of coolant. In addition, for each experimental point, the fixed cutting parameters include cutting speed of 26 m/s, depth of cut 0.015 mm, velocity of workpiece at 12 m/min and crossing traverse at 6 mm/stroke.

3 Results and Discussion

Surface roughness when grinding is presented in Table 4. After using Minitab 16 software to analyze test results, the analytical results are presented as shown in Tables 5, 6 and Figs. 3, 4 for Tectyl cool 290MC; and Tables 6, 7 and Figs. 5, 6 for Emulsion (Tables 8, 9).

Based on these results, the regression equation that presents the relationship between surface roughness and the flow and concentration of the each lubricant, is shown (Eq. 1), with the determination coefficient of 98.27% for Tectyl cool 290MC

Table 4. Experimental results

No	Coding value		Actual value		Ra (μm)	
	F	C	F (litre/min)	C (%)	Tectyl cool 290MC	Emulsion
1	−1	−1	1.8	3	0.598	0.658
2	1	−1	3.2	3	0.59	0.708
3	−1	1	1.8	5	0.578	0.694
4	1	1	3.2	5	0.476	0.571
5	−1.41	0	1.52	4	0.593	0.771
6	1.41	0	3.48	4	0.527	0.632
7	0	−1.41	2.5	2.59	0.618	0.742
8	0	1.41	2.5	5.41	0.518	0.544
9	0	0	2.5	4	0.428	0.514
10	0	0	2.5	4	0.424	0.509
11	0	0	2.5	4	0.414	0.455
12	0	0	2.5	4	0.453	0.544
13	0	0	2.5	4	0.407	0.488

Table 5. Results of variance analysis when using Tectyl cool 290MC

Analysis of Variance for Ra

Source	DF	Seq SS	Adj SS	Adj MS	F	P
Regression	5	0.074704	0.074704	0.014941	79.40	0.000
Linear	2	0.014651	0.014651	0.007325	38.93	0.000
F	1	0.005170	0.005170	0.005170	27.47	0.001
C	1	0.009481	0.009481	0.009481	50.39	0.000
Square	2	0.057844	0.057844	0.028922	153.70	0.000
F*F	1	0.023336	0.030694	0.030694	163.12	0.000
C*C	1	0.034508	0.034508	0.034508	183.39	0.000
Interaction	1	0.002209	0.002209	0.002209	11.74	0.011
F*C	1	0.002209	0.002209	0.002209	11.74	0.011
Residual Error	7	0.001317	0.001317	0.000188		
Lack-of-Fit	3	0.000078	0.000078	0.000026	0.08	0.965
Pure Error	4	0.001239	0.001239	0.000310		
Total	12	0.076021				

(Eq. 2), and 91.73% for Emulsion. These equations are the basis for selecting the flow and concentration of lubricants for each specific requirement of the value of surface roughness with each type of lubricant.

$$R_a = 0.42521 - 0.0359F - 0.04861C + 0.13262F^2 + 0.14062C^2 - 0.04672FC \quad (1)$$

$$R_a = 0.50204 - 0.04755F - 0.06721C + 0.1917F^2 + 0.1332C^2 - 0.08599FC \quad (2)$$

Based on the results in Table 4, the values of surface roughness of workpiece in two case of lubricant are shown in Fig. 7.

Table 6. Regression model information when using Tectyl cool 290MC

Response Surface Regression: Ra versus F, C

The analysis was done using coded units.

Estimated Regression Coefficients for Ra

Term	Coef	SE Coef	T	P
Constant	0.42521	0.006135	69.313	0.000
F	-0.03590	0.006849	-5.241	0.001
C	-0.04861	0.006849	-7.098	0.000
F*F	0.13262	0.010384	12.772	0.000
C*C	0.14062	0.010384	13.542	0.000
F*C	-0.04672	0.013636	-3.426	0.011

S = 0.0137176 PRESS = 0.00249317

R-Sq = 98.27% R-Sq(pred) = 96.72% R-Sq(adj) = 97.03%

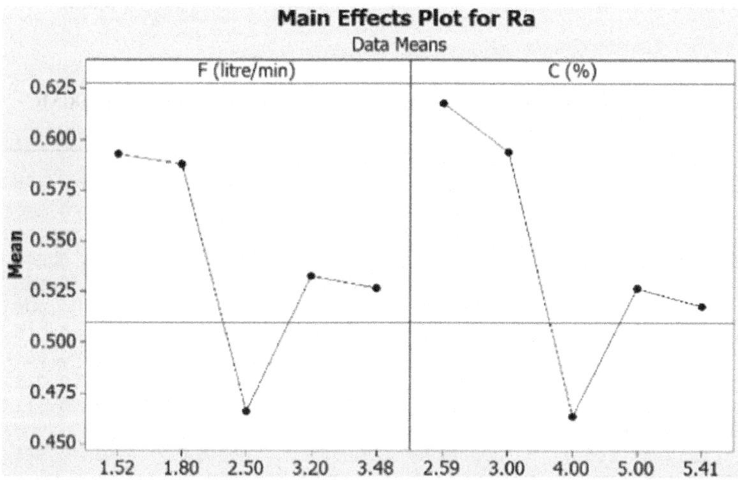

Fig. 3. Graph on influence of parameters F and C on Ra when using Tectyl cool 290MC

Based on Eqs. 1 and 2, the problem of optimizing the objective function of surface roughness was solved to determine the optimal value of concentration and flow in each type of lubricant. In the encoded form of the input parameters, the optimal problem is written as in expression (3).

$$\begin{cases} R_a = f(F, C) \rightarrow min \\ -1.41 \leq F, C \leq 1.41 \\ R_a > 0 \end{cases} \quad (3)$$

The optimal value in encoded form, numerical form of flow and concentration of lubricant; the surface roughness value of the workpiece are shown in Table 7.

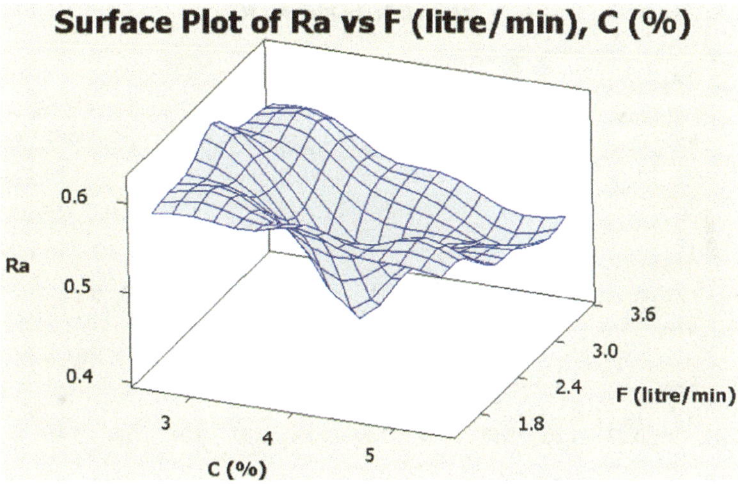

Fig. 4. Graph on interactive influence of F and C on Ra when using Tectyl cool 290MC

Table 7. Results of variance analysis when using Emulsion

Analysis of Variance for Ra						
Source	DF	Seq SS	Adj SS	Adj MS	F	P
Regression	5	0.119789	0.119789	0.023958	15.54	0.001
Linear	2	0.027192	0.027192	0.013596	8.82	0.012
F	1	0.009071	0.009071	0.009071	5.88	0.046
C	1	0.018121	0.018121	0.018121	11.75	0.011
Square	2	0.085115	0.085115	0.042557	27.60	0.000
F*F	1	0.054152	0.064130	0.064130	41.59	0.000
C*C	1	0.030963	0.030963	0.030963	20.08	0.003
Interaction	1	0.007482	0.007482	0.007482	4.85	0.063
F*C	1	0.007482	0.007482	0.007482	4.85	0.063
Residual Error	7	0.010793	0.010793	0.001542		
Lack-of-Fit	3	0.006431	0.006431	0.002144	1.97	0.261
Pure Error	4	0.004362	0.004362	0.001091		
Total	12	0.130583				

From the above Tables and Figures, for two types of lubricant, it can be shown that:

- The parameters of the flow and concentration of lubricant as well as the interaction between them have a significant influence on surface roughness, in which the influence of solution concentration is greater effect than that of the flow.
- The influence of flow and concentration of lubricant on surface roughness is quite complicated. When the flow of the solution and the solution concentration is increased, the surface roughness tends to decrease. However, if the values of these two parameters continue increasing, the surface roughness tends to increase.

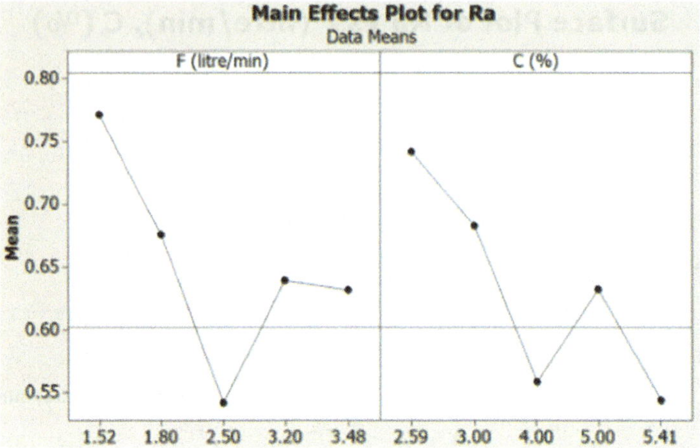

Fig. 5. Graph on influence of parameters F and C on Ra when using Emulsion

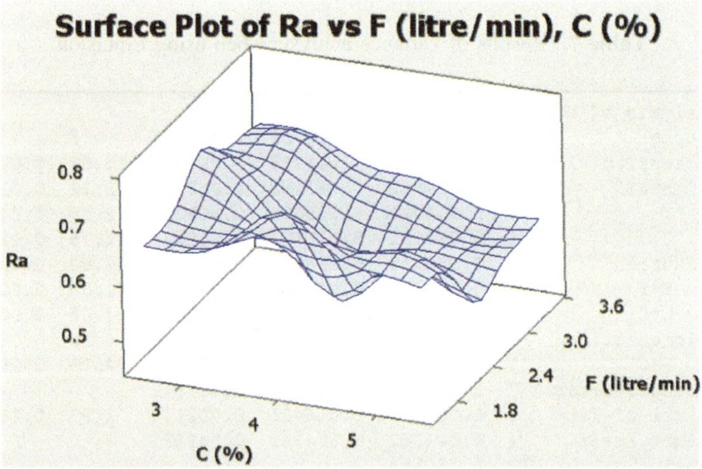

Fig. 6. Graph on interactive influence of F and C on Ra when using Emulsion

- In the same processing condition, when Tectyl cool 290MC is used, the value of surface roughness is smaller than when Emulsion oil is used.
- For Tectyl cool 290MC, the optimal values of the flow and concentration of lubricant are 2.62 (L/min) and 4.2 (%), respectively, when the processed surface roughness reaches 0.41 (μm). And for Emulsion, the optimal values of the flow and concentration of lubricant are 2.63 (L/min) and 4.32 (%), respectively, when the processed surface roughness reaches 0.49 (μm).

Table 8. Regression model information when using Emulsion

Response Surface Regression: Ra versus F, C

The analysis was done using coded units.

Estimated Regression Coefficients for Ra

Term	Coef	SE Coef	T	P
Constant	0.50204	0.01756	28.589	0.000
F	-0.04755	0.01960	-2.426	0.046
C	-0.06721	0.01960	-3.428	0.011
F*F	0.19170	0.02973	6.449	0.000
C*C	0.13320	0.02973	4.481	0.003
F*C	-0.08599	0.03903	-2.203	0.063

S = 0.0392671 PRESS = 0.0525510

R-Sq = 91.73% R-Sq(pred) = 59.76% R-Sq(adj) = 85.83%

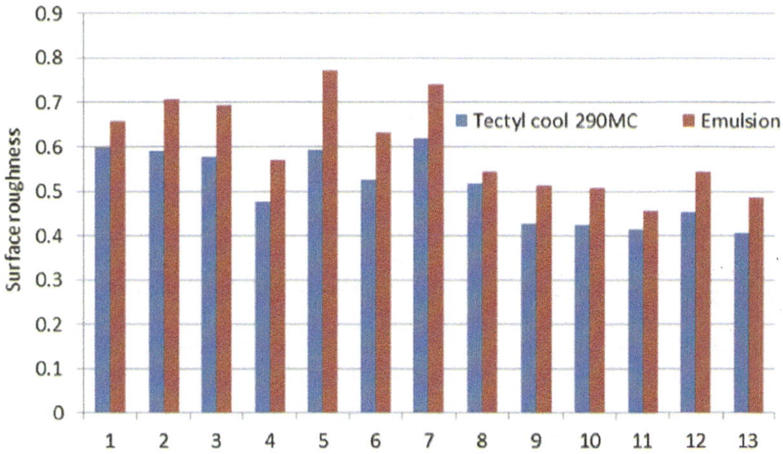

Fig. 7. Surface roughness in two types of lubricant

Table 9. The optimal value of the parameters

Type of lubricant	Decoded value		Numerical value		R_a (μm)
	F	C	$F(litre/min)$	$C(\%)$	
Tectyl cool 290MC	0.17	0.20	2.62	4.20	0.41
Emulsion	0.19	0.32	2.63	4.32	0.49

4 Conclusion

Based on the results in this research, when surface grinding of SKD11 steel with 36A60LV grinding wheel, lubricants Tectyl cool 290MC and Emulsion, some conclusions are given as follows:

- The influence of flow and concentration of lubricant on surface roughness is quite complicated. When the flow of the solution and the solution concentration is increased, the surface roughness tends to decrease. However, if the values of these two parameters continue increasing, the surface roughness tends to increase.
- When Tectyl cool 290MC lubricant is used, the value of surface roughness is smaller than when Emulsion is used.
- In fine grinding conditions, with Tectyl cool 290MC lubricant, if the values of the flow and concentration of lubricant are 2.62 (L/min) and 4.2 (%) respectively, the surface roughness will reach the smallest value at 0.41 (μm).
- For lubricant is Emulsion, if the values of the flow and concentration of lubricant are 2.63 (L/min) and 4.32 (%) respectively, the surface roughness will reach the smallest value at 0.49 (μm).

Acknowledgements. The work described in this paper was supported by Ha Noi University of Industry (https://www.haui.edu.vn/vn).

References

1. Malkin, S., Guo, C.: Grinding Technology - Theory and Applications of Machining with Abrasives. Industrial Press, New York (2008)
2. Hung, L.X., Pi, V.N., Tung, L.A., Khiem, V.H.: A Study on Coolant Parameters in Internal Grinding of 9CrSi Harden Steel, Science and technology project at University level. Thai Nguyen University (2018)
3. Choi, H.Z., Lee, S.W., Jeong, H.D.: A comparison of the cooling effects of compressed cold air and coolant for cylindrical grinding with a CBN wheel. J. Mater. Process. Technol. **111**, 265–268 (2001)
4. Xiao, K.Q., Zhang, L.C.: The effect of compressed cold air and vegetable oil on the subsurface residual stress of ground tool steel. J. Mater. Process. Technol. **178**, 9–13 (2006)
5. Tu, H.X., Pi, V.N., Jun, G.: A study on determination of optimum parameters for lubrication in external cylindrical grinding base on taguchi method. Key Eng. Mater. **796**, 97–102 (2019)
6. Tu, H.X., Jun, G., Hien, B.T., Hung, L.X., Tung, L.A., Pi, V.N.: Determining optimum parameters of cutting fluid in external grinding of 9CrSi steel using Taguchi technique. Int. J. Mech. Eng. (SSRG-IJME) **5**(6), 1–5 (2018)
7. Kiyak, M., Cakir, O., Altan, E.: A study on surface roughness in external cylindrical grinding. In: 12th International Scientific Conference – Achievements in Mechanical & Materials Engineering (AAME), pp. 459–462 (2003)
8. Choi, H.Z., Lee, S.W., Kim, D.J.: Optimization of cooling effect in the grinding with mist type coolant. Korea Institute of Industrial (2001)

9. Trung, D.D.: Research on the influence of lubricant on surface quality when grinding SUJ bearing steel with Al_2O_3 and CBN grinding wheel on surface grinder, Master thesis, Thai Nguyen University of Technology (2010)
10. Cuong, N., Man, N.D.: Research on choosing cutting parameters and technological measures for improvement of economic and technical efficiency when grinding 3X13 stainless steel with Hai Duong grinding wheel, Science research project at ministerial level. Thai Nguyen University (2009)
11. Fredj, N.B., Sidhom, H., Braham, C.: Ground surface improvement of the austenitic stainless steel AISI304 using cryogenic cooling. Surf. Coat. Technol. **200**, 4846–4860 (2006)
12. Thu, N.T.: Research on the influence of cooling-lubrication technology on the machined surface quality when grinding 9XC tempered steel through with Hai Duong grindstone, Master thesis. Thai Nguyen University of Technology (2015)
13. Webster, J.A, Cui, C: Flow rate and jet velocity determination for design of a grinding cooling system. Technical Papers Supplement of the First International Machining and Grinding Conference, Dearborn, Michigan, pp. 345–356 (1995)
14. Stephenson, D.J., Jin, T.: Physical basics in grinding. In: European Conference on Grinding, Aachen, pp. 13–21 (2003)
15. Myers, R.H., Montgomery, D.C., Anderson-Cook, C.M.: Response Surface Methodology: Process and Product Optimization Using Designed Experiments, 3rd edn. Wiley, Hoboken (2009)

Investigations of Defects in Inverted Organic Solar Cells

Jordan Goilard[1], Kai Xue[1], Cédric Renaud[2], P. Y. Chen[3], Sheng-Hsiung Yang[3], and Thien-Phap Nguyen[1(✉)]

[1] Institut des Matériaux Jean Rouxel, University of Nantes, CNRS, Nantes, France
Thien-Phap.nguyen@cnrs-imn.fr
[2] Laplace University of Toulouse, Toulouse Cedex 9, France
[3] Institute of Lighting and Energy Photonics, National Chiao Tung University, Hsinchu, Taiwan

Abstract. The utilization of inverted structures in organic solar cells (OSCs) has been demonstrated to provide higher efficiency and stability as compared to standard structure devices. The improvement of the cell performance is thought to be linked to the electron and hole transporting layers (ETL and HTL), which also play key role in preventing the cell from extrinsic degradation. However, from the defect point of view, the presence of these layers can introduce new sources of charge carrier trapping, and therefore can impact on the long term stability and electrical property of the solar cells. In this work, we report results on investigations of defects in inverted solar cells using blends of poly (hexylthiophene) (P3HT) and 6,6-phenyl-C61-butyric acid methyl ester (PCBM) as the absorbing layer, while zinc oxide (ZnO) was used as the ETL. The defects in devices were determined by the charge based deep level spectroscopy (Q-DLTS). The results indicated new defect states in inverted OSCs as compared to the standard P3HT:PCBM device. Defects of energy in the range of 10–470 meV have been determined by the charge peak corresponding to the high relaxation time domain, assigning to the heterojunction bulk of the cells. Additional traps observed through the onset of a charge peak in the low relaxation time domain have a low energy level and are assigned to interface defects. These defects are supposed to originate from the zinc oxide contact and may affect the stability of the solar cells in operation.

Keywords: Organic solar cells · Inverted structures · Defects · Interface

1 Introduction

At present, organic solar cells (OSCs) represent the most attractive technology for sustainable energy production of low cost by their ease to fabricate modules of large surface on flexible or rollable substrates. Intensive investigations to improve their performance have been undertaken by worldwide research work and a threshold for 10% conversion efficiency is recently reached and passed. It can be observed that progress in reduction of production costs as well as in efficiency of OSCs has been much faster than in their stability improvement. Several approaches have been

K.-U. Sattler et al. (Eds.): ICERA 2019, LNNS 104, pp. 448–454, 2020.
https://doi.org/10.1007/978-3-030-37497-6_51

proposed for obtaining a better stability of devices under operation. One of the promising approaches for an efficient protection of devices against moistures consists of changing the device architecture by using the inverted structure cells.

The structure of the inverted cells is typically: ITO/hole blocking layer (HBL)/ active layer/electron blocking layer (EBL)/metal electrode. ITO remains a good transparent conducting oxide but usually is covered by a layer of metal oxide such as zinc oxide (ZnO) or titania (TiO_2), whose work function allows a fine electronic level matching with the conduction band of the polymer of the active blend [1]. The influence of these interfacial layers on the performance of inverted OSCs has been examined by different groups [2, 3]. In this work, we reports on defect investigation results obtained in inverted solar cells using a poly(3-hexylthiophene-2,5-diyl) (P3HT): phenyl-C61-butyric acid methyl ester (PCBM) blend as an active layer. The current-voltage (I-V) characteristic of the cells was performed and the defect parameters were determined by using the charge based deep level spectroscopy (Q-DLTS). By analyzing the defect parameters in inverted devices and by comparing them to those obtained in standard ones, we discuss the modifications in the trap characteristics and from the discussion results, we propose a possible scenario of process of defect formation in inverted devices.

2 Experimental

2.1 Materials and Preparation of Devices

PEDOT:PSS aqueous solution (CleviosTM P VP AI 4083) was purchased from Heraeus Precious Metals GmbH & Co. KG. P3HT and PCBM were purchased from FEM Technology Co., Inc. Other chemicals including reagents and solvents were purchased from Alfa Aesar and used without further purification.

The preparation of ZnO nanorod arrays and fabrication of inverted devices were referred to the previous literature [4]. The device structure is ITO/ZnO/P3HT: PCBM/PEDOT/WO_3/Au. It should be noted that we did not fabricate devices having a symmetrical structure to that of the inverted cells because of the high temperature treatment of the ZnO layer, which would alter the organic films. Therefore, standard cells were fabricated with a conventional structure of ITO/PEDOT:PSS/P3HT: PCBM/Ca/Al [5].

2.2 Electrical Characterization

Photovoltaic characteristics of the devices were measured with a Keithley 2400 source measurement unit under simulated AM1.5G illumination from a 300 W solar simulator (Oriel 9600, 150 W) with an intensity of 100 mW/cm^2. For studying the current density-voltage-temperature (J-V-T) characteristics in the dark, the solar cells were mounted in an Oxford cryostat, the temperature of which was controlled by an ITC controller. The Q-DLTS measurements were performed using an automated system, ASMEC-06, supplied by InOmTech Inc, the sample being mounted in a cryostat working in the temperature range from 77 to 320 K.

The key points of the Q-DLTS measurements can be summarized as follows [6]. In order to determine the trap levels in devices, the spectra were recorded with a constant charging voltage ΔV and at a constant temperature T using different charging times t_C in the range of 0.5 ms–5 s. Depending on t_C, the trap filling and hence the intensity of the Q-DLTS peaks varies. Short charging times allow low-density traps to be filled and with increasing t_C, they are entirely filled. The corresponding Q-DLTS peaks will be then maximized and saturated. By the same process, charge peak of high density traps will be saturated by using long charging times. Thus, it is possible to determine the relaxation times τ_M of different trap levels by identifying the Q-DLTS peak maxima from the spectra recorded with different charging times at a given charging voltage and a given temperature. Knowing τ_M one can resolve a spectrum into components. The trap parameters including activation energy E_T, capture cross-section σ and trap density N_T can be evaluated from the measurements of the Q-DLTS spectra as a function of temperature.

3 Results and Discussion

3.1 Current-Voltage Characteristic

The structure of the inverted organic solar cell and its J-V characteristic are shown in Fig. 1. Under illumination, the inverted cell exhibits an average short circuit current (J_{SC}), open circuit voltage (V_{OC}), and fill factor (FF) of 14.9 mA/cm^2, 0.54 V and 41%, respectively (Fig. 1). The power conversion efficiency of the cell is 3.31%. The photovoltaic performance of the devices is reproducible from different cells, which were fabricated under the same conditions.

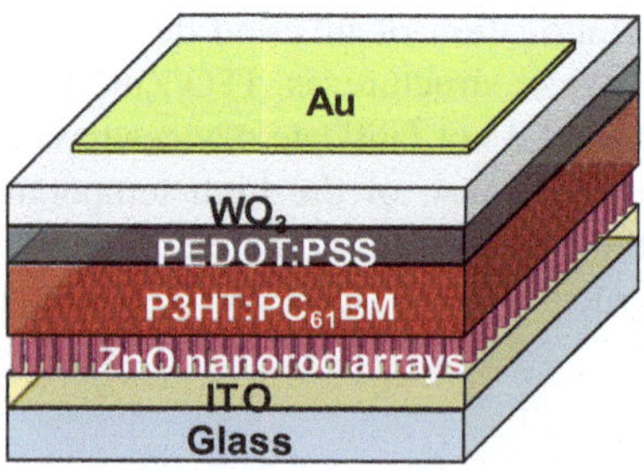

(a)

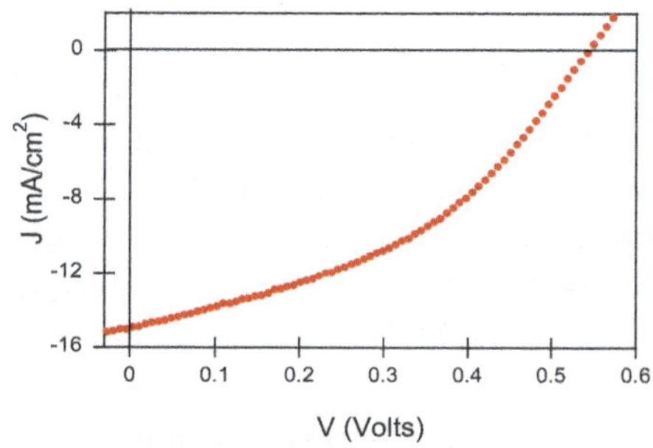

(b)

Fig. 1. (a) Structure of the inverted organic solar cell and (b) J-V characteristic of the device under AM1.5G illumination.

3.2 Analysis of QDLTS Spectra

Figure 2a show the Q-DLTS spectra of an inverted OSC at T = 300 K with an applied voltage ΔV = +1 V and different charging times. The spectra show two apparent peaks.

A peak labeled 1, is localized in the low relaxation time region and is apparently independent of the charging time when the temperature T is fixed. A second peak labeled 2 is localized in the high relaxation time region. Peak 2 is similar to that observed in the Q-DLTS spectra of a standard OSC [5]. When the relaxation time increases, its intensity increases and its peak shifts to higher relaxation times. Indeed, the apparent shift of the peak corresponds to the growth of one trap level of the device, which becomes progressively dominant in the measurement conditions (T = 300 K, ΔV = +1 V).

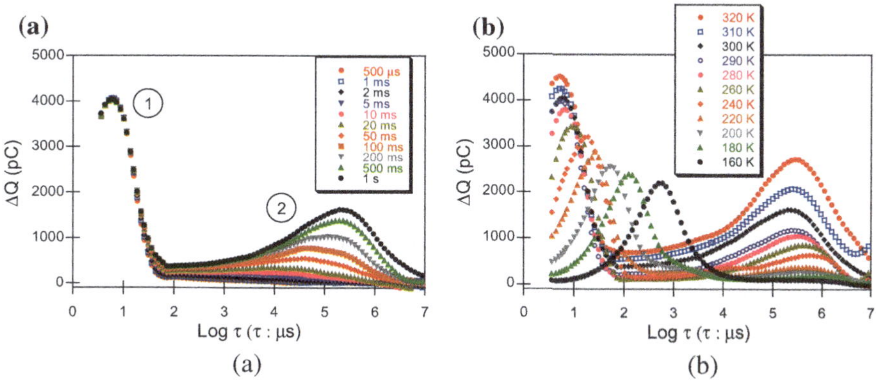

Fig. 2. Q-DLTS spectra of an inverted OSC : (a) at T = 300 K with an applied voltage ΔV = +1 V and different charging times, (b) with an applied voltage ΔV = +1 V, a charging time t_C = 1 s and at different temperatures of 160–320 K.

For the determination of the activation energy of the traps, measurements of the relaxation times at different temperatures of the devices have been carried out within the range of 160–320 K and are shown in Fig. 2b. We observe two different kinetic behaviors of the Q-DLTS peaks. With increasing temperatures, the relaxation time of peak 1 decreases quickly while the shift of peak 2 is much slower. A fast variation of the relaxation times with the temperature variation suggests that defects are shallow ones.

By analyzing the spectra recorded with short charging times, we can determine the position of the trap peaks and use it to calculate the relaxation times τ_M of the trapped charges. The QDLTS spectrum of each trap level is then computed by using the technique described in the previous work [5]. Figure 3 shows the resolved QDLTS spectrum of the inverted cell recorded at T = 300 K with an applied voltage of ΔV = +1 V and a charging time of t_C = 1 s. Traps from A2 to F2 have been found in the standard cells [5].

In Figs. 4 and 5, the detailed Q-DLTS spectra for both peaks 1 and 2 at different temperatures and the corresponding Arrhenius plots are shown. Finally, using this protocol for each temperature of the sample over the range from 160 to 320 K, measurements of a set of relaxation times at different temperatures have been carried out.

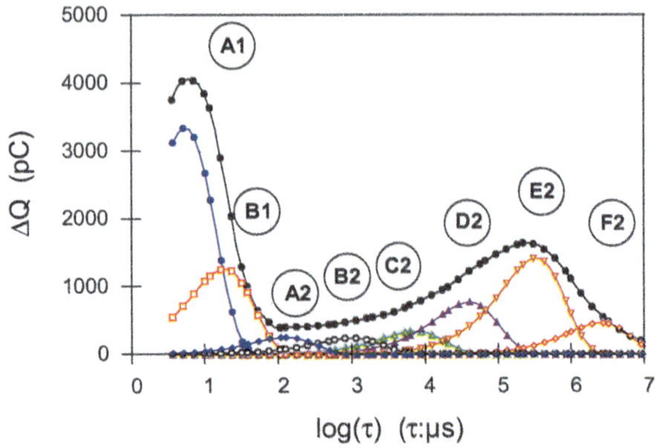

Fig. 3. Resolution of the Q-DLTS spectrum using the following parameters: T = 300 K, $\Delta V = +1$ V, $t_C = 1$ s.

Using these data, we determine the trap activation energy by plotting the Arrhenius graph. The trap parameters of all the components including the capture cross section and the trap density are summarized in Table 1.

Basing upon the above analysis, we suggest the following process of formation of defects in inverted solar cells using ZnO nanorod arrays as an electron collection layer. Defects detected in the P3HT:PCBM active layer have not been affected by the use of the inverted structure. These defects are the same as those found in standard cells, that is, they are created in both the polymer and the fullerene, and have not been modified by the use of additional layer on both electrode contacts (polyfluorene on the HBL side and PEDOT:PSS on the EBL side). Additional defects (A1 and B1) are mainly introduced by the ZnO arrays, which capture the electrons during their transport. The trap parameters of these defects are similar to those reported in ZnO based devices [7].

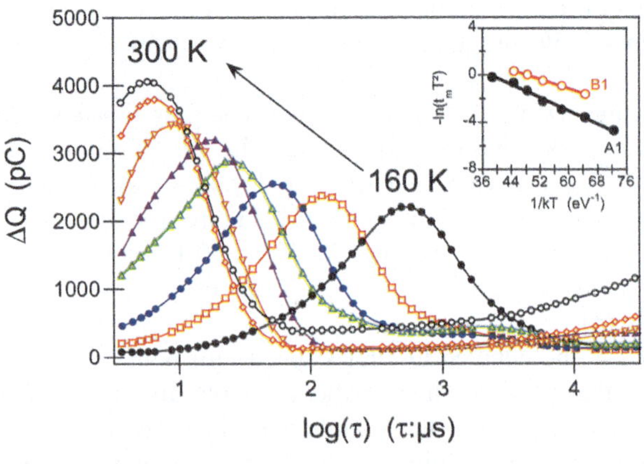

Fig. 4. Q-DLTS spectra and Arrhenius graph for peak 1.

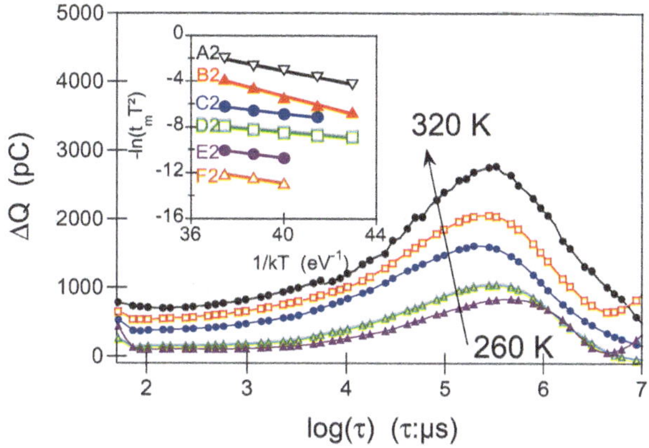

Fig. 5. Q-DLTS spectra and Arrhenius graph for peak 2.

Table 1. Trap parameters in inverted diode structures for $\Delta V = +1$ V.

Peak	E_T (eV)	σ (cm^2)	N_T (cm^{-3})
A1	0.10	6×10^{-20}	4.3×10^{17}
B1	0.15	4×10^{-20}	1.6×10^{17}
A2	0.36	1×10^{-16}	3.5×10^{16}
B2	0.47	1×10^{-17}	3.3×10^{16}
C2	0.20	3×10^{-21}	4.5×10^{16}
D2	0.09	4×10^{-20}	9.9×10^{17}
E2	0.26	2×10^{-21}	1.8×10^{17}
F2	0.31	1×10^{-21}	6.3×10^{16}

4 Conclusion

We have investigated defects in inverted organic solar cells using P3HT:PCBM blend as an active layer by using the Q-DLTS technique. In addition to the traps created in the blend active layer, we have observed new defects having weak relaxation times at room temperature. These defects have been assigned to traps in ZnO nanorod arrays which are native defects and not impurities introduced on the film surface. Since ZnO buffer player plays a key role in the performance of inverted solar cells, intensive investigations have focused on the improvement of its structure aiming at enhancing its stability and at reducing defects.

References

1. Lattante, S.: Electron and hole transport layers: their use in inverted bulk heterojunction polymer solar cells. Electronics **3**, 132–164 (2014)
2. Espinosa, N., Dam, H.F., Tanenbaum, D.M., Andreasen, J.W., Jørgensen, M., Krebs, F.C.: Roll-to-roll processing of inverted polymer solar cells using hydrated vanadium (V) oxide as a PEDOT:PSS replacement. Materials **4**, 169–182 (2011)
3. Bovill, E., Scarratt, N., Griffin, J., Yi, H., Iraqi, A., Buckley, A.R., Kingsley, J.W., Lidzey, D. G.: The role of the hole-extraction layer in determining the operational stability of a poly carbazole:fullerene BHJ photovoltaic device. Appl. Phys. Lett. **106**, 073301 (2005)
4. Chen, W.C., Chen, P.Y., Yang, S.H.: Solution-processed hybrid light emitting and photovoltaic devices comprising zinc oxide nanorod arrays and tungsten trioxide layers. AIMS Mater. Sci. **4**, 551–560 (2017)
5. Nguyen, T.P., Renaud, C., Reisdorffer, F., Wang, L.: Degradation of PCBM:P3HT organic photovoltaic cells and structure changes as determined by defect investigation. J. Energy Photonics **2**, 021013 (2012)
6. Nguyen, T.P.: Defects in organic electronic devices. Phys. Status Solidi (a) **205**, 162–166 (2008)
7. Auret, F.D., Goodman, S.A., Legodi, M.J., Meyer, W.E., Look, D.C.: Electrical characterization of vapor-phase-grown single-crystal ZnO. Appl. Phys. Lett. **80**, 1340–1342 (2002)

Kinematic Analysis of a Resonant Flexible-Wing Nano Air Vehicle Using a Bond Graph Approach

Le Anh Doan[1](✉), Thanh Nghi Ngo[2], Nhu Thanh Vo[2], and Phuoc Vinh Dang[2]

[1] Department of Mechanical Engineering, The University of Danang-University of Technology and Education, 48, Cao Thang Street, Danang, Vietnam
doanleanh024@gmail.com

[2] Department of Mechanical Engineering, The University of Danang-University of Science and Technology, 54, Nguyen Luong Bang Street, Danang, Vietnam

Abstract. In recent decades, the prospect of exploiting the exceptional flying capacities of insects has prompted much research on the elaboration of flapping-wing nano air vehicles (FWNAV). However, when designing such a prototype, designers have to wade through a vast array of design solutions that reflect the wide variety of flying insects to identify the correct combination of parameters that meets their requirements. To alleviate this burden, the purpose of this paper is to develop a suitable tool to analyze the kinematics of a resonant flexible-wing nano air vehicle. The proposed tool uses a Bond Graph formalism because it is well suited to simulating multi-physical systems. Moreover, the prototype studied combines two resonant vibration modes – bending and twisting – to reproduce insect wing kinematics. This could be considered as the key to optimize the generated lift.

Keywords: Flapping wing nano air vehicle · Power · Energy · Bond graph · Flexible structure

1 Introduction

People have always been fascinated by the impressive aerial capacities of small birds and insects such as hovering, landing on sloping and inverted surfaces, flying backward, and recovering after a shock [1, 2]. In recent decades, much effort has been put into developing small-sized flying vehicles such as flapping-wing micro air vehicles (FWMAV) [3, 4] and flapping-wing nano air vehicles (FWNAV) [5, 6]. However, developing a flying vehicle with a wingspan of just a few millimeters and a very low weight of a few milligrams is a challenge. The weight and size limits imposed are in contrast to the huge amount of power required to overcome aerodynamic drag, the inertia of the constantly oscillating wing and accompanying mechanisms. Fortunately, the solution exists in nature, since insects such as flies, bees, wasps, and beetles use, mechanical resonance to counteract these problems [7]. This design concept – inspired

K.-U. Sattler et al. (Eds.): ICERA 2019, LNNS 104, pp. 455–461, 2020
https://doi.org/10.1007/978-3-030-37497-6_52

by the resonant thorax of these dipteran insects – is usually achieved by introducing elasticity into the micro and nano air vehicle structures. The elastic element can be any of the following components: helical spring [8, 9], virtual spring [10] or compliant link [11]. The aim, therefore, is to develop a model to analyze the performance of a flexible vehicle capable of operating at resonance. The model itself is original as it is a distributed-parameter model and is based on a flexible micro-structure. The vehicle studied is a micromachined FWNAV – named "object volantmimantl'insecte" (OVMI) or insect-mimicking flying object – powered by a single electromagnetic actuator to vibrate the structure made primarily of flexible materials [6].

2 OVMI Dynamic Bond Graph Model

2.1 Prototype Description

The FWNAV prototype studied here is a 3D skeleton composed mainly of SU-8 photoresist multilayers (Fig. 1). A very thin parylene layer (0.4 μm) based wing membrane was deposited on the skeleton veins resulting in wings 22 mm in length.

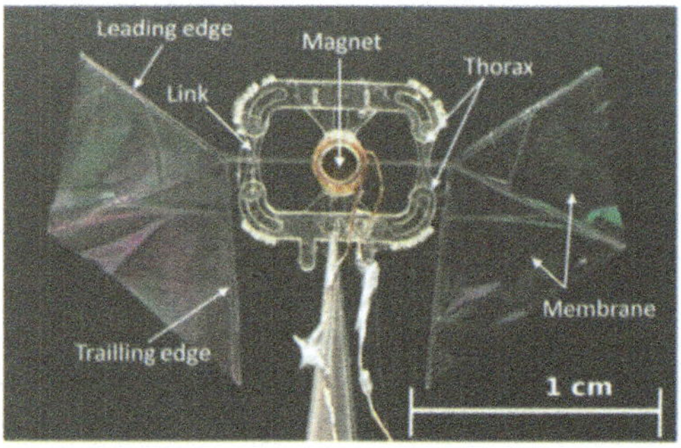

Fig. 1. OVMI prototype with wings and electromagnetic actuator with a total mass of 22 mg and a wingspan of 22 mm.

A made in-house coil glued to the thorax and a magnet positioned in the center of the tergum act as an electromagnetic actuator to generate and maintain the vibration. Weighing 22 mg in total, excluding the electronic board and battery, this prototype is within the dimensional range of insects. Kinematics of this flexible structure is defined by the combination of two elementary movements: a flapping motion and a twisting motion.

2.2 OVMI Word Bond Graph

Figure 2 presents the Word Bond Graph corresponding to the technological level of the model. Here, the global OVMI system is broken down into three subsystems: the generator, the electromagnetic actuator, and the wings. The first subsystem corresponds to a signal generator that supplies a sine wave to drive the electromagnetic actuator, whereas the last subsystem, named "Wings", actually includes the entire flexible skeleton. The Word Bond Graph has the "bonds" representing the bi-directional exchange of physical energy. Each bond depicts is a pair of power variables called flow and effort. For example, the bond next to the generator block represents the flow of electrical energy and the corresponding power variables are the current (i_i) and the voltage (U_i), the product of which is power (P_{in}).

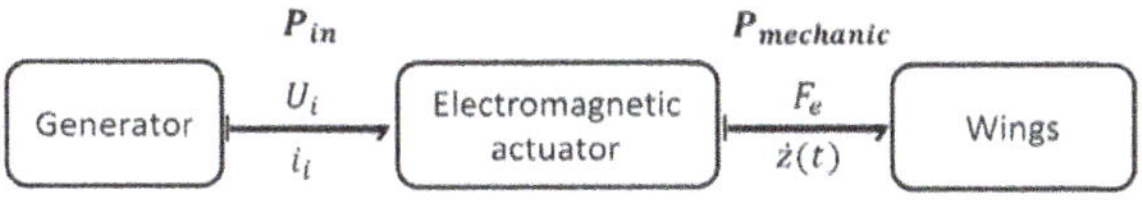

Fig. 2. Word Bond Graph of the prototype.

2.3 BG Model

Bond Graph models of the generator and the electromagnetic actuator are already developed in our previous work [12]. The originality is therefore on the model of the flexible skeleton. The "Wings" model actually includes the wings and all the beams that form the flexible structure but does not include membrane since its influence is decoupled from the kinematic part and represented only by an aerodynamic effect. This assumption is accepted as the membrane thickness is very small compared to the veins.

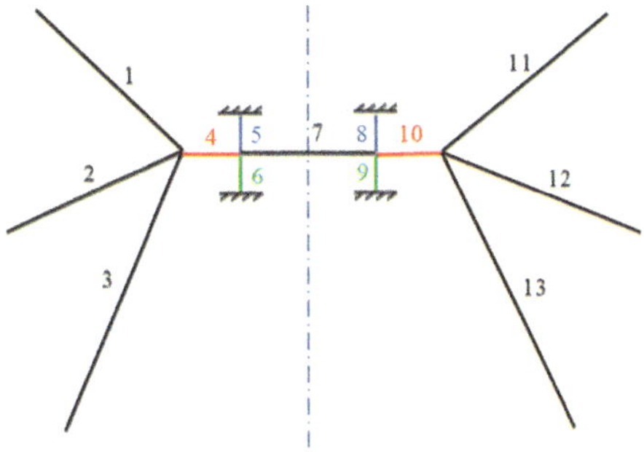

Fig. 3. Diagram of the "Wings" skeleton; the colors used to distinguish between vicinal beams.

As shown in Fig. 3, the flexible skeleton is represented by a set of thirteen Euler Bernoulli beams. The bending and twisting vibration in each beam can be expressed by the two following partial differential equations:

$$EI_i w_i''''(x_i,t) + \rho S_i \ddot{w}_i(x_i,t) = 0; GJ_i\theta_i''(x_i,t) - \rho J_i \ddot{\theta}_i(x_i,t) = 0 \tag{1}$$

where $w_i(x_i,t)$ and $\theta_i(x_i),t$ are respectively the transverse displacement (due to bending) and the twisting angle (the rotation of the cross section around the beam axis) of beam i, at axial position x_i and time t. The space and time derivatives are denoted by $w_i' = \partial w_i/\partial x_i$ and $\dot{w}_i = \partial w_i/\partial t$. E and ρ are respectively the Young's modulus and the density of the material (SU-8 photoresist) and $G = E/(2(1+\upsilon))$ is the shear modulus; S_i, Ii and Ji are the cross section's area, bending and twisting moments of inertia of beam i, respectively. The dynamics of the structure is computed by classical vibration mode expansion, that writes:

$$w_i(x_i,t) = \sum_{k=1}^{N} \phi_{i(n)}(x_i)q_n(t); \theta_i(x_i,t) = \sum_{k=1}^{N} \gamma_{i(n)}(x_i)q_n(t) \tag{2}$$

Where, $q_n(t)$ is the n-th. modal coordinate and $\phi_{i(n)}(x_i)$, $\gamma_{i(n)}(x_i)$ are n-th. mode shape of beam i at x_i (in bending and twisting, respectively). The modes $\left(\omega_n, \phi_{i(n)}(x_i), \gamma_{i(n)}(x_i)\right)$ (with ω_n being the n-th. natural angular frequency) are computed by using (1) along with some specific boundary conditions between the beams of the skeleton. Among the resonant modes, the two symmetric modes (modes 1 and 3) are more relevant to this work as they define directly two elementary motions of the wing including bending and twisting motions, named bending mode and twisting mode, respectively. To adapt the modal model to the Bond Graph formalism, each of the working modes is represented by a mass-spring-damper system [13]. The equivalent spring stiffness k_n and mass m_n of mode number n are determined by inserting (2) in (1) and using the orthogonal properties of the modes. One obtains the following integrals taken along all beams of the skeleton:

$$
\begin{aligned}
\mathrm{m_n} &= \sum_{i=1}^{13} \int_0^{L_i} \left[\rho S_i \phi_{i(n)}^2(x) + \rho J_i \gamma_{i(n)}^2(x)\right] dx \\
k_n &= \sum_{i=1}^{13} \int_0^{L_i} \left[E I_i \phi_{i(n)}^{''''}(x) \phi_{i(n)}(x) + G J_i \gamma_{i(n)}^{''}(x) \gamma_{i(n)}(x)\right] dx
\end{aligned}
\tag{3}
$$

Where, L_i is the length of the ith beam. According to classical vibration theory [14], the modal coordinates $q_n(t)$ satisfy the following ordinary differential equation:

$$
m_{\mathrm{n}} \ddot{q}_{\mathrm{n}}(t) + R_{\mathrm{n}} \dot{q}_{\mathrm{n}}(t) + k_{\mathrm{n}} q_{\mathrm{n}}(t) = \phi_{7(\mathrm{n})}\left(\frac{L_7}{2}\right) F_e \tag{4}
$$

where $\phi_{7(\mathrm{n})}\left(\frac{L_7}{2}\right)$ is the bending modal amplitude at the midpoint of beam number seven when mode n is excited. F_e acts in the middle of this beam, and in the center of the actuator. (4) has the form of Newton's second law. As we are interested in bending and twisting modes, only two sets are employed: (m_1, k_1, R_1) and (m_3, k_3, R_3). The damping values R_n are derived from the quality factor discussed in the next section.

In the Bond Graph formalism, force balance equations as in (4) can be represented by a 1-junction. The spring and the mass are characterized by a C-element (potential energy) and an I-element (kinetic energy), respectively, whereas the R-element depicts damping (R_n) (Fig. 4).

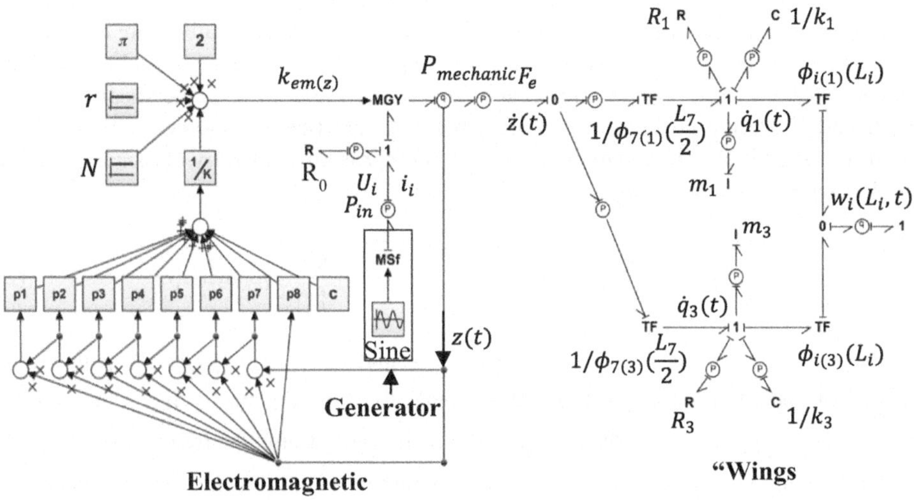

Fig. 4. Global OVMI Bond Graph model.

2.4 Parameters Estimation

For the generator, a current of 300 mA was chosen. This value was calculated and verified experimentally to made the visualization of the wing movement without burning the coil or breaking the prototype. Concerning the electromagnetic actuator, a polynomial of degree 8 representing $B_{interp}(z)$ is obtained from Finite Element analysis as provided in [15]. The equivalent stiffness and mass corresponding to each resonant mode could be theoretically derived from (4). The equivalent damping, including the damping effect of the aeroelastic forces, is however evaluated experimentally through the impact of air pressure on the dynamic behavior of the prototypes. The frequency response functions (FRFs) at specific points on the prototype can thus be established for different pressures. These FRFs make it possible to evaluate aeroelastic damping at different pressures by identifying the bandwidth at -3 dB below the resonant peak level.

3 Kinetic Simulation and Operating Modes Identification

By employing the global model, we searched for the existence of operating modes using the frequency response. The frequency response shown in Fig. 5 is deduced from the transfer function presenting the ratio between the wing motion at the free end of the leading edge (beam 11) and the electrical current (i_i). The motion (1) (in Fig. 5) is always the synthesis of bending (2) and twisting (3) mode shapes. In Fig. 5a), local peaks of a magnitude of (1) occur close to the resonant frequencies of (2) and (3) where their phases intercept at 90°, as depicted in Fig. 5b).

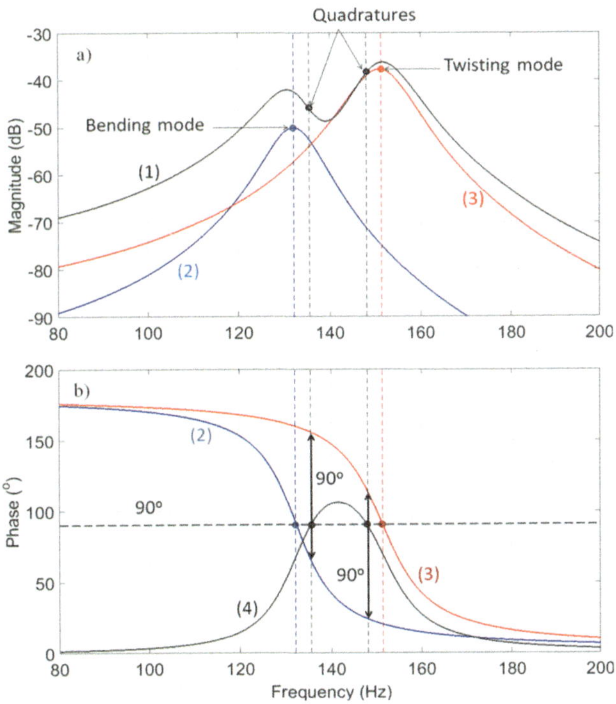

Fig. 5. Simulated Bond Graph amplitude and frequency response phase of the prototype

The peaks are defined as bending and twisting modes. Quadrature is found when observing the phase shift between (2) and (3). Looking at Fig. 5b, there are two frequencies (135.5 Hz and 148 Hz) at which the phase difference (4) is 90°. These correspond to the kinematics with the bending and twisting motions in quadrature.

To sum up, bending and twisting modes devote big wing motion since found at resonances while quadrature modes bring a phase difference of 90° which is desired for standard insect-wing kinematic.

4 Conclusion

In this work, a Bond Graph-based dynamic model for an FWNAV has been successfully developed. The one proposed in this work is original since it is based on a completely flexible skeleton. The input parameters are determined by either from experiment or simulation. From the global model, four operating modes have been identified, two at resonance and the other two promising to bring an insect-wing kinematic.

Acknowledgment. This research is funded by Funds for Science and Technology Development of University of Technology and Education under project number T2019-06-116.

References

1. Evangelista, C., Kraft, P., Dacke, M., Reinhard, J., Srinivasan, M.V.: The moment before touchdown: landing manoeuvres of the honeybee Apis mellifera. J. Exp. Biol. **213**(2), 262–270 (2010)
2. Card, G., Dickinson, M.H.: Visually mediated motor planning in the escape response of drosophila. Curr. Biol. **18**(17), 1300–1307 (2008)
3. Karasek, M., Preumont, A.: Robotic hummingbird: design of a control mechanism for a hovering flapping wing micro air vehicle. Ph.D. thesis University (2014)
4. de Croon, G.C.H.E., Perçin, M., Remes, B.D.W., Ruijsink, R., De Wagter, C.: The DelFly. Springer, Dordrecht (2016)
5. Teoh, Z.E., Fuller, S.B., Chirarattananon, P., Prez-Arancibia, N.O., Greenberg, J.D., Wood, R.J.: A hovering flapping-wing microrobot with altitude control and passive upright stability. In 2012 IEEE/RSJ International Conference on Intelligent Robots and Systems, Vilamoura-Algarve, Portugal, pp. 3209–3216 (2012)
6. Faux, D., Thomas, O., Cattan, E., Grondel, S., Doan, L.A.: Two modes resonant combined motion for insect wings kinematics reproduction and lift generation. EPL Europhys. Lett. **121**(6), 66001 (2018)
7. Bolsman, C.T., Palsson, B., Goosen, H., Schmidt, R., van Keulen, F.: The use of resonant structures for miniaturizing FMAVs. In: 3rd US-European Competition and Workshop on Micro Air Vehicle & European Micro Air Vehicle Conference and Flight Competition, Toulouse, France (2007)
8. Zhang, J., Cheng, B., Roll, J.A., Deng, X., Yao, B.: Direct drive of flapping wings under resonance with instantaneous wing trajectory control. In: 2013 IEEE International Conference on Robotics and Automation (ICRA), pp. 4029–4034 (2013)

9. Hines, L., Colmenares, D., Sitti, M.: Platform design and tethered flight of a motor-driven flapping-wing system. In: 2015 IEEE International Conference on Robotics and Automation (ICRA), pp. 5838–5845 (2015)
10. Roll, J.A., Cheng, B., Deng, X.: An electromagnetic actuator for high-frequency flapping-wing microair vehicles. IEEE Trans. Robot. **31**(2), 400–414 (2015)
11. Lau, G.-K., Chin, Y.-W., Goh, J.T.-W., Wood, R.J.: Dipteran-insect-inspired thoracic mechanism with nonlinear stiffness to save inertial power of flapping-wing flight. IEEE Trans. Robot. **30**(5), 1187–1197 (2014)
12. Doan, L.A., Faux, D., Dupont, S., Cattan, E., Grondel, S.: Modeling and simulation of the vertical take off and energy consumption of a vibrating wing nano air vehicle. In: Mechatronics/17th International Conference REM, pp. 123–128 (2016)
13. Borutzky, W.: Bond Graph Methodology. Springer, London (2010)
14. Meirovitch, L.: Fundamental of Vibration. Waveland (2000)
15. Bontemps, A.: Prototypage d'un Objet Volant Mimant l'Insecte. Ph.D. thesis University of Valenciennes (2012)

Manufacturing Cost of Robot Structures with Tolerance Calculated on the View of Kinetic Response and that of Technology

Thang Nguyen Huu, Khanh Duong Quoc, Thuy Le Thi Thu(✉), and Long Pham Thanh

Thai Nguyen University of Technology, Thai Nguyen, Vietnam
hanthuyngoc@tnut.edu.vn

Abstract. Robot-type structures have many different methods to calculate tolerances, the results of these methods are not the same and therefore the manufacturing costs and processing technology are not identical. This paper presents two points of view to calculate different robot structure tolerances and then show reasonable methods to apply in practice. The basis of calculating tolerances from the view of structural kinetic response and the view of manufacturing technology is presented. Calculated results are documented on the same robot with the same required accuracy. Then the comparison of manufacturing costs will be made to identify the more suitable option. The results of this paper are suggestions for calculating similar structures effectively for those interested.

Keywords: Tolerance · Robot structure · Kinematic response · View of technology · Cost

1 Introduction

Industrial robots have an increasingly important role in industrial applications. The more precise manipulators are, the higher the product quality and the higher the price of these products are. Therefore, how to manufacture robots that meet the requirements of quality and price is always designers' interest.

The inaccuracy at the end manipulator can be caused by errors of kinetics, dynamics, friction, wear, vibration, surroundings etc. in which the kinetic factor plays a critical role and takes more than 70% of the total errors [1].

Previous studies mentioned cost-based method to design robot kinematics tolerances. Kim [2] presented the problem of finding the optimal tolerances of robot link and joint parameters with minimal manufacturing cost. Meanwhile, the cost optimization model was established using pseudo-boolean program. Rout and Mittal [3] used evolutionary optimization techniques to select optimal parameters and tolerance simultaneously to minimize the manufacturing cost. Kinematic tolerances (links, joints), kinetics (mass, moment etc.) are assumed to have known initial values.

Meanwhile, several studies presented the evaluation of kinetic parameter tolerance (links, joints) to show which parameters have a greater influence on the end effector difference. Weill and Shani [1] developed a model to evaluate the influence of

K.-U. Sattler et al. (Eds.): ICERA 2019, LNNS 104, pp. 462–470, 2020.
https://doi.org/10.1007/978-3-030-37497-6_53

components' geometric error on position and direction errors of the end effector R and to identify which component has a greater influence. The model is developed and implemented on a computer SILICON-GRAPHIC in language C. Ting et al. [4, 5] analyzed and evaluated the effect of joint clearance on the direction and position errors of linkages and manipulators. Kim et al. [6] use advanced first-order second moment (AFOSM) method to determine the influence of link and joint tolerances of the end effector of robot, and verified with Monte Carlo simulation method. Taguchi method was applied by Fattah Hanafi Sheikhha to determine how parameters (link tolerance, joint clearance) affect the accuracy of its end effector [7]. These more influential links will be the first tolerance adjustment when changing the accuracy. Tolerance on the major influences will only need minor changes to have the desired impact to the overall accuracy. Therefore, according to this calculation, there may be links with large nominal dimensions but very small tolerances and vice versa. This is of course not beneficial in terms of manufacturing, the links with large nominal dimensions and small tolerances will require more machining costs. The method proposed in this paper can solve this problem, as the links will keep the same tolerance and nominal dimension ratio. Therefore, the technology of manufacturing will be the first priority when calculating, the tolerance is not adjusted on the more influential effect but on all links of the chain. The relationship between accuracy, tolerance and cost is shown in Sect. 3 to calculate and compare the price. The problem becomes clear when determining which is more effective from an economic perspective.

2 Calculation Principles of Chain Structure Tolerance

2.1 Describing the End-Point Kinematic Quality of the Structure

The concept of accuracy is often used to describe the quality of end-point kinematics. Assuming that the end point is the center of the sphere (see Fig. 1), the error radius is given. The end points achieved are not exceeded the above-mentioned sphere, the robot is considered to ensure kinematic accuracy. The error radius is determined by:

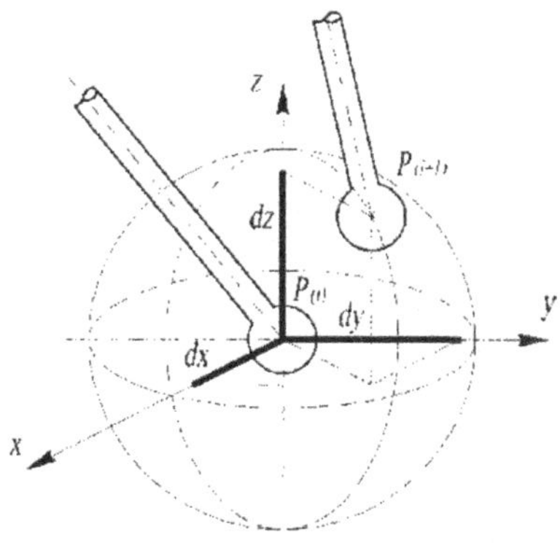

Fig. 1. Allowance error sphere to describe accuracy

$$\delta = \sqrt{d_x^2 + d_y^2 + d_z^2} \quad (1)$$

Every end point that falls into this sphere ensures accuracy and vice versa.

2.2 Method to Calculate Tolerance Based on Kinematics

There are two tolerant components that need to be calculated (size a_i and d_i in the Denavit-Hartenberg (D-H) table), how to calculate this tolerance is based on the viewpoint of setting limit states as shown in Fig. 2:

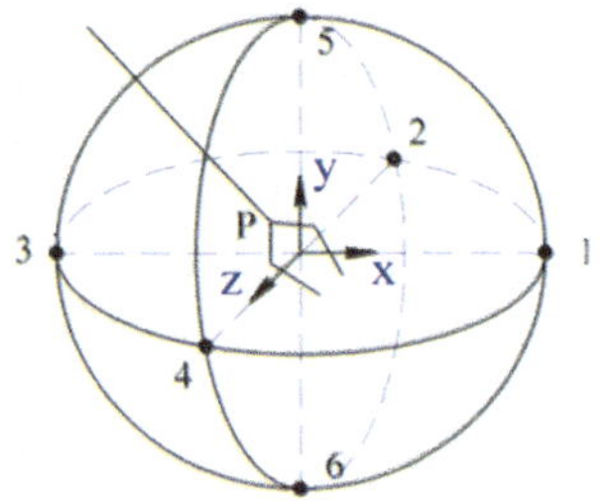

Fig. 2. Six limited positions to calculate the structural tolerances on the sphere

Suppose that the kinetic equations of robots are Eq. (2):

$$f(x, y, z, a_i, d_i, q_i) = p_j \quad (2)$$

If the tolerance of generalized variable q_i and dimensional tolerances D-H including a_i and d_i are presented, Eq. (2) is rewritten as Eq. (3):

$$f(x, y, z, a_i \pm \delta a_i, d_i \pm \delta d_i, q_i \pm \delta q_i) = p_j \pm \delta \quad (3)$$

To reduce the degree of Eq. (3), consider subproblem of Eqs. (4) and (5):

$$f(x, y, z, a_i \pm \delta a_i, d_i \pm \delta d_i, q_i) = p_j \pm \delta \quad (4)$$

$$f(x, y, z, a_i, d_i, q_i \pm \delta q_i) = p_j \pm \delta \quad (5)$$

The problem (4) is only counted for dimensional tolerances D-H and the generalized variable is considered not to have motor errors.

The problem (5) only considers the generalized variable tolerance and D-H dimensional tolerance is not mentioned. Equations (4) and (5) are solved by giving (x, y, z) as the coordinates of the sphere's center as shown in Fig. 2, move p_i from points 1 to 6 in turn to obtain a corresponding kinematic equation system. D-H dimensions are considered as variables in Eq. (4), generalized coordinates are considered as variables in Eq. (5). These equations are solved by Trung et al. [8]. $j = 1 \div m$ are survey points selected in advance with sufficiently large numbers, the calculation results will be as follows.

Equation (4) uses generalized variables without errors of the resolution of the motor, the D-H dimensional difference can be calculated as shown in Fig. 3.

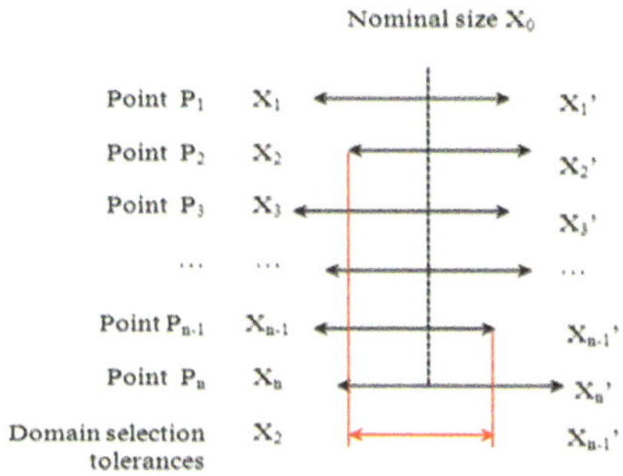

Fig. 3. Summary of tolerance domain of D-H dimension after solving the problem (4)

Two adjacent points p_{j+1} and p_j are solution of Eq. (5) and are coordinates in the joint space:

$$\begin{aligned} p_j &= (q_1, q_2, \ldots, q_6)^{(j)} \\ p_{j+1} &= (q_1, q_2, \ldots, q_6)^{(j+1)} \end{aligned} \tag{6}$$

Each joint can be calculated as completing a movement given by Eq. (7):

$$\begin{cases} \delta q_1 = q_1^{(j+1)} - q_1^{(j)} \\ \vdots \\ \delta q_6 = q_6^{(j+1)} - q_6^{(j)} \end{cases} \tag{7}$$

Since the two problems Eqs. (4) and (5) are separated from the problem Eq. (3), this is an incompatible transformation due to the nonlinear system Eq. (3). The solution of the tolerance domain of D-H dimension on Fig. 3 and the generalized variable tolerance in Eq. (7) need to be checked again by Eq. (3). If all of the falling points satisfy Eq. (8), the calculation is complete, otherwise reduce the width of the re-check tolerance domain.

$$\begin{aligned} & f(\mathrm{x}, \mathrm{y}, \mathrm{z}, \mathrm{a}_i \pm \delta \mathrm{a}_i, \mathrm{d}_i \pm \delta \mathrm{d}_i, \mathrm{q}_i \pm \delta \mathrm{q}_i) \leq \mathrm{p}_j \pm \delta \\ & i = 1 \div n \\ & j = 1 \div m \end{aligned} \tag{8}$$

This step needs to be programmed to check all possible combinations when assembling the machine parts together. In this calculation, the calculated results will always have ratios $k_i = \frac{\delta_d^i}{d_i}$ different between linkages. This is the factor because it affects the technology when manufacturing.

2.3 Method to Calculate Tolerance Due to Technological Point of View

The manipulator is presented as Fig. 4.

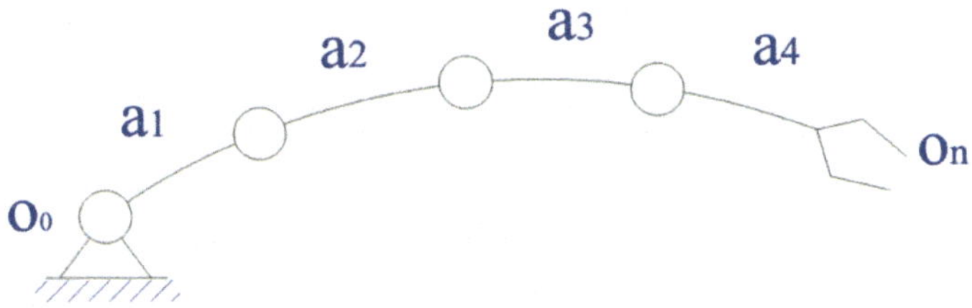

Fig. 4. Reach of robot structure

a_i with $i = 1 \div n$ is the nominal dimension of each link;
Spherical radius of the desired error of the end point is $\delta(mm)$;
The largest reach of the manipulator is given by:

$$R = \sum_{i=1}^{n} a_i \tag{9}$$

k is a non-dimensional quantity representing accuracy:

$$k = \frac{\delta}{R} \tag{10}$$

This ratio is used to set the initial approximation value of component tolerance for each link in the series:

$$(\mathrm{k} = \frac{\delta_1}{a_1}; \mathrm{k} = \frac{\delta_i}{a_i}; ..; \mathrm{k} = \frac{\delta_n}{a_n}) \tag{11}$$

In which δ_i, a_i is the tolerance of the ith link and its nominal size in that order respectively, a_i is known value so δ_i is calculated by k taken from Eq. (10). Assigning links with the same ratio of tolerances and nominal sizes creates favorable conditions for mechanical processing.

These values, if accepted, are the tolerances of the corresponding nominal dimensions meaning that the random endpoint position p_i satisfies the following constraint:

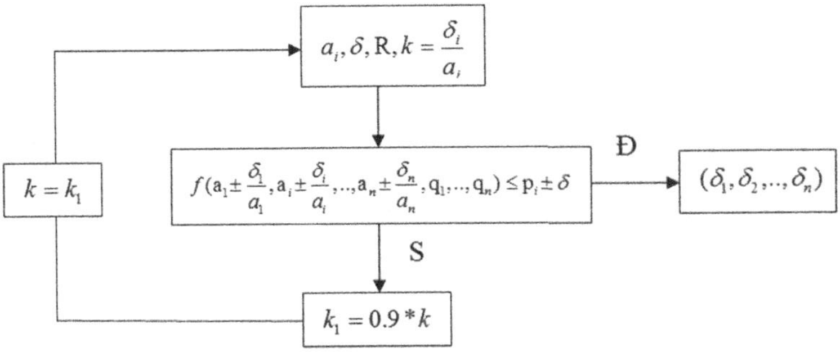

Fig. 5. Diagram to determine tolerance of component links

$$f(a_1 \pm \frac{\delta_1}{a_1}, a_i \pm \frac{\delta_i}{a_i}, \ldots, a_n \pm \frac{\delta_n}{a_n}, q_1, \ldots, q_n) \leq p_i \pm \delta \tag{12}$$

Where p_i is any i^{th} testing point in the robot's working area, (12) is the forward kinetic equation of the robot. There are two possibilities when examining the constraint Eq. (12), as shown in Fig. 5. The purpose of this process is to find a reasonable value k_i instead of the value k given by Eq. (10) at which the following two constraints satisfy simultaneously:

- The links with the same ratio of tolerances and nominal sizes create favorable conditions for mechanical processing;
- k_i is the maximum value that satisfies Eq. (12) so that the tolerance can be maximum to reduce manufacturing costs;

Calculation results ensure that the received tolerance is always proportional to the nominal dimensions, which is different from the method presented in Sect. 2.2.

3 Calculation Basis of Manufacturing Cost According to the Tolerance

When machining, the lower the accuracy level (high accuracy), the more tasks need to process. After each task, the accuracy will gradually increase in proportion to the cost of investment. The accuracy itself can be looked up from the technology manuals if the nominal dimension (the dimension D-H consists of a_i and d_i) and the calculation tolerance are known as shown in Sects. 2.2 and 2.3;

Another way to calculate the cost is to use the relationship between machining cost and tolerance as shown in Fig. 6.

It is necessary to determine the base price of the workpiece (A_i), this is the price formed by the volume of materials used, the cost of machine tools, jigs…

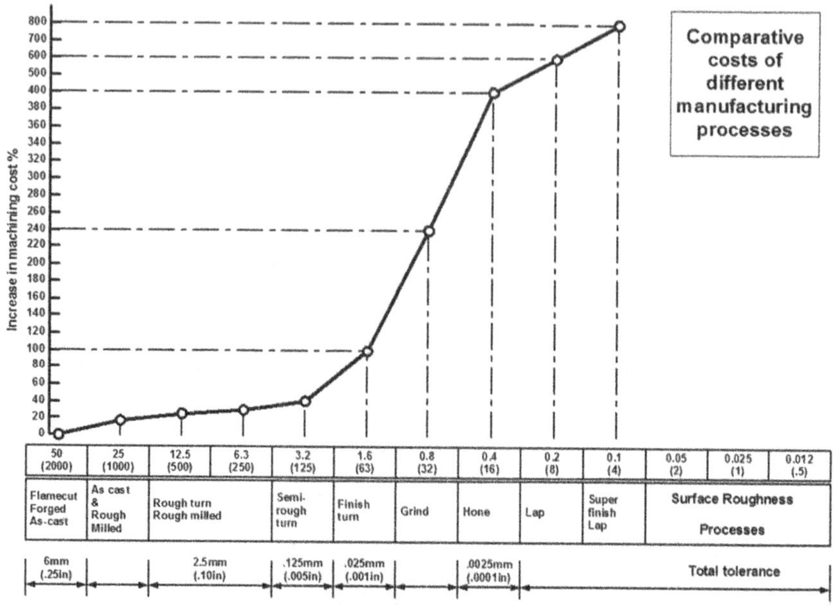

Fig. 6. Relationship between cost and tolerance of machined parts [9]

The second part is the value determined by the tolerance (or precision level, B_i) to adjust the cost due to the requirements of the given tolerance.

The production cost will be calculated as:

$$G = \sum_{i=1}^{n} A_i B_i \quad i = 1 \div n \tag{13}$$

Where: n is the degrees of freedom of the manipulator;

A_i is determined by the market price corresponding to the machine part with basic tolerance (H or h according to Vietnam Standards);

B_i is determined by the tolerance of the workpiece based on the diagram in Fig. 6;

It is possible to use this method to compare the cost of the two structures with the same mechanisms and nominal dimensions, with different component tolerances as in this paper. Component tolerances, although allocated differently because of the calculation point of view, only compare manufacturing costs if these two mechanisms have the same endpoint accuracy.

4 Case Study

A robot is given as in Fig. 7:

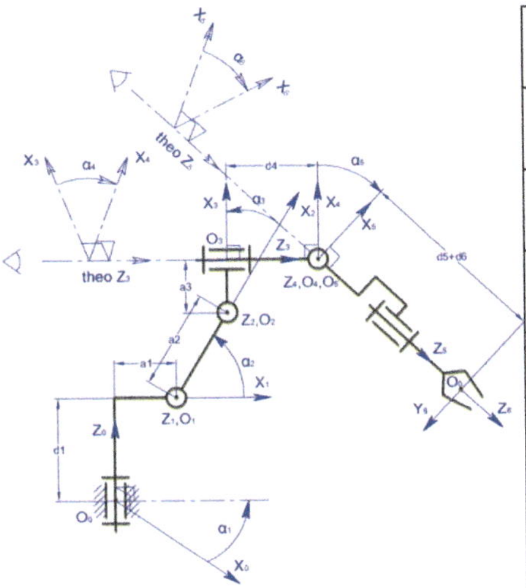

a. Typically six rotary joints robot

Joint	R_z	T_z	T_x	R_x
1	(α_1)	d_1	a_1	908
2	(α_2)	0	a_2	0
3	(α_3)	0	a_3	908
4	(α_4)	d_4	0	-908
5	(α_5)	0	0	908
6	(α_6)	d_5+d_6	0	0

b. Table of robot's DH dimensions

Fig. 7. Robot and kinetic parameters

Dimensions are (mm):

$$d1 = 335,\ a1 = 75,\ a2 = 270,\ a3 = 90,\ d4 = 295,\ d5 + d6 = 80$$

Calculate tolerance to get the endpoint accuracy with radius $\delta = 0.5(\text{mm})$;

Calculated results by kinematic response method (method 1) and technological point of view (method 2) as shown in Table 1.

Table 1. Cost results calculated by the two proposed methods

Nominal dimensions	Reference cost A_i	Tolerance (method 1)	Cost B_i (method 1)	Tolerance (method 2)	Cost B_i (method 2)
$d_1 = 335$	3 500 000	±0.150	4900 000	±0.28	4 830 000
$a_1 = 75$	900 000	±0.100	1 350 000	±0.06	1 440 000
$a_2 = 270$	2850 000	±0.150	3990000	±0.22	3 876 000
$a_3 = 90$	1 050 000	±0.100	2 075 000	±0.07	1 659 000
$d_4 = 295$	3 100 000	±0.100	4 650 000	±0.24	4 340 000
$d_6 = 80$	950 000	±0.100	1 425 000	±0.06	1 567 000
In total			**18 .390 .000**		**17 .712. 000**

The end result shows that keeping a fixed ratio between tolerance and nominal dimensions has higher economic efficiency than dynamic response. In this case, the 2nd method gets the same ratio with all links. The cost difference depends on the specific price, but setting component tolerance gives positive results.

This is especially true when processing large series, bulk products, the efficiency is generalized with the number of machine parts in the series. Therefore, in order to optimize economic efficiency, it is necessary to pay attention to the method of calculating tolerance so as to achieve good results in both economy and technology.

5 Conclusion

A technical design has to pay attention to both kinetics and kinematics features and manufacturing technology. When technological properties are ensured, the manufacturing process can be done by conventional equipment, vice versa, the parts may not be machined by conventional methods (due to narrow tolerance) or the cost will be too high if machining with special methods.

This paper provides two methods of tolerance calculation based on two different points of view and how to evaluate the economic efficiency and technology of each method. It allows to create reasonable designs in both economic and technical aspects. It can be seen that only when the technology achieves its rationality, the economy is guaranteed indirectly.

References

1. Weill, R., Shani, B.: Assessment of accuracy of robots in relation with geometrical tolerances in robot links. Ann. CIRP **40**(Vei 88), 395–399 (1991)
2. Kim, S.-H.: The optimal tolerance design for kinematic parameters of a robot. J. Des. Manuf. Autom. **1**(4), 269–282 (2001)
3. Rout, B.K., Mittal, R.K.: Simultaneous selection of optimal parameters and tolerance of manipulator using evolutionary optimization technique. Struct. Multidiscip. Optim. **40**, 513–528 (2010)
4. Ting, K., Zhu, J., Watkins, D.: The effects of joint clearance on position and orientation deviation of linkages and manipulators. Mech. Mach. Theory **35**, 391–401 (2000)
5. Zhu, J., Ting, K.: Uncertainty analysis of planar and spatial robots with joint clearances. Mech. Mach. Theory **35**, 1239–1256 (2000)
6. Kim, J., Song, W.J., Kang, B.S.: Stochastic approach to kinematic reliability of open-loop mechanism with dimensional tolerance. Appl. Math. Model. **34**(5), 1225–1237 (2010)
7. Sheikhha, F.H., Akbarzadeh, A.: Effect of link tolerance and joint clearance on end-effector positioning of the 3-PSP manipulator using Taguchi method. Appl. Mech. Mater. **798**, 20–24 (2015)
8. Trung, T.T., Guang, L.W., Long, P.T.: Tolerance design of robot parameters using generalized reduced gradient algorithm. Int. J. Mech. Mater. Eng. **5**(2), 96–105 (2017)
9. http://mca.valuedrivendesign.co.uk/help/Finish.aspx

Microstructure and Permeability of Anisotropic Open-Cell Foams

Van Hai Trinh(✉)

Le Quy Don Technical University (LQDTU), Hanoi 10000, Vietnam
hai.tv@lqdtu.edu.vn

Abstract. Solid foams within open-cell structure have been used widely in mechanical, thermal and acoustical applications due to their functional properties such as lightweight, high surface area-to-volume ratio. The present paper investigates numerically the effect of microstructural properties of anisotropic random foams on the permeability. To this regard, we first employ molecular dynamics simulation to generate the Representative Volume Element (RVE) based on Voronoi tessellation, the RVEs presenting the structure of random foams are then used to compute the foam permeability behavior by solving the viscous problem. The obtained results reveal that the finite element computations agree well with both reference analytical model and experimental data. The anisotropy degree has a significant effect on the permeability of foams.

Keywords: Anisotropic · Open-cell · Random foam · Microstructure · Permeability

1 Introduction

The problems raised in transport phenomena in natural and man-made media have been attracted to the growing attention of both industrialists and scientists from many engineering disciplines (e.g., geophysics, acoustics, thermal engineering). In which, understanding of permeability behavior of porous material provides the fundamentals (i.e., fluid flow and heat transfer) for industrial and environmental processes [1]. From various studies over several decades, it can be stated that the permeability is a highly variable parameter according to the basic morphological parameters such as pore size and solid volume fraction [1–6]. Due to the importance of permeability, such parameter of various porous media (e.g., foam [4–6], granular [1, 2], fibrous [7]) is characterized by numerous approaches: analytical model [3], semi-empirical law [8], experimental [4–6] or numerical [1, 2] curve-fitting.

Analytical models even for highly idealized morphologies often require adjustable factors or in the range of geometrical parameters. Similarly, performing a large number of experiments on various series of real samples is a hard task in terms of consumption. Contrary, numerical approach as testing on virtual material samples allows us to design structures having derived functional properties.

In this paper, we employ the numerical method for predicting the permeability behavior of isotropy and anisotropy open-cell foams.

K.-U. Sattler et al. (Eds.): ICERA 2019, LNNS 104, pp. 471–476, 2020.
https://doi.org/10.1007/978-3-030-37497-6_54

2 Microstructure Modeling of Random Open-Cell Foams

In a Voronoi pattern, the partitioning is based on a set of seed points distributed in a model space where each cell is defined by all points that are closer to one particular seed point than to any others. Mathematically, given a set S of N points in a three-dimensional space, the process of associating all the locations of the space into polyhedral regions with the closest point of S is called Voronoi partitioning process. The polyhedral regions are called cells. The union of all the cells is then referred to as a Voronoi diagram. Theoretically, a Voronoi diagram may be constructed in any dimensional space. A cellular foam model based on Voronoi partitions of 3D space is built as follows [9]. First, a set of N nuclei (seed points) is given in a three-dimensional finite space R^3. For each nucleus, let cell V_i be the region consisting of all locations in the space which are closer to P_i than any other nucleus P_j (with $j \neq i$), a cell V_L corresponding to seed point P_i is defined as:

$$V_L(\boldsymbol{x}_i) = \left\{ \boldsymbol{x} \in R^3 \mid \|\boldsymbol{x} - \boldsymbol{x}_i\| \leq \left\|\boldsymbol{x} - \boldsymbol{x}_j\right\| \; \forall i \neq j \right\} \tag{1}$$

where $\boldsymbol{x}_i$ and $\boldsymbol{x}_j$ are respectively the coordinates of seed points P_i and P_j.

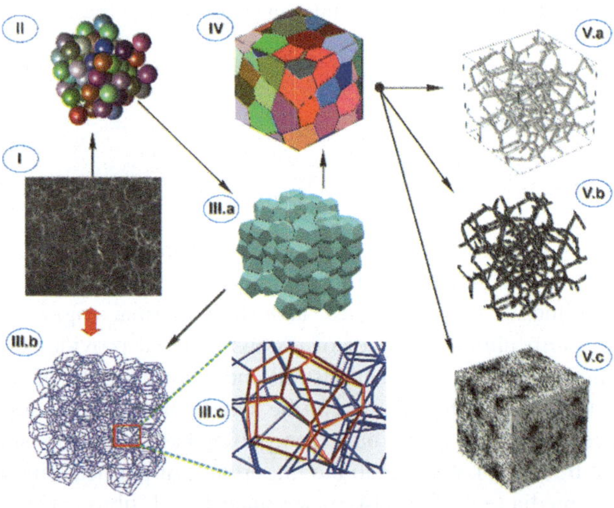

Fig. 1. Schematic diagram of major steps in the procedure for generating random open-cell foam structures: SEM image of real foam (I); assembly of randomly close sphere packing (II); Voronoi cell pattern (III.a) reconstructed structure of open foam with cylindrical structs (III.b) and detail of a typical polyhedron (III.c); various periodic REVs and the FE meshes (IV, V).

Each seed point is surrounded by a cell that contains all points in space that are closer to this particular seed point than to any other. Consequently, cell walls will appear centrally aligned on, and perpendicular to, lines that fictively connect two neighbor seed points. Cell edges appear wherever cell walls intersect and cell vertices appear where cell edges intersect (see Fig. 1(III)). The result is strictly convex cells with flat faces.

An ordered set of seed points can be used for creating ordered and regular structures, for instance, structures made of Kelvin or Weaire-Phelan pattern. Voronoi algorithm generates around each seed a convex polyhedral unit cell made of vertices, joined by edges delimiting planar faces, which connect neighbor cells. Finally, a foam skeleton is completely established (see Fig. 1(III.a)). The corresponding finite mesh models of skeleton and pore domain are graphed in Fig. 1(V.c) and 1(V.c). From this, both isotropic and anisotropic foam material is studied by the virtual REV. The anisotropic virtual REV is construction by elongating and compressing in x-directions and z-direction by a factor R_A and $1/R_A$ respectively while the y-direction is kept (Fig. 2).

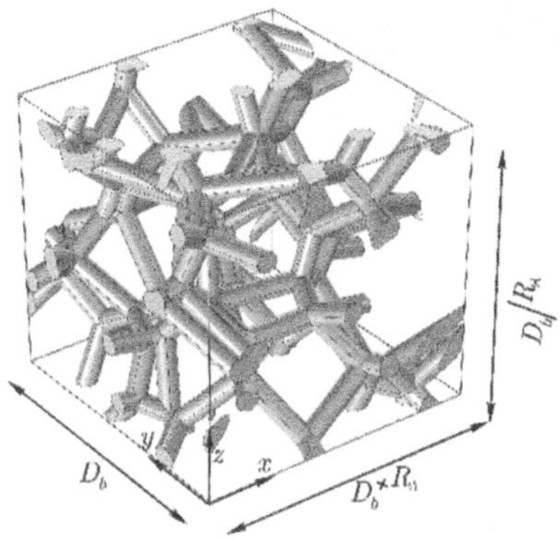

Fig. 2. REV configuration with an introduction of the ratio of anisotropy R_A.

3 Permeability Predicting of Anisotropic Foams

The low Reynolds number flow of an incompressible Newtonian fluid is governed by the usual Stokes equations in the fluid phase [10]:

$$\eta \Delta \mathbf{v} - \nabla p = -\boldsymbol{G} \text{ with } \nabla.\mathbf{v} = 0 \text{ in } \Omega_f \tag{2}$$

where $\boldsymbol{G} = \nabla p^{\mathrm{m}}$ a macroscopic pressure gradient acting as a source term. $\mathbf{v}$, p, and η denote respectively the velocity, pressure, and viscosity of the fluid. The velocity field $\mathbf{v}$ satisfies the no-slip condition ($\mathbf{v} = 0$) at the fluid-solid interface, $\partial\Omega$.

According to the Darcy law, fluid permeability tensor $\boldsymbol{K}$ is determined from the average velocity vector over the whole microstructure and the macroscopic pressure gradient as $\langle \mathbf{v} \rangle = (1/\eta)\, \boldsymbol{K} \cdot \boldsymbol{G}$, this could be expressed in three-dimensional space,

$$\begin{bmatrix} \langle x \rangle \\ \langle y \rangle \\ \langle z \rangle \end{bmatrix} = -\frac{1}{\mu} \begin{bmatrix} K_{xx} & K_{xy} & K_{xz} \\ K_{yx} & K_{yy} & K_{yz} \\ K_{zx} & K_{zy} & K_{zz} \end{bmatrix} \begin{bmatrix} G_x \\ G_y \\ G_z \end{bmatrix} \tag{3}$$

Finally, nine components (Kij) of the permeability tensor could be estimated by solving Eq. (3) within three independent steps with a pressure gradient that is applied only the considered direction (e.g., by applying the gradient G = [Gx 0 0]T in the x-direction, three left coefficients Kxx, Kyx and Kzx of the permeability tensor is deduced, see Fig. 3 for the obtained solution fields of local permeability components).

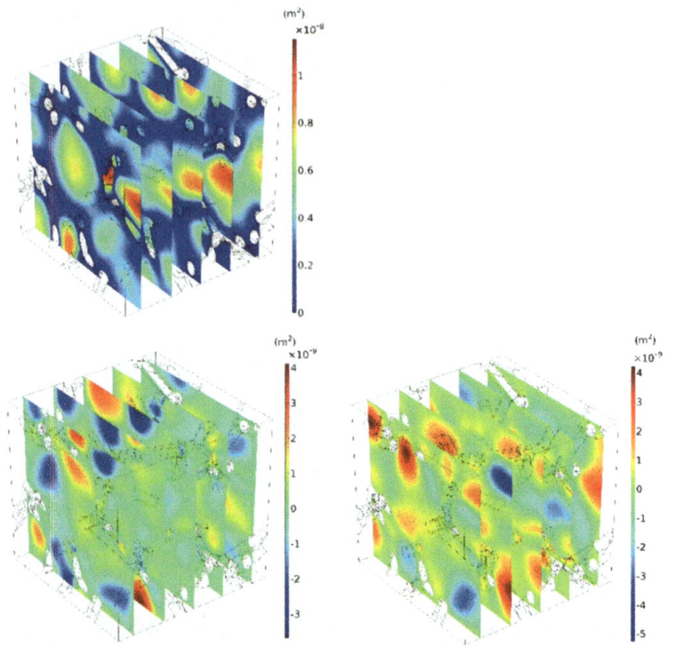

Fig. 3. The solution fields of local permeability components correspond to the applied gradient $\boldsymbol{G} = [G_x\ 0\ 0]^T$. The results are for: $x\eta/G_x$ (upper), $y\eta/G_x$ (lower, left) and $z\eta/G_x$ (lower, right).

4 Results and Discussion

In this section, a foam structure configuration proposed in Ref. [2] is used to validate our numerical procedure (e.g., based the perfect Kelvin pattern with a cell size of 4 mm and a porosity of 0.786). At the same time, the effect of anisotropy level on the regular foam permeability is investigated. Firstly, applying the numerical procedure presented in Sect. 3, we obtain a fully symmetric permeability tensor of isotropy structure ($R_A = 1$) as $\boldsymbol{K} = \mathrm{diag}(8.32, 8.32, 8.32) \times 10^{-8}$ (m^2). As shown in Fig. 4, the numerical obtained results of the permeability tensor is in very good agreement with the reference data proposed in Ref. [2], which validates our finite element model, the present smooth curve shows also the better stable and convergence property of our numerical model. In term of permeability, it is seen that (i) for $R_A \in [1, 1.6]$, within the increase of R_A, the x-direction permeability increases, while the x-direction and y-direction permeability decrease; (ii) for $R_A \in [1.6, 3]$, all permeability components decrease due to the increase of the specific surface area or the effect of strut shape [2].

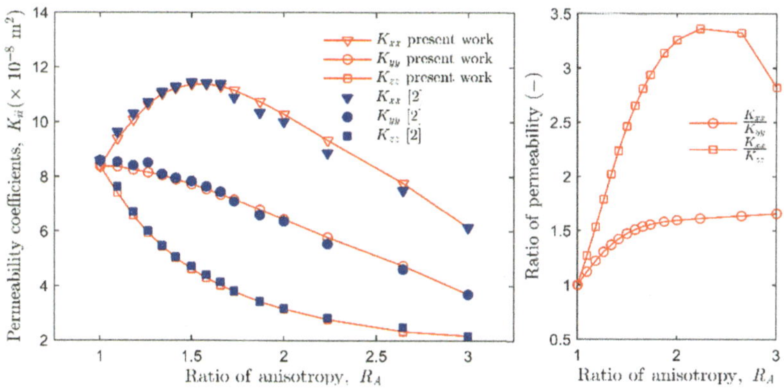

Fig. 4. Permeability tensors diagonal components (left) and their relative ratio (right) versus ratio of anisotropy.

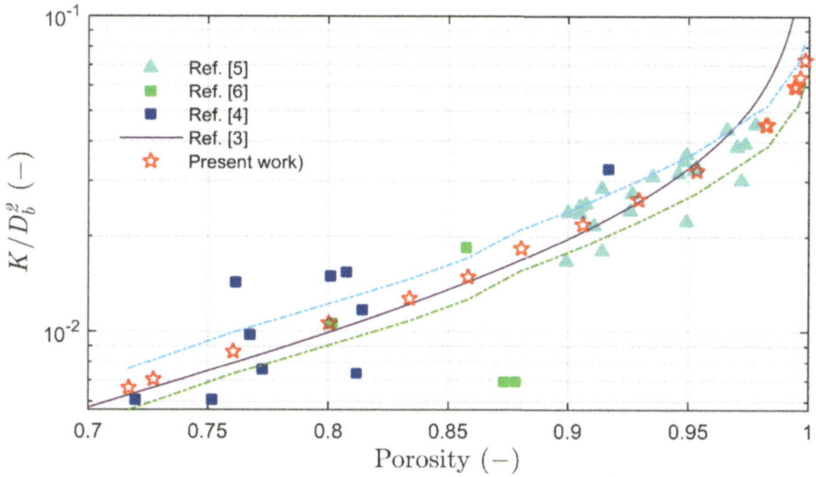

Fig. 5. Dependence of permeability on porosity and pose size of open-cell foams.

Figure 5 compares the permeability of random foam materials obtained from different approaches, it can be seen from the numerical present results (star makers) that our computations are in good agreement both with analytical model [3] and experimental data [4–6] of the static permeability in the whole porosity ranging from 0.70 to 0.99. Interestingly, compared to the analytical law, our permeability predictions show a better fitting with measurements for highly porous foams (i.e., with porosity > 96%). By considering the anisotropic foam materials having the relative ratio (e.g., 15%) of their permeability components, it is seen from the permeability zone created by two dash lines (for +15% and −15%) that the proposed numerical modeling of random open-cell foam within an anisotropic factor enables to capture the static permeability behavior of real structures.

5 Conclusion

This paper presents the microstructure-based approach to characterize numerically the relationship between local geometrical information and permeability behavior of random anisotropic open-cell foams. The proposed method is validated by its accuracy and robustness in compared with existing experimental data and analytical model in terms of determining components of permeability tensors for both regular and irregular foam structures. As demonstrated, the fluctuation of experimental permeability of real foams can be explained by the effects of the ratio of anisotropic. Our numerical modeling of random open-cell foam within an anisotropic factor enables to capture the static permeability behavior of real structures in a wide range of porosity.

The potential application of this work relies on not only modeling the foamy permeability parameter but also other important transport behaviors of mass and heat.

References

1. Boutin, C., Geindreau, C.: Estimates and bounds of dynamic permeability of granular media. J. Acoust. Soc. Am. **124**, 3576–3593 (2008)
2. Jobic, Y., Kumar, P., Topin, F., Occelli, R.: Determining permeability tensors of porous media: a novel 'vector kinetic' numerical approach. Int. J. Multiph. Flow **110**, 198–217 (2019)
3. Yang, X., Lu, T.J., Kim, T.: An analytical model for permeability of isotropic porous media. Phys. Lett. A **378**, 2308–2311 (2014)
4. Garrido, G.I., Patcas, F., Lang, S., Kraushaar-Czarnetzki, B.: Mass transfer and pressure drop in ceramic foams: a description for different pore sizes and porosities. Chem. Eng. Sci. **63**, 5202–5217 (2008)
5. Bhattacharya, A., Calmidi, V., Mahajan, R.: Thermophysical properties of high porosity metal foams. Int. J. Heat Mass Transf. **45**, 1017–1031 (2002)
6. Richardson, J., Peng, Y., Remue, D.: Properties of ceramic foam catalyst supports: pressure drop. Appl. Catal. A **204**, 19–32 (2000)
7. Hunt, M., Tien, C.: Effects of thermal dispersion on forced convection in fibrous media. Int. J. Heat Mass Transf. **31**, 301–309 (1988)
8. Doutres, O., Atalla, N., Dong, K.: Effect of the microstructure closed pore content on the acoustic behavior of polyurethane foams. J. Appl. Phys. **110**, 064901 (2011)
9. Okabe, A.: Spatial Tessellations. Wiley Online Library (1992)
10. Avellaneda, M., Torquato, S.: Rigorous link between fluid permeability, electrical conductivity, and relaxation times for transport in porous media. Phys. Fluids A **3**, 2529–2540 (1991)

Multiexciton Properties in CdSe Core and CdSe/CdSe$_{1-x}$S$_x$ Tetrapod Nanostructures Under Pulse Wave Optical Pumping

Nguyen Thi Luyen[1], Nguyen Xuan Ca[1], and Pham Minh Tan[2(✉)]

[1] Faculty of Physics and Technology,
Thai Nguyen University of Science, Thai Nguyen, Vietnam
[2] Faculty of Fundamental Sciences,
Thai Nguyen University of Technology, Thai Nguyen, Vietnam
tanpm@tnu.edu.vn

Abstract. In this research, the detail photoluminescence (PL) spectra of CdSe core and CdSe/CdSe$_{1-x}$S$_x$ tetrapod nanostructures (NSs) were studied by using different pump-photon energies at 488 nm and 337.1 nm with continuous-wave and pulsed excitation regimes, respectively. Using the excitation source at a wavelength of 488 nm, only one exciton emission peak of NSs is observed. In the case of pulsed excitation, the clear emission features were observed by both exciton and charged biexcitons, as well as the higher-order multi-exciton, which significantly reduced the rate of Auger recombination.

Keywords: CdSe nanocrystal · Core/shell nanocrystal · Tetrapod nanostructures

1 Introduction

Semiconductor nanocrystals (NCs) have many interesting optoelectronic properties due to quantum confinement effect, making them promising materials in a wide range of applications, such as biological imaging [1, 24, 25], photovoltaics [2, 24, 26], photodetectors [3], light-emitting diodes [4] and lasing [5, 6, 26]. In many applications, the NC is excited under modest conditions of one absorbed photon per nanocrystal (NC) or small numbers of electrically injected charge carriers. Under low power, the NC relevant excitation is the single exciton [7, 8], while multiexciton can be generated with strong power [9, 10]. This generation multiple excitons from the absorption of one high energy photon ($h\nu > 2Eg$) is termed as multiexciton generation (MEG). These multiexcitons release to a single-exciton state in a sequential manner through the Auger recombination process in which one exciton is recombined by transferring its energy to another charge carrier. MEG observed in different nanostructures such as core/shell [11, 12, 24, 25], nanorod [13, 14], nanotetrapod CdSe/CdS [14–18], however, the nature of multiexciton in these structure is explained that dual excitons emission: low energy photoluminescence peak is attributed to the exciton emission of CdSe core, while high energy photoluminescence peak shown the exciton emission of CdS arm or shell. Specifically, Lutich [14] showed that PL spectra of CdSe/CdS tetrapod (TP) at

K.-U. Sattler et al. (Eds.): ICERA 2019, LNNS 104, pp. 477–486, 2020.
https://doi.org/10.1007/978-3-030-37497-6_55

different pump frequencies with longer arms have dual multiexciton emission (diameter 4.0 nm of CdSe core; length 28 and 55 nm of CdS arm). Due to the reduction of Auger recombination in TP with large volume and filling space, the high energy peak of hole-electron recombination in CdS arm was appeared. Masser [16] demonstrated that TP with long CdS arm, holes not only located in CdSe core but also trapped in CdS arm, resulted in dual multiexciton emission of CdSe/CdS TP. Strong Coulomb interaction of hole and electron in CdS arm, hole in CdSe core closed CdS arm and emitted exciton in CdS arm. Choi [17] also observed dual multiexciton emission in CdSe/CdS TP, photoluminescence peak at low energy related to recombination exciton in CdSe core, while at high energy due to emission from recombination of carriers at interface core/arm. In this case, the energy structure of CdSe/CdS TP is considered as molecules energy determined by LUMO and HOMO levels. The transition from LUMO to HUMO states corresponds to direct exciton emission in CdSe core, whereas the latter transition from LUMO+1 to HOMO states corresponds to spatially indirect recombination at interface core/arm. Besides, dual thermalization excitons emission [18] were shown in PL spectra of single CdSe/CdS TP at low 100 K temperature. The strain-induced interfacial barrier generated exciton recombination at higher energy. Mishra [15] showed dual multiexciton emission in CdSe/CdS TP, higher energy peak origins recombination carriers in CdS arm at high pump. The partial state-filling in CdSe core and hole-trap filling effect in the CdS arm are conditional for observation dual emission.

In this work, with an interesting approach, we clearly explain the multiexciton emission in CdSe/CdSe1-xSx TP nanoheterostructure. The different pump excitation including continuous laser and pulse source was used. On the basis of the spectral position of emission peaks, sole emission by single excitons under excitation independent of pump level was presented, while the pulsed excitation at 337.1 nm pump can produce both excitons and charged biexcitons, as well as triexcitons at high pump levels.

2 Experimentals

2.1 Materials

All chemical reagents and solvents used without further purification were purchased from commercial suppliers. Cadmium oxide (CdO, tech.98%) and selenium powder (Se, tech. 99%) were purchased from Alfa Aesar. Oleic acid (OA, tech. 90%), 1-Octadecene (ODE, tech.90%) and trioctylphosphine (TOP, tech.90%) were purchased from Aldrich. All moisture/air-sensitive chemicals were stored under nitrogen atmosphere.

2.2 Tetrapod $CdSe/CdSe_{1-x}S_x$ Synthesis

$CdSe/CdSe_{1-x}S_x$TP are synthesized by a wet chemical method in one pot. CdSe core (seed) are synthesized in octadecene (ODE) using a mixture of cadmium oxide (CdO), oleic acid (OA) and trioctylphosphineselenide (TOPSe) as precursors. To synthesize the CdSe/$CdSe_{1-x}S_x$ TP, the TOPS precursor is injected into a hot reaction mixture that contains CdSe cores. The preparation of CdSe cores and then $CdSe_{1-x}S_x$ arms was sequentially performed at 280 and 200 °C, respectively. After that, the samples were purified through centrifugation and redissolved in toluene for spectroscopic measurements.

2.3 Optical Measurements

The continuous photonic excitation at 488 nm was obtained with a 12 W continuous-wave Argon laser. The pulsed photonic excitation at 337.1 nm was obtained with a pulsed nitrogen laser (Laser Photonics LN 1000, 1.4 mJ energy per pulse, pulse width 0.6 ns) with a maximum power of the order of 50 MW/cm^2. The visible emitted light from the sample, collected by an optical fiber located at 20 mm perpendicular to the surface on the same side as the excitation was analyzed with a Jobin-Yvon Spectrometer HR460 and nitrogen-cooled Si Charged-Coupled Device (CCD) detector. This setup has a resolution of 0.3 nm/point for a slit $\leq$ 30 μm. The decays were analyzed by a PM Hamamatsu R5600U and a scope Tektronix TDS 784A with a time constant of the order of 1 ns.

2.4 Transmission Electron Microscopy (TEM)

TEM images of CdSe core and CdSe/$CdSe_{1-x}S_x$ TP were obtained using a JEM 1010 microscope (Jeol) operating at an acceleration voltage of 80 kV with samples mounted on a carbon-coated copper grid.

3 Results and Discussion

Figure 1(a) and (b) display the TEM images of CdSe core and CdSe/$CdSe_{1-x}S_x$ TP samples, respectively. The diameter of the CdSe core was estimated from TEM images to about 8 nm (Fig. 1(a)) and $CdSe_{1-x}S_x$ with long arms about 15 nm (Fig. 1(b)). XRD measurement can provide the important formation about the crystallographic phase of NC. Figure 1(c) show the powder XRD patterns of CdSe core and CdSe/$CdSe_{1-x}S_x$ TP, respectively.

It is known that CdSe semiconductors exist in two different crystalline structures at ambient temperature: hexagonal symmetry (Wz) and cubic symmetry (Zb). They often present a polytypism between the two phases due to low internal energy differences (<20 meV per atom) and small structural variations [19]. The diffraction patterns of CdSe core shows (111), (110), (112) reflections at 2θ angles of 25.5, 42.7, and 49.8°, respectively while XRD patterns of CdSe/CdSe1-xSx TP confirmed the formation of Wz-CdSe1-xSx arms. The peak around 25° in the diffraction patterns of CdSe/CdSe1-xSx TP is a convolution of (111), (101), (100) reflections and broader than the (110), (112) reflection at 42, 49°, respectively.

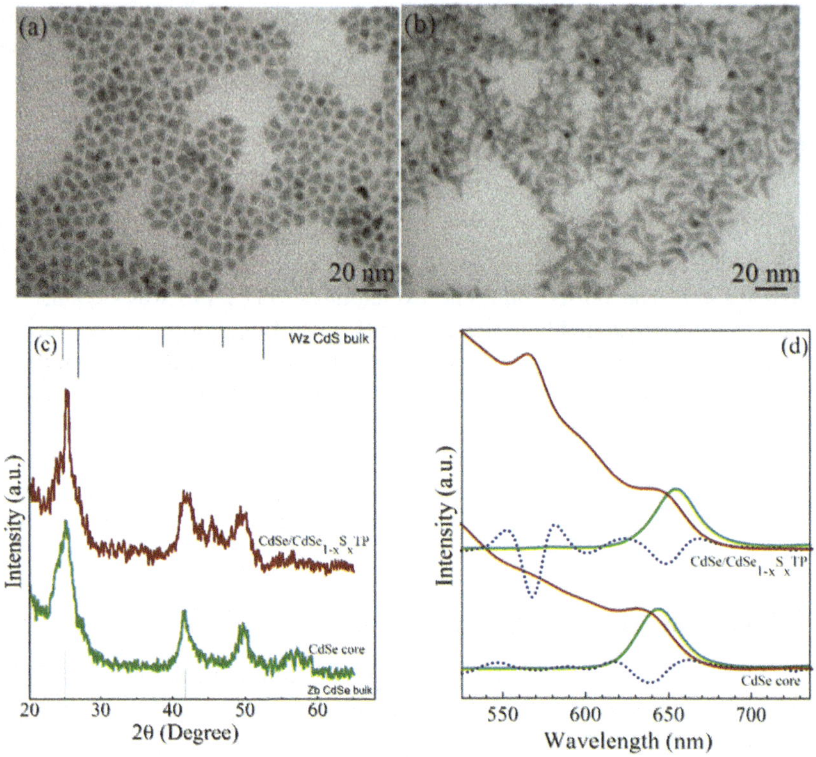

Fig. 1. (a, b) TEM images, (c) XRD diagram, (d) the absorption (wine solid line), the second derivative of absorption (navy dot line) and the PL (olive solid line) spectrum of CdSe core and CdSe/$CdSe_{1-x}S_x$ TP, respectively.

The absorption and PL spectra of CdSe core and CdSe/CdSe1-xSx TP (Fig. 1(d)) show the first excitonic (X) peaks associated at 638 and 648 nm, respectively. Delocalization of the electron wave functions into CdSe1-xSx arms results in a red shift of X peaks in CdSe/CdSe1-xSx TP. The new peak at 567 nm (absorption spectra of CdSe/CdSe1-xSx TP) is related to the CdSe1-xSx arms.

3.1 Continuous-Wave 488 nm Excitation: Observation of Excitons

Figure 2 shows the PL spectra of CdSe core and CdSe/CdSe1-xSx TP as a function of laser power at 488 nm, continuous excitation in relative at room temperature. We observe a little variation of the spectra which present only one emission peak. The respective photon energy is below the absorption onset of CdSe1-xSx arms and therefore, it corresponds to the situation when charge carriers are photoinjected directly into the CdSe core. The emission is centered at 642 nm with a FWHM of 31 nm, which shows a relatively homogeneous size of the different quantum dots.

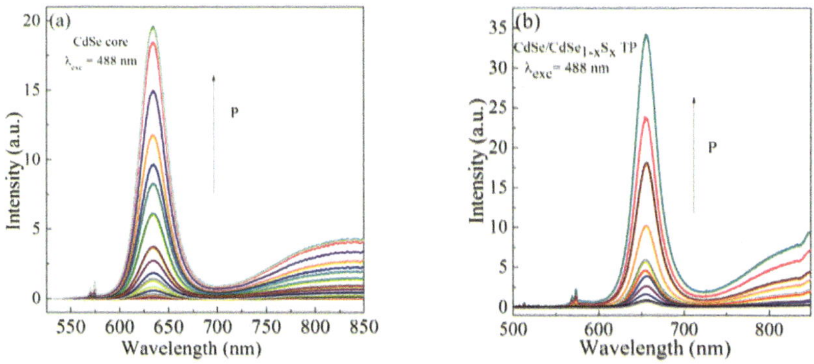

Fig. 2. PL spectra as a function of power excitation at 488 nm using continuous excitation at room temperature of CdSe core (1 mW to 620 mW) and CdSe/CdSe$_{1-x}$S$_x$ TP (3 mW to 1050 mW).

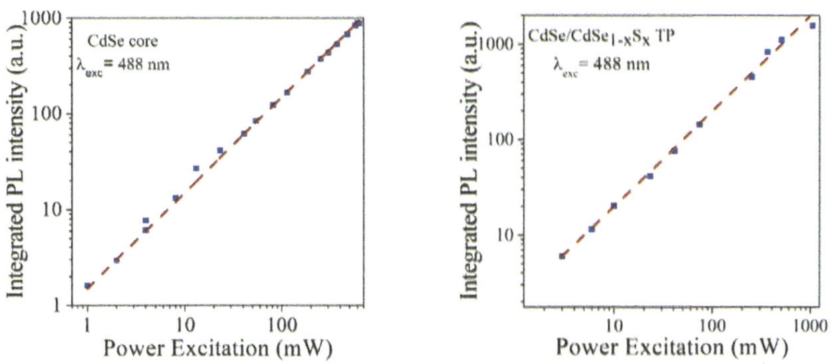

Fig. 3. The integrated PL intensity of the excitons peaks as obtained from Lorentzian fits, as a function of the excitation power of CdSe core and CdSe/CdSe$_{1-x}$S$_x$ TP, respectively.

Figure 3 shows the variation of the integrated PL intensity as a function of excitation power with continuous laser Ar at 488 nm for exciton emission of CdSe core and CdSe/CdSe1-xSx TP, respectively. The integrated PL intensity of the exciton emission varies linearly with the excitation power. We have a variation of the integrated PL intensity of the exciton emission with a slope of 1 when we only obtain the exciton due to the excitation power is not sufficient to create the multi-excitons.

3.2 Pulsed Wave 337.1 nm Excitation: Observation of High Order Multi-excitons

We plot in Fig. 4(a, b) the PL spectra of CdSe core and CdSe/CdSe$_{1-x}$S$_x$ TP as a function of the wavelength λ for various excitation powers using pulsed wave 337.1 nm, respectively.

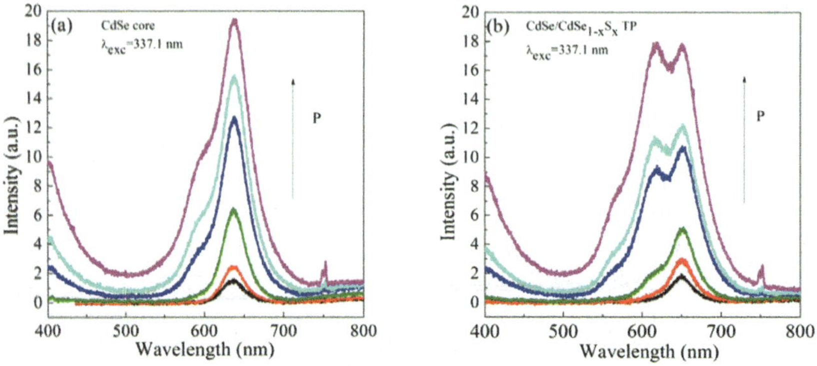

Fig. 4. PL spectra of: (a) CdSe core and (b) CdSe/CdSe$_{1-x}$S$_x$ TP in solution under pulsed 337.1 nm excitation, with excitation powers ranging from 1 to 120 mW.

For CdSe core, at lower excitation power, the PL spectra is a single peak, at higher excitation power, a second peak appears at a lower wavelength. Such as CdSe core, the PL spectra of CdSe/CdSe1-xSx TP at lower excitation power shows only one peak. However, a second peak appears very soon as the excitation power is increased. For keep increasing of excitation powers, a third fluorescence peak is observed. We fit the PL spectra by a sum of two and three Lorentzian dependences on λ with reasonable success for CdSe core and CdSe/CdSe1-xSx TP, respectively. Using the Lorentzian fit, we determine the integrated PL intensity of each peak, which is plotted as a function of the excitation power in Fig. 5.

The integrated PL intensity of the peaks exhibits an increase as a function of the excitation power, which can be fit by a power-law dependence. The obtained exponents for CdSe core sample are 0.44 for the first peak and 0.96 for the second peak. We thus have found two peaks with different relative intensities as a function of the excitation power, but constant position; this indicates that each peak corresponds to a specific multiparticle state. The single peak observed at lower pump power is attributed to the exciton emission and noted X. A second peak appears at higher pump power. It is located about 40 nm below the first peak, corresponding to a 130 meV at higher energy. This peak cannot be attributed to the biexciton emission: in a type I nanocrystal, the biexciton energy is lower than the exciton energy, due to attractive interactions between the two excitons [20]. However, the intensity of this peak scales as the square of the intensity of the exciton peak. Following the authors who observed a similar peak at 60 to 330 meV above the exciton peak [21], we attribute this peak to the emission from the "charged biexciton" state: three electrons and two holes, resulting from two-photon absorptions in a negatively-charged nanocrystal. More specifically, the charged-biexciton state is described as two 1S and one 1P electron and two 1S holes; the charged-biexciton emission (noted XX*) corresponds to the transition from the 1P electron state to the 1S hole state. Ionization of nanocrystals is a well-documented phenomenon, and is believed to explain fluorescence blinking [22]; it occurs when one charge from an exciton is trapped in a defect outside the nanocrystal so that the other charge remains in the nanocrystal core. Ionization is thus expected to be especially important for our nanocrystals as they have no passivating shell. The energy difference between the X and XX* bands is 130 meV for our samples with 8 nm diameter; it is the same order of magnitude but quite larger ~50 meV value which can be extrapolated from the data in ref. [21], possibly because the confinement is larger in our CdSe nanocrystals than in the CdSe/ZnS nanocrystals. The excitation power dependence of our X and XX* bands is a power law with respective exponents ~0.5 and ~1, while the respective exponents are 1 and 2 [21].

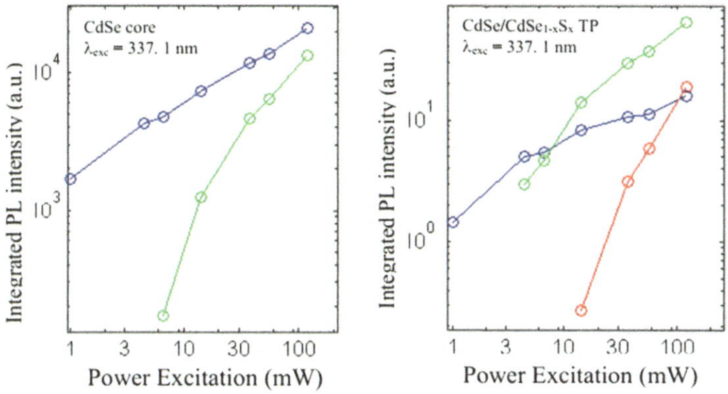

Fig. 5. The integrated PL intensity of the first, second and third peaks (respectively blue, green and red), as obtained from Lorentzian fits, as a function of the excitation power.

We explain this by the fact that we are in a regime of higher pump power (of the order of 10 mJ/cm^2/pulse as compared to 3 mJ/cm^2/pulse [21]; the beginning of this regime is observed in the high-excitation parts [20, 21]. The same experiment is performed for CdSe/CdSe$_{1-x}$S$_x$ TP sample. The similar results are obtained, with power-law exponents. The dependence of the area of the peak on the excitation power is fit by a power law, with respective exponents 0.46, 0.93 and 2.0. The origin of this behavior should be the same as for the CdSe core and involve some multi-excitonic states. We can assume that the two first bands correspond to the X and XX* emissions like for CdSe core sample. The third peak could be attributed to the TX emission [20, 23].

Moreover, the subtraction of the exciton spectrum obtained under excitation with the Xe lamp makes to observe in addition the emission XX*, TX, which increases with the excitation density (Fig. 6).

It had been difficult to highlight this emission XX during the decomposition by Lorentzian. We should try to resume this decomposition to clarify this point. In addition, the decomposition procedure seems more reliable to follow the evolution of the different emissions as a function of the excitation density.

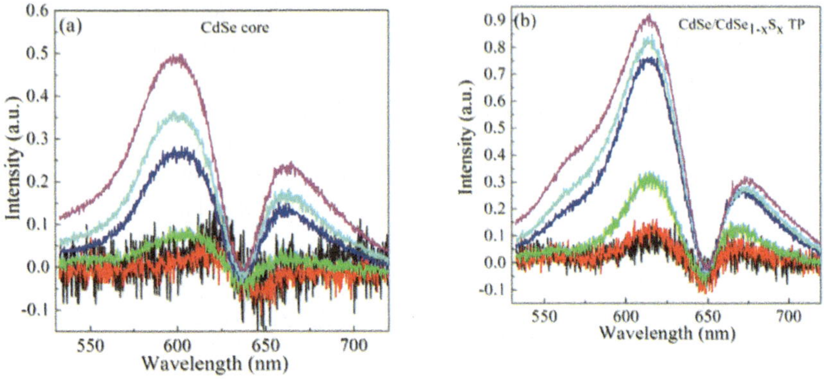

Fig. 6. The subtraction PL spectra of CdSe core, $CdSe/CdSe_{1-x}S_x$ TP, respectively.

Figure 7 shows the influence of excitation power on the peak decay curves of CdSe core and CdSe/CdSe1-xSx TP, respectively. In order to better discern the first part of declines, we consider only the first 20 ns. If we first look at the beginning of the decline curves characterized by a very fast decline corresponding to the resolution limit of our equipment of the order 1.5 ns, this part increases in importance with the excitation density and this more strongly for TP compared to core. This is consistent with a greater contribution of multiexciton emissions with excitation density. Moreover, if we consider the decline corresponding to the core for the lowest excitation density which almost exclusively the exciton emission is obtained, the decline although non-exponential presents a first part that it can be characterized by a time of several ns, corresponding to the exciton emission in the presence of defects. For a decline of CdSe core about 23 ns is observed, consistent with the decline of the exciton.

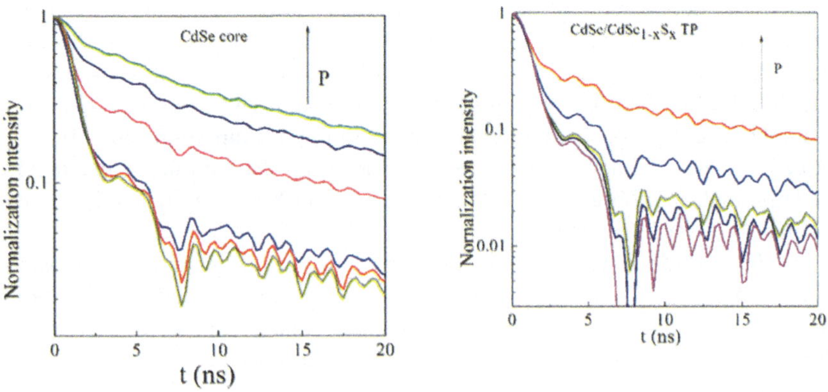

Fig. 7. The photoluminescence decay curves in a solution of CdSe core and $CdSe/CdSe_{1-x}S_x$ TP.

4 Conclusion

In this work, the continuous excitation at 488 nm with the Ar laser was used. As a result, only the exciton emission (X) was obtained, even for the highest available powers. The higher-order exciton emission including charged biexciton and triexciton was obtained by using pulsed excitation power at 337.1 nm. Furthermore, the subtraction PL spectra of CdSe/CdSe$_{1-x}$S$_x$TP is more visible than XX*, TX. The faster photoluminescence decay curves of multiexciton shown are especially important for application in future solar cell technologies.

Acknowledgements. This work is supported by Thai Nguyen University of Technology (Grant No. T2019-B18).

References

1. Michalet, X., Pinaud, F.F., Bentolila, L.A., Tsay, J.M., Doose, S., Li, J.J., Sundaresan, G., Wu, A.M., Gambhir, S.S., Weiss, S.: Quantum dots for live cells, in vivo imaging, and diagnostics. Science **307**, 538–544 (2005)
2. Singh, A., Gunning, R.D., Ahmed, S., Barrett, C.A., English, N.J., Garatead, J.-A., Ryan, K. M.: Controlled semiconductor nanorodassembly from solution: influence of concentration, charge and solvent nature. J. Mater. Chem. **22**, 1562–1569 (2012)
3. Bae, W.K., Kwak, J., Lim, J., Lee, D., Nam, M.K., Char, K., Lee, C., Lee, S.: Multicolored light-emitting diodes based on all-quantum-dot multilayer films using layer-by-layer assembly method. Nano Lett. **10**, 2368–2373 (2010)
4. Bawendi, M.G., Caruge, J.M., Halpert, J.E., Wood, V., Bulovic, V.: Colloidal quantum-dot light-emitting diodes with metal-oxide charge transport layers. Nat. Photonics **2**, 247–250 (2008)
5. Klimov, V.I., Mikhailovsky, A.A., Xu, S., Malko, A., Hollingsworth, J.A., Leatherdale, C. A., Eisler, H.-J., Bawendi, M.C.: Optical gain and stimulated emission in nanocrystal quantum dots. Science **290**, 314–317 (2000)
6. Klimov, V.I., Ivanov, S.A., Nanda, J., Achermann, M., Bezel, I., McGuire, J.A., Piryatinski, A.: Single-exciton optical gain in semiconductor nanocrystals. Nature **447**, 441–446 (2007)
7. Kambhampati, P.: Hot exciton relaxation dynamics in semiconductor quantum dots: radiationless transitions on the nanoscale. J. Phys. Chem. C **115**, 22089–22109 (2011)
8. Kambhampati, P.: Unraveling the structure and dynamics of excitons in semiconductor quantum dots. Acc. Chem. Res. **441**, 1–13 (2011)
9. Mondal, N., De, A., Samanta, A.: All-inorganic perovskite nanocrystal assisted extraction of hot electrons and biexcitons from photoexcited CdTe quantum dots. Nanoscale **10**, 639–645 (2018)
10. Klimov, V.I.: Spectral and dynamical properties of multiexcitons in semiconductor nanocrystals. Annu. Rev. Phys. Chem. **58**, 635–673 (2007)
11. Galland, C., Brovelli, S., Bae, W.K., Padilha, L.A., Meinardi, F., Klimov, V.I.: Dynamic hole blockade yields two-color quantum and classical light from dot-in-bulk nanocrystals. Nano Lett. **13**, 1321–1328 (2013)
12. Brovelli, S., Bae, W.K., Galland, C., Giovanella, V., Meinardi, F., Klimov, V.I.: Dual-color electroluminescence from dot-in-bulk nanocrystals. Nano Lett. **14**, 486–494 (2014)

13. Eshet, H., Grünwald, M., Rabani, E.: The electronic structure of CdSe/CdS core/shell seeded nanorods: type-I or quasi-type-II? Nano Lett. **13**, 5880–5885 (2013)
14. Lutich, A.A., Mauser, C., Como, E.D., Huang, J., Vaneski, A., Talapin, D.V., Rogach, A.L., Feldmann, J.: Multiexcitonic dual emission in CdSe/CdS tetrapods and nanorods. Nano Lett. **10**, 4646–4650 (2010)
15. Mishra, N., Orfield, N.J., Wang, F., Hu, Z., Krishnamurthy, S., Malko, A.V., Casson, J.L., Htoon, H., Sykora, M., Hollingsworth, J.A.: Using shape to turn off blinking for two-colour multiexciton emission in CdSe/CdS tetrapods. Nat. Commun. **8**, 1–9 (2017)
16. Mauser, C., Como, E.D., Baldauf, J., Rogach, A.L., Huang, J., Talapin, D.V., Feldmann, J.: Spatio-temporal dynamics of coupled electrons and holes in nanosize CdSe-CdS semiconductor tetrapods. Phys. Rev. B **82**, 1–4 (2010)
17. Choi, C.L., Li, H., Olson, A.C.K., Jain, P.K., Sivasankar, S., Alivisatos, A.P.: Spatially indirect emission in a luminescent nanocrystal molecule. Nano Lett. **11**, 2358–2362 (2011)
18. Shafran, E., Borys, N.J., Huang, J., Talapin, D.V., Lupton, J.M.: Indirect exciton formation due to inhibited carrier thermalization in single CdSe/CdS nanocrystals. J. Phys. Chem. Lett. **4**, 691–697 (2013)
19. Mahler, B., Lequeux, N., Dubertret, B.: Ligand-controlled polytypism of thick-shell CdSe/CdS nanocrystals. J. Am. Chem. Soc. **13**, 953–959 (2010)
20. Sitt, A., Sala, F.D., Menagen, G., Banin, U.: Multiexciton engineering in seeded core/shell nanorods: transfer from type-I to quasi-type-II regimes. Nano Lett. **9**, 3470–3476 (2009)
21. Achermann, M., Hollingsworth, J.A., Klimov, V.I.: Multiexcitons confined within a subexcitonic volume: spectroscopic and dynamical signatures of neutral and charged biexcitons in ultrasmall semiconductor nanocrystals. Phys. Rev. B **68**, 1–5 (2003)
22. Efros, A.L., Rosen, M.: Random telegraph signal in the photoluminescence intensity of a single quantum dot. Phys. Rev. Lett. **78**, 6–9 (1997)
23. Oron, D., Kazes, M., Shweky, I., Banin, U.: Multiexciton spectroscopy of semiconductor nanocrystals under quasi-continuous-wave optical pumping. Phys. Rev. B **74**, 1–6 (2006)
24. Ca, N.X., Hien, N.T., Tan, P.M., Phan, T.L., Thanh, L.D., Do, P.V., Bau, N.Q., Lien, V.T. K., Van, H.T.: Tunable dual emission in type-I/type-II CdSe/CdS/ZnSe nanocrystals. J. Alloy. Compd. **791**, 144–151 (2019)
25. Lien, V.T.K., Tan, P.M., Hien, N.T., Hoa, V.X., Chi, T.T.K., Truong, N.X., Oanh, V.T.K., Thuy, N.T.M., Ca, N.X.: Tunable photoluminescent Cu-doped CdS/ZnSe type-II core/shell quantum dots. J. Lumin. **215**, 116627 (2019)
26. Ca, N.X., Hien, N.T., Luyen, N.T., Tan, P.M.: Near-infrared emitting type-II CdTe/CdSe core/shell nanocrystals: synthesis and optical properties. In: Fujita, H., et al. (eds.) Proceedings of the International Conference, ICERA 2018. LNNS, vol. 63, pp. 398–407. Springer Nature, Switzerland, AG (2019)

Multi-response Optimization of WEDM Process Parameters of Inconel 718 Alloy Using TGRA Method

T. Muthuramalingam[1](✉), L. Ganesh Babu[2], K. Sridharan[3], T. Geethapriyan[4], and K. P. Srinivasan[5]

[1] Department of Mechatronics Engineering, SRM Institute of Science and Technology, Kattankulathur, Kancheepuram, India
muthu1060@gmail.com

[2] Department of Mechatronics Engineering, TISHK International University, Erbil, Iraq

[3] Department of Mechanical Engineering, Madha Engineering College, Chennai, India

[4] Department of Mechanical Engineering, SRM Institute of Science and Technology, Kattankulathur, Kancheepuram, India

[5] Department of Engineering Design, Visteon Electronics India Pvt Ltd, Chennai, India

Abstract. It is very important and complexity to find the optimum values of wire EDM process parameters and contribution of each parameter to attain the better performance characteristics. In this study, an attempt has been made to optimize those parameters while machining the Inconel 718 alloy. Since the process involves more one than one response parameter, it is essential to carry out the multi-response optimization methodology. The experiments have been conducted with different levels of input factors such as pulse on time, pulse off time, servo voltage and wire tension based on Taguchi L_9 orthogonal table. Wire EDM optimal process parameter has been identified using grey relational analysis and significant parameter has been determined by analysis of variance. The experimental results have indicated that the multi-response characteristic such as material removal rate and surface roughness can be improved effectively through grey relational analysis.

Keywords: WEDM · Inconel · Grey relational analysis

1 Introduction

Wire electrical discharge machining (WEDM) process has grown up in recent years for manufacturing the complex dies and press tools. The mechanism of WEDM process involves the erosion of material due higher temperature with applying the electrical pulse between tool and work piece separated by a dielectric medium. Ramamurthy et al. explained about the wire EDM process in detailed manner with the researches on recent developments to improve the process performance [1]. Mahapatra and Patnaik developed the mathematical models to optimize the WEDM process parameters using

K.-U. Sattler et al. (Eds.): ICERA 2019, LNNS 104, pp. 487–492, 2020.
https://doi.org/10.1007/978-3-030-37497-6_56

the non-linear regression method [2]. Liao and Hunag found that the table feed and pulse on time have a significant influence on metal remove rate [3]. Kuriakose and Shanmugam suggested that the optimization of WEDM process parameters has been very much essential and optimized those parameters by using non-dominated sorting genetic algorithm [4]. Tosun et al. developed empirical model between machining parameter using multi objective optimization technique [5]. Data envelopment analysis and research (DEAR) method can be used for deriving the multi criteria decision making (MCDM) in any process [6]. However, the accuracy of the methodology still needs an improvement [7]. Hence Taguchi-grey relational analysis (TGRA) can be utilized for the better parameters analysis and optimization [8].

From the literatures, it has been clearly understood that the optimal process parameters have highly contributed on determining machinability in WEDM Process. There are fewer literatures available to obtain multi–response optimizing parameter in WEDM process on machining Inconel 718 alloy using Taguchi grey relational analysis Method. Hence Taguchi method and grey relational analysis based optimization has been proposed to optimize the multi-response involved in WEDM in the present study.

2 Experimental Methodology

The experiments have been performed on four axis ELECTRIKA sprint cut wire EDM machine. Pulse on time, pulse off time, servo voltage and wire tension have been chosen as input process parameters with different levels as shown in the Table 1. The peak current 180 A, wire speed 4 mm/min, servo feed 2060 mm/min, high dielectric pressure 2 bar have been kept as constant throughout experiments. Brass wire with 0.25 mm diameter has been used in the experiments. Due to its need in the manufacturing industries, Nickel based super alloy i.e. Inconel 718 alloy has been used as the work piece. The performance measures have been taken as material removal rate (MRR) and surface roughness (R_a) to evaluate the machinability [6]. MRR has been computed by finding the volume difference of the work piece before and after the machining process and it is denoted by mm^3/min. surface roughness has been obtained using Mitutayo portable Surftester SJ-201P with the cutoff length of 0.8 mm. In this experiment L_9 orthogonal array (OA) has been selected based on Taguchi design of experiments. The selection of process variable for L_9 orthogonal array is shown in the Table 2. Taguchi method is a quality analysis to evaluate the results obtained. This analysis can be achieved by signal to noise ratio (S/N) and this ratio depends on the quality characteristics of the process to be optimized. In WEDM process, the main objective is to maximize the MRR and minimize the R_a. The steps involved in Fig. 1.

Table 1. Factor and level of process parameter

Factor	Factor	Level 1	Level 2	Level 3
A	Pulse on time (μs)	110	120	130
B	Pulse off time (μs)	30	40	50
C	Servo voltage (V)	40	60	80
D	Wire tension (kg)	5	7	9

Calculation of S/N ratio:

- Larger the better characteristics

$$\frac{S}{N}\ ratio = -10\log\frac{1}{A}\sum\frac{1}{Y_c^2}$$

- Smaller the better characteristics

$$\frac{S}{N}\ ratio = -10\log\frac{1}{A}\sum Y_c^2$$

A - Number of experimental replication

Y_c - Response of nth trial of ith dependent level.

Calculation of normalized S/N ratio:

- Larger the better characteristics

$$Z_c = \frac{Y_c - \min(Y_c)}{\max(Y_c) - \min(Y_c)}$$

- Smaller the better characteristics

$$Z_c = \frac{\max(Y_c) - Y_c}{\max(Y_c) - \min(Y_c)}$$

Z_{ni} - Normalized of nth trial of ith dependent level.

Computation of grey co-efficient (GC):

$$GC_{ni} = \frac{(\pounds_{min} + \mu\pounds_{max})}{(\pounds_{ni} + \mu\pounds_{max})}$$

$\pounds_{ni}$ - Grey co-efficient for n^{th} trial of ith dependent response

μ - Distinctive co efficient which has value from 0 to 1

Computation of grey relation grade (G_n):

$$G_n = \frac{1}{M}\sum GC_{ni}$$

M - Number of the performance measures

Fig. 1. Steps involved in TGRA method

Table 2. Orthogonal table L_9 for response

Trial	A	B	C	D	MRR (mm^3/min)	R_a (μm)
1	1	1	1	1	58.54	3.19
2	1	2	2	2	79.68	2.34
3	1	3	3	3	87.56	3.50
4	2	1	2	3	98.97	3.98
5	2	2	3	2	104.78	4.19
6	2	3	1	1	97.55	3.90
7	3	1	3	2	108.56	4.34
8	3	2	1	3	104.35	6.36
9	3	3	2	1	159.16	4.17

3 Results and Discussion

Table 3 shows S/N ratio and normalized S/N ratio with their grey scale values for MRR and surface roughness where considered as HB type for data processing in the grey relation analysis. The value of distinguishing coefficient ζ was taken as 0.5 [8]. The higher grey relation grade the better is the better multi performance characteristics in EDM process. Table 4 shows the response table of the average grey relation grade for each level of the parameters. It has been inferred as the optimal combination for process parameter as pulse on time (level 3), pulse off time (level 3), servo voltage (level 1) and wire tension (level 3). Hence the values of optimum parameter have been found as pulse on time (110 μs), pulse off time (50 μs), servo voltage (40 V) and wire tension (9 kg).

Table 3. Computation of grey scale analysis

No	S/N MRR	N MRR	S/N Ra	N Ra	GC_{MRR}	GC_{Ra}	GC
1	35.34905	0	−10.0758	0.309909	0.3333	0.4201	0.3767
2	38.02699	0.308244	−7.38432	0	0.4195	0.3333	0.3764
3	38.84612	0.402528	−13.0643	0.402669	0.4556	0.4557	0.4557
4	39.91007	0.525001	−11.9977	0.531204	0.5128	0.5161	0.5145
5	39.79611	0.582036	−12.4443	0.582627	0.5447	0.5450	0.5449
6	41.18066	0.510555	−11.8213	0.510893	0.5053	0.5055	0.5054
7	40.71340	0.617466	−12.7498	0.617804	0.5666	0.5668	0.5667
8	40.36985	0.577927	−16.0691	1	0.5423	1	0.7723
9	44.03668	1	−12.4027	0.577837	1	0.5422	0.7711

The confirmation test has to be conducted after the identification of optimal input process parameters to analyze the accuracy of the optimal combination. In this test, the trial has been conducted in EDM process with the obtained optimal combination of

Table 4. Response table of the average grey relational grade

Factor	Control factor	Average grey relational grade by factor level			MAX-MIN
		Level 1	LEVEL 2	Level 3	
A	Pulse on time	0.4029	0.5216	0.7034	0.3008
B	Pulse off time	0.4860	0.5645	0.5774	0.0914
C	Servo voltage	0.5540	0.5515	0.5224	0.0316
D	Wire tension	0.5511	0.4960	0.5808	0.0848

input process parameters. The predicted grey relational grade (G_A) with optimal parameters can be computed by Eq. 1 as follows

$$G_A = G_B + \sum^{q} (G_{\epsilon} - G_B) \tag{1}$$

where G_B is the total mean Grey relational grade, G_C is the optimal mean Grey relational grade for each level of factor and q is the number of input process parameters. The Grey relational grade value has been improved by 2.8% from the predicted mean grade.

The difference between maximum and minimum value (max-min) of the grey grade is indicated in Table 4. The maximum of max-min value indicates that it is the effective factor affecting the multi performance characteristic the maximum of max-min value is 0.3008. The corresponding control factor is the pulse on time and has stronger effect on multi performance characteristics. The order of importance of the controllable factor in sequence A (pulse on time) B (pulse off time) C (wire tension) D (servo voltage) factor A is the most controllable factor in WEDM process for the multi performance characteristic. Since pulse on time is the duration of the spark energy delivered, it possesses higher significance on performance measures. Pulse off time indicates duration of flushing process; it possesses considerable significance on performance measures. Wire tension has higher influence on surface roughness owing to its importance on determining distance across the machining zone in WEDM process.

4 Conclusion

In the present study, an attempt has been made to optimize those parameters while machining the Inconel 718 alloy using TGRA method for enhancing the multiple performance characteristic such as MRR and R_a in WEDM process. From the experimental results, the following conclusions have been made.

- The values of optimum parameter have been found as pulse on time (110 μs), pulse off time (50 μs), servo voltage (40 V) and wire tension (9 kg).
- The confirmation has also verified the optimal process parameters.
- Pulse on time has higher significance on performance measures owing to its importance on duration of the spark energy delivered in WEDM process.

References

1. Ramamurthy, A., Sivaramakrishnan, R., Muthuramalingam, T., Venugopal, S.: Performance analysis of wire electrodes on machining Ti-6Al-4V alloy using wire electrical discharge machining process. Mach. Sci. Technol. **19**(4), 577–593 (2015)
2. Mahapatra, S.S., Patnaik, A.: Optimization of wire electrical discharge machining (WEDM) process parameters using Taguchi method. Int. J. Adv. Manu. Tech. **34**, 911–925 (2007)
3. Liao, Y.S., Huang, J.T.: A study on the machining parameters optimization of wire electrical discharge machining. J. Mater. Process. Tech. **71**, 487–493 (1997)
4. Kuriakose, S., Shunmugam, M.S.: Multi-objective optimization of wire – electro discharge machining process by non-dominated sorting genetic algorithm. J. Mater. Process. Tech. **170**, 133–141 (2005)
5. Tosun, N., Cogun, C., Tosun, G.: A study on kerf and material removal rate in wire electrical discharge machining based on Taguchi method. J. Mater. Process. Tech. **152**, 316–322 (2004)
6. Manoj, M., Jinu, G.R., Muthuramalingam, T.: Multi response optimization of AWJM process parameters on machining TiB_2 particles reinforced Al7075 composite using Taguchi- DEAR methodology. Silicon **10**(5), 2287–2293 (2018)
7. Muthuramalingam, T., Vasanth, S., Vinothkumar, P., Geethapriyan, T., Mohamed, R.M.: Multi criteria decision making of abrasive flow oriented process parameters in abrasive water jet machining using Taguchi-DEAR methodology. Silicon **10**(5), 2015–2021 (2018)
8. Geethapriyan, T., Kalaichelvan, K., Muthuramalingam, T.: Multi performance optimization of electrochemical micro-machining process surface related parameters on machining Inconel 718 using Taguchi-grey relational analysis. La Metallurgia Italiana **2016**(4), 13–19 (2016)

Multi-responses Optimization of Process Parameters in Die-Sinking EDM Process on SKD11 Steel Using PSI Based Taguchi Method

Nguyen Huu Phan[1], Nguyen Van Duc[1], Pham Van Bong[1], T. Muthuramalingam[2], Ngoc Pi Vu[3], and Le Xuan Hung[3](✉)

[1] Hanoi University of Industry, Hanoi, Vietnam
[2] SRM Institute of Science and Technology, Kattankulathur, India
[3] Thai Nguyen University of Technology, Thai Nguyen, Vietnam
lexuanhung@tnut.edu.vn

Abstract. Electrical discharge machining (EDM) is a very common method used in mould making with complex shape. The optimizing multi-objective process parameters and weight selection for each quality indicator in EDM is very complex. In the present study, an attempt has been made to perform multi-objective optimization in EDM using Taguchi - Preference selection index. Current (I), voltage (V), pulse on time (T_{on}) and pulse off time (T_{of}) has been considered as process parameters to optimize the performance measures such as material removal rate (MRR), tool wear rate (TWR) and surface roughness (R_a) under L_{25} Taguchi orthogonal matrix based experimental design. From the experimental investigation, the optimal process parameters on machining SKD11 steel has been identified as I (5 A), V (70 V), T_{on} (50 μs) and T_{of} (18 μs). It has also been observed that the proposed method can significantly reduce the experimental time and cost.

Keywords: EDM · PSI · Taguchi

1 Introduction

EDM is a non-traditional machining method which has been used to fabricate complex shaped moulds. This method can machine higher strength materials which is difficult to perform with traditional machining methods. Since the processing capacity and quality of EDM is low compared with other non-conventional machining process, it is important to optimize the process parameters owing to its wide range of values. in this technology. Hence the adoptation of multi response optimization of such process has been attracting the attention of technical experts. Many solutions have been used to solve the multi-objective optimization problem in EDM. Taguchi method is commonly combined with other methods to solve the multi-objective optimization problem. The research results have shown that combining Taguchi with Grey Relational Analysis (GRA), TOPSIS, PCA and Fuzzy can optimize multiple objectives simultaneously in EDM. Taguchi - GRA has been used to simultaneously optimize the MRR, TWR and

K.-U. Sattler et al. (Eds.): ICERA 2019, LNNS 104, pp. 493–499, 2020.
https://doi.org/10.1007/978-3-030-37497-6_57

overcut in μ-EDM [1–3]. The accuracy of TGRA method is appropriate. It has been inferred that Taguchi - TOPSIS could give higher efficiency of multi-objective optimization than Taguchi - GRA [4, 5]. TOPSIS is a simple, understandable and powerful method on solving multiple performance measures problem [6]. It is the solution that allows access to the multi-objective optimization problem to enhance the quality indicators in μ-EDM by TOPSIS method [7]. The optimal results of the productivity, surface quality and machining accuracy of AISI - 304 in μ-EDM were determined by TOPSIS method [8]. The experimental results showed that the proposed multi-objective optimization can provide better accuracy in EDM. However, the determination the value of the weights of the quantitative indicators is quite difficult. The Performance Selection Index (PSI) method can optimize the multiple performance measures in machining methods [9]. This method does not require any determination of the weight of the targets. Hence the optimal problem can be solved in a much simple way. Hence the present investigation has been performed to optimize the process parameters on machining SKD11 steel using PSI method along with Taguchi method.

2 Experimental Methodology

2.1 Selection of Machining Conditions

SKD11 die steel has been commonly used to make cold stamping molds with the size of the workpiece is 15 × 15 × 20 mm in the present study. The cylindrical copper (Cu) electrode has been chosen as tool electrode with the diameter of 10 mm and length of 35 mm. The CNC-AG40L die-sinking electrical discharge machining (Sodick, Inc. USA) has been used to perform the machining experiments under HD-1 oil as the dielectric medium. The weight of the workpiece and electrode before and after machining is determined by electronic scale AJ 203 (Shinko Denshi Co. LTD - Japan) with the accuracy of 0.001 g. Surface roughness (SR) has been computed by SJ-210 (Mitutoyo - Japan) with cutoff length of 0.8 cm. The measurements have been taken three times and the average has been considered as measured values in the present study. In the present study, four technological parameters (U, I, T_{on} and T_{of}) have been selected under five different levels as shown in Table 1.

Table 1. Process parameters

Process parameters	Unit	Symbol	Levels				
			1	2	3	4	5
Current	A	I	1	2	3	4	5
Voltage	V	U	30	40	50	60	70
Pulse on time	μs	Ton	18	25	37	50	75
Pulse off time	μs	Tof	9	12	18	25	37

2.2 Experimental Design with Taguchi-PSI Method

In this study, L_{25} orthogonal matrix has been used as per Taguchi based design of experiments. The combination of Taguchi–PSI method was used to simultaneously optimize the three output criterias in EDM: MRR, TWR and R_a. The steps are as follows:

Step1: Sort the selected criterias in matrix.
Step 2: In solving multi-objective problems, it is required to make the values of the indicators have to be dimensionless.
Step 3: The average of the standardized criteria is calculated.
Step 4: Optional variation values between the values of each indicator are calculated.
Step 5: Determine the deviation of the value of the priority level relative to each criteria.
Step 6: In this step of the PSI method, the overall priority value is determined for each criteria.
Step 7: The preferred selection index is calculated for each experiment.
Step 8: Based on the value of the priority index to rank.

3 Results and Discussion

The various machining experiments have been conducted as per the design of experiments which have been framed in the previous section. The experimental results values along with standardized values have been computed and tabulated as shown in Table 2. The values of conversion parameters and priority indicators is shown in Table 2. It has been found that the experiment number 25 could provide the larger value of index based on the calculation of PSI index as shown in Fig. 1. Hence the corresponding optimal technological parameters of EDM process among the chosen process variables on machining SKD11 steel has been identified as I (5 A), V (70 V), T_{on} (50 μs) and T_{of} (18 μs), Fig. 2.

The PSI results have been compared with the results of TOPSIS and GRA methods to assess the relevance of PSI in multi-objective optimization of EDM with quality criteria including MRR, TWR and R_a as shown in Table 3. The results show that the optimal results of different multi-objective decision methods could be different. Table 4 shows quality improvement comparison among all the optimization algorithms with respect to experiment number 1. Hence it is necessary to have an assessment of the optimal efficiency of methods that determine multiple objectives to find the optimal solution in the present study. The number of steps for solving the optimal problem by PSI is less than that with TOPSIS and GRA. It is also not necessary to determine the weight value of quality indicators in PSI. Hence it is easier to solve the multi-objective optimization problem compared with other algorithms.

Table 2. Values of conversion parameters and priority indicators

Exp. No	PSI								TOPSIS	GRA	
	I	U	T_{on}	T_{of}	MRR (mg/min)	TWR (mg/min)	R_a (μm)	θj	Ranking		
1	1	30	18	9	1.45	0.25	0.71	0.031	23	6	20
2	1	40	25	12	0.98	0.15	0.59	0.029	24	12	16
3	1	50	37	18	1.02	0.12	0.67	0.036	19	16	12
4	1	60	50	25	0.72	0.11	0.64	0.035	21	20	9
5	1	70	75	37	0.74	0.03	0.80	0.042	11	17	17
6	2	30	25	18	9.08	0.17	1.32	0.039	14	5	14
7	2	40	37	25	6.42	0.10	1.24	0.044	10	22	2
8	2	50	50	37	5.37	0.03	1.24	0.047	7	24	22
9	2	60	75	9	3.50	0.03	1.07	**0.046**	9	25	**1**
10	2	70	18	12	6.75	0.22	1.26	0.035	22	2	11
11	3	30	37	37	12.23	0.18	1.54	0.040	13	8	21
12	3	40	50	9	4.38	0.05	0.98	0.054	2	7	23
13	3	50	75	12	3.80	0.05	1.08	0.052	4	9	19
14	3	60	18	18	14.48	0.38	1.48	0.036	17	10	10
15	3	70	25	25	12.73	0.30	1.57	**0.035**	20	**1**	5
16	4	30	50	12	9.13	0.10	1.08	0.049	5	21	3
17	4	40	75	18	9.50	0.10	1.36	0.046	8	3	4
18	4	50	18	25	34.74	1.26	2.27	0.038	16	15	25
19	4	60	25	37	31.25	0.92	2.70	0.036	18	14	8
20	4	70	37	9	16.30	0.25	1.66	0.039	15	13	6
21	5	30	75	25	17.60	0.10	1.55	0.052	3	18	15
22	5	40	18	37	41.70	2.20	2.70	0.028	25	11	7
23	5	50	25	9	25.90	0.60	1.81	0.040	12	23	13
24	5	60	37	12	25.30	0.20	1.70	0.047	6	4	24
25	5	70	50	18	26.30	0.10	1.86	**0.056**	**1**	19	18

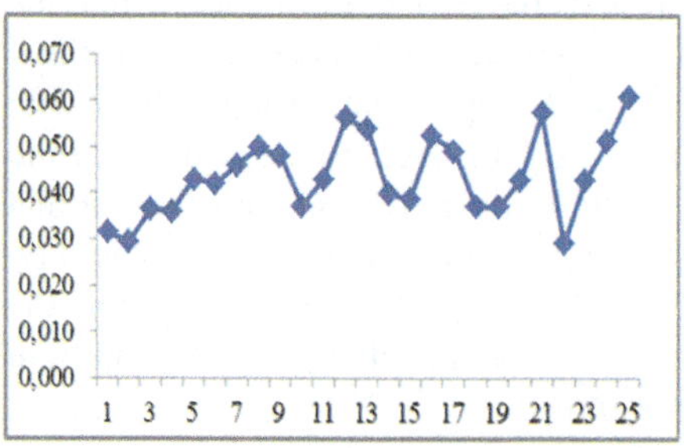

Fig. 1. Ranking index of PSI

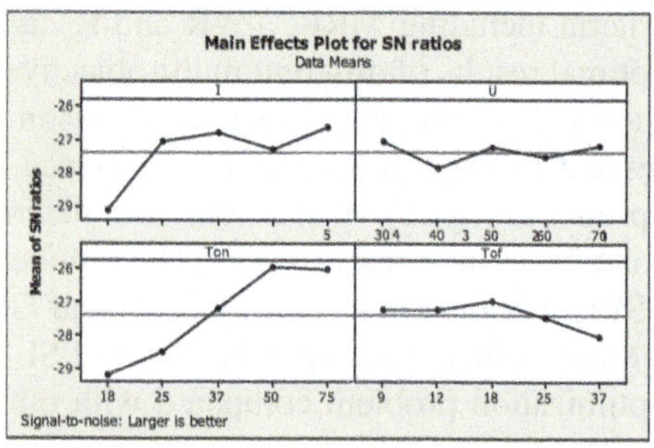

Fig. 2. Influence of machining conditions on the S/N ratio of θj.

Analysis of Machined Surface at Optimal Condition: The craters of machining surfaces in EDM are randomly distributed as shown in Fig. 3. It has been observed that the size of the craters is quite uniform (13 −30 μm). The bottom of the spherical craters can increase the ability to store lubricating oil on the machining surface. This could enhance the wear resistance of the molds. Figure 4 shows that microscopic cracks appear on the machining surface. The size and depth of cracks has been observed as small (≤ 0.7 μm). It negatively affects the fatigue strength of the mold surfaces. The white layer is formed on the machined surface layer as shown in Fig. 5. The depth of cracks is approximately equal to the thickness of the white layer [10]. The white layer formed after EDM has a direct effect on the surface quality of machining. This layer needs to be reduced by finishing methods [11]. The thickness of the white layer of the machined surface at optimal conditions has shown that the thickness of the surface layer should be removed by finishing could be around 10.286 μm.

Table 3. Quality improvement compared to exp. No.1

Optimization methods	Overall performance of exp. No. 1	Optimal results					Quality improvement (%)
		The best experimental order	The value of the quality indicator				
			MRR	TWR	R_a	Overall performance	
PSI	0.031	25th	26.30	0.10	1.86	0.056	80.65
TOPSIS	0.517	15th	12.73	0.30	1.57	0.575	11.22
GRA	0.181	9th	3.50	0.03	1.07	0.247	36.46

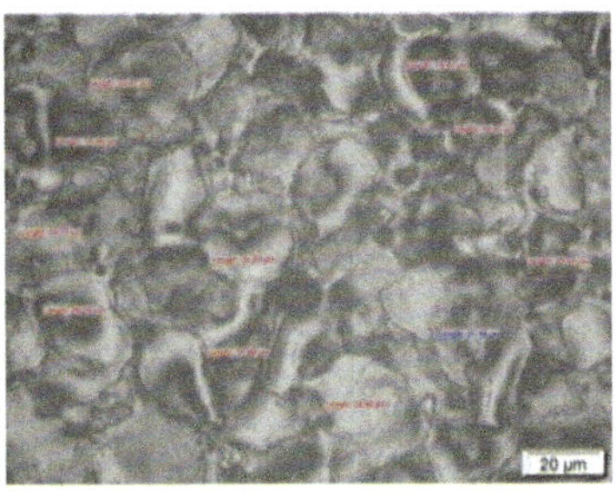

Fig. 3. Carters on machined surface

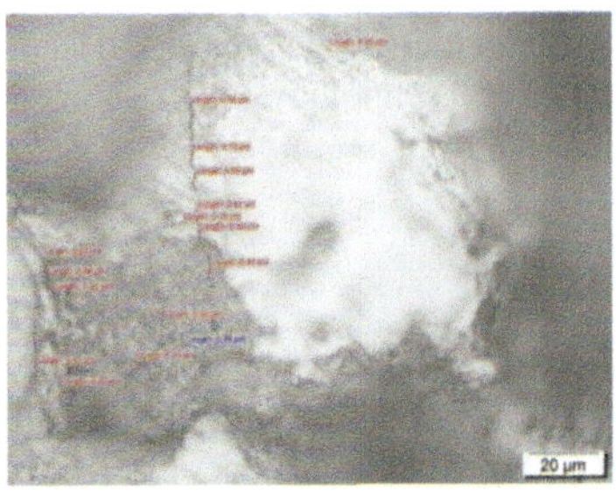

Fig. 4. Cracks on machined surface

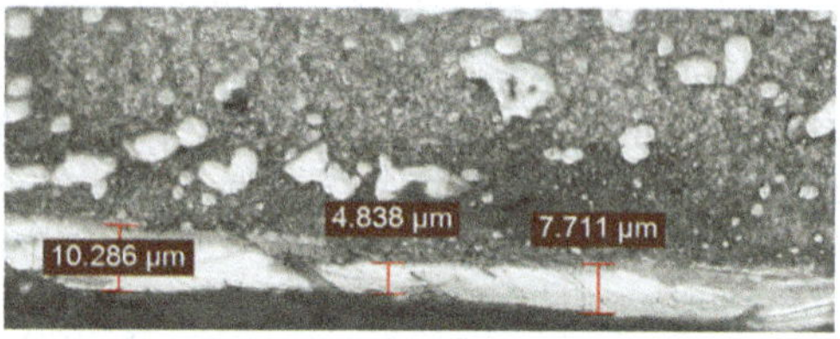

Fig. 5. White layer thickness of machined surface

4 Conclusions

In this study, the technological parameters of die sinking EDM for SKD11 steel were optimized by PSI, the results are as follows:

- The optimal technology parameters are U = 70 V, Ton = 50 μs, I = 5A, Tof = 18 μs. The optimal value of the indicators are MRR = 26 mm^3/min, TWR = 0,1 mm^3/min and R_a = 1,86 μm.
- Optimal results with PSI have been compared to that of TOPSIS and GRA, and optimal efficiency by PSI is the largest.
- The surface layer quality at optimum conditions has also been analyzed, and it is the basis for making recommendations for further finishing methods.
- Taguchi - PSI combination to solve multi-objective problems will help solve this problem short and easy, in addition, it can save time and cost in the implementation process.
- PSI is a very effective solution to solve the multi-objective optimization problem in the field of EDM in particular and other machining technologies in general.

Acknowledgements. The work described in this paper was supported by Thai Nguyen University of Technology.

References

1. Vijay, K.M., Man, S.A.: Micro-EDM multiple parameter optimization for Cp titanium. Int. J. Adv. Manuf. Technol. **89**, 897–904 (2017)
2. Durairaj, M., Sudharsun, D., Swamynathan, N.: Analysis of process parameters in wire EDM with stainless steel using single objective Taguchi method and multi objective grey relational grade. Procedia Eng. **64**, 868–877 (2013)
3. Pragadish, N., Pradeep, K.M.: Optimization of dry EDM process parameters using grey relational analysis. Arab. J. Sci. Eng. **41**, 4383–4390 (2016)
4. Dastagiri, M., Srinivasa, R.P., Valli, P.M.: TOPSIS, GRA methods for parametric optimization on wire electrical discharge machining (WEDM) process. In: Design and Research Conference (AIMTDR–2016). College of Engineering (2016)
5. Prabhu, S., Vinayagam, B.K.: Multiresponse optimization of EDM process with nanofluids using TOPSIS method and genetic algorithm. Arch. Mech. Eng. **63**(1), 45–71 (2016)

6. Gadakh, V.S.: Parametric optimization of wire electrical discharge machining using TOPSIS method. Adv. Prod. Eng. Manag. **7**(3), 157–164 (2012)
7. Manivannan, R., Pradeep, K.M.: Multi-attribute decision making of cryogenically cooled Micro-EDM drilling process parameters using TOPSIS method. Mater. Manuf. Process. **32** (2), 209–215 (2017)
8. Manivannan, R., Pradeep, K.M.: Multi-response optimization of Micro-EDM process parameters on AISI304 steel using TOPSIS. J. Mech. Sci. Technol. **30**, 137–144 (2016)
9. Dušan, P., Miloš, M., Miroslav, R., Valentina, G.: Application of the performance selection index method for solving machining MCDM problems. Mech. Eng. **15**, 97–106 (2017)
10. Muthuramalingam, T.: Measuring the influence of discharge energy on white layer thickness in electrical discharge machining process. Measurement **131**, 694–700 (2019)
11. Lee, L.C., Lim, L.C., Narayanan, V., Venkatesh, V.C.: Quantification of surface damage of tool steels after EDM. Int. J. Mach. Tools Manuf. **28**(4), 359–372 (1988)

New Vibration Model to Analyze the Correlation of Components in the Washing Machine Suspension System

Nguyen Thi Hoa(✉), Ngo Nhu Khoa, and Nguyen Thi Bich Ngoc

Thai Nguyen University of Technology, Thai Nguyen, Vietnam
hoanguyen@tnut.edu.vn

Abstract. This paper presents the new vibration model of the horizontal washing machine (HWM) using a nonlinear geographic of the suspension system with nonlinear friction force - velocity (F-V) relationship of dampers obtained from experiment. Furthermore, using the dissipated energy function of dampers (Ed) criteria, some results of the effect of the spring stiffness on the dampers' loading level are achieved. The simulation was done in Matlab/Simulink® environment. The results were then compared with experimental validation results to determine the suitability of the model. The conclusions can provide some available evidences for the design and improvement of nonlinear dampers.

Keywords: Dynamic modeling · Washing machine · Friction damper · Vibration

1 Introduction

For washing machines, the common cause of vibration and noise is due to centrifugal force generated by the distribution of unbalanced laundry in the drum when it rotates at high speed. The collision between the tub and the frame can cause vibrations to the cabinet, even the movement of the machine on the floor. To reduce the oscillation amplitude of the cabinet, a solution that many researchers used is investigating in damping configurations of the HWM suspension system.

Turkay et al. [1] presented a nonlinear time variant rigid body dynamic model of the suspension system of an HWM using Newton-Euler method, programmed for simulation, and assessed experimentally. The results demonstrated the validity of the dynamic model and user-friendly simulation code to predict tub displacements and, sliding and jumping tendencies of the cabinet. Thus, the mechanical performance of a suspension system configuration of an HWM can be assessed for a prototype design by undertaking a parametric investigation. They [2] proposed various formulations for suspension design optimization using grid and sequential quadratic programming optimization methods. It is concluded that minimization of the maximum orbit displacement or minimization of a weighted multi-objective function subject to the resistive stepping force constraint are suitable formulations. Moreover, the actual design of the test washing machine is verified for its optimality.

K.-U. Sattler et al. (Eds.): ICERA 2019, LNNS 104, pp. 500–511, 2020.
https://doi.org/10.1007/978-3-030-37497-6_58

Zhou et al. [3] studied analytically dynamic characteristics of a generalized suspension system. The paper showed that the length and inclined angle of the strings have significant effects on dynamic behavior of the suspension system. In another words, cables have great influence on free vibrations of the suspension system, whereas the dynamic behavior of the suspension system under crowd excitation has not been considered.

Kim et al. [4] had proposed and compared several useful damping models for multibody dynamic simulation. For frictional damping, they conducted a new modified stick-transition velocity model of frictional damping instead of a Coulomb model and this produced the best results. Whereas, YalçJn and Erol [5] developed a semi-active vibration control method to cope with the dynamic stability problem of an HWM based on adjusting the maximum force values produced by the semi-active suspension elements. Erol, in [6], stated that the side panel vibrations are affected by the forces that are transmitted from the oscillation group, the damping and material properties of the housing, the connection between the ground and the housing, and from the connections between the springs and friction dampers and the housing. Their model the focus of the attention was details such as flexible components and the noise characteristics of the machine. Deriving a dynamic model is the first step of the design improvement cycle and further work is required to validate the model and parameterize the system variables so that it can be recast as a well-defined optimization problem.

Several authors had used parametric approach to obtain the effect of suspension system especially, spring and damper coefficients and then validated by simulation. Ma et al. [7], based on ADAMS/View and ADAMS/Vibration platform, studied the simulation of the system in time domain and frequency domain and the effect of the key variables parameters (spring stiffness, damping coefficient) of vibration system. Bascetta et al. [8] characterized how the control performance is influenced by each mechanical, electrical and control components. A model of the machine including a description of the electrical aspects involved in the torque generation, as well as the transmission chain, the load, the spring–damper system connecting the tub to the cabinet, and the velocity estimator, is discussed. While [9] developed a model of the machine that includes an accurate description of the electrical aspects involved in the torque generation, as well as of the transmission chain, the load and the spring-damper system connecting the tub to the cabinet. The model was then used to identify the main mechanical parameters to simplify the tuning process. Boyraz, Gündüz [10] presented a 2D dynamic model of an HWM regarding the rotation plane in order to examine the vibration characteristics of the spin-cycle and improve the design proposing a new optimization scheme based on Genetic Algorithms (GA). The dynamical model is numerically simulated then validated by using experimental vibration data. The measurements are performed using piezo-transducers and a novel measurement scheme is used to obtain displacement values.

Although there are several researches on suspension parameters, their effect on vibration of HWMs was not fully considered. Our study aims to develop the new vibration model and based on the new model, as close to reality as possible, to evaluate the effects of the spring stiffness to the other components in the HWM's suspension system. The nonlinear vibration model is numerically simulated in Matlab/Simulink environment and evaluated by using experimental vibration.

2 Dynamic Model Derivation for Vibration Behavior

In this section, the model assumptions are given and then the model of HWM in static and dynamic state are presented.

In order to obtain a model which is simple enough to comprehend and for later uses, however, detailed enough to represent the real dynamics of HWM, some assumptions are included in the model. The vibration block which includes drum and motor driver rotating freely in a tub is assumed to move on x-y plane only. The movement on z-axis and rotations are restricted in this model. The cabinet of the HWM is assumed to be rigid and stable on the ground. Any effect of complex motion is not considered in the model. The unbalanced mass (m), suspension group are assumed to be in the middle plane of the vibration block. The real direction variables of springs and dampers are applied into the model to improve the reality of the model. Therefore, the nonlinear geographic dynamic model should be used instead of the linear geographic model as in [11] and [12]. The spinning speeds of the drum is set as realistic in the experiments. Based on the basic assumptions, the physical model in static state can be shown in Fig. 1 and in the dynamical state with movement on x and y positive directions is shown in Fig. 2.

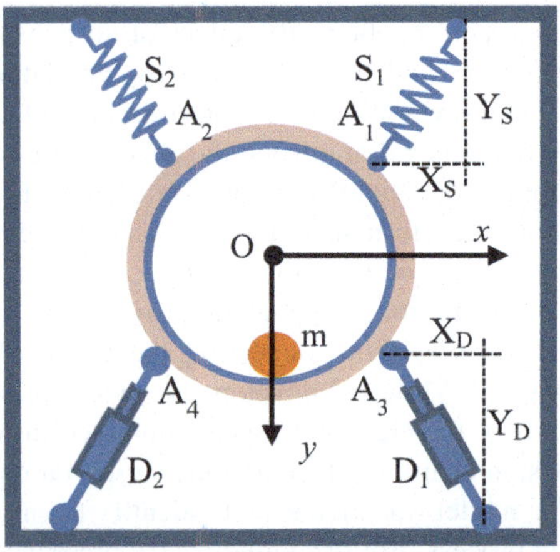

Fig. 1. HWM physical model of the static state

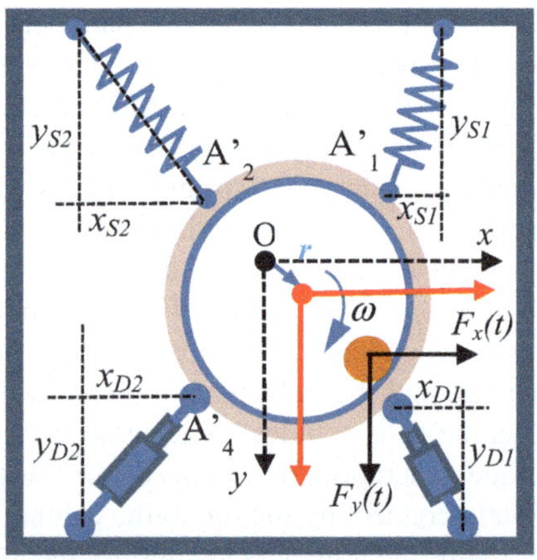

Fig. 2. Dynamical state of the HWM

Same as the 2D generation vibration models, the Equation of Motions (EOMs) of the HWM shown in Fig. 1 can be written as in (1) for x and y axes.

$$M\ddot{x} = F_x(t) - F_{Dx} - F_{Sx}; \qquad M\ddot{y} = F_y(t) - F_{Dy} - F_{Sy} \tag{1}$$

Where:

+M is mass of the tub;

+$F_x(t)$ and $F_y(t)$ are the external force components caused by unbalanced mass have tangential and normal components given by (2) and shown in Fig. 2:

$$\begin{cases} F_n = mR\omega^2 \quad F_t = mR\dot{\omega} \\ F_x(t) = F_n \sin(\omega t) - F_t \cos(\omega t) \\ F_y(t) = -F_n \cos(\omega t) - F_t \sin(\omega t) \end{cases} \quad (2)$$

+F_{Sx} and F_{Sy} are resultant spring forces in x and y directions. For springs, the dynamic forces are resulted from the dynamic length changes and are formulated by using Fig. 2. As the spring rotation angles attached to the tub are changing, the spring force components are nonlinear. Derivation of these forces are:

- The right spring (S_1):

$$\begin{cases} L_{S1} = \sqrt{(X_S - x)^2 + (Y_S + y)^2} \\ \Delta L_{S1} = L_{S1} - L_S \quad F_{S1} = k * \Delta L_{S1} \\ sin\theta_{S1} = \frac{X_S - x}{L_{S1}} \quad cos\theta_{S1} = \frac{Y_S + y}{L_{S1}} \\ F_{S1x} = F_{S1} * sin\theta_{S1} \quad F_{S1y} = F_{S1} * cos\theta_{S1} \end{cases} \quad (3)$$

- The left spring (S_2):

$$\begin{cases} L_{S2} = \sqrt{(X_S + x)^2 + (Y_S + y)^2} \\ \Delta L_{S2} = L_{S2} - L_S \quad F_{S2} = k * \Delta L_{S2} \\ sin\theta_{S2} = \frac{X_S + x}{L_{S2}} \quad cos\theta_{S2} = \frac{Y_S + y}{L_{S2}} \\ F_{S2x} = F_{S2} * sin\theta_{S2} \quad F_{S2y} = F_{S2} * cos\theta_{S2} \end{cases} \quad (4)$$

The resultant spring forces F_{Sx} and F_{Sy} are:

$$\begin{cases} F_{Sx} = F_{LSx} - F_{RSx} \\ F_{Sy} = F_{LSy} + F_{RSy} \end{cases} \quad (5)$$

Where: $L_S = \sqrt{X_S^2 + Y_S^2}$

+The F_{Dx} and F_{Dy} are resultant friction forces in x and y directions of the dampers/shock absorbers. In general, the shock absorbers are modeled as nonlinear elements with a high order force–velocity curve. The shock absorbers movement of suspension system is shown in Fig. 3 with the velocity vectors, force components and the relevant equations are given in Eqs. (6)–(8), in where *nold* and *nord* is number of the left and the right dampers, respectively. The damper force - velocity relationship of dampers in HWMs: F_D (V) = $C_2.V^2 + C_1.V + C_0$, where $C_2 = -14.305$, $C_1 = 40.734$ and $C_0 = 59.243$ [12].

- The right damper, D_1:

$$\begin{cases} L_{D1} = \sqrt{(X_D - x)^2 + (Y_D - y)^2} \\ sin\theta_{D1} = \frac{X_D - x}{L_{D1}} \quad cos\theta_{D1} = \frac{Y_D - y}{L_{D1}} \\ V_{D1} = -V_x sin\theta_{D1} - V_y cos\theta_{D1} \quad F_{D1} = \sum_{i=0}^{n} C_i (V_{D1})^i \\ F_{D1x} = F_{D1} * sin\theta_{D1} \quad F_{D1y} = F_{D1} * cos\theta_{D1} \end{cases} \tag{6}$$

- The left damper, D_2:

$$\begin{cases} L_{D2} = \sqrt{(X_D + x)^2 + (Y_D - y)^2} \\ sin\theta_{D2} = \frac{X_D + x}{L_{D2}} \quad cos\theta_{D2} = \frac{Y_D - y}{L_{D2}} \\ V_{D2} = V_x sin\theta_{D2} - V_y cos\theta_{D2} \quad F_{D2} = \sum_{i=0}^{n} C_i (V_{D2})^i \\ F_{D2x} = F_{D2} * sin\theta_{D2} \quad F_{D2y} = F_{D2} * cos\theta_{D2} \end{cases} \tag{7}$$

Finally, the resultant damping forces F_{Dx} and F_{Dy} are:

$$\begin{cases} F_{Dx} = nold * F_{D2x} - nord * F_{D1x} \\ F_{Dy} = -nold * F_{D2y} - nord * F_{D1y} \end{cases} \tag{8}$$

The simulation model of the vibration system model under the effect of unbalanced mass when the HWM operates in rinse stage at some design speed range is programmed in MATLAB/SIMULINK environment. The full model is given in Fig. 4, all the relationships of forces calculated in Eqs. (2)–(8) are coded in the V2N_Func

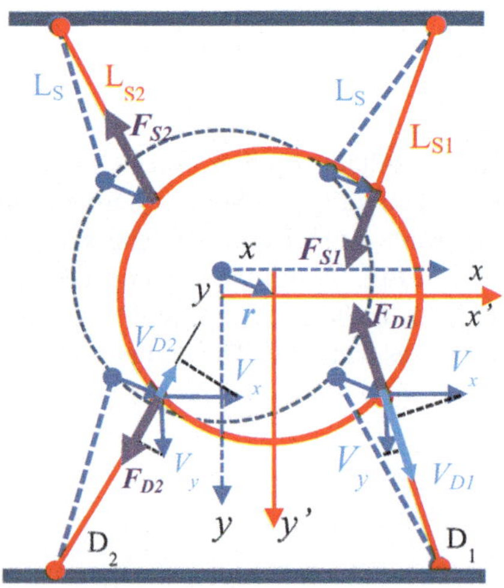

Fig. 3. Dynamic displacement vector ***r***, length and force for springs, velocity and force components for dampers

function block as an m-file. The function blocks get the needed parameters including mass values (*M*, *m*), table of increase of coordinates which are geometric locations of suspension elements in x and y axes (X_S, Y_S; X_D, Y_D), and coefficients such as stiffness of springs, damping coefficients which are also in a m-file.

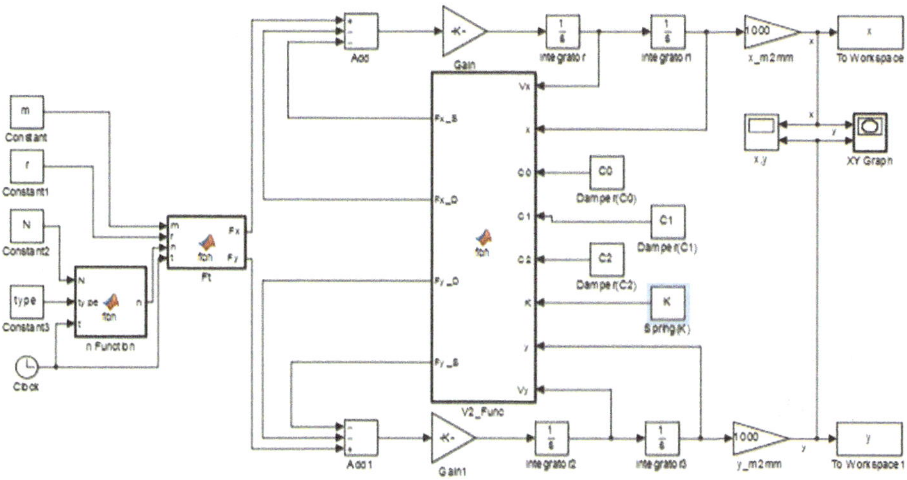

Fig. 4. Simulink diagram for the HWM vibration mathematical model

3 Experimental Model of the Damper and Measurements

3.1 Experimental Set-Up for Damper Testing

Dampers used in HWMs are non-Coulomb friction. HWMs use friction dampers as a low-cost solution to absorb vibration energy transposing to HWM's case. For an ideal friction damper, the friction coefficient is constant, but experimental results show that it strongly depends on velocity. In this study, to determine the characteristics of a damper, a damping test was performed by using the Electro-Servo tensile/compress testing machine, as shown in Fig. 5. In the experimental set-up, the linear displacement system include a servo drive model ASD-A04 21 LA and a ball screw model RCP2-SA7C-I-56P-

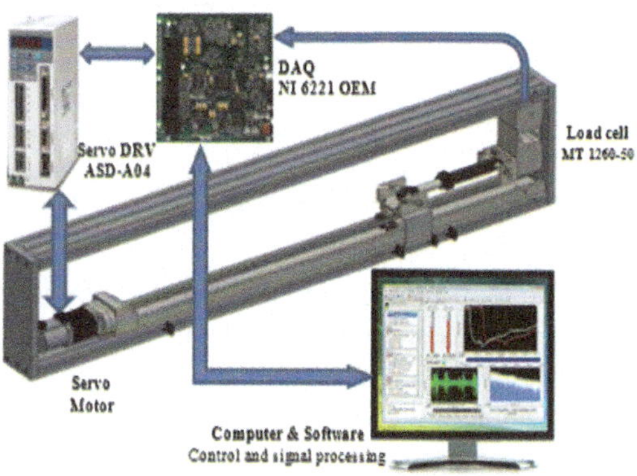

Fig. 5. Experimental setup of the damping test system

16-800-P1-M; a Mettler Toledo loadcell, model MT 1260-50 is used to measure friction force on the damper; The data acquisition board NI-6221 OEM is used to control linear displacement, velocity, force data acquisition and processes via computer software.

3.2 Validation of the Vibration Model

The numerical results of the vibration model and the data of real system are compared to validate the derived dynamic system model. In this study, vibration characteristics of an HWM are measured using a specific setup shown in Fig. 8. The experimental system was built base on LG WD 8990TDS HWM model with cabinet removal and all the other parts of the machine were installed on a stable steel structure. The unbalance mass of 620 g was attached into the inside drum. Two Linear Variable Differential Transformer (LVDT - RDP DCT 1000A models, stroke ±25 mm, output ±5 V, linearity error of full scale ±0.5%) sensors were used to measure the horizontal and vertical displacement on the middle of tub; 03 loadcells (MT 1260-50) measure dynamic forces on the dampers. The analog data from the LVDTs and loadcells amplified by Signal Conditioning modules (NI SCC-SG04) are converted to numerical data by Data acquisition board (DAQ, NI-USB 6251) via I/O Connector Block NI SCC-68. And finally, a computer running NI Signal Express were used for real-time computation and interfacing with the experimental hardware via the data acquisition card (Fig. 6).

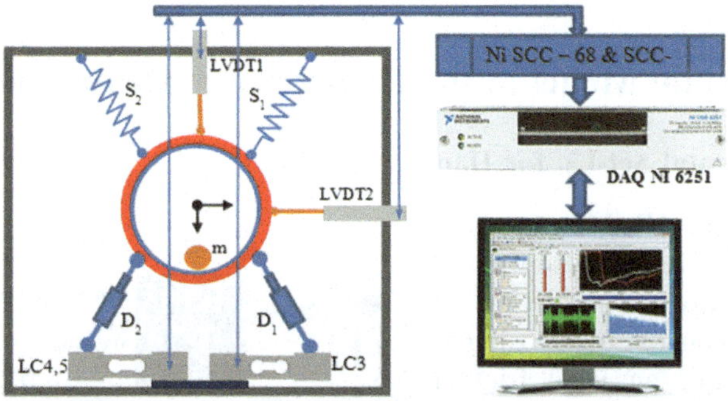

Fig. 6. Experimental setup of the vibration test system

The numerical simulation for the vibration model is carried out with the real HWM parameters as shown in the Table 1 (Fig. 7).

The measurement application built on the Ni Signal Express software allows direct observation and evaluation of displacement as well as the force at the links in the spin modes. Experimental and simulation results for nonlinear frictions are shown in Table 2.

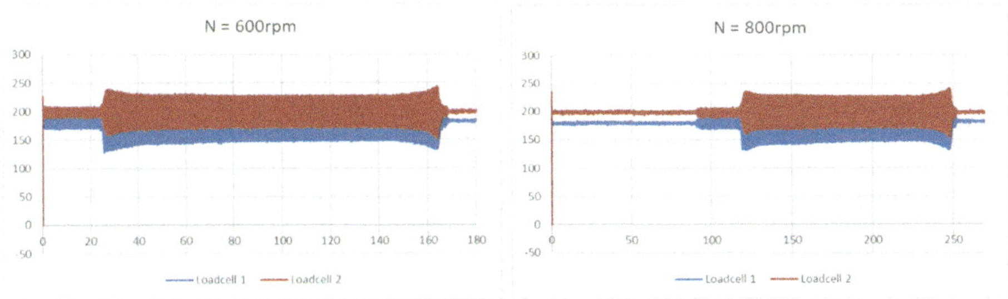

Fig. 7. Loadcell 1, 2 – Data of forces at springs in full time

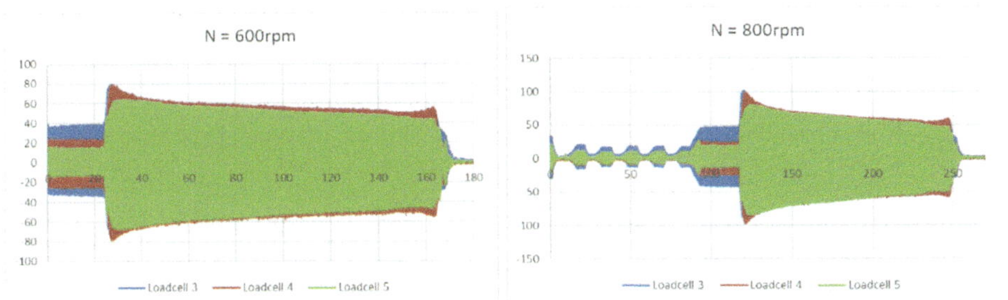

Fig. 8. Loadcell 3, 4, 5 - Data of forces at dampers in full time

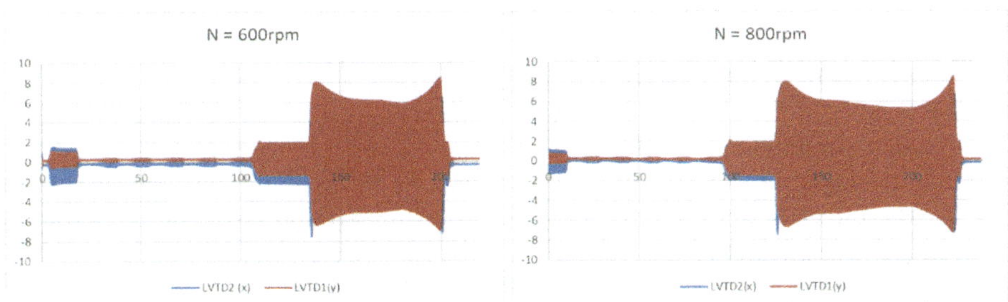

Fig. 9. LVDT 1, 2 - Displacement in directions x (LVDT 2) and y (LVDT 1)

As shown in Table 2, the different between modelling results of the new model and experimental data, especially damper force components are smaller than that of the linear geographic dynamic model in [11]. It means the nonlinear geographic dynamic model is closer with real model than linear geographic dynamic model. And the new model with the simulation programs can be used to study the effect of parameters on the vibration properties of the HWM's suspension system (Fig. 9).

Table 1. Nomenclature

Symbol	Definition	Parameters of LG WD 8990TDS
M	Drum total mass (kg)	32.5
m	The unbalanced mass (kg)	0.62
R	The eccentric or the inner radius of the drum (m)	0.24
K_0	Stiffness coefficients of springs (N/m)	5500
X_S, Y_S	The geometric locations of the springs on the tub (m)	0.03, 0.165
X_D, Y_D	The geometric locations of the dampers on the tub (m)	0.11, 0.165
N	The angular speed of the laundry rotation (rpm)	600 and 800

Table 2. Displacements and forces at the links in spin modes

Contents	N = 600 rpm				N = 600 rpm			
	Previous work [11]	Present work	Experimental	% different	Previous work [11]	Present work	Experimental	% different
x_amp (mm)	4.201	**4.2118**	4.302	2	4.520	**4.451**	4.197	−6
y_min (mm)	5.087	**5.0971**	5.364	5	5.006	**5.037**	4.855	−4
y_max (mm)	−5.092	**−5.184**	−5.603	7	−5.025	**−5.053**	−4.798	−5
y_amp (mm)	5.089	**5.140**	5.4835	6	5.0162	**5.045**	4.8265	−5
± Fd_R (N)	66.639	**64.018**	53.86	−19	66.639	**65.045**	60.705	−7
± Fd_L (N)	133.278	**130.064**	113.34	−15	133.278	**130.672**	124.11	−5

4 Results and Discussion

The vibration amplitude in x and y directions and dissipated energy of dampers are two criteria to evaluate the correlation of components in the washing machine suspension system. The effect of spring stiffness to vibration of tub and working conditions of dampers are investigated in this paper.

Based on original spring stiffness coefficient ($K_0 = 5500$ N/m), the effect of the varied spring stiffness ($K = hs * K_0$) to the vibration behavior of the tub at 600 RPM and 800 RPM rotation speeds, which corresponds to different spring stiffness, from 2750 N/m to 11000 N/m ($hs = 0.5\ldots2$) can be determined. The results are shown in the Table 3, Figs. 10 and 11. Where, Table 3 shows the vibration amplitude in x and y directions of the tub; the spring force trend in x direction and the dissipated energy curve of the dampers are shown in Figs. 10 and 11, respectively.

As shown as in the Table 3, the effect of the spring stiffness to the vibration amplitude is not very significant. The results are matched with the conclusion presented by Boyraz [10]. In fact, the increase of spring stiffness would lead to the increase of spring force in x direction, as shown in Fig. 10. The results are risk to the walking washing machine phenomena.

Table 3. The x, y displacement amplitude in varied stiffness of the springs

Spring stiffness K (N/m)	N = 600 rpm		N = 800 rpm	
	x_amp (mm)	y_amp (mm)	x_amp (mm)	y_amp (mm)
2750	4.807902	4.064379	4.769564	4.306678
3300	4.808268	4.123037	4.799942	4.335449
3850	4.812694	4.155131	4.77899	4.353648
4400	4.827736	4.166295	4.778179	4.379116
4950	4.823409	4.213992	4.830347	4.422338
5500	4.824765	4.250668	4.813985	4.45912
6050	4.856429	4.283621	4.826483	4.466985
6600	4.844538	4.310384	4.813823	4.484559
7150	4.853528	4.363002	4.84892	4.515382
7700	4.854632	4.400646	4.835484	4.52792
8250	4.88368	4.415938	4.809856	4.553591
11000	4.897716	4.628447	4.848597	4.670641

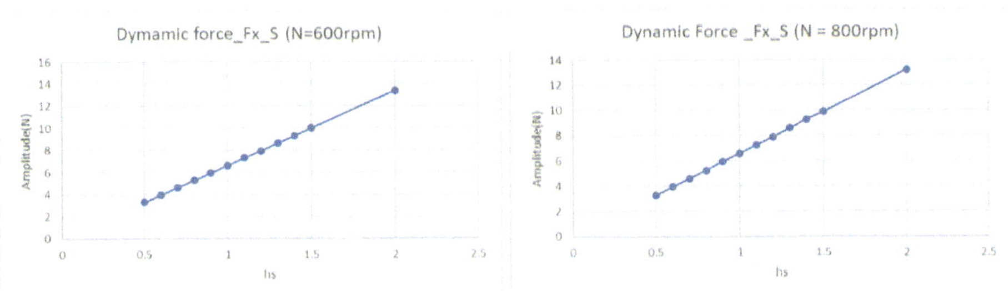

Fig. 10. The spring forces in x direction in varied stiffness of the springs via *hs*

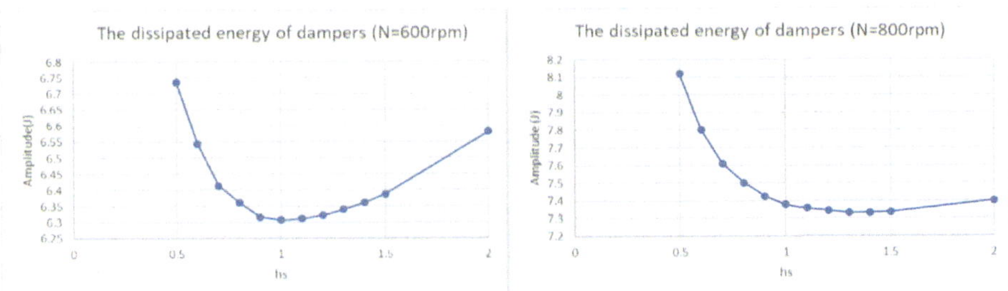

Fig. 11. The dissipated energy curve of the dampers in varied *hs*

However, the spring stiffness effects the most to the loading level of the dampers, which is indicated by dissipated energy criteria. In which, the dissipated energy function Ed [12] is calculated by:

$$E_d = \int_{t0}^{t} [F_{LD}(t).x_{LD}(t) + F_{RD}(t).x_{RD}(t)]dt$$

Where, $F_{LD}(t)$, $F_{RD}(t)$ are dynamic friction forces in the left and right dampers, and $x_{LD}(t)$, $x_{RD}(t)$ are dynamic displacements of the left and right damper cylinders.

As show as in the Fig. 11, the Ed curve has minimum at $hs = 1$ when the tub rotates at 600 RPM and $hs = 1.3$ at the 800 RPM. In fact, the dampers will be much durable if the dissipated energy is small. So, these results almost matched with the original spring stiffness from the HWM LG WD 8990TDS producers.

5 Conclusion

The new vibration model is developed, with the varied springs and dampers directions are considered. The new model is proved much more closed with the real vibration model comparing to previous linear geographic model. By using the *Ed* criteria to examine the effect of the spring stiffness to the loading level of the dampers, the results are matching with the origin parameters of the producer in realty. Based on the new model with simulation program established with new criteria - the dissipated energy function of the dampers, the studies about HWM vibration behaviors can be further studied.

References

1. Türkay, O.S., Sümer, I.T., Tuğcu, A.K., Kiray, B.: Modeling and experimental assessment of suspension dynamics of a horizontal – axis washing machine. J. Vib. Acoust. **120**, 534–543 (1998)
2. Türkay, O.S., Kiray, B., Tugcu, A.K., Sümer, İ.T.: Formulation and implementation of parametric optimisation of a washing machine suspension system. Mech. Syst. Sign. Process. **9**(4), 359–377 (1995). https://doi.org/10.1006/mssp.1995.0029
3. Zhou, D., Ji, T.: Dynamic characteristics of a generalized suspension system. Int. J. Mech. Sci. **50**, 30–42 (2008)
4. Kim, H.-J., Yoo, W.-S., Ok, J.-K., Kang, D.-W.: Parameter identification of damping models in multibody dynamic simulation of mechanical systems. Multibody Syst. Dyn. **22**, 383–398 (2009). https://doi.org/10.1007/s11044-009-9163-5
5. YalçJn, B.C., Erol, H.: Semiactive vibration control for horizontal axis washing machine. Shock Vib. **2015**, 10 (2015). https://doi.org/10.1155/2015/692570
6. Öztürk, M.E., Erol, H.: On the dynamics of a washing machine with flexible components. Noise Control Eng. J. **58**(6), 572–590 (2010)
7. Ma, X., Hu, F., Liu, J.: Dynamic characteristic simulation of drum washing machine rigid-flexible coupling model. Int. J. Control Autom. **8**(5), 167–176 (2015). https://doi.org/10.14257/ijca.2015.8.5.16
8. Bascetta, L., Rocco, P., Zanchettin, A.M., Magnani, G.: Velocity control of a washing machine: a mechatronic approach. Mechatronics **22**, 778–787 (2012)

9. Bascetta, L., Magnani, G., Rocco, P.: Mechatronic analysis of the velocity control of a washing machine. In: Proceedings of the 2009 IEEE International Conference on Mechatronics, Malaga, Spain, April 2009
10. Boyraz, P., Gündüz, M.: Dynamic modeling of a horizontal washing machine and optimization of vibration characteristics using Genetic Algorithms. Mechatronics **23**, 581–593 (2013)
11. Khoa, N.N., Hoa, N.T., Ngoc, N.T.B.: Numerical modeling and experimental study on vibration of a horizontal washing machine. In: ICERA 2018. LNNS, vol. 63, pp. 415–424 (2019). https://doi.org/10.1007/978–3-030-04792-4_54
12. Khoa, N.N., Hoa, N.T., Ngoc, N.T.B.: The effect of damper Configurations on the vibration of horizontal washing machines. In: ICERA 2018. LNNS, vol. 63, pp. 298–308 (2019). https://doi.org/10.1007/978-3-030-04792-4_40

Nonlinear Backstepping-Sliding Mode Control of Electro-Hydraulic Systems

Tung Lam Nguyen[1], Hong Quang Nguyen[2(✉)], Manh Cuong Nguyen[1], Van Manh Tran[1], Danh Huy Nguyen[1], and Anh Duc Nguyen[3]

[1] Hanoi University of Science and Technology, Hanoi, Vietnam
[2] Thai Nguyen University of Technology, Thai Nguyen, Vietnam
quang.nguyenhong@tnut.edu.vn
[3] Thai Nguyen University, Thai Nguyen, Vietnam

Abstract. A speed control of an electro-hydraulic system is investigated in the paper. The nonlinear model of the electro-valve driven hydraulic system is developed. A systematic control approach using backstepping-sliding mode control which is designed with can regulate the system to the desired speed. The system stability is proven, and several simulations are given to illustrate the effectiveness of the control.

Keywords: Electro-hydraulic · Backstepping · Sliding mode control

1 Introduction

Thanks to the ability of handling heavy tasks, hydraulic systems are widely employed in various applications [1]. Many researches focus on position and tracking problems of the hydraulic systems in the quest of seeking suitable control schemes for better control performances [2–4]. In [5], the authors consider the position control of the hydraulic actuator taking into account system dead-zone, control input saturation, discharge coefficient and friction. A PID control is designed and reinforced by particle swarm optimization, the proposed control is validated through numerical simulations and experiments. With an idea of exploiting simplicity of pulse-width-modulated on/off valve accompanied with a feedback linearization control to stabilize the supply pressure, the authors successfully construct a nonlinear position control for a hydraulic system [6]. In [7], the authors suggest a fuzzy logic controller whose parameters are tuned using particle swarm optimization. The analogous technique of using optimization can be found in [8]. Other nonlinear control approaches to the hydraulic system can be listed in [9–11].

The mentioned researches either need accurate system information or relatively comprehensive understandings about the system. In actual situation, hydraulic system parameters are not easy to identify correctly. The paper proposes a backstepping based position control of the hydraulic system backed with sliding mode control for the sake of system robustness. The system stability and performance are proven analytically and validated numerically.

K.-U. Sattler et al. (Eds.): ICERA 2019, LNNS 104, pp. 512–519, 2020.
https://doi.org/10.1007/978-3-030-37497-6_59

2 System Description and Dynamical Modeling

Two stages servo valve consists of three main parts: the electrical torque motor, the hydraulic amplifier, and the valve spool assembly. The dynamics of the valve spool with no noticeable decline in accuracy in a wide range of frequencies can be described through the first order transfer function between the valve opening area A_{sv} and control input u.

$$\tau.\dot{A}_{sv} + A_{sv} = K_{sv}.u \tag{1}$$

where K_{sv} is the servo valve gain and τ_{sv} is the servo valve time constant. The constants mentioned can be determined for by certain tests. Due to the fact that the input of the valve is an electric current but the interface card output is in the form of an electric voltage, it is in common to use a current to voltage converter.

For an ideal critical center, the servo valve with a matched and symmetric orifice the input/output flow rate from the servo valve through the orifices (assuming negligible leakage) can be expressed in the following form:

$$Q_L = C_d A_{sv} \sqrt{\frac{P_s - p_L sign(A_{sv})}{\rho}} \tag{2}$$

where $p_L = p_{C1} - p_{C2}$ is a load pressure or pressure difference between both chambers, $p_S = p_{C1} + p_{C2}$ is the supply pressure and Q_L is the load flow. Assuming no external leakage, QL can be considered as the average flow in each path $Q_L = \frac{Q_{C1}+Q_{C2}}{2}$, Q_{C1} and Q_{C2} are flow rates to and from the servo valve.

$$\frac{V_0}{2\beta}\dot{p}_L = C_d A_{sv} \sqrt{\frac{p_s - p_L sign(A_{sv})}{\rho}} - D_m\dot{\Theta} - C_L p_L \tag{3}$$

where β and V_0 are, respectively, the fluid bulk modulus and the oil under compression in one chamber of the actuator. D_m and C_L represent the actuator volumetric displacement and total leakage coefficient, respectively. By applying Newton's second law for the rotary motion of a hydraulic actuator and neglecting the Coulomb's frictional torque:

$$J_T\ddot{\Theta} = D_m p_L - B.\dot{\Theta} + T_L \tag{4}$$

Combining Eqs. (1) ÷ (4), the third-order nonlinear system that describe the system dynamics can be derived as:

$$\begin{aligned} \ddot{\Theta} &= -a_1\dot{\Theta} + a_2 p_L + a_3 \\ \dot{p}_L &= -a_4\dot{\Theta} - a_5 p_L + a_6(\sqrt{P_s - p_L})A_{sv} \\ \dot{A}_{sv} &= -a_7 A_{sv} + a_8 u \end{aligned} \tag{5}$$

where $a_1 = \frac{B}{J_t}$, $a_2 = \frac{D_m}{J_t}$, $a_3 = \frac{T_L}{J_t}$, $a_4 = \frac{2\beta D_m}{V_0}$

$$a_5 = \frac{2\beta}{V_0} C_L,\ a_6 = \frac{2\beta}{V_0\sqrt{\rho}} C_d,\ a_7 = \frac{1}{\tau_{sv}} \text{ and } a_8 = \frac{K_{sv}}{\tau_{sv}}$$

3 Nonlinear Control Design

This section addresses the problem of designing a controller which provides asymptotic stability of the operating point of interest. Assuming that the full state information is available, the underlying technique for solving this problem is SMC-backstepping. The first and second tracking error is defined as:

$$\begin{aligned} e_1 &= \dot{\Theta} - \dot{\Theta}_d \\ e_2 &= p_L - p_{Ld} \\ e_3 &= A_{sv} - A_{svd} \end{aligned} \tag{6}$$

Define the first candidate of control Lyapunov candidate function

$$V_1 = \frac{1}{2} e_1^2 \tag{7}$$

Then p_{Ld} is chosen as virtual control to drive e_1 to zero so that $\dot{\Theta}$ can track $\dot{\Theta}_d$.

$$p_{Ld} = \frac{1}{a_2}\left(a_1\dot{\Theta} + \ddot{\Theta}_d - a_3 - \xi_1 e_1\right) \tag{8}$$

$$\dot{V}_1 = e_1\dot{e}_1 = e_1\left(-a_1\dot{\Theta} + a_2 e_2 + a_2 p_{Ld} + a_3 - \ddot{\Theta}_d\right) \tag{9}$$

With such selection, we achieve the following form of augmented Lyapunov function:

$$\dot{V}_1 = a_2 e_1 e_2 - \xi_1 e^2 \tag{10}$$

The dynamics error e_1 can be described

$$\dot{e}_1 = \ddot{\Theta} - \ddot{\Theta}_d = -a_1\dot{\Theta} + a_2 p_L + a_3 - \ddot{\Theta}_d \tag{11}$$

Then p_{Ld} is chosen as virtual control to drive e_1 to zero, by substituting (6) into (11):

$$\dot{e}_1 = -a_1\dot{\Theta} + a_2(e_2 + p_{Ld}) + a_3 - \ddot{\Theta}_d \tag{12}$$

We obtain the virtual signal p_{Ld} as

$$p_{Ld} = \frac{1}{a_2}\left(a_1\dot{\Theta} + \ddot{\Theta}_d - a_3 - \xi_1 e_1\right) \tag{13}$$

Step 2: The virtual control error e_2 can be described by

$$\dot{e}_2 = \dot{p}_L - \dot{p}_{Ld} \tag{14}$$

By substituting (6) into (14):

$$\dot{e}_2 = -a_4\dot{\Theta} - a_5 p_L + \left(a_6\sqrt{P_s - p_L}\right)A_{sv} - \dot{p}_{Ld} \tag{15}$$

Define the second Lyapunov candidate function:

$$V_2 = V_1 + \frac{1}{2}e_2^2 \tag{16}$$

Differentiating V_2 gets:

$$\dot{V}_2 = \dot{V}_1 + e_2\dot{e}_2 \tag{17}$$

$$\dot{V}_2 = -\xi_1 e_1^2 + e_2(-\xi_2 e_2) + a_6\sqrt{P_s - x_2}e_2 e_3 \tag{18}$$

$$\dot{V}_2 = a_2 e_1 e_2 - \xi_1 e_1^2 + e_2\left[-a_4\dot{\Theta} - a_5 p_L + a_6\sqrt{P_s - p_L}A_{sv} - \dot{p}_{Ld}\right] \tag{19}$$

From (6), it is straightforward to show that:

$$\dot{V}_2 = a_2 e_1 e_2 - \xi_1 e_1^2 + e_2\left[-a_4\dot{\Theta} - a_5 p_L + a_6\sqrt{P_s - p_L}A_{svd} + a_6\sqrt{P_s - p_L}e_3 - \dot{p}_{Ld}\right] \tag{20}$$

Similarly, we define the sliding surface as:

$$s = \lambda(A_{sv} - A_{svr}) \tag{21}$$

Differentiating s yields

$$\dot{s} = \lambda(-a_7 A_{sv} + a_8 u - \dot{A}_{svr}) \tag{22}$$

The signal control can be computed as:

$$u = u_s + u_e \tag{23}$$

where u_e is the equivalent control and u_s is the switching control. By differentiating s with respect to time t in (6), letting $\dot{s} = 0$, and substituting (6) into it, the equivalent control is formulated as

$$u_e = \frac{1}{a_8 \lambda}\left(a_7 A_{sv} + \dot{A}_{svr}\right) \tag{24}$$

$$u_s = -\frac{k}{a_8 \lambda} s - \frac{\eta}{a_8 \lambda} sat(s) \tag{25}$$

Where k and η are positive constants. Define the Lyapunov candidate function as

$$V_s = \frac{1}{2} s^2 \tag{26}$$

Differentiating both sides of (26) gives:

$$\dot{V}_s = s\dot{s} \tag{27}$$

Then $$\frac{dV_s}{dt} = s[-\text{ks} - \eta sat(s)] = -ks^2 - \eta|s| \leq 0 \tag{28}$$

Equation (28) ensuring the sliding mode is reachable in finite time. In the subsequent time interval, the system trajectory moves along the sliding surface and converges to the coordinate origin constructed by the intermediate variable s and its derivative.

4 Simulation Studies

In this section, numerical simulation results are presented to verify the accuracy of the proposed control approach for the hydraulic rotary actuator. The results are shown in two cases: the system working without any external noises and the system affected by the disturbance. Different reference speed forms are set for the system to validate the robustness of the designed algorithm.

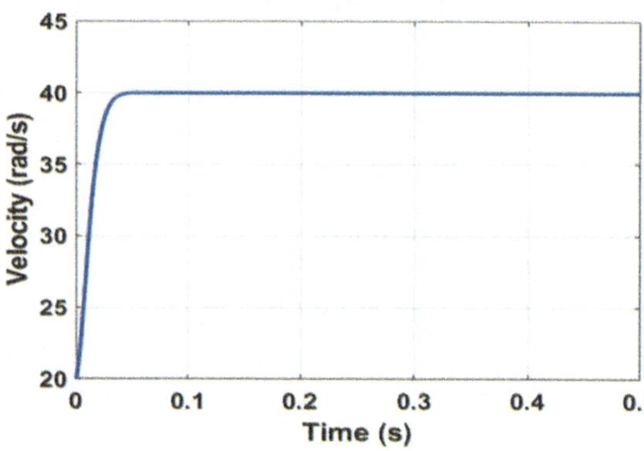

Fig. 1. Actual velocity with fixed reference

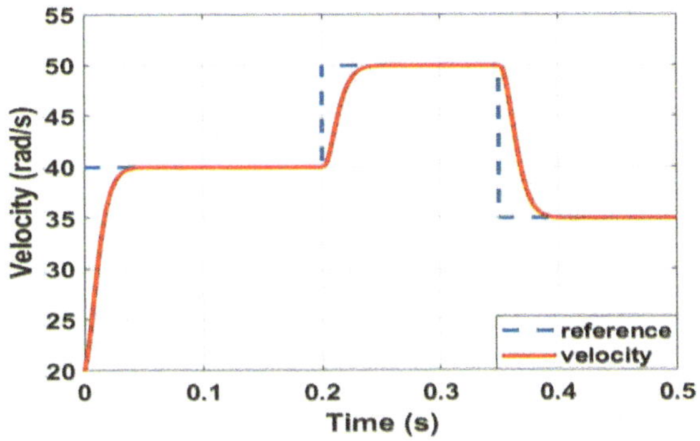

Fig. 2. Actual velocity with variable reference

Under the ideal condition, assuming that all system's parameters are clearly known, the performance of the electrohydraulic velocity servo system is demonstrated in Figs. 1 and 2. The reference speed is chosen as $\dot{\Theta} = 40\,(\mathrm{rad/s})$ and the actual value can be seen in Fig. 1. For the variation of the reference velocity with time, the proposed controller still ensures the high tracking quality of the actual velocity in Fig. 2. The speed response is in an acceptable range while guarantees the stability of the system while guarantees the stability of the system.

Commonly, in actual application, the system is always affected by the unknown disturbances, so that in this case, at the time 0.2 s, the simulation condition is under an external noise. The purpose of this simulation is to verify the robustness of the controller.

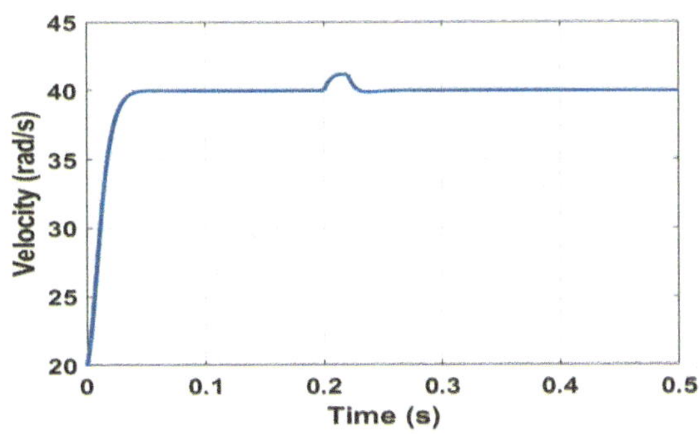

Fig. 3. Actual velocity with noise

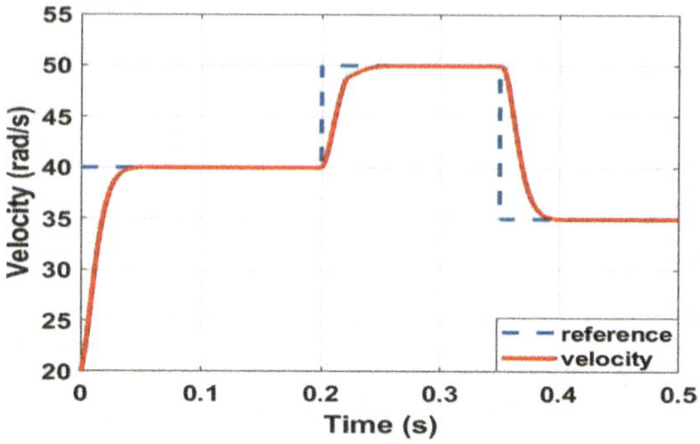

Fig. 4. Velocity response with system disturbance

It can be seen that in Figs. 3 and 4, the proposed controller has the ability to reject external noise after a short period of time (Table 1).

Table 1.

$J_t = 3.4 \times 10^{-3}\,\mathrm{kgm}^2$	$D_m = 0.75 \times 10^{-6}\ \mathrm{m}^3/\mathrm{rad}$
$B = 1.25 \times 10^{-6}\ \mathrm{Nms/rad}$	$C_L = 9.5 \times 10^{-12}\ \mathrm{m}^5/\mathrm{Ns}$
$\beta = 0.35 \times 10^9\ \mathrm{Pa}$	$V_0 = 2.75 \times 10^{-5}\ \mathrm{m}^3$
$C_d = 0.65$	$T_L = 0.7\ \mathrm{Nm}$
$K_{sv} = 4.23 \times 10^{-7}\ \mathrm{m}^2/\mathrm{V}$	$\tau_{sv} = 0.001\ \mathrm{s}$

5 Conclusions

In the paper, a backstepping-sliding mode-based control for electro-hydraulic system is introduced. Hydraulic nonlinearities are taken care of using the systematic approach. System uncertainties and disturbances are handled thanks to the existence of sliding mode control. The system is proven to be stable in Lyapunov's sense. Several numerical studies are included to show the effectiveness of the proposed control.

References

1. Aly, A.A.: Model reference PID control of an electro-hydraulic drive. Int. J. Intell. Syst. Appl. **4**(11), 24–32 (2012)
2. Najafi, F., Fathi, M., Saadat, M.: Dynamic modelling of servo pneumatic actuators with cushioning. Int. J. Adv. Manuf. Technol. **42**(7–8), 757–765 (2009)
3. Soon, C.C., Ghazali, R., Jaafar, H.I., Hussien, S.Y.S., Rozali, S.M., Rashid, M.Z.A.: Position tracking optimization for an electro-hydraulic actuator system. J. Telecommun. Electron. Comput. Eng. **8**(7), 1–6 (2016)

4. Kim, H.M., Park, S.H., Song, J.H., Kim, J.S.: Robust position control of electro-hydraulic actuator systems using the adaptive back-stepping control scheme. Proc. Inst. Mech. Eng. Part I J. Syst. Control Eng. **224**(6), 737–746 (2010)
5. Ye, Y., Yin, C.B., Gong, Y., Zhou, J.J.: Position control of nonlinear hydraulic system using an improved PSO based PID controller. Mech. Syst. Sig. Process. **83**, 241–259 (2017)
6. Rannow, M.B., Li, P.Y.: On/off valve based position control of a hydraulic cylinder, no. April, pp. 141–149 (2009)
7. Wonohadidjojo, D.M., Kothapalli, G., Hassan, M.Y.: Position control of electro-hydraulic actuator system using fuzzy logic controller optimized by particle swarm optimization. Int. J. Autom. Comput. **10**(3), 181–193 (2013)
8. Essa, M.E.S.M., Aboelela, M.A.S., Hassan, M.A.M.: Position control of hydraulic servo system using proportional-integral-derivative controller tuned by some evolutionary techniques. JVC/J. Vibr. Control **22**(12), 2946–2957 (2016)
9. Mihajlov, M., Nikolić, V., Antić, D.: Position control of an electro-hydraulic servo system using sliding mode control enhanced by fuzzy PI controller. Mech. Eng. **1**(9), 1217–1230 (2002)
10. Abd Rahman, R.: Dynamical adaptive backstepping-sliding mode control for servo-pneumatic positioning applications: controller design and experimental evaluation, no. February (2016)
11. Yao, J., Jiao, Z., Ma, D.: Extended-state-observer-based output feedback nonlinear robust control of hydraulic systems with backstepping. IEEE Trans. Ind. Electron. **61**(11), 6285–6293 (2014)

Nonlinear Design of Adaptive Controllers for Bilateral Teleoperation Systems with Variable Time Delays

Dang Ngoc Trung[1(✉)], Dao Phuong Nam[2], and Do Trung Hai[1]

[1] Thai Nguyen University of Technology, Thai Nguyen, Vietnam
trungcsktd@tnut.edu.vn
[2] Hanoi University of Science and Technology, Hanoi, Vietnam

Abstract. Controlling Bilateral Teleoperation systems in the presence of time-varying delays and disturbances presents a significant problem. This paper presents robust sliding mode control for slave side and traditional input state stability (ISS) controller for master side with the effects of external forces. Furthermore, synchronous control law based on wave variable and passivity based control in the presence of variable time delays. The wave variable theory enables us to consider the two-side BT systems as a unified circuit system. The attraction region in closed systems is estimated by using theoretical analysis. Offline simulation results are also implemented to verify the efficiency of the proposed controllers.

Keywords: Bilateral Teleoperation systems · Robust control · Wave form variables

1 Introduction

Bilateral Teleoperation (BT) Systems are the human control systems which enable human to interact with the remote environment based on a double-sided robot system [1, 10, 15]. They have been extensively applied in industrial applications such as robotic telesurgery, rehabilitation [2]. Due to the distance between local and remote robots, time delays appearing from communication are a problem affecting stability of whole BT systems [3, 4].

In order to deal with communication delay, a lot of methods have been presented to cope with this challenge [6–10]. In the work of [5, 6, 9], the authors proposed the constant time delay based wave variables theory to consider BT systems as two port network circuit with long-lines. It enables us to employ the passivity based control law for these BT systems. However, several disadvantages have been mentioned, such as the wave variable needs to be computed in time interval, instability of whole BT systems due to the variable time delay. By using the passivity based theory, the transmission of wave variable as well as scattering [5, 6] has been handled to stabilize BT systems under constant time delay. These works were extended to Time domain passive approach (TDPA) for the case of variable time delays [8, 11, 12, 14]. However, the control objective has been mentioned also being position control problem.

K.-U. Sattler et al. (Eds.): ICERA 2019, LNNS 104, pp. 520–533, 2020.
https://doi.org/10.1007/978-3-030-37497-6_60

Furthermore, dynamic uncertainties are also the obstacle which has been extensively considered [3–5]. In order to handle the teleoperators under time delays and uncertainties, the adaptive synchronous control law has been proposed [14]. The synchronizing signal was mentioned to tackle the dynamic uncertainties [13] based on classical Lyapunov theoretical analysis. The motion/force adaptive control for uncertain teleoperator has been presented in [1] based on Jacobian matrix and certainty equivalence. The optimal control approach was considered in [2] for both sides in BT systems by considering the linearized model and the equivalent variational method.

Thanks to the effectiveness of being more robust to uncertainties, the robust adaptive sliding mode control for BT systems is developed to handle the dynamic uncertainties in this paper. Furthermore, the wave variable based synchronization controller is considered for the drawback of time-varying delays. Numerical examples are implemented to demonstrate the efficiency of the proposed controller for nonlinear BT systems.

2 Problem Formulation

Consider the BT systems including the operated robot (slave robot) and controlled robot (master robot) being considered to the dual robot n - Degree of Freedom with the interaction of robot and environment, robot and human by Euler - Lagrange equation [3]:

$$\begin{aligned} M_m(q_m)\ddot{q}_m + C_m(q_m, \dot{q}_m)\dot{q}_m + g_m(q_m) = \tau_h - \tau_m \\ M_s(q_s)\ddot{q}_s + C_s(q_s, \dot{q}_s)\dot{q}_s + g_s(q_s) = \tau_s - \tau_e \end{aligned} \tag{1}$$

where m, s are master and slave side, respectively, $\ddot{q}_i$, $\dot{q}_i$, $q_i \in R^n (i = m, s)$ are accelerator, velocity, position of robot respectively. $M_i(q_i) \in R^{n\times n}$ is inertia matrix, $C_i(\dot{q}_i, q_i)$ is the specific component Coriolis, τ_h, τ_e are force (torque) of supervisor and work place, τ_m, τ_s are force (torque) of controller with the assumption of perfect actuators.

Some important well-known properties were mentioned as follows [3]:

1. The inertia matrix is lower and upper bounded, i.e.,

$$0 < \lambda_m\{M_i\}I \leq M_i(q_i) \leq \lambda_m\{M_i\}I < \infty.$$

2. The relation between Coriolis and inertia matrix is described:

$$\dot{M}_i(q_i) = C_i(q_i, \dot{q}_i) + C_i^T(q_i, \dot{q}_i)$$

3. The Coriolis matrices are bounded as follows:
   $\forall q_i, \dot{q}_i \in R^n \exists k_{c_i} \in R_{>0}$ so that $|C_i(q_i, \dot{q}_i)\dot{q}_i| \leq k_{c_i}|\dot{q}_i|^2$.
4. The dynamics are linearly parameterizable:

$$M_i(q_i)\ddot{q}_i + C_i(q_i, \dot{q}_i)\dot{q}_i + g_i(q_i) = Y_i(q_i, \dot{q}_i, \ddot{q}_i)\theta_i$$

The control objective is to find a new wave transformation and the equivalent synchronization control for BT systems under variable time delay. Besides, the robust sliding mode control is proposed to ensure the tracking problem on master and slave sides.

3 Robust Sliding Mode Control Law for BT Systems

In this chapter, the robust SMC Law is proposed for slave side and traditional SMC scheme is presented for master side of BT systems.

On the slave side, by using the following slave controller:

$$\tau_s = \mathbf{M}_s(q_s)\mathbf{u}_s + \mathbf{C}_s(q_s, \dot{q}_s)\dot{q}_s + \mathbf{G}_s(q_s) + \tau_b \tag{2}$$

We obtain the relation:

$$\mathbf{M}_s(q_s)\ddot{q}_s = \mathbf{M}_s(q_s)\mathbf{u}_s - \Delta_e \tag{3}$$

Let's design the sliding surface as:

$$\mathbf{S}_s = \dot{e}_s + \mathbf{\Gamma}_p e_s + \mathbf{\Gamma}_I \int_o^t e_s(\xi)d\xi \tag{4}$$

This work proposed the following Lemma:

Lemma 1: The controller is described:

$$\mathbf{u}_s = \mathbf{u}_{seq} + \mathbf{K}_s\mathbf{S}_s + \mathbf{A}sgn\mathbf{S}_s \tag{5}$$

$$\text{with}\,\mathbf{u}_{seq} = \ddot{q}_{sd} + \mathbf{\Gamma}_p\dot{e}_s + \mathbf{\Gamma}_I e_s \tag{6}$$

$$\mathbf{S}_s \in \mathbb{R}^n;\ \mathbf{\Gamma}_I = diag[\gamma_{I1}\ \ \gamma_{I2}\cdots\gamma_{In}] \tag{7}$$

$$\gamma_{pi} > 0;\ \gamma_{Ii} > 0,\ i = 1,\ 2,\ \ldots,\ n;\ \mathbf{\Gamma}_p = diag\left[\gamma_{p1}\ \ \gamma_{p2}\cdots\gamma_{pn}\right] \tag{8}$$

ensure that the state trajectory of (3) converges to sliding surface (4).

On the master side, based on the following master controller:

$$\tau_m = \mathbf{M}_m(q_m)\left[\ddot{q}_{dm} + \mathbf{K}_p e_m + \mathbf{K}_D\dot{e}_m\right] + \mathbf{C}_m(q_m, \dot{q}_m)\dot{q}_m + \mathbf{G}_m(q_m) + \hat{\tau}_{Nm} \tag{9}$$

we obtain the closed system as follows:

$$\mathbf{M}_m(q_m)\left[\ddot{q}_m - \ddot{q}_{md} - \mathbf{K}_p e_m - \mathbf{K}_D\dot{e}_m\right] = \tau_{op} - \hat{\tau}_e^* + \hat{\tau}_{Nm} - \tau_{Nm} \tag{10}$$

Using the intermediate variable: $\Delta_{Nm} = \tau_{Nm} - \hat{\tau}_{Nm}$

$$\mathbf{X}_m = \begin{bmatrix} e_m \\ \dot{e}_m \end{bmatrix}, \ \mathbf{A} = \begin{bmatrix} 0 & \mathbf{I}_m \\ -\mathbf{K}_p & -\mathbf{K}_D \end{bmatrix}, \ \mathbf{B} = \begin{bmatrix} 0 \\ \mathbf{M}_m^{-1}(q_m) \end{bmatrix} = \begin{bmatrix} 0 \\ \mathbf{B}_2(q_m) \end{bmatrix} \tag{11}$$

$$\mathbf{B}_2(q_m) = \mathbf{M}_m^{-1}(q_m)$$

We imply:

$$\dot{\mathbf{X}}_m = \mathbf{A}\mathbf{X}_m + \mathbf{B}\left[-\tau_{op} + \hat{\tau}_e^* + \Delta_{Nm}\right] \tag{12}$$

With the following assumptions:

$$\begin{aligned} &\left\|\tau_{op}\right\|_\infty = \sup_t\left|\tau_{op}(t)\right| = \eta_{op}; \ \left\|\hat{\tau}_e^*\right\|_\infty = \sup_t\left|\hat{\tau}_e^*(t)\right| = \eta_e \\ &\left\|\Delta_{Nm}\right\|_\infty = \sup_t\left|\Delta_{Nm}(t)\right| = \delta_{Nm}; \ \max_{1\le i\le n}\sum_{j=1}^{n}\left|b_{ij}(q_m)\right| \le \sigma_b \end{aligned} \tag{13}$$

Lemma 2: The closed system (12) is ISS stability with the attractor being determined as follows:

$$\mathbb{O} = \left\{\mathbf{X}_m \in \mathbb{R}^{2n} : \ \|\mathbf{X}_m\| < 2\sigma_p\sigma_b\eta_v\right\} \tag{14}$$

$$\eta_v = \eta_{op} + \eta_e + \delta_{Nm} \tag{15}$$

σ_p is the upper bound of norm $\|(.)\|$ of matrix $\mathbf{P}_m \in \mathbb{R}^{2nx2n}$:

$$\mathbf{A}^T\mathbf{P}_m + \mathbf{P}_m\mathbf{A} = -\mathbf{I} \tag{16}$$

In order to consider the influence of time delays, this work is considered not only the two above lemmas but also some following errors:

$$\begin{aligned} \tilde{\tau}_e(t) = \ \hat{\tau}_e^*(t-T); \ q_{sd}(t) = q_m(t-T); \\ \dot{q}_{sd}(t) = \dot{q}_m(t-T); \ q_{md}(t) = q_s(t-T); \ \dot{q}_{md}(t) = \dot{q}_s(t-T) \end{aligned} \tag{17}$$

$$\begin{aligned} e_m(t) &= q_{md}(t) - q_m(t) = q_s(t-T) - q_m(t) \\ \dot{e}_m(t) &= \dot{q}_{md}(t) - \dot{q}_m(t) = \dot{q}_s(t-T) - \dot{q}_m(t); \\ e_s(t) &= q_{sd}(t) - q_s(t) = q_m(t-T) - q_s(t) \\ \dot{e}_s(t) &= \dot{q}_{sd}(t) - \dot{q}_s(t) = \dot{q}_m(t-T) - \dot{q}_s(t) \end{aligned} \tag{18}$$

Therefore, the equivalent controllers under time delays for both sides are described as follows:

Theorem 1: Considering the BT systems (1) under time delays. The equivalent controllers are established as the two above lemmas. The closed system, which is obtained from these controllers, is ISS stable with the attractors as follows:

$$\mathbb{O}_t = \left\{ \mathbf{X}_t = \left[\mathbf{X}_m^T \ \mathbf{X}_s^T\right]^T \in \mathbb{R}^{4n} \right\} \tag{19}$$

$$\begin{aligned} &\|\mathbf{X}_m\| < \tfrac{g+\sqrt{g^2+4\rho}}{2};\ g = 2\|\mathbf{P}_m\|.\|\mathbf{B}\|.\|\tau_v\|;\ \rho = \rho_m + \rho_s + \rho_{\tau_e};\ \rho_m = \rho_{m1} + \rho_{m2} \\ &\|\mathbf{X}_s\| = 0 \end{aligned}$$

$$\rho_{m1} \geq \left\| \sum_{i=1}^{\infty} \frac{\partial^i q_m^T(t-T) q_m(t-T)}{\partial t^i} T^i \right\|;\ \rho_{m2} \geq \left\| \sum_{i=1}^{\infty} \frac{\partial^i \dot{q}_m^T(t-T) \dot{q}_m(t-T)}{\partial t^i} T^i \right\| \tag{20}$$

$$\rho_s = \rho_{s1} + \rho_{s2};\ \rho_{s1} \geq \left\| \sum_{i=1}^{\infty} \tfrac{\partial^i q_s^T(t-T) q_s(t-T)}{\partial t^i} T^i \right\|$$

$$\rho_{s2} \geq \left\| \sum_{i=1}^{\infty} \tfrac{\partial^i \dot{q}_s^T(t-T) \dot{q}_s(t-T)}{\partial t^i} T^i \right\|;\ \rho_{\tau_e} \geq \left\| \sum_{i=1}^{\infty} \tfrac{\partial^i \hat{\tau}_e^{*T}(t-T) \hat{\tau}_e^*(t-T)}{\partial t^i} T^i \right\|$$

Remark 1: The proof of this Theorem absolutely implemented based on the following Lyapunov candidate function:

$$\begin{aligned} \mathbf{V} = \mathbf{V}_m + \mathbf{V}_S + \int_{t-T}^{t} \left[q_m^T(\xi) q_m(\xi) + \dot{q}_m^T(\xi) \dot{q}_m(\xi)\right] d\xi \\ + \int_{t-T}^{t} \left[q_s^T(\xi) q_s(\xi) + \dot{q}_s^T(\xi) \dot{q}_s(\xi)\right] d\xi + \int_{t-T}^{t} \hat{\tau}_e^{*T}(\xi) \hat{\tau}_e^*(\xi) d\xi \end{aligned} \tag{21}$$

4 Synchronization Control of BT Systems Under Time-Varying Delays

Scattering being specific power variable of system is transmitted and communication between dual robot. The master robot is able to send signal u_m and receive v_m, the slave robot obtains u_s and directs v_s (see Fig. 1).

Proposition 1: The proposed wave variable:

$$u_m = \frac{bF_m' + B\dot{x}_m'}{\sqrt{2bB}};\ u_s = \frac{bF_s' + B\dot{x}_s'}{\sqrt{2bB}};\ v_m = \frac{B\dot{x}_m' - bF_m'}{\sqrt{2bB}};\ v_s = \frac{B\dot{x}_s' - bF_s'}{\sqrt{2bB}} \tag{22}$$

Parameters b, B are virtual independence of double transmission line, $F'_m, \dot{x}'_m$ and $\dot{x}'_s, F'_s$ import and export master/slave scattering. The power of the system is calculated with variable time delay:

$$P = \frac{1}{2}\int_{t-T}^{t} \left(u_m(\tau)^T u_m(\tau) + v_s(\tau)^T v_s(\tau)\right) d\tau = \frac{1}{2}\left(u_m^T u_m - u_s^T u_s + v_s^T v_s - v_m^T v_m\right) \quad (23)$$

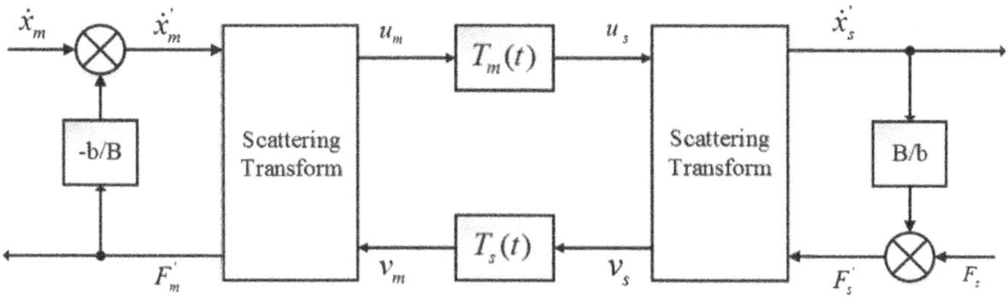

Fig. 1. Equivalent diagram of BTS

The notion is that the dual robot system and communication are considered as one union together with the notion about the circuit theory. Considering communication as that wave is transmitted in long–lines, system can lose energy because of wave reflection which is part of transmitted signal and appear in received signal. Wave reflection is unnecessary; therefore, that must to be eliminated. The proposed method in this paper places $\dot{x}'_m, F'_m$ backwards $\dot{x}_m, F_m$.

The proposed wave transformation with wave reflection elimination:

$$\dot{x}'_m = \dot{x}_m - \frac{b}{B}F'_m;\ F'_s = F_s + \frac{B}{b}\dot{x}'_s \quad (24)$$

The new wave transformation will be changed when replacing eliminated wave-reflection (24) on (22).

$$u_m = \frac{B\dot{x}_m}{\sqrt{2bB}};\ u_s = \frac{bF_s + 2B\dot{x}'_s}{\sqrt{2bB}};\ v_m = \frac{B\dot{x}_m - 2bF'_m}{\sqrt{2bB}};\ v_s = \frac{-bF_s}{\sqrt{2bB}} \quad (25)$$

We definitely achieve intermediate variable control from (25).

$$\begin{aligned} \dot{x}_m &= \sqrt{\frac{2b}{B}}u_m & F'_m &= \sqrt{\frac{B}{2b}}(u_m - v_m) \\ F_s &= -\sqrt{\frac{2B}{b}}v_s & \dot{x}'_s &= \sqrt{\frac{b}{2B}}(u_s + v_s) \end{aligned} \quad (26)$$

We realize (26) that two transmitted variables u_m, v_s only depend on one variable. The fact that time delay is random signal, but in this paper it is considered as variable time function.

$$u_s(t) = u_m(t - T_1(t)); v_m(t) = v_s(t - T_2(t)) \tag{27}$$

The power of system is calculated as following (28):

$$\begin{aligned} P = F_m^{\prime T} \dot{x}_m - F_s^T \dot{x}_s^{\prime} &= u_m^T(u_m - v_m) + (u_s + v_s)^T v_s \\ &= \frac{1}{2}u_m^T u_m - u_m^T v_m + \frac{1}{2}v_m^T v_m + \frac{1}{2}u_s^T u_s + u_s^T v_s + \frac{1}{2}v_s^T v_s \\ &+ \frac{1}{2}u_m^T u_m - \frac{1}{2}v_m^T v_m - \frac{1}{2}u_s^T u_s + \frac{1}{2}v_s^T v_s \\ &= \left(\frac{1}{2}u_m^T u_m - \frac{1}{2}v_m^T v_m\right)^2 + \left(\frac{1}{2}u_s^T u_s + \frac{1}{2}v_s^T v_s\right)^2 \\ &- \frac{1}{2}\dot{T}_1(t)u_s^T u_s - \frac{1}{2}\dot{T}_2(t)v_m^T v_m + \frac{d}{dt}\int_{t-T_1(t)}^{t} \frac{1}{2}u_m^T(\tau)u_m(\tau)d\tau \\ &+ \frac{d}{dt}\int_{t-T_2(t)}^{t} \frac{1}{2}v_s^T(\tau)v_s(\tau)d\tau = P_{diss} + \frac{dE}{dt} \end{aligned} \tag{28}$$

Derivative of energy system and power of dissipation are considered as:

$$\frac{dE}{dt} = \frac{d}{dt}\int_{t-T_1(t)}^{t} \frac{1}{2}u_m^T(\tau)u_m(\tau)d\tau + \frac{d}{dt}\int_{t-T_2(t)}^{t} \frac{1}{2}v_s^T(\tau)v_s(\tau)d\tau \tag{29}$$

$$\begin{aligned} P_{diss} = &\left(\frac{1}{2}u_m^T u_m - \frac{1}{2}v_m^T v_m\right)^2 + \left(\frac{1}{2}u_s^T u_s + \frac{1}{2}v_s^T v_s\right)^2 \\ &- \frac{1}{2}\dot{T}_1(t)u_s^T u_s - \frac{1}{2}\dot{T}_2(t)v_m^T v_m \end{aligned} \tag{30}$$

The dissipation of system is separated to become the master and slave dissipation including m-sides and s-sides, respectively:

$$\begin{aligned} P_{diss}^m &= \left(\frac{1}{2}u_m^T u_m - \frac{1}{2}v_m^T v_m\right)^2 - \frac{1}{2}\dot{T}_2(t)v_m^T v_m \\ P_{diss}^s &= \left(\frac{1}{2}u_s^T u_s + \frac{1}{2}v_s^T v_s\right)^2 - \frac{1}{2}\dot{T}_1(t)u_s^T u_s \end{aligned} \tag{31}$$

This paper must apply passive theory, but the variable time delay is being discussed these days because it causes damage to the stability of the system. TDPA proposed by Hannaford was able to support the system to become passive, including calculation and controller. The mission of calculation is to examine satisfied requirements and modify coefficients of controller to the system to obtain stability.

The condition of absolute derivative time-delay no more than one is applied in previous papers but will be eliminated in this paper. The absolute derivative time-delay must be limited $|\dot{T}_{1,2}| < \varepsilon$.

Proposition 2: The controller is selected as follows:

$$F_m = F_m^{'} + \kappa_m \dot{x}_m;\ \dot{x}_s = \dot{x}_s^{'} + \kappa_s F_s \tag{32}$$

Proposition 3: The calculation is employed to regulate two coefficients of the proposed controller:

$$P_{cal}^{m} = -\frac{1}{2}\varepsilon v_m^T v_m;\ P_{cal}^{s} = -\frac{1}{2}\varepsilon u_s^T u_s \tag{33}$$

The power of the system is calculated by new variable following two ports network:

$$\begin{aligned} P = F_m^T \dot{x}_m - F_s^T \dot{x}_s &= \left(F_m^{'} + \kappa_m \dot{x}_m\right)^T \dot{x}_m - F_s^T\left(\dot{x}_s^{'} + \kappa_s F_s\right) \\ &= \left(F_m^{'T} \dot{x}_m - F_s^T \dot{x}_s^{'}\right) + \kappa_m \dot{x}_m^T \dot{x}_m - \kappa_s F_s^T F_s \\ &= \left(P_{diss}^{m} + P_{diss}^{s} + \frac{dE}{dt}\right) + \kappa_m \dot{x}_m^T \dot{x}_m - \kappa_s F_s^T F_s \\ &= \left(P_{diss}^{m} + \kappa_m \dot{x}_m^T \dot{x}_m\right) + \left(P_{diss}^{s} - \kappa_s F_s^T F_s\right) + \frac{dE}{dt} \\ &= P_{diss}^{m'} + P_{diss}^{s'} + \frac{dE}{dt} \end{aligned} \tag{34}$$

Applying (33) to (34) to demonstrate the stability of system:

$$\begin{aligned} P_{diss}^{m'} &= P_{cal}^{m} + \kappa_m \dot{x}_m^T \dot{x}_m + \left(\frac{1}{2}u_m^T u_m - \frac{1}{2}v_m^T v_m\right)^2 + \frac{1}{2}\left(\varepsilon - \dot{T}_2(t)\right) v_m^T v_m \\ P_{diss}^{s'} &= P_{cal}^{s} - \kappa_s F_s^T F_s + \left(\frac{1}{2}u_s^T u_s + \frac{1}{2}v_s^T v_s\right)^2 + \frac{1}{2}\left(\varepsilon - \dot{T}_1(t)\right) u_s^T u_s \end{aligned} \tag{35}$$

In this paper, the system needs to satisfy a passivity condition proposed by M. Spong to become stable, in that the derivative of the energy of the system must not be more than the power of that, which derives power of dissipation should be more than zero. Because the two last terms in (35) are more than zero, the coefficient of controller

κ_m and κ_s are set to the two first terms in (35) to be eliminated with that $\dot{x}_m^T\dot{x}_m$ and $F_s^T F_s$ differ zero. The coefficients of controller are selected as follows:

$$\kappa_m = -P_{cal}^m\left(\dot{x}_m^T\dot{x}_m\right)^{-1};\ \kappa_s = P_{cal}^s\left(F_s^T F_s\right)^{-1} \tag{36}$$

The system will be passive because of positive power of dissipation $P_{diss}^{m'}$ and $P_{diss}^{s'}$:

The previous papers must apply assumption absoluteness of derivative time-delay no more than one to ignore cases $\dot{x}_m^T\dot{x}_m$ or $F_s^T F_s$ being zero. This paper employs two calculations to consider these two cases.

Where $\dot{x}_m^T\dot{x}_m = 0$

$$\begin{aligned} P &= -F_s^T\dot{x}_s = -F_s^T\left(\dot{x}_s' + \kappa_s F_s\right) = \left(u_s + v_s\right)^T v_s - \kappa_s F_s^T F_s \\ &= \frac{d}{dt}\int\limits_{t-T_1(t)}^{t} u_m^T(\tau)u_m(\tau)d\tau + \left(1 - \dot{T}_1(t)\right)\left(u_s^T u_s\right) + u_s^T v_s + v_s^T v_s - \kappa_s F_s^T F_s \end{aligned} \tag{37}$$

The new derivative of energy and power of dissipation follow (37).

$$\frac{dE}{dt} = \frac{d}{dt}\int\limits_{t-T_1(t)}^{t} u_m^T(\tau)u_m(\tau)d\tau \tag{38}$$

$$\begin{aligned} P_{diss} &= \left(1 - \dot{T}_1(t)\right)\left(u_s^T u_s\right) + u_s^T v_s + v_s^T v_s - \kappa_s F_s^T F_s \\ &= \left(u_s^T u_s + u_s^T v_s + v_s^T v_s\right) + (\varepsilon - T_1(t))u_s^T u_s + \left(2P_{cal}^s - \kappa_s F_s^T F_s\right) \end{aligned} \tag{39}$$

Employing passive standard - Spong to select coefficient of controller κ_s:

$$\kappa_s = 2F_{cal}^s\left(F_s^T F_s\right)^{-1} = -\varepsilon\frac{b}{2B}\left(u_s^T u_s\right)\left(v_s^T v_s\right)^{-1} \tag{40}$$

Where $F_s^T F_s = 0$

$$\begin{aligned} P = F_m^T\dot{x}_m &= \left(F_m' + \kappa_m\dot{x}_m\right)^T\dot{x}_m = u_m^T(u_m - v_m) + \kappa_m\dot{x}_m^T\dot{x}_m \\ &= \frac{d}{dt}\int\limits_{t-T_2(t)}^{t} v_s^T(\tau)v_s(\tau)d\tau + \left(1 - \dot{T}_2(t)\right)v_m^T v_m \\ &\quad - u_m^T v_m + u_m^T u_m + \kappa_m\dot{x}_m^T\dot{x}_m \end{aligned} \tag{41}$$

The new derivative of energy and power of dissipation follow (42).

$$\frac{dE}{dt} = \frac{d}{dt}\int\limits_{t-T_2(t)}^{t} v_s^T(\tau)v_s(\tau)d\tau \tag{42}$$

$$\begin{aligned} P_{diss} &= \left(1 - \dot{T}_2(t)\right) v_m^T v_m - u_m^T v_m + u_m^T u_m + \kappa_m \dot{x}_m^T \dot{x}_m \\ &= \left(v_m^T v_m - u_m^T v_m + u_m^T u_m\right) + \left(\varepsilon - \dot{T}_2(t)\right) \dot{x}_m^T \dot{x}_m \end{aligned} \tag{43}$$

The similar method directs coefficient of controller κ_m:

$$\kappa_m = -2P_{cal}^m \left(\dot{x}_m^T \dot{x}_m\right)^{-1} = \varepsilon \frac{B}{2b} \left(v_m^T v_m\right) \left(u_m^T u_m\right)^{-1} \tag{44}$$

Diagram (see Fig. 2) presents a new Scattering transformation, an elimination of

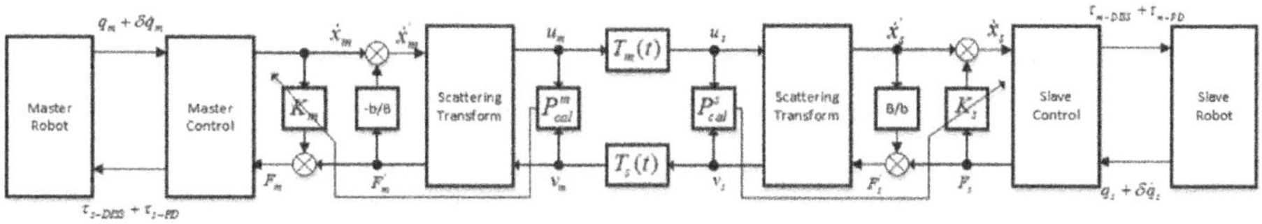

Fig. 2. Scattering with TDPA

wave reflection and TDPA applying to dual robot system.

Signal $\dot{x}_m(t)$ is sent from the master robot to the slave robot with position and velocity, in contrast, slave robot's position and velocity are transmitted to that with signal $F_s(t)$ to establish PD-like controller

$$\begin{aligned} \dot{x}_m(t) &= q_m(t) + \delta \dot{q}_m(t) \\ F_s(t) &= q_s(t) + \delta \dot{q}_s(t) \end{aligned} \tag{45}$$

Force/Torque applying the slave robot includes two components $\tau_s = \tau_{s-DISS} + \tau_{s-PD}$ which mean dissipation and PD-like controller respectively:

$$\begin{aligned} \tau_{s-PD} &= \kappa_s (r_m(t - T_1(t)) - r_s(t)) \\ \tau_{s-PD} &= \kappa_s (q_m(t - T_1(t)) - q_s(t)) + \kappa_s \delta (\dot{q}_m(t - T_1(t)) - \dot{q}_s(t)) \end{aligned} \tag{46}$$

$$\tau_{s-DISS} = -B_s \dot{q}_s(t) \tag{47}$$

Force/Torque impact on the master robot is the same as that on the slave robot:

$$\begin{aligned} \tau_{m-PD} &= \kappa_m (r_s(t - T_2(t)) - r_m(t)) \\ \tau_{m-PD} &= \kappa_m (q_s(t - T_2(t)) - q_m(t)) + \kappa_m \delta (\dot{q}_s(t - T_2(t)) - \dot{q}_m(t)) \end{aligned} \tag{48}$$

$$\tau_{m-DISS} = B_m \dot{q}_m(t) \tag{49}$$

If the dual robots are passive, they will be stable. Considering Euler-Lagrange equation of master robot without gravity:

$$\begin{aligned} & M_m(q_m)\ddot{q}_m + C_m(\dot{q}_m, q_m)\dot{q}_m = \tau_h - (\tau_{m-PD} + \tau_{DISS}) \\ & M_m(q_m)\ddot{q}_m + (C_m + B_m)(\dot{q}_m, q_m)\dot{q}_m = \tau_h - \tau_{m-PD} \\ & [M_m(q_m)\ddot{q}_m + (C_m + B_m)(\dot{q}_m, q_m)\dot{q}_m](q + \delta\dot{q}) = (\tau_h - \tau_{m-PD})(q + \delta\dot{q}) \end{aligned} \tag{50}$$

It can be seen that right side of (50) is power of robot; therefore, the rest must be sum of derivative of positive store energy and positive function. The left side of (50):

$$\frac{dSE}{dt} = M_m(q_m\ddot{q}_m + \delta\dot{q}_m\ddot{q}_m) + \frac{M_m}{\delta}q_m\dot{q}_m + M\ddot{q}_m^2 \tag{51}$$

$$\frac{dSE}{dt} = \frac{M_m}{\delta}(q_m\dot{q}_m + \delta q_m\ddot{q}_m + \delta\ddot{q}_m^2 + \delta^2\dot{q}_m\ddot{q}_m) + \left((C_m + B_m) - \frac{M_m}{\delta}\right)q_m\dot{q}_m \tag{52}$$

From (40) and (41) we can derive

$$SE = \frac{M_m}{2\delta}\left[(q_m + \delta\dot{q}_m)^2 + (M_m^{-1}\delta(C_m + B_m) - 1)q_m^2\right] \tag{53}$$

$$PF = [(C_m + B_m)\delta - M_m]\dot{q}_m^2 \tag{54}$$

Proposition 4: δ can be chosen to SE and PF to become positive:

$$(C_m + B_m)\delta - M_m > 0 \tag{55}$$

$$\delta > 0 \tag{56}$$

5 Offline Simulation Results

Consider the BT systems as described in the three tables (Tables 1, 2 and 3):

With the time delay being 2 s, the simulation result for the proposed controller in chapter 3 are illustrated (see Figs. 3 and 4). It can be seen that the proposed system is able to reach the desired trajectories.

Table 1. Robot parameters

Parameters	l_1; l_2 (m)	r_1 and r_2 (m)	m_1 and m_2 (kg)	I_1 and I_2 (kg.m^2)
Value	0.2 (m)	0.1 and 0.082	0.72 and 0.48	0.0050191 and 0.0031088

Table 2. Master side parameters

Parameters	τ_{op}	τ_{Nm}	K_P	K_D	G_m
Value	$\begin{bmatrix} 8\sin 0.5t \\ 3\sin 0.2t \end{bmatrix}$	$\begin{bmatrix} 0.03\sin 2.5t \\ 0.012\sin t \end{bmatrix}$	[5 0; 0 5]	[100 0; 0 100]	[0; 0]

Table 3. Slave side parameters

Parameters	P	Γ_I	K_s	A	τ_{Ns}	τ_e
Value	1	0.25	[10 0; 0 10]	[0.005 0; 0 0.005]	$\begin{bmatrix} 0.5\sin 2.5t \\ 0.3\sin 4t \end{bmatrix}$	$\begin{bmatrix} 2\sin 0.5t \\ 1\sin 0.2t \end{bmatrix}$

The proposed controllers in chapter 4 are definitely demonstrated by simulation with 2-DOF(revolution) dual robot without gravity (see Figs. 5 and 6). Master and slave robot have parameters manipulators' massive and length respectively:

$$m_m = \begin{bmatrix} 4 \\ 1 \end{bmatrix}(kg) \qquad l_m = \begin{bmatrix} 0.3 \\ 0.2 \end{bmatrix}(m) \quad m_s = \begin{bmatrix} 6 \\ 2 \end{bmatrix}(kg) \qquad l_s = \begin{bmatrix} 0.4 \\ 0.3 \end{bmatrix}(m)$$

The initial state of dual robot:

$$\dot{q}_{m0} = \begin{bmatrix} 0 \\ 0 \end{bmatrix}(rad/s) \qquad q_{m0} = \begin{bmatrix} 0.5 \\ 0.3 \end{bmatrix}(rad)$$
$$\dot{q}_{s0} = \begin{bmatrix} 0 \\ 0 \end{bmatrix}(rad/s) \qquad q_{s0} = \begin{bmatrix} 0.5 \\ 0.3 \end{bmatrix}(rad)$$

The BTS can be recognized easily as stability even with the appearance of human force or the influence of the environment.

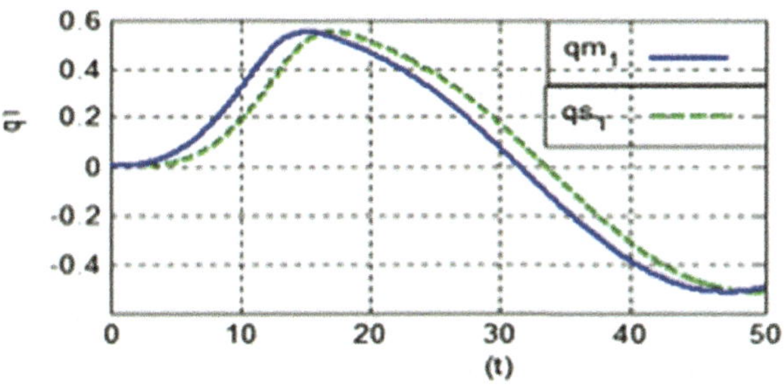

Fig. 3. The response of the first joint

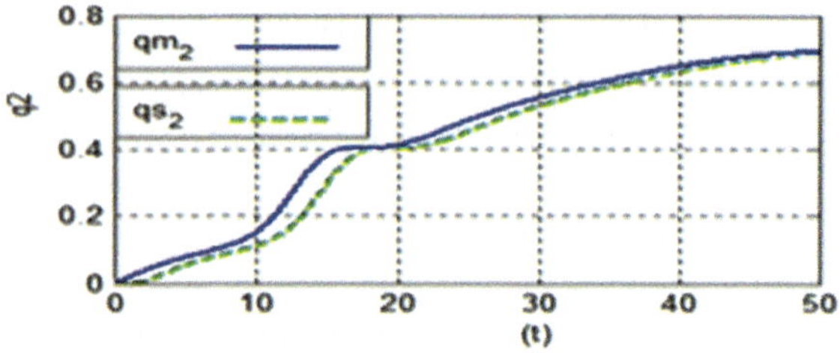

Fig. 4. The response of the second joint

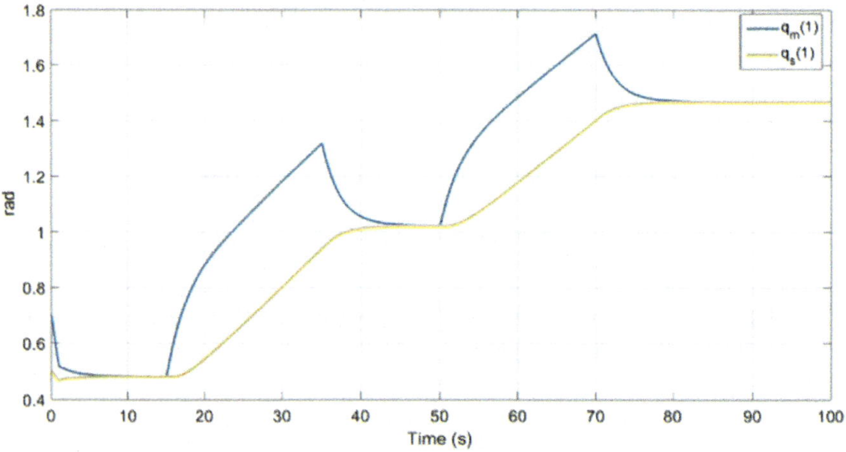

Fig. 5. The response of the robust SMC for BT systems in the first join

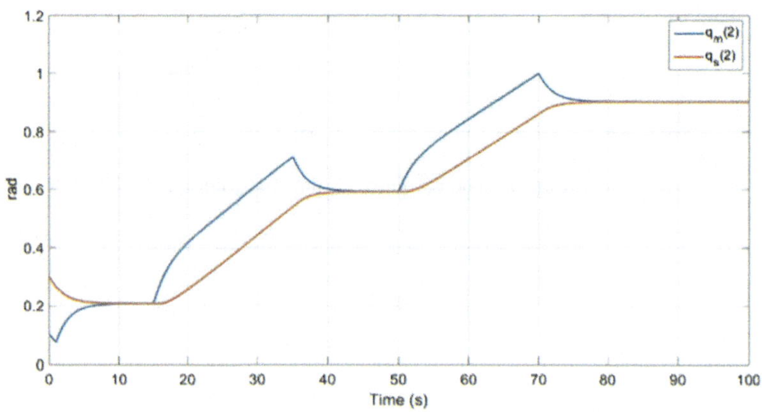

Fig. 6. The response of the robust SMC for BT systems in the second join

6 Conclusion

This paper addresses the problem of control design for BT systems by utilizing the robust SMC law and wave variables. The robust SMC is mentioned in the slave side of BT systems. It is able to ensure the tracking control objective in the issue of constant time delay. In order to tackle the challenge of time-varying delays, the new wave variables is proposed to obtain the equivalent master & slave control algorithm. Offline simulation demonstrates the efficiency of these proposed controller.

References

1. Trung, D.N., Hai, D.T., Nam, D.D.: The constructing of robust adaptive controller for the remotely operated system with uniform communication delays. VietNam Autom. Today (16), P69–76 (2016). ISSN 1859-0551
2. Nam, D.P., Vu, T.A., Thiem, T.D., Thiet, N.H.: Optimal control for bilateral teleoperation system with variational method. In: 2016 International Conference on Biomediacl Engineering, Hanoi, pp. 130–136. IEEE (2016)
3. Tung, N.A., Binh, N.T., Anh, T.H., Nam, D.P., Dong, N.M.: Synchronization control of bilateral teleoperation systems by using wave variable method under varying time delay. In: 2017 International Conference on System Science and Engineering, Hanoi, pp. 521–525. IEEE (2017)
4. Anh, T.H., Tung, N.A., Binh, N.T., Nam, D.P., Van Tu, V.: An adaptive control law against time – varying delays in bilateral teleoperation systems. In: 2017 International Conference on System Science and Engineering, Hanoi, pp. 542–547. IEEE (2017)
5. Niemeyer, G., Siotine, J.J.E.: Stable adaptive teleoperation. IEEE Oceanic Eng. **16**(1), 99–110 (1991)
6. Anderson, R., Spong, M.: Bilateral control of teleoperators with time delay. IEEE Trans. Autom. Control **34**(5), 494–501 (1989)
7. Nuno, E., Basanez, L., Ortega, R., Spong, M.: Position tracking for nonlinear teleoperators with variable time delay. Int. J. Robot. Res. **28**(7), 895–910 (2009)
8. Chawda, V., et al.: Position synchronization in bilateral teleoperation under time - varying communication delays. IEEE/ASME Trans. Mechatron. **20**, 245–253 (2015)
9. Anderson, R.J., Spong, M.: Asymptotic stability for force reflecting teleoperators with time delay. Int. J. Robot. Res. **11**(2), 135–149 (1992)
10. Lee, D., Spong, M.: Passive bilateral teleoperation with constant time delay. IEEE Trans. Rob. **22**(2), 269–281 (2006)
11. Sun, D., et al.: Wave-variable-based passivity control of four-channel nonlinear bilateral teleoperation system under time delays. IEEE/ASME Trans. Mechatron. **21**, 1–13 (2016)
12. Ryu, J.H., Kwon, D., Hannaford, B.: Stable teleoperation with time domain passivity control. IEEE Trans. Robot. Autom. **20**(2), 365–373 (2004)
13. Nuiio, E., Ortega, R., Basaiiez, L.: An adaptive controller for nonlinear bilateral teleoperators. Automatica J I FAC **46**(1), 155–159 (2010)
14. Chopra, N., Spong, M.W., Lozano, R.: Synchronization of bilateral teleoperators with time delay. Automatica **44**(8), 2142–2148 (2008)
15. Liu, Y.C., Khong, M.H.: Adaptive control for nonlinear teleoperators with uncertain kinematics and dynamics. IEEE/ASME Trans. Mechatron. **20**(5), 2550–2562 (2015)

Optimal Planning Model for Grid-Connected Micro-grids Considering Uncertainties of Renewable Sources, Loads and Electrical Price

Vu Van Thang(✉) and Nguyen Hien Trung

Faculty of Electrical Engineering, Thai Nguyen University of Technology, Thai Nguyen, Vietnam
{thangvvhtd, nguyenhientrung}@tnut.edu.vn

Abstract. This study proposes a planning framework for grid-connected micro-grids to determine the power, technology, types and invested time of the renewable sources during planning period of project. The uncertainty of parameters is modeled by the probability distribution functions, then the clustering technique is utilized to divide into different states and integrated in an optimal model. A mixed-integer programming model with objective is based on life cycle cost of project including investment cost of renewable sources, operation cost, energy cost and emission cost of micro-grid. Constraints are created to balance energy and limit power of sources in each scenario. The simulation result by GAMS/CPLEX for the test micro-grid shows effects of the model with uncertainty parameters because of reducing life cycle cost and emission cost. Moreover, the change of parameters simulated in model that is similar to the practical parameters enhances the accuracy and effectiveness of the project.

Keywords: Life cycle cost · Micro-grid · Planning · Renewable sources · Uncertainty

1 Introduction

Micro-grid is a new type of the electric energy system comprising the low voltage grid with distributed energy sources to supply energy for consumers [1]. The distributed energy sources or distributed generators are defined as an electric power source connected directly to the distribution network or on the customer site of the network with small size [2, 3]. Many technologies of the distributed generators, such as micro-turbine, gas turbine, reciprocating engines, combustion turbine and renewable sources (RS) which comprise wind turbine, photovoltaic, bio-mass, geo-thermal, tidal and hydro-power are introduced and applied in reality. Among these sources, the photovoltaic (PV) and wind turbine (WT) are widely used because they have great potential: they are built in modules with short installing time, they reduce output emissions of pollutants, lower operation and maintenance cost and reduce investment risks.

K.-U. Sattler et al. (Eds.): ICERA 2019, LNNS 104, pp. 534–545, 2020.
https://doi.org/10.1007/978-3-030-37497-6_61

The micro-grids can be built by two structures including the grid-connected and islanded mode [1, 4]. In islanded mode, the RS should be in balance with demand of loads, thus the reliability of the micro-grids reduces because of the uncertainty and faults of RS. In contrast, the demand for loads is simultaneously supplied from either renewable sources or utility grid through the point of common coupling (PCC) in the grid-connected mode. Therefore, the modern optimization methods can be applied in the grid-connected micro-grids planning and operating to improve efficiency, economics, and resiliency.

Many studies have been made to plan, design and operate the micro-grids [3, 4]. In studies [5, 6], the models analyzing optimal operation and planning for the grid-connected micro-grids are introduced. The objective function is the total net present cost of investment project including cost for the capital, operation and maintenance (O&M), cost for reducing pollutant emissions and purchasing energy from a utility grid. Similarly, the effect of RS in islanded micro-grids is also evaluated in study [7] that the objective function is minimizing of net present worth of micro-grid. The result shows the positive effects of RS consisting of higher net present worth and greater profit. The structures and planning models of direct current and alternating current micro-grids are proposed in study [8] with the objective of minimizing the total cost of the planning project.

As previously seen, the RS has been successfully studied and applied in planning the micro-grids. The objective mostly utilized is the total net present cost of investment project. The different lifetime and uptime of equipment are not considered, thus the application of these models has significant error due to nonconformity of computed parameters with reality. Hence, the planning and designing models of the energy systems, which utilize the objective function the life cycle cost (LCC) have been introduced in the recent years. The LCC is defined as a technique, which enables comparative cost assessments to be made over a specified period, taking into account all relevant economic factors, both in terms of initial costs and future operational cost [9]. The design options for energy systems with LCC analyzed show that the implementing LCC analysis in planning and design time can develop more sustainable energy system in long-term. Similarly, the study [10] proposed an optimal model with the objective function being LCC to find the economical optimal size of BESS. The LCC of model includes cost for investment, O&M, recycling or disposal of the equipment in system.

Additionally, the output power of RS and load profile randomly changes depending on properties of the primary energy sources (wind speed or solar radiation) and weather and becomes more challenging in planning and operating the micro-grids. Therefore, the many uncertainty modeling techniques and methods of parameters are proposed in optimal planning and operating model of the energy systems [3, 11]. An optimal operation model of the micro-grid considered uncertainty of RS is presented in study [12] to minimize real power loss and voltage deviation. Similarly, the optimal micro-grid expansion planning or demand side management models under uncertainty of the RS, demand of loads and electrical price is proposed in studies [13] and [14]. The objective function consists of the capital costs of the system, the fuel supply costs, and O&M costs. The results lead to the observations that models provide the tools to design and plan micro-grids with the various advantages and disadvantages of each micro-grid configuration. The study [15] also presents a planning framework for the isolated multi-energy micro-grids with objective including the investment and O&M cost of RS. Besides, this model adds the costs of unsupplied loads and the CO_2 emissions of micro-grid. However, the previous micro-grid planning models only separately consider the lifetime and uptime of the RS by objective function minimizing the LCC with determined parameters or integrate the uncertainty parameters in optimal models but the investment of the RS often is assumed in the first year and operates during planning period. Besides, the rated power of the RS in fact comes in many sizes and is discrete values.

Therefore, an optimal planning framework proposed in this study aims at addressing above disadvantages at the same time. The optimal rated power is jointly selected together with technology of the RS. This study considers not only the lifetime and uptime of RS by the objective function LCC but also the rated power of them with discrete values by binary variables. The uncertainty of parameters is integrated in model then the all states are computed and thus increase in the accuracy of planning results. The optimal planning problem that is a mixed-integer problem is programmed in GAMS environment solved with CPLEX solver [16].

The rest of paper is organized as follows: Sect. 2 presents the structure of micro-grid and mathematics model of problem. Section 3 discusses the numerical results and Sect. 4 presents the conclusive remarks and few insights for future work.

2 Planning Framework for Grid-Connected Micro-grids

2.1 Structure of Grid-Connected Micro-grids

A grid-connected micro-grid often consists of a low voltage grid and distributed generators connected to utility grid through the PCC. In recent years, the various technologies of the RS have been successfully developed and applied in distributed systems and micro-grids because of improving the efficiency, reliability and reducing investment cost [2]. In which, the PV and WT are widely used because they have great potential built in modules [3, 17]. Hence, these RS are selected to supply energy for loads and the structure of grid-connected micro-grids is proposed as Fig. 1 [17].

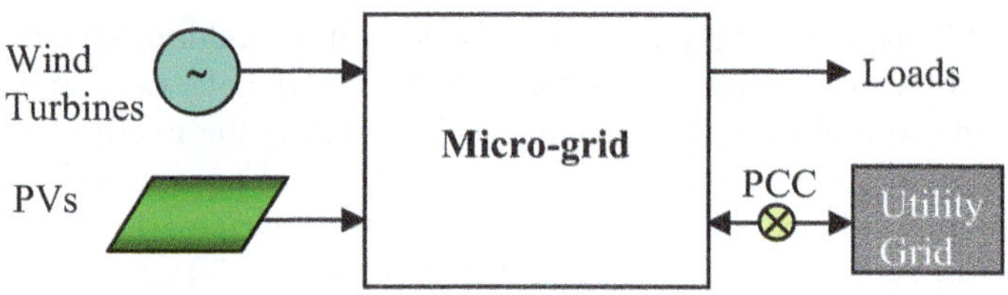

Fig. 1. The structure of grid-connected micro-grid

In grid-connected micro-grid, the two scenarios that can be operated are an autonomous operation, called the islanded mode, or operation in conjunction with the utility grid, called the grid-connected mode. Either the RS or the utility grid jointly supplies the energy for loads in the grid-connected mode. On the contrary, the only RS is operated to supply energy for load in the islanded mode. Consequently, the RS can not meet the demand for loads, the reliability is not guaranty and this is the disadvantage of grid-connected micro-grid. However, this structure can optimize the electric energy purchased from utility gird and investing RS and thus improves the effectiveness of the micro-grid.

2.2 Modeling Uncertainty of Parameters

The output power of RS, the energy price and electrical demand often change random and are uncertain parameters. These uncertainties are often modeled by the probability distribution functions (PDF) such as beta, rayleigh, weibull and normal [11]. Then, the k-mean clustering technique is applied to divide stochastic parameters into different states because the PDFs are the continuous functions. In each state, there is a specific value with the related probability.

The probability of solar irradiance is expressed by beta PDF as Eq. (1) with the I_{ir} is the solar irradiance, μ is mean and σ is the standard deviation of the stochastic variable [13, 18].

$$\begin{aligned} & f_b(I_{ir}) = \begin{cases} \frac{\Gamma(\alpha+\beta)}{\Gamma(\alpha)\Gamma(\beta)}.I_{ir}^{(\alpha-1)}.(1-I_{ir})^{(\beta-1)} & if\, 0 \le I_{ir} \le 1, \alpha, \beta \\ 0 & else \end{cases} \\ & \beta = (1-\mu).\left(\frac{\mu.(1+\mu)}{\sigma^2} - 1\right); \qquad \alpha = \frac{\mu.\beta}{1-\mu} \end{aligned} \tag{1}$$

In each state, a value of solar irradiance related with probability is determined and hence the output power is computed as expression (2) in which the $k_{F.s}$ is the fill factor and the n is the number of modules [19, 20]. The I_s, U_s are the generated current and voltage of the PV module at state s. Besides, the ISC and UOC are short circuit current and open circuit voltage, the UMPP and IMPP are current and the voltage at the maximum power point, respectively. The k_U, k_I are the coefficients depending on temperature, voltage and current of PV. The θ_A, θ_N are the ambient and nominal operating temperature of module PV.

$$\begin{aligned} & P_s^{pv}(I_{ir}) = n.k_{F.s}.U_s(I_{ir}).I_s(I_{ir}) \\ & k_{F.s} = \frac{U_{MPP.s}.I_{MPP.s}}{U_{OC}.I_{SC}} \\ & U_s(I_{ir}) = U_{OC} - k_U.\left(\theta_A + I_{ir.s}.\frac{\theta_N-20}{0.8}\right) \\ & I_s(I_{ir}) = I_{ir.s}.\left[I_{SC} + k_I.\left(\theta_A + I_{ir.s}.\frac{\theta_N-20}{0.8} - 25\right)\right] \end{aligned} \tag{2}$$

Similarly, the probability of wind speed v also represented under rayleigh PDF which is a special case of weibull PDF as expression (3) with the shape index k = 2 and the scale index c = 1.128 vm. The value of wind speed related with probability in each state is determined and the output power of WT is computed as Eq. (4) [21]. Where, the P_s^{wt} and P_r^{wt} are output and rated power of WT. The v_{ci}, v_{cr}, v_{co} and v_m are the cut-in speed, rated speed, cut-off speed and mean wind speed of WT, respectively.

$$f_r(v) = \left(\frac{2v}{c^2}\right) \exp\left[-\left(\frac{v}{c}\right)^2\right] \quad (3)$$

$$P_s^{wt}(v) = \begin{cases} 0 & v_s \leq v_{ci} \text{ or } v_{co} \leq v_s \\ P_r^{wt} \cdot \frac{v_s - v_{ci}}{v_r - v_{ci}} & v_{ci} \leq v_s \leq v_r \\ P_r^{wt} & v_r \leq v_s \leq v_{co} \end{cases} \quad (4)$$

The probability of loads and electrical price are modeled by normal PDF as present at Eq. (5) in which the μ is mean, σ is the standard deviation and σ^2 is variance of the distribution [21, 22]. The x is stochastic value of loads or electrical price in competitive market. The clustering technique is also used to divide stochastic loads and energy price into different states with specific value and probability in each state.

$$P(X = x|\mu, \sigma^2) = f(x) = \frac{1}{\sqrt{2\pi\sigma}} \exp\left[-\frac{(x-\mu)^2}{2\sigma^2}\right] \quad (5)$$

A matrix is utilized to integrate all possible states of loads, electrical price and out power of RS as Eq. (6) [19, 23]. In which, the Cs is the matrix that enumerates the possible values of parameters, the $\lambda_s\{C_s\}$ is one-column matrix that represents probability corresponding with the matrix Cs.

$$\begin{aligned} &M = \{C_s, \lambda_s\{C_s\}\} \\ &\lambda_s\{C_s\} = \lambda_s^{\rho}\{\rho_s\}.\lambda_s^{l}\{P_s^{l}\}\,.\,\lambda_s^{wt}\{P_s^{wt}\}.\lambda_s^{pv}\{P_s^{pv}\}; S = n_s^{\rho}.n_s^{l}.n_s^{wt}.n_s^{pv} \end{aligned} \quad (6)$$

In Eq. (6), S is the sum of individual states of matrix M which is computed by the product the possible states of parameters. ρ_s is the electricity price in state s. P_s^l is the load value in state s. $\lambda_s^{\rho}, \lambda_s^{l}, \lambda_s^{wt}$ and λ_s^{pv} are probabilities of electrical price, loads and output power of WT and PV, respectively. $n_s^{\rho}, n_s^{l}, n_s^{wt}$ and n_s^{pv} are number of states of electrical price, loads and output power of the RS, respectively.

2.3 Mathematics Model

Optimal planning model for grid-connected micro-grids considering uncertainties of RS, loads and electrical price includes objective and constraints. The objective function which is the LCC of the project during planning period converted to the base year by discount rate r is proposed in Eq. (7) [17, 24]. In which, $C_{I.t}^{RS}$ is investment cost and $C_{OM.t}^{RS}$ is operation and maintenance of RS at planning year t. $C_{E.t}^{G}$ is cost for purchasing energy from a utility grid and $B_{Emi.t}$ is benefit of reducing emission taxes due to low emissions of RS. Final, $B_{Res.t}$ is residual value of equipment at the end of planning time.

$$J = Min \sum_{t=1}^{T} \frac{1}{(1+r)^t} \left(C_{I.t}^{RS} + C_{OM.t}^{RS} + C_{E.t}^{G} - B_{Emi.t} - B_{Res.t}\right) \qquad \forall t \in T \quad (7)$$

The investment cost of the RS is computed by Eq. (8). In this cost function, C_k^{RS} is capital cost and $P_{k,t}^{RS}$ is selected power variable of the RS with technology k. K is total number of technologies of the RS considered for selection.

$$C_{I.t}^{RS} = \sum_{k=1}^{K} C_k^{RS}.P_{k,t}^{RS} \qquad \forall t \in T \tag{8}$$

The O&M cost of the RS at planning year t depends on out power $P_{k,t,s}^{RS}$ and probability λ_s of state s as Eq. (9). Where, $\rho_{OM.k}^{RS}$ is the coefficient for determining the O&M cost based on the output power of RS, technology k. The out power of RS at state s is reflected by the coefficient $k_{k,h}^{RS}$ and rated power which is selected to invest in a year t, $P_{k,t}^{RS}$.

$$C_{OM.t}^{RS} = 8760 \sum_{k=1}^{K} \sum_{s=1}^{S} \lambda_s.\rho_{OM.k}^{RS}.P_{k,t,s}^{RS}; \quad P_{k,t,s}^{RS} = k_{k,s}^{RS}.P_{k,t}^{RS} \quad \forall s \in S, t \in T, k \in K \tag{9}$$

Similarly, the cost for energy purchased from utility grid also depends on probability and power which receives through the PCC at state s and year t, $P_{t,s}^{G}$, as Eq. (10). In which, $\rho_{E.s}$ is market electrical price which is computed by maximum price $\rho_{E.s}^{\max}$ and factor k_s^e in each state s.

$$C_{E.t}^{G} = 8760 \sum_{s=1}^{N_S} \lambda_s.\rho_{E.s}.P_{t,s}^{G}; \rho_{E.s} = k_s^e.\rho_{E.s}^{\max} \qquad \forall t \in T, s \in S \tag{10}$$

The energy supplied by RS will substitute energy that is received from utility grid by traditional energy sources and thus the emission of micro-grids reduces. Hence, the benefit due to decrease emission taxes is computed as Eq. (11) [25], where ξG and ξRS are emission coefficient of traditional and renewable sources, respectively. β is emission tax probably enforced by the government.

$$B_{Emi.t} = \beta.(\xi_G - \xi_{RS}).8760.\sum_{k=1}^{K} \sum_{s=1}^{S} P_{k,t,s}^{RS} \quad \forall t \in T \tag{11}$$

The residual value of equipment at the end of project is computed by expression (12). Where, $T_{C.k}^{RS}$ is the life cycle and $t_{I.k}^{RS}$ is investment year of the RS.

$$B_{Res.t} = \frac{T_{C.k}^{RS} - t_{I.k}^{RS}}{T_{C.k}^{RS}} C_{I.t}^{RS} \quad \forall t \in T \tag{12}$$

Balancing energy for each state s of micro-grid is guaranteed by constraint (13) with $P_{t,s}^{l}$ is load at state s in a year t. Besides, P_t^l is the load at the year t, k_s^l is a factor of load in state s and k_d is annual growth factor of load at the year t.

$$P^G_{t,s} + \sum_{k=1}^{N_K} P^{RS}_{k,t,s} = P^l_{t,s}; \quad P^l_{t,s} = k^l_s.P^l_t; \quad P^l_t = P^l_{t-1}.k_d \qquad \forall t \in T, s \in S \tag{13}$$

The rated power of the RS is discrete values thus the binary variables are integrated in model to determine optimal technology corresponding with rated power of the RS as presented in Eq. (14) [19]. In which, $\alpha_{i.k}$ is binary variable and $P^{RS}_{r.k.j}$ is rated power of technology k, type j,

$$P^{RS}_{k,t} = \sum_{j=1}^{n} \alpha_{k,j} P^{RS}_{r.k.j} \qquad \forall t \in T, k \in K \tag{14}$$

The power of the RS that can select to invest is limited by potential of primary energy sources. Because of that, the constraint (15) is utilized to guarantee the invested power of RS is feasible in fact. Similarly, the power that micro-grid can be received from a utility grid is limited by power of PCC. In which, $P^{RS}_{k.max}$ is the maximum rated power of each technology that the RS can be invested and P^G_{max} is maximum capacity of equipment of PCC. In addition, the energy of micro-grid, which receives from a utility grid, is unlimited because of large capacity of utility grid.

$$P^{RS}_{k,t} \le P^{RS}_{k.max} \qquad \forall t \in T, k \in K \tag{15}$$

$$P^G_{t,s} \le P^G_{max} \qquad \forall t \in T, s \in S \tag{16}$$

As seen above, the optimal model is mixed-integer problem thus the GAMS/CPLEX solver is utilized to find out the optimal solution. The simulation result for test micro-grid is presented below.

3 Simulation Result and Discussion

3.1 Parameters and Assumptions of Test Micro-grid

The loads of micro-grid assumed at base year are 1000 kW, the annual growth factor kd is 5%. Similarly, the electrical price is 15 cents/kWh and the annual rise factor of it is 2%. The stochastic parameters, which are conceptualized under the pdfs, are divided by clustering technique into states. The 11 states corresponding with 11 levels of the specific value and related probability is selected to guarantee a reasonable balance between accuracy and computed speed shown in Table 1 [12, 23].

The aggregation all states of uncertainty parameters made by the matrix M includes 14641 scenarios. There are several extreme scenarios in operation the micro-grid. In scenarios from the first to 11th, the out power of the RS is zero and thus all of loads is supplied from utility, whereas the out power of RS is maximum in scenario 14641 and the received power from utility is minimum.

The rated powers of the RS are discrete values with maximum rated power of PV and WT assumed as Table 2 are 300 kW and 250 kW, respectively [5, 26, 27]. Similarly, the capital cost, O&M cost, CO2 emission coefficient and lifetime of the RS are also presented in this table. Besides, the CO2 emission coefficient of traditional sources in a utility grid is assumed 0.68 kg/kWh and the emission tax probably enforced by the government is 4.12 $/ton [28]. The planning period is 10 years and discount rate is 10%.

Table 1. Uncertainty parameters of micro-gird

State	PV		WT		Load		Electrical price	
	k_s^{pv}	λ_s^{pv}	k_s^{wt}	λ_s^{wt}	k_s^{l}	λ_s^{l}	k_s^{e}	λ_s^{e}
1	0	0.518	0	0.154	0.30	0	0.3	0.059
2	0.1	0.003	0.1	0.125	0.35	0.033	0.35	0.023
3	0.2	0.006	0.2	0.118	0.40	0.047	0.4	0.069
4	0.3	0.015	0.3	0.106	0.45	0.091	0.45	0.092
5	0.4	0.029	0.4	0.096	0.51	0.163	0.5	0.116
6	0.5	0.049	0.5	0.085	0.59	0.163	0.55	0.131
7	0.6	0.071	0.6	0.072	0.65	0.165	0.65	0.132
8	0.7	0.091	0.7	0.059	0.71	0.165	0.75	0.215
9	0.8	0.104	0.8	0.047	0.77	0.106	0.8	0.115
10	0.9	0.090	0.9	0.036	0.85	0.056	0.9	0.026
11	1	0.025	1	0.102	1	0.011	1	0.022

Table 2. The parameters of RS

Technologies of RS	Rated power of type j - $P_{r.k.j}^{RS}$ (kW)	C_k^{RS} ($/kW)	$\rho_{OM.k}^{RS}$ ($/kWh)	ξ_{RS} (kg/kWh)	Lifetime (year)
PV	50; 75; 100; 150; 200; 250; 300	1500	0.025	0.02	20
WT	25; 50; 100; 125; 150; 200; 250	1800	0.035	0.02	20

3.2 Calculation Result

The feasibility of the proposed model and the efficiency of the RS are investigated in three cases. The first one, the deterministic parameters are utilized to compute the micro-grid planning with RS (case A). The capacity factors of the PV and WT are assumed 0.27 and 0.41, respectively. The average load factor is 0.62 and average electrical price is about 9.15 cent/kWh. In the second case (case B), the stochastic parameters are considered but the RS are not integrated in micro-grid and hence all demand of loads is only supplied by a utility grid. Either the RS or the stochastic parameters of micro-grid are simultaneously considered in the final case (case C).

The simulation result determines the technology, type, power and invested time of the RS in case A and case C shown on Table 3. Total power of WT and PV selected on both cases are equal 450 kW corresponding to 45% of the peak load of micro-grid, in which the rated power of WT is 150 kW and the rated power of WT is 300 kW. However, the invested time of the RS in case C is earlier than case A because it takes full advantage of the RS in states which have high electrical price with large probability. In case C, the WT and PV are selected at the 3rd year and 4th compared with at 5th and 6th years in case A, respectively.

Table 3. Invested decision for RS

Technologies of RS	Case A			Case C		
	Types of RS	Rated power, kW	Invested year	Types of RS	Rated power, kW	Invested year
PV	7	300	6	7	300	4
WT	5	150	5	5	150	3

Investing the RS and considering to uncertainty of parameters improved the efficiency of micro-grids verified due to reduce the expected value of LCC as well as emission as shown in Table 4. The expected value of the LCC during planning time in case C reduces about $0.19 M compared with case A corresponding to 4.49%. This is due to the consideration to change of the electricity price, loads and out power of the RS in states with different value and probability. Similarly, the efficiency of the RS in case C is also shown when the LCC of case C decreases $0.2 M corresponding to 4.95% compared with case B.

Table 4. Comparison of LCC and emission in cases of project

Cases	Total life cycle cost, M$	Electrical energy purchased from utility gird, 10^6 kWh	CO2 Emission, 10^3 tons
Case A	4.23	61.53	41.98
Case B	4.24	65.88	44.80
Case C	4.04	57.89	39.53

Besides, the electrical energy purchased from utility grid also reduces when the RS are installed in both cases, the deterministic and stochastic parameters. Comparing between cases C and B shown the reduced electrical energy is about 7.99×10^6 kWh corresponding to 12.13%. Therefore, CO_2 emission is decreased to 5.27×10^3 tons and the emission tax is reduced about $21,712.4 during planning period.

As seen above, the RS have highly effective in planning the grid-connected micro-grid because of reducing the LCC and emission. The stochastic parameters considered in optimal model greatly enhance the accuracy of calculation result and effectiveness of invested project. Moreover, the RS also improve the reliability of micro-grids, which can still provide a portion of power for loads when the utility grid is a fault. However, the loads cannot be supplied in several extreme scenarios (Fig. 2).

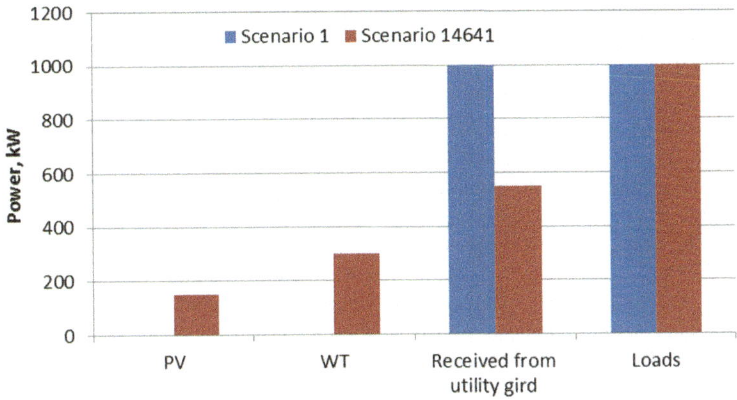

Fig. 2. The power of micro-grid in several extreme scenarios

4 Conclusion

An optimal planning framework for grid-connected micro-grids is proposed to select the optimal power, technology, type and planning time of the RS. The model considers not only the lifetime, uptime and the rated power of the RS with discrete values by binary variables integrated in the objective function LCC but also the uncertainty of parameters. Therefore, the accuracy of computed results and effectiveness of the RS in invested project are improved and more appropriate to the actual equipment.

The simulation results show the feasibility and suitability of the proposed model for planning the grid-connected micro-grid considering the RS and uncertainty of parameters. The RS can significantly improve the effectiveness of the micro-grid because of reduction of LCC and emission cost. The uncertain parameters examined in model lead to the investment of the RS 2 years earlier than the deterministic parameters. Hence, LCC of project reduces by 4.95% and CO_2 emission also decreases by 5.84%. However, the solutions to guarantee the energy supply for loads when the utility grid is fault should be further studied in next studies.

Acknowledgments. This work is supported in part by Thai Nguyen University of Technology (TNUT) with the finance coming from another project.

References

1. Lopes, J.A.P.A., Madureira, G., Moreira, C.C.L.M.: A view of micro-grids. WIREs Energy Environ. **2**, 86–103 (2013)
2. Elmubarak, E.S., Ali, A.M.: Distributed generation: definitions, benefits, technologies & challenges. Int. J. Sci. Res. **5**(7), 1941–1948 (2016)
3. Zuboa, R.H.A., Mokryani, G., Rajamani, H.S., Aghaei, J., Niknam, T., Pillai, P.: Operation and planning of distribution networks with integration of renewable distributed generators considering uncertainties: a review. Renew. Sustain. Energy Rev. **72**, 1177–1198 (2017)

4. Mumtaz, F., Bayram, I.S.: Planning, operation, and protection of micro-grids: an overview. Energy Procedia **107**, 94–100 (2017)
5. Hafez, O., Bhattacharya, K.: Optimal planning and design of a renewable energy based supply system for micro-grids. Renew. Energy **45**, 7–15 (2012)
6. Su, W., Yuan, Z., Chow, M.Y.: Micro-grid planning and operation: solar energy and wind energy. In: 2010 IEEE Power and Energy Society General Meeting, USA (2010)
7. Qamar, S.B., Janajreh, I.: Renewable energy sources for isolated self-sufficient micro-grids: comparison of solar and wind energy for UAE. Energy Procedia **103**, 413–418 (2016)
8. Lotfi, H., Khodaei, A.: AC versus DC micro-grid planning. IEEE Trans. Smart Grid **8**(1), 296–304 (2017)
9. Ristimäki, M., Säynäjoki, A., Heinonen, J., Junnila, S.: Combining life cycle costing and life cycle assessment for an analysis of a new residential district energy system design. Energy **63**, 168–179 (2013)
10. Marchi, B., Pasetti, M., Zanoni, S.: Life cycle cost analysis for BESS optimal sizing. Energy Procedia **113**, 127–134 (2017)
11. Ehsana, A., Yanga, Q.: State-of-the-art techniques for modelling of uncertainties in active distribution network planning: a review. Appl. Energy **239**, 1509–1523 (2019)
12. Meena, N.K., Swarnkar, A., Gupta, N., Niazi, K.R.: Stochastic Volt/Var planning and operation for microgrids with renewables. Int. J. Electr. Energy **4**, 159–164 (2016)
13. Zhenga, Y., Jenkins, B.M., Kornblutha, K., Kendall, A., Træholt, C.: Optimization of a biomass-integrated renewable energy microgrid with demand side management under uncertainty. Appl. Energy **230**, 836–844 (2018)
14. Mishraa, S., Bordinc, C., Tomasgard, A., Palua, I.: A multi-agent system approach for optimal microgrid expansion planning under uncertainty. Electr. Power Energy Syst. **109**, 696–709 (2019)
15. Ehsana, A., Yanga, Q.: Scenario-based investment planning of isolated multi-energy microgrids considering electricity, heating and cooling demand. Appl. Energy **235**, 1277–1288 (2019)
16. GAMS/SCENRED User Guide (2009). http://www.gams.com/
17. Thang, V.V., Trung, N.H.: Optimal planning of renewable sources in micro-grids based on life cycle cost. In: ICERA 2018. Lecture Notes in Networks and Systems (LNNS 63), vol. 63, pp. 440–447 (2018)
18. Ying, Y., Wu, Y., Su, Y., Fu, R., Liang, X., Xu, H.: Dispatching approach for active distribution network considering PV generation reliability and load predicting interval. J. Eng. **2017**(13), 2433–2437 (2017)
19. Thang, V.V., Ha, T.: Optimal siting and sizing of renewable sources in distribution system planning based on life cycle cost and considering uncertainties. AIMS Energy **7**(2), 211–226 (2019)
20. Mohamed, E.A., Hegazy, Y.G.: A novel probabilistic strategy for modeling photovoltaic based distributed generators. Int. Sch. Sci. Res. Innov. **9**(8), 904–907 (2015)
21. Dolatabadi, A., Mohammadi-Ivatloo, B.: Stochastic risk-constrained scheduling of smart energy hub in the presence of wind power and demand response. Appl. Therm. Eng. **123**, 40–49 (2017)
22. Jadidbonab, M., Dolatabadi, A., Mohammadi-Ivatloo, B., Abapour, M., Asadi, S.: Risk-constrained energy management of PV integrated smart energy hub in the presence of demand response program and compressed air energy storage. IET Renew. Power Gener. **13**(6), 998–1008 (2019)
23. Atwa, Y.M., El-Saadany, E.F.: Probabilistic approach for optimal allocation of wind based distributed generation in distribution systems. IET Renew. Power Gener. **5**, 79–88 (2011)

24. Liu, W., Niu, S., Xu, H.: Optimal planning of battery energy storage considering reliability benefit and operation strategy in active distribution system. J. Mod. Power Syst. Clean Energy **5**(2), 177–186 (2017)
25. Santos, F., Fitiwi, D.Z., Bizuayehu, A.W., Asensio, M., Contreras, J., Cabrita, M.P., Catalao, J.P.S.: Impacts of operational variability and uncertainty on distributed generation investment planning: a comprehensive sensitivity analysis. IEEE Trans. Sustain. Energy **8**, 855–869 (2017)
26. Ioannou, A., Fuzuli, G., Brennan, F., Yudha, S.W., Angus, A.: Multi-stage stochastic optimization framework for power generation system planning integrating hybrid uncertainty modeling. Energy Econ. **80**, 760–776 (2019)
27. Liu, Z., Wen, F., Ledwich, G.: Optimal siting and sizing of distributed generators in distribution systems considering uncertainties. IEEE Trans. Power Delivery **26**(4), 2541–2551 (2011)
28. Deng, S., Wu, Q., Jing, Z., Wu, L., Wei, F., Zhou, X.: Optimal capacity configuration for energy hubs considering part-load characteristics of generation units. Energies **10**(1966), 1–29 (2017). https://doi.org/10.3390/en10121966

Optimization of Exchanged Grinding Wheel Diameter for Minimum Cost in External Grinding

Nguyen Anh Tuan[1], Le Hong Ky[2], Tran Thi Hong[3], Nguyen Van Cuong[4], Luu Anh Tung[5], Nguyen Thanh Tu[5], Hoang Xuan Tu[5], and Ngoc Pi Vu[5(✉)]

[1] University of Economic and Technical Industries, Ha Noi, Vietnam
[2] Vinh Long University of Technology Education, Vinh Long, Vietnam
[3] Nguyen Tat Thanh University, Ho Chi Minh City, Vietnam
[4] University of Transport and Communications, Ha Noi, Vietnam
[5] Thai Nguyen University of Technology, Thai Nguyen, Vietnam
vungocpi@tnut.edu.vn

Abstract. Grinding cost is an essential factor in a grinding process. In external cylindrical grinding, there exist different input technology parameters which have significant influences on the grinding costs. This paper presents a study on optimization of exchanged grinding wheel diameter to minimize the grinding cost in the external grinding stainless steel. The work is concerned with the influences of seven input grinding process parameters including the initial grinding wheel diameter, the grinding wheel width, the wheel life, the radial grinding wheel wear per dress, the total depth of dressing cut, the machine tool hourly rate, and the grinding wheel cost on the optimum exchanged grinding wheel diameter in the external grinding process based on the optimization of grinding cost. In addition, the effect of the interactions between the input grinding parameters was also evaluated. Finally, the regression equation for computing the optimum exchanged grinding wheel diameter was introduced. The proposed model can be further applied effectively for the external grinding process.

Keywords: External grinding · Cost optimization · Exchanged grinding wheel diameter, optimum diameter

1 Introduction

During the mechanical processing, grinding is constantly the final processing method that decides the final surface features of the part [1, 2]. Therefore, numerous researches have focused on optimization problems of a grinding process. Different optimum grinding process parameters were investigated in [3–16]. The optimum internal grinding cycles were proposed in [4]. Studies to optimise process parameters to achieve the desired stock removal in a compliant Abrasive Belt Grinding process [5] or to get the minimal surface roughness in surface grinding process [6] were carried out. Based on minimizing grinding time and maximizing the volume of material removed rate,

K.-U. Sattler et al. (Eds.): ICERA 2019, LNNS 104, pp. 546–556, 2020.
https://doi.org/10.1007/978-3-030-37497-6_62

optimum grinding process parameters including wheel and machine parameters were proposed in [7–9]. The other research was also concerned with the optimum grinding parameters for grinding table punches on CNC milling machine to minimize grinding time [10]. Relevant to the CBN grinding wheel is also the work on analysis of wear for the grinding process of aeronautical alloys [11].

On the other hand, the influence of dressing on internal cylindrical grinding [12] or external grinding [13] was also considered for grinding process optimization. In recent years, multi-dressing parameters including the dressing feed rate, the coarse dressing depth, the coarse dressing times, the fine dressing depth, the fine dressing times and the dressing number without depth of cut for maximum material removal rate were optimized in order to maximize material removed rate [14]. The influence of dressing parameters on the topography, morphology and characteristics of the surface of different vitrified abrasive wheels was proposed by Pamer [15]. In addition, there have been several studies on determining optimum parameters for lubrication in internal [16] or external [17] grinding processes.

Regarding the optimization of cost for grinding processes, several previous researches presented the cost function [18–26]. Based on these cost optimization studies, a method to determine the optimum exchanged grinding wheel diameter for the grinding processes was reported. However, there has not been a study for determining the optimum exchanged diameter when external grinding stainless steel.

Therefore, this paper presents a study on optimization of exchanged grinding wheel diameter to minimize the manufacturing cost in the external grinding stainless steel. Based on the cost function analysis of the external grinding process, the influences of seven input grinding process parameters including the initial grinding wheel diameter (D_0), the grinding wheel width $\left(W_{gw}\right)$, the wheel life (T_w), the radial grinding wheel wear per dress $\left(W_{pd}\right)$, the total depth of dressing cut (a_{ed}), the machine tool hourly rate (C_{mh}), and the grinding wheel cost (C_{gw}) on the optimum exchanged grinding wheel diameter $(D_{e,op})$ in the external cylindrical grinding process were taken into account. The screening experiments were conducted to evaluate the influences of the grinding parameters on the optimum exchanged grinding wheel diameter for the external grinding process. By using regression analysis, a mathematical model for calculating the optimum diameter was established.

2 Cost Function of External Grinding

This section presents the procedure for determining the manufacturing cost function in the external grinding. This is the theoretical basis to determine the optimal value of $D_{e,op}$ to minimize the cost of the external grinding process. Based on the manufacturing time t_s (h), the machine tool hourly rate C_{mh} (USD/h) and the grinding wheel cost per part $C_{gw,p}$ (USD/part), the manufacturing single cost per piece in the external grinding C_{sin} is computed as follows:

$$C_{sin} = t_s \cdot C_{mh} + C_{gw,p} \tag{1}$$

The manufacturing time t_s including the auxiliary time can be expressed as follows:

$$t_s = t_c + t_{sp} + t_{lu} + t_{d,p} + t_{cw,p} \tag{2}$$

Where $t_c, t_{sp}, t_{lu}, t_{d,p}$ and $t_{cw,p}$ are the grinding time (h), the spark-out time (h), the time for loading and unloading workpiece (h), the dressing time per part (h), and the time for changing a grinding wheel per work-piece, respectively. $t_c, t_{d,p}$ and $t_{cw,p}$ can be calculated as the following equations:

$$t_c = l_w\, a_{e,tot}/(v_{fa} f_r) \tag{3}$$

$$t_{d,p} = t_d/n_{p,d} \tag{4}$$

$$t_{cw,p} = t_{cw}/n_{p,w} \tag{5}$$

In Eq. (3), l_w (mm), $a_{e,tot}$ (mm), v_{fa} (mm/min), and f_r (mm/double stroke) are the length of part, the total depth of cut, the axial speed, and the radial wheel feed, respectively. v_{fa} and f_r are determined as follows [27]:

$$v_{fa} = 0.2557\, d_w^{0.0749}\, n_w^{1.1093}\, b_s^{0.9841} \tag{6}$$

$$f_r = f_{r,tab} \cdot c_1 \cdot c_2 \cdot c_3 \cdot c_4 \tag{7}$$

Here, d_w, n_w, and b_s are the workpiece diameter, the workpiece speed, and the grinding wheel width, respectively. For grinding tool steel, n_w can be computed by [27]:

$$n_w = 9.9985.2\, d_w^{-0.966} \tag{8}$$

In Eq. (7), $f_{r,tab}$ (mm/double stoke) is the tabled radial wheel feed. The coefficient c_1 is determined by the workpiece material and tolerance grade tg; c_2 is calculated by the grinding peripheral speed v_s (m/second) and the grinding wheel diameter d_s; the coefficient c_3 is characterized for measurement and $c_3 = 0.8$ with using a micrometer for measurement [27]; c_4 is the coefficient related to the ratio of the workpiece length to its diameter and $c_4 = 1$ if the ratio is smaller than 10 [27]. The tabled radial wheel feed $f_{r,tab}$ and the coefficients c_1, c_2 are expressed as the equations below [27].

$$f_{r,tab} = 150.3605\, d_w^{-0.9331}\, v_{fa}^{-0.4922}\, (2\,a_e)^{0.2194} \tag{9}$$

$$c_1 = 0.017\, tg^{1.6935} \tag{10}$$

$$c_2 = 0.0032\, v_s^{0.7174}\, d_s^{0.4984} \tag{11}$$

Where, a_e is the total depth of cut (mm) and tg is the required tolerance grade.

In Eqs. (4) and (5) above, t_d and t_{cw} are the dressing time and the time (h) for changing a grinding wheel (h), respectively. The number of parts per dress $n_{p,d}$ and the total number of parts ground by a grinding wheel $n_{p,w}$ can be expressed as:

$$n_{p,d} = \frac{T_w}{t_c} \tag{12}$$

$$n_{p,w} = \frac{(D_{s,0} - D_{s,e})n_{p,d}}{2(W_{pd} + a_{ed,ges})} \tag{13}$$

Where, T_w is the wheel life (h), $D_{s,o}$ is the initial grinding wheel diameter (mm), $D_{s,e}$ is the exchanged grinding wheel diameter (mm), W_{pd} is the radial grinding wheel wear per dress (mm/dress), and $a_{ed,ges}$ is the total depth of dressing cut (mm).

From Eq. (1), the machine tool hourly rate C_{mh} (USD/h) includes the cost of wages, overhead, and maintenance and the grinding wheel cost per part $C_{gw,p}$ (USD/part). Based on the grinding wheel cost per piece C_{gw} (USD/piece) and the total number of parts ground by a grinding wheel $n_{p,w}$ as shown in Eq. (12), $C_{gw,p}$ can be displayed as:

$$\mathrm{C_{gw,p}} = \frac{\mathrm{C_{gw}}}{\mathrm{n_{p,w}}} \tag{14}$$

Based on the cost function C_{sin} in Eq. (1), it is indicated that the exchanged grinding wheel diameter affects the cost of external grinding process. For a given value of the input parameters i.e., D_O = 500 (mm), a_e = 0.1 *(mm)*, C_{mh} = 10 *(USD/h)*, C_{wg} = 30 *(USD/h)*, n_{pd} = 35, a_{ed} = 0.1, T_w = 20, W_{pd} = 0.02 *(mm/dress)*, the cost function C_{sin} depends on the variable of the exchanged grinding wheel diameter D_e as shown in Fig. 1. It is observed that the grinding cost reaches the minimum value at $D_{e,op} = 475$(mm). This optimum value of exchanged grinding wheel diameter $D_{e,op}$ is much larger than the conventional exchanged grinding wheel diameter. Therefore, the objective function for the grinding cost depending on the exchanged grinding wheel diameter D_e can be demonstrated as follows:

$$Minimize\, C_{sin} = \mathrm{f}(D_e) \tag{15}$$

With the constraint as:

$$\mathrm{D}_{e,min} \leq D_e \leq \mathrm{D}_{e,max} \tag{16}$$

Based on the value of the selected input parameters i.e., the initial grinding wheel diameter D_O, the grinding wheel width W_{gw}, the wheel life T_w, the radial grinding wheel wear per dress W_{pd}, the total depth of dressing cut a_{ed}, the machine tool hourly rate C_{mh}, and the grinding wheel cost C_{gw}, the optimum exchanged grinding diameter is

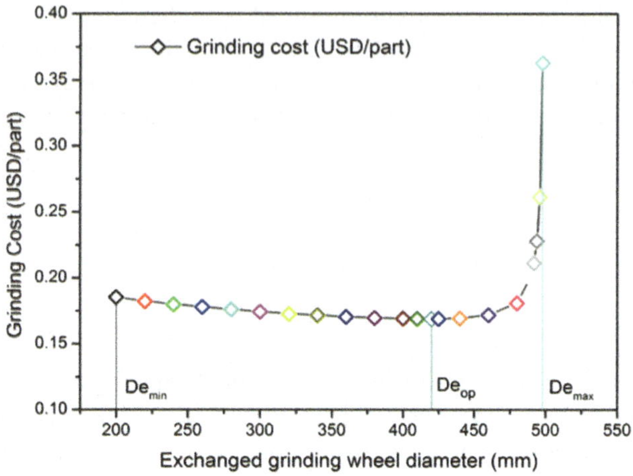

Fig. 1. Grinding cost versus exchanged grinding wheel diameter

determined to minimize the grinding cost. Therefore, the function of the optimum diameter $D_{e,op}$ can be written as follows:

$$D_{e,op} = f\left(D_0, W_{gw}, T_w, W_{pd}, a_{ed}, C_{mh}, C_{gw}\right) \quad (17)$$

3 Experimental Methodology

In this section, the experimental work is presented to analyse the influences of the input parameters i.e., the initial grinding wheel diameter D_0, the grinding wheel width W_{gw}, the wheel life T_w, the radial grinding wheel wear per dress W_{pd}, the total depth of dressing cut a_{ed}, the machine tool hourly rate C_{mh}, and the grinding wheel cost C_{gw} on the optimum diameter $D_{e,op}$. The input parameters and their value range are shown in Table 1.

Table 1. Input parameters and their value range.

Factor	Code	Unit	Low	High
Initial grinding wheel diameter	D_0	mm	250	500
Grinding wheel width	W_{gw}	mm	20	60
Radial grinding wheel wear per dress	W_{pd}	mm	0.01	0.03
Total depth of dressing cut	a_{ed}	mm	0.1	0.3
Wheel life	T_w	min.	10	30
Machine tool hourly rate	C_{mh}	USD/h	4	12
Grinding wheel cost	C_{gw}	USD/p.	10	60

To investigate the effects of the input parameters on the optimum diameter $D_{e,op}$, the screening experiment was conducted. The purpose of using the experiment method is to: (1) determine the main factor which has the largest effect on the optimum diameter $D_{e,op}$, (2) evaluate the level of the influence of the parameters on the optimum diameter $D_{e,op}$, (3) estimate the effect of the interaction of the parameters, and (4) establish the regression equation which shows the relation between the optimum diameter $D_{e,op}$ and the significant effect parameters.

After carrying out experiments and collecting results, the data are analysed and processed by using Minitab 18 software. The establishment of the experimental variables and the installation of their value range are designed with 2-level full factorial design. From the experimental installation with the seven input parameters as shown in Table 1, the total of test runs is 128. Based on the values of the input parameters, the optimum diameter value $D_{e,op}$ at each test run is determined as shown in Table 2.

Table 2. Experimental plan and the result values of Dop.

Std order	Run order	Center Pt	D_0	B_{gw}	a_{ed}	T_w	W_{pd}	$C_{m,h}$	C_{gw}	D_{op}
56	1	1	500	60	0.03	0.1	30	12	10	484.21
15	2	1	250	60	0.03	0.3	10	4	10	212.97
19	3	1	250	60	0.01	0.1	30	4	10	235.12
110	4	1	500	20	0.03	0.3	10	12	60	430
53	5	1	250	20	0.03	0.1	30	12	10	239
115	6	1	250	60	0.01	0.1	30	12	60	230.35
...										
124	127	1	500	60	0.01	0.3	30	12	60	454.17
89	128	1	250	20	0.01	0.3	30	4	60	203.32

4 Results and Discussion

This section describes the influences of the input grinding parameters and the interactions between these parameters on the optimum exchanged grinding wheel diameter. In addition, a mathematical model for calculating the optimum diameter $D_{e,op}$ was introduced by using regression analysis.

4.1 Effect of the Factors on the Optimum Exchanged Grinding Wheel Diameter

Based on the analysis of variance (ANOVA), Fig. 2 shows the main effect of each factor on the optimum diameter $D_{e,op}$. The relation between the optimum exchanged grinding wheel diameter and each factor is described as the straight lines in Fig. 2. Based on the analysis of the slope angles of the straight lines in Fig. 2, it can be clearly seen that the initial grinding wheel diameter D_0 has the largest effect on the optimum diameter $D_{e,op}$. Meanwhile, the influence of the other factors is smaller than that of the initial grinding wheel. In addition, the grinding wheel width W_{gw} does not affect the optimum diameter $D_{e,op}$.

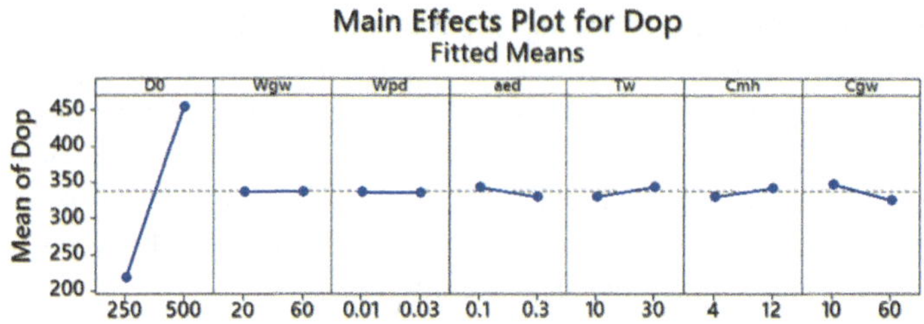

Fig. 2. Main effects plot on the optimum exchanged grinding wheel diameter $D_{e,op}$

Figure 3 shows the Normal Plot of the standardized effects to investigate the effect trend of the factors and the interactions of the parameters on the optimum diameter $D_{e,op}$. It is recognizable that the distribution of the standardized effects for most of the factors is close to the reference line (see the red line in Fig. 3). For the positive effect of the three factors i.e., the initial grinding wheel diameter D_0 (A), the wheel life T_w (C), the machine tool hourly rate C_{mh} (F) and the four interactions i.e., FG $(C_{mh} * C_{gw})$, CG $(T_w * C_{gw})$, AC $(D_0 * T_w)$, AF $(D_0 * C_{mh})$, the optimum diameter D_{op} increases when the values of these factors and interactions change from the low value to the high value. Conversely, the negative effects of the three factors including the radial grinding wheel wear per dress W_{pd} (D), the total depth of dressing cut a_{ed} (E), the grinding wheel cost C_{gw} (G) and the three interactions including AE $(D_0 * a_{ed})$, EG $(a_{ed} * C_{gw})$, AG $\left(D_0 * C_{gw}\right)$ show that the optimum diameter D_{op} increases when the values of these factors and interactions decrease. Remarkably, the distribution of the standardized effects indicates that the optimum diameter D_{op} is tremendously affected by the initial grinding wheel diameter D_0 whereas the grinding wheel width W_{gw} (B) does not affect the optimum diameter.

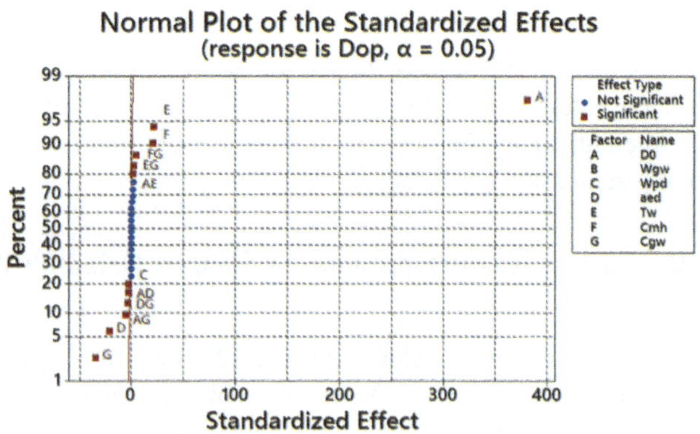

Fig. 3. Normal plot for the optimum diameter $D_{e,op}$

Figure 4 shows the Pareto chart of the standardized effects for determining the magnitude and the importance of the effect on the optimum exchanged grinding diameter $D_{e,op}$. As presented in this figure, the influence magnitude of the grinding process parameters is arranged from the lowest value to the highest value. The largest influence factor on the optimum diameter is the initial grinding wheel D_0 (the factor A in Fig. 4). The influence is gradually reduced in the following order of the grinding process parameters: D_0, C_{gw}, a_{ed}, T_w, C_{mh}, $C_{mh} * C_{gw}$, $D_0 * C_{gw}$, $a_{ed} * C_{gw}$, $T_w * C_{gw}$, $D_0 * a_{ed}$, $D_0 * T_w$ and W_{pd}. Especially, as noted above, the optimum exchanged diameter is not affected by the grinding wheel width W_{gw} (the factor B in Fig. 4).

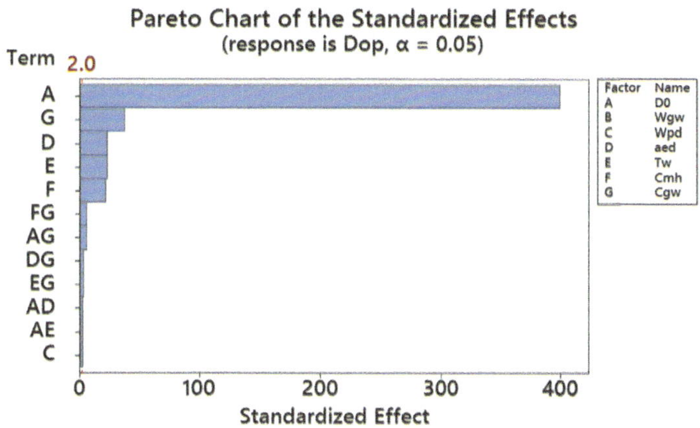

Fig. 4. Pareto chart of the standardized effects on the optimum diameter $D_{e,op}$

4.2 Regression Analysis

Table 3 shows the estimated effects and the coefficients of the regression function after neglecting the insignificant effects on $D_{e,op}$. It is realized that the input parameters including the initial grinding wheel D_0, the wheel life T_w, the radial grinding wheel wear per dress W_{pd}, the total depth of dressing cut a_{ed}, the machine tool hourly rate C_{mh}, the grinding wheel cost C_{gw} and the interactions including $D_0 * T_w, D_0 * a_{ed}$, $D_0 * C_{mh}, D_0 * C_{gw}, T_w * C_{gw}$, $a_{ed} * C_{gw}, C_{mh} * C_{gw}$ have significant effects on the response $D_{e,op}$ because their P-values are smaller than the significant level ($\alpha = 0.05$). Based on that, the regression equation for calculating the optimum diameter $D_{e,op}$ can be expressed as follows:

$$\begin{aligned} D_{e,op} = & -15.75 + 0.96541D_0 - 68.6W_{pd} - 32.9a_{ed} + 0.331T_w + 1.029C_{mh} \\ & - 0.3816C_{gw} - 0.0556D_0 * a_{ed} + 0.000550D_0 * T_w - 0.000493D_0 * C_{gw} \\ & - 0.376a_{ed} * C_{gw} + 0.00372T_w * C_{gw} + 0.01606C_{mh} * C_{gw} \end{aligned} \quad (18)$$

The correlation coefficients shown in Table 3, i.e. R2 = 0.99.95, adj-R2 = 0.99.94 and pred-R2 = 0.9993 are approximate to 1, which proves that the model in Eq. (18)

Table 3. Variance analysis of regression after elimination of the insignificant factors

Coded Coefficients

Term	Effect	Coef	SE Coef	T-Value	P-Value	VIF
Constant		337.769	0.297	1138.38	0.000	
D0	237.004	118.502	0.297	399.39	0.000	1.00
Wpd	-1.373	-0.686	0.297	-2.31	0.022	1.00
aed	-13.384	-6.692	0.297	-22.55	0.000	1.00
Tw	13.340	6.670	0.297	22.48	0.000	1.00
Cmh	12.733	6.367	0.297	21.46	0.000	1.00
Cgw	-21.945	-10.972	0.297	-36.98	0.000	1.00
D0*aed	-1.391	-0.695	0.297	-2.34	0.021	1.00
D0*Tw	1.375	0.687	0.297	2.32	0.022	1.00
D0*Cgw	-3.083	-1.541	0.297	-5.20	0.000	1.00
aed*Cgw	-1.878	-0.939	0.297	-3.17	0.002	1.00
Tw*Cgw	1.858	0.929	0.297	3.13	0.002	1.00
Cmh*Cgw	3.213	1.606	0.297	5.41	0.000	1.00

Model Summary

S	R-sq	R-sq(adj)	R-sq(pred)
3.35688	99.93%	99.92%	99.91%

presents a good fit to the data. Therefore, it can be used to determine the optimum exchanged grinding wheel diameter for the external grinding stainless steel.

5 Conclusion

In this paper, the optimum exchanged grinding wheel diameter in the external grinding process was determined to minimize the grinding cost. The influences of technological parameters including the initial grinding wheel diameter D_O, the grinding wheel width W_{gw}, the wheel life T_w, the radial grinding wheel wear per dress W_{pd}, the total depth of dressing cut a_{ed}, the machine tool hourly rate C_{mh}, and the grinding wheel cost C_{gw} on the optimum exchanged grinding wheel diameter were also investigated. The findings are drawn as follows:

- The cost function for external grinding was established;
- The optimum exchanged grinding wheel diameter was determined to minimize the grinding cost;
- The influence of the seven grinding parameters on the optimum diameter were investigated. The initial grinding wheel had the largest effect on the optimum diameter;
- The influence of the interactions between the parameters was also examined;
- Based on the analysis regression, the fitting model for computing the optimum diameter was introduced for the external grinding process.

Acknowledgments. The work described in this paper was supported by Thai Nguyen University of Technology for a scientific project.

References

1. Rowe, W.B.: Principle of Modern Grinding Technology. William Andrew, Boston (2009)
2. Malkin, S., Guo, C.: Grinding Technology: Theory and Applications of Machining with Abrasives. Industrial Press, New York (2008)
3. Li, G.F., Wang, L.S., Yang, L.B.: Multi-parameter optimization and control of the cylindrical grinding process. J. Mater. Process. Technol. **129**, 232–236 (2002)
4. Pereverzev, P.P., Akintseva, A.V.: Automatic cycles multi-parametric optimization of internal grinding, Procedia Eng. **129**, 121–126 (2015). International Conference on Industrial Engineering
5. Pandiyan, V., Caesarendra, W., Tjahjowidodo, T., Praveen, G.: Predictive modelling and analysis of process parameters on material removal characteristics in abrasive belt grinding process. Appl. Sci. **7**(4), 363 (2017). https://doi.org/10.3390/app7040363
6. Luu, A.T., Vu, N.P., Do, T.T.H., Le, X.H., Banh, T.L.: A study on optimization of surface roughness in surface grinding 9CrSi tool steel by using Taguchi method. In: Fujita, H., et al. (eds.) ICERA 2018. LNNS, vol. 63, pp. 100–108 (2019). https://doi.org/10.1007/978-3-030-04792-4_15
7. Gupta, R., Shishodia, K.S., Sekhon, G.S.: Optimization of grinding parameters using enumeration method. J. Mater. Process. Technol. **112**, 63–67 (2001)
8. Wen, X.M., Tay, A.A.O., Nee, A.Y.C.: Micro-computer-based optimization of the surface grinding process. J. Mater. Process. Technol. **29**(1–3), 75–90 (1992)
9. Mekala, K., Chandradas, J., Chandrasekaran, K., Kannan, T.T.M., Ramesh, E., Narasing Babu, R.: Optimization of cylindrical grinding parameters of austenitic stainless steel rods (AISI 316) by Taguchi method. Int. J. Mech. Eng. Rob. Res. **2**, 208 (2014)
10. Vu, N.-P., Nguyen, Q.-T., Tran, T.-H., Le, H.-K., Nguyen, A.-T., Luu, A.-T., Nguyen, V.-T., Le, X.-H.: Optimization of grinding parameters for minimum grinding time when grinding tablet punches by CBN wheel on CNC milling machine. Appl. Sci. **9**, 957 (2019)
11. Vidal, G., Ortega, N., Bravo, H., Dubar, M., González, H.: An analysis of electroplated cBN grinding wheel wear and conditioning during creep feed grinding of aeronautical alloys. Metals, 1–24 (2018)
12. Daneshia, A., Jandaghia, N., Tawakoli, T.: Effect of dressing on internal cylindrical grinding. Procedia CIRP **14**, 37–41 (2014)
13. Tu, H.X., Thao, L.P., Hong, T.T., Nga, N.T.T., Trung, D.D., Gong, J., Pi, V.N.: Influence of dressing parameters on surface roughness of workpiece for grinding hardened 9XC tool steel. In: IOP Conference Series: Materials Science and Engineering, vol. 542, Conference 1. https://doi.org/10.1007/978-3-030-04792-4_15
14. Le, X.H., Tran, T.H., Luu, A.T., Nguyen, T.T.N., Vu, N.P.: Optimum dressing parameters for maximum material removal rate when internal cylindrical grinding using Taguchi method. Int. J. Mech. Eng. Technol. **9**, 123–129 (2018)
15. Palmer, A., Ghadbeigi, H., Novovic, D., Curtis, D.: An experimental study of the effects of dressing parameters on the topography of grinding wheels during roller dressing. J. Manuf. Process. **31**, 348–355 (2018)
16. Hung, L.X., Lien, V.T., Pi, V.N., Long, B.T.: A study on coolant parameters in internal grinding of 9CrSi steel. Mater. Sci. Forum **950**, 24–31 (2019)
17. Tu, H.X., Pi, V.N., Jun, G.: A study on determination of optimum parameters for lubrication in external cylindrical grinding base on Taguchi method. Key Eng. Mater. **796**, 97–102 (2019)
18. Pi, V.N., Hung, L.X., Tung, L.A., Long, B.T.: Cost optimization of internal grinding. J. Mater. Sci. Eng. B **6**, 291–296 (2016)

19. Hung, L.X., Pi, V.N., Tung, L.A., Tu, H.X., Jun, G., Long, B.T.: Determination of optimal exchanged grinding wheel diameter when internally grinding alloy tool steel 9CrSi. In: IOP Conference Series: Materials Science and Engineering, vol. 417, 012–026 (2018)
20. Hung, L.X., Ky, L.H., Hong, T.T., Dung, H.T., Lien, V.T., Tung, L.A., Long, B.T., Pi, V.N.: A study on cost optimization of internal cylindrical grinding. Int. J. Mech. Eng. Technol. (IJMET) **10**(1), 414–423 (2019)
21. Tran, T.-H., Le, X.-H., Nguyen, Q.-T., Le, H.-K., Hoang, T.-D., Luu, A.-T., Banh, T.-L., Ngoc-Pi, V.: Optimization of exchanged grinding wheel diameter for minimum grinding cost in internal grinding. Appl. Sci. **9**(7), 1363 (2019). https://doi.org/10.3390/app9071363
22. Le, X.H., Vu, N.P., Luu, A.T., Tu, H.X., Jun, G., Banh, T.L.: Determination of optimum exchanged grinding wheel diameter when internal grinding alloy tool steel 9CrSi. In: IOP Conference Series: Materials Science and Engineering, vol. 417, conference 1 (2018). https://doi.org/10.1088/1757-899X/417/1/012026
23. Pi, V.N., Tung, L.A., Hung, L.X., Ngoc, N.V.: Experimental determination of optimum exchanged diameter in surface grinding process. J. Environ. Sci. Eng. A **6**, 85–89 (2017)
24. Tran, T.-H., Luu, A.-T., Nguyen, Q.-T., Le, H.-K., Nguyen, A.-T., Hoang, T.-D., Le, X.-H., Banh, T.-L., Ngoc-Pi, V.: Optimization of exchanged grinding wheel diameter for surface grinding tool steel based on the cost analysis. Metals **9**(4), 448 (2019). https://doi.org/10.3390/met9040448
25. Pi, V.N., The, P.Q., Khiem, V.H., Huong, N.N.: Cost optimization of external cylindrical grinding. Appl. Mech. Mater. **312**, 982–989 (2013)
26. Tu, H.X., Le, G.J., Hung, L.X., Tung, L.A., Pi, V.N.: Calculation of optimum exchanged grinding wheel diameter when external grinding tool steel 9CrSi. Int. J. Mech. Eng. Robot. Res. **8**(1), 59–64 (2019)
27. Kozuro, L.M., Panov, A.A., Remizovski, E.I., Tristosepdov, P.S.: Handbook of Grinding. Publish Housing of High-education, Minsk (1981). (in Russian)

Optimization of Manufacturing Time in Internal Grinding

Le Xuan Hung[1], Le Hong Ky[2], Tran Thi Hong[3], Nguyen Van Cuong[4], Do Duc Trung[5], Nguyen Huu Phan[5], Luu Anh Tung[1], and Ngoc Pi Vu[1](✉)

[1] Thai Nguyen University of Technology, Thai Nguyen, Vietnam
vungocpi@tnut.edu.vn
[2] Vinh Long University of Technology Education, Vinh Long, Vietnam
[3] Nguyen Tat Thanh University, Ho Chi Minh City, Vietnam
[4] University of Transport and Communications, Hanoi, Vietnam
[5] Faculty of Mechanical Engineering, Hanoi University of Industry, No. 298, Cau Dien Street, Bac Tu Liem District, Hanoi, Vietnam

Abstract. This paper introduces an optimization study on determination of optimum replaced grinding wheel diameter in internal grinding process. In this study, the minimum manufacturing time was chosen as the objective of optimization problem. Also, several grinding parameters containing the initial grinding wheel diameter, the radial grinding wheel wear per dress, the total depth of dressing cut and the wheel life were taken into the investigation. In addition, to evaluate the influences of these input factors on the optimum replaced wheel diameter, a simulation experiment was conducted. Lastly, a regression model was presented for determination of the optimum replaced wheel diameter.

Keywords: Grinding · Internal grinding · Manufacturing time · Manufacturing time optimization · Replaced grinding wheel diameter · Tool steel

1 Introduction

Grinding is an abrasive machining process commonly used for producing finished or semi-finished parts [1]. It enables high accuracy and close tolerances to be kept for size, shape and surface quality [2]. Therefore, research to improve the quality and the productivity of grinding process has been the object of numerous researchers [3–16]. Grinding process has been investigated in several grinding types including surface grinding [5, 7], external grinding [9, 16], and internal grinding [13, 14]. In addition, researchers have sought many ways to improve the quality, the performance and reduce the manufacturing cost of the grinding process such as finding the optimum coolant conditions [14–17], suggesting the optimum dressing parameters [9, 11, 12, 18], improving the material removal rate and the productivity of grinding [19, 20] or using optimum replaced grinding wheel diameter for getting the minimum manufacturing cost [21–30]. Nonetheless, no proposals have been made to determine the optimum replaced wheel diameter for the minimum manufacturing time.

K.-U. Sattler et al. (Eds.): ICERA 2019, LNNS 104, pp. 557–565, 2020.
https://doi.org/10.1007/978-3-030-37497-6_63

This paper introduces a study on finding the optimum replaced wheel diameter for minimizing the manufacturing time in internal grinding tool steel. In the study, the effects of the input process parameters containing the initial grinding wheel diameter (D_0), the total depth of dressing cut (t_{sd}), the radial grinding wheel wear per dress (D_{max}) and the wheel life (T_d) on the optimum diameter (d_{op}) were examined by conducting a simulation experiment. Also, a regression equation to calculate the optimum replaced diameter was proposed.

2 Methodology

2.1 Manufacturing Time Analysis

In internal grinding, the manufacturing time t_s can be determined by:

$$t_s = t_c + t_{lu} + t_{sp} + t_{d,p} + t_{cw,p} \tag{1}$$

In which, $t_c, t_{lu}, t_{sp}, t_{d,p}$ and $t_{cw,p}$ are the grinding time (h), the part loading and unloading time (h), the spark-out time (h), the dressing time per part (h), and the grinding wheel replacement time per part, respectively. $t_c, t_{d,p}$ and $t_{cw,p}$ can be calculated as the following equations:

$$t_c = l_w a_{e,t} / (v_{fa} \cdot f_r) \tag{2}$$

$$t_{d,p} = t_d / n_{p,d} \tag{3}$$

$$t_{cw,p} = t_{cw} / n_{p,w} \tag{4}$$

In Eq. (2), l_w(mm), $a_{e,t}$(mm), v_{fa}(mm/min), and f_r(mm/double stroke) are the part length, the total grinding depth, the axial speed, and the radial wheel feed, respectively. In addition, when grinding stainless steel, v_{fa} and f_r are determined by the following regression equations [31]:

$$v_{fa} = 22.88 \cdot B_{gw}^{0.9865} \cdot d_w^{0.0821} \cdot S_{rg}^{-2.9833} \cdot n_w^{1.2471} \tag{5}$$

$$f_r = f_{r,tab} \cdot c_1 \cdot c_2 \cdot c_3 \cdot c_4 \tag{6}$$

In which, n_w is the workpiece speed; For grinding stainless steel, n_w is calculated as follows [31]:

$$n_w = 1255.8 \cdot d_w^{-0.3491} \tag{7}$$

$f_{r,tab}$ is the tabled radial wheel feed (mm/stroke) which is found by [27]:

$$f_{r,tab} = 30.2944 \cdot a_{e,t}^{0.567} \cdot v_{fa}^{-0.9693} \cdot d_{w}^{-0.1269} \tag{8}$$

In Eq. (6), c_1, c_2, c_3 and c_4 are the coefficients for determining the radial wheel feed (Table 1); B_{gw} is the grinding wheel width; d_w is the workpiece diameter; S_{rg} is the surface roughness grade;

Table 1. Coefficients for determining the radial wheel feed

Coefficient depends on	Code	Value	Reference
Workpiece material and tolerance grade (for work material is tool steel)	c_1	$c_1 = 0.0857 \cdot tg^{1.2767}$	[31]
Grinding wheel diameter	c_2	$c_2 = 0.5657 \cdot d_s^{0.153}$	[31]
Measurement type	c_3	1 (using micrometer)	[32]
Ratio of length to diameter of workpiece	c_4	$c_4 = 1.0642 \cdot R_{ld}^{-0.5079}$	[31]

In Eq. (4), $n_{p,w}$ is the total number of parts ground by a grinding wheel which is calculated as follows:

$$n_{p,w} = (D_0 - d_e) \cdot n_{p,d} / [2(D_{\max} + t_{sd})] \tag{9}$$

Where, D_0 is the initial diameter of the grinding wheel (mm); d_e is the replacement diameter of the wheel (mm); $D_{\max}$ is the radial grinding wheel wear per dress (mm/dress); t_{sd} is the total depth of dressing cut (mm); $n_{p,d}$ is the number of parts per dress:

$$n_{p,d} = T_d / t_c \tag{10}$$

2.2 Optimization Problem

From Sect. (2.1), it is found that when the grinding wheel replacement diameter $d_{s,e}$ is large, f_r is in large values (see Eq. (6) and c_2 in Table (1)) and it leads to the manufacturing time decline. However, the large value of the replacement diameter will result in a small number of parts ground by a grinding wheel (Eq. (9), and therefore, an increase of the manufacturing time. It is obvious that there will be an optimum value of the replacement diameter of the wheel at which the machining time is the smallest.

Consequently, the optimization problem to calculate the optimum replacement diameter for the minimum manufacturing time can be expressed by:

$$Minimize\ t_s = \mathrm{f}(d_{s,r}) \tag{11}$$

subject to

$$d_{s,rmin} \le d_{s,r} \le d_{s,rmax} \tag{12}$$

2.3 Experimental Work

A simulation experiment, a 2-level full factorial design, was prepared with the aim to evaluate the influences of the input parameters on the optimum replaced wheel diameter. In the experiment, four input parameters including the initial grinding wheel diameter (D_0), the total depth of dressing cut (tsd)) the wheel life (Td) and the radial grinding wheel wear per dress (D_{max}) were taken for the evaluation (Table 2). Consequently, a number of $2^4 = 16$ experimental runs were conducted by a computing program. The experimental plan and the output results (the optimum replaced grinding wheel diameters d_{op}) are described in Table 3.

Table 2. Input parameters

Factor	Code	Unit	Low	High
Initial grinding wheel diameter	D_0	mm	5	30
Total depth of dressing cut	tsd	mm	0.01	0.03
Wheel life	Td	min	10	30
Radial grinding wheel per dress	D_{max}	mm	0.01	0.03

Table 3. Experimental plans and result values of d_{op}

StdOrder	RunOrder	CenterPt	Blocks	D_0	t_{sd}	T_d	D_{max}	d_{op}
1	1	1	1	10	0.01	10	0.01	8.58
5	2	1	1	10	0.01	30	0.01	9.1
7	3	1	1	10	0.03	30	0.01	9.03
9	4	1	1	10	0.01	10	0.03	8.58
12	5	1	1	30	0.03	10	0.03	27.24
...	...	...	...	...	...	...	...	...
13	15	1	1	10	0.01	30	0.03	9.1
6	16	1	1	30	0.01	30	0.01	28.41

3 Results and Discussion

The analysis of the input parameter influences on the optimum replaced grinding wheel diameter is featured by the graph of parameters' main effect plotted in Fig. 1. It is interpreted that the optimum replaced wheel diameter dop is largely influenced by the initial grinding wheel diameter D_0, whereas it is totally independent upon the total depth of dressing cut tsd and the radial grinding wheel wear per dress (D_{max}). However, the wheel life Td has a slight effect on this factor.

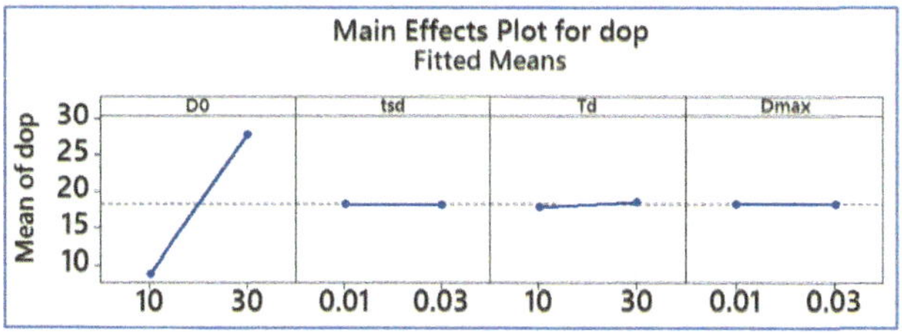

Fig. 1. Main effects plot on d_{op}

The Normal Plot of the standardized effects is shown in Fig. 2. It is discernible that similar to the comment in Fig. 1, the optimum replaced wheel diameter depends strongly on the initial wheel diameter (factor A). Besides, the initial wheel diameter, the wheel life (factor C) and the interactions AC and BC have positive standardized effects. That means if these factors grow, the optimum diameter increases. Also, the total depth of dressing cut (factor B) and the interaction AB are the negative affecting factors. If these values ascend, the optimum diameter discends.

Figure 3 shows the Pareto chart which describes the effects of the input factors on the optimum replaced wheel diameter. It is recognized from this chart that the bars symbolising factor A (the initial wheel diameter), factor C (the wheel life), factor B (the total depth of dressing cut) and the interaction AC cross the reference line. Those signals mean these factors are statistically significant at the 0.05 level with the response model.

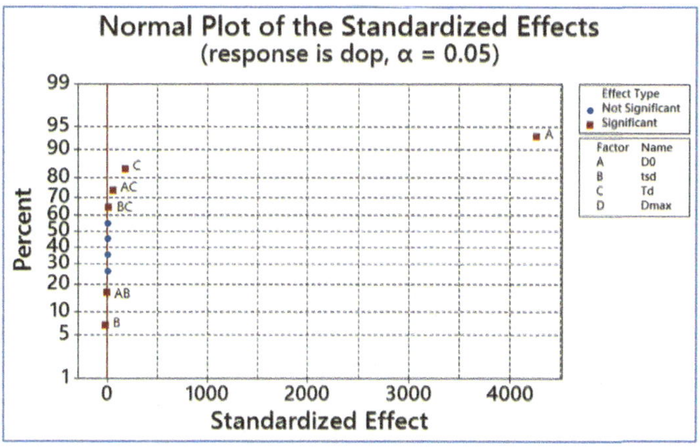

Fig. 2. Normal plot for the optimum diameter d_{op}

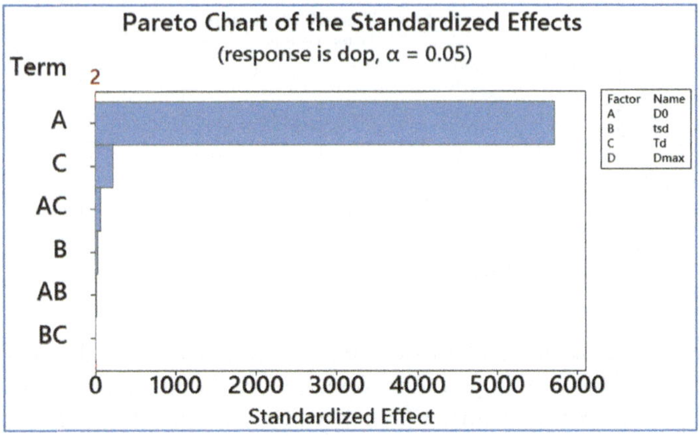

Fig. 3. Pareto chart of the standardized effects on d_{op}

Table 4. Estimated effects and coefficients for d_{op}

Coded Coefficients

Term	Effect	Coef	SE Coef	T-Value	P-Value	VIF
Constant		18.3200	0.0017	10992.00	0.000	
D0	19.05000	9.52500	0.00167	5715.00	0.000	1.00
tsd	-0.13000	-0.06500	0.00167	-39.00	0.000	1.00
Td	0.77000	0.38500	0.00167	231.00	0.000	1.00
D0*tsd	-0.04000	-0.02000	0.00167	-12.00	0.000	1.00
D0*Td	0.23000	0.11500	0.00167	69.00	0.000	1.00
tsd*Td	0.03000	0.01500	0.00167	9.00	0.000	1.00

Model Summary

S	R-sq	R-sq(adj)	R-sq(pred)
0.0066667	100.00%	100.00%	100.00%

The estimated effects and coefficients for d_{op}- is exhibited in Table 4 with the negligence of the insignificant effects. It is found from the figure that the initial wheel diameter D_0, the total depth of dressing cut t_{sd}, the wheel life T_d and the interactions $D_0 * t_{sd}$, $D_0 * T_d$ and $t_{sd} * T_d$ have significant effects on the response (the P-values lower than 0.05) (Fig. 4). Consequently, the regression model to find the optimum diameter d_{op} can be defined by:

$$\begin{aligned} d_{op} &= -0.93 + 0.9335 \cdot D_0 - 5.5 \cdot t_{sd} + 0.0125 \cdot T_d \\ &\quad -0.2 \cdot D_0 \cdot t_{sd} + 0.00115 \cdot D_0 \cdot T_d + 0.15 \cdot t_{sd} \cdot T_d \end{aligned} \tag{13}$$

From Table 4 it is drawn that all of the values of R-sq and R-sq(adj) are 100%, which signifies that Eq. (13) fits very well with the experimental data. Therefore, the Equation can be used to calculate the optimum replaced wheel diameter when internal grinding tool steel.

4 Conclusions

An optimization study on determination of the optimum replaced grinding wheel diameter when internal grinding tool steel was conducted. In the study, the manufacturing time consumed when internal grinding process was analysed. Also, it was chosen as the objective of the optimization problem. Moreover, a simulation experiment was planned and conducted to evaluate the effects of the input parameters on the optimum replaced wheel diameter. Finally, from the experimental results, a regression equation to find the optimum replaced wheel diameter was suggested. As the equation is explicit, the optimum replaced diameter for internal grinding process can be calculated simply and effectively.

Acknowledgments. The work described in this paper was supported by Thai Nguyen University of Technology for a scientific project.

References

1. Malkin, S., Guo, C.: Grinding Technology: Theory and Applications of Machining with Abrasives. Industrial Press, New York (2008)
2. Rowe, W.B.: Principle of Modern Grinding Technology. William Andrew, Norwich (2009)
3. Chatterjee, S., Rudrapati, R., Kumarpal, P., Nandi, G.: Experiments, analysis and parametric optimization of cylindrical traverse cut grinding of aluminium bronze. Mater. Today Proc. **5** (2), 5272–5280 (2018). Part 1
4. Vu, N.-P., Nguyen, Q.-T., Tran, T.-H., Le, H.-K., Nguyen, A.-T., Luu, A.-T., Nguyen, V.-T., Le, X.-H.: Optimization of grinding parameters for minimum grinding time when grinding tablet punches by CBN wheel on CNC milling machine. Appl. Sci. **9**, 957 (2019)
5. Tung, L.A., Pi, V.N., Ha, D.T.T., Hung, L.X., Banh, T.L.: A study on optimization of surface roughness in surface grinding 9CrSi tool steel by using Taguchi method. In: Fujita, H., et al. (eds.) ICERA 2018. LNNS, vol. 63, pp. 100–108 (2019). https://doi.org/10.1007/978-3-030-04792-4_15
6. Gupta, R., Shishodia, K.S., Sekhon, G.S.: Optimization of grinding parameters using enumeration method. J. Mater. Process. Technol. **112**, 63–67 (2001)
7. Wen, X.M., Tay, A.A.O., Nee, A.Y.C.: Micro-computer-based optimization of the surface grinding process. J. Mater. Process. Technol. **29**(1–3), 75–90 (1992)
8. Mekala, K., Chandradas, J., Chandrasekaran, K., Kannan, T.T.M., Ramesh, E., Babu, R.N.: Optimization of cylindrical grinding parameters of austenitic stainless steel rods (AISI 316) by Taguchi method. Int. J. Mech. Eng. Robot. Res. **3**, 208–215 (2014)
9. Tu, H.X., Thao, L.P., Hong, T.T., Nga, N.T.T., Trung, D.D., Gong, J., Pi, V.N.: Influence of dressing parameters on surface roughness of workpiece for grinding hardened 9XC tool steel. In: IOP Conference Series: Materials Science and Engineering, vol. 542 (2019). https://doi.org/10.1088/1757-899X/542/1/012008

10. Vidal, G., Ortega, N., Bravo, H., Dubar, M., González, H.: An analysis of electroplated cBN grinding wheel wear and conditioning during creep feed grinding of aeronautical alloys. Metals **8**, 1–24 (2018)
11. Daneshia, A., Jandaghia, N., Tawakoli, T.: Effect of dressing on internal cylindrical grinding. Procedia CIRP **14**, 37–41 (2014)
12. Le, X.H., Tran, T.H., Luu, A.T., Nguyen, T.T.N., Vu, N.P.: Optimum dressing parameters for maximum material removal rate when internal cylindrical grinding using Taguchi method. Int. J. Mech. Eng. Technol. **9**, 123–129 (2018)
13. Gupta, R., Shishodia, K.S., Sekhon, G.S.: Optimization of grinding process parameters using enumeration method. J. Mater. Process. Technol. **112**(1), 63–67 (2001)
14. Alberdi, R., Sanchez, J.A., Pombo, I., Ortega, N., Barrenetxea, D.: Strategies for optimal use of fluids in grinding. Int. J. Mach. Tools Manuf **51**(6), 491–499 (2011)
15. Hung, L.X., Liên, V.T., Pi, V.N., Long, B.T.: A study on coolant parameters in internal grinding of 9CrSi steel. Mater. Sci. Forum **950**, 24–31 (2019)
16. Tu, H.X., Jun, G., Hien, B.T., Hung, L.X., Tung, L.A., Pi, V.N.: Determining optimum parameters of cutting fluid in external grinding of 9CrSi steel using Taguchi technique. SSRG Int. J. Mech. Eng. **5**(6), 1–5 (2018). https://doi.org/10.14445/23488360/IJME-V5I6P101
17. Tu, H.X., Pi, V.N., Jun, G.: A study on determination of optimum parameters for lubrication in external cylindrical grinding base on Taguchi method. Key Eng. Mater. **796**, 97–102 (2019)
18. Wegener, K., Hoffmeister, H.-W., Karpuschewski, B., Kuster, F., Rabiey, M.: Conditioning and monitoring of grinding wheels. CIRP Ann. **60**(2), 757–777 (2011)
19. Yadav, H.S., Shrivastava, R.K.: Effect of process parameters on surface roughness and MRR in cylindrical grinding using response surface method. Int. J. Eng. Res. Technol. **3**(3) (2014)
20. Agarwal, S.: Optimizing machining parameters to combine high productivity with high surface integrity in grinding silicon carbide ceramics. Ceram. Int. **42**(5), 6244–6262 (2016)
21. Pi, V.N., The, P.Q., Khiem, V.H., Huong, N.N.: Cost optimization of external cylindrical grinding. Appl. Mech. Mater. **312**, 982–989 (2013)
22. Tu, H.X., Jun, G., Hung, L.X., Tung, L.A., Pi, V.N.: Calculation of optimum exchanged grinding wheel diameter when external grinding tool steel 9CrSi. Int. J. Mech. Eng. Robot. Res. **8**(1), 59–64 (2019)
23. Hung, L.X., Pi, V.N., Tung, L.A., Tu, H.X., Jun, G., Long, B.T.: Determination of optimal exchanged grinding wheel diameter when internally grinding alloy tool steel 9CrSi. In: IOP Conference Series: Materials Science and Engineering, vol. 417, pp. 012–026 (2018)
24. Pi, V.N., Hung, L.X., Tung, L.A., Long, B.T.: Cost optimization of internal grinding. J. Mater. Sci. Eng. B **6**, 291–296 (2016)
25. Hung, L.X., Ky, L.H., Hong, T.T., Dung, H.T., Lien, V.T., Tung, L.A., Long, B.T., Pi, V.N.: A study on cost optimization of internal cylindrical grinding. Int. J. Mech. Eng. Technol. (IJMET) **10**(1), 414–423 (2019)
26. Tran, T.-H., Le, X.-H., Nguyen, Q.-T., Le, H.-K., Hoang, T.-D., Luu, A.-T., Banh, T.-L., Ngoc-Pi, Vu: Optimization of exchanged grinding wheel diameter for minimum grinding cost in internal grinding. Appl. Sci. **9**(7), 1363 (2019). https://doi.org/10.3390/app9071363
27. Hung, L.X., Pi, V.N., Tung, L.A., Xuan, T.H., Jun, G., Long, B.T.: Determination of optimum exchanged grinding wheel diameter when internal grinding alloy tool steel 9CrSi. In: IOP Conference Series: Materials Science and Engineering, vol. 417 (2018). https://doi.org/10.1088/1757-899X/417/1/012026
28. Pi, V.N., Tung, L.A., Hung, L.X., Ngoc, N.V.: Experimental determination of optimum exchanged diameter in surface grinding process. J. Environ. Sci. Eng. A **6**, 85–89 (2017)

29. Hoang, T.D., Tran, T.H., Cuong, N.V., Le, H.K., Nga, N.T.T.: An optimization study on surface grinding stainless steel. Int. J. Eng. Technol. **7**(4), 6621–6625 (2018). https://doi.org/10.14419/ijet.v7i4.29442
30. Tran, T.-H., Luu, A.-T., Nguyen, Q.-T., Le, H.-K., Nguyen, A.-T., Hoang, T.-D., Le, X.-H., Banh, T.-L., Vu, N.-P.: Optimization of exchanged grinding wheel diameter for surface grinding tool steel based on the cost analysis. Metals **9**(4), 448 (2019). https://doi.org/10.3390/met9040448
31. Pi, V.N., Khiem, V.H., Cuong, P.T.: Building formulas for calculation of cutting regime for external cylindrical grinding. Vietnam Mech. Eng. J. (in Vietnamese), 12, 18–23 (2012)
32. Kozuro, L.M., Panov, A.A., Remizovski, E.I., Tristosepdov, P.S.: Handbook of Grinding. Publish Housing of High-education, Minsk (1981). (in Russian)

Optimization of Manufacturing Time in Surface Grinding

Luu Anh Tung[1], Tran Thi Hong[2], Nguyen Van Cuong[3], Le Hong Ky[4], T. Muthuramalingam[5], Do Duc Trung[6], Le Xuan Hung[1], and Ngoc Pi Vu[1(✉)]

[1] Thai Nguyen University of Technology, Thai Nguyen, Vietnam
vungocpi@tnut.edu.vn
[2] Nguyen Tat Thanh University, Ho Chi Minh City, Vietnam
[3] University of Transport and Communications, Ha Noi, Vietnam
[4] Vinh Long University of Technology Education, Vinh Long, Vietnam
[5] SRM Institute of Science and Technology, Kattankulathur, India
[6] Hanoi University of Industry, No. 298, Cau Dien Street, Bac Tu Liem district, Hanoi, Vietnam

Abstract. The objectives of this research are to determine the optimum values of grinding wheel parameters. An optimization problem with manufacturing time as the objective function was solved. The initial wheel diameter, the wheel life, the radial grinding wheel wear per dress and the total depth of dressing cut were considered as the input factors in the present study. Moreover, their effects on the optimum replaced wheel diameter were explored based on the implementation of a simulation experiment. The obtained results showed that the optimum replaced wheel diameter can be calculated by employing the proposed regression equation.

Keywords: Grinding · Surface grinding · Manufacturing time · Manufacturing time optimization · Replaced grinding wheel diameter

1 Introduction

Grinding is one of finished machining methods commonly used in mechanical processing. It determines the surface quality and the precision of machining parts [1, 2]. Therefore, a number of studies have focused on optimization problems of a grinding process to increase the grinding performance or to reduce the grinding cost etc. For example, various studies were carried out on optimal grinding process parameters [3–14]. To increase the grinding quality, a proposed methodology was introduced in [15]. In addition, the influence of the process parameters on the surface roughness and the material removal rate when external grinding SAE 8620 steel was reported in [16]. Studies were carried out with different types of grinding, such as abrasive belt grinding [3], grinding by CBN wheel on CNC milling machines [4], external grinding [9, 16], surface grinding [5, 7], or internal grinding [13, 14], etc. Besides, to optimize the dressing process, the influences of dressing parameters on the surface quality were investigated in external grinding [9] or internal grinding [11, 12]. Also, the effects of

K.-U. Sattler et al. (Eds.): ICERA 2019, LNNS 104, pp. 566–574, 2020.
https://doi.org/10.1007/978-3-030-37497-6_64

dressing parameters on the topography of different vitrified abrasive wheel surface were evaluated in [13]. Besides, the optimization problem to determine optimum lubrication conditions in internal grinding [14] or external grinding [17, 18] has been carried out.

Concerning the cost optimization in grinding processes, different studies have been done [19–28]. From the results of these studies, the models to calculate the optimum replaced grinding wheel diameter for different types of grinding processes were proposed for external grinding [19, 20], internal grinding [21–25] or surface grinding [26, 28]. However, the above discussions reveal that there is still a lack in research on manufacturing time optimization to determine the optimum replaced wheel diameter.

Thus, this paper introduces an optimization study to find optimum replaced grinding wheel diameter to minimize the manufacturing time in surface grinding stainless steel. In the study, the effects of several input grinding parameters including the initial grinding wheel diameter (D_0), the wheel life (T_w), the radial grinding wheel wear per dress $\left(W_{pd}\right)$ and the total depth of dressing cut (a_{ed}) on the optimum replaced grinding wheel diameter $\left(D_{e,op}\right)$ in the surface cylindrical grinding process were investigated by designing and conducting a simulation experiment. In addition, a regression model to determine the optimum replaced wheel diameter was proposed.

2 Methodology

2.1 Calculating Manufacturing Time

In surface grinding, the manufacturing time t_s can be determined by:

$$t_s = t_c + t_{lu} + t_{sp} + t_{d,p} + t_{cw,p} \tag{1}$$

In which, t_c, t_{lu}, t_{sp}, $t_{d,p}$ and $t_{cw,p}$ are the grinding time (h), the loading and unloading time (h), the spark-out time (h), the dressing time per part (h), and the grinding wheel replacing time per part, respectively.

For surface grinding, t_{sp}, $t_{d,p}$ and $t_{cw,p}$ can be found by the following equations:

$$t_{sp} = l_c w_c / \left(1000.v_w.v_f.N_t\right) \tag{2}$$

$$t_{d,p} = t_d / n_{p,d} \tag{3}$$

$$t_{cw,p} = t_{cw} / n_{p,w} \tag{4}$$

In the above equations, t_d is the dressing time; t_{cw} is the replacing grinding wheel time (h). $n_{p,d}$ is the number of parts per dress and $n_{p,w}$ is the total number of parts which are ground by a grinding wheel:

$$n_{p,d} = T_w / t_c \tag{5}$$

$$n_{p,w} = \frac{\left(d_{s,0} - d_{s,e}\right) n_{p,d}}{2\left(W_{pd} + a_{ed}\right)} \tag{6}$$

Where, T_w is the life of wheel (h), $d_{s,o}$ is the initial grinding wheel diameter (mm), $d_{s,e}$ is the replaced diameter of the grinding wheel (mm), W_{pd} is the radial grinding wheel wear per dress (mm/dress), and a_{ed} is the total dressing depth (mm). In surface grinding, the grinding time t_c is calculated by:

$$t_c = l_c.\mathrm{w}_c.a_{e,tot}/\left(1000.v_w.v_f.f_d.N_t\right) \tag{7}$$

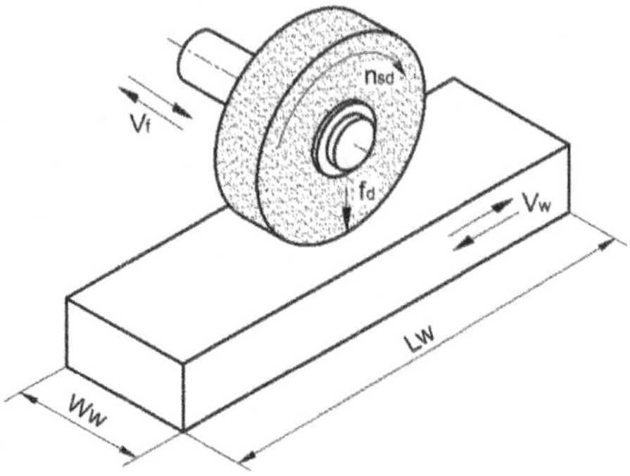

Fig. 1. Schematic representation of surface grinding

In which, l_c is the calculated grinding length (mm); $l_c = l_w + (20\ldots30)$, where l_w is the workpiece length (mm); w_c is the calculated grinding width (mm); $w_c = w_w + w_{gw} + 5$, where w_w is the workpiece width (mm) (Fig. 1) and w_{gw} is the width of grinding wheel (mm); $a_{e,tot}$ is the total depth of cut (mm); v_w is the work speed (m/s); v_f is the work feed rate (mm/min); f_d is the downfeed (mm/pass) (Fig. 1), and N_t is the number of parts which is ground at the same time; v_w, v_f and f_d are determined as follows:

With the workpiece material as carbon steel, the work speed v_w can be calculated by [29]:

$$v_\mathrm{w} = 0.0598.HRC^{1.4} \tag{8}$$

The work feed rate $\mathrm{v_f}$ is determined as [29]:

$$v_f = 46.\mathrm{w}_{\mathrm{gw}}^{0.983}/N_{Ra}^{2.44} \tag{9}$$

Where, $\mathrm{N_{Ra}}$ is the required roughness grade number; $\mathrm{w_{gw}}$ is the grinding wheel width (mm).

The downfeed $\mathrm{f_d}$ is found by the following equation [30]:

$$f_d = f_{d,t}.c_1.c_2.c_3 \tag{10}$$

In which, $f_{d,t}$ is the tabulated downfeed (mm/pass); c_1, c_2 and c_3 are the coefficients.

If the workpiece material is stainless steel, the tabulated downfeed $f_{d,t}$ and c_1 can be determined by [29]:

$$f_{d,t} = 0.649 \cdot a_{e,tot}^{0.651} \cdot v_f^{-0.985} \quad (11)$$

$$c_1 = 1.87 \cdot tg^{0.477} \quad (12)$$

The coefficient c_2 is calculated as [28]:

$$c_2 = 0.0292.d_s^{0.5151} / D_W^{0.4949} \quad (13)$$

Where, d_s is the grinding wheel diameter and D_W is the workpiece density on the machine table.

The coefficient c_3 depends on the grinding machine age; $c_3 = 1$ if the age is less than 10 years; $c_3 = 0.85$ if the age ranges from 10 to 20 years and $c_3 = 0.7$ if the age is more than 20 years [30].

2.2 Optimization Problem

It can be found from the above analyses that with the increase of the replaced grinding wheel diameter $d_{s,e}$, the manufacturing time decreases because the downfeed increases (see Eqs. (10), (13) and (1)). On the other hand, when the replaced grinding wheel diameter increases, the number of parts ground by a grinding wheel (Eq. (6)) is reduced. It causes the time to change the grinding wheel increase (Eq. (4)), and as a result, the manufacturing time increases (Eq. (1)).

Hence, the optimization problem to determine the optimum replaced grinding wheel diameter for getting the minimum manufacturing time is described as

$$Minimize\ t_s = \mathrm{f}(d_{s,r}) \quad (14)$$

$$\text{subject to } d_{s,r\,min} \leq d_{s,r} \leq d_{s,r\,max} \quad (15)$$

2.3 Experimental Work

To investigate the effects of grinding parameters on the optimum replaced wheel diameter, a simulation experiment was designed and conducted. In the experiment, four input parameters containing the initial grinding wheel diameter (D_0), the radial grinding wheel wear per dress (W_{pd}), the wheel life (T_w), and the total depth of dressing cut (a_{ed}) were selected (Table 1) and planned with a 2-level factorial design. Therefore, a number of $2^4 = 16$ test runs were performed by a computing program. Table 2 shows the experimental plan and the output results (the optimum replaced wheel diameters d_{op}).

Table 1. Input parameters

Factor	Code	Unit	Low	High
Initial grinding wheel diameter	D_0	mm	250	500
Total depth of dressing cut	a_{ed}	mm	0.1	0.2
Wheel life	T_w	min.	10	30
Radial grinding wheel wear per dress	W_{pd}	mm	0.01	0.03

3 Results and Discussion

To analyse the influences of the input factors on the response (the optimum replaced wheel diameter), the graph of main effect of each input factor is described in Fig. 2. It is reported from the graph that the optimum replaced wheel diameter $D_{s,rop}$ depends powerfully on the initial wheel diameter D_0. However, it does not depend on the radial grinding wheel wear per dress (W_{pd}) and is slightly influenced by the total depth of dressing cut a_{ed} and the wheel life T_w. This relationship can be seen more clearly in Pareto chart (Fig. 3).

Table 2. Experimental plans and the result values of d_{op}

StdOrder	RunOrder	CenterPt	Blocks	D_0	a_{ed}	T_w	W_{pd}	d_{op}
8	1	1	1	500	0.2	30	0.01	486.94
15	2	1	1	250	0.2	30	0.03	240.46
3	3	1	1	250	0.2	10	0.01	236.32
2	4	1	1	500	0.1	10	0.01	485.6
16	5	1	1	500	0.2	30	0.03	486.35
12	6	1	1	500	0.2	10	0.03	479.45
...	...	...	...	...	...	...	...	...
14	15	1	1	500	0.1	30	0.03	489.66
1	16	1	1	250	0.1	10	0.01	239.94

From the chart, among four input factors, only three bars representing factor A (the initial wheel diameter), factor B (the total depth of dressing cut) and factor C (the wheel life) cross the reference line. That means these factors are statistically significant at the 0.05 level for the model of optimum replaced wheel diameter.

Figure 4 presents the Normal Plot of the standardized effects. It has been found from the figure that the initial wheel diameter (factor A) is the most significant affecting factor on the optimum replaced diameter. Also, the initial wheel diameter (factor A), the wheel life (factor C) and the interactions AC and BC have positive standardized effects. When these factors increase, the optimum replaced diameter grows. Besides, the total depth of dressing cut (factor B), the radial grinding wheel wear per dress (factor D) and the interaction AB have negative standardized effects. That means when they rise, the optimum diameter reduces.

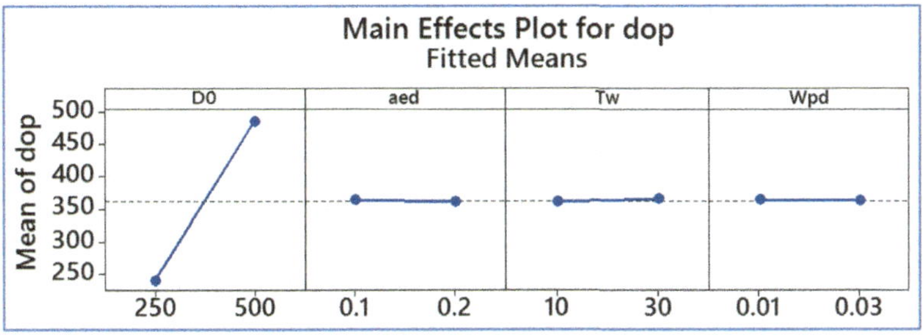

Fig. 2. Main effects plot on d_{op}

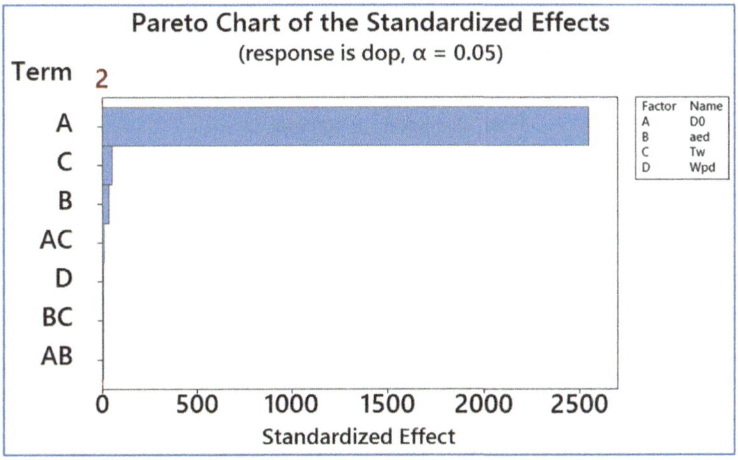

Fig. 3. Pareto chart of the standardized effects on d_{op}

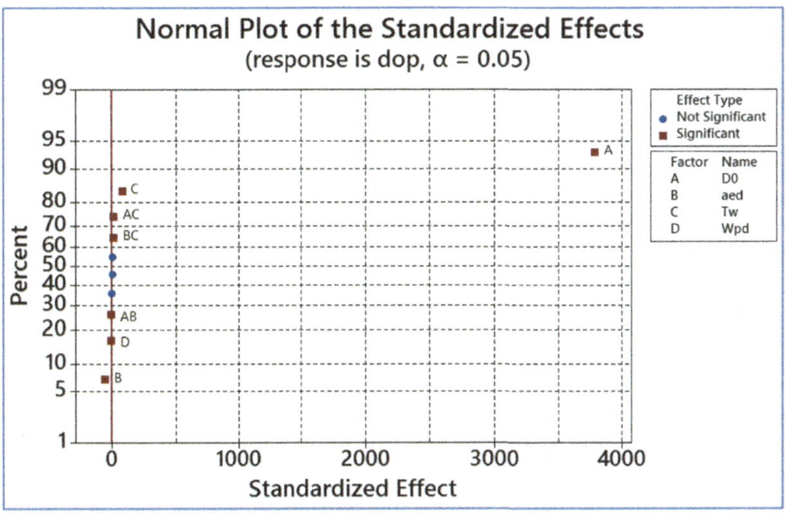

Fig. 4. Normal plot for the optimum diameter d_{op}

Table 3. Estimated effects and coefficients for d_{op}

Coded Coefficients

Term	Effect	Coef	SE Coef	T-Value	P-Value	VIF
Constant		362.607	0.048	7520.12	0.000	
D0	245.586	122.793	0.048	2546.61	0.000	1.00
aed	-3.6013	-1.8006	0.0482	-37.34	0.000	1.00
Tw	4.9963	2.4981	0.0482	51.81	0.000	1.00
Wpd	-0.7362	-0.3681	0.0482	-7.63	0.000	1.00
D0*aed	-0.6638	-0.3319	0.0482	-6.88	0.000	1.00
D0*Tw	0.9188	0.4594	0.0482	9.53	0.000	1.00
aed*Tw	0.7013	0.3506	0.0482	7.27	0.000	1.00

Model Summary

S	R-sq	R-sq(adj)	R-sq(pred)
0.192873	100.00%	100.00%	100.00%

For finding the significant effects on the optimum replaced wheel diameter, the insignificant effects were disregarded. Therefore, it is noted that the initial wheel diameter D_0, the total depth of dressing cut a_{ed}, the wheel life T_w, the radial grinding wheel wear per dress $\left(W_{pd}\right)$ and the interactions $D_0 * a_{ed}$ $D_0 * T_w$ and $a_{ed} * T_w$ are the significant affecting factors on the response (with the P-values lower than 0.05) (Table 3). As a result, the regression equation to determine the optimum diameter d_{op} can be described as:

$$
\begin{aligned}
d_{op} = {} & -2.757 + 0.98296 \cdot D_0 - 30.12 \cdot a_{ed} + 0.0068 \cdot T_w - 36.81 \cdot W_{pd} - \\
& - 0.0531 \cdot D_0 \cdot a_{ed} + 0.000368 \cdot D_0 \cdot T_w + 0.7013 \cdot a_{ed} \cdot T_w
\end{aligned} \tag{16}
$$

It is found from Fig. 4 that Eq. (16) fits very well with the data of the experiment. This is because all of the values of adj-R2 and pred-R2 are 100% (Table 3). Thus, this equation can be used to find the optimum replaced wheel diameter when surface grinding stainless steel.

4 Conclusions

In this paper an optimization study to determine the replaced grinding wheel diameter when surface grinding of stainless steel was presented. In the study, the minimum manufacturing time was selected as the objective of the optimization problem. Also, the influences of several process parameters on the optimum replaced wheel diameter were investigated by designing and performing a simulation experiment. Moreover, a regression equation to calculate the optimum replaced diameter was proposed. As the equation is explicit, the optimum replaced wheel diameter when surface grinding stainless steel can be determined easily and conveniently.

Acknowledgments. The work described in this paper was supported by Thai Nguyen University of Technology for a scientific project.

References

1. Rowe, W.B.: Principle of Modern Grinding Technology. William Andrew, New York (2009)
2. Malkin, S., Guo, C.: Grinding Technology: Theory and Applications of Machining with Abrasives. Industrial Press, New York (2008)
3. Pandiyan, V., Caesarendra, W., Tjahjowidodo, T., Praveen, G.: Predictive modelling and analysis of process parameters on material removal characteristics in abrasive belt grinding process. Appl. Sci. **7**(4), 363 (2017). https://doi.org/10.3390/app7040363
4. Vu, N.-P., Nguyen, Q.-T., Tran, T.-H., Le, H.-K., Nguyen, A.-T., Luu, A.-T., Nguyen, V.-T., Le, X.-H.: Optimization of grinding parameters for minimum grinding time when grinding tablet punches by CBN wheel on CNC milling machine. Appl. Sci. **9**(5), 957 (2019)
5. Luu, A.T., Vu, N.P., Do, T.T.H., Le, X.H., Banh, T.L.: A study on optimization of surface roughness in surface grinding 9CrSi tool steel by using Taguchi method. In: Fujita, H., et al. (eds.) ICERA 2018, LNNS 63, pp. 100–108 (2019). https://doi.org/10.1007/978-3-030-04792-4_15
6. Gupta, R., Shishodia, K.S., Sekhon, G.S.: Optimization of grinding parameters using enumeration method. J. Mater. Process. Technol. **112**, 63–67 (2001)
7. Wen, X.M., Tay, A.A.O., Nee, A.Y.C.: Micro-computer-based optimization of the surface grinding process. J. Mater. Process. Technol. **29**(1–3), 75–90 (1992)
8. Mekala, K., Chandradas, J., Chandrasekaran, K., Kannan, T.T.M., Ramesh, E., Babu, R.N.: Optimization of cylindrical grinding parameters of austenitic stainless steel rods (AISI 316) by Taguchi method. Int. J. Mech. Eng. Robot. Res. **3**(2), 208 (2014)
9. Tu, H.X., Thao, L.P., Hong, T.T., Nga, N.T.T., Trung, D.D., Gong, J., Pi, V.N.: Influence of dressing parameters on surface roughness of workpiece for grinding hardened 9XC tool steel. In: IOP Conference Series: Materials Science and Engineering, vol. 542, no. 1. https://doi.org/10.1088/1757-899X/542/1/012008
10. Vidal, G., Ortega, N., Bravo, H., Dubar, M., González, H.: An analysis of electroplated cBN grinding wheel wear and conditioning during creep feed grinding of aeronautical alloys. Metals **8**(5), 1–24 (2018)
11. Daneshia, A., Jandaghia, N., Tawakoli, T.: Effect of dressing on internal cylindrical grinding. Procedia CIRP **14**, 37–41 (2014)
12. Le, X.H., Tran, T.H., Luu, A.T., Nguyen, T.T.N., Vu, N.P.: Optimum dressing parameters for maximum material removal rate when internal cylindrical grinding using Taguchi method. Int. J. Mech. Eng. Technol. **9**, 123–129 (2018)
13. Palmer, A., Ghadbeigi, H., Novovic, D., Curtis, D.: An experimental study of the effects of dressing parameters on the topography of grinding wheels during roller dressing. J. Manuf. Process. **31**, 348–355 (2018)
14. Hung, L.X., Liên, V.T., Pi, V.N., Long, B.T.: A study on coolant parameters in internal grinding of 9CrSi steel. Mater. Sci. Forum **950**, 24–31 (2019)
15. Kumar, J.P., Indhirajith, B., Thiruppathi, K.: Quality improvement of a grinding operation using process capability studies. Int. J. Mech. Eng. Robot. Res. **1**(1), 76–90 (2012)

16. Yadav, H.S., Shrivastava, R.K.: Effect of process parameters on surface roughness and MRR in cylindrical grinding using response surface method. Int. J. Eng. Res. Technol. **3**(3) (2014)
17. Tu, H.X., Pi, V.N., Jun, G.: A study on determination of optimum parameters for lubrication in external cylindrical grinding base on Taguchi method. Key Eng. Mater. **796**, 97–102 (2019)
18. Tu, H.X., Jun, G., Hien, B.T., Hung, L.X., Tung, L.A., Pi, V.N.: Determining optimum parameters of cutting fluid in external grinding of 9CrSi steel using Taguchi technique. SSRG Int. J. Mech. Eng. **5**(6), 1–5 (2018). https://doi.org/10.14445/23488360/IJME-V5I6P101
19. Pi, V.N., The, P.Q., Khiem, V.H., Huong, N.N.: Cost optimization of external cylindrical grinding. Appl. Mech. Mater. **312**, 982–989 (2013)
20. Tu, H.X., Jun, G., Hung, L.X., Tung, L.A., Pi, V.N.: Calculation of optimum exchanged grinding wheel diameter when external grinding tool steel 9CrSi. Int. J. Mech. Eng. Robot. Res. **8**(1), 59–64 (2019)
21. Hung, L.X., Pi, V.N., Tung, L.A., Tu, H.X., Jun, G., Long, B.T.: Determination of optimal exchanged grinding wheel diameter when internally grinding alloy tool steel 9CrSi. In: IOP Conference Series: Materials Science and Engineering, vol. 417, pp. 012–026 (2018)
22. Pi, V.N., Hung, L.X., Tung, L.A., Long, B.T.: Cost optimization of internal grinding. J. Mater. Sci. Eng. B **6**, 291–296 (2016)
23. Hung, L.X., Ky, L.H., Hong, T.T., Dung, H.T., Lien, V.T., Tung, L.A., Long, B.T., Pi, V.N.: A study on cost optimization of internal cylindrical grinding. Int. J. Mech. Eng. Technol. (IJMET) **10**(1), 414–423 (2019)
24. Tran, T.-H., Le, X.-H., Nguyen, Q.-T., Le, H.-K., Hoang, T.-D., Luu, A.-T., Banh, T.-L., Ngoc-Pi, V.: Optimization of replaced grinding wheel diameter for minimum grinding cost in internal grinding. Appl. Sci. **9**(7), 1363 (2019). https://doi.org/10.3390/app9071363
25. Le, X.H., Vu, N.P., Luu, A.T., Tu, H.X., Jun, G., Banh, T.L.: Determination of optimum exchanged grinding wheel diameter when internal grinding alloy tool steel 9CrSi. In: IOP Conference Series: Materials Science and Engineering, vol. 417, no. 1 (2018). https://doi.org/10.1088/1757-899X/417/1/012026
26. Pi, V.N., Tung, L.A., Hung, L.X., Ngoc, N.V.: Experimental determination of optimum exchanged diameter in surface grinding process. J. Environ. Sci. Eng. **6**, 85–89 (2017)
27. Hoang, T.D., Tran, T.H., Cuong, N.V., Le, H.K., Nga, N.T.T.: An optimization study on surface grinding stainless steel. Int. J. Eng. Technol. **7**(4), 6621–6625 (2018). https://doi.org/10.14419/ijet.v7i4.29442
28. Tran, T.-H., Luu, A.-T., Nguyen, Q.-T., Le, H.-K., Nguyen, A.-T., Hoang, T.-D., Le, X.-H., Banh, T.-L., Ngoc-Pi, V.: Optimization of replaced grinding wheel diameter for surface grinding based on a cost analysis. Metals **9**(4), 448 (2019). https://doi.org/10.3390/met9040448
29. Pi, V.N., Khiem, V.H., Cuong, P.T.: Building formulas for calculation of cutting regime for external cylindrical grinding. Vietnam Mech. Eng. J. (in Vietnamese) (12), 18–23 (2012)
30. Kozuro, L.M., Panov, A.A., Remizovski, E.I., Tristosepdov, P.S.: Handbook of Grinding. Publish Housing of High-education, Minsk (1981). (in Russian)

Optimization of the Infrared Assisted Heat Pump Drying Operation of White Leg Shrimp Using Particle Swarm Optimization

Thi Thom Hoang[1], Thi-Nguyen Nguyen[2(✉)], and Nhu Chinh Le[1]

[1] Nha Trang University, Nha Trang, Vietnam
[2] Thai Nguyen University of Technology, Thai Nguyen, Vietnam
thaonguyen491988@tnut.edu.vn

Abstract. In this paper, a meta-heuristic, namely particle swarm optimization (PSO) is investigated to optimize the infrared (IR)-assisted heat pump drying operation of White leg shrimp in order to ensure relatively shorter drying time with satisfactory product quality. The work aims to evaluate the multi-objective function based on the influence of technological factors, such as final product temperature, rehydration ratio, and final moisture content. The proposed method is implemented under the optimal drying conditions of (10–25) mm sample thickness, (30–60) mm sample distance from the IR heater, (55–65) °C IR temperature, and (IR)-assisted heat pump drying time of (5.5–6.5) h. The experimental results show the completed product with the following minimum values: the final product temperature of 12.24 °C, the rehydration ratio of 2.33%, and the final moisture content of 1.51%, demonstrate the effectiveness and high efficiency of the developed method.

Keywords: Particle swarm optimization · Multi-objective optimization · (IR)-assisted heat pump drying · White leg shrimp

1 Introduction

Due to low fat and calorie content, White leg shrimps are one of very healthy choices of food. They provide not only a high level of vitamin B_{12}, zinc, iodine, phosphorus, potassium, selenium, iron; but also a trace of calcium, magnesium, and sodium [1]. Thus, it is very important to store White leg shrimp by drying processes in order to increase their shelf life. Normally, shrimp drying is mostly composed of boiling shrimp with salt water and then they are dried using sun drying [1]. However, this technique has some drawbacks, such as weather uncertainties, mixing with dust and other materials. The superheated steam drying is introduced to control parameters of drying process better [2]. In paper [3], the vacuum freeze-drying of shrimp using a plate heater was employed, but it also consumes a long time to reach the desired product quality. The authors in paper [4] proposed the heat pump drying method, which makes the shorter drying time with higher product quality. The main disadvantage of this method is based on the structure of the foodstuffs and the intensity of heat pump.

K. U. Sattler et al. (Eds.): ICERA 2019, LNNS 104, pp. 575–581, 2020.
https://doi.org/10.1007/978-3-030-37497-6_65

In this paper, a novel method based upon PSO technique is developed to simultaneously optimize three parameters of drying process, including: the final product temperature, the rehydration ratio, and the final moisture content. The work is employed under the optimal parametric conditions, such as sample thickness, distance from the IR heater, IR temperature, and (IR)-assisted heat pump-drying time. The proposed method not only enhances drying efficiency (shorter drying time and better product quality), but also decreases the operation cost.

2 Problem Description

The determination of the (IR)-assisted heat pump drying mode is shown as Fig. 1, in which X_1, X_2, X_3 and X_4 are IR heater temperature, the distance between sample and IR heater, (IR)-assisted heat pump - drying time, and sample thickness, respectively. Meanwhile, p_1, p_2, p_3 are the outputs, including final product temperature, rehydration ratio, and final moisture content of the (IR)-assisted heat pump-dried products.

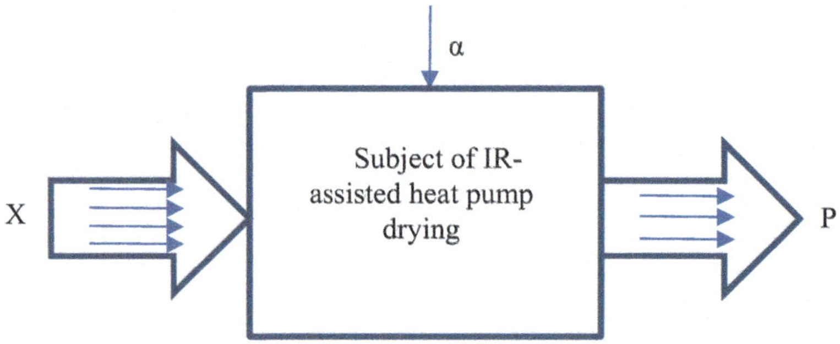

Fig. 1. Diagram of (IR)-assisted heat pump drying process

It can be clearly seen from Fig. 1, to simultaneously determine the minimal value of all these output factors, it is necessary to solve the multi-objective optimization problem. In other words, we should find the minimum value for the set of m one-objective optimization problems:

$$f_{jmin} = f_j(X_1^j, X_2^j, \ldots, X_n^j) = Min f_j(X_1, X_2, \ldots, X_n) \tag{1}$$

$$X = \{X_i\} = (X_1, X_2, \ldots, X_n) \in \Omega_X \tag{2}$$

$$j = 1 \div m; \; i = 1 \div n \tag{3}$$

In this paper, the PSO technique is applied for the multi-objective optimization for the (IR)-assisted heat pump drying process of White leg shrimp. Integrated with the heat and mass transfer model of the (IR)-assisted drying operation, the obtained results are used to design the optimal dryer of White leg shrimp.

3 Application of MOPSO for (IR)-Assisted Heat Pump Drying Process of White Leg Shrimps

3.1 PSO

PSO is a population-based optimization approach, which was firstly introduced by Kennedy and Eberhard in 1995 [5]. It is inspired by social and cooperative behavior displayed by various species to fill their needs in the search space. The algorithm is guided by personal experience (Pbest), overall experience (Gbest) and the present movement of the particles to decide their next positions in the search space. Mathematically, updated positions of each particle in the search space can be expressed using the two following equations:

$$V_{p,q}^{k+1} = w \times V_{p,q}^{k} + c_1 r_1 (Pbest_{p,q}^{k} - X_{p,q}^{k}) + c_2 r_2 (Gbest_{q}^{k} - X_{p,q}^{k}) \tag{4}$$

$$X_{p,q}^{k+1} = X_{p,q}^{k} + V_{p,q}^{k+1} \tag{5}$$

In Eq. (4), $\mathrm{Pbest}_{p,q}^{k}$ represents personal best q^{th} component of pth individual, whereas Gbest_{q}^{k} represents q^{th} component of the best individual of the population up to iteration k. Figure 2 shows the search mechanism of PSO in a multidimensional search space. More details about the basic concept of PSO can be found in [6]

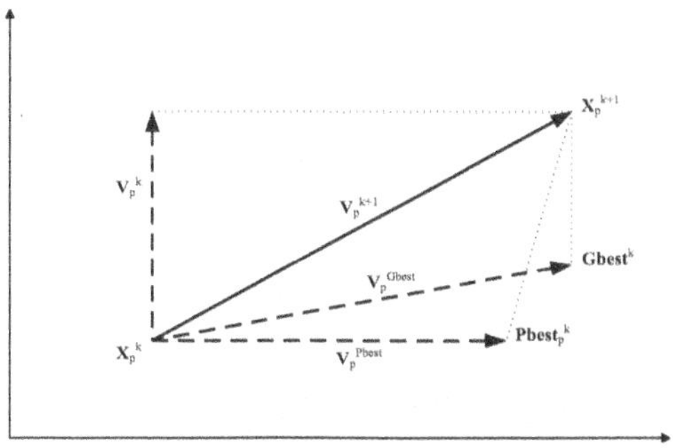

Fig. 2. The PSO search mechanism p^{th} particle at k^{th} iteration

3.2 Application of MOPSO for (IR)-Assisted Heat Pump Drying Process of White Leg Shrimps

For multi-objective problems, the objective is to find not one "global best" solution. For this, an archive of non-dominated solutions is kept, where all non-dominated solutions found at each iteration are stored. The MOPSO steps are applied for (IR)-assisted heat pump drying operation as follows:

Step 1: Determine distance and sample thickness. Measure the temperature by sensors in the machine.

Step 2: Design a general second-order polynomial response surface model using Response Surface Methodology [7]:

$$P_k = \beta_{ko} + \sum_{i=1}^{3} \beta_{ki} X_i + \sum_{i=1}^{3} \beta_{kii} X_i^2 + \sum_{i=1}^{3} \sum_{j=1}^{3} \beta_{kii} X_i X_j + \alpha \qquad (6)$$

Step 3: Initialize the swarm and archive

Step 4: For each particle in the swarm:

Select leader from the archive
Update velocity and position using Eqs. (4) and (5)

Step 5: Update the archive of non-dominated solutions

Step 6: Repeat

4 Results and Discussions

In this paper, the proposed MOPSO algorithm is performed on the White leg shrimp materials whose weight is (40÷50) prawns/pound and their size coefficient is 10. These materials are blanched at 60 °C in 30 s, and then are processed minimally by peeling off the shells and cutting off the heads. The (IR)-assisted heat pump drying is implemented by a machine at Nha Trang University as shown in Fig. 3.

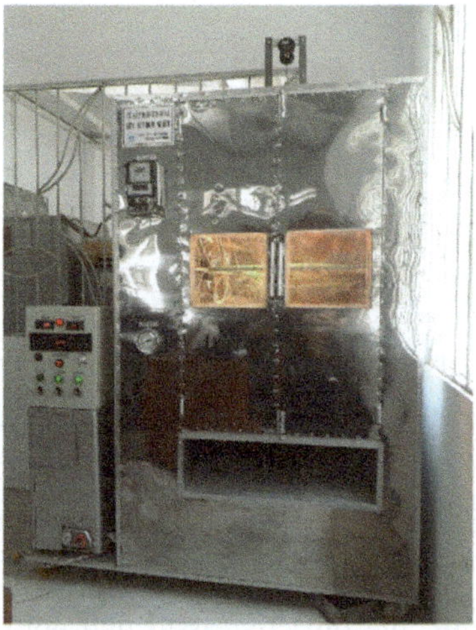

Fig. 3. The (IR)-assisted heat pump drying system at Nha Trang University

Normally, the set point temperature of the IR radiator (T_p) is maintained within the range of 35–65 °C with the distance range (D) of 30–60 mm. The tested kind of White leg shrimp with the thickness range of 3-10 mm during the drying time range (t) of

5.5–6.5 h. However, to demonstrate the efficiency of the proposed method, the drying process was implemented under the following conditions: sample thickness of 10 mm, distance from the IR heater of 60 mm, IR temperature of 65 °C, and (IR)-assisted heat pump-drying time of 6.37 h. Meanwhile, (p_{1min}, p_{2min}, and p_{3min}) are the minimum values the final product temperature of 12.24 °C, the rehydration ratio of 2.33%, and the final moisture content of 1.51%, respectively. These targets of the (IR)-assisted heat pump drying process are obtained by using the MOPSO as shown in Table 1. Furthermore, it can be clearly seen from Table 1 that these results are compared to those obtained by the Infrared-Assisted Freeze Drying method to demonstrate the effectiveness of the proposed method.

It is observed that the targets of the (IR)-assisted heat pump drying process in case of using the MOPSO method is better as compared to the IR-Assisted Freeze Drying technique.

Table 1. The optimal results of the (IR)-assisted heat pump drying process using different techniques

Target	p_{1min} (kWh)	p_{2min} (%)	p_{3min} (%)
IR-freeze drying	13.34	2.78	4.08
IR-Heat pump drying + MOPSO	12.24	2.33	1.51

Table 2 shows the experimental results of the (IR)-assisted heat pump drying process of White leg shrimp at the optimal drying test above: sample thickness of 10-mm, distance from the IR heater of 60-mm, IR temperature of 65 °C, and (IR)-assisted heat pump-drying time of 6.37 h.

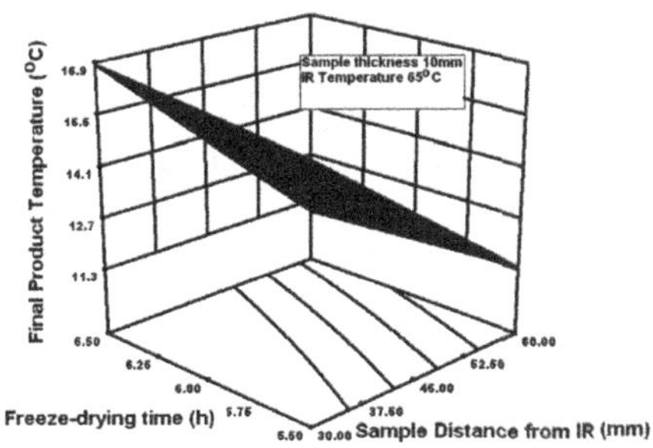

Fig. 4. Response surface plot of final product temperature as a function of freeze-drying time and sample to IR distance.

Table 2. The experimental results of the (IR)-assisted heat pump drying process

Target	p_{1min} (kWh)	p_{2min} (%)	p_{3min} (%)
Heat pump drying + MOPSO	12.24	2.33	1.51
Experimental results	12.52	2.41	1.58

From Table 2 it can be observed that the obtained results by using the proposed MOPSO algorithm for the (IR)-assisted heat pump drying process are suitable with the experimental results. Furthermore, Fig. 4 reveals the effects of sample distance from the IR heater and freeze-drying time on the final product temperature T_p. It can be observed from Fig. 5 that at the lower value of D (30 mm), with an increase in t from 5.5 to 6.5 h, T_p was found to reach around 16.9 °C. Figure 5 shows a picture of White leg shrimp obtained from the experimental (IR)-assisted heat pump drying process.

Fig. 5. The (IR)-assisted heat pump dried White leg shrimp of experimental process

5 Conclusion

In this paper, a novel MOPSO is investigated to solve the three-objective optimization problem (p_1, p_2, p_3) determined with four optimal technological parameters: sample thickness, distance from the IR heater, IR temperature, and (IR)-assisted heat pump - drying time. The obtained results show the proposed MOPSO for the (IR)-assisted heat pump drying operation of White leg shrimp is capable of enhancing product quality as well as decreasing drying time and operating cost. Furthermore, a comparison between the MOPSO and IR-Assisted Freeze Drying technique is employed to demonstrate the effectiveness and high efficiency of the developed method.

References

1. Jain, D., Pathare, P.B.: Study the drying kinetics of open sun drying of fish. J. Food Eng. **78**, 1315–1319 (2007)
2. Mujumdar, A.S.: Handbook of Industrial Drying. Marcel-Dekker Inc., New York (1987)
3. Mitra, P., Meda, V.: Optimization of microwave-vacuum drying parameters of Saskatoon berries using response surface methodology. Drying Technol. **27**(10), 1089–1096 (2009)
4. Donsi, G., Ferrari, G., Matteo, D.I.: Utilization of combined processes in (IR)-assisted heat pump-drying of shrimps. Food Bioprod. Process. **79**(3), 152–159 (2001)
5. Cho, M.Y., Hoang, T.T.: Feature selection and parameters optimization of SVM using particle swarm optimization for fault classification in power distribution systems. Comput. Intell. Neurosci. (2017)

6. Kennedy, J., Eberhart, R., Shi, Y.: Swarm Intelligence. Morgan Kaufmann Publishers Inc., San Francisco (2001)
7. Chakraborty, R., Mukhopadhyay, P., Bera, M., Suman, S.: Infrared-assisted freeze drying of tiger prawn: parameter optimization and quality assessment. Dry. Technol. **29**(5), 508–519 (2013)

Parameter Optimization of Pseudo-Rigid-Body 3R Model for Modeling Large Deflection of Compliant Mechanisms

Thi Thanh Nga Nguyen[1(✉)], David Schoenen[2], Mathias Hüsing[2], and Burkhard Corves[2]

[1] Faculty of Mechanical Engineering, Thai Nguyen University of Technology, Thai Nguyen, Vietnam
nguyennga@tnut.edu.vn
[2] Institute of Mechanism Theory, Machine Dynamics and Robotics, RWTH Aachen University, Aachen, Germany
{schoenen, huesing, corves}@igmr.rwth-aachen.de

Abstract. Pseudo-Rigid-Body (PRB) models provide a simple method for simulating large deflection of flexible links for compliant mechanisms. The PRB model is useful to analyze and synthesize compliant mechanisms. Therefore, increasing the accuracy of the PRB model for modeling flexible links is a necessary problem. This paper presents an optimization of parameters of PRB 3R model for modeling large deflection of flexible beams in order to increase the accuracy and working range of the model. The objective function is established based on the difference of the tip locus between flexible beams and the PRB model. The results demonstrate that the PRB 3R model obtains a high accuracy and a large working range.

Keywords: Compliant mechanisms · Flexible beams · Pseudo-Rigid-Body model

1 Introduction

Compliant mechanisms are commonly used to transform motion, force, and energy. They have many advantages compared to traditional mechanisms, such as reduced parts and assembly time, less friction and wear, improved precision, and easy miniaturization. However, analyzing compliant mechanisms is complicated since flexible links may have large deflection and depend on material properties and geometry [1].

The Pseudo-Rigid-Body (PRB) model is thus used for modeling flexible links in order to analyze, design and simulate compliant mechanisms. The research proposed the PRB model for flexible beam subjected to arbitrary end loads with an inflection point in the beam [2]. The application of the PRB model for large deflection of the flexible beam with an inflection point was presented in [3, 4]. The PRB 1R model with two rigid links and a torsion spring for modeling large deflection of flexible beams was proposed [5, 6]. For further researches of this model, improving parameters for this model, i.e., the characteristic radius parameter and the spring stiffness was presented in

K.-U. Sattler et al. (Eds.): ICERA 2019, LNNS 104, pp. 582–588, 2020.
https://doi.org/10.1007/978-3-030-37497-6_66

[7, 8]. The PRB 3R model with four rigid links and three torsion springs for a Steerable Ablation Catheter was also shown in [9]. In this research, however, the parameters of the model have not yet been carefully evaluated in specific cases i.e., a pure moment and pure end force. Therefore, this paper presents an optimization of parameters by minimizing the error of the tip locus between the flexible beam and the model in order to improve the accuracy of the tip locus and increase the working range of the model.

2 Pseudo-Rigid-Body 3R Model for Large Deflection

This section presents the formulation for large deflection of a flexible beam suspected to an end force and an end moment. The PRB 3R model is then proposed for modeling flexible beams.

2.1 Formulation of Large Deflection Beam

Considering large deflection of a flexible beam under an end force F and end moment M, the deflection at the tip point of the beam is defined by the coordinates a, b and the slope angle θ_0 as shown in Fig. 1a. The Bernoulli-Euler beam equation can be expressed as [10, 11].

$$M = EI\frac{d\theta}{ds} = EI\frac{d^2y/dx^2}{\left(1 + (dy/dx^2)\right)^{3/2}} \tag{1}$$

The normalized tip coordinates can be determined from Eq. (1) as follows

$$\begin{aligned} \alpha &= \frac{1}{\sqrt{2}}\int_0^{\theta_0} \frac{d\theta}{\sqrt{[\cos(\theta + \phi) - \cos(\theta_0 + \phi)] + \mathrm{k}}} \\ \frac{a}{l} &= \frac{1}{\sqrt{2}\alpha}\int_0^{\theta_0} \frac{cos\theta d\theta}{\sqrt{[\cos(\theta + \phi) - \cos(\theta_0 + \phi)] + \mathrm{k}}} \\ \frac{b}{l} &= \frac{1}{\sqrt{2}\alpha}\int_0^{\theta_0} \frac{sin\theta d\theta}{\sqrt{[\cos(\theta + \phi) - \cos(\theta_0 + \phi)] + \mathrm{k}}} \end{aligned} \tag{2}$$

where,

$$\alpha^2 = Fl^2/EI; \quad \beta = M_0 l/EI; \quad \mathrm{k} = \beta^2/2\alpha^2. \tag{3}$$

Here, α^2 and β are characterized for forces and moments acting on the beam end, and k is the load ratio.

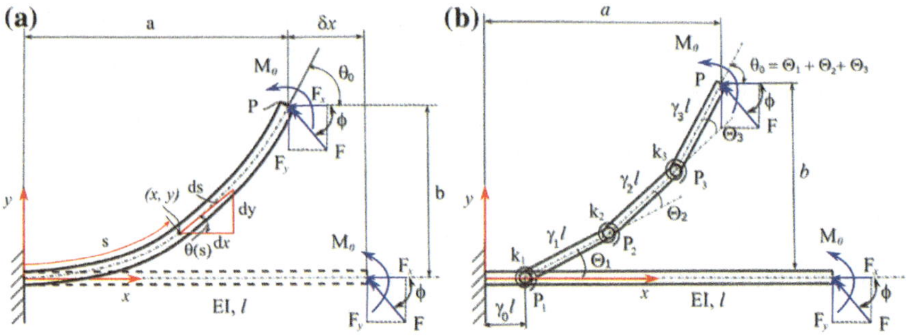

Fig. 1. (a) Cantilever beam subjected to an end load and end moment; (b) Pseudo-Rigid-Body 3R model for large deflection of a cantilever beam

2.2 Pseudo-Rigid-Body 3R Model for Large Deflection

The PRB 3R model is used for modeling large deflection of a flexible link within compliant mechanisms as found in [9]. The model contains four PRB links with the dimensions $\gamma_0 l, \gamma_1 l, \gamma_2 l$, and $\gamma_3 l$, respectively. These links are connected by three pin joints and torsion springs (see Fig. 1b). For given PRB angles $\Theta_1, \Theta_2, \Theta_3$, the tip locus of the PRB 3R model can be easily computed by forward kinematics as

$$\begin{aligned} P_x &= [\gamma_0 + \gamma_1 \cos\Theta_1 + \gamma_2 \cos(\Theta_1 + \Theta_2) + \gamma_3 \cos(\Theta_1 + \Theta_2 + \Theta_3)].l \\ P_y &= [\gamma_1 \sin\Theta_1 + \gamma_2 \sin(\Theta_1 + \Theta_2) + \gamma_3 \sin(\Theta_1 + \Theta_2 + \Theta_3)].l \end{aligned} \tag{4}$$

where, $\gamma_0, \gamma_1, \gamma_2$, and γ_3 are characteristic radius parameters of the PRB model and they fulfill the condition $\gamma_0 + \gamma_1 + \gamma_2 + \gamma_3 = 1$.

In order to resist the deflection of the flexible link, the non-dimensional torsion spring constants called the stiffness coefficients $k_{\Theta_i} (i = 1, 2, 3)$ are characterized to compute the torsion springs. For given external forces (F, M_0, ϕ), the procedure for calculating the stiffness coefficients is based on the torques at the three pin joints and the torques of the springs as found in [6]. From this, the normalized and non-dimensional equation can be expressed as

$$\begin{bmatrix} k_{\Theta_1}\Theta_1 \\ k_{\Theta_2}\Theta_2 \\ k_{\Theta_3}\Theta_3 \end{bmatrix} = [J] \begin{bmatrix} \alpha^2\cos\phi \\ \alpha^2\sin\phi \\ \beta \end{bmatrix}. \tag{5}$$

Here, matrix J can be written as

$$[J] = \begin{bmatrix} \gamma_1 s_1 + \gamma_2 s_{12} + \gamma_3 s_{123} & \gamma_1 c_1 + \gamma_2 c_{12} + \gamma_3 c_{123} & 1 \\ \gamma_2 s_{12} + \gamma_3 s_{123} & \gamma_2 c_{12} + \gamma_3 c_{123} & 1 \\ \gamma_3 s_{123} & \gamma_3 c_{123} & 1 \end{bmatrix} \tag{6}$$

With

$$\begin{aligned} s_1 &= \sin\Theta_1;\ s_{12} = \sin(\Theta_1 + \Theta_2);\ s_{123} = \sin(\Theta_1 + \Theta_2 + \Theta_3); \\ c_1 &= \cos\Theta_1;\ c_{12} = \cos(\Theta_1 + \Theta_2);\ c_{123} = \cos(\Theta_1 + \Theta_2 + \Theta_3) \end{aligned} \tag{7}$$

From Eq. (5), the stiffness coefficients k_{Θ_i} are determined.

3 Determination of Characteristic Radius Parameters

In order to determine the value of the parameters, $\gamma_i (i = 0, 1, 2, 3)$, the difference of the tip point between the beam and the model must be minimized. Figure 2 shows the error of the tip locus of the model compared to the beam.

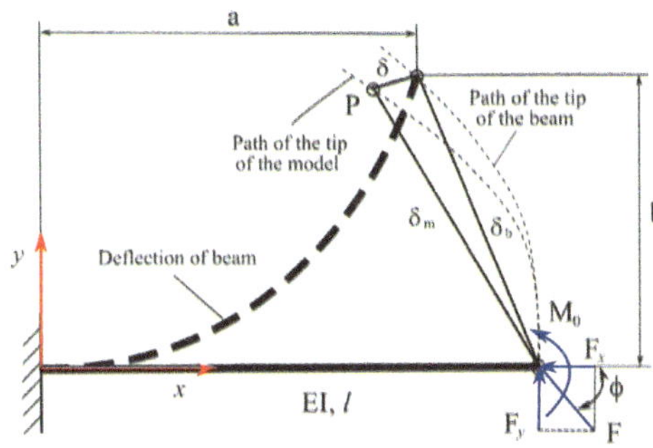

Fig. 2. The difference of the tip between the beam and the model

Therefore, the objective function depending on the parameters γ_i can be written as

$$\min f = f(\gamma_i), \quad for\, i = 0, 1, 2, 3 \tag{8}$$

subject to $0.05 < \gamma_i < 0.5$ and $\gamma_0 + \gamma_1 + \gamma_2 + \gamma_3 = 1$.

The objective function f is calculated from the error

$$f = \delta / \delta_b. \tag{9}$$

Here, the difference between two related points that are one on the path of the beam and the other on the path of the model δ can be calculated by

$$\delta = \sqrt{(P_x - a)^2 + (P_y - b)^2}, \tag{10}$$

where P_x and P_y are shown in Eq. (4). They depend on values of the characteristic radius parameters γ_i. As shown in Eq. (2), a and b are coordinates of the tip locus of the beam deflection under acting external loads.

In Eq. (9), δ_b can also be computed as

$$\delta_b = \sqrt{(l-a)^2 + (b)^2}. \tag{11}$$

The optimum characteristic parameters $\gamma_i (i = 0, 1, 2, 3)$ are obtained by minimizing the error function f in Eq. (9).

4 Result and Discussion

This section presents two main examples: modeling the large deflection of compliant links with a pure moment and a pure vertical force. These examples are compared with the PRB 1R model.

Pure Moment. The optimal characteristic radius parameters γ_i are determined by minimizing the objective function in Eq. (9), and the stiffness coefficients are then computed from Eq. (5) as follows: $\gamma_0 = 0.1, \gamma_1 = 0.4, \gamma_2 = 0.38, \gamma_3 = 0.12$ and $k_{\Theta_1} = 3.3333$, $k_{\Theta_2} = 2.4881$, $k_{\Theta_3} = 3.3548$.

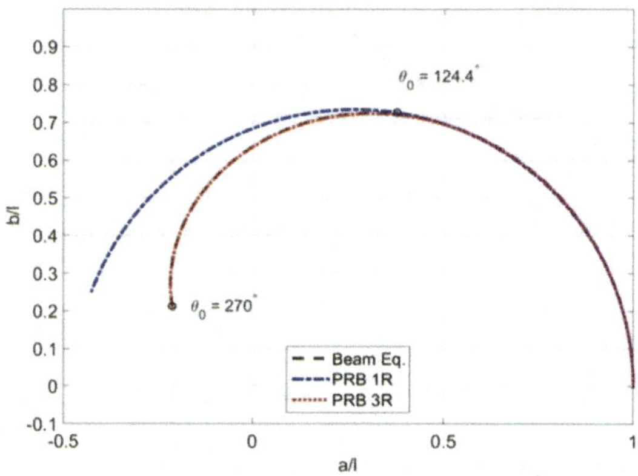

Fig. 3. Comparison of the tip locus between the beam and PRBM in case of pure moment

For the PRB 1R model, the parameters are $\gamma_0 = 0.7346$ and $k_{\Theta_1} = 1.5164$ [6]. Figure 3 shows the normalized tip locus of the flexible beam, the PRB 1R, and the PRB 3R model with the end moment load. It can be observed from Fig. 3 that the tip locus of the PRB 3R model (red line) is very close to the tip locus of the flexible beam (black line). The PRB 1R (blue line) can be accurately modeled in the range $\theta_0 = [0, 124.4^\circ]$ for the tip locus of the flexible beam [6]. The tip locus of the PRB 3R model is more accurate than the PRB 1R model not only in the interval $\theta_0 = [0, 124.4^\circ]$ but also in the whole range $\theta_0 = [0, 270^\circ]$. For this, Fig. 4 a presents the error function f in the interval $\theta_0 = [0, 270^\circ]$. As seen that the error is reduced when the slope angle θ_0

increases, and the accuracy of the 3R model extremely increases in the interval $\theta_0 = [200^0, 270^\circ]$. In addition, the accuracy of the slope angle is very high by using the PRB 3R model for modeling flexible beams as shown in Fig. 4b.

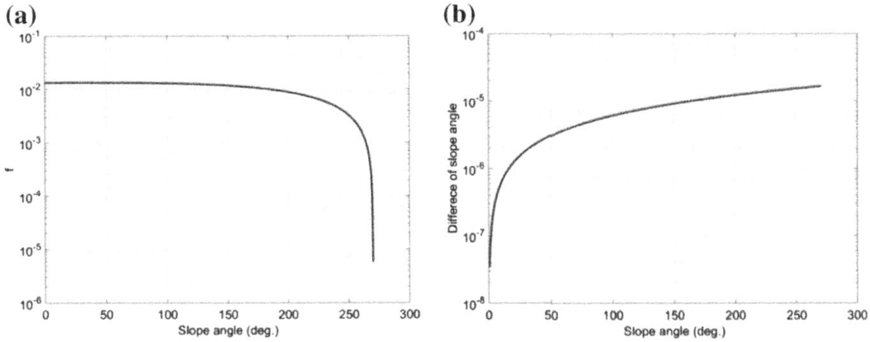

Fig. 4. (a). Displacement error of the tip locus of PRB 3R model compared to the beam; (b). Deviation of the slope angle of the flexible beam and the PRB 3R model

Pure Force. A vertical force F acts on the beam end. For this case, the characteristic radius parameters and the stiffness coefficients are also obtained by minimizing Eq. (9) and solving Eq. (5), respectively. The optimal parameters are derived as follows: $\gamma_0 = 0.1$, $\gamma_1 = 0.38$, $\gamma_2 = 0.42$, $\gamma_3 = 0.1$ and $k_{\Theta_1} = 3.7066$, $k_{\Theta_2} = 3.2530$, $k_{\Theta_3} = 1.6149$.

Figure 5 presents the tip locus of the flexible beam (black dashed line), the PRB 1R (blue line), and the PRB 3R model (red line). In Fig. 5, the working range of the PRB 1R model is in $[0, 77^\circ]$ [5]. With the PRB 3R model, the working range of the model is larger than the PRB 1R model i.e., $[0, 90^\circ]$. Moreover, the tip locus of the PRB 3R model is more accurate than the PRB 1R model.

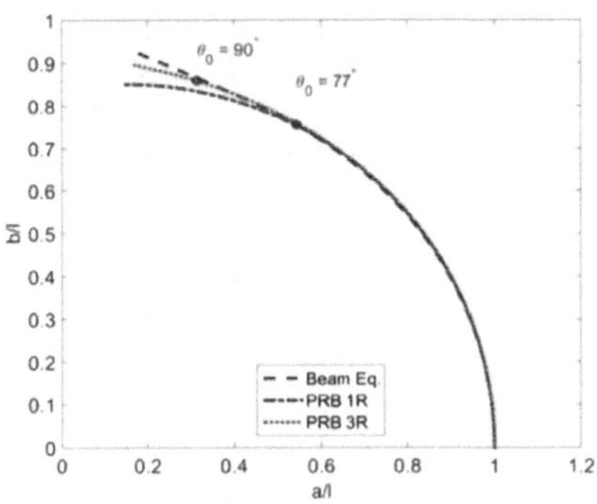

Fig. 5. Comparison of the tip locus between the beam and PRBM in case of pure force

5 Conclusion

This paper proposed a methodology for optimizing parameters of the Pseudo-Rigid-Body 3R model for modeling the deflection of compliant links. The characteristic radius and stiffness coefficients are determined by minimizing the error function of the tip point between the beam and the model. The results show that using the PRB 3R model for the flexible beam increases the accuracy of the tip locus. Moreover, the working range of the 3R model is much larger than the PRB 1R model. For this reason, the Pseudo-Rigid-Body 3R model is not only used to model the large deflection of compliant links but also used to analyze kinematics and dynamics of compliant mechanisms. Furthermore, this model can be useful in an initial design of compliant mechanisms.

Acknowledgements. The work described in this paper was supported by Thai Nguyen University of Technology for a scientific project.

References

1. Cao, L.: On advancing the topology optimization technique to compliant mechanisms and robots. Thesis of Doctor Philosophy. In: Department of Mechanical Engineering, University of Saskatchewan (2015)
2. Kimball, C., Tsai, L.-W.: Modeling of flexural beams subjected to arbitrary end loads. J. Mech. Des. **124**(2), 223–235 (2002)
3. Midha, A., Bapat, G.S., Mavanthoor, A., Chinta, V.: Analysis of a fixed-guided compliant beam with an inflection point using the pseudo-rigid-body model concept. J. Mech. Robot. **7** (2015)
4. Lyon, S.M., Howell, L.L., Roach, G.M.: Modeling flexible segments with force and moment end loads via the pseudo-rigid-body model. In: Proceedings of the ASME Dynamics and Control Division. Orlando: Proceedings of the ASME International Mechanical Engineering Congress & Exposition, pp. 883–890 (2000)
5. Howell, L.L., Midha, A.: Parametric deflection approximations for end-loaded, large-deflection beams in compliant mech. J. Mech. Des. **117**, 156–165 (1995)
6. Howell, L.L.: Compliant Mechanisms. Wiley, Hoboken (2000)
7. Pauly, J., Midha, A.: Improved pseudo-rigid-body model parameter values for end-force-loaded compliant beams. In: Design and Engineering Technical Conferences and Computes and Information in Engineering Conference, pp. 1513–1517 (2004)
8. Howel, L.L., Midha, A., Norton, T.W.: Evaluation of equivalent spring stiffness for use in a pseudo-rigid-body model of large-deflection compliant mechanisms. J. Mech. Eng. Des. **118**, 126–131 (1996)
9. Mahta, K., Rajni, V.P.: A pseudo-rigid-body 3R model for a steerable ablation catheter. In: IEEE International Conference on Robotics and Automation (ICRA), pp. 4427–4432 (2013)
10. Banerjee, A., Bhattacharya, B., Mallik, A.K.: Large deflection of cantilever beams with geometric non-linearity: analytical and numerical approaches. Int. J. Non-Linear Mech. **43** (5), 366–376 (2008). https://doi.org/10.1016/j.ijnonlinmec.2007.12.020
11. Saxena, A.A., Kramer, S.N.: A simple and accurate method for determining large deflections in compliant mechanisms subjected to end forces and moments. J. Mech. Des. **120**, 392–400 (1998)

Pigeon-Inspired Optimization for Node Location in Wireless Sensor Network

Trong-The Nguyen[1,2], Jeng-Shyang Pan[1], Thi-Kien Dao[1(✉)], Tien-Wen Sung[1], and Truong-Giang Ngo[2]

[1] Fujian Provincial Key Lab of Big Data Mining and Applications, Fujian University of Technology, Fuzhou 350118, Fujian, China
jspan@cc.kuas.edu.tw, jvnthekien@gmail.com, twsung@foxmail.com

[2] Department of Information Technology, Haiphong Private University, Haiphong, Vietnam
{vnthe, giangnt}@hpu.edu.vn

Abstract. Wireless Sensor Network (WSN) refers to a network of devices that can communicate the information gathered from a monitored field through wireless links. As a critical technology of WSN, the localization algorithm plays a vital role in improving node location accuracy and network efficiency. A hybrid Pigeon Inspired Optimization (PIO) with a typical localization model is proposed to solve the problem of node localization in WSN. The self-learning idea of PIO and speed formula are combined to improve exploring and exploiting agents of PIO. Fitness function for optimization is mathematically modeled based on analysis Pareto distances. The simulation results compared with the other approaches in the literature, e.g., the improved particle swarm optimization (PSO) and the cuckoo search (CS) show that the proposed method effectively improves the location accuracy of nodes and reduces the cumulative error caused by success positioning nodes.

Keywords: Wireless sensor network · Pigeon-inspired Optimization · Location accuracy

1 Introduction

The fast development of the field of the computer hardware and software, e.g., the integrated circuits (IC) and Information technology (IT) has caused having more and more devices in cheap and small size such as sensor nodes of Wireless sensor work (WSN) [1]. WSN refers to a network of devices that can communicate the information gathered from a monitored field through wireless links [2–4]. The applications of WSN have been widely applied in a variety of fields of industry, traffic control, healthcare, and home automation [5, 6]. However, the sensor nodes are limited in computing capability and storage capacity [7]. Due to its unique interdisciplinary nature, it has become a key technology with excellent development potential in the century. It has the advantages of low cost, small size, and long transmission distance [8]. It is widely employed for environment monitoring or tracking in areas of military, industry, and healthcare [5]. One of the critical technologies in WSN is the positioning that plays an

K.-U. Sattler et al. (Eds.): ICERA 2019, LNNS 104, pp. 589–598, 2020.
https://doi.org/10.1007/978-3-030-37497-6_67

essential role in tracking and monitoring WSN applications [7, 9]. The positioning method, such as range-based and range-free that has a wide range of its forms, e.g., capturing required location information. The improvements of optimization have been put forward to aiming at the shortcomings of various localization algorithms. The genetic algorithm (GA) was applied to improve the localization algorithm in wireless sensor networks [10]; The DV-distance method was used to estimate node position in WSN by applying PSO algorithm that showed better localization effect [11]; Cuckoo Search CS) algorithm for the nodes localization in WSN [12]; The DV-Hop algorithm was used to improve positioning accuracy effectively [13]; Establish the relationship model between the sensor node and neighbor node, and to calculate the node position based on Flower algorithm [14]..

The node information is coded into the matrix and applied to node location, which improves the positioning accuracy to a certain extent [15]. Firefly algorithm is applied to the location of wireless sensor networks, and the purpose of locating unknown nodes is quickly achieved [16]. Those mentioned methods have achieved good improving positioning accuracy in WSN. However, efficiency performance is not high in a complex environment.

In this paper, sensor nodes are carried out to optimize localization by combining a new algorithm PIO with localization model. The adaptive value calculation approach based on Pareto distance grading is suggested to reasonably distribute the fireflies. The optimized pigeon speed and position update formula improve the positioning accuracy. The simulation results show that the IPO can effectively improve the positioning accuracy under various conditions.

2 Node Localization Problem in WSN

Among the classical non-ranging algorithms, DV-Hop algorithm has attracted much attention because of its simplicity and high accuracy [11, 13]. This algorithm often uses the least square method to calculate the node coordinates. However, the positioning accuracy will be affected by the cumulative error. For solving the problem effectively, a node location model is figured out, and the possible errors are analyzed as follows.

$$\begin{cases} d_1 = \sqrt{(x_1 - x)^2 + (y_1 - y)^2} \\ d_2 = \sqrt{(x_2 - x)^2 + (y_2 - y)^2} \\ \vdots \\ d_n = \sqrt{(x_n - x)^2 + (y_n - y)^2} \end{cases} \tag{1}$$

where (x, y) is unknown node coordinates and coordinates of (x_i, y_i) is node coordinates; d_i is the distance between the reference node and the unknown node; $i = 1, 2, \ldots, n$. Equation (1) is expressed as a linear equation as follows.

$$A \times X = b \tag{2}$$

where matrix of A is expressed as follows.

$$A = -2 * \begin{bmatrix} (x_1 - x_n) & (y_1 - y_n) \\ (x_2 - x_n) & (y_2 - y_n) \\ \vdots & \vdots \\ (x_{n-1} - x_n) & (y_{n-1} - y_n) \end{bmatrix}, \tag{3}$$

The vector X is formulated as:

$$X = \begin{bmatrix} x \\ y \end{bmatrix} \tag{4}$$

The coefficient b is stated as:

$$b = \begin{bmatrix} x_1^2 - x_n^2 + y_1^2 - y_n^2 - d_1^2 + d_n^2 \\ x_2^2 - x_n^2 + y_2^2 - y_n^2 - d_2^2 + d_n^2 \\ \vdots \\ x_{n-1}^2 - x_n^2 + y_{n-1}^2 - y_n^2 - d_{n-1}^2 + d_n^2 \end{bmatrix} \tag{5}$$

The above formulas are constructed theory under ideal conditions; however, in the practice applications that would have noise or not correct all in distance measurement. Let N be an error vector that represents an n − 1 dimensional vector, Eq. (2) is written as $A \times X + N = b$, so, N should be minimized to make X more accurate.

$$X = \left(A^T A\right)^{-1} A^T b \tag{6}$$

where b is an active factor in the solution of X and it in turns d_n would be determined to b. If the value of d_n is large, the least square method used to calculate the coordinates of specific nodes can not be applied effectively. In order to solve this problem, the location problem in the environment with large errors is transformed into a function constrained optimization problem. The above should be rewritten as follow:

$$\begin{cases} d_1^2 = (x - x_1)^2 + (y - y_1)^2 \\ d_2^2 = (x - x_2)^2 + (y - y_2)^2 \\ \cdots \\ d_n^2 = (x - x_n)^2 + (y - y_n)^2 \end{cases} \tag{7}$$

The formula for distance measurement error between nodes is as follows:

$$|r_i - d_i| < \varepsilon_i \tag{8}$$

where ε_i is the ranging error between nodes r_i is the actual distance between the reference node i and the unknown node in the network. $i = 1, 2, \ldots, n$ as follow:

$$\begin{cases} d_1^2 - \varepsilon_1^2 \le (x - x_1)^2 + (y - y_1)^2 \le d_1^2 + \varepsilon_1^2 \\ d_2^2 - \varepsilon_2^2 \le (x - x_2)^2 + (y - y_2)^2 \le d_2^2 + \varepsilon_2^2 \\ \cdots \\ d_n^2 - \varepsilon_n^2 \le (x - x_n)^2 + (y - y_2)^2 \le d_2^2 + \varepsilon_2^2 \end{cases} \tag{9}$$

$$f_i = \sum_{i=1}^{n} \sqrt{(x - x_i)^2 + (y - y_i)^2 - d_i^2} \tag{10}$$

where f_i represents the ranging error between the reference node and the unknown node. Therefore, to (x, y), Define the target function as:

$$f(x, y) = \sum_{i=1}^{n} \sqrt{(x - x_i)^2 + (y - y_i)^2 - d_i^2} \tag{11}$$

The smaller the value of $f(x, y)$ is, the closer the coordinate value of the solution is to the actual value. In this way, the problem of network node location is transformed into multi-dimensional constrained optimization equation. The global optimal pigeon swarm algorithm is used to find out the location of the node and gradually reduce the difference until the optimal value is obtained.

3 Pigeon-Inspired Optimization (PIO)

PIO simulated the homing behavior of pigeons [17] that are capable of homing. The homing behavior of the pigeon group can be summarized as the optimization problem of finding the best value in a certain region. The map, compass, and landmark operators have simulated the influence on the homing behavior of pigeon groups. PIO has the characteristics of simple principle, few adjustment parameters, and easy to be implemented. PIO has obvious advantages such as relatively simple calculation, relatively strong robustness.

Map and Compass Operator

In simulated flying pigeons, the position direction information is updated according to the sense direction of the surrounding magnetic field that recorded as $X_i = [x_{i1}, x_{i2}, .., x_{iD}$ and $V_i = [v_{i1}, v_{i2}, .., v_{iD}$, i = 1, 2, …, Np. NP is the total number of pigeons,

The number of row solutions, D denotes the dimension. The position and velocity of each pigeon group at the t iteration are updated as follows:

$$V_i(t) = V_i(t-1) * e^{-Rt} + rand * \left(X_g - X_i(t-1)\right) \tag{12}$$

$$X_i(t) = X_i(t-1) + V_i(t) \tag{13}$$

where R is the map, rand is the random number of [0, 1], Xg indicates that the optimal position of the whole group of pigeons at t time is obtained by selecting the individuals with the best fitness value after comparing the position of the whole group of pigeons, which is similar to the direction pointed by the compass. After a certain number of iterations, the Xi is transferred to the landmark operator to continue the iteration.

Landmark Operators

In the late flight of pigeon herd, the pigeon flock is located through familiar buildings in the area during the process of approaching the end point. If there are familiar buildings, the pigeons can fly directly to the end. If not, they will fly with familiar pigeons. Each iteration halved the number of pigeons, ranked the first half of the population as the current population, abandoned the other half of the poor, and looked for the center of the new population, X_c, as a new reference direction, the position updating formula is as follows:

$$N_p(t) = \frac{N_p(t-1)}{2} \tag{14}$$

$$X_c(t) = \frac{\sum X_i(t) * fitness(X_i(t))}{N_p(t) * \sum fitness(X_i(t))} \tag{15}$$

$$X_i(t) = X_i(t-1) + rand * (X_c(t) - X_i(t-1)) \tag{16}$$

$fitness(X_i(t))$ is used to measure the advantages and disadvantages of the solution, constructed by objective function. rand is a random number between [0, 1]. After the number of iterations reaches the preset maximum number of cycles, the landmark operator stops the operation and outputs the X_g optimal solution.

4 Node Localization Based on PIO

4.1 Position Correction Factor

The position correction factor is suggested in the pigeon group in the landmark operator search stage to improve the selection formula of the optimal position in the late stage of the pigeon breeding search, and to optimize the positioning accuracy of the pigeon group. A parameter ω represents the position of pigeons

$$\delta = \left(\mu_{\text{best}} - \omega \cdot \frac{(\mu_{\text{best}} - \mu_{\text{worse}})}{\text{N}}\right) \cdot \frac{1}{N_p} \tag{17}$$

μ_{best} is represented as the best pigeon position, $\mu_{best} = N.P_1$, μ_{worse} represents the position of the least ideal pigeon. $\mu_{worse} = N.p_N$, suppose $\sum_{i=1}^{N} p_i = 1$ and $p_i \geq 1$. $1 \leq \mu_{best} \leq 2$ and $\mu_{best} + \mu_{worse} = 2$. The position correction factor brings into Eq. (18):

$$X_c(t) = \frac{\sum X_i(t) * fitness(X_i(t))}{N_p(t) * \sum fitness(X_i(t))} \cdot \delta \tag{18}$$

After the position correction factor, half of the better-fitted pigeons do not fly only to the middle of the population but can identify the individuals with the best fitness values in the new community and flight.

4.2 Fitness Function

The fitness value directly determines the position of pigeons. In this paper, each individual in the pigeon group is compared to a random solution of an unknown node. The formula of fitness function is:

$$fitness(x, y) = \sum_{i=1}^{n} \sqrt{(x - x_i)^2 + (y - y_i)^2 - d_i^2} \tag{19}$$

In the form, (x, y) is the coordinates of unknown nodes, (x_i, y_i) is the coordinates of the i reference node, di is the distance between the reference node and the unknown node. The node positioning research is carried out, and the position problem of the node to be tested is determined by combining the PIO with the above three-point positioning model for solving the constraint optimization problem. The exact position of the node is figured out by the intelligent search of PIO as following specific steps:

Step 1: Establish the positioning model mentioned above, and use the PIO to solve the optimization problem transformed by the model.
Step 2: Randomly initialize the pigeon speed V_i, X_i, set the maximum number of iterations G.
Step 3: Pareto distance was used to calculate the fitness of pigeon population.
Step 4: Combine pigeon speed formula in map and compass operator stage, update pigeon group position X_i.
Step 5: Introduce the position correction factor δ to update the pigeon position X_i in the position update formula of the landmark operator search stage.
Step 6: If the termination condition is met, the algorithm ends and the global best value is output. This value is an approximation of the optimal position of the node to be tested. Otherwise go to the third step and increase the number of iterations by one.

5 Simulation Results

The location of the unknown nodes is estimated by applying PIO for minimizing the objective function in Eq. (17). The obtained result outcomes of the proposed method are compared with the other methods in the literature, e.g., Particle Swarm Optimization (PSO) [11] and Cuckoo Search (CS) [12] methods. There are N unknown nodes (N = 100, 200, 300) and M reference nodes (M = 10, 20) randomly deployed in a region size of Z × Z (Z = 100, 200, 300) sensing field. Figure 1 shows the GUI of the parameter setting of the location area, the anchor node ratio, the node density, the communication radius. The unknown nodes are colored with yellow and anchor nodes are a blue color.

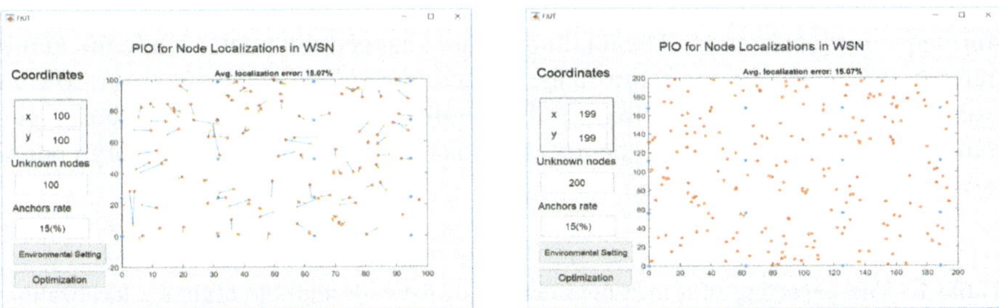

a (x=100, y =100) b (x=199, y =199)

Fig. 1. Node localization in WSN with the parameter setting of the location area, the anchor node ratio, the node density, the communication radius.

A variable of influence on positioning accuracy is used to measure ranging error of estimating positions. The specific formula of average error is to measure the accuracy of node locations as follows:

$$E_r = \frac{\sum_{i=1}^{N} \sqrt{(\overline{x_i} - x_i)^2 + (\overline{y_i} - y_i)^2}}{N \times R}, \tag{20}$$

where $\overline{x_i}, \overline{y_i}$ are the estimated coordinates of node i, (x_i, y_i) are the estimated coordinates, the number of unknown nodes is N, and the communication radius is R. Figure 2 depicts the error fluctuation curve of three algorithms after optimizing the positioning nodes. It is clearly seen that the error fluctuation curve ratio of the proposed method is the smallest approach.

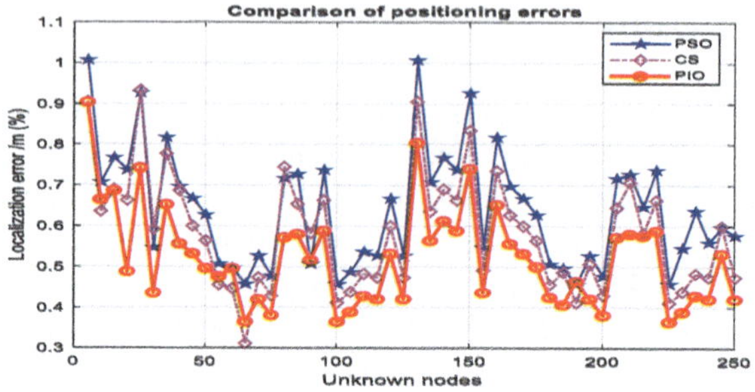

Fig. 2. Comparison of positioning errors of the proposed method with PSO and CS methods.

Table 1 shows the effects of a measured density of network nodes for accuracy performance for localization. The total nodes are changed differently while the number of referencing node and the radio range of communication radius is fixed 20% and 25 m. It is clearly seen that increasing the number of anchors and the radio range of communication usually makes the localization accuracy more satisfactory, but the incurring cost is increased more.

Table 1. The impacting of a measured density of network nodes to accuracy localization

Methods	Total number of nodes						
	80	100	120	140	160	180	200
FA	34.20%	26.30%	21.00%	20.00%	19.20%	18.50%	18.00%
pFA	31.20%	25.30%	20.00%	19.50%	19.20%	17.50%	17.00%
PIO	**26.80%**	**24.80%**	**19.00%**	**19.00%**	18.60%	**16.90%**	16.50%

Figure 3 shows the error curve of a comparison of average positioning errors for three methods. The effect of location area on average positioning accuracy with the increase of the area of the location area. It can be seen that the number of nodes does little effect on the errors when it is over 150. Therefore, in the actual node positioning process, the relationship between the area and the node density is measured effectively.

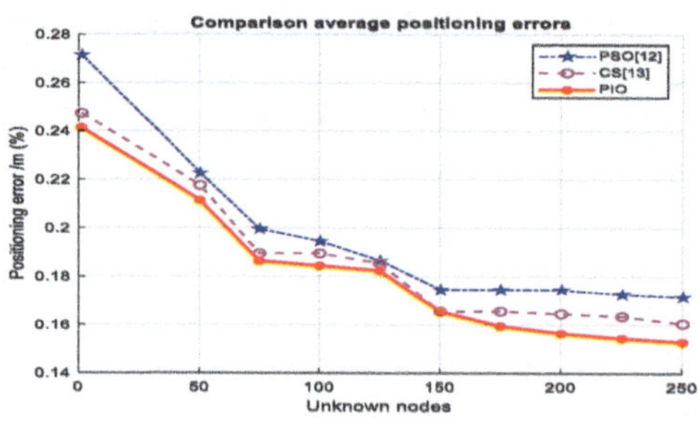

Fig. 3. Comparison of the average positioning errors for three methods of PIO, PSO, and CS.

It is obviously seen that the localization errors obtained by the proposed methods were lower than the errors generated by PSO and CS.

Table 2 displays the related positioning errors with the referencing node while fixing the radio range of the communication radius to 25 m. The errors decrease as the referencing node rate increases due to the increase of anchor nodes around unknown nodes resulting in increasing localization accuracy.

Table 2. The effect of the rate for positioning errors with various referencing node rate.

Approaches	Referencing nodes rates					
	5%	10%	15%	20%	25%	30%
CS [12]	32%	27%	23%	21%	20%	20%
PSO [11]	27%	22%	20%	20%	18%	18%
PIO	**24%**	**21%**	**19%**	**19%**	**18%**	**17%**

It is clearly seen that the outputs of optimization of the proposed method for positioning node accuracy are better than the other methods, e.g., CS and PSO.

6 Conclusion

In this paper, a solution to the node localization based on Pigeon-inspired Optimization (PIO) was proposed. The transformed positioning model into node location problem in Wireless Sensor Network (WSN) was figured out for the objective function by combining PIO. The adaptive fitness value based on Pareto distance was used to construct the model of the positioning node. Correcting the location factor is introduced in the updating pigeon positions formula to make it effectively search for the optimal solution. The simulation results compared with the other methods in the literature, e.g., Particle Swarm Optimization (PSO) and Cuckoo Search (CS) method show that the proposed method effectively improves more the node location accuracy and effectively reducing the positioning error than the PSO and CS methods.

References

1. Gungor, V.C., Lu, B., Hancke, G.P.: Opportunities and challenges of wireless sensor networks in smart grid. IEEE Trans. Ind. Electron. (2010). https://doi.org/10.1109/TIE.2009.2039455
2. Nguyen, T.-T., Dao, T.-K., Kao, H.-Y., Horng, M.-F., Shieh, C.-S.: Hybrid particle swarm optimization with artificial bee colony optimization for topology control scheme in wireless sensor networks. J. Internet Technol. **18**, 743–752 (2017). https://doi.org/10.6138/jit.2017.18.4.20150119
3. Nguyen, T.-T., Pan, J., Dao, T.: An improved flower pollination algorithm for optimizing layouts of nodes in wireless sensor network. IEEE Access **7**, 75985–75998 (2019). https://doi.org/10.1109/ACCESS.2019.2921721

4. Pan, J.-S., Nguyen, T.-T., Dao, T.-K., Pan, T.-S., Chu, S.-C.: Clustering formation in wireless sensor networks: a survey. J. Netw. Intell. **02**, 287–309 (2017)
5. García-hernández, C.F., Ibargüengoytia-gonzález, P.H., García-hernández, J., Pérez-díaz, J. A.: Wireless sensor networks and applications: a survey. J. Comput. Sci. **7**, 264–273 (2007). https://doi.org/10.1109/MC.2002.1039518
6. Nguyen, T.-T., Pan, J.-S., Dao, T.-K.: A compact bat algorithm for unequal clustering in wireless sensor networks (2019). https://doi.org/10.3390/app9101973
7. Nguyen, T.-T., Pan, J.-S., Chu, S.-C., Roddick, J.F., Dao, T.-K.: Optimization localization in wireless sensor network based on multi-objective firefly algorithm. J. Netw. Intell. **1**, 130–138 (2016)
8. Pan, J.-S., Nguyen, T.-T., Chu, S.-C., Dao, T.-K., Ngo, T.-G.: Diversity enhanced ion motion optimization for localization in wireless sensor network. J. Inf. Hiding Multimedia Signal Process. **10**, 221–229 (2019)
9. Nguyen, T.-T., Pan, J.-S., Dao, T.-K.: A novel improved bat algorithm based on hybrid parallel and compact for balancing an energy consumption problem (2019). https://doi.org/10.3390/info10060194
10. Peng, B., Li, L.: An improved localization algorithm based on genetic algorithm in wireless sensor networks. Cogn. Neurodyn. **9**, 249–256 (2015). https://doi.org/10.1007/s11571-014-9324-y
11. Low, K.S., Nguyen, H.A., Guo, H.: A particle swarm optimization approach for the localization of a wireless sensor network. In: IEEE International Symposium on Industrial Electronics (2008). https://doi.org/10.1109/ISIE.2008.4677205
12. Goyal, S., Patterh, M.S.: Wireless sensor network localization based on cuckoo search algorithm. Wirel. Pers. Commun. **79**, 223–234 (2014). https://doi.org/10.1007/s11277-014-1850-8
13. Chuang, P.J., Wu, C.P.: Employing PSO to enhance RSS range-based node localization for wireless sensor networks. J. Inf. Sci. Eng. **27**, 1597–1611 (2011)
14. Pan, J.-S., Dao, T.-K., Pan, T.-S., Nguyen, T.-T., Chu, S.-C., Roddick, J.F.: An improvement of flower pollination algorithm for node localization optimization in WSN. J. Inf. Hiding Multimedia Signal Process. **08**, 500–509 (2017)
15. Nguyen, T.-T., Thom, H.T.H., Dao, T.-K.: Estimation localization in wireless sensor network based on multi-objective grey wolf optimizer (2017). https://doi.org/10.1007/978-3-319-49073-1_25
16. Sai, V.-O., Shieh, C.-S., Nguyen, T.-T., Lin, Y.-C., Horng, M.-F., Le, Q.-D.: Parallel firefly algorithm for localization algorithm in wireless sensor network. In: Proceedings - 2015 3rd International Conference on Robot, Vision and Signal Processing, RVSP 2015 (2016). https://doi.org/10.1109/RVSP.2015.78
17. Duan, H., Qiao, P.: Pigeon-inspired optimization: a new swarm intelligence optimizer for air robot path planning. Int. J. Intell. Comput. Cybern. **7**, 24–37 (2014)

Pneumatically-Driven Micropump Using Active Check-Valve for Liquid Transportation

Gia Thinh Bui[1(✉)], Nhu-Tung Nguyen[2], and Jr-Lung Lin[3]

[1] Hai Phong University, Hai Phong 180000, Vietnam
thinhbg@dhhp.edu.vn
[2] Hanoi University of Industry, Hanoi, Vietnam
[3] I-Shou University, Kaohsiung City 84001, Taiwan

Abstract. In this paper, we designed a micro-pump using active check-valve for sample transportation. The micro-pump is designed as a moving structure using Polydimethylsiloxane (PDMS) membrane with an active valve deformed pneumatically to generate a transporting of the sample streams to the detected region. Theoretical analyses and numerical simulations were employed to determine the deformation mechanism. Hence, the numerical calculations were also used to design the optimum actuated membrane. Variations of the generated flow rate with the pneumatic frequencies, air pressures and operated cases were analyzed. The larger flow rates are obtained for the basic parameters such as driving frequencies and actuated air pressures depending on location of active check-valve. Additionally, three different design features are employed to alternately induce the flow field and to generate an effectively transporting performance. The operation of the case III can be achieved more efficient flow rate than the one of the single-side one. The development of micro-pump is a crucial component in many microfluidic applications.

Keywords: Microfluidics · Micro-pump · Active check-valve · Moving structure

1 Introduction

Nowadays, the applications for biochemical analysis [1–3], drug delivery [4], DNA test [5], and nucleic acid synthesis [6], etc almost use micromachining technologies which have been successfully introduced to provide means to miniaturize microfluidic. The speedy evolution of biological-microelectromechanical systems (Bio-MEMS) is one of the most dramatic progresses in microfluidic applications. The biomedical samples can be delivered for testing and analysis in an integrated fashion, hence, these applications have huge benefits in dramatically reducing many steps of sample handling, processing and improving data quality as well as sensitivity. Besides, with capabilities in quantitation and specific analysis, it will also help to save the overall cost and time of the measurement. In addition, the microfluidics and Bio-MEMS have been successful combination of basic microfluidic components which include microchannels, microvalves, micropumps, micromixers and microreactors with various novel sensors and detection platforms. Because of the using of sample delivery, manipulation in the microfluidic devices and

K.-U. Sattler et al. (Eds.). ICERA 2019, LNNS 104, pp. 599–606, 2020.
https://doi.org/10.1007/978-3-030-37497-6_68

systems, the micropumps become the most important components among them. Therefore, the most difficult task is to create the efficient micropump in microstructures for the applications of the miniaturized biomedical systems.

In principle, micropumps may be briefly classified as either non-mechanical or mechanical depending on movable or fixed ability of components, respectively. With simple structures without moving parts, the performance of non-mechanical pumps mostly depends on fluids and surface properties. Recently, peristaltic micropumps actuated pneumatically provide an effective means of driving fluids in microchannels and can be readily integrated with bio-sensing chips. It was successfully capable of executing ELISA (Enzyme-linked immunosorbent assay) tests in an automated fashion [7]. Later, a rotary pneumatic peristaltic pump has exhibited effective functions of transport and mixed the samples by using four-controlled membranes in a circular channel [8]. Presently, serpentine-shape (S-shape) pneumatic peristaltic micropump provides an improved pumping rate and is controlled by a single electromagnetic valve (EMV) switch [9] for the purpose of transportation. Owing to this standpoint, the above-mentioned pneumatic peristaltic micropumps for detecting or manipulating microfluids have been integrated in bio-sensing chips. However, they are not only to be fabricated by three PDMS layers (i.e., form the air chamber), but also to damp the pumping rate at the highly driving frequencies due to plastic properties of PDMS material.

In this study, the micropump using the pneumatically-actuated PDMS membrane with an active check valve as a flexible structure will be discussed in detail as follows. Theoretical and numerical models are employed to predict for membrane deformation mechanisms. Consequently, the proposed actuated membrane design is optimized by numerical simulations. Moreover, we used the design parameters of applied air pressure and operating frequency to investigate the transporting effect.

2 Theoretical and Numerical Methods

The dimensionless maximum deformation w_0 due to pressure loading is expressed as below [10].

$$\frac{w_0}{a} = \left(\frac{4}{\pi}\right)^2 \left[\frac{(1-\nu^2)\left(\beta\frac{b}{a}+\gamma\lambda\right)}{4\left(\alpha\frac{b}{a}+(9-\alpha)\lambda+\frac{2}{\lambda}+\frac{9}{\lambda^3}\right)}\frac{p_a}{Eh}\right]^{\frac{1}{3}} \tag{1}$$

Here, P and h are applied pressure and membrane thickness, respectively. $\alpha = \frac{128}{5\pi}$, $\beta = \frac{\pi^2}{4}$, $\gamma = 2 - \frac{\pi^2}{4}$, $\nu = 0.5$, and $\lambda = 3$. w_0 is the z-directional maximum displacement; a and b are the short-side (longitudinal) and long-side (transversal) lengths, respectively. Equation (1) can be theoretically calculated to obtain the larger deformation of the pneumatically actuation membrane for the different pneumatic pressures.

In this case, Eq. (1) can be simplified by

$$\frac{w_0}{a} = 0.551\left[\frac{pa}{Eh}\right]^{\frac{1}{3}} \tag{2}$$

This study aimed to investigate the transporting performance of the developed micro-pump by establishing the optimum design by utilizing the active check-valve structures activated pneumatically to transport the bio-sample. A numerical simulation was applied on the design of micro-pumps and investigation of deformation of PDMS membrane. The deformation was simulated numerically using a commercial code (CFD-ACE+, CFD-RC, USA). In the simulations, the moving boundary condition was calculated by using the stress module as well as the deformation module. The moving boundary of the membrane was discretely separated to allow smooth motion. The deformed grid of the moving boundary was accomplished using the auto-remesh function in the deformation process. Dense grids were used in the regions of the moving boundary, where the deformation was induced by the buried side chamber. The mechanical properties of PDMS are listed as: the density ρ = 970 kg/m^3; Young's modulus E = 740 kPa and Poisson's ratio ν = 0.5.

3 Design Principle

Figure 1 shows a schematic diagram of the proposed micropump. The proposed micro-pump composes of three parts: primary moving structure with an active check-valve, auxiliary moving structures and a microfluidic channel, respectively. The primary PDMS moving element is not activated by the external air pressure, the active check-valve is vertical with the bottom substrate to prevent the liquid from flowing. When the PDMS moving element is activated, the active check-valve is lifted-up due to the deflection of moving structure to generate the pumping rate of the fluid. Similarly, the auxiliary moving membrane is deflected by external pressure to avoid the fluids compressed by the primary membrane to pass through. When the air is exported from the air chamber, the active check valve will be recovered the original position, i.e., it is still vertical with the substrate to block the forward of fluids from flowing backward. Therefore, the liquids can be smoothly transported in series of operation by the proposed micropump.

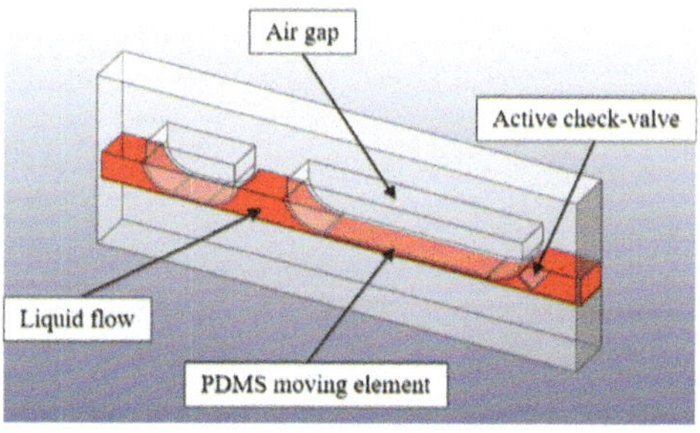

Fig. 1. Schematic illustration shows the actuation mechanism of the micropump.

The pneumatic pump comprises two PDMS layers and one glass-based layer. The proposed micropump includes a pneumatically driving layer, microchannel/active check valve layer and glass substrate. Figure 2 shows a schematic dimension of micropump. The dimensions of auxiliary moving membrane are 2.5 mm and 2.5 mm for the length and width, respectively. The one of primary moving membranes are 7.5 mm and 2.5 mm for the length and the width, respectively. The thick-ness of two moving membrane is 500 μm. The connected air channel is designed to be 0.1 mm width. The dimension of fluidic channel is 2.0 mm for width, 17.5 mm for length, and 1.0 mm for depth. And the distances from an active check valve to outlet are 1.8, 2.0, 2.2, and 2.4 mm, respectively. Figure 3 shows the completed micropump photograp hafter fabricating: (a) top-side view and (b) up-side down view.

3.1 Membrane Deflection Simulation

Figure 4 (a)–(d) shows the applied on the primary PDMS moving membrane at the external pressures of 5 psi (34.4 kPa), 10 psi (69.0 kPa), 15 psi (103.4 kPa), and 20 psi (137.8 kPa), respectively. As we expected, the membrane deflection increases with the increasing air pressure. A maximum deflection of membrane of 0.9 mm is obtained at an applied air pressure of 10 psi (69.0 kPa). In the design principle of the proposed micropump, the higher active check valve lift-up performance will obtain the higher pumping rate. Conversely, the lower active check valve lift-up performance will obtain the lower pumping rate. This numerical calculation is reasonable agreement with the experimental observations since an applied pressure higher to 10 psi (68.0 kPa) can push the PDMS moving membrane to the glass substrate.

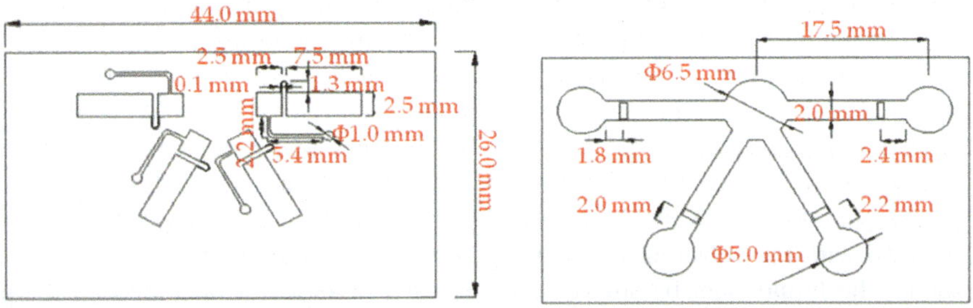

Fig. 2. Schematic dimension illustrations of top layer and bottom layer of the pneumatic pump.

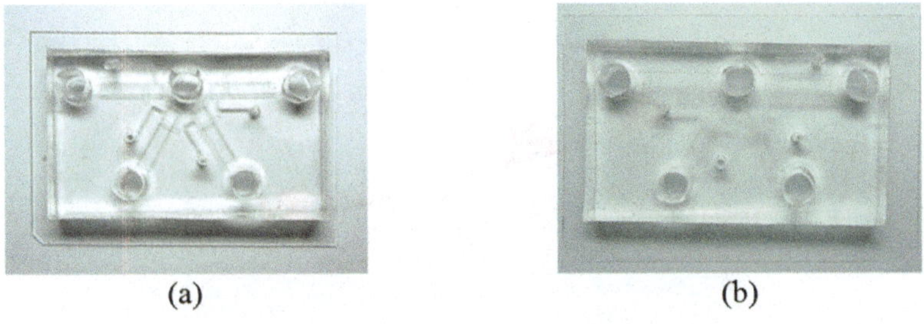

Fig. 3. (a) Top-side view and (b) up-side down view photographs of completed micropump.

4 Results and Discussion

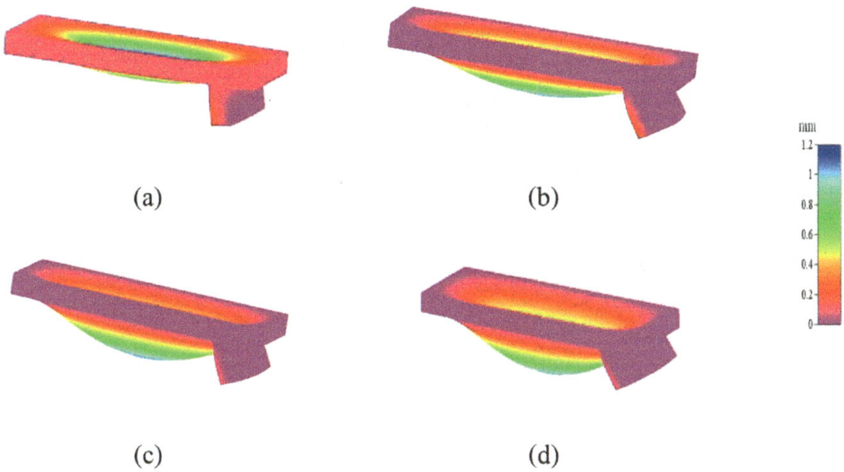

Fig. 4. Deflection of the PDMS membrane with an active check valve with the different air pressures of (b) 5 psi, (c) 10 psi, (d) 15 psi, (e) 20 psi, respectively.

4.1 Pumping Effect

Operation of External Pressure

The flow rate influenced by the applied air pressure at different driving frequency is shown in Fig. 5. During the test process, the operating frequency is varied sequentially to test the effect of the driving frequency. In general, the flow rate increases along with driving frequency (at a constant pneumatic pressure). However, the maximum flow rate at a constant applied pressure is limited by the fill and releases times of the compressed air. If the operating frequency is too high, the air chamber cannot be completely filled and released, and the flow rate will not increase, but start to fall, as shown in Fig. 5. Specifically, the optimum operational frequencies of air pressures at 5 psi (34.4 kPa), 10 psi (69.0 kPa) and 20 psi (136.1 kPa) are found to be 30, 26 and 24 Hz, respectively. Experimental data also shows that the higher driving pressures are, the higher flow rates are for the applied pressures of 5 psi (34.4 kPa), 10 psi (69.0 kPa), and 20 psi (138 kPa). A maximum flow rate of 532 μL/min is obtained at a driving frequency of 24 Hz and at an applied air pressure of 20 psi (136.1 kPa). This observation is reasonable because a higher than 10 psi applied pressure can force the PDMS membrane to completely touch down to the glass substrate. It can also explain the same flow rates appearing at the pressure of 10 psi and 20 psi. In this study, the proposed applied pressure of valve is 20 psi to block the fluids from passing through the microchannel.

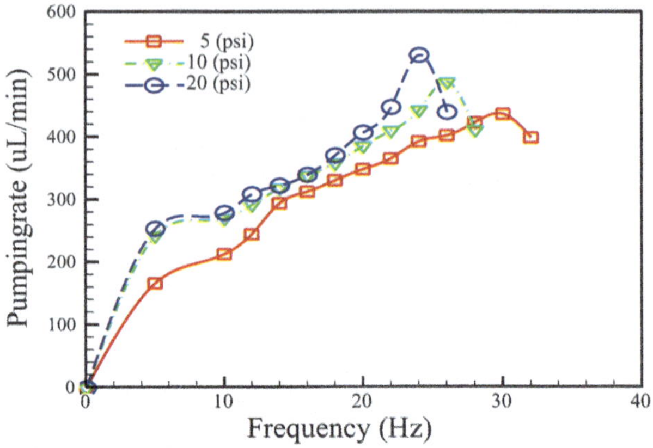

Fig. 5. Relationship of the pumping rate and the applied air pressure at different driving frequencies.

4.2 Locations of Active Check-Valve

It was also tested for the influence of the different location of active check valve to pumping rate. The distance from active check valve to outlet of channel with 1.8 (Case 1), 2.0 (case 2) and 2.2 mm (case 3), respectively, was tested with a constant pneumatic pressure of 10 psi. Figure 6 plots the relationship between the pumping rate and the operational frequency as a function of the different location of active check valve. It can be seen that the first pumping rate increases along with frequency, and certainly also increases along with distance of active check valve. However, the optimum operational frequency, f, i.e. the frequency corresponding to the maximum pumping rate, reduces as the distance of active check valve increases. Specifically, the optimum operational frequencies for the channels with distance of check valve at 1.8, 2.0 and 2.2 mm are found to be 26, 24 and 26 Hz, respectively. Conversely, the maximum pumping rate increases with an increasing distance of active check valve. Apparently, the maximum pumping rates obtained for the distance of check valve at 1.8, 2.0 and 2.2 mm are 296, 373 and 488 μL/mm, respectively. This observation is reasonable since the peristaltic driving effect is enhanced when the distance of active check valve increases. However, the time required to rise the distance of active check valve depends on the complete release of the compressed air. Therefore, the optimum driving frequency reduces when the distance of active check valve increases.

The relationship between the pumping rate and the different location of active check valve was also experimentally explored. Figure 7 shows the maximum flow rates at the optimal driving frequency (26 Hz) for different locations of valve of channel, i.e. the distance from active check valve to the outlet of reservoir. Again, the membrane thickness is 500 μm and the applied air pressure is 10 psi (69.0 kPa). The results indicate that the micropump can be operated for various location of active check valve. In general, the higher distance of active check-valve is, the higher pumping rate is due to more stroke volume is squeezed by the defected PDMS membranes. However, the maximum flow rate experimentally obtained is at distance of 600 μm. The flow rates

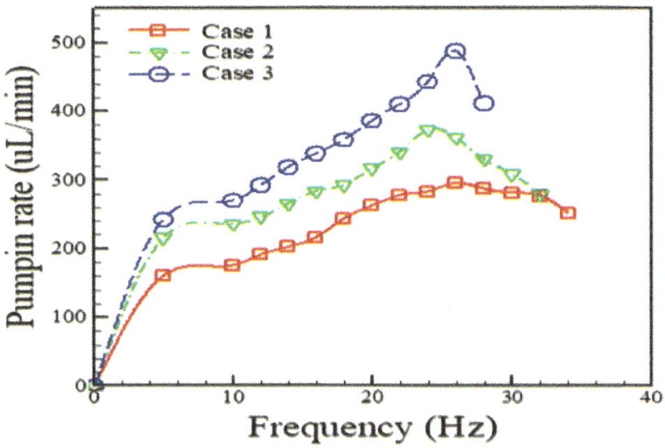

Fig. 6. Relationship between pumping rate and operational frequency as a function of different location of active check valve for a pneumatic pressure of 10 psi.

will increase along with the distance of active check-valve. However, it will decrease when the distance is higher than 600 μm. The smaller distance of active check-valve obtaines the lower pumping rates of the proposed micropump. Conversely, the longer distance of active check valve obtaines higher pumping rates. The reason is that the active check valve is lifted up by the deflected moving structure. The smaller distance of active check valve causes a smaller lift to let fluids pass through the microchannel. Certainly, the longer distance will interrupt the deflection of moving structure. Therefore, it exists an optimum distance of active check valve to achieve a higher pumping rate. The data in Fig. 7 are fitted to a curve line, which can be found the optimum location of active check valve to be 600 μm (i.e., the distance is 2.2 mm) for this study.

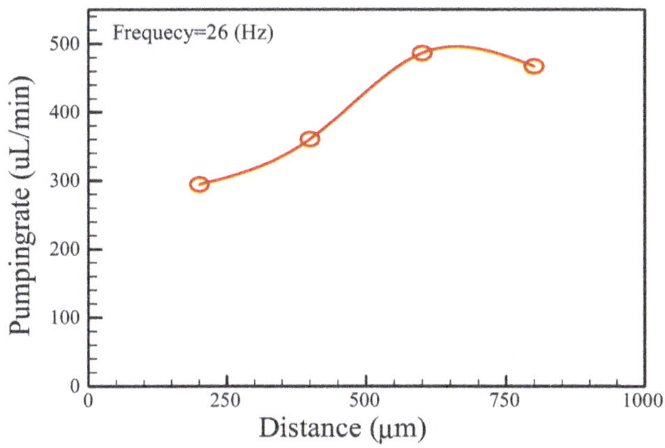

Fig. 7. Thepumping rates of the micropumps with different locations of active check valve of channels under the optimal driving frequency of 26 Hz.

5 Conclusion

This study has proposed a novel pneumatic pump which is based on the peristaltic-like actuation effect caused by the sequential deflection of membrane with an active check valve. The pneumatic micropump improves pumping rates and is determined by the pneumatic pressure, the operational frequency, the location of active check valve. The performance of the pneumatic pump has been thoroughly investigated for the various effects. The optimum parameters of air pressure, driving frequency, and location of active check valve are experimentally 10 psi, 26 Hz, and 600 μm, respectively. The maximum pumping rate has been found to be 488 μL/min. Furthermore, the proposed micro-pump could be crucial for the transportation of the microfluidics and could be incorporated with the micro-total-analysis-systems.

References

1. Laser, D.J., Santiago, J.G.: A review of micropumps. J. Micromech. Microeng. **14**, 35–64 (2004)
2. Fu, A.Y., Chou, H.P., Spence, C., Frances, H.A., Stephen, R.Q.: An integrated microfabricated cell sorter. Anal. Chem. **74**, 2451–2457 (2002)
3. Fu, A.Y., Spence, C., Scherer, A., Frances, H.A., Stephen, R.Q.: A microfabricated fluorescence-activated cell sorter. Nat. Biotechnol. **17**, 1109–1111 (1999)
4. Maillefer, D., Gamper, S., Frehner, B., Balmer, P.: A high-performance silicon micropump for disposable drug delivery systems. In: Proceedings of IEEE Micro Electro Mechanical Systems (MEMS), pp. 413–417 (2001)
5. Hong, J.W., Studer, V., Hang, G., Anderson, W.F., Quake, S.: A nanoliter-scale nucleic acid processor with parallel architecture. Nat. Biotechnol. **22**(4), 435–439 (2004)
6. Liu, J., Enzelberger, M., Quake, S.: A nanoliter rotary device for polymerase chain reaction. Electrophoresis **23**, 1531–1536 (2002)
7. Wang, C.H., Lee, G.B.: Automatic bio-sampling chips integrated with micropumps and microvalves for multiple disease detection. Biosens. Bioelectron. **21**(3), 419–425 (2005)
8. Tseng, H.Y., Wang, C.H., Lin, W.Y., Lee, G.B.: Membrane-activated microfluidic rotary devices for pumping and mixing. Biomed. Microdevices **9**(4), 545–554 (2007)
9. Wang, C.H., Lee, G.B.: Pneumatically-driven peristaltic micropumps utilizing serpentine-shape channels. J. Micromech. Microeng. **16**, 341–348 (2006)
10. Chiou, C.-H., Yeh, T.-Y., Lin, J.-L.: Deformation analysis of a pneumatically-activated polydimethylsiloxane (PDMS) membrane and potential micro-pump applications. Micromachines **6**, 216–229 (2015)

Prediction of Surface Roughness in Turning with Diamond Insert

Do Duc Trung[1], Nguyen Nhu Tung[1], Nguyen Hong Son[2], Tran Thi Hong[3], Nguyen Van Cuong[4], Vu Nhu Nguyet[5], and Ngoc Pi Vu[5](✉)

[1] Faculty of Mechanical Engineering, Hanoi University of Industry, No. 298, Cau Dien Street, Bac Tu Liem District, Hanoi, Vietnam
[2] Center for Mechanical Engineering, Hanoi University of Industry, No. 298, Cau Dien Street, Bac Tu Liem District, Hanoi, Vietnam
[3] Center of Excellence for Automation and Precision Mechanical Engineering, Nguyen Tat Thanh University, Ho Chi Minh City, Vietnam
[4] Faculty of Mechanical Engineering, University of Transport and Communications, Ha Noi City, Vietnam
[5] Faculty of Mechanical Engineering, Thai Nguyen University of Technology, Thai Nguyen City, Vietnam
vungocpi@tnut.edu.vn

Abstract. Prediction of the surface quality of a manufacturing process has a very important role. Therefore, in this paper, a study on the prediction of surface roughness in turning process with the use of diamond insert was presented. Based on the results of previous studies, a new surface roughness model was formulated. In the model, the effects of the depth of cut, the nose radius of the insert, the feed rate, and the characteristics of the workpiece material were taken into account. The surface roughness calculated from the model has been successfully verified by an experiment. Hence, this proposed model can be used to predict the surface roughness in turning processes using diamond inserts.

Keywords: Surface roughness · Surface roughness model · Turning process · Diamond insert

1 Introduction

In turning as well as in other machining processes, the quality of the processing is evaluated through many factors, of which the surface roughness is one of the most important elements. This is because the surface roughness has a great influence on the working quality and the life of machine parts. As a result, numerous researches have been carried out on modelling surface roughness.

Until now, the surface roughness in turning have been calculated by different methods. Abburi and Dixit applied the artificial neutron network to predict surface roughness [1]. Buldum et al. [2], Costes [3] predicted the surface roughness by spectrum analysis of the machining surface and Fourier transform. An Adaptive Neuro-Fuzzy Inference System was used by Reddy to predict the surface roughness [4]. The above

K.-U. Sattler et al. (Eds.): ICERA 2019, LNNS 104, pp. 607–612, 2020.
https://doi.org/10.1007/978-3-030-37497-6_69

mentioned methods often give high accuracy to predict the surface roughness, however, it is complicated to apply them to practice, which reveals their limitations.

In another method, the surface roughness has been found based on the theoretical of machining processes. Boothroyd and Knight [5] determined the value of the undulating surface height R_{th} by the following equation:

$$R_{th} = 1000\frac{f^2}{18r\sqrt{3}} \approx 1000\frac{0.0321f^2}{r} \tag{1}$$

Where, f is the feed rate (mm/rev); r is the nose radius (mm).

Ståhl [6] proposed the Eq. (2) to determine the undulating surface height when f < 2*r:

$$R_{th} \approx 770\left[1 - \frac{\frac{f}{2r}}{arcsin\left(\frac{f}{2r}\right)}\right]r \tag{2}$$

Besides, Groover [7] determined the undulating surface height as

$$R_{th} \approx 1000\frac{f^2}{32 \cdot r} \tag{3}$$

The relationship between the undulating surface height and axial feed rate and nose radius can also be explained by the following equation [8].

$$R_{th} = 1000\frac{f^2}{8 \cdot r} \tag{4}$$

Moreover, the undulating surface height is calculated by [9]:

$$R_{th} = 1000\frac{f_I^2}{24 \cdot r_I} \tag{5}$$

In Eq. (5), f_I is the axial feed rate (inch/rev); r_I is the nose radius (inch).

From the above equations, is found that only two process parameters including the axial feed rate and the nose radius of the insert were taken into the models for determining the undulating surface height. In practice, the undulating surface height depends not only on the mentioned parameters but also on many other process parameters such as the depth of cut, the characteristics of workpiece material, and so on. Therefore, the surface roughness calculated from these equations may lead to inaccurate results. To clarify this observation, it is assumed that the turning process is implemented with an axial feed rate of 0.097 (mm/rev), the tool nose radius of 0.4 (mm), the values of R_{th} calculated by these Eqs. (1) to (5) are 0.76 (μm), 1.89 (μm), 0.74 (μm), 2.94 (μm), and 0.98 (μm), respectively.

From the above analyses, to increase the accuracy of the surface roughness prediction, the influences of many input parameters on the surface roughness such as the axial feed rate, the tool nose radius, the depth of cut and the characteristics of workpiece material need to be considered. This issue will be described in detail in the next section of this study.

2 Prediction of Surface Roughness

In turning process, the surface roughness is determined by the following equation [10]:

$$R_a = 0.2566 R_{th} \tag{6}$$

Where, R_{th} is the undulating surface height. When turning with diamond insert, R_{th} is determined by [11]:

$$R_{th} = \frac{f^2}{8r} + \frac{h_{Dmin}}{2}\left(1 + \frac{r \cdot h_{Dmin}}{2}\right) + k_1 r_n \frac{H}{E} k_2 \tag{7}$$

Where, r_n is the radius of main cutting edge; H is the Vickers hardness of workpiece material; E is the Young modulus of workpiece material; h_{Dmin} is the minimum undeform chip thickness which can be found by [12]:

$$h_{Dmin} = c \cdot r_n \tag{8}$$

In which, c is a constant, c = 0.3–0.4 and it is chosen as 0.35. k_1 is a coefficient which depends on the radius of the tool nose and the main angle of the cutter (or the insert); $k_1 = 12$ when the cutter nose radius is greater than 0.9 (mm) and $k_1 = 10$ when the cutter nose radius is less than or equal to 0.9 (mm) [11]. k_2 is a coefficient that considers the effect of maximum undeform chip thickness and radius of the main cutting edge on R_{th}; k_2 is determined by the following equations [11]:

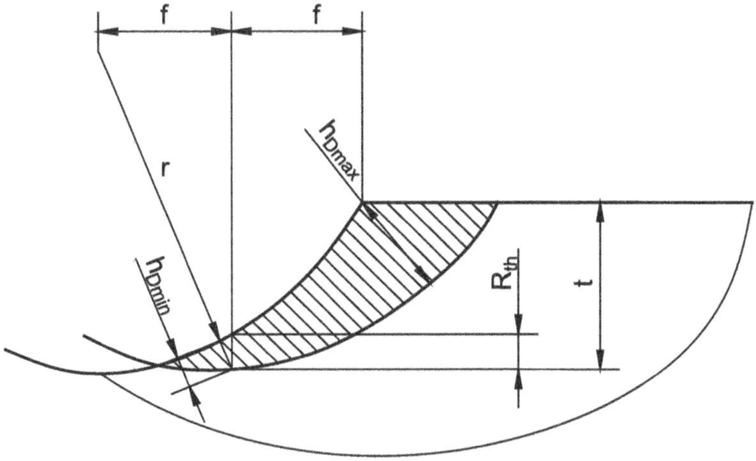

Fig. 1. The undeform chip thickness in turning process

$$k_2 = 0.8 + 0.002356\left(\frac{h_{Dmax}}{r_n} - 1\right) \text{ khi } \frac{h_{Dmax}}{r_n} \geq 1 \quad (9)$$

$$k_2 = 0.8 - 0.2356\left(\frac{h_{Dmax}}{r_n} - 1\right) \text{ khi } \frac{h_{Dmax}}{r_n} < 1 \quad (10)$$

Wherein, h_{Dmax} is the maximum undeform chip thickness which is determined by (see Fig. 1):

$$h_{Dmax} = r - \sqrt{r^2 + f^2 - 2f\sqrt{2 \cdot r \cdot t - t^2}} \quad (11)$$

In which, t is the depth of cut.

3 Results and Discussion

Equation (6) used was to calculate the surface roughness when turning CK 60 plain carbon steel by using the diamond insert. The input parameters used in the calculation are shown in Table 1. Then, the calculated results were compared to the measured values of surface roughness in the experimental study in [13]. Figure 2 presents the surface roughness received from predicted model and from the experiment in [13]. From the figure, it is clear that the calculated surface roughness is very close to the measured one. That means the model results and experimental results are in good agreement. Therefore, the proposed model for calculation of the surface roughness can be used to predict the surface roughness in turning process when using diamond inserts.

Table 1. The input parameters of surface roughness calculation

Parameters	Symbol	Value	Unit	Reference
Workpiece material	CK 60 plain carbon steel	–	–	
Tool nose radius	r	0.4/0.8/1.2/1.6	mm	[13]
Main cutting edge radius	r_n	29	nm	
Constant	c	0.35	–	
Main tilt angle	–	30	degree	
Feed rate	f	0.1/0.15/0.18/0.23/0.3	mm/v	
Depth of cut	t	0.3	mm	
Viker hardness	H	2.2	Gpa	[14]
Young modulus	E	117	Gpa	[15]

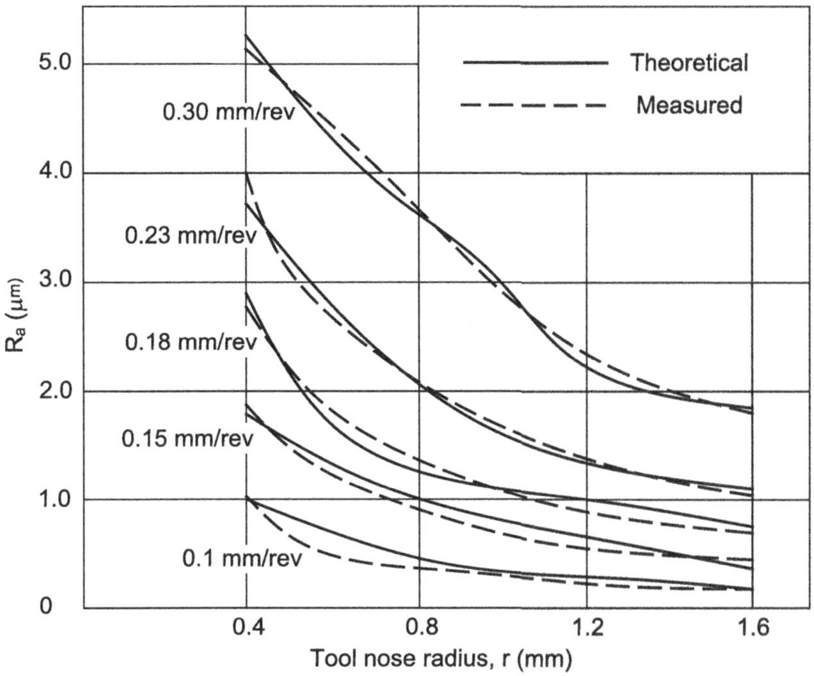

Fig. 2. Comparison of predicted and measured surface roughness

4 Conclusions

In this study, the combination of theoretical and experimental methods was performed to determine the surface roughness in turning process with the use of diamond inserts. From the study results, several conclusions can be drawn as follows:

- A predicted equation to calculate the surface roughness was proposed. Many input parameters including the axial feed rate, the tool nose radius, the cutting depth and the characteristics of workpiece material were considered;
- The predicted values of surface roughness are very close to the measured values which were found from a previous study;
- The proposed model of the surface roughness can be used to predict the surface roughness in turning process when using diamond inserts.

Acknowledgements. The work described in this paper was supported by Ha Noi University of Industry and Thai Nguyen University of Technology.

References

1. Abburi, N.R., Dixit, U.S.: A knowledge-based system for the prediction of surface roughness in turning process. Robot. Comput.-Integr. Manuf. **22**, 363–372 (2006)
2. Buldum, B., Şik, A., Akdagli, A., Biçer, M.B.: ANN surface roughness prediction of AZ91D magnesium alloys in the turning process. Mater. Test. **59**, 916–920 (2017)
3. Costes, J.-P.: A predictive surface profile model for turning based on spectral analysis. J. Mater. Process. Technol. **213**(1), 94–100 (2013)
4. Sidda Reddy, B., Suresh Kumar, J., Vijaya Kumar Reddy, K.: Prediction of surface roughness in turning using adaptive neuro-fuzzy inference system. Jordan J. Mech. Ind. Eng. **3**(4), 252–259 (2009)
5. Boothroyd, G., Knight, W.A.: Fundamental of Machining and Machine Tool. Marcel Dekker, New York (1989)
6. Ståhl, J.-E., Schultheiss, F., Hägglund, S.: Analytical and experimental determination of the Ra surface roughness during turning. Procedia Eng. **19**, 349–356 (2011)
7. Groover, M.P.: Fundamentals of Modern Manufacturing. Prentice Hall, Upper Saddle River (1996)
8. http://www.mitsubishicarbide.com/en
9. https://www.custompartnet.com/
10. Grzesik, W.: A revised model for predicting surface roughness in turning. Wear **194**, 143–148 (1996)
11. Zong, W.J., Huang, Y.H., Zhang, Y.L., Sun, T.: Conservation law of surface roughness in single point diamond turning. Int. J. Mach. Tools Manuf. **84**, 58–63 (2014)
12. Wang, H.X.: Research on establishing theoretical model of surface microtopography and experiment in ultraprecision turning (Ph.D. dissertation), Harbin Institute of Technology (2002)
13. Petros, G.: Petropoulos: the effect of feed rate and of tool nose radius on the roughness of oblique finish turned surface. Wear **23**, 299–310 (1973)
14. http://vietmetal.com.vn/bang-tra-do-cung-hrc-hrb-hb-hv-cua-kim-loai-thep.html
15. https://en.wikipedia.org/wiki/Young%27s_modulus

Research on Determining the Limited Roll Angle of Vehicle

Anh Nguyen Tuan[1(✉)] and Binh Hoang Thang[2]

[1] Mechatronic-Automobile Department, Hanoi University of Business and Technology, 29A/124 Vinh Tuy, Hai Ba Trung, Ha Noi, Viet Nam
taae177@gmail.com
[2] Automotive Engineering Department, Hanoi University of Science and Technology, 1 Dai Co Viet, Hai Ba Trung, Ha Noi, Viet Nam

Abstract. The rollover of the vehicle often occurs when the vehicle steers at a high speed. The limit of rollover phenomenon is expressed through the value of the roll angle (φ_x), this value is a multivariate function depending on many input parameters. This research establishes the dynamic vehicle model to determine the value of the limited roll angle of the vehicle when steering. The results of the study show that as the velocity or height increases, the value of the limited roll angle tends to decrease almost linearly. At the high-speed range, the instantaneous roll angle should not exceed 8.0^0.

Keywords: Roll angle of the vehicle · Rollover phenomenon · Limited roll angle · Dynamic vehicle model

1 Introduction

1.1 Lateral Instability on Vehicle

Lateral stability of the vehicle when moving is one of the extremely important factors determining the safety of the drivers and passengers. The vehicle is considered lateral instability when faced with a lateral slip or rollover, in which rollover is assessed to have a very dangerous angle [1].

1.2 The Cause of Rollover

When the vehicle steers, under the effect of centrifugal force, it will cause a change in the position of the center of gravity, thus appearing the roll moment M_R [2, 3].

Assume that M_{AR} is the anti-roll moment of the suspension, according to [4] if:

$M_{AR} < M_R < M_{AR,\ max}$: The body of the vehicle is tilted.
$M_R > M_{AR,\ max}$: The vehicle is rollover.

The warning sign of the rollover is the phenomenon of one side of the wheel is separated from the road surface [5]. The wheel lifted off from the road also means that the corresponding suspension of that wheel is not working. At this time, the vehicle's anti-roll moment will dramatically reduce while the roll moment continuously rises,

K.-U. Sattler et al. (Eds.): ICERA 2019, LNNS 104, pp. 613–619, 2020.
https://doi.org/10.1007/978-3-030-37497-6_70

which will cause the vehicle to roll over quickly. Vehicles with increased dimensions and weights are known to be at higher risk of rollover [6, 7].

1.3 The Roll Angle of the Vehicle

Roll angle (φ_x) is a value that characterizes the lateral instability of the vehicle. This value is a multivariate function depending on many factors such as the velocity (v), the distance from the center of gravity to the roll axis (h), the steering angle (δ) [8, 9]. It is difficult to determine exactly the limit at which the vehicle is roll over. This research focused on simulates limited roll angle under ideal conditions.

$$\varphi_x = f(v, h, \delta) \quad (1)$$

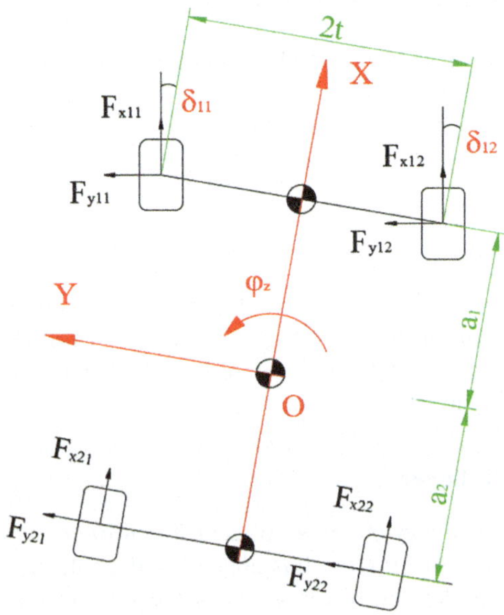

Fig. 1. Double track dynamic vehicle model

2 Dynamic Vehicle Model

2.1 Double Track Dynamic Vehicle Model

Assuming that the vehicle is moving on a flat road and does not include the influence of external forces (Fig. 1).

With: $F_{y1} = F_{y11} + F_{y12}$ $F_{y2} = F_{y21} + F_{y22}$

In the case of a steering with a constant velocity, the steering angle is small. The matrix describing the vehicle's moving state on the road plane is of the form:

$$\left(\dot{v}_y\ \ddot{\varphi}_z\right) = \left(F_{y1}\ F_{y2}\right)\begin{pmatrix} \frac{1}{M} & \frac{a_1}{I_z} \\ \frac{1}{M} & \frac{-a_2}{I_z} \end{pmatrix} - v_x\left(\dot{\varphi}_z\ 0\right)$$

2.2 Dynamic Vehicle Model of 7 DOF

With:

$$F_1 = F_{C11} + F_{K11} + F_{C21} + F_{K21}\quad F_2 = F_{C12} + F_{K12} + F_{C22} + F_{K22}$$

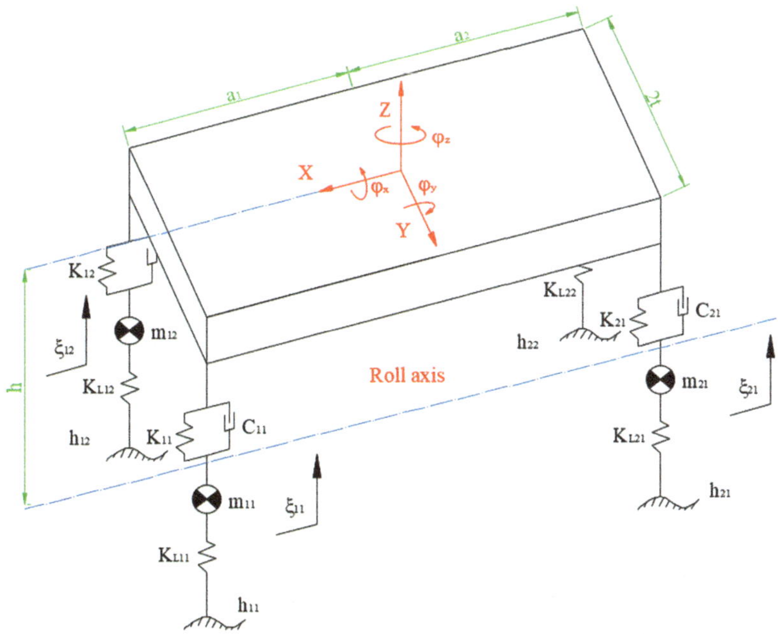

Fig. 2. Dynamic vehicle model of 7 DOF

The matrix describing the vertical displacement and the roll angle of the sprung mass is of the form (Fig. 2):

$$(\ddot{z}\ \ddot{\varphi}_x) = (F_1\ F_2)\begin{pmatrix} \frac{1}{m} & \frac{t}{I_x+mh^2} \\ \frac{1}{m} & -\frac{t}{I_x+mh^2} \end{pmatrix} + \left(0\ \frac{M_R - M_{SB}}{I_x+mh^2}\right)$$

Where:

M_{SB}: Moment of stabilizer bar, K_{SB}: Torsional stiffness of stabilizer bar, i: Ratio between the torsional stiffness of the vehicle and the torsional stiffness of the bar.

$$M_{SB} = K_{SB}\varphi_x i \tag{2}$$

M_R: Roll moment:

$$M_R = (a_y cos^2\varphi_x + g sin\varphi_x)mh \tag{3}$$

In order to be able to identify the above matrix, use the corresponding link equations.

According to [10, 11], vertical displacement of the unsprung mass will be as follows:

$$m_{ij}\ddot{\xi}_{ij} = F_{KTij} - F_{Cij} - F_{Kij} \tag{4}$$

The elastic force of the suspension system:

$$F_{Kij} = K_{ij}(\xi_{ij} - z \pm b\varphi_x) \tag{5}$$

The damping force of the suspension system:

$$F_{Cij} = C_{ij}(\dot{\xi}_{ij} - \dot{z} \pm b\dot{\varphi}_x) \tag{6}$$

The elastic force of the tire:

$$F_{KTij} = K_{Tij}(h_{ij} - \xi_{ij}) \tag{7}$$

3 Simulation of Vehicle

3.1 Input Parameters

For the parameters of the vehicle refer to the following Table 1:

Table 1. Input parameters

Symbol	Description	Value
m	Sprung mass	1800 (kg)
m_{ij}	Unsprung mass	50 (kg)
a_1	Distance from the center of gravity to front axle	1.2 (m)
a_2	Distance from the center of gravity to rear axle	1.6 (m)
t	Half of track width	0.8 (m)
h	Distance from the center of gravity to roll axis	0.8 ÷ 1 (m)

Assume that the vehicle moves steadily at velocities v = {120; 130; 140; 150} km/h and the steering angle as shown in Fig. 3.

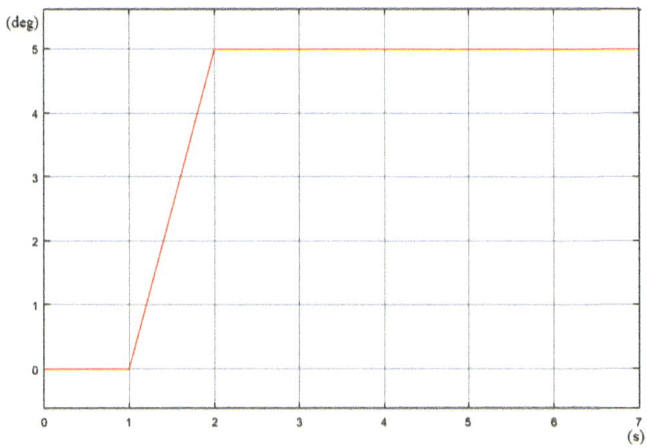

Fig. 3. Steering angle

3.2 The Limited Roll Angle of the Vehicle

With the steering angle as shown in Fig. 3, the graph shows the limited roll angle depending on the velocity (v) and the distance from the center of gravity to the roll axis (h) is shown as in Fig. 4.

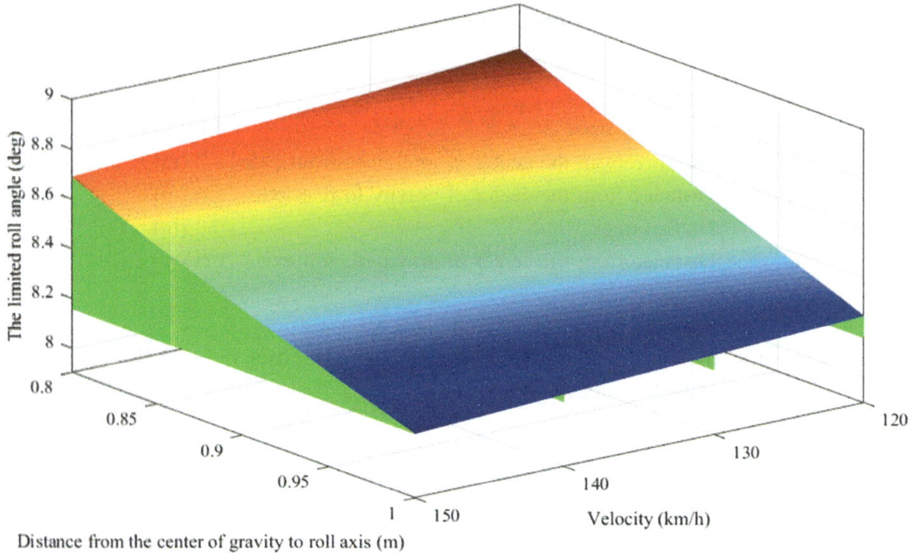

Fig. 4. Limited roll angle.

Based on the graph in Fig. 4, it can be seen that:

At the same height, if the velocity increases, the limited roll angle will decrease. At the same velocity, if the height increases, the limited roll angle will decrease.

3.3 Establishing Rollover State Function

The graph in Fig. 4 shows the dependency between limited roll angle, height, and velocity under defined conditions. Assume that the values above in an inclined plane are limited to the scope of the survey, the inclined plane equation describes the dependence of the limited roll angle on the height and velocity as follows:

$$A_1 v + A_2 h + A_3 \varphi_x + A_4 \approx 0 \tag{8}$$

Where: A_1, A_2, A_3, A_4 is the coefficient.

With a fixed input as a steering angle, the parameters of height and velocity can change, the limited roll angle of the vehicle is considered being a function depending on the two variables above.

$$\varphi_x \approx \alpha_1 v + \alpha_2 h + \alpha_3 \tag{9}$$

Where:

$$\alpha_1 = -\frac{A_1}{A_3};\ \alpha_2 = -\frac{A_2}{A_3};\ \alpha_3 = -\frac{A_4}{A_3} \tag{10}$$

Using parameters and simulation conditions as mentioned above, the limited roll angle of the vehicle in this research can be approximately calculated as follows:

$$\varphi_x \approx 11.46 - 0.0043v - 2.65h \tag{11}$$

Note: Eq. (11) is only used in the simulation limit (v $\geq$ 120 km/h, h1 $\geq$ 0.8 m).

4 Conclusions

The limited roll angle depends on velocity, steering angle, and distance from the center of gravity to the roll axis. As the value of input parameters increases, the value of the limited roll angle tends to decrease. Each vehicle has different parameters will give different results, in general, the value of limited roll angle of most vehicles (SUV) will be similar to the graph Fig. 4.

The model has neglected the influence of external factors and simplified the survey process, so the results may be different from the actual, but the difference is not large. To ensure safety during moving, the roll angle of the vehicle should not exceed 8.0^0.

References

1. Anh, N.T., Tran, T.T., Binh, H.T., Nam, P.H., Dung, L.T.: The study on the method of calculating and designing the stabilizer bar on the vehicle using solidworks software. Viet Nam Mech. Eng. J. **7**(12), 0866–7056 (2018)
2. Aleksander, H.: Rollover stability index including effects of suspension design. SAE technical papers, pp. 0148–7191 (2002, in press)
3. Trai, N.K., Hoan, N.T., Hai, H.H., Huong, P.H., Chuong, N.V., Hoang, T.M.: Structure Vehicle. Bach Khoa Ha Noi Publishing, Ha Noi (2010)
4. Anh, N.T.: The study on dynamic vehicle model equipped active stabilizer bar. Master thesis. Ha Noi University of Science and Technology, Ha Noi (2019)
5. Cruz, P., Echaveguren, T., Gonzalez, P.: Estimation of heavy vehicle rollover potential using reliability principles. Revista Ingenieria de Construction **32**(1), 0718–5073 (2017)
6. Phanomchoeng, G., Rajamani, R.: New rollover index for detection of tripped and un – tripped rollovers. In: 50th IEEE Conference on Decision and Control and European Control, pp. 7440–7445. IEEE Publishing, Florida (2011)
7. Anh, N.T., Binh, H.T.: Study on the influence of vehicle's dimension to rollover. Int. J. Sci. Eng. Investig. **8**(5), 2251–8843 (2019)
8. Muniandy, V., Mohdsamin, P., Jamaluddin, H., Rahman, R.A., Abubakar, S.A.: Double anti-roll bar hardware-in-loop experiment for the active anti-roll control system. J. VibroEng. **19**(4), 1392–8716 (2016)
9. Fei, X., Fengchong, L., Jiqing, C., Yunjiao, Z.: The study for anti-rollover performance based on fishhook and J turn simulation. In: 3rd International Conference on Material, Mechanical and Manufacturing Engineering, pp. 2084–2093. Atlantis Press Publishing, Guangzhou (2015)
10. Hassaan, G.A., Mohammad, N.A.: Vehicle dynamics response to road hump using 10 degrees of freedom full-car model. Int. J. Comput. Tech. **2**(1) (2015). ISSN 2394-2231
11. Huong, V.V., Dung, N.T., Khanh, D.N., Phuc, D.H.: Dynamic Vehicle. Viet Nam Education Publishing, Ha Noi (2014)

Research on Dynamic Modelling for Hydraulic Power Automotive Steering Systems with Nonlinear Friction

Nguyen Xuan Tuan[1], Huyen T. Dinh[2(✉)], and Nguyen Van Bang[2]

[1] Faculty of Automobile Technology, Hanoi University of Industry, Hanoi, Vietnam
tuannx@haui.edu.vn
[2] Faculty of Mechanical Engineering, University of Transport and Communications, Hanoi, Vietnam
huyentdinh@utc.edu.vn

Abstract. The paper introduces a comprehensive mathematical model for power steering systems mounted on cars. Dynamic modelling for the steering system is studied in consideration of nonlinear friction forces in the system such as friction in the steering column, friction in the steering mechanism (gear and rack), friction between tires with road. The dynamic modelling for both the mechanism subsystems and the hydraulic subsystem is provided. Simulations were conducted to show the accuracy of the research model which can support further research on design for safer and more efficient driving systems.

Keywords: Dynamic modeling · Hydraulic power steering · Steering system dynamics

1 Introduction

The paper investigates a comprehensive mathematical model for rack – pinion gear drive steering system which is powered by hydraulic power assistant system in cars. The gear-bar steering system is widely used on many cars due to its advantages such as: less linkage, high mechanical efficiency, and ease in fabrication. This steering system has a gear transmission which reduces the driving force slightly; however, the transmitter is only limited to the appropriate gear ratio (reducing the steering force slightly will increase the steering wheel rotation angle). The driving force is increased when the load impacting on the guide wheel increases, or when the vehicle goes on difficult road sections therefore the current driving systems are equipped with a power assistant steering system to increase steering performance and reduce driving power for the driver. Hydraulic power steering systems have currently been playing a main role in steering assistance on the market because of their higher assistant force, higher reliability and lower cost. Hydraulic power type is widely used because it has many advantages such as larger forces, higher reliability and lower cost.

The construction of mathematical modelling and dynamics simulation is an effective and inexpensive method widely used to research, analyze and improve steering performance for cars. Several models of no assistant power rack and pinion

K.-U. Sattler et al. (Eds.): ICERA 2019, LNNS 104, pp. 620–627, 2020.
https://doi.org/10.1007/978-3-030-37497-6_71

gear steering systems have been developed by researchers. A steering model with free-play and dry friction was used in the vehicle dynamics simulation [1]. Steering system parameters were analyzed and optimized in a steering system model in [2]. However, these models often ignore the effect of friction forces on the system and are not suitable for describing the dynamics of power steering systems which are much more complicated than the no assistant power systems.

While the previous steering models had different emphases and advantages, an integrated model of hydraulic power steering system which includes the detailed dynamic modelling for both the mechanism subsystems and the hydraulic subsystem is presented in this paper. This paper also considers the influence of friction forces in the system including: friction in the steering column, friction in the steering mechanism (gear and rack), friction impacting on the rack, between friction the guide wheel with suspension, friction between the tire with the road … In addition, the study also considers the soft link between the guiding wheel and the suspension, the elasticity of the guide wheel modeled by springs and dampers.

2 Mechanism Subsystem Model

The model of research steering system includes: Steering wheel, steering axles, hydraulic rotating valves, pinion gears, pumps, cylinders, hydraulic hoses, link rod structure, wheels and tires. During the driving, the rotating movement between the valve shaft and the valve covered due to the driver's driving will open and close the hydraulic valve. The hydraulic flow generated by the hydraulic pump is directed to one side of the plunger in the hydraulic cylinder, since the piston is directly connected to the gear bar the hydraulic pressure on the piston rod will force it through. Link bar mechanism acts on two guide wheels resulting in reduced driver's driving force on the steering wheel.

The diagram of the system is shown in Fig. 1, in which the steering axle and the twist bar are modeled as the spring-damper systems, the stiffness of the gear, the bar, the blow bar are also considered in the article. The torsion angle of the torsion control valve is the main link between the mechanical system and the hydraulic system (indicated by the dashed line in Fig. 1). During a steering process, the rotational displacement between the sleeve and spool of the rotary spool valve caused by the driver steer input directs the hydraulic force to one side of the piston in the hydraulic cylinder. Since the piston is directly connected with the rack, the hydraulic force on the piston helps the driver steer the two front wheels. In this paper, we also consider both the direction and horizontal displacement of the guide wheel and the symbols of these degrees of freedom are denoted as steering angles θ_{FW-L}, θ_{FW-R} and horizontal displacements x_{FW-L}, x_{FW-R} respectively of left and right wheels. The hydraulic cylinder attached to the body is modeled by spring and damping system. This study also considers the soft link between the guide wheel and the body through independent suspensions modeled by spring-damper system.

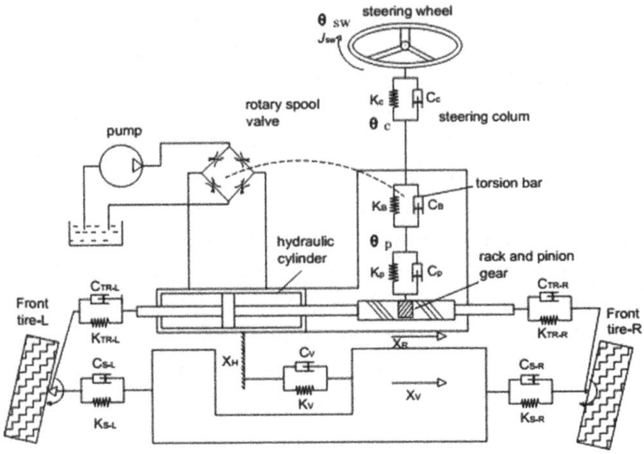

Fig. 1. The diagram of steering system model, including the connection with front suspensions.

The steering model in Fig. 1 consists of 12 degrees of freedom and is represented by 12 differential equations of motion from (1) to (12) as follows. The differential equation of motion of the steering wheel rim is as follows:

$$J_{sw}\ddot{\theta}_{sw} + C_c\left(\dot{\theta}_{sw} - \dot{\theta}_c\right) + K_c(\theta_{sw} - \theta_c) = T_{sw}, \quad (1)$$

where θ_{sw}, θ_c are rotation angles of the steering wheel rim and steering column; K_c, C_c are stiffness coefficient and viscous resistance of the steering column, respectively, J_{sw} denotes the moment of inertial of the steering wheel rim, T_{sw} is driving torque. The differential equation of motion of the steering column is represented by Eq. (2) as:

$$J_c\ddot{\theta}_c - C_c\left(\dot{\theta}_{sw} - \dot{\theta}_c\right) + C_B\left(\dot{\theta}_c - \dot{\theta}_p\right) - K_c(\theta_{sw} - \theta_c) + K_B\left(\theta_c - \theta_p\right) = -T_{FR-C}, \quad (2)$$

where θ_p is rotation angle of the gear, K_B, C_B are stiffness and viscous resistance coefficient of the helical rod, J_c is the moment of inertial of the steering column, T_{FR-c} is the friction torque in the steering column. The differential equation of motion of gears is represented by Eq. (3) as follows

$$J_p\ddot{\theta}_p - C_B\left(\dot{\theta}_c - \dot{\theta}_p\right) + C_p\left(\dot{\theta}_p - \frac{1}{N}\dot{x}_r\right) - K_B(\theta_c - \theta_p) + K_p\left(\theta_p - \frac{1}{N}x_r\right) = -T_{FR-p} \quad (3)$$

with J_p: moment of inertia of the gear, N: the gear ratio between rack and pinion, x_r: the displacement of the rack, K_p, C_p: stiffness coefficient and viscous resistance of gear,

T_{FR-P}: torque friction of rack and pinion. The differential equation of motion of hydraulic cylinder and piston is represented by Eq. (4):

$$m_H \ddot{x}_H + C_V(\dot{x}_H - \dot{x}_V) + K_V(x_H - x_V) = F_{FR-H} - F_B, \tag{4}$$

with: m_H: the mass of hydraulic cylinder-piston assemblies, K_V, C_V: the stiffness coefficient and viscous resistance of cylinder-piston assembly with body, x_H: the movement of cylinder-piston, T_{FR-H}: the friction torque in the cylinder-piston assembly, F_{FR-H}: is the friction force between piston and hydraulic cylinder and F_B: the hydraulic force acting on the piston (this force is generated from the hydraulic power unit studied in the following section). The differential equation of movement of the rack is represented by Eq. (5):

$$\begin{aligned} m_R \ddot{x}_r - \tfrac{C_p}{N}\left(\dot{\theta}_P - \tfrac{1}{N}\dot{x}_r\right) + C_{TR-L}\left(\dot{x}_r - l\dot{\theta}_{FW-L}\right) + C_{TR-R}\left(\dot{x}_r - l\dot{\theta}_{FW-R}\right) - \tfrac{K_p}{N}\left(\theta_P - \tfrac{1}{N}x_r\right) \\ + K_{TR-L}(x_r - l\theta_{FW-L}) + K_{TR-R}(x_r - l\theta_{FW-R}) = F_B - F_{FR-H} \end{aligned} \tag{5}$$

where m_R is the rack (with the piston and tie rods) mass, K_{TR-L}, C_{TR-L}, K_{TR-R}, C_{TR-R} denotes the stiffness coefficient and viscous resistance of the left and right tie rods, respectively, l is the distance from the end of tie rod to the center of tire twisting, θ_{FW-L}, θ_{FW-R} are the rotation angles of left and right guide wheels. The differential equations of rotation of the left and right guide wheels are shown by Eqs. (6) and (7):

$$\begin{aligned} J_{FW-L}\ddot{\theta}_{FW-L} - lC_{TR-L}\left(\dot{x}_r - l\dot{\theta}_{FW-L}\right) + C_{Ti-L}\left(\dot{\theta}_{FW-L} - \dot{\theta}_{CP-L}\right) \\ -lK_{TR-L}(x_r - l\theta_{FW-L}) + K_{Ti-L}(\theta_{FW-L} - \theta_{CP-L}) = -T_{FL-K} \end{aligned} \tag{6}$$

$$\begin{aligned} J_{FW-R}\ddot{\theta}_{FW-R} - lC_{TR-R}\left(\dot{x}_r - l\dot{\theta}_{FW-R}\right) + C_{Ti-R}\left(\dot{\theta}_{FW-R} - \dot{\theta}_{CP-R}\right) \\ -lK_{TR-R}(x_r - l\theta_{FW-R}) + K_{Ti-R}(\theta_{FW-R} - \theta_{CP-R}) = -T_{FR-K} \end{aligned} \tag{7}$$

with: J_{FW-L}, J_{FW-R}: moments of inertia of left and right guide wheels, K_{Ti-L}, C_{Ti-L}, K_{Ti-R}, C_{Ti-R}: the stiffness and viscous resistance of the left and right tires (noted that each tire is modeled by a spring system and damping), T_{FL-K}, T_{FR-K}: the friction torques applied to the left and right tires, θ_{CP-L}, θ_{CP-R}: the rotation angles of the left and right tires in the vertical direction. The dynamic equations of left and right tires are shown by Eqs. (8) and (9):

$$J_{CP-L}\ddot{\theta}_{CP-L} - C_{Ti-L}\left(\dot{\theta}_{FW-L} - \dot{\theta}_{CP-L}\right) - K_{Ti-L}(\theta_{FW-L} - \theta_{CP-L}) = -T_{FL-G} \tag{8}$$

$$J_{CP-R}\ddot{\theta}_{CP-R} - C_{Ti-R}\left(\dot{\theta}_{FW-R} - \dot{\theta}_{CP-R}\right) - K_{Ti-R}(\theta_{FW-R} - \theta_{CP-R}) = -T_{FR-G} \tag{9}$$

where J_{CP-L}, J_{CP-R} are the moments of inertia of contact patches of left and right tires with road surface, T_{FL-G}, T_{FR-G} are the rotation resistance torques which are calculated by the Pacejka formula as $T_{FL-G} = K_{FL-G}\theta_{FW-L} + C_{FL-G}\dot{\theta}_{FW-L}$, $T_{FR-G} = K_{FR-G}\theta_{FW-R} + C_{FR-G}\dot{\theta}_{FW-R}$. The differential equations of motion of horizontal displacements of the left and right tires are represented by Eqs. (10) and (11):

$$\begin{aligned} & m_{FW-L}\ddot{x}_{FW-L} + C_{S-L}(\dot{x}_{FW-L} - \dot{x}_V) + C_{T-LA}\dot{x}_{FW-L} + K_{S-L}(x_{FW-L} - x_V) \\ & \quad + K_{T-LA}x_{FW-L} = 0 \end{aligned} \tag{10}$$

$$\begin{aligned} & m_{FW-R}\ddot{x}_{FW-R} + C_{S-R}(\dot{x}_{FW-R} - \dot{x}_V) + C_{T-LA}\dot{x}_{FW-R} + K_{S-R}(x_{FW-R} - x_V) \\ & \quad + K_{T-LA}x_{FW-R} = 0 \end{aligned} \tag{11}$$

with: m_{FW-L}, m_{FW-R}: the masses of left and right guide wheels, x_{FW-L}, x_{FW-R} are the horizontal displacements of left and right guide wheels, K_{S-L}, C_{S-L}, C_{S-R}, K_{S-R} are the stiffness coefficient and viscous resistance of the left and right suspensions. The differential equation of movement of the front body is as follows

$$\begin{aligned} & m_V\ddot{x}_V - C_V(\dot{x}_H - \dot{x}_V) - C_{S-L}(\dot{x}_{FW-L} - \dot{x}_V) - C_{S-R}(\dot{x}_{FW-R} - \dot{x}_V) - K_V(x_H - x_V) \\ & \quad - K_{S-L}(x_{FW-L} - x_V) - K_{S-R}(x_{FW-R} - x_V) = 0 \end{aligned} \tag{12}$$

where x_V: the displacement of the front half of vehicle and m_V: the mass of the front half of vehicle.

3 Hydraulic Subsystem Model

Hydraulic power steering system model includes a hydraulic pump, a rotating valve assembly, a piston and a cylinder as shown in Fig. 2:

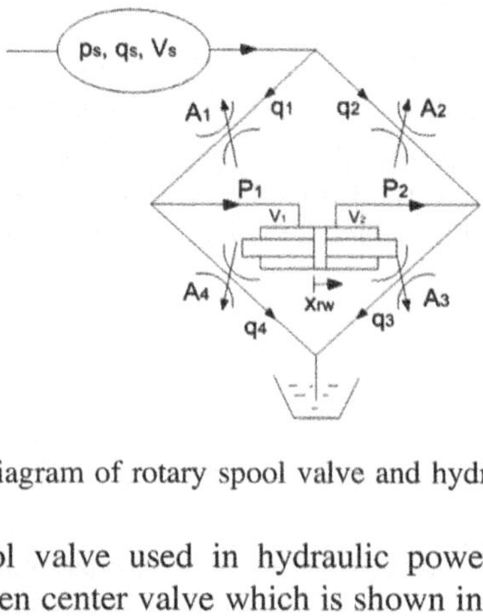

Fig. 2. The diagram of rotary spool valve and hydraulic cylinder.

A typical rotary spool valve used in hydraulic power steering system can be modeled as a four-way open center valve which is shown in Fig. 2. The rotating valve assembly consists of torsion bar, inner valve, and outer valve cover. When the driver turns the steering wheel, the twist bar is twisted, the valve rotates relatively to the outer shell to open the valve which increases the hydraulic pressure acting on one side of the piston while limiting the flow of hydraulic the other side. The pressure difference between the two sides of the piston will create the force. According to the results published in [3], the power assistant force F_b is calculated as follows

$$F_b = A_p(P_1 - P_2) \tag{13}$$

where P_1 P_2 are the fluid pressures in the cylinder chamber on the left and right sides of the cylinder, which are calculated according to the following differential equations [4]:

$$\dot{P}_s = \frac{\beta}{V_s}\left[Q_s - A_1 C_d\sqrt{\frac{2}{\rho}|P_s - P_2|}sign(P_s - P_2) - A_2 C_d\sqrt{\frac{2}{\rho}|P_s - P_1|}sign(P_s - P_1)\right] \tag{14}$$

$$\dot{P}_2 = \frac{\beta}{A_p\left(\frac{L}{2} - x_r\right)}\left[A_1 C_d\sqrt{\frac{2}{\rho}|P_s - P_2|}sign(P_s - P_2) - A_2 C_d\sqrt{\frac{2}{\rho}|P_s - P_1|}sign(P_s - P_1) + A_p\dot{x}_r\right] \tag{15}$$

$$\dot{P}_1 = \frac{\beta}{A_p\left(\frac{L}{2} - x_r\right)}\left[A_2 C_d\sqrt{\frac{2}{\rho}|P_s - P_1|}sign(P_s - P_1) - A_1 C_d\sqrt{\frac{2}{\rho}|P_s - P_2|}sign(P_s - P_2) + A_p\dot{x}_r\right] \tag{16}$$

In (14)–(16), A_p is the piston surface area, A_1, A_2 are open areas of the left and right orifices of the valve. According to [5], A_1, A_2 are estimated as $A_1 = \frac{1}{2}(\theta_c - \theta_p) + \frac{3}{4}$ if $\theta_c - \theta_p \leq -1$, $A_1 = 3(\theta_c - \theta_p) + 4$ if $\theta_c - \theta_p \geq -1$, with θ_c, θ_p are introduced in (2) and (3), and $A_2 = A_O - A_1$; C_d is the flow discharge coefficient, and ρ is the density of the fluid; L is the cylinder length, Q_s is the supply flow rate.

4 Simulation Results

The paper uses Matlab/Simulink software to simulated the dynamics of the steering system with the main parameters of mechanism subsystems and hydraulic system are as follows [3, 4]: J_{sw} = 0.0344 Nms^2/rad; C_c = 0.36042 Nm/(rad/s); K_c = 400000 Nm/rad; J_C = 0.00344 Nms^2/rad; C_B = 0.36042 Nm/(rad/s); K_B = 2 Nm/rad; J_p = 0.00344 Nms^2/rad; C_p = 0.36042 Nm/(rad/s); K_p = 400000 Nm/rad; N = 0.01 m; m_H = 3 kg; C_V = 4000 Nm/(rad/s); K_V = 1000 Nm/rad; m_R = 0.5 kg; l = 0.2 m; $C_{TR-L} = C_{TR-R}$ = 1450 Nm/(rad/s); $K_{TR-L} = K_{TR-R}$ = 300000 Nm/rad; $C_{Ti-L} = C_{Ti-R}$ = 4000 Nm/(rad/s); $K_{Ti-L} = K_{Ti-R}$ = 1000 Nm/rad; $J_{FW-L} = J_{FW-R}$ = 1.8 Nms^2/rad; $J_{CP-L} = J_{CP-R}$ = 1.8 Nms^2/rad; $m_{FW-L} = m_{FW-R}$ = 6 kg; $C_{S-L} = C_{S-R}$ = 10000 Nm/(rad/s); $K_{S-L} = K_{S-R}$ 28000 Nm/rad; m_V = 70 kg; the constants as: $\beta = 7 * 10^8$; $V_s = 1.02 * 10^{-4}$; $Q_s = 0.15 * 10^{-3}$ m^3/s; C_d = 0.61; A_p = 8.26.10^{-4} m^2; L = 0.8 m; ρ = 800 Kg/m^3; P_0 = 0.

To simulate the dynamics of the steering system and easily compare the simulated front wheel angles with the desired front wheel angles, the driving torque T_{sw} is chosen so that the steering wheel angle θ_{sw} is obtained as depicted in Fig. 3, then the desired front wheel angles are calculated from dividing the steering wheel angle θ_{sw} for the corresponding steering ratio. In this simulation, the corresponding steering ratio equals to 20. The hydraulic assist force applied to the steering system is plotted in Fig. 4. Figure 5 compares to the result of the simulated front wheel angle to the desired front wheel angle. It is clearly that there is an acceptably small difference between the desired front wheel angle and the simulated front wheel angle, whereas the maximum of angle differences is 0.47° and the RMS of the angle difference is 0.19°.

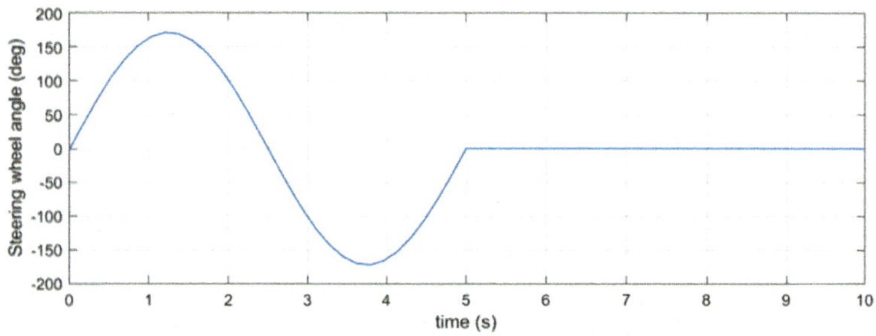

Fig. 3. Simulation of the steering wheel angle

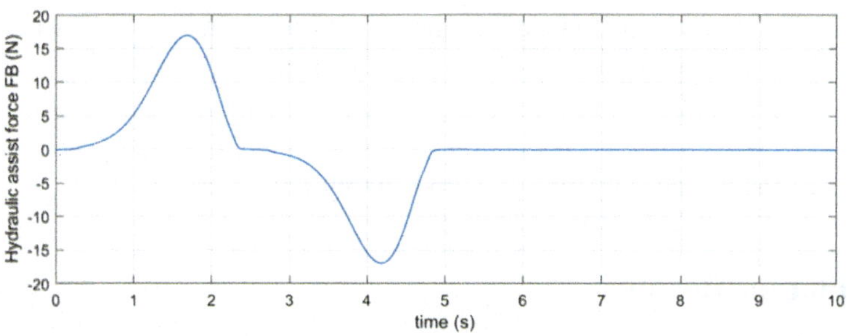

Fig. 4. Simulation of hydraulic assist force

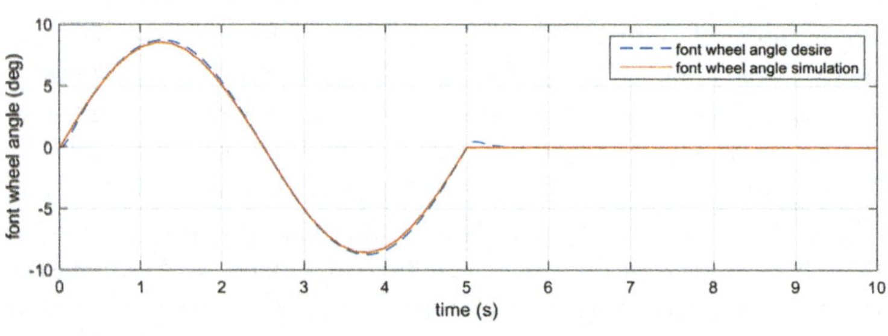

Fig. 5. Comparison between the desire and simulated front wheel angles.

5 Conclusion

A highly nonlinear hydraulic power steering system model has been developed. The mechanism and hydraulic subsystems are modeled in detail and then integrated into the completed system model kinetically. The numerical simulations have been conducted to investigate the dynamics of the steering systems, where the maximum difference between the desired and simulated front wheel angles is 0.47°. This simulation results show the accuracy of the dynamics model.

References

1. Lozia, Z., Zardecki, D.: Vehicle dynamics simulation with inclusion of free play and dry friction in steering system. SAE technical paper series 2002-01-0619 (2002)
2. Taheri, S., Kazemi, R., Tabatabai, M.: Analysis and optimization of vehicle steering system. SAE technical paper series 981113 (1998)
3. DellAmico, A., Krus, P.: Modeling, simulation, and experimental investigation of an electrohydraulic closed-center power steering system. IEEE-ASME Trans. Mechatron **20**, 2452–2462 (2015)
4. Tuấn, N.X., Trần, V.N., Bang, N.: Xây dựng mô hình hệ thống lái Steer by wire điện tử - thủy lực. Tạp chí Cơ khí, số đặc biệt (2017)
5. Yih, P.: Steer-by-wire: implications for vehicle handling and safety. Ph.D. dissertation, Stanford University (2005)

Robust Model Predictive Control Based Kinematic Controller for Nonholonomic Wheeled Mobile Robotic Systems

Dao Phuong Nam[1], Nguyen Hong Quang[2(✉)], Dao Cu Hung Phi[1], Tran Nam Anh[1], and Dinh Lam Bao[1]

[1] Hanoi University of Science and Technology, Hanoi, Vietnam
[2] Thainguyen University of Technology, Thai Nguyen, Vietnam
quang.nguyenhong@tnut.edu.vn

Abstract. A wheel mobile robot (WMR) is a remarkable mobile robot system with unknown parameters, or unpredictable and irregular features, external disturbances and it is hard to implement a control design. Due to the nonholonomic constraint's description, this work considers the combination between the model separation technique and model predictive control (MPC) approach to obtain the robust control design. Thanks to the advantage of this proposed solution, the external disturbances will be neglected in kinematic model. Hence, the MPC kinematic control design only cares about the under-actuated description. The stability of the whole system is determined via theoretical analysis of intermediate terminal controller. The results of offline simulation demonstrate the good performances of the proposed controller for a WMR.

Keywords: Robust model predictive control (RMPC) · Wheel mobile robot (WMR) · Cascade control design · Trajectory tracking control

1 Introduction

It is determined that designing control schemes for wheel mobile robots (WMRs), which has widely prospective applications in many important fields (search and rescue, space exploration, mine clearance, hospital task), is popular to researchers throughout the world. Due to the nonholonomic constraints and under-actuated property, a WMR can be separated into two subsystems to employ a cascade tracking and stabilization control law [1–3]. This separation technique can be absolutely extended for a WMR in presence of unknown wheel slip [4]. The outer loop control design implemented for underactuated subsystem in WMR, namely the kinematic model, based on traditional Lyapunov technique after using trigonometric transform technique [5, 6]. However, the condition of constant velocities need to be employed in [5, 6] and it was handled by the second transform in [7, 8]. The fully-actuated subsystem, namely dynamic model was implemented by robust adaptive nonlinear control based on Backstepping technique in presence of external disturbances and parametric uncertainties [7, 8]. Furthermore, the above cascade controllers are guaranteed to achieve the whole system stability [1–3, 7, 8], which is the extension of the work in [5, 6]. The authors in [9] also continue

K.-U. Sattler et al. (Eds.): ICERA 2019, LNNS 104, pp. 628–635, 2020.
https://doi.org/10.1007/978-3-030-37497-6_72

to implement cascade control design with the different approach of uncertain nonlinearities based on two time-scale filtering blocks. However, the condition that external disturbance does not vary slowly is necessary to use. The closed system stability was determined via theoretical analysis.

In recent years, the model predictive control (MPC) has been known as an appropriate control law in the situation of state/input constraints as well as guaranteeing many control objectives. The main distinguishing feature between MPC and classical nonlinear control algorithms is that a control input sequence being the solution of optimization problem, is only determined at each time instant [10]. Furthermore, although much more objectives will be developed under MPC technique rather than traditional control techniques, it is essential to consider the stability effectiveness of closed systems. In [10], the MPC law was proposed for approximate linear kinematic subsystem without considering the disturbances. The stability as well as tracking control problem was also mentioned by using cost function plays role of Lyapunov candidate function. This work continues to extend for H infinity problem under linear matrix inequalities (LMIs) technique in [11]. The influence of disturbances was handled in [12] by using separated disturbance observer and the tracking control problem was not cared about. In the view point of discrete time systems, the approximate linear model of WMR was employed to implement the MPC design with finite time horizon [13]. LMIs were considered to find the solution of optimization problem at each instant time. The tube based MPC was an important solution to implement MPC law for disturbed nonlinear system [14, 16]. The feasible region, feasibility problem for discrete time systems were mentioned in [14, 16] under the LMIs technique. A different approach of Tube-based MPC for constrained unicycle robots was proposed without the disturbance observer, LMIs [16]. The proposed robust MPC is the combination between tube-MPC algorithm for nominal systems obtaining by neglecting the disturbances and additional feedback law [17]. In order to consider the stability of this RMPC, feasibility as well as feasible region, terminal function were mentioned. In order to consider the stability problem, the intermediate terminal state controller as well as terminal state region was mentioned [17]. The robust MPC was developed for holonomic manipulators without separation technique, based on approximate linearization solution [18]. This brief address RMPC for WMRs based on analysis of nonholonomic model separation with the advantage of eliminating disturbances and uncertainties in kinematic model. The tracking control problem as well as stability was ensured by proposed robust MPC under appropriate tube set. It means that theoretical analysis of closed systems' stability was considered. The results of offline simulation point out good effectiveness of proposed RMPC law. The rest of the paper is structured as follows. In Sect. 2, we briefly describe problem statements as well as the problem of model separation. In Sect. 3, we demonstrate that RMPC law for WMRs absolutely was achieved by optimization algorithm. In Sect. 4, we provide offline simulation to validate the theoretical results. Finally, in Sect. 5, we give some remarkable problems and future directions.

2 Problem Statements and Model of WMRs

As known in traditional Euler-Lagrange theory, the WMRs can be described by the following dynamic equations (1–4) and parameter table (Table 1):

$$\begin{cases} M(q)\ddot{q} + C(q,\dot{q})\dot{q} + B(q)(F(\dot{q}) + \tau_d) = B(q)\tau + J^T(q)\lambda \\ J(q)\dot{q} = 0 \end{cases} \tag{1}$$

where $q = [x \quad y \quad \theta]^T$ is the joint variables vector. Describing in more details, x and y, θ are denoted the position coordinates, the orientation angle of the WMR with respect to the x axis, respectively.

Table 1. Parameter table [14]

Parameter	Meaning	Unit
b	Half of distance between two wheels	m
r	Radius of wheels	m
m_r	Mass of platform	kg
m_w	Mass of wheels	kg
I_{rz}	Moment of inertia of platform about z axis	kgm^2
I_{wz}	Moment of inertia of wheels about z axis	kgm^2
I_{wy}	Moment of inertia of wheels about wheel axis	kgm^2
x	Coordinate in Decartes frame	m
y	Coordinate in Decartes frame	m
θ	Heading angle of WMR	rad
θ_r, θ_l	Angular displacement of wheels	rad
v	Velocity of WMR	m/s

$M(q) \in \Re^{3\times3}; C(q,\dot{q}) \in \Re^{3\times3}; B(q) \in \Re^{3\times2}$ are the system inertia matrix, the centripetal, corriolis torques vector, the input transformation matrix, respectively.

In previous work [1, 2, 7, 8], in order to find the appropriate controller, we continue considering the separation of WMRs model due to the non-holonomic property in the following proposition.

Proposition 1. Thanks to the existence of solutions in the equation $S^T(q)J^T(q) = 0$ where $J(q)$ is the term of constraint force. There exists a vector η satisfying $\dot{q}(t) = S(q).\eta$. Implement the time derivative of above equation we obtain:

$$\dot{q} = \dot{S}(q)\eta + S(q)\dot{\eta} \tag{2}$$

Therefore, we achieve the two subsystems of WMRs described in (1), as follows:

$$\dot{q} = S(q)\eta \tag{3}$$

$$D(q)S(q)\dot{\eta} + C_1(q,\dot{q})\eta = B(q)\tau \tag{4}$$

Remark 1: It can be seen that the viewpoint of two above subsystems is the starting point of cascade robust adaptive control law in the work of [1–4]. However, it is determined that this technique only enable us to implement trajectory tracking control because of eliminating of the Lagrangian constraint coefficient (3, 4). In order to implement the motion/force control problem, the different separation techniques were considered in [15].

Remark 2: The above separation technique has following advantages:

- The inner subsystem (dynamic model) (4) is a fully-actuated being affected by unknown parameters, external disturbances.
- The outer subsystem, namely the kinematic model (3), is only under-actuated systems without uncertainties as well as external disturbances.

The control objective is to find the input control η by using the model predictive control (MPC) to guarantee not only the optimization problem but also the stability by means that the trajectory tracking control target.

3 Tube-MPC Control Design

After realizing the model separation as above contents, this work focused on finding the tube-MPC as following steps:

Because of the existence of trigonometric term in (3), the tracking error model of kinematic model will be transformed as described (5):

$$\begin{bmatrix} x_e(t) \\ y_e(t) \\ \theta_e(t) \end{bmatrix} = \begin{bmatrix} \cos\theta & \sin\theta & 0 \\ -\sin\theta & \cos\theta & 0 \\ 0 & 0 & 1 \end{bmatrix} \begin{bmatrix} x_r(t) - x(t) \\ y_r(t) - y(t) \\ \theta_r(t) - \theta(t) \end{bmatrix} \tag{5}$$

where $[x_r(t), y_r(t), \theta_r(t)]^T$ is the desired trajectory vector.

The dynamic of tracking error model of kinematic model can be considered as follows:

$$\frac{d}{dt}\begin{bmatrix} x_e(t) \\ y_e(t) \end{bmatrix} = \begin{bmatrix} 0 & \omega \\ -\omega & 0 \end{bmatrix} \begin{bmatrix} x_e(t) \\ y_e(t) \end{bmatrix} + \begin{bmatrix} -\upsilon + (\upsilon_r)\cos\theta_e \\ \upsilon_r \sin\theta_e \end{bmatrix} \tag{6}$$

$$\frac{d}{dt}\begin{bmatrix} x_e(t) \\ y_e(t) \end{bmatrix} = \begin{bmatrix} 0 & \omega \\ -\omega & 0 \end{bmatrix} \begin{bmatrix} x_e(t) \\ y_e(t) \end{bmatrix} + u_e(t) \tag{7}$$

Where $u_e(t)$ is given by

$$u_e(t) = \begin{bmatrix} -\upsilon + (\upsilon_r)\cos\theta_e \\ \upsilon_r \sin\theta_e \end{bmatrix} \tag{8}$$

This work continued to consider the MPC problem for the kinematic model (6) based on the following algorithm:

Algorithm 1:

1. At time $t_k = k\delta$, solving the Optimization Problem as follows:

$$\begin{aligned} &\min_{u(\tau|t_k)} J(x_e, y_e, u_e) = \int_{t_k}^{t_k+T} L(x_e(\tau|t_k), y_e(\tau|t_k), u_e(\tau|t_k))d\tau \\ &\qquad + g(x_e(t_k+T|t_k), y_e(t_k+T|t_k)); T = N\delta \\ &\min_{u(\tau|t_k)} J(x_e, y_e, u_e) = \\ &\int_{t_k}^{t_k+T} \left\{ \left[x_e^T(\tau|t_k), y_e^T(\tau|t_k)\right] Q \begin{bmatrix} x_e(\tau|t_k) \\ y_e(\tau|t_k) \end{bmatrix} + u_e^T(\tau|t_k) R u_e(\tau|t_k) \right\} d\tau \\ &+ \frac{1}{2}\left[x_e^T(\tau|t_k), y_e^T(\tau|t_k)\right] \begin{bmatrix} x_e(\tau|t_k) \\ y_e(\tau|t_k) \end{bmatrix} \end{aligned} \tag{9}$$

$$\begin{aligned} &u(\tau|t_k) \in U_{tube} = \left\{ [\upsilon, \omega]^T : \frac{|\upsilon|}{a} + \frac{|\omega|}{b} \le \lambda_{tube} = \frac{1}{\sqrt{2}} \right\} \\ &s.t.\ x_e(t_k+T|t_k)\ \&\ y_e(t_k+T|t_k) \\ &\qquad \in \Omega_{tube} = \left\{ [x_e, y_e]^T : k_1|x_e| + k_2|y_e| \le a\lambda_{tube} \right\} \end{aligned} \tag{10}$$

where k_1, k_2 are selected from the matrix Q, R in cost function as described in Remark 3.

2. Compute the following control input signal $[\upsilon, \omega]^T$ from the solution u_e of above optimization problem based on (8).
3. Apply this control input $[\upsilon, \omega]^T$ to the real system during the sampling time interval $[t_k, t_{k+1})$.
4. Updating the time instant $t_k \to t_{k+1}$ and come back step 1.

Remark 3: In order to ensure the tracking control system under Algorithm 1, this work need to consider the terminal region Ω for the terminal controller $u(\tau|t_{k+1}) = u^*(\tau|t_{k+1})$(in Lemma 1 [17]), for any $\tau \in (t_k+T, t_{k+1}+T]$, satisfying that if $x_e(t_k+T|t_k)\&y_e(t_k+T|t_k) \in \Omega$ and implementing this terminal controller, it obtains that:

$$
\begin{aligned}
& x_e(\tau|t_k) \& y_e(\tau|t_k) \in \Omega, \\
& u(\tau|t_k) = \begin{bmatrix} \upsilon(\tau|t_k) \\ \omega(\tau|t_k) \end{bmatrix} \in U, \\
& \frac{d}{dt} g(x_e(\tau|t_k), y_e(\tau|t_k)) + L(x_e(\tau|t_k), y_e(\tau|t_k), u_e(\tau|t_k)) \leq 0.
\end{aligned} \tag{11}
$$

Theorem 1: The control input under Algorithm 1 guarantees the asymptotical tracking control problem.

Remark 4: The key idea of Proof is as follows: Based on the Lyapunov Candidate function being the minimum cost function in Algorithm 1.

Remark 5: The proposed Algorithm 1 is different from the work in [17] because of neglecting the disturbances as well as unknown parameters in kinematic model by using the matrix $S^T(q)$. Furthermore, thanks to the view point of (7), the MPC was proposed without linearization technique as described in [10].

4 Offline Simulation Results

The simulation is demonstrated by Matlab software for kinematic model of a WMR under the parameters as follows:

$$\begin{array}{l} v_r = 0.015(m/s) \\ \omega_r = 0.04(rad/s) \end{array} \text{ and } \xi_r(0) = \begin{bmatrix} 0 & 0 & \pi/3 \end{bmatrix}^T$$

$$\xi_f(0) = \begin{bmatrix} 0.4 & -0.2 & -\pi/2 \end{bmatrix}^T$$

$$T = 1; \delta = 0.1$$

$$
\begin{aligned}
& P = diag\{0.4, 0.4\}; R = diag\{0.2; 0.2\}; \\
& \tilde{k}_1 = \tilde{k}_2 = 1.2;
\end{aligned}
$$

$\lambda_{\text{tube}} = 0.6636, \Omega_{tube} = \{\tilde{p}_e : |\tilde{x}_e| + |\tilde{y}_e| < 0.0542\}$, $P_{fe} = \{p_{fe}(t) : |p_{fe}| \leq [0.0017 \quad 0.0017]^T\}$. The results show the good behavior as described in following figurations (Figs. 1 and 2):

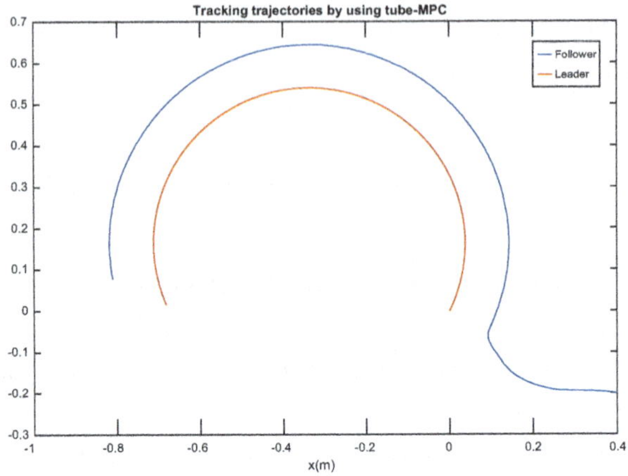

Fig. 1. The trajectory of a WMR

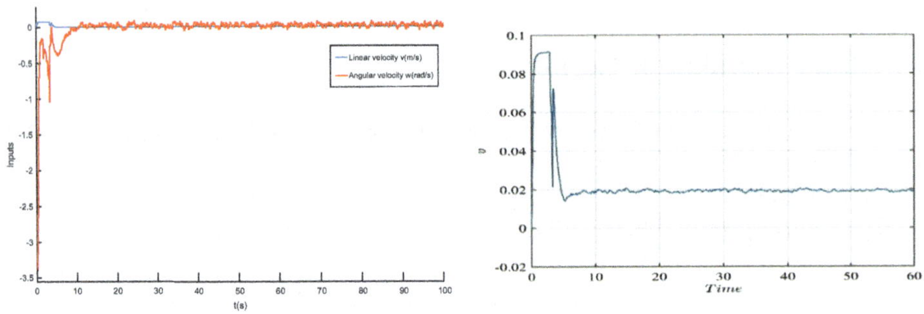

Fig. 2. The control inputs of a WMR

5 Conclusion

We introduce the simple MPC kinematic control scheme after implementing the whole system separation technique. The advantage of this proposed solution is to move the external disturbances and uncertainties into the dynamic model. Therefore, the kinematic control design only cares about the under-actuated description. The online proposed MPC approach for kinematic model ensures the tracking control problem under intermediate terminal controller.

References

1. Nam, D.P., Hung, N.M., Ty, N.T., Van Hau, P., Long, N.T.: A cascade controller for tracking and stabilization of wheeled mobile robotic systems. In: Proceedings of the 5th IEEE International Conference on System Science and Engineering (ICSSE), Ho Chi Minh City, pp. 305–309, July 2017

2. Van Hau, P., Nam, D.P., Ha, N.T., Thanh, P.T., Hai, H.T., Hanh, H.D.: Asymptotic stability of the whole tractor-trailer control system. In: Proceedings of the 5th IEEE International Conference on System Science and Engineering (ICSSE), Ho Chi Minh City, pp. 444–449, July 2017
3. Binh, N.T., Tung, N.A., Nam, D.P., Quang, N.H.: An adaptive backstepping trajectory tracking control of a tractor trailer wheeled mobile robot. Int. J. Control Autom. Syst. **17**(2), 465–473 (2019)
4. Van Tinh, N., Thuong, H.T., Tuan, P.M., Nam, D.P.: A Gaussian wavelet network-based robust adaptive tracking controller for a wheeled mobile robot with unknown wheel slips. Int. J. Control (2019). https://doi.org/10.1080/00207179.2018.1458156
5. Yue, M., Hu, P., Sun, W.: Path following of a class of non-holonomic mobile robot with underactuated vehicle body. IET Control Theory Appl. **4**(10), 1898–1904 (2010)
6. Asif, M., Khan, M.J., Cai, N.: Adaptive sliding mode dynamic controller with integrator in the loop for nonholonomic wheeled mobile robot trajectory tracking. Int. J. Control **87**(5), 964–975 (2014)
7. Huang, J., Wen, C., Wang, W., Jiang, Z.-P.: Adaptive stabilization and tracking control of a nonholonomic mobile robot with input saturation and disturbance. Syst. Control Lett. **62**, 234–241 (2013)
8. Huang, J., Wen, C., Wang, W., Jiang, Z.-P.: Adaptive output feedback tracking control of a nonholonomic mobile robot. Automatica **50**, 821–831 (2014)
9. Sun, W., Tang, S., Gao, H., Zhao, J.: Two time – scale tracking control of nonholonomic wheeled mobile robots. IEEE Trans. Control Syst. Technol. **24**(6), 2059–2069 (2016)
10. Dongbing, G., Huosheng, H.: Receding horizon tracking control of wheeled mobile robots. IEEE Trans. Control Syst. Technol. **14**(4), 743–749 (2006)
11. Chen, H., Miao-Miao Ma, H., Wang, Z.-Y.L., Cai, Z.-X.: Moving horizon H infinity tracking control of wheeled mobile robots with actuator saturation. IEEE Trans. Control Syst. Technol. **21**(2), 449–457 (2009)
12. Yu, S., Guo, Y., Meng, L., Ting, Q., Chen, H.: MPC for path following problems of wheeled mobile robots. Int. Fed. Autom. Control **51**(20), 247–252 (2018)
13. Pitanga, J.R., Araújo, H.X., Conceição, A.G., Oliveira, G.H.: Stable model-based predictive control for wheeled mobile robots using linear matrix inequalities. Int. Fed. Autom. Control **48**(19), 033–038 (2015)
14. Nam, D.P., Quang, N.H., Van Huong, N.: A new approach of a tube based output feedback model predictive control: control design for 2D overhead crane. In: Proceedings in ICERA: International Conference on Engineering Research and Applications, pp. 52–62 (2018)
15. Phuong, N.D., Quang, H.T., Nhat, A.D., Hong, Q.N.: Cascade motion/force control strategy of nonholonomic wheeled mobile robotic systems. In: 5th International Conference on Mechatronics and Robotics Engineering, Rome, Italy, pp. 69–72, February 2019
16. Phuong, N.D., Hong, Q.N., Minh, D.N.: Tube based robust model predictive control for disturbed nonlinear systems via solving linear matrix inequalities. In: 5th International Conference on Mechatronics and Robotics Engineering, Rome, Italy, pp. 118–122, February 2019
17. Sun, Z., Dai, L., Liu, K., Xia, Y., Johansson, K.H.: Robust MPC for tracking constrained unicycle robots with additive disturbances. Automatica **90**, 172–184 (2018)
18. Sun, T., Pan, Y., Zhang, J., Yu, H.: Robust model predictive control for constrained continuous-time nonlinear systems. Int. J. Control **91**(2), 359–368 (2018)

Scanning Noise Evaluation Based on 3D Mesh Model

Nguyen Duy Minh Phan(✉)

The University of Danang-University of Technology and Education, Danang 550 000, Vietnam
pndminh@ute.udn.vn

Abstract. Nowadays, with the evolution of new technologies, the use of 3D scanning systems is becoming more prevalent in the industry. These systems allow the acquisition of a large amount of data on the geometric representation of surfaces in a relatively short time. As the acquired data is usually exploited for many purposes such as reverse engineering or manufactured part inspection, the quality of point cloud or scanning quality has become a critical issue. The scanning noise is a common indicator to evaluate the scanning quality of the acquired data. In this paper, we propose a method to evaluate the scanning noise based on 3D mesh model. 3D point clouds obtained from the scanning system are aligned with 3Dmesh model of the scanned part and the noise is then calculated for each triangular facet. This noise is qualified in terms of quality by comparing with a threshold value of noise obtained from the sensor qualification protocol.

Keywords: Quality indicators · Scanning noise · Scanning system · Laser-plane sensor · Mesh

1 Introduction

3D scanning consists of the acquisition of 3D point clouds by means of a scanning system, to obtain a numeric representation of the geometry of the part. The 3D scanning system usually includes three different elements specifically the acquisition system, the displacement system and the treatment system [1] (Fig. 1).

The acquisition system allows acquiring data from the surface by using the 3D optical sensors. The most used optical sensors in the field of inspection are triangulation laser sensors because they offer high accuracy and relatively low cost compared to other systems. The displacement system (CMM, Robot, Machine-tool, …) allows moving the sensor to perform the part scanning. The treatment system is used to obtain coherent geometric data (3D coordinates generally) from the raw data provided by the sensor and/or the displacement system. Most of the work concentrates on the topic of scan path planning for automated part inspection [2–5], few of them propose the optimal scan path to obtain an expected scanning data quality. The scanning data is generally evaluated through the scanning noise by measuring the reference surface [6, 7]. The reference surface is generally a plane. The definition of the scanning noise is given by Hoppe et al., [8]. Given a set of n points M_i acquired by the scanning and the points P_i considered as

K.-U. Sattler et al. (Eds.): ICERA 2019, LNNS 104, pp. 636–641, 2020.
https://doi.org/10.1007/978-3-030-37497-6_73

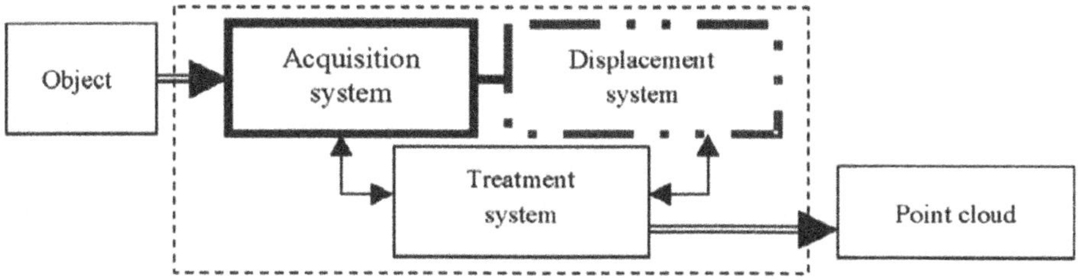

Fig. 1. 3D scanning system [1].

the orthogonal projections of the points M_i on the ideal surface associated by the least squares criteria (Fig. 2), the distance of each scanned point to the ideal surface e_i is defined by $e_i = M_i - P_i$. It is possible to calculate the standard deviation σ of the distribution of e_i in point cloud data. The scanning noise δ is calculated in the Eq. (1):

$$\delta = k.\sigma = k.\sqrt{\frac{1}{n}.\sum_{i=1}^{n}(e_i - \overline{e_i})^2} \quad (1)$$

where k is the coverage factor (k is generally set to 1), $\overline{e_i}$ is the mean value of the distance e_i, n number of points in the point cloud.

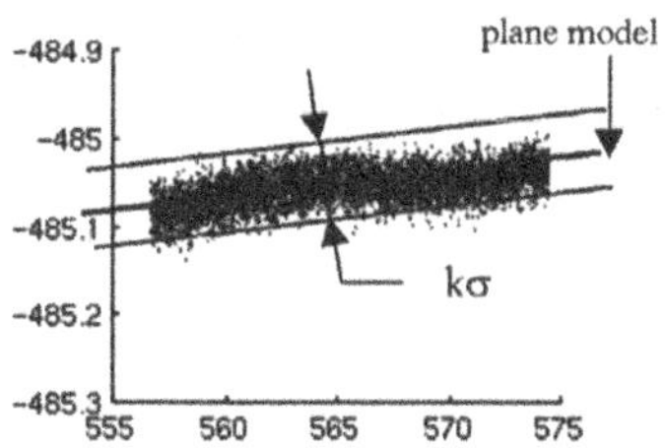

Fig. 2. Noise representation [8]

From the definition of the scanning noise in [8], this paper deals with the evaluation of point cloud quality through the scanning noise indicator. The paper is organized as follows. First, a protocol of the sensor qualification proposed in [6] is carried out to identify the threshold value of the noise for scanning quality. The second part of this paper presents a method to evaluate the scanning noise based on the 3D mesh model.

2 Sensor Qualification

Many studies have emphasized the importance of the scanning distance and the view angle on the scanning noise, that strongly influences the scanning quality [6, 9, 10]. The scanning quality is assured when the scanning noise is below a threshold that is given by the user in the considered application. In this part, we propose to use the protocol described in [6] to determine the threshold of scanning noise of the laser-plane sensor in function of the scanning distance and the view angle. Scanning noise is evaluated for different scanning distances and various view angles as shown in Fig. 3. The evaluation of scanning noise in function of the scanning distance is performed on the reference plane, allowing changing the scanning distance by moving the sensor normal to the surface within the field of view (FOV) of the sensor (Fig. 3a). The FOV of the laser-plane corresponds to the zone of the laser-plane that is visible by the camera. The FOV is defined by the height H and the width L. The scanning noise corresponding to each scanning distance is calculated from the point cloud obtained.

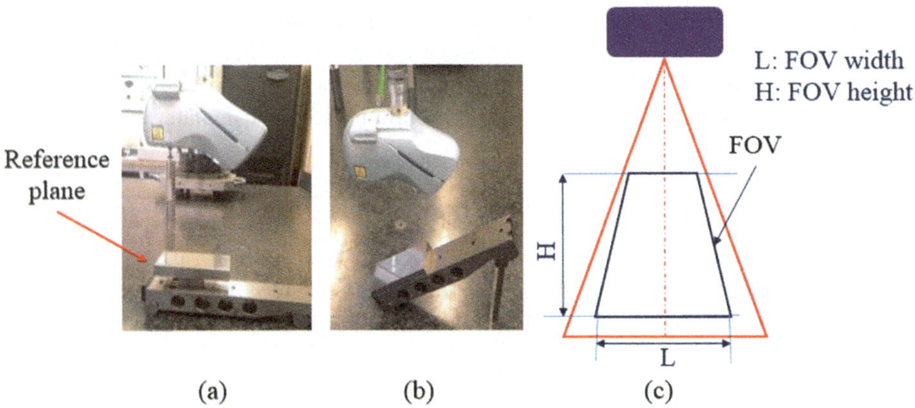

Fig. 3. Scanning noise evaluation in function of the distance (a) and the view angle (b), FOV of laser-plane (c)

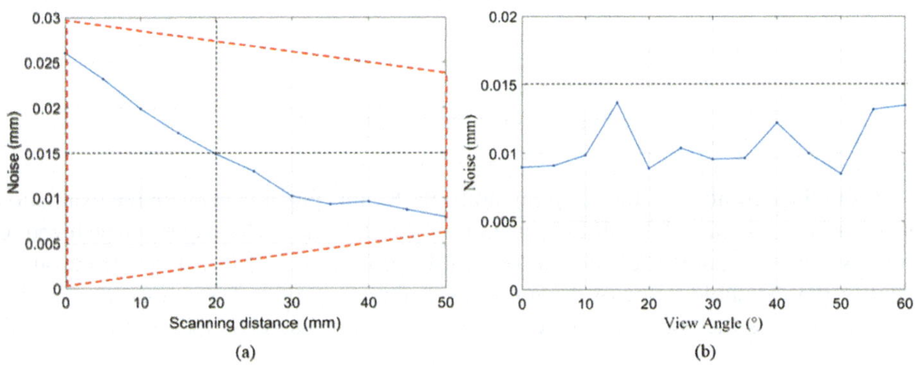

Fig. 4. Noise in function of the scanning distance (a) and the view angle (b)

The influence of the view angle is evaluated by scanning the reference plane at a constant distance while varying the view angle (Fig. 3b). The view angle is defined as the angle between the normal vector at the measured point and the laser plane. A sinus ruler is used to vary the view angles (Fig. 3b). For this study, the sensor we are using is a laser-plane sensor Zephyr KZ25 (Fig. 3). It consists of a laser source that projects a laser-plane on the surface to be scanned, and a CCD camera that observes the laser-plane. The scanning noise in function of the scanning distance shows a decrease of the noise from the bottom to the top of the FOV (Fig. 4a). On the other hand, we can point out that the surface is visible by the sensor if the view angle is less than 60° and the noise is less than 0.015 mm for the view angle range from [0°; 60°]. We consider the noise value of 0.015 mm as a threshold for the scanning quality, the scanning distance is determined by [20; 50] mm in FOV. This threshold value is used for evaluating the scanning noise of point cloud obtained from the actual scanning.

3 Scanning Noise Evaluation

The actual scanning gives a point cloud that is aligned to a CAD model as a stl format. The stl format is used in this study since it stores information about 3D models such as vertices, edges and the facets. Therefore, we can get the additional properties of mesh surface like the normal vector to the facet.

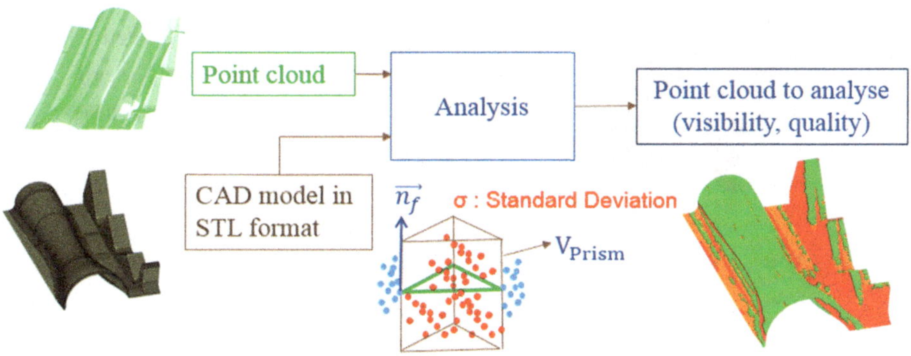

Fig. 5. Point cloud qualification approach

For each facet, a triangular prism is created, the base of which is the triangle defining the facets of the surface. The set of numerical points belonging to the prism volume V_{Prism} with an infinite height corresponds to the actual scanned facet (Fig. 5). For each triangle, the geometrical deviations between the scanned points and the facet are calculated. The associated standard deviation accounts for the actual scanning noise as defined in Eq. (1). A facet is *not-seen* if there are no scanned points in V_{Prism} and its color is red. A facet is *well-seen* if the noise is less than the threshold value 0.015 mm and its color is green. Conversely, if the noise of a facet is greater than 0.015 mm, this facet is tagged *poorly-seen* and set to the orange color.

Application. The noise evaluation is carried out with the actual scanning data obtained from a scanning system. This system consists of a Coordinate Measuring Machine (CMM) equipped with a motorized indexing head allowing the sensor to be oriented in discrete angular orientations. On the CMM, the sensor orientations are given by the two rotation angles around a horizontal axis (A) and a vertical axis (B) (Fig. 6a).

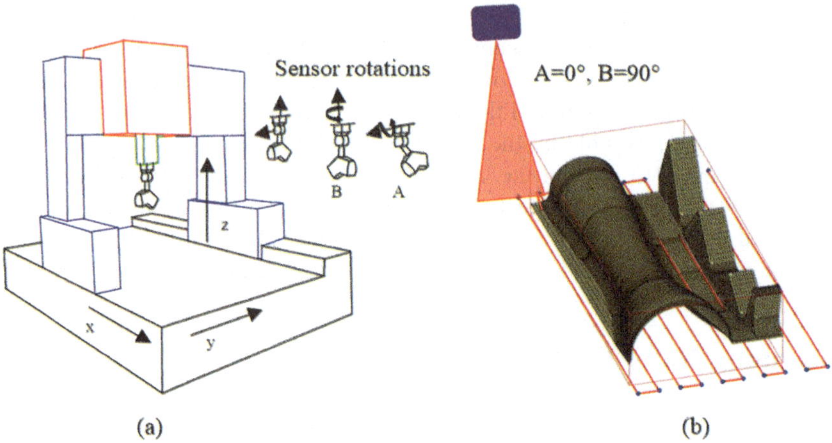

Fig. 6. (a) Scanning system [11] (b) Scan path

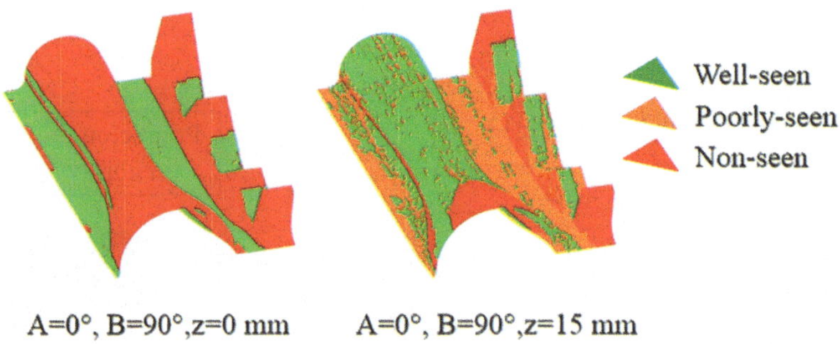

Fig. 7. Results for scanning noise evaluation

The different zigzag scan paths for various scanning distances are used to scan the test part. The results associated with the orientation (A = 0°; B = 90°) of the sensor and two different distances (z = 0 and z = 15 mm) are discussed in this paper. The orientation (A = 0°; B = 90°) corresponds to the axis of the laser beam in the perpendicular direction to the horizontal. The scan path z = 0 mm corresponds to the altitude of the zigzag scan path shown in Fig. 6b. Then, the scan path z = 15 mm is generated by shifting the path z = 0 mm vertically by 15 mm. The results of scanning noise evaluation are shown on Fig. 7. It is clear to identify the change in the visibility and quality of scanning areas from z = 0 mm to z = 15 mm. Some areas of the surface are visible only for the path z = 15 mm but not for the path z = 0 mm. The lowest scanning area for the path z = 15 mm is nearly orange (poorly-seen) (Fig. 7), since the scanning distance for this area is within the bottom of FOV which gives the scanning noise greater than the threshold value of noise. The results of scanning noise evaluation on the actual point cloud are therefore relevant to the results obtained from the sensor qualification described previously. However, it is easy to observe that there are some orange facets located in the green zones. This may come from the overlap between two consecutive passes of the scan path.

4 Conclusion

This presented paper proposes a method to evaluate the scanning noise of the acquired data based on the 3D mesh model. The noise is calculated for each facet of the mesh surface and then is qualified in function of quality by comparing with a threshold value of noise obtained from the sensor qualification protocol. The noise evaluation enables to validate the scan path in terms of the scanning quality. This might help to find the optimal scan path to obtain the expected scanning quality.

References

1. Lartigue, C., Bourdet, A., Contri, P.: Digitised point quality in relation with point exploitation. Measurement **32**(3), 193–203 (2002)
2. Phan, N.D.M., Quinsat, Y., Lartigue, C.: Optimal scanning strategy for on-machine inspection with laser-plane sensor. Int. J. Adv. Manuf. Technol., 1–14 (2019)
3. Son, S., Park, H., Lee, K.H.: Automated laser scanning system for reverse engineering and inspection. Int. J. Mach. Tools Manuf. **42**(8), 889–897 (2002)
4. Phan, N.D.M., Quinsat, Y., Lavernhe, S., Lartigue, C.: Scanner path planning with the control of overlap for part inspection with an industrial robot. Int. J. Adv. Manuf. Technol. **98**, 629–643 (2018)
5. Phan, N.D.M, Quinsat, Y., Lartigue, C.: Defining scanning trajectory for on-machine inspection using a laser-plane scanner. In: Advances on Mechanics, Design Engineering and Manufacturing II, pp. 349–358. Springer (2019)
6. Audfray, N.: Uneapprocheglobale pour la métrologie 3D automatique multi-systemes. Ph.D. thesis, Ecolenormalesupérieure de Cachan- ENS Cachan (2012)
7. Phan, N.D.M., Quinsat, Y., Lartigue, C.: Simulation of laser-sensor digitizing for on-machine part inspection. In: Advances on Mechanics, Design Engineering and Manufacturing, pp. 301–311. Springer (2017)
8. Hoppe, H., DeRose, T., Duchamp, T., McDonald, J., Stuetzle, W.: Surface reconstruction from unorganized points. ACM (1992)
9. Lartigue, C., Quinsat, Y., Mehdi-Souzani, C., Zuquete-Guarato, A., Tabibian, S.: Voxel-based path planning for 3D scanning of mechanical parts. Comput.-Aided Des. Appl. **11**, 220–227 (2014)
10. Mahmud, M., Joannic, D., Roy, M., Isheil, A., Fontaine, J.-F.: 3D part inspection path planning of a laser scanner with control on the uncertainty. Comput.-Aided Des. **43**, 345–355 (2011)
11. Mehdi-Souzani, C., Thiebaut, F., Lartigue, C.: Scan planning strategy for a general digitized surface. J. Comput. Inf. Sci. Eng. **6**, 331–339 (2006)

Simulation Vibration in the Automotive Powertrain

Thanh Quang Nguyen(✉)

Hanoi University of Industry, 298 Cau Dien, Hanoi, Vietnam
nguyenthanhquang@haui.edu.vn

Abstract. Vibration of the Automotive powertrain system affects not only the efficiency of the powertrain but also the noise-vibration-harshness (NVH) quality of the overall vehicle. In some severe cases, parts of the powertrain can be damaged by the excessive stress caused by vibration and particular resonance.

This paper presents a simulation result of modeling vibration of the powertrain system. The model takes into account of the static and dynamic force from the engine, from the clutch, from the transmission and active wheels. Using finite element methods (FEM), the harmonic responses of the powertrain in different loading scenarios are analyzed. From the numerical simulation results, references for several methods will be proposed to optimize the performance of the powertrain system.

Keywords: Powertrain · Vibration · Dynamic analysis

1 Introduction

Powertrain (PWT) performs the function of transferring and adjusting the torque from the engine to the vehicle wheels moving at different speeds corresponding to different load modes. During that process, twisting vibrations will occur and vibration and noise will appear. When torsional resonance phenomenon appears, it will cause damage to details in PWT. The cause of fluctuations is due to sudden changes in static and dynamic loads in PWT. These sudden changes are affected by the structure parameter and also by the assembly mode with gaps between the gears, bearings and joints in PWT.

From theoretical research results by using 3-D design method and Matlab Simulink software to simulate powertrain and experiment, it shows that structural parameters affect load in the powertrain. When the tension of the clutch spring increases, the load on the clutch is increased. The clutch closing time increases (the driver releases the clutch slowly), the car moves more smoothly because the engine torque and clutch torque do not reach the maximum value. The oscillation of the torque on the cardan shaft depends on the tilt angle and the shaft length. For large angle axes, vibrations with higher frequencies and amplitudes will occur [1]. By Lagerberg, Egardt by PWT simulation study is a closed loop that is oscillated by gaps. Experimental results also show the influence of design parameters on torsional vibration frequency of the system [2]. The authors Templin, and Egardt used the Linear Quadratic Regulator (LQR) optimal control theory to control the gap according to damping in PWT. Control algorithm has controlled the origin causing torsional vibration. Based on a gap change process,

K.-U. Sattler et al. (Eds.): ICERA 2019, LNNS 104, pp. 642–647, 2020.
https://doi.org/10.1007/978-3-030-37497-6_74

an optimized controller is used to limit vibration [3]. A real-time PWT simulation study (Real Time) has used the theory of multi-object kinetics. Clusters in the system are things that link together. The study has developed a model of input load from the motor related to the load of the road surface during the driving process, thus simulating the reactions of torsional oscillations occurring in PWT to provide the optimal vehicle operation mode [4]. Another study analyzed the effect of stiffness of clutch assemblies, gearboxes, axles and wheels to oscillation of powertrain. The authors used modal analysis and FFT method to simulate the oscillation of powertrain with 1-D modeling analysis of truck. The results show that vehicle vibrations should be investigated deeply with modal analysis and necessary countermeasures can be taken to prevent unwanted vibration values, for this reason clutch vibration levels at different stiffness values were compared with each other and mode shapes were found to illustrate the change [5].

The finite element method (FEM) and Ansys Workbench software have been used in this study, and the harmonic responses of the powertrain in different loading scenarios are analyzed.

2 Methology of Research

The study uses proprietary analysis methods (modal and harmonic analysis) to describe the kinetics and dynamics of the powertrain, thereby analyzing torsional vibrations in the system. The excitation force parameters include engine torque and torque from active wheels.

The linear kinetic equation of a mechanical system of form Eq. (1).

$$M\dot{x} + C\dot{x} + Kx = F \tag{1}$$

where:

M is mass matrix, K is stiffness matrix, C is the elastic matrix, F is the external force vector, x is the displacement vector in the general coordinate system.

Defined the modal matrix denoted by Φ as a column vector that describes the individual oscillator vectors Ω^2 of the system, form I matrix and cross matrix.

$$\Phi' M \Phi = I \tag{2}$$

$$\Phi' K \Phi = \Omega^2 \tag{3}$$

The relation between the displacement vector x and the vectors varies separately and is modalized according to the z-axis for torsional oscillation (z-axis is the vertical of PWT).

$$\ddot{x} = \Phi\ddot{z};\ \dot{x} = \Phi\dot{z}; x = \Phi z \tag{4}$$

Thus can represent Eq. (1) to the form (5)

$$\ddot{z} + C_{\Phi}\dot{z} + \Omega^2 z = F_{\Phi} \quad (5)$$

with:

$$C_{\Phi} = \Phi' C \Phi \quad (6)$$

$$F_{\Phi} = \Phi F \quad (7)$$

C_{Φ} is a cross-matrix when we substitute approximately equal to the elastic matrix C of the system.

3 Simulation of Vibration

3.1 Building 3-D PWT Model

The 3-D model is constructed in accordance with the actual shape and size of LF3070G1 light truck PWT with 3 tons load as shown in Fig. 1, on which the engine and clutch assemblies are described by crankshaft and connecting flywheel. Continued with the gearbox, the shaft transmits open and active bridges.

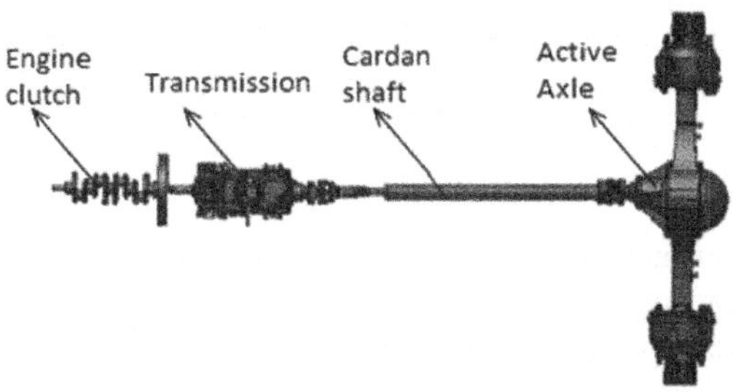

Fig. 1. The model of 3-D powertrain system of light truck LF3070G1

3.2 Boundary and Loading Conditions

The 3-D PWT model is updated into the Ansys Workbench software when surveyed at each gear of transmission. Figure 2 shows the updated results when surveyed at gear speed 5. Using the input conditions of the problem including:

Cylindrical boundary condition, position of bearing of intermediate axis. Two constraints fixed. Load conditions include: Maximum power of engine QLZ6012 is 81 kW and 320 Nm torque. Torque is the driving force of the engine placed on the active shaft of the gearbox at point B. Torque on the secondary shaft box. The number at hand 5 is 180 Nm (point E) is the active moment on the shaft of the propeller shaft (point G). At the top position under the axle cluster (point H) is the active torque on pineapple gears (point J) and 320 Nm. The barrier moments on the two active wheels are referred to the two outer ends of the right shaft (point C) of 22000 Nm, on the left half axis (point D) is 19000 Nm, Fig. 2.

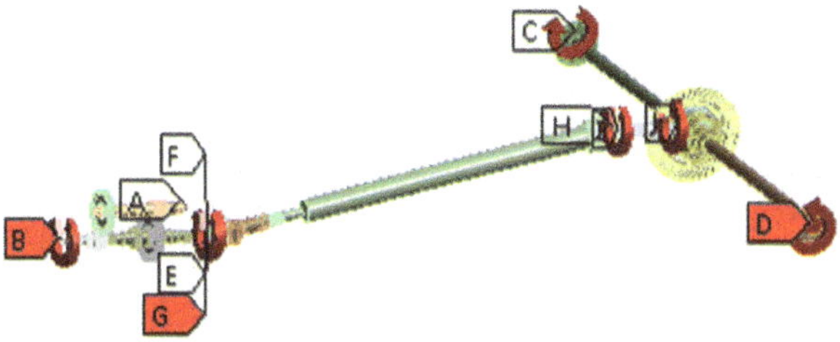

Fig. 2. Model of boundary and loading on the powertrain

4 Results of Simulation

Solving Eq. (5) with excitation force FΦ in Eq. (7) include static load and dynamic load. Static loading is of Engine computed torque, torque of clutch mechanism, torque of transmission, torque of cardan shaft, torque of rear axle [1]. Dynamic loading comes from the torque of active wheels. The vehicle is considered to run on a straight path, the road profile at active wheels can be expressed in terms of the one-dimensional wave [6]. We consider two speed levels of vehicle that are 40 km/h and 60 km/h and from there will get the same results to consider for other speeds of the vehicles. Loading of powertrain system on the gearbox is moment 1 (or cardan shaft), on the right half axle is moment 2 and on the left half axle is moment 3. These moment values will change with time and speeds of the vehicle and usually determine by experiments [7]. An experimental result of the research team performed on the light truck LF3070G1 is shown in Fig. 3 when the vehicle runs at 40 km/h and in Fig. 4 with the vehicle running at 60 km/h speeds. On the two half axes, the torque is relatively symmetrical, while the gearbox is fluctuated. At 40 km/h the maximum moment value on the gearbox is larger than the case where the car runs at higher speeds. Due to varying loads and different speeds should powertrain vibration with different frequencies and amplitudes. If the excitation force frequency is nearly to the vibration frequency of the powertrain, resonance will occur, when the system vibration very strongly and may cause damage to the powertrain. Therefore, it is important to determine the excitation force and vibration analysis of powertrain to avoid these resonant frequencies.

Some simulation results vibration with the time domain of powertrain of light trucks LF3070G1 shown in Fig. 5 when the vehicle runs at 40 km/h and in Fig. 6 when the vehicle runs at 60 km/h. It is clear that when the vehicle runs at low speed, the amplitude of vibration is greater than the case of high speed vehicles. As for choosing a reasonable movement speed for a vehicle, the powertrain has a low vibration amplitudes range and has little effect on the smoothness of the vehicle's movement.

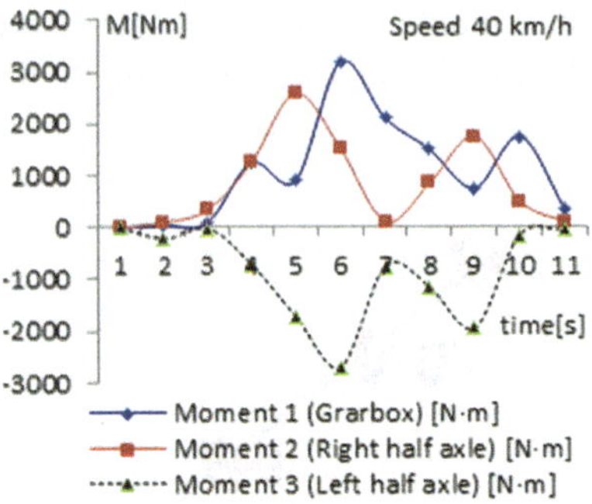

Fig. 3. Loading at 40 km/h

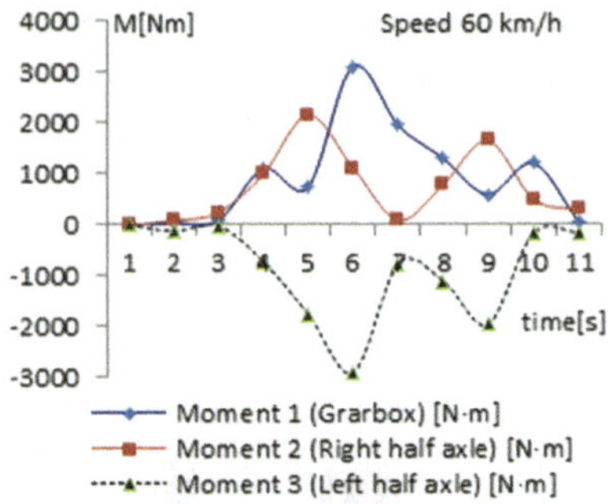

Fig. 4. Loading at 60 km/h

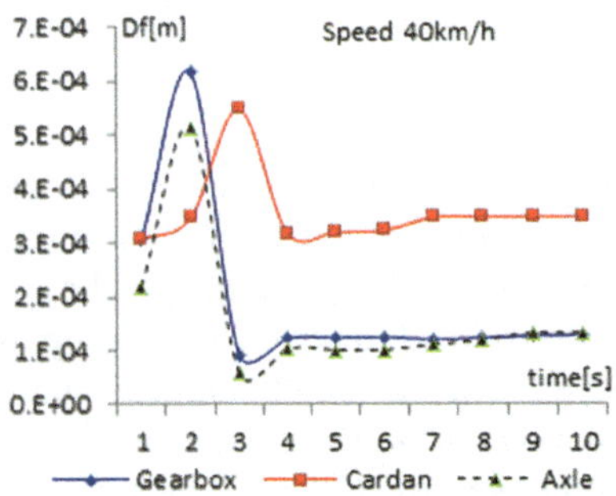

Fig. 5. Deformation at 40 km/h

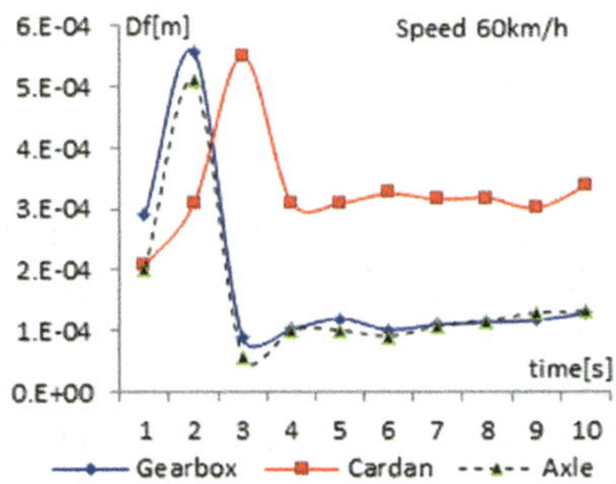

Fig. 6. Deformation at 60 km/h

We get the values shown in Table 1 of the deformation, equivalent elastic strain and (von-Mises) stress.

Table 1. Harmonic analysis of powertrain

	Object name	Total deformation (m)	Equivalent elastic strain (m/m)	Equivalent (von-Mises) stress (Pa)
Gearbox	Minimum	0	4.6226e–008	7600.5
	Maximum	2.3564e–004	8.0933e–003	1.4615e+009
Cardan	Minimum	0	6.9786e–008	5840.5
	Maximum	2.4126e–004	2.4918e–002	4.4823e+009
Active axle	Minimum	0	8.1038e–008	7350.2
	Maximum	1.9506e–003	8.2612e–003	1.3979e+009

The cardan shaft with maximum deformation is 2.4126e–004 (m) and maximum equivalent elastic strain is 2.4918e–002 (m/m). So, if the cardan shaft is imbalanced there will be a large vibrations of the powertrain. The equivalent (von-Mises) stress maximum at the gearbox and active axle because excitation force of engine and road surface.

5 Conclusion

Using Finite Element Method and Ansys Workbench software to simulate the powertrain can perform vibration analysis in real time through the results of harmonic and transient vibration analysis of the system. On the updated torque model, the angular velocity of powertrain creates the torque for active wheels and analyzes the reactions in powertrain. Excitation forces from engines, clutches, gearboxes, cardan shaft, and differential active axle are all major components of the model.

Developing the model will complement the system's damping system to overcome the disadvantages of vibration and increasing the durability of the system.

References

1. Nguyen, Q.T.: Analysis of static loading in automotive powertrain system. In: Fujita, H., et al. (Eds.) Advances Engineering Research and Application. Springer Nature Switzerland AG 2019: ICERA 2018. LNNS, vol. 63, pp. 1–7 (2019). https://doi.org/10.1007/978-3-030-04792-4_24
2. Lagerberg, A., Egardt, B.: Estimation of backlash in automotive powertrains – an experimental validation. In: IFAC Advances in Automotive Control Salerno, Italy (2004)
3. Templin, P., Egardt, B.: A powertrain LQR-torque compensator with backlash handling. Oil Gas Sci. Technol. Rev. IFP Energies Nouvelles **66**(4), 645–654 (2011). https://doi.org/10.2516/ogst/2011147
4. Yao, Z., Mousseau, R., Kao, B.G.: A Powertrain Model for Real-Time Vehicle Simulation, DSC North America 2003 Proceedings, Dearborn, Michigan, 8–10 October 2003. ISSN 1546-5071
5. Genç, M.O., Budak, B., Kaya, N.: Modelling and vibration analysis of powertrain system. Int. J. Automot. Sci. Technol. **2**(10), 17–25 (2018)
6. Nguyen, T.Q.: Finite Element Analysis in Automobile Chassis Design, Applied Mechanics and Materials, vol. 889, pp. 461–468. Trans Tech Publications, Switzerland (2019). https://doi.org/10.4028/www.scientific.net/AMM.889.461. ISSN: 1662-7482
7. Bai, J., Wu, X., Gao, F., Li, H.: Analysis of Powertrain loading dynamic characteristics and the effects on fatigue damage. Appl. Sci. **7**, 1027 (2017). https://doi.org/10.3390/app7101027
8. Galvagno, E., Velardocchia, M., Vigliani, A.: Torsional oscillations in automotive transmissions: experimental analysis and modelling. Shock Vibr. (2016). http://dx.doi.org/10.1155/2016/5721960. Article ID 5721960

Sliding Mode Control for a Pneumatic Servo System with Friction Compensation

Xuan Bo Tran[1(✉)], Van Lai Nguyen[1,2], Nam Chung Nguyen[1], Duc Thinh Pham[1], and Van Lanh Phan[1]

[1] Hanoi University of Science and Technology, Hanoi, Vietnam
bo.tranxuan@hust.edu.vn
[2] Hanoi University of Mining and Geology, Hanoi, Vietnam

Abstract. In this study, we propose a new control approach for the position tracking of a pneumatic cylinder. The proposed control approach is based on a multiple surface sliding (MSS) control in combination with a LuGre model-based friction observer. A pneumatic servo system using a pneumatic cylinder and two proportional flow control valves is considered. The mathematical model of the system is built to match the design of the multiple-surface sliding control. In the model, the friction force is modelled by the LuGre friction model. The analyzation of absolute stability of the system is based on Lyapunov stability standard. Experimental evaluation with sinusoidal reference inputs at different amplitudes and frequencies is carried out to show good tracking performance and robustness of the proposed control approach.

Keywords: Pneumatic cylinder · Multiple surface sliding control · LuGre model

1 Introduction

Pneumatic systems are widely used in automatic systems because they are cheap, clean, easy to assemble and maintain. With available air source, they can provide a good force/weight ratio comparing to mechanical and electric driving systems [1]. However, the pneumatic systems are well known as high nonlinearity and are subjects to modelling uncertainties [2] due to the compressibility of the air, nonlinear properties of the proportional valve and friction in the pneumatic actuators.

In order to deal with the nonlinear characteristics and modelling uncertainties, nonlinear control strategies or advanced control strategies need to be applied [3]. Several nonlinear control strategies have been applied to control position of the pneumatic systems such as adaptive control [4], adaptive control combined with neural network [5], and sliding mode control [6]. In these control strategies, the main effort of the controllers is to compensate for the systematic nonlinearity and the modelling uncertainties of the system while the friction force in the pneumatic actuators are often considered negligible or existent in bounded uncertain parameters.

Friction that always exists in pneumatic actuators can cause limit-cycles phenomena and stick-slip motion and deteriorate the system control performance. The design of a controller coupled with friction compensation has the potential to improve the control

K.-U. Sattler et al. (Eds.): ICERA 2019, LNNS 104, pp. 648–656, 2020.
https://doi.org/10.1007/978-3-030-37497-6_75

performance of the system [7]. To design the controller with friction compensation, it is necessary to obtain a suitable dynamic friction model for the pneumatic actuators. Several dynamic friction models have been proposed so far [8–10] and among them, the LuGre model [8] is most widely utilized in control applications [11]. This friction model can observe basic friction properties in most mechanical systems. In addition, it is quite suitable for theoretical calculations and easy to implement.

The nonlinear control strategies combined with friction compensation based on LuGre friction model for the pneumatic systems have been carried out by some authors. Particularly, Raul et al. [12] used a cascade controller in combination with friction compensation based on LuGre friction model, Karim et al. [13] applied an output-feedback controller based on a LuGre friction closed-loop observer dynamics and Sobczyk et al. [14] used a continuous version of the LuGre friction model applied to the adaptive control. The experimental evaluations of the methods indicate a significant improvement over the control methods without friction compensation. However, to the best of our knowledge, currently no research has applied the sliding control scheme combined with friction compensation based on the friction model to evaluate the ability of control performance for the pneumatic systems. In this study, we apply the multiple - surface sliding control method [6] in combination with friction compensation using LuGre friction model [8] to control the position of a pneumatic cylinder. A pneumatic servo system using a pneumatic cylinder and two proportional flow control valves will be considered. The mathematical model of the system is derived to adapt the design of a multiple-surface sliding controller. Absolute stability of the system based on Lyapunov stability standard is proved. Experimental evaluations of sinusoidal reference inputs with variable amplitudes and frequencies are presented to indicate the significantly improved performances of the proposed controller.

2 System Model

The servo pneumatic system considered in this investigation is shown in Fig. 1. In the system, a pneumatic double-acting cylinder is used, and the piston motion is controlled by two proportional flow control valves. As shown later in the experimental setup in Sect. 4, if the valve control inputs u_1 or u_2 vary from 2.5 to 5 V, the valves will push air into the cylinder chamber (the building pressure case) and if the valve inputs u_1 or u_2 vary from 0 to 2.5 V, the valves will release air into the atmosphere (the exhausting pressure case). Therefore, by combining signals between u_1 and u_2 of the two valves with the control law u of the system as shown in Eq. (1), the extending and retracting motions of the piston can be obtained.

$$\begin{aligned} u_1 &= 2.5 + u \\ u_2 &= 2.5 - u \end{aligned} \tag{1}$$

In Fig. 1, x is the piston position, p_1 and p_2 are respectively the pressures in the cylinder chambers, p_s is the supply pressure, and A_1 and A_2 are respectively the piston areas.

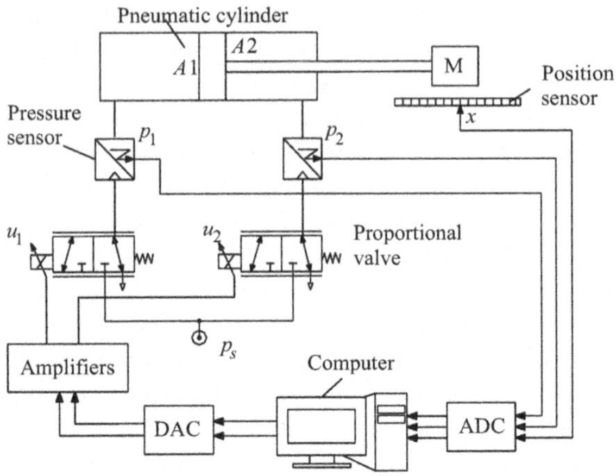

Fig. 1. Schematic of the pneumatic servo system

By defining $x_1 = x,\ x_2 = \dot{x}_1$, and $x_3 = p_1A_1 - p_2A_2$, we can derive the following stated equation from the system dynamics.

$$\begin{aligned} \dot{x}_1 &= x_2 \\ \dot{x}_2 &= \frac{1}{M}x_3 - \frac{F_r}{M} \\ \dot{x}_3 &= F(\mathbf{x}, t) + G(\mathbf{x}, t)u \end{aligned} \tag{2}$$

Where, $\mathbf{x} = [x_1 \quad x_2 \quad x_3]$ is state-vector, M is the load mass and

$$\begin{aligned} F(\mathbf{x}, t) &= -\frac{k\mathrm{A}_1^2 p_1 x_2}{V_1} - \frac{k\mathrm{A}_2^2 p_2 x_2}{V_2} \\ G(\mathbf{x}, t) &= k\sqrt{kRT_s}\left[\begin{array}{l} \frac{1+\mathrm{sgn}(u)}{2}\left(\frac{\gamma_{1b} p_s K_{V1} \mathrm{A}_1}{V_1} + \frac{\gamma_{2e} p_2 K_{V2} \mathrm{A}_2}{V_2}\right) \\ + \frac{1-\mathrm{sgn}(u)}{2}\left(\frac{\gamma_{1e} p_1 K_{V2} \mathrm{A}_1}{V_1} + \frac{\gamma_{2b} p_s K_{V1} \mathrm{A}_2}{V_2}\right) \end{array}\right] \end{aligned} \tag{3}$$

In Eq. (3), V_1 and V_2 refer respectively to the volumes of the cylinder chambers 1 and 2, R is the gas constant, k is the specific heat ratio, T_s is the temperature of the supply source, K_{V1} and K_{V2} are respectively the valve gains in the building and exhausting pressure cases, and γ_{ib} and γ_{ie} are respectively the modifying factors in the building and exhausting pressure cases.

The term F_r in Eq. (2) is the friction force that is described by the LuGre model proposed by Canudas et al. [8]. The LuGre model is given by the following equations.

$$\frac{dz}{dt} = v - \frac{\sigma_0 z}{g(v)} v \tag{4}$$

$$g(v) = F_c + (F_s - F_c)e^{-(v/v_s)^n} \tag{5}$$

$$F_r = \sigma_0 z + \sigma_1 \frac{dz}{dt} + \sigma_2 v \tag{6}$$

Where, z is the mean deflection of the elastic bristle, σ_0 is the stiffness of the elastic bristle, $g(v)$ is the Tribeck function, F_s is the static friction force, F_c is the Coulomb friction force, v_s is the Stribeck velocity, n is the exponent that affects the slope of the Stribeck curve, σ_1 is the micro-viscous friction coefficient and σ_2 is the viscous friction coefficient.

3 Controller Design

It is shown in Eqs. (2) to (6) that the relationship between the control input u and the position output x_1 of the pneumatic piston is highly nonlinear. Therefore, the purpose here is to design a controller so that the piston position x_1 can track the desired trajectory x_{1d} well. In this study, we propose a controller that is based on the multiple-surface sliding control (MSS control) in [6] in combination with a friction observer. The control diagram for the system is shown in Fig. 2. In order to apply the multiple-surface sliding control, we define three sliding surfaces s_i, i = 1, 2, 3 as follows:

$$s_i = x_i - x_{id} \tag{7}$$

where, x_{id} is the desired trajectories of x_i. Obviously, x_{1d} is the desired position of the pneumatic piston and x_{2d}, x_{3d} are selected by

$$\begin{aligned} x_{2d} &= \dot{x}_{1d} - k_1 \frac{s_1}{\phi_1} \\ x_{3d} &= M\dot{x}_{2d} + \hat{F}_r - k_2 \frac{s_2}{\phi_2} \end{aligned} \tag{8}$$

where ϕ_1 and ϕ_2 are respectively the boundary layer thickness of the surfaces s_1 and s_2. k_1 and k_2 are constants and are determined so that the x_1 and x_2 converge asymptotically to the desired trajectories x_{1d} and x_{2d}, respectively. $\hat{F}_r$ is the friction force that is estimated by the friction observer based on the LuGre friction model as follows [8]:

$$\hat{F}_r = \sigma_0 \hat{z} + \sigma_1 \frac{d\hat{z}}{dt} + \sigma_2 x_2 \tag{9}$$

with $\hat{z}$ is the estimated mean deflection of the elastic bristle and is given by

$$\frac{d\hat{z}}{dt} = x_2 - \frac{x_2 \sigma_0 \hat{z}}{g(x_2)} - k_0 s_1 \tag{10}$$

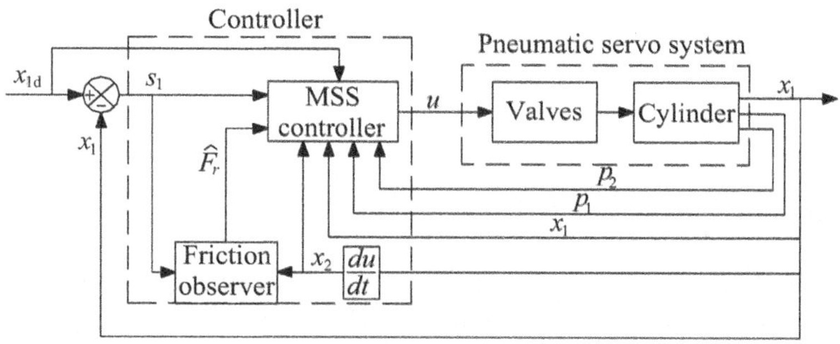

Fig. 2. Diagram of the proposed controller

Where, $g(x_2)$ is calculated by Eq. (5) and k_0 is constant and is determined so that the mean deflection z converges a symptotically to the estimated mean deflection $\hat{z}$.

The control law u of the proposed controller is proposed as follows:

$$u = \frac{1}{\hat{G}}\left(\dot{x}_{3dk} - k_3 \frac{s_3}{\phi_3} - F(\mathbf{x}, t)\right) \tag{11}$$

where $\hat{G}$ is the known nominal value of $G(\mathbf{x}, t)$, $\dot{x}_{3dk}$ is the known part of $\dot{x}_{3d}$, ϕ_3 is the boundary thickness of the sliding surface s_3, and k_3 is constant and is determined so that the force x_3 converges asymptotically to the desired force x_{3d}.

It is noted in the proposed controller that two revisions are made to the MSS controller [6]: first, the term $\hat{F}_r$ is added in Eq. (8) to compensate for the friction force in calculating the desired force x_{3d} and second, the term $F(\mathbf{x}, t)$ is added in Eq. (11) in order to reduce the oscillation in the measured signal of the control law.

The analyzation of the stability property of the closed-loop system is based on Lyapunov stability method [3]. To ensure the stability of the proposed control system, the following conditions need to be satisfied:

$$\begin{aligned}
&k_1 = \phi_2 + \eta_1 \\
&k_2 = M_{\max}\left[\tfrac{1}{M_{\min}}\phi_3 + \tfrac{1}{M_{\min}}2F_s + \eta_2\right] \\
&k_4 = \tfrac{\hat{G}}{G_{\min}}\left[\left(1 + \tfrac{G_{\max}}{\hat{G}}\right)F_{\max} + \left(1 + \tfrac{G_{\max}}{\hat{G}}\right)\dot{x}_{3\text{dkmax}} + \dot{x}_{3\text{dumax}} + \eta_3\right] \\
&k_0 < \tfrac{2F_s\sigma_0}{M_{\min}g(x_2)} \\
&\phi_1 \le \tfrac{\phi_2}{|s_2|}|s_1| \\
&\eta_1 > \tfrac{\phi_1}{2}
\end{aligned} \tag{12}$$

where η_1, η_2 and η_3 are positive constants, $M_{\min}$ and $M_{\max}$ are respectively the minimum and maximum load masses, $G_{\min}$ and $G_{\max}$ are respectively the minimum and maximum values of $G(\boldsymbol{x}, t)$, $F_{\max}$ is the maximum value of $F(\boldsymbol{x}, t)$, and $\dot{x}_{3\text{dumax}}$ is the maximum value of the unknown part of $\dot{x}_{3\text{d}}$.

4 Results and Discussion

The experimental pneumatic servo system for evaluation of the proposed control method is shown in Fig. 3. It includes a pneumatic cylinder (SMC, CM2L25-300) with internal diameter of 0.025 m, rod diameter of 0.01 m and piston stroke of 0.3 m respectively. Two proportional flow control valves (SMC, VEF3121) that can supply a flow rate up to 720 l/min with a rated voltage of 5 V were used. If the valve inputs vary from 2.5 to 5 V, the valves will push air into the cylinder chamber and if the valve inputs vary from 0 to 2.5 V, the valves will release air into the atmosphere. The position of the piston was measured with a position sensor (Novotechnik, LWH0300). Measuring accuracy of the displacement sensor is less than 0.5% F.S. Two pressure sensors (SMC, PSE540) with a measuring accuracy less than 1% F.S were used to measure the pressures in the cylinder chambers. The supply pressure p_s and the atmosphere pressure p_{atm} were set at 5 bar and 0.1 bar, respectively. The position signal and the pressure signals were read via a personal computer through a 12 bits Analog to Digital converter (ADC) and the computer sent the control signals of the two valve through a 12 bits Digital to Analog converter (DAC) (Advantage, USB4711). Two amplifiers (SMC, VEA250) were used to convert the voltage signals to the current signals of the valves. The program for data acquisition was done by using Microsoft visual C++. The measured signals were recorded at the interval of 1.1 ms.

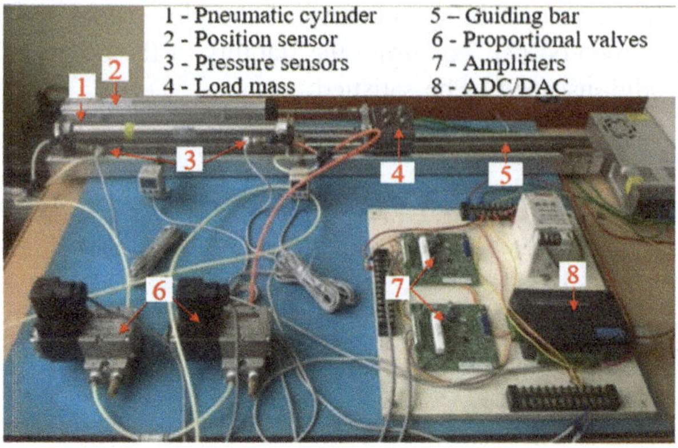

Fig. 3. Photo of the experimental test setup

Control performance of the proposed controller is evaluated in this study by sinusoidally desired inputs with different amplitudes and frequencies. In addition, the control performance of the proposed controller is also compared to that of the MSS controller [6] at the same conditions of the desired inputs. Figure 4 shows the control performance of the proposed controller and of the MSS controller for a sinusoidal input$x_{1d} = 0.15 + A\sin(2\pi ft)$ with $A = 0.1$ m and $f = 0.1$ Hz. The parameters of the proposed controller were chosen as $\phi_1 = 0.001$, $\phi_2 = 0.01$, $\phi_3 = 0.5$, $\eta_1 = \eta_2 = \eta_3 = 1$, and the parameter of the friction force observer was chosen as $k_3 = 20$. It is shown in Fig. 4a for the MSS controller that the actual position of the piston x_1 tracks the desired input x_{1d} with a relatively good accuracy. The maximum tracking error obtained in the steady state shown in Fig. 4b is about 0.008 mm (4% of the maximum crossed distance); the large errors appear mainly at high velocities of the piston. However, as shown in Fig. 4c, a much better tracking ability of the proposed controller than that of the MSS controller can be obtained. The maximum tracking error in the steady state of the proposed controller is significantly reduced to 0.005 mm (2.5% of the maximum crossed distance) in Fig. 4d. In addition, it can be calculated that the root means square error (RMS error) is 2.1% with the proposed controller and is 3.7% with the MSS controller. More comparisons of the peak error and of the RMS error between the proposed controller and the MSS controller for sinusoidally desired inputs with different amplitudes at a constant frequency $f = 0.1$ Hz and for sinusoidally desired inputs with different frequencies at a constant amplitude $A = 0.1$ m are shown in Tables 1 and 2, respectively. It shows that the percentage rates of both the peak and the RMS errors with the maximum crossed distance increase while the amplitude reduces and the frequency rises. However, in all cases, the tracking performances of the proposed controller are better than those of the MSS controller.

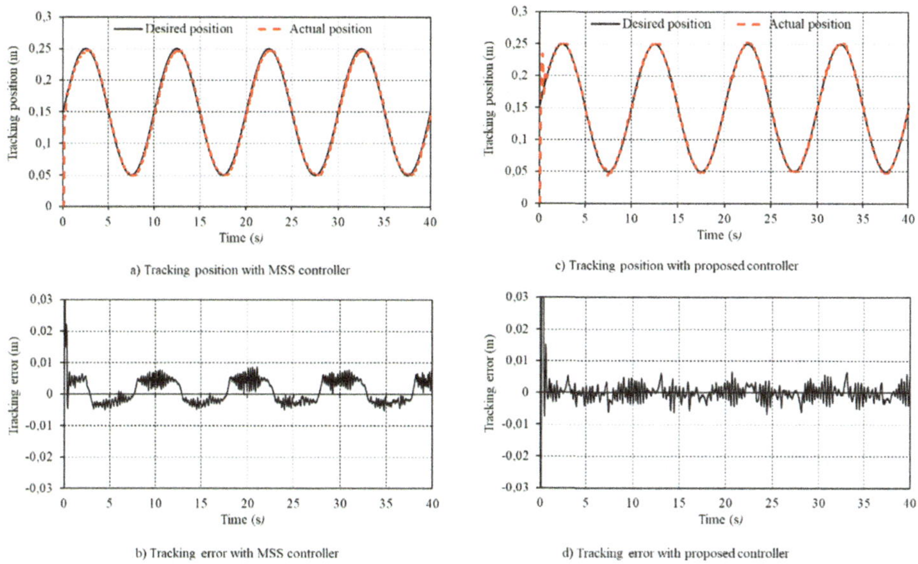

Fig. 4. Comparison of the control performances between the proposed control and the MSS control with a desired input $x_{1d} = 0.15 + 0.1\sin(2\pi ft)$, $f = 0.1$ Hz.

Table 1. The peak error and RMS error with different amplitudes of the sinusoidally desired input at a constant frequency $f = 0.1$ Hz

Amplitude A (mm)	Peak error (mm)		RMS error (mm)	
	Proposed control	MSS control	Proposed control	MSS control
10	2.5	5	1.2	2.5
25	4.5	6	1.4	2.7
50	4.8	7	1.6	3.0
75	5	7	1.8	3.2
100	5	8	2.1	3.7

Table 2. The peak error and RMS error with different amplitudes of the sinusoidally desired input at a constant amplitude $A = 0.1$ m

Frequency of desired input (Hz)	Peak error (mm)		RMS error (mm)	
	Proposed control	MSS control	Proposed control	MSS control
0.1	5	8	2.1	3.7
0.25	13	15	4.6	6.3
0.5	16	23	6.4	10.4
0.75	20	24	9.4	10.8
1	28	30	11.9	12.6

5 Conclusions

A new nonlinear controller for the position control performance of a pneumatic cylinder has been presented. The controller is based on the multiple surface sliding (MSS) control combining with the model based-friction force compensation. For the MSS control strategy, three surfaces of the position, velocity and the force generated from the cylinder pressures are considered. The friction force is estimated by an observer based on the LuGre model. The experimental evaluations of the proposed controller and of the MSS controller under different conditions of the sinusoidal reference input have shown significant improvements of the proposed controller.

Acknowledgements. This research was funded by the Hanoi University of Science and Technology (HUST) under project number T2018-PC-042.

References

1. Krivts, I.L., Krejnin, G.V.: Pneumatic Actuating Systems for Automatic Equipment, Structure and Design. Routledge, Taylor & Francis Group LLC (2006)
2. Ali, H.I., Noor, S.B.M., Bashi, S.M., Marhaban, M.H.: A review of pneumatic actuators (modeling and control). Aust. J. Basic Appl. Sci. **3**, 440–454 (2009)
3. Slotine, J.J.E., Li, W.: Applied Nonlinear Control. Prentice-Hall, New Jersey (1991)
4. McDonell, B.W., Bobrow, J.E.: Adaptive tracking control of an air powered robot actuator. Trans.-Am. Soc. Mech. Eng. J. Dyn. Syst. Meas. Control. **115**, 427–433 (1993)
5. Tanaka, K., Yamada, Y., Satoh, T., Uchibori, A., Uchikado, S.: Model reference adaptive control with multi-rate type neural network for electro-pneumatic servo system. In: Proceedings of IEEE International Conference on Control Applications, pp. 1716–1721. IEEE (1999)
6. Tsai, Y.C., Huang, A.C.: Multiple-surface sliding controller design for pneumatic servo systems. Mechatronics **18**, 506–512 (2008)
7. Armstrong, H.B., Dupont, P., Canudas, C.W.: A survey of models analysis tools and compensation methods, for the control of machines with friction. Automatica **30**(7), 1083–1138 (1994)
8. Canudas, C.W., Olsson, H., Aström, K.J., Lischinsky, P.: A new model for control of systems with friction. IEEE Trans. Automat. Contr. **40**(3), 419–425 (1995)
9. Lampaert, V., Swevers, J., Al-Bender, F.: Modification of the Leuven integrated friction model structure. IEEE Trans. Autom. Control **47**(4), 683–687 (2002)
10. Tran, X.B., Hafizah, N., Yanada, H.: Modeling of dynamic friction behaviors of hydraulic cylinders. Mechatronics **22**(1), 65–75 (2012)
11. Freidovich, L., Robertsson, A., Shiriaev, A., Johansson, R.: LuGre-model-based friction compensation. IEEE Trans. Control Syst. Technol. **18**(1), 194–200 (2010)
12. Raul, G., Eduardo, C.P., Edson, R.D., Antônio, C.V.: Cascade controlled pneumatic positioning system with LuGre model-based friction compensation. J. Braz. Soc. Mech. Sci. Eng. **28**(1), 48–57 (2005)
13. Karim, K., Pascal, B., Louis, A.D.: LuGre model-based friction compensation and positioning control for a pneumatic actuator using multi-objective output-feedback control via LMI optimization. Mechatronics **19**, 535–547 (2009)
14. Sobczyk, M.R., Gervini, V.I., Perondi, E.A., Cunha, M.A.B.: A continuous version of the LuGre friction model applied to the adaptive control of a pneumatic servo system. J. Franklin Inst. **353**, 3021–3039 (2016)

Smooth and Time Optimal Trajectory Planning for Industrial Robot Using a Single Polynomial

Minh-Tuan Nguyen[1,3](✉) and Jin-Huang Huang[2]

[1] Mechanical and Aeronautical Engineering, Feng Chia University, No. 100 Wenhwa Rd., Seatwen, Taichung 40724, Taiwan, ROC
tuanctm7@gmail.com

[2] Department of Mechanical and Computer Aided Engineering, Feng Chia University, No. 100 Wenhwa Rd., Seatwen, Taichung 40724, Taiwan, ROC

[3] Faculty of Mechanical Engineering, Hung Yen University of Technology and Education, Dan Tien Commune, Khoai Chau, Hung Yen 160000, Vietnam

Abstract. This paper proposes a method for smoothing and optimizing trajectory in robotics. The trajectory in the joint space with the constraints of velocity, acceleration and jerk is interpolated by a single polynomial and the traveling time is then minimized by sequential quadratic programming algorithm. The proposed approach is performed on an industrial robot and results are compared with another planning method. The results show that the single polynomial makes the trajectory simple because only one interpolation polynomial is used while the continuity conditions of the trajectory, velocity, acceleration and jerk are simultaneously satisfied. Additionally, the optimized traveling time is also obtained.

Keywords: Trajectory planning · Single polynomial · Time optimal trajectory · Sequential quadratic programming

1 Introduction

The trajectory refers to the displacement, velocity and acceleration of a robot in movement. The requirements of trajectory planning should be able to control the velocity and acceleration of the robot during movement, and the movement space is always within the allowable range of each joint movement. The running trajectory also needs to be smooth and continuous. There are two approaches to generating such trajectories: joint space and Cartesian space. Joint space is simpler and more convenient than Cartesian space, so it is often used in trajectory planning. In joint space, trajectories have been mostly constructed by several interpolation functions such as polynomial [1–3], radial basis function [4], B-spline [5], trigonometric spline [6], etc. In addition, in light of the kinematic characteristics of the robot, the velocity, acceleration and jerk of each joint should be limited to the allowable range. Therefore, it is necessary to optimize the trajectory, not only to ensure smooth operation of each joint arm, but also to reduce traveling time and jerk and improve the efficiency of the robot as well. The first proposal in the literature to improve the efficiency of robots is the

K.-U. Sattler et al. (Eds.): ICERA 2019, LNNS 104, pp. 657–663, 2020.
https://doi.org/10.1007/978-3-030-37497-6_76

minimum-time trajectory planning. Such a planning has become an important area in trajectory optimization. The purpose of minimum-time trajectory planning is to plan the shortest traveling time trajectory according to the given points, and also to satisfy the boundary constraints [1, 2, 4–10]. Aribowo [1] reached an implementation of cubic spline optimization with free via points for reducing motion time, combined with input shaping for suppressing vibration. By using a 5th-order polynomial function to interpolate several given points of each joint of the robot, Zhang [2] applied genetic algorithm to optimize time trajectory planning. These researches focused on using many different interpolation functions to establish motion and optimization in trajectory planning. However, the limitation is that the equation of motion is complicated and takes a long time to calculate.

In this study, a new approach for generating smooth joint trajectories for robot manipulators by using a single polynomial interpolation technique is proposed. The practical importance of the proposed work is that single polynomial can be easily applied for generating smooth reference trajectories with continuous derivatives for any order of derivation within the interpolation interval. Therefore, the velocity, acceleration and jerk of motion is continuous over time. In addition, the motion equation of robot is simpler because only one single polynomial used. The traveling time of robot is also minimized by using sequential quadratic programming. This article is organized as follows: Sect. 2 describes the problem statement in the trajectory planning; Sect. 3 presents the simulation results and Sect. 4 summarizes the article.

2 Problem Statement

2.1 Trajectory Construction by Single Polynomial

A single polynomial is used to interpolate trajectory points in joint space. The equations of the joint angle, velocity, acceleration and jerk of the nth-order polynomial versus time t are given as:

$$\begin{aligned} q(t) &= a_0 + a_1 t + a_2 t^2 + \ldots + a_n t^n \\ v(t) &= \dot{q}(t) = a_1 + 2a_2 t + 3a_3 t^2 + \ldots + n a_n t^{n-1} \\ a(t) &= \ddot{q}(t) = 2a_2 + 6a_3 t + 12a_4 t^2 + \ldots + (n-1) n a_n t^{n-2} \\ j(t) &= \dddot{q}(t) = 6a_3 + 24a_4 t + \ldots + (n-2)(n-1) n a_n t^{n-3} \end{aligned} \tag{1}$$

where a_0, a_1, …,a_n are polynomial coefficients.

The boundary conditions come from the initial point, via points and end point (Fig. 1):

$$\begin{aligned} &q(0) = q_i, \quad \dot{q}(0) = 0, \quad \ddot{q}(0) = 0, \quad \dddot{q}(0) = 0 \\ &q(t_f) = q_f, \quad \dot{q}(t_f) = 0, \quad \ddot{q}(t_f) = 0, \quad \dddot{q}(t_f) = 0 \\ &q(t_i) = q_{vi}, \quad i = 1 \sim m-2 \end{aligned} \tag{2}$$

m is the number of given points.

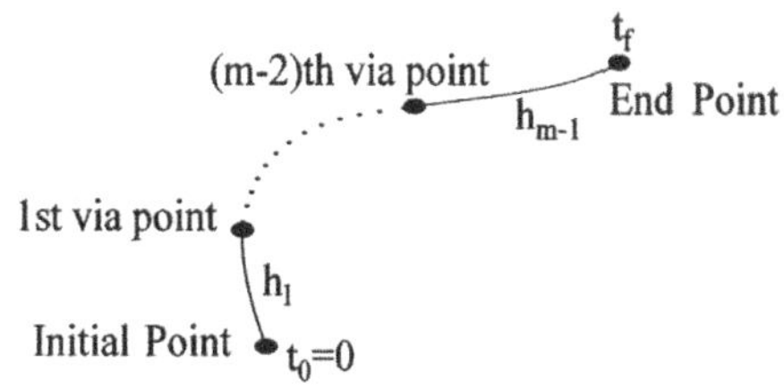

Fig. 1. Motion trajectory of robot manipulator with via points

2.2 Formulation of the Optimization Problem

The optimization problem is how to set up the time intervals between each pair of adjacent knots so that the traveling time is minimized. The objective function to be minimized is defined as

$$f = T = \sum_{i=1}^{m-1} h_i \tag{3}$$

where T is the traveling time and h_i is the time intervals between each pair of via points.

The constraints for the maximum angular velocity, acceleration and jerk of each joint j of the robot are given by

$$|\dot{q}_j(t)| \leq \dot{q}_j^{\max}; \quad |\ddot{q}_j(t)| \leq \ddot{q}_j^{\max}; \quad |\dddot{q}_j(t)| \leq \dddot{q}_j^{\max} \tag{4}$$

$\dot{q}_j(t), \ddot{q}_j(t), \dddot{q}_j(t)$ are velocity, acceleration and jerk of joint j at time t, respectively; $\dot{q}_j^{\max}, \ddot{q}_j^{\max}, \dddot{q}_j^{\max}$ represent the maximum velocity, acceleration and jerk of joint j in turn.

2.3 Sequential Quadratic Programming (SQP)

SQP [11] is an algorithm for solving nonlinear optimization problems (NLP). It can handle any degree of nonlinearity including nonlinear constraints. The method dates back to 1963 and was developed and refined in the 1970's. SQP combines two fundamental algorithms for solving nonlinear optimization problems: an active set method and Newton's method. In this study, we applied the SQP methodology to NLP of the form

$$\begin{aligned} &\text{minimize } f(x) \\ &\text{over } x \in R_{+}^{n} \\ &\text{subject to } g(x) \leq 0 \end{aligned} \tag{5}$$

where f is the objective function and g is the inequality constraint.

3 Results and Discussions

The industrial manipulator PUMA560 with six degrees of freedom (n = 6) is considered in the problem. Each joint trajectory has three spaced knots (m = 4). The joint space via point angles are given in Table 1 and the kinematic limits are reported in Table 2.

Table 1. The via points for trajectory planning

Via points	Joint 1 (°)	Joint 2 (°)	Joint 3 (°)	Joint 4 (°)	Joint 5 (°)	Joint 6 (°)
1	−10	20	15	150	30	120
2	60	50	100	100	110	60
3	20	120	−10	40	90	100
4	55	35	30	10	70	25

Table 2. Kinematic limits of the robot joints

Joint	1	2	3	4	5	6
Velocity (°/s)	100	95	100	150	130	110
Acceleration (°/s^2)	60	60	75	70	90	80
Jerk (°/s^3)	60	66	85	70	75	70

With 10 boundary conditions (4 from initial point, 4 from final point and 2 from via points, respectively), a 9th-order single polynomial is required for equation of joint angle. According to Eq. (1), we can get equations of velocity, acceleration and jerk of joint j (j = 1–6) that are its derivatives over time:

$$\begin{array}{l} q_j(t) = a_{j0} + a_{j1}t + a_{j2}t^2 + \ldots + a_{j9}t^9 \\ v_j(t) = \dot{q}_j(t) = a_{j1} + 2a_{j2}t + 3a_{j3}t^2 + \ldots + 9a_{j9}t^8 \\ a_j(t) = \ddot{q}_j(t) = 2a_{j2} + 6a_{j3}t + 12a_{j4}t^2 + \ldots + 72a_{j9}t^7 \\ j_j(t) = \dddot{q}_j(t) = 6a_{j3} + 24a_{j4}t + \ldots + 504a_{j9}t^6 \end{array} \tag{6}$$

The objective function to be minimized is

$$f = T = \sum_{i=1}^{m-1} h_i = h_1 + h_2 + h_3 \tag{7}$$

The SQP algorithm with nonlinear constraints was used to minimize the objective function from initial point [10]. The obtained time intervals are h_1 = 3.002 s, h_2 = 2.617 s, h_3 = 3.097 s, and T = 8.716 s. The convergence of SQP algorithm is displayed in Fig. 2.

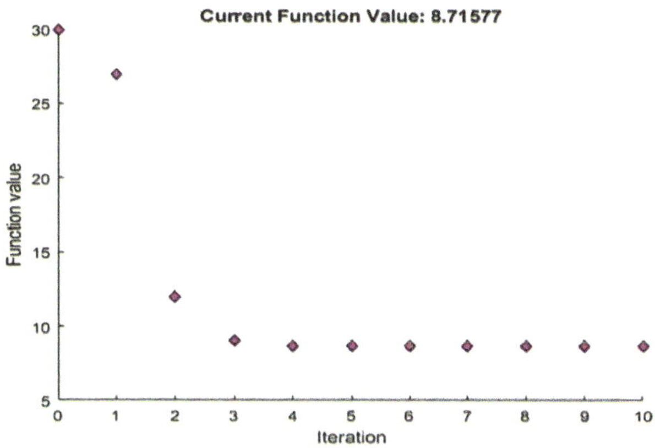

Fig. 2. Convergence history of SQP algorithm

From the optimal time T obtained, the equation of six joint angles versus time t are expressed (Eq. (8)):

$$\begin{aligned}
q_1(t) &= -10 + 10.1801t^4 - 6.5805t^5 + 1.6923t^6 - 0.2159t^7 + 0.0137t^8 - 0.0003t^9 \\
q_2(t) &= 20 - 2.3867t^4 + 2.5387t^5 - 0.8480t^6 + 0.1276t^7 - 0.0090t^8 + 0.0002t^9 \\
q_3(t) &= 15 + 14.5039t^4 - 9.6166t^5 + 2.5069t^6 - 0.3223t^7 + 0.0205t^8 - 0.0005t^9 \\
q_4(t) &= 150 - 4.2324t^4 + 2.3969t^5 - 0.5697t^6 + 0.0694t^7 - 0.0043t^8 + 0.0001t^9 \\
q_5(t) &= 30 + 8.5501t^4 - 5.0404t^5 + 1.1954t^6 - 0.1421t^7 + 0.0085t^8 - 0.0002t^9 \\
q_6(t) &= 120 - 10.5458t^4 + 7.1250t^5 - 1.8998t^6 + 0.2495t^7 - 0.0162t^8 + 0.0004t^9
\end{aligned} \tag{8}$$

Figures 3, 4, 5 and 6 demonstrate the variation curves of the angle, velocity, acceleration, jerk of six joints, respectively. It is seen from the figures that the joint angle, velocity, acceleration and jerk curves are smooth and continuous, hence ensuring the smooth operation of the robot.

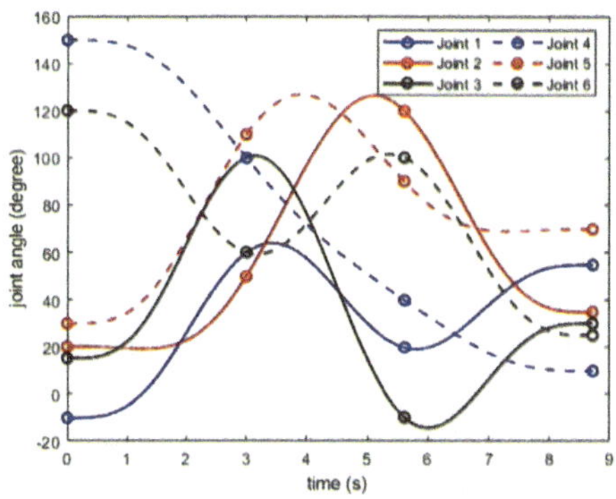

Fig. 3. Angle of six joints

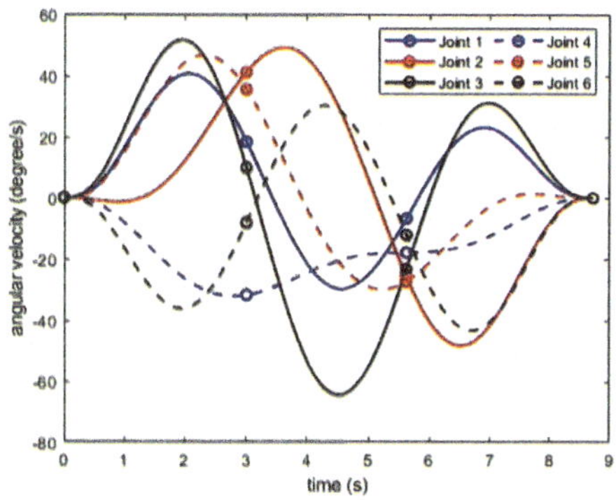

Fig. 4. Angular velocity of six joints

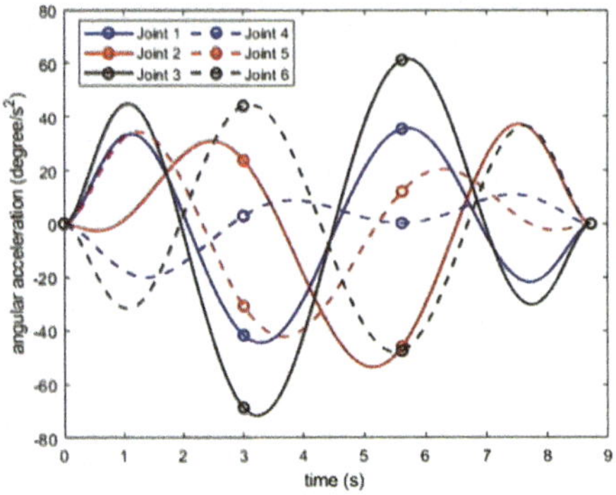

Fig. 5. Angular acceleration of six joints

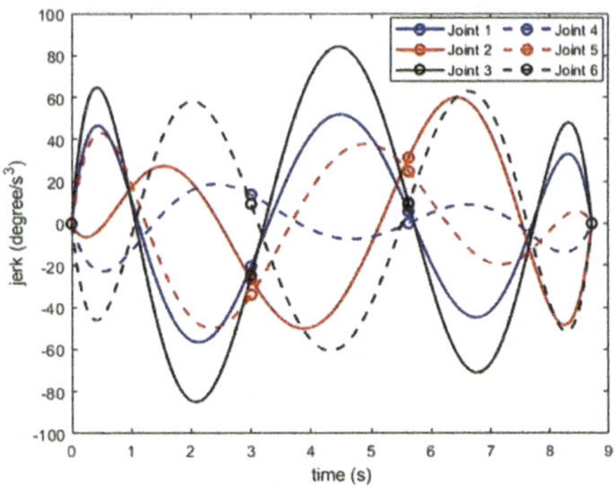

Fig. 6. Jerks of six joints

Table 3 shows the maximum values of velocity, acceleration and jerk of six joints after the optimization and compared to the results from 5th-order B-spline [5]. As shown in Table 3, the travelling time is less than the traveling time of the 5th-order B-spline used, which was 4.2%. Jerks of joint 1, 2, 4, 5 are lower than those by the 5th-order B-spline with 4.3%, 3.7%, 14.0%, 9.9%, respectively. Jerks of joint 3 and joint 6 are larger than those by the 5th-order B-spline with 2.5%, 4.5%, respectively.

Table 3. Maximum values of velocity, acceleration and jerk recorded during a point-to-point task

Kinematic parameters	Joint					
	1	2	3	4	5	6
Single 9th-order Polynomial	T = 8.717 s					
Velocity (°/s)	40.83	49.32	64.50	31.78	46.81	43.22
Acceleration (°/s^2)	44.39	53.56	71.71	19.71	42.15	48.07
Jerk (°/s^3)	56.49	59.97	84.97	22.53	49.90	62.71
5th-order B-Splines	T = 9.097 s					
Velocity (°/s)	39.27	47.05	50.65	29.15	43.35	41.98
Acceleration (°/s^2)	42.67	53.04	70.53	17.60	35.75	48.19
Jerk (°/s^3)	59.00	62.28	82.92	26.20	55.40	60.02

4 Conclusions

In order to obtain the smooth and an optimal time trajectory of robot manipulator, a methodology has proposed in this work. The method adopts a single polynomial interpolation to construct the trajectory in the joint space and minimizes traveling time using SQP and is simulated on an industrial robot. Analysis results show that the single polynomial interpolation satisfies the continuity of velocity, acceleration and jerk with kinematic constraints, and hence the motion trajectory is smooth with optimal time. Future work will be devoted to conducting experiments to verify the accuracy of the proposed method.

Acknowledgement. This research was supported by the Ministry of Science and Technology of Taiwan under Contract Nos. MOST 107-2221-E-035-074-MY3 and MOST 107-2218-E-035-016.

References

1. Aribowo, W., Terashima, K.: Cubic spline trajectory planning and vibration suppression of semiconductor wafer transfer robot arm. Int. J. Autom. Technol. **8**(2), 265–274 (2014)
2. Zhang, J., Meng, Q., Feng, X., Shen, H.: A 6-DOF robot-time optimal trajectory planning based on an improved genetic algorithm. Robot. Biomim. **5**(3), 1–7 (2018)
3. Williams II, R.L.: Simplified robotics joint-space trajectory generation with a via point using a single polynomial. J. Robot. **2013**, 1–7 (2013)
4. Chettibi, T.: Smooth point-to-point trajectory planning for robot manipulators by using radial basis functions. Robotica **37**(3), 539–559 (2018)
5. Huang, J., Hu, P., Wu, K., Zeng, M.: Optimal time-jerk trajectory planning for industrial robots. Mech. Mach. Theory **121**, 530–544 (2018)
6. Simon, D., Isik, C.: Optimal trigonometric robot joint trajectories. Robotica **9**, 379–386 (1991)
7. Egerstedt, M., Martin, C.F.: Optimal trajectory planning and smoothing splines. Automatica **37**, 1057–1064 (2001)
8. Lu, S., Zhao, J., Jiang, L., Liu, H.: Solving the time-jerk optimal trajectory planning problem of a robot using augmented lagrange constrained particle swarm optimization. Math. Probl. Eng. **2017**, 1–10 (2017)
9. Lu, S., Zhao, J., Jiang, L., Liu, H.: Time-jerk optimal trajectory planning of a 7-DOF redundant robot. Turk. J. Electr. Eng. Comput. Sci. **25**, 4211–4222 (2017)
10. Fang, Y., Hu, J., Liu, W., Shao, Q., Qi, J., Peng, Y.: Smooth and time-optimal S-curve trajectory planning for automated robots and machines. Mech. Mach. Theory **137**, 127–153 (2019)
11. Northwestern University Homepage. https://optimization.mccormick.northwestern.edu/index.php/Sequential_quadratic_programming. Accessed 07 June 2015

State of Charge Estimation for Lithium-Ion Battery Using Sigma-Point Kalman Filters Based on the Second Order Equivalent Circuit Model

Nguyen Vinh Thuy[1(✉)] and Nguyen Van Chi[2]

[1] Electrical Engineering Faculty, Thai Nguyen University of Technology, Thai Nguyen, Vietnam
nguyenvinhthuy-tdh@tnut.edu.vn

[2] Institute of High-Technology Research and Development for Industry, Thai Nguyen University of Technology, Thai Nguyen, Vietnam

Abstract. Today Lithium-Ion battery (LiB) is very widely used in vehicular applications including Hybrid-Electric Vehicle (HEV), Plug-in Hybrid-Electric Vehicle (PHEV), Extended-Range Electric Vehicle (E-REV), and Electric Vehicle (EV). The improved discharge and charge efficiency, the longer life span, high energy density and the ability to deep cycle while maintaining power are the typical advantages of LiB. The most significant feature of vehicular applications powered by LiB which needs to be considered is the process of charging and discharging suddenly concerning to acceleration and breaking. The state of charge (SoC) estimation is an important problem related to energy and power control in the operation process of electric vehicles. In this paper, the SoC estimation method of LiB based on Sigma-point Kalman filter (SPKF) is proposed. The results based on the real data indicate that using the second-order equivalent circuit model of LiB can increase the accuracy of the SoC estimation compared with methods using first-order model.

Keywords: Lithium-Ion batteries · State of charge · Sigma-point Kalman filter · Battery management system · The equivalent circuit model

1 Introduction

The use of LiB in portable devices like mobile phones, the instruments and especially the electric vehicle has grown significantly in recent years. The LiB has many advantages including high energy density, the longer life span, self-discharge is much lower than that of other rechargeable cells such as Ni-Cad and NiMH forms, low maintenance and no requirement for priming. By the LiB's advantages, the LiB is used widely in the vehicular applications, especially in HEV, P-HEV, E-REV, and EV in the world. To make the LiB battery module with desired larger power and voltage, the LiB cells are connected by series and parallel structure [1, 2].

One of the most important problems of the vehicular applications using LiB is the power and energy control, this function is conducted by a system called battery management system (BMS). The BMS has some basic functions including the voltage

K.-U. Sattler et al. (Eds.): ICERA 2019, LNNS 104, pp. 664–678, 2020.
https://doi.org/10.1007/978-3-030-37497-6_77

control, over-charge and over-discharge protection, communicating with peripheral devices, SoC estimation, charge balance, etc. [1]. The SoC is an unmeasured parameter related to the lithium concentration stoichiometry. Therefore, it must be estimated by other measured quantities from each battery such as discharge current, battery terminal voltage, and temperature. When knowing the SoC, we can infer the remaining power of the LiB.

There are many SoC estimation algorithms in the research literature including the Coulomb counting method, intrinsic parameter direct measurement method, computer intelligence method, and model-based method [3]. The Coulomb counting method is the simplest but the estimation error is accumulated by the counting time increasing, so this method requires the additional error correction. The intrinsic parameter direct measurement method includes the open-circuit voltage (OCV), impedance spectrum, electromotive force (EMF), and internal resistance (IR) [4, 5]. The SoC can be inferred directly from through the mapping relationship between the SoC and the battery intrinsic parameters but the estimation accuracy relying excessively on the accuracy of the mapping relationship between SoC and intrinsic parameters of battery is a disadvantage.

To estimate the SoC more exactly, in recent years the model-based SoC estimation method is used more. This method can be divided into two categories: physics-based models and equivalent circuit models [6]. The capability of estimation for SoC in the long term of operation and the LiB inside parameters are the advantages of the physics-based models, but the complex model regarding to a partial differential equation is its disadvantage.

The equivalent circuit model-based SoC estimation method mainly includes the Kalman filter algorithm, the Gauss–Hermitian orthogonal filter algorithm. Although it can not reflect upon the inside states of the LiB, the advantages of this method are simpler SoC calculations and voltage and current estimation capability. The first-order model of the equivalent circuit model is mostly used for the LiB with the applications that have no requirements for charging and discharging suddenly like portable electronic devices.

When coming to the LiB applications requiring the large power to charge and discharge suddenly, the first order models can not describe fully the dynamic characteristics of the LiB. Another side of estimation SoC based on the Kalman filters and Extended Kalman filters is linearized by the nonlinear characteristic sampled by sample times, therefore the output distributions calculated by that linearized characteristic are less exact [7–9].

This paper uses the second-order equivalent circuit model in order to reflect better upon the dynamics of LiB in case of charge and discharge suddenly, continuously. To estimate more exactly the dynamics of LiB, the estimation method based on sigma-point is used. The measurement noise, the dynamic uncertainties are taken into account, mapped via the LiB's nonlinear characteristics, therefore the LiB's SoC is estimated more accurately. The method's advantage is proofed by the results based on real data.

This paper is organized as follows. In Sect. 2, the second-order LiB model is used and model parameters are identified with test data. In Sect. 3, the SoC estimation based on the model introduced in Sect. 1 is conducted, this SoC estimation uses the Sigma-point Kalman filter. Section 4 are some conclusions and future works.

2 Second-Order Equivalent Circuit Model and Parameter Identification of LiB

2.1 Second-Order Equivalent Circuit Model of LiB

Consider the second-order equivalent circuit model of LiB as described in the following Fig. 1.

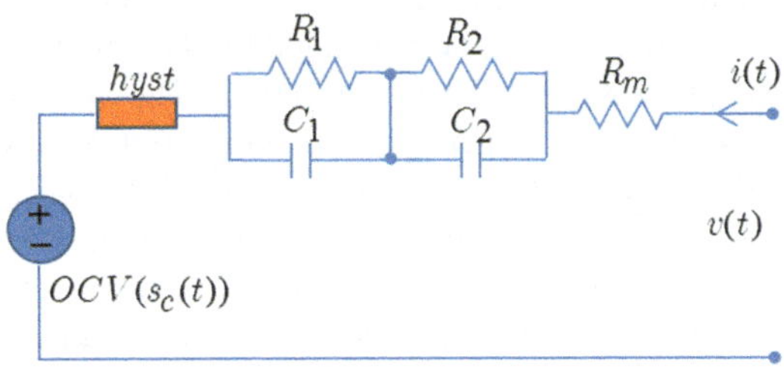

Fig. 1. Model structure of a Lithium-Ion battery.

Where: $OCV(s_c(t))$ is open-circuit voltage; *hyst* shows hysteresis voltage; $v(t)$ is the battery terminal voltage; R_m represents internal cell resistor; (R_1, C_1), (R_2, C_2) represent two parallel resistor-capacitor pairs from the circuit, demonstrating the process of battery transient dynamics when changing the load current.

We define SoC as $s_c(t)$ of a cell, we have:

$$s_c(t) = s_c(t_0) - \frac{1}{Q}\int_{t_0}^{t} i(t)dt \tag{1}$$

In discrete time, if we assume that current is constant over sampling interval Δt, we have:

$$s_c(k+1) = s_c(k) - \frac{i(k)\eta(k)\Delta t}{Q} \tag{2}$$

in which $\eta(k)$ is called "coulombic efficiency"; Q is total capacity of a cell.

Diffused resistor current:

$$\begin{aligned} i_{R_1}(k+1) &= \exp\left(-\frac{\Delta t}{R_1C_1}\right)i_{R_1}(k) + \left(1 - \exp\left(-\frac{\Delta t}{R_1C_1}\right)\right)i(k) \\ i_{R_2}(k+1) &= \exp\left(-\frac{\Delta t}{R_2C_2}\right)i_{R_2}(k) + \left(1 - \exp\left(-\frac{\Delta t}{R_2C_2}\right)\right)i(k) \end{aligned} \tag{3}$$

Hysteresis voltage:

$$hyst(k+1) = \exp\left(-\left|\frac{i(k)\eta(k)\Delta t}{Q}\right|\right)hyst(k) - \left(1 - \exp\left|\frac{i(k)\eta(k)\Delta t}{Q}\right|\right)\operatorname{sgn}(i(k)) \quad (4)$$

The circuit contains two parallel resistor-capacitor pairs. We can define vector valued:

$$i_R(k+1) = \left[\begin{array}{c|c} \exp\left(-\frac{\Delta t}{R_1C_1}\right) & 0 \\ \hline 0 & \exp\left(-\frac{\Delta t}{R_2C_2}\right) \end{array}\right] i_R(k) + \left[\begin{array}{c} 1-\exp\left(-\frac{\Delta t}{R_1C_1}\right) \\ \hline 1-\exp\left(-\frac{\Delta t}{R_2C_2}\right) \end{array}\right] i(k) \quad (5)$$

Let:

$$\boldsymbol{A}_{RC} \triangleq \left[\begin{array}{c|c} \exp\left(-\frac{\Delta t}{R_1C_1}\right) & 0 \\ \hline 0 & \exp\left(-\frac{\Delta t}{R_2C_2}\right) \end{array}\right], \boldsymbol{B}_{RC} \triangleq \left[\begin{array}{c} 1-\exp\left(-\frac{\Delta t}{R_1C_1}\right) \\ \hline 1-\exp\left(-\frac{\Delta t}{R_2C_2}\right) \end{array}\right], \quad (6)$$

$$\boldsymbol{A}_H(k) = \exp\left(-\left|\frac{\eta_k(k)i_k(k)\gamma(k)\Delta t}{Q}\right|\right)$$

We have:

$$\begin{bmatrix} s_c(k+1) \\ i_R(k+1) \\ hyst(k+1) \end{bmatrix} = \left[\begin{array}{c|c|c} 1 & 0 & 0 \\ \hline 0 & \boldsymbol{A}_{RC} & 0 \\ \hline 0 & 0 & \boldsymbol{A}_H(k) \end{array}\right] \begin{bmatrix} s_c(k) \\ i_R(k) \\ hyst(k) \end{bmatrix} + \left[\begin{array}{c|c} -\frac{\eta_k(k)\Delta t}{Q} & 0 \\ \hline \boldsymbol{B}_{RC} & 0 \\ \hline 0 & 1-\boldsymbol{A}_H(k) \end{array}\right] \begin{bmatrix} i(k) \\ \operatorname{sgn}(i(k)) \end{bmatrix} \quad (7)$$

If defining $\underline{z}(k) = [\, s_c(k) \quad i_R(k) \quad hyst(k) \,]^T$, we have the second- order equivalent circuit model of LiB as the following equation:

$$\underline{z}(k+1) = \mathbf{A}(i(k))\underline{z}(k) + \mathbf{B}(i(k))i(k) \quad (8)$$

in which:

$$\mathbf{A}(i(k)) = \left[\begin{array}{c|c|c} 1 & 0 & 0 \\ \hline 0 & A_{RC} & 0 \\ \hline 0 & 0 & A_H(k) \end{array}\right], \ \mathbf{B}(i(k)) = \left[\begin{array}{c|c} -\frac{\eta_k(k)\Delta t}{Q} & 0 \\ \hline B_{RC} & 0 \\ \hline 0 & 1-A_H(k) \end{array}\right] \quad (9)$$

The output equation is:

$$v(k) = OCV(s_c(k), T_k) + M_0.s(k) + M.hyst(k) - \sum_i R_i i_{R_i}(k) - R_m i(k) \tag{10}$$

Defined the matrices:

$$\mathbf{C}(k) = [0 \quad -R_1 \quad -R_2 \quad M], \quad \mathbf{D}(k) = -R_0 \tag{11}$$

We have:

$$v(k) = OCV(s_c(k), T_k) + \mathbf{C}(k)\underline{z}(k) + \mathbf{D}(k)i(k) \tag{12}$$

So finally, we have a second- order LiB model as follows:

$$\begin{cases} \underline{z}(k+1) = \mathbf{A}(i(k))\underline{z}(k) + \mathbf{B}(i(k))i(k) \\ v(k) = OCV(s_c(k), T_k) + \mathbf{C}(k)\underline{z}(k) + \mathbf{D}(k)i(k) \end{cases} \tag{13}$$

In the model (13), the nonlinear characteristic is the relationship between the OCV and SoC, the matrices $\mathbf{A}(i(k))$ and $\mathbf{B}(i(k))$ both depend on the model's input that is the current $i(k)$, the parameters in the matrices R_1, C_1, R_2, C_2, M_0, M, η, γ, R_0, $OCV(s_c(k),\ T_k)$ all depend on the operating temperature of LiB.

2.2 Parameter Identification of LiB

To estimate the SoC, all parameters in the model (13) need to be determined by using the practice data for particular LiB. The practice data are collected by some charging and discharging scenarios. The range of temperature is −25 °C to 45 °C, the reference temperature is 25 °C. The LiB SAMSUNG INR18650 is used for collecting practice data. The algorithm to determine all parameters in the model (13) is based on the algorithm presented in the material [6] and some modifications conducted by us, the algorithm is implemented in the Matlab script environment. The identified parameters are shown in Table 1 and Fig. 2.

Figure 2 describes the variation of the parameters R_1, C_1, R_2, C_2, M_0, M, η, γ, R_0 by the temperatures. Figure 3 shows the relationship between OCV versus SoC. In the Fig. 4, to verify the accuracy of the estimation we compare test voltage and estimated voltage by the second-order model with identified parameters, the error between two those voltages are about 0.47%.

Table 1. The values of the identified parameters R_1, C_1, R_2, C_2, M_0, M, η, γ, R_0

Parameters	T (^{0}C)	R_0 (Ω)	R_1 (Ω)	R_1–C_1 (s)	R_2 (Ω)	R_2–C_2 (s)
Values	−25	0.2207	0.0945	1.7418	0.00011	1.1960
	−15	0.1292	0.0539	6.0148	0.00016	0.3413
	−5	0.0652	0.0249	3.6797	0.00015	2.5840
	5	0.0325	0.0189	27.4782	0.0023	3.4605
	15	0.0185	0.0076	5.2395	0.00012	0.3681
	25	0.0113	0.0048	10.2956	0.0004	0.8575
	35	0.0081	0.0048	15.8687	0.0004	0.9339
	45	0.0061	0.0048	20.0195	0.0004	0.9553

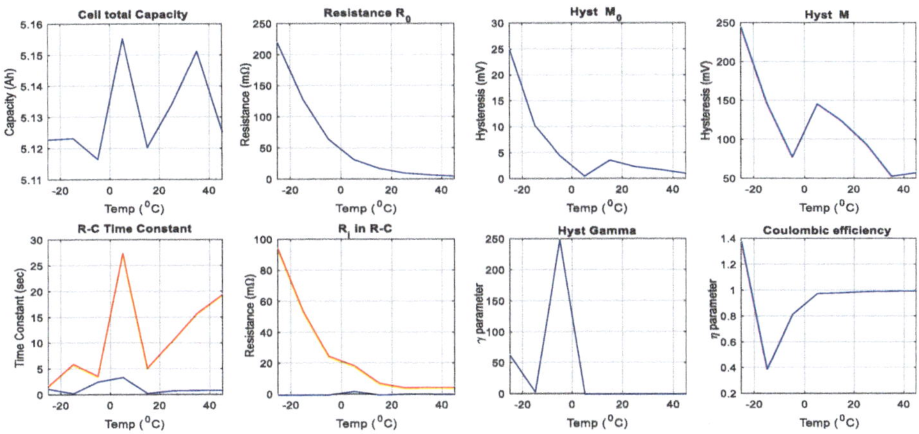

Fig. 2. The relationship between identified parameters and temperature

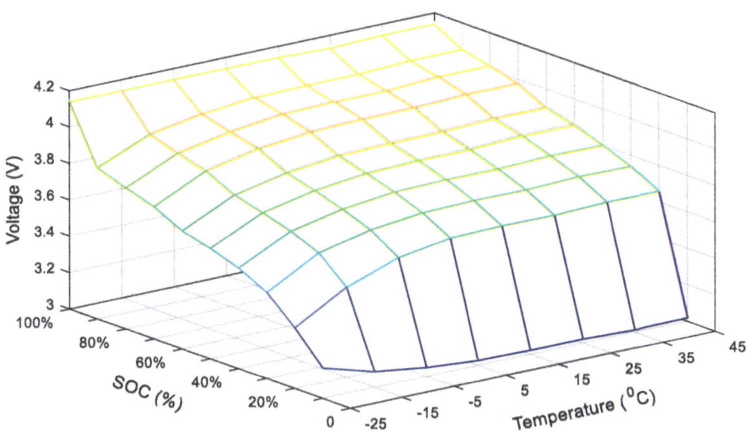

Fig. 3. Relationship between OCV versus SOC

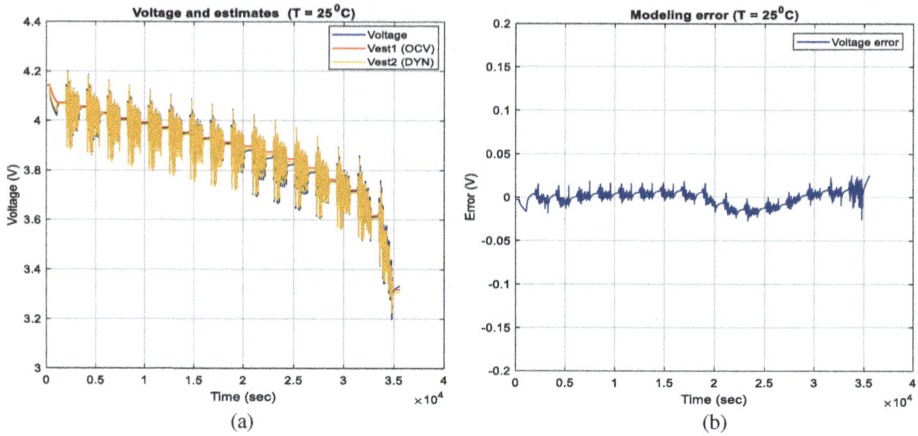

Fig. 4. Comparing test voltage vs estimated voltage: (a) test voltage vs estimated voltage, (b) voltage error

3 SoC Estimation Using Sigma-Point Kalman Filters

The SPKF known as one of the state estimation methods has some advantages better than EKF by application ranges, this method estimates more accurately the states for the second-order system or higher meanwhile calculating the complex is the same as EKF. SPKF is more easily implemented as no requirement for the Jacobian matrix calculation. For the EKF, the system's nonlinear dynamic is linearized at the sampling time, the input distributions are mapped via the linearized dynamic function to calculate the output distributions, the output distributions is less accurate if there is high non-linear of the system dynamics. For the SPKF, the whole of the input distributions are mapped through the nonlinear dynamic functions to get the output distribution so that the distribution of the system's nonlinear dynamic is calculated more accurately than EKF [10].

Consider the sample time chosen small enough, the Eq. (13) is described in discrete form (14), the system (14) is taken into account with the system noises $\underline{w}(k)$ and measurement noise $n(k)$:

$$\begin{cases} \underline{z}(k+1) = \mathbf{A}(i(k))\underline{z}(k) + \mathbf{B}(i(k))i(k) + \underline{w}(k) \\ v(k) = OCV(s_c(k), T_k) + \mathbf{C}(k)\underline{z}(k) + \mathbf{D}(k)i(k) + n(k) \end{cases} \tag{14}$$

The algorithm for estimation SoC (the state vector $\underline{z}(k)$) using Sigma-point Kalman filters at the sampling time k is conducted by two main steps:

- Step 1

- Form the augmented state vector at the sampling time $k-1$ based on the estimated states at the sampling time $k-1$ and the vector of mean values of the system noise $\bar{w}$ and measure noise $\bar{n}$:

$$\hat{\underline{z}}_{k-1}^{a,+} = \left[\left(\hat{\underline{z}}_{k-1}^{a,+}\right)^T \quad \bar{w} \quad \bar{n} \right]^T \tag{15}$$

- Form the estimated covariance matrix of the estimation error at the previous step including the covariance matrix of the estimation error, covariance matrix of the noise error and covariance matrix of the measured error:

$$\boldsymbol{C}_{\tilde{z},k-1}^{a,+} = diag\left(C_{\tilde{z},k-1}^{+} \quad C_{\tilde{w}} \quad C_{\tilde{n}} \right) \tag{16}$$

- The augmented sigma-point matrix containing $p+1$ sigma-points is established as (17), $\chi_{k-1}^{a,+}$ is split into state portion $\chi_{k-1}^{z,+}$, process noise portion $\chi_{k-1}^{w,+}$ and sensor noise portion $\chi_{k-1}^{n,+}$:

$$\chi_{k-1}^{a,+} = \left[\hat{\underline{z}}_{k-1}^{a,+}, \hat{\underline{z}}_{k-1}^{a,+} + \lambda\sqrt{C_{\tilde{\underline{z}},k-1}^{a,+}}, \hat{\underline{z}}_{k-1}^{a,+} - \lambda\sqrt{C_{\tilde{z},k-1}^{a,+}} \right] \tag{17}$$

- Calculate priori sigma-points at the sampling time k by using the (14) and the augmented sigma point matrix at the sampling time $k-1$, for the sigma-point i estimation of the priori sigma-points is calculated by the following equation:

$$\chi_{k,i}^{z,-} = \mathbf{A}(i(k))\chi_{k-1}^{a,+} + \mathbf{B}(i(k))i(k) + \chi_{k-1,i}^{w,+} \tag{18}$$

- The priori state estimation is calculated by using the priori sigma points mapped via the system dynamic $\chi_{k1,i}^{a,-}$ as:

$$\hat{\underline{z}}_k^- = E[\mathbf{A}(i(k-1))\hat{\underline{z}}(k-1) + \mathbf{B}(i(k-1))i(k-1) + \underline{w}(k-1)] = \sum_{i=0}^{p} \alpha_i^{(m)} \chi_{k,i}^{x,-} \tag{19}$$

– The covariance matrix of the estimation errors is updated by the equation:

$$\begin{aligned} C_{\tilde{z},k}^{-} &= \sum_{i=0}^{p} \alpha_i^{(c)} \left(\chi_{k,i}^{z,-} - \hat{z}_k^{-}\right)\left(\chi_{k,i}^{z,-} - \hat{z}_k^{-}\right)^T \\ &= \alpha_0^{(c)} \left(\chi_{k,0}^{x,-} - \hat{z}_k^{-}\right)\left(\chi_{k,0}^{z,-} - \hat{z}_k^{-}\right)^T \\ &\quad + \sum_{i=1}^{p} \left(\sqrt{\alpha_i^{(c)}}\left(\chi_{k,i}^{z,-} - \hat{z}_k^{-}\right)\right)\left(\sqrt{\alpha_i^{(c)}}\left(\chi_{k,i}^{z,-} - \hat{z}_k^{-}\right)^T\right) \\ &= \alpha_0^{(c)} \left(\chi_{k,0}^{z,-} - \hat{z}_k^{-}\right)\left(\chi_{k,0}^{z,-} - \hat{z}_k^{-}\right)^T + \chi_a \chi_a^T \end{aligned} \tag{20}$$

– The estimated output distribution is calculated by using $\chi_{k,i}^{z,-}$:

$$\zeta_{k,i} = OCV(s_c(k-1), T_k) + \mathbf{C}(k-1)\chi_k^{z,-} + \mathbf{D}(k-2)i(k-1) + \chi_{k-1,i}^{v,+} \tag{21}$$

– The estimated output of the suspension system at the sampling time k is estimated by the following equation:

$$\begin{aligned} \hat{v}(k) &= E\left[OCV(s_c(k-1), T_k) + \mathbf{C}(k-1)\chi_k^{z,-} + \mathbf{D}(k-2)i(k-1) + \chi_{k-1,i}^{v,+}\right] \\ &\approx \sum_{i=0}^{p} \alpha_i^{(m)} \zeta_{k,i} \end{aligned} \tag{22}$$

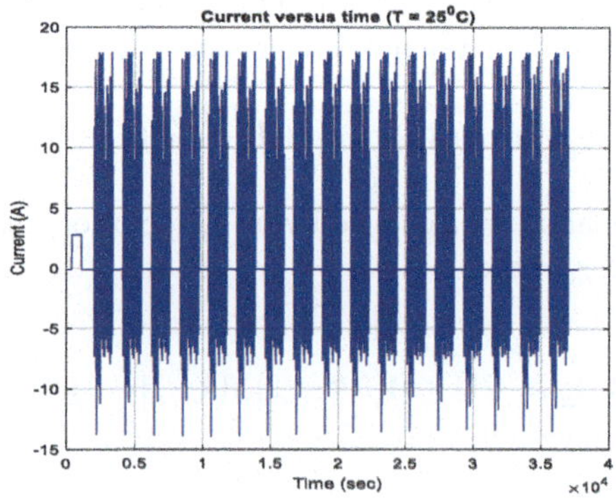

Fig. 5. The scenario of current applied to the LiB in the test

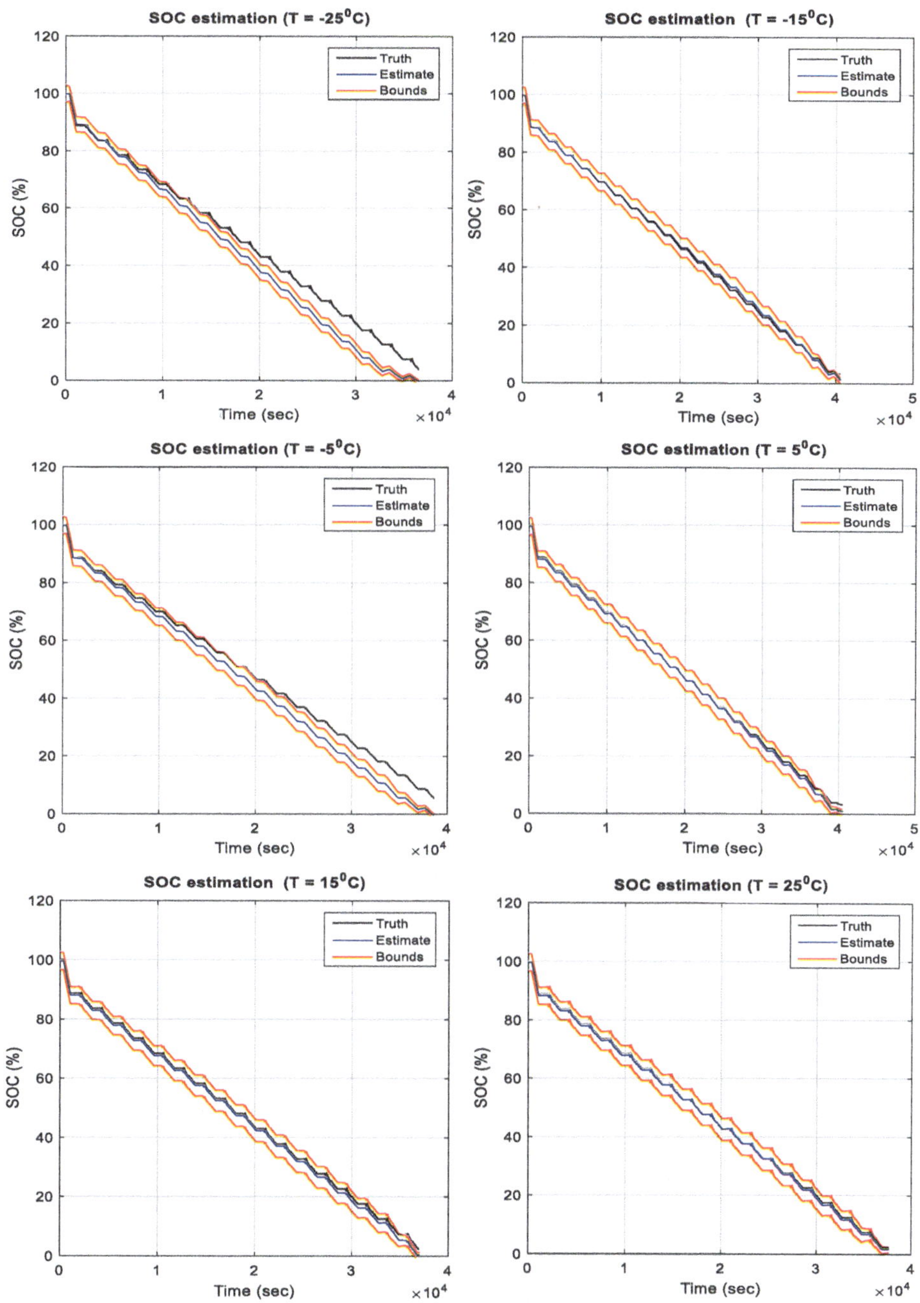

Fig. 6. The SoC estimation in comparing with practice SoC for the test temperatures

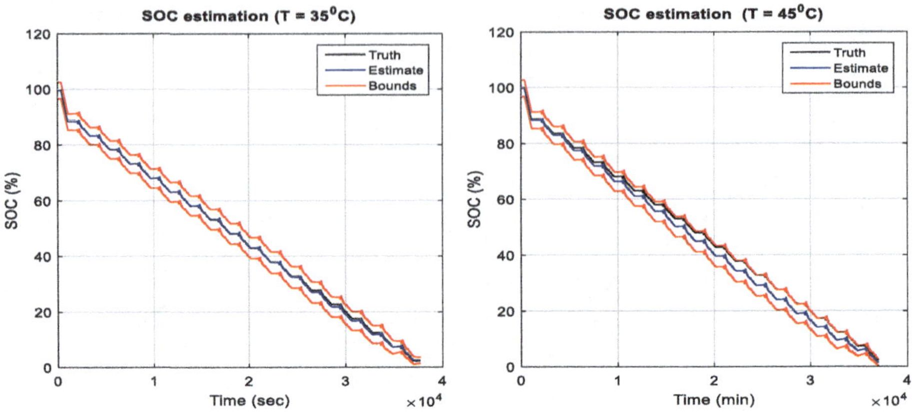

Fig. 6. *(continued)*

- Step 2

– Update the covariance matrices of the output errors at the sampling time k and the output errors and state estimation errors, these matrices depend on the sigma-points and matrix $\psi_{k,i}$ (calculated by Eq. 18) and also $\hat{\underline{z}}_k^-$, $\hat{v}(k)$:

$$C_{\tilde{v},k} = \sum_{i=0}^{p} \alpha_i^{(c)} \left(\zeta_{k,i} - \hat{v}(k)\right)\left(\zeta_{k,i} - \hat{v}(k)\right)^T \tag{23}$$

$$C_{\tilde{z}\tilde{v},k}^{-} = \sum_{i=0}^{p} \alpha_i^{(c)} \left(\chi_{k,i}^{z,-} - \hat{\underline{z}}_k^-\right)\left(\zeta_{k,i} - \hat{v}(k)\right)^T \tag{24}$$

– Calculate the estimation gain matrix as:

$$\mathbf{G}_k = \frac{C_{\tilde{z}\tilde{v},k}^{-}}{C_{\tilde{v},k}} \tag{25}$$

– Update the priori estimation states calculated at the step 1 by taking into account the output errors by following equation, we have the estimation state as:

$$\hat{\underline{z}}_k^+ = \hat{\underline{z}}_k^- + G_k(v(k) - \hat{v}(k)) \tag{26}$$

– Update the covariance matrix of the estimation errors:

$$C_{\tilde{z},k}^{+} = C_{\tilde{z},k}^{-} - G_k C_{\tilde{v},k} G_k^T \tag{27}$$

– Back to the step 1 for the next sampling time $k + 1$.

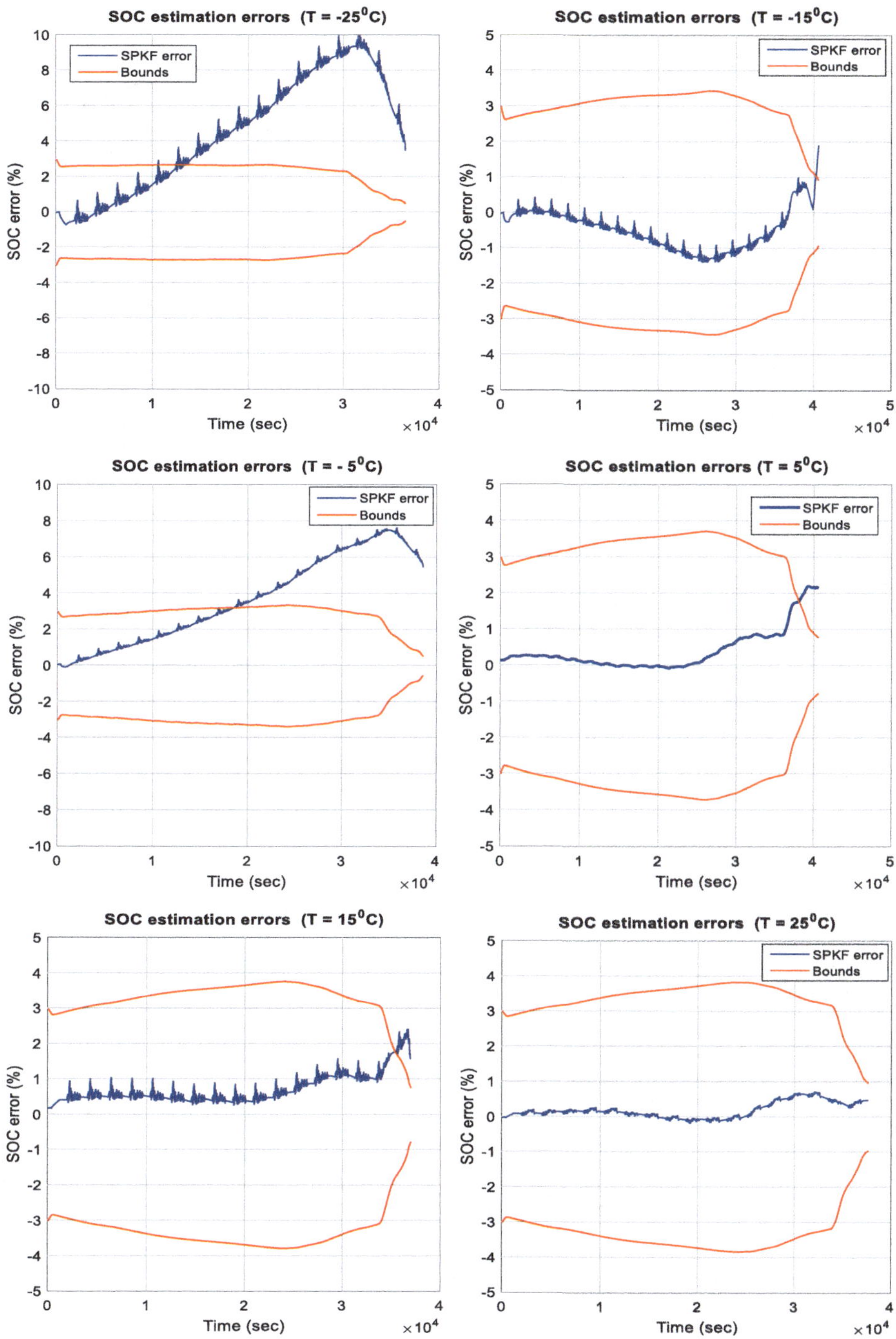

Fig. 7. The SoC estimation error

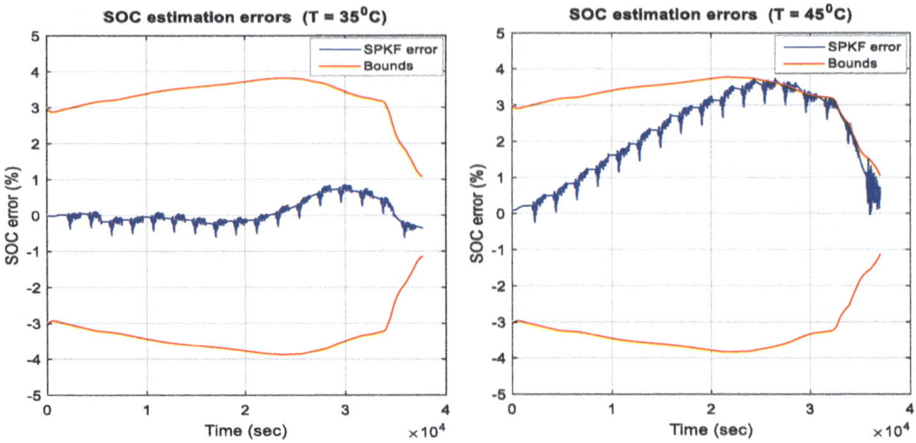

Fig. 7. *(continued)*

To compare the SoC estimation results using the second-order model based on sigma point Kalman filter, the scenarios of charge and discharge with current shown in Fig. 5 are applied. In those scenarios, LiB is charged and discharged consequently, discharged amount is more than the charged amount.

Figure 6 describes the SoC estimation in comparing with practice SoC in the temperature range −25 °C, −15 °C, −5 °C, 5 °C, 15 °C, 25 °C, 35 °C and 45 °C. We see that the SoC estimation error is small for SoC >60%, in which estimation error is the smallest in the range 5 °C to 35 °C. The smaller SoC is (<10%), the more estimation error increases. Figure 7 depicts the SoC estimation error varying by SoC decreasing from 100% to 0% for several operating temperatures.

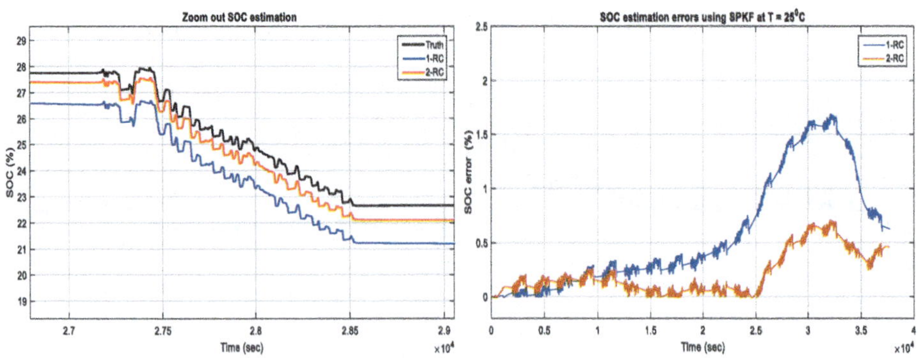

Fig. 8. The comparison between SoC estimation and SoC estimation error at 25 °C for two models of LiB

Table 2 shows the comparison between means of SoC estimation errors (%) for two cases of the first and second- order model of LiB. From Table 2, we conclude that SoC estimation error for the second-order model is smaller than SoC estimation error for the first-order model. Figure 8 shows the comparing SoC estimation at 25 °C for two models of LiB.

Table 2. The comparison between means of SoC estimation errors for two LiB's models

Temp (^{0}C)	−25	−15	−5	5	15	25	35	45
Second order model (%)	5.49	0.73	4.38	0.72	0.86	0.30	0.34	2.47
First order model (%)	6.25	2.60	4.77	1.82	1.77	0.78	0.69	2.62

4 Conclusions

In this study, the method SoC estimation using the algorithm of Sigma-point Kalman filters is proposed. The second-order model of LiB is used in order to describe more exactly the SoC dynamics for the charging and discharging suddenly. The model parameters are determined by using some appropriate charging and discharging scenarios. The range of test temperature is from −25 °C to 45 °C, the reference temperature is 25 °C. The algorithm of Sigma-point Kalman filters is applied to the second-order model of LiB. The results based on the practice data indicates that using the second-order equivalent circuit model of LiB can increase the accuracy of the SoC estimation when compared with methods using the first-order model.

Using the second-order equivalent circuit model and using Sigma-point Kalman filters to estimate SoC for LiB module, in which cells connected in parallel and serial are our future works.

References

1. Andrea, D.: Battery management systems for large Lithium-Ion battery packs. Artech House, pp. 22–110 (2010)
2. Nishi, Y.: Lithium ion secondary batteries; past 10 years and the future. J. Power Sources **100**(1–2), 101–106 (2001)
3. Chang, W.-Y.: The state of charge estimating methods for battery: a review. ISRN Appl. Math., 7 (2013). Article ID 953792, http://dx.doi.org/10.1155/2013/953792
4. Bundy, K., Karlsson, M., Lindbergh, G., Lundqvist, A.: An electrochemical impedance spectroscopy method for prediction of the state of charge of a nickel-metal hydride battery at open circuit and during discharge. J. Power Sources **72**, 118–125 (1998)
5. Holger, B., Oliver, B., Stephan, B., de Doncker, R.W., Fricke, B., Hammouche, A., Linzen, D., Thele, M., Sauer, D.U.: Impedance measurements on lead–acid batteries for state-of-charge, state-of-health and cranking capability prognosis in electric and hybrid electric vehicles. J. Power Sources **144**, 418–425 (2005)
6. Gregory, L.P.: Battery Management Systems, Volume I: Battery Modeling. Artech House (2015)

7. He, W., Williard, N., Chen, C., Pecht, M.: State of charge estimation for electric vehicle batteries using unscented kalman filtering. Microelectron. Reliab. **53**, 840–847 (2013)
8. Lee, J., Nam, O., Cho, B.H.: Li-ion battery SoC estimation method based on the reduced order extended Kalman filtering. J. Power Sources **174**, 9–15 (2007)
9. Zhang, C.P., Jiang, J.C., Zhang, W.G., Sharkh, S.M.: Estimation of state of charge of lithium-ion batteries used in HEV using robust extended Kalman filtering. Energies **5**(4), 1098–1115 (2012)
10. Ritter, B., Mora, E., Schlicht, T., Schild, A., Konigorski, U.: Adaptive sigma-point Kalman filtering for wind turbine state and process noise estimation. J. Phys. Conf. Ser. 1037 (2018). Control and Monitoring
11. Rudolph, V.M., Eric, W.: Sigma-point Kalman filters for probabilistic inference in dynamic state-space models. OGI School of Science & Engineering Oregon, Health & Science University Beaverton, Oregon, 97006, USA (2003)
12. Gregory, L.P.: Sigma-point Kalman filtering for battery management systems of LiPB-based HEV battery packs. Part 1: Introduction and state estimation. J. Power Sources **161**(2), 1356–1368 (2006)

Strain Effect on Hysteresis Loop of $PbTiO_3$ Bulk

Do Van Truong[1,2(✉)], Tran The Quang[1,4], Nguyen Hoang Linh[1], Nguyen Van Hoi[1], and Vuong Van Thanh[3]

[1] Department of Mechatronics, Hanoi University of Science and Technology, Hanoi, Vietnam
truong.dovan@hust.edu.vn
[2] International Research Center for Computational Materials Science, Hanoi University of Science and Technology, Hanoi, Vietnam
[3] Department of Design of Machinery and Robot, Hanoi University of Science and Technology, Hanoi, Vietnam
[4] Department of Basic Engineering, Faculty of Technology, Hanoi, Thai Binh, Vietnam

Abstract. Ferroelectric material has been popularly used in micro/nano electronics devices. In order to elucidate the strain effect on the hysteresis loop, simulation using a shell model based on the potential functions fitting from ab initio is used to investigate the effect of the uniaxial strain, the iso-strain and the shear strain on the hysteresis loop of the $PbTiO_3$ bulk. The obtained results show that the hysteresis loop is expanded under the tension and shrunk under the compression of the uniaxial strain. Contrarily, the hysteresis loop is shrunk under the tension and expanded under the compression of the iso-strain. Besides, the hysteresis loop investigated is unchanged under the tension and compression of the shear strain. The obtained results expectantly provide useful guidelines for the development of ferroelectric devices in the future.

Keywords: Ferroelectric hysteresis loop · $PbTiO_3$ · Shell model · Strain effect

1 Introduction

Ferroelectric materials have been widely used in micro/nano electronics devices as FRAM [1], sensors, actuators, transducers [2], field-effect transistors [3], tunnel junctions [4] owing to the special switch of ferroelectric polarization vectors in each unit cell. Ferroelectrics are characterized by a hysteresis loop relating between the polarization and the applied electric field. In the real structures, ferroelectrics are usually grown on a substrate or sandwiched between the electrodes. Because of the lattice mismatch between the ferroelectric layer and the substrate (or the electrode), misfit strains appear in the ferroelectric layer, make changes of the remnant polarization and the coercive field, and affect the working reliability of the devices.

By different methods, studies have pointed out that the misfit strains affect the remnant polarization, the coercive field as well as the hysteresis loop. Wang *et al.* [5] investigated the strain effect of $PbTiO_3$ thin film is on the polarization switching by a

K.-U. Sattler et al. (Eds.): ICERA 2019, LNNS 104, pp. 679–685, 2020.
https://doi.org/10.1007/978-3-030-37497-6_78

phase-field model. The obtained results show that the remanent polarization and the coercive field increase with the compressive strain, while those decrease with the tensile strain. Zhang *et al.* [6] used a molecular dynamics study have proved that a tensile strain increases the polarization of the nanowire while a compressive strain decreases it. Shimada et al. [7] has conducted the effect of the terminated surfaces on the remnant polarization in $PbTiO_3$ nanowires by ab initio study and revealed that the remnant polarization increases linearly with the axial tensile strain in the PbO-terminated nanowires. For the TiO_2-terminated nanowires, the paraelectric and ferro-electricity phases appear, but the paraelectric prevails under a small tensile strain. Although numerous studies have been considered the effect of strain on the ferro-electricity, most of researches [8–10] have focused on the uniaxial strain and the iso-strain, and only few studies [11, 12] have investigated the shear strain. Moreover, the ferroelectric thin film and wire have intensely studied while the ferroelectric bulk has basically mentioned. In order to elucidate the effect of the strain on the hysteresis loop, the present study aims to the $PbTiO_3$ bulk. The focus is to put on the uniaxial strain, iso-strain and shear strain. A shell model [13] as a simulation tool is used to perform the research work.

2 Computational Details

2.1 Shell Model

In the shell model, each atom is assumed to contain two charged particles as a positive core q_c and a negative shell q_s. The electronic polarization happens owing to the displacement between the core and the shell. Interaction between the core-core, shell-shell, core-shell in an atom and among atoms is expressed by the potential functions. The potential functions [13] as the long-range V_{LR}, the short-range V_{SR}, and the core-shell spring V_{CS} potentials are used, which are defined as

$$V_{LR} = \frac{q_i q_j}{4\pi\varepsilon_o}\frac{1}{r_{ij}} \tag{1}$$

$$V_{SR} = \left\{Aexp\left(-\frac{r}{\rho}\right) - \frac{C}{r^6}\right\}f_c \tag{2}$$

$$V_{CS} = \frac{1}{2}k_2 r^2 + \frac{1}{24}k_4 r^4 \tag{3}$$

here r_{ij}, q_i and ε_o indicate the distance between *i* and *j* particles, the charge of core or shell, and the dielectric constant in vacuum, respectively. f_c is the function that limits the distance of interaction among ions. A, ρ, C, k_2, k_4 are the parameters of the potential functions. In this study, the shell model is used via GULP packet [14]. The potential functions obtained by Shimada *et al.* [13] are used. Table 1 shows the detail of the parameters of the potential functions.

2.2 Electronic Polarization

The local electronic polarization P can be evaluated by [15].

$$P = \frac{1}{v}\sum_{m}\frac{1}{w_m}q_m(r_m - r_A) \tag{4}$$

here v indicates the volume of the unit cell. q_m, r_m are respectively the charge and the position of atom m^{th}. r_A is the corresponding reference position to the core of atom A and the remaining 29 particles. The weight factors w_m are set to 1/8 for Pb, 1 for Ti and 1/2 for O.

Table 1. Parameters of potential functions of the shell model for $PbTiO_3$ [13]

Species	q_c (e)	q_s (e)	k_2 (eV Å^{-2})	k_4 (eV Å^{-4})
Pb	+5.49586	−3.63322	154.8713	22416.67
Ti	+19.36901	−16.27952	8829.4096	1928581.70
O	+2.54843	−4.19917	180.9134	6 945.78
Interactions	A (eV)	ρ (Å)	C (eV Å^6)	
Pb–O	2538.4110	0.300698	2.61676	
O–O	1698.6653	0.271756	61.84354	
Ti–O	2555.2075	0.278391	2.25557	
Pb–Ti	387.7316	0.39495	223.24409	

2.3 Models and Simulation Procedure

Figure 1*(a)* depicts the unit cell of tetragonal $PbTiO_3$ bulk with the lattice constants of a = 3.894 Å and c = 4.076 Å [13]. Figure 1*(b)*, *(c)*, *(d)* show the models of $PbTiO_3$ under the uniaxial strain, iso-strain and shear strain. The atomic structures are fully relaxed under the free stress condition at the applied electric field E of 0 V/Ang. Then, the strains (uniaxial strain, iso-strain, shear strain) are applied stepwise with an increment of 4% in both tension and compression. Finally, in each strain step, E is applied step by step with an increment of 0.05 V/Ang.

3 Results and Discussions

In this study, we only focus on the z direction polarization and stress component because of ferroelectric materials mainly applied in the z direction. Besides, the hysteresis loop is only conducted at the temperature of 0 °K. Based on the potential functions [13], the remanent polarization obtained is 86.8 μC/cm^2, compared with P_r = 79.8 μC/cm^2 obtained by Resta R. [16] and P_r = 82.0 μC/cm^2 by Ho Nyung Lee et al. [17]. Thus, the current study gives a relative consistent result compared with the results obtained from theory [16, 17], which shows the reliability of the shell model.

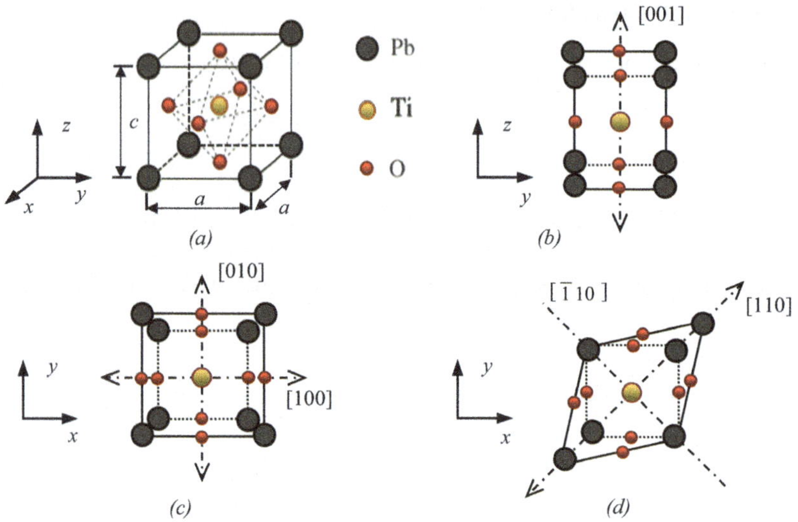

Fig. 1. Unit cell of tetragonal $PbTiO_3$ bulk *(a)*, the cell under uniaxial strain *(b)*, the cell under iso-strain *(c)*, the cell under shear strain *(d)*.

Figure 2*(a)* shows the hysteresis loop of $PbTiO_3$ bulk under the uniaxial strain ($\varepsilon_{zz} \neq 0$). The hysteresis loop is expanded under the tension of 4%. The remanent polarization P_r increases 1.18 times and the coercive field E_e increases 1.74 time compared with that at the uniaxial strain of 0%. These rates decrease to 0.68 and 0.87 times at the compression of −4%. The stored energy as the area inside the closed curve of the hysteresis loop is arranged in order at the strains of 4%, 0% and −4%. The increase of the remanent polarization P_r explained is due to shift the atoms far from the ideal positions (paraelectric phase), especially the shift of the atom Ti (Eq. (4)).

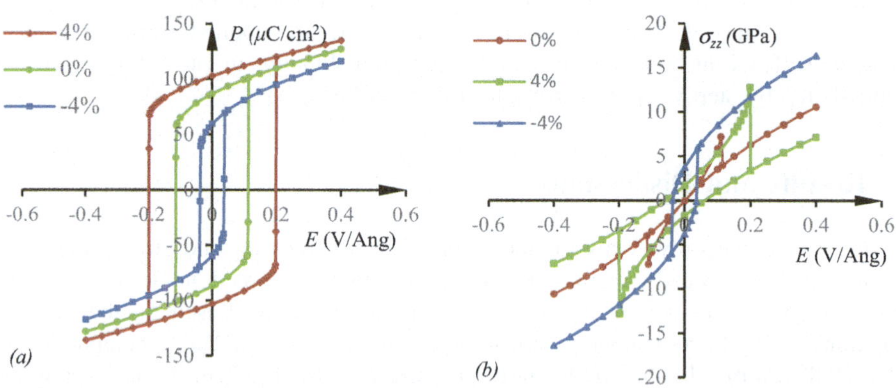

Fig. 2. Hysteresis loops of $PbTiO_3$ under the uniaxial strains of 0% and ±4% *(a)*; Relationship between the normal stress σ_{zz} and the applied electric field under the uniaxial strains of 0% and ±4% *(b)*.

Therefore, the coercive field E_c also needs increasing to back the atoms to the ideal positions. For the unit cells under the compression, the remanent polarization P_r and the coercive field E_c decrease, and these decreases can be explained in the opposite way mentioned above. Figure 2*(b)* depicts the relationship between the normal stress σ_{zz} and the electronic filed *E*. Stress behavior is also a closed loop. The maximum stress appears in the compression of −4%, and is lager 1.55 time than that in the strain of 0%.

Figure 3*(a)* và *(b)* show the hysteresis loop and the relationship of $\sigma_{zz} - E$ under the iso-strain ($\varepsilon_{xx} = \varepsilon_{yy} \neq 0$), respectively. Contrary to the uniaxial strain, the hysteresis loop is expanded under the compression and shrunk under the tension. Under the iso-strain, Ti is shifted largely in the *z* direction, and the shift causes the increase of the remanent polarization P_r and the coercive field E_c. The maximum stress appears in the compression and is reached at the coercive field E_c. However, this value is only 0.76 of the largest value in the case of the uniaxial strain.

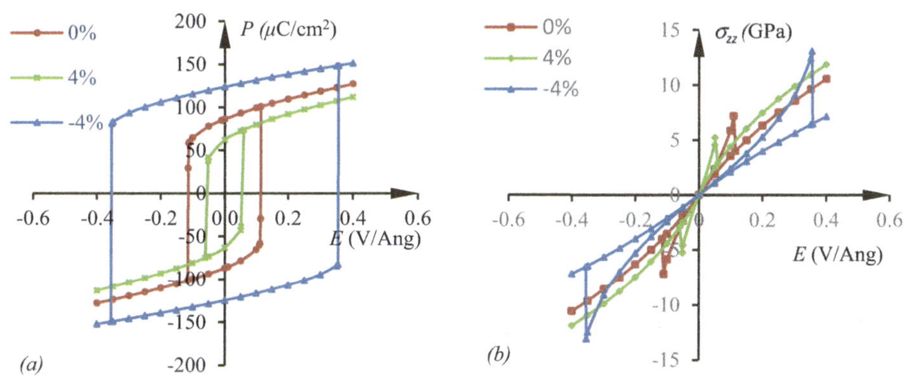

Fig. 3. Hysteresis loops of $PbTiO_3$ under the isostrains of 0% and ±4% *(a)*; Relationship between the normal stress $\boldsymbol{\sigma}_{zz}$ and the applied electric field under the isostrains of 0% and ±4% *(b)*.

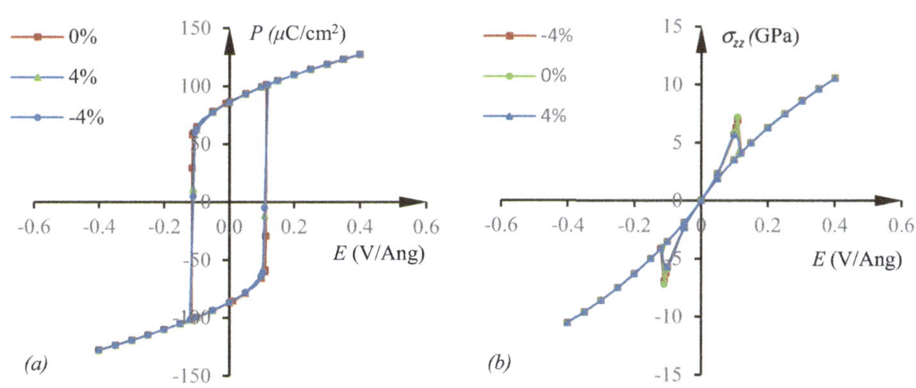

Fig. 4. Hysteresis loops of $PbTiO_3$ under the shear strains of 0% and ±4% *(a)*; Relationship between the normal stress $\boldsymbol{\sigma}_{zz}$ and the applied electric field under the shear strains of 0% and ±4% *(b)*.

In addition to the uniaxial strain and the iso-strain, the shear strain ($\varepsilon_{xy} = \varepsilon_{yx} \neq 0$), is also interested in and expect to have the effects on the hysteresis loop and the relationship of $\sigma_{zz} - E$. However, the results obtained in Fig. 4*(a)* and *(b)* show that the shear strain does not significantly affect the hysteresis loop and the relationship of $\sigma_{zz} - E$. This can be explained that the cells are only deformed in the *xy*-plane, the atoms do not shift in the *z* direction, especially Ti atoms, and thus P_r (calculated by Eq. (4)) is almost constant in this case.

4 Conclusions

In this study, the simulation is performed by using the shell model based on the potential functions fitting from ab initio study to investigate the effect of the uniaxial strain, the iso-strain and the shear strain on the hysteresis loop of the $PbTiO_3$ bulk. The obtained results could be summarized below.

The hysteresis loop is expanded under the uniaxial strain. The remanent polarization P_r increases 1.18 times and the coercive field E_c increases 1.74 time compared with that at the uniaxial strain of 0%. These rates are decreased to 0.68 and 0.87 times at the compression of −4%. The greatest stress appears in the compression of −4%, and is lager 1.55 time than that in the strain of 0%.

Under the iso-strain, the hysteresis loop is changed, the stress behavior is a closed loop. The maximum stress appears under the compression of −4%, and is reached at the coercive field E_c. This value is only 0.76 of the largest value in the case of the uniaxial strain.

Under the shear strain, the hysteresis loop changes insignificantly under the shear strains of −4%, 0% and 4%. The stresses obtained are not different among strains.

Acknowledgments. This research was funded by the Vietnam's National Foundation for Science and Technology Development (NAFOSTED) with No.107.02-2016.18.

References

1. Scott, J.F., Paz de Araujo, C.A.: Ferroelectric memories. Science **246**, 1400 (1989)
2. Chen, W.J., Zheng, Y., Wang, B.: Vortex domain structure in ferroelectric nanoplatelets and control of its transformation by mechanical load. Sci. Rep. **2**, 796 (2012)
3. Tsymbal, E.Y., Kohlstedt, H.: Tunneling across a ferroelectric. Science **313**, 181 (2016)
4. Julian, D.B., John, Y.Z., Mikhail, Y.T.: Predictive modelling of ferroelectric tunnel junctions. npj Comput. Mater. **2**, 16009 (2016)
5. Wang, J., Xia, Y., Chen, L.Q., Shi, S.Q.: Effect of strain and deadlayer on the polarization switching of ferroelectric thin film. J. Appl. Phys. **110**, 114111 (2011)
6. Zhang, Y., Hong, J., Liu, B., Fang, D.: Strain effect on ferroelectric behaviors of BaTiO3 nanowires: a molecular dynamics study. Nanotechnology **21**, 015701 (2010)
7. Shimada, T., Tomoda, S., Kitamura, T.: First-principles study on ferroelectricity at PbTiO3 surface steps. J. Phys.: Condens. Matter **22**, 355901 (2010)

8. Shimada, T., Wang, X., Tomoda, S., Martin, P., Elsässer, C., Kitamura, T.: Coexistence of rectilinear and vortex polarizations at twist boundaries in ferroelectric PbTiO3 from first principles. Phys. Rev. B **83**, 094121 (2011)
9. Marton, P., Shimada, T., Kitamura, T., Elsässer, C.: First-principles study of the interplay between grain boundaries and domain walls in ferroelectric PbTiO3. Phys. Rev. B **83**, 064110 (2011)
10. Wang, X., Tomoda, S., Shimada, T., Kitamura, T.: Local suppression of ferroelectricity at PbTiO3 surface steps: a density functional theory study. J. Phys.: Condens. Matter **24**, 045903 (2012)
11. Simon, W.K., Akdogan, E.K., Safari, A.: Anisotropic strain relaxation in $(Ba_{0.6}Sr_{0.4})$ TiO_3 epitaxial thin films. J. Appl. Phys. **97**, 103530 (2005)
12. Yan, W.S., Zhang, R., Xie, Z.L., Xiu, X.Q., Zheng, Y.D., Liu, Z.G., Xu, S.: Impact of effective shear strain on the equilibrium phases and polarization states of PbTiO3 thin film. Appl. Phys. Lett. **95**, 222901 (2009)
13. Shimada, T., Wakahara, K., Umeno, Y., Kitamura, T.: Shell model potential for PbTiO3 and its applicability to surfaces and domain walls. J. Phys.: Condens. Matter **20**, 325225 (2008)
14. Gale, J.D., Rohl, A.L.: The general utility lattice program (GULP). Mol. Simul. **29**, 341 (2003)
15. Sepliarsky, M., Cohen, R.E.: First principles based atomistic modeling of phase stability in PMN-xPT. J. Phys.: Condens. Matter **23**, 435902 (2011)
16. Resta, R.: Manifestations of Berry's phase in molecules and condensed matter. J. Phys.: Condens. Matter **12**, R107 (2000)
17. Lee, H.N., Nakhmanson, S.M., Chisholm, M.F., Christen, H.M., Rabe, K.M., Vanderbilt, D.: Suppressed dependence of polarization on epitaxial strain in highly polar ferroelectrics. Phys. Rev. Lett. **98**(21), 217602 (2007)

Studying an Active Anti-roll Bar Control System for Tractor-Semi Trailer Vehicles

Van Tan Vu[1(✉)] and Duc Tien Bui[2]

[1] Department of Automotive Mechanical Engineering, University of Transport and Communications, Hanoi, Vietnam
vvtan@utc.edu.vn

[2] Faculty of Mechanical Engineering, Thuyloi University, Hanoi, Vietnam
buiductien.ckoto@tlu.edu.vn

Abstract. Today, articulated vehicles are widely used in the world because they benefit the society. However, accidents related to articulated vehicles tend to be fatal, one of the main reasons is the instability of articulated vehicles. This paper is concerned with the dynamic modeling of tractor–semi trailer vehicles. A simplified linear model is proposed, which includes the roll and yaw motions, based on the assumption that the longitudinal velocity is constant. The simulation results in a cornering and a double lane change manoeuver show that the stability of tractor-semi trailer depends on the stiffness and damping coefficients and the forward velocity. Therefore, in order to improve the vehicle roll stability, three types of active anti-roll bar system for the tractor-semi trailer are proposed: the first and second systems are equipped only for tractor and trailer respectively, while the last one can be used for both kinds of vehicles.

Keywords: Vehicle dynamics · Articulated vehicles · Active anti-roll bar · Roll stability · Rollover

1 Introduction

Figure 1 illustrates the two types of common heavy vehicles, which are: a single unit truck and articulated vehicles. The main types of articulated vehicles are the tractor - semi trailer, the truck - center axle trailer with one articulation, the truck - full trailer with two articulations and the Longer Combination Vehicles (LCVs). Compared to conventional trucks, articulated vehicles provide such advantages as reduction in fuel consumption per ton per kilometer leading to less emission, decrease of transport costs thanks to increase of the transport volume and payload of one delivery time, and road protection due to reduced axle loads.

However, it has been well documented that accidents related to articulated vehicles are deadly serious and linked to roll stability. As defined, roll stability refers to the ability of a vehicle to overcome overturning moments generated during cornering and lane changing to avoid an obstacle. Although being relatively rare events, rollovers are usually deadly accidents when they occur. The three major contributing factors to rollover accidents are side wind gusts, abrupt steering and braking manoeuvers by the driver. In all these cases, roll stability loss is caused when the tyre-road contact force of

K.-U. Sattler et al. (Eds.): ICERA 2019, LNNS 104, pp. 686–692, 2020.
https://doi.org/10.1007/978-3-030-37497-6_79

the wheels on one side of the vehicle becomes zero. Technically, heavy vehicles have relatively high centers of mass and narrow track widths (maximum of 2.5 m) and can lose roll stability at moderate levels of lateral acceleration. Hence, minimizing the effects of low roll stability is a challenging mission. In order to deal with the possibilities of rollovers, it is necessary to develop an electronic stability function for articulated vehicles.

The dynamics of tractor semi trailer and tractor full-trailer combinations were summarized in [1] and [2]. In [3] and [4], the authors reviewed the dynamics of long combination vehicles. Publications from the University of Michigan Transportation Research Institute are among the most comprehensive general reviews of heavy vehicle dynamics [5–7]. The roll dynamics of heavy vehicles in cornering maneuvers are much more relevant to vehicle safety than those of automobiles [6].

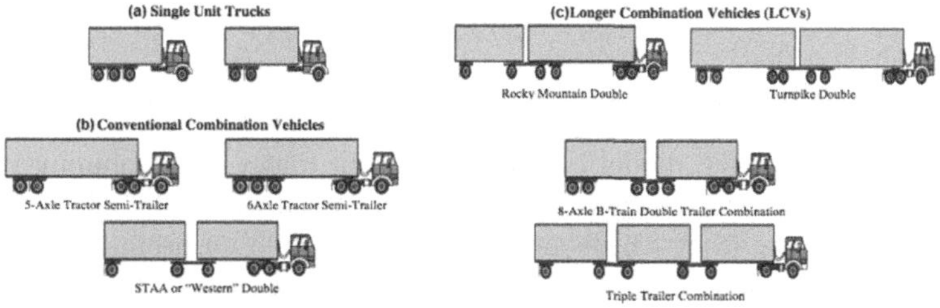

Fig. 1. Truck combinations (a: Single vehicles; b: Conventional articulated vehicles; c: Longer Combination Vehicles)

The contributions of this paper are the following: (1) A yaw - roll model of 9 degrees of freedom is used for studying the dynamics of a tractor - semi trailer, by considering the vertical and lateral displacement. (2) With the simulation results of roll stability of tractor- semi trailer, an active anti-roll bar is proposed. It also allows the assessment of the effects of the induced normalized load transfer. (3) The simulation results show that with the passive suspension system, vehicle rollover occurs when the forward velocity is 38 km/h in a double lane change manoeuvre. Therefore, we propose three solutions in order to apply the active anti-roll bar system on tractor - semi trailer vehicles.

2 Vehicle Modeling

2.1 Modeling of a Tractor – Semi Trailer

Figure 2 describes a simplified model that captures the essential tractor - semi trailer dynamics. In this model, the car body is assumed to be perfectly rigid. The vehicle body has the roll axis with distance *r* upwards from the ground. Pitching and bouncing motions have only a small effect on the roll and handling behavior of the vehicle and so can be neglected. The effects of aerodynamic inputs (wind disturbances) and road inputs are also ignored. Disturbances of the roll model are produced from the steering angle. The dynamic equations of the vehicle are formulated by equating the change of momentum (or, in the rotational case, the moment of momentum) with the sum of external forces (or moments) acting on the system. The motion is described using a coordinate system (x′, y′, z′) fixed in the vehicle. The roll axis is replaced by an x′ axis parallel to the ground, and the z′ axis passes downward through the center of mass of the tractor - semi trailer.

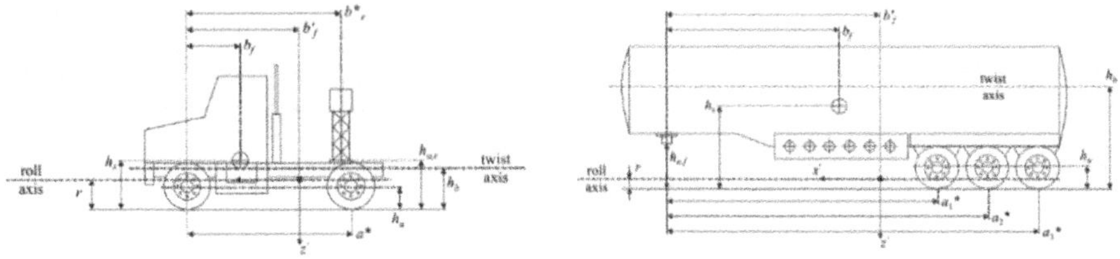

Fig. 2. Yaw – roll model of tractor - semi trailer.

Figure 2 illustrates the model of a tractor - semi trailer by combining two parts: tractor and semi trailer, in which $m_{s,i}$ is the sprung mass, $m_{uf,i}$ is the unsprung mass at the front including the front wheels and axle, and $m_{ur,i}$ the unsprung mass at the rear with the rear wheels and axle, I_{xi} is the roll moment of inertia, I_{zi} is the yaw moment of inertia of tractor or semi trailer. F_C is the lateral force at the coupling. The subscripts f and r denote the front and rear axle of the tractor or trailer, respectively. The subscripts i (1 and 2) denote the tractor and semi trailer, respectively. Some symbols and parameters of the model are shown in [8].

2.2 Dynamic Equations of a Tractor – Semi Trailer

In the vehicle modelling, the differential equations of the Yaw – Roll dynamics of the tractor - semi trailer model, that are the lateral dynamics, the yaw moment, the roll moment of the sprung mass, the roll moment of the front and rear unsprung masses of tractor are formalized in Eqs. (1) to (5). The Eqs. (6) to (8) show the dynamic motions of the semi trailer, which are the yaw moment of the sprung mass, the roll moment of the sprung mass, the roll moment of the rear unsprung mass. The kinematic constraint at the vehicle coupling is described by Eq. (9).

$$m_1 v(\dot{\beta}_1 + \dot{\psi}_1) - \mathrm{m}_{s1} h\, \dot{\phi}_1 = F_{yf} + F_{yr} - F_C \tag{1}$$

$$I_{xz1} \cdot \ddot{\phi}_1 - I_{zz1} \cdot \ddot{\psi}_1 = F_{yr1}.dl_{r1} - F_{yf1}.dl_{f1} - b'_{r1}.F_C - k_{\psi 1}(\psi_1 - \psi_2) \tag{2}$$

$$\begin{aligned} &- \mathrm{m}_{s1}.a_y.h_{s1} - \mathrm{m}_{s1} g.\phi_1.h_{s1} + I_{xx1} \cdot \ddot{\phi}_1 - I_{xz1} \cdot \ddot{\psi}_1 + k_{f1}(\phi_1 - \phi_{tf1}) + l_{f1}(\dot{\phi}_1 - \dot{\phi}_{tf1}) \\ &+ k_{r1}(\phi_1 - \phi_{tr1}) + l_{r1}(\dot{\phi}_1 - \dot{\phi}_{tr1}) + u_{f1} + u_{r1} + (h_{ar1} - r_1) F_C - k_{\phi 1}(\phi_1 - \phi_2) = 0 \end{aligned} \tag{3}$$

$$\begin{aligned} & -F_{yf1}.r_1+(r_1-h_{uf1}).m_{uf1}.v(\dot{\beta}_1+\dot{\psi}_1)+k_{tf1}.\phi_{tf1}+(r_1-h_{uf1}).m_{uf1}.g.\phi_{tf1} \\ & -k_{f1}(\phi_1-\phi_{tf1})-l_{f1}(\dot{\phi}_1-\dot{\phi}_{tf1})-u_{f1}=0 \end{aligned} \tag{4}$$

$$\begin{aligned} & -F_{yr1}.r_1+(r_1-h_{ur1}).m_{ur1}.v(\dot{\beta}_1+\dot{\psi}_1)+k_{tr1}.\phi_{tf1}+(r_1-h_{uf1}).m_{ur1}.g.\phi_{tr1} \\ & -k_{r1}(\phi_1-\phi_{tr1})-l_{r1}(\dot{\phi}_1-\dot{\phi}_{tr1})-u_{r1}=0 \end{aligned} \tag{5}$$

$$I_{xz2}.\ddot{\phi}_2-I_{zz2}.\ddot{\psi}_2=F_{yr2}.dl_{r2}+b'_{f2}.F_C-k_{\psi1}(\psi_1-\psi_2) \tag{6}$$

$$\begin{aligned} & I_{xx2}.\ddot{\phi}_2-I_{xz2}.\ddot{\psi}_2-\mathrm{m}_{s2}.a_y.h_{s2}-\mathrm{m}_{s2}g.\phi_2.h_{s2}+k_{r2}(\phi_2-\phi_{tr2}) \\ & +l_{r2}(\dot{\phi}_2-\dot{\phi}_{tr2})+u_{r2}-(h_{af2}-r_2)F_C+k_{\phi1}(\phi_1-\phi_2)=0 \end{aligned} \tag{7}$$

$$\begin{aligned} & -F_{yr2}.r_2-(h_{ur2}-r_2).m_{ur2}.v(\dot{\beta}_2+\dot{\psi}_2)-(h_{ur2}-r_2).m_{ur2}.g.\phi_{tr2} \\ & -k_{r2}(\phi_2-\phi_{tr2})-l_{r1}(\dot{\phi}_2-\dot{\phi}_{tr2})-u_{r2}+k_{tr2}.\phi_{tr2}=0 \end{aligned} \tag{8}$$

$$\beta_1-\beta_2-\frac{(r_1-h_{ar1})}{v}\dot{\phi}_1+\frac{(r_2-h_{af2})}{v}\dot{\phi}_2+\frac{b'_{r1}}{v}\dot{\psi}_1-\frac{b'_{f2}}{v}\dot{\psi}_2+\psi_1-\psi_2=0 \tag{9}$$

The lateral tyre forces $F_{y,i}$ in the direction of velocity at the wheel ground contact points are modeled by using linear stiffness coefficients as:

$$\begin{aligned} F_{yfi} &= \mu C_f \alpha_f \\ F_{yri} &= \mu C_r \alpha_r \end{aligned} \tag{10}$$

With the tyre side slip angles:

$$\alpha_{f1}=-\beta_1+\delta_f-\frac{l_{f1}\dot{\psi}_1}{v} \qquad \alpha_{r1}=-\beta_1+\frac{l_{r1}\dot{\psi}_1}{v} \qquad \alpha_{r2}=-\beta_2+\frac{l_{r2}\dot{\psi}_2}{v} \tag{11}$$

The motion differential equations from (1) to (11) can be rewritten in the LTI state-space representation as:

$$\begin{cases} \dot{x}=Ax+B_1w+B_2u \\ z=Cx+D_1w+D_2u \end{cases} \tag{12}$$

With the state vector: $x=[\beta_1\ \dot{\psi}_1\ \phi_1\ \dot{\phi}_1\ \phi_{tf1}\ \phi_{tr1}\ \phi_2\ \dot{\phi}_2\ \beta_2\ \dot{\psi}_2\ \phi_{tr2}]^T$, The exogenous disturbance: $w=[\delta]^T$, The control input: $u=[u_{f1}\ u_{r1}\ u_{r2}]^T$, The output vector: z = x.

2.3 The Control Objectives for Vehicle Roll Stability

In order to evaluate the vehicle roll stability of articulated vehicles using the active anti-roll bar system, we would like to minimize the normalized load transfers of the three axles R_i: $|R_i| = \left|\frac{\Delta F_{zi}}{F_{zi}}\right| < 1$, where ΔF_{zi} is lateral load transfer and F_{zi} total axle load. If R_i takes the value ± 1 then the inner wheel in the bend lifts off, so the rollover occurs.

3 Simulation Results Analysis

In this section, the simulation results of the yaw - roll model of a tractor - semi trailer are shown in the time domain. A double lane change is used in this study as the common disturbance for studying the roll stability. The double lane change maneuver has a 2 (m) path deviation over 10 0 (m). The size of the path deviation is chosen to test a real obstacle avoidance in an emergency [9]. Figure 3 shows the time response of the steering angle, the roll angle of sprung mass, the normalized load transfer at two axles of the tractor when the vehicle velocity is considered at 45 km/h.

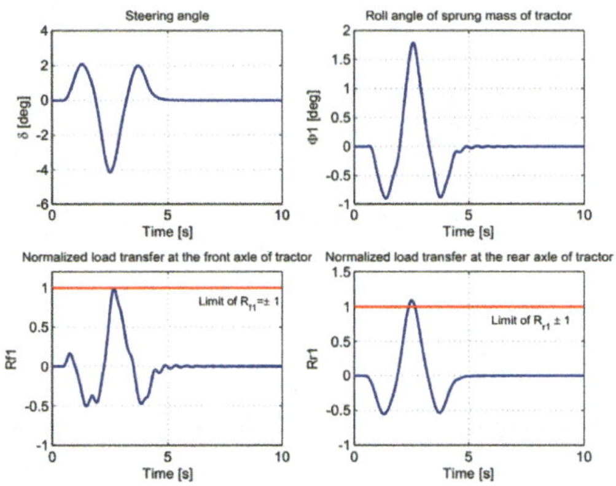

Fig. 3. Time responses when the vehicle is in a double lane change maneuver.

We can see that the value of the normalized load transfers at 2.3 s exceeds 1 so that the inner wheels lift off, and vehicle rollover occurs. The simulation results also show that when the forward velocity is 38 km/h, the normalized load transfers reach 1, this means that the roll stability can't be maintained with the speed higher of more than 38 km/h. Of course, these values depend on the steering angle and its speed. However, the forward velocity (38 km/h) is quite low with a standard steering behavior.

4 Proposal of Active Anti-roll Bar System for a Tractor – Semi Trailer

From the simulation results in Sect. 3, we can see that the passive anti-roll bar system is not good enough for the roll stability of the tractor - semi trailer in emergency situations. Hence, in order to improve the vehicle roll stability, here we have three following proposals:

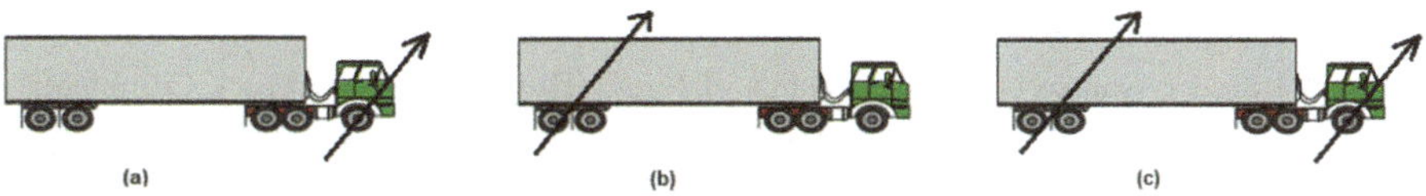

Fig. 4. The proposal methods to equip the active anti-roll bar system for tractor semitrailer.

The first one: an active anti-roll bar system is equipped for the only tractor (Fig. 4a). In this case, the cost for the active anti-roll bar system is reduced. Additionally, the interchangeability of tractors is quite convenient.

The second one: an active anti-roll bar system is equipped for the semitrailer (Fig. 4b). Because the semitrailer is exchangeable, the stability of articulated vehicles is significantly improved.

The third one: two active anti-roll bar systems are equipped for the tractor and semitrailer (Fig. 4c). It is clear that the stability of articulated vehicles in this case is the highest. In particularly, when the tractor and semitrailer work together for long periods.

5 Conclusions

We would like to emphasize that the use of the active anti-roll bar system on the tractor - semi trailer is a new approach to reduce accidents. The simulation results show the importance to equip the active anti-roll bar system for the tractor - semi trailer in order to enhance its roll stability. Hence, we proposed a yaw-roll model of 9 degrees of freedom with three solutions for applying this system to this type of vehicle. The dynamic equations of the tractor - semi trailer in this study are useful for applying the advanced control methods in further studies.

References

1. Nalecz, A.G., Genin, J.: Dynamic stability of articulated vehicles. Int. J. Veh. Des. **5**(4), 417–426 (1984)
2. Vlk, F.: Lateral dynamics of commercial vehicle combinations – a literature survey. Veh. Syst. Dyn. **11**(5), 305–324 (1982)

3. Segel, L., Ervin, R.D.: The influence of tire factors on the stability of trucks and truck-trailers. Veh. Syst. Dyn. **10**(1), 39–59 (1981)
4. Ellis, J.R.: Vehicle Handling Dynamics. Mechanical Engineering Publications, London (1994)
5. Fancher, P.S., Mathew, A.: A vehicle dynamics handbook for single and articulated heavy trucks. Technical report UMTRI-86-37, University Michigan Transportation Research Institute, Ann Arbor, MI, USA (1987)
6. Fancher, P.S., Ervin, R.D., Winkler, C.B., Gillespie, T.D.: A factbook of the mechanical properties of the components for single-unit and articulated heavy trucks. Technical report UMTRI-86-12, University of Michigan Transportation Research Institute, Ann Arbor, MI, USA (1986)
7. Segel, L. (ed.): Course on the Mechanics of Heavy-Duty Trucks and Truck Combinations. Surfers Paradise, Qld, Australia. University of Michigan Transportation Research Institute (1988)
8. Sampson, D.J.M.: Active roll control of articulated heavy vehicles. Ph.D. thesis (2000)
9. Gaspar, P., Bokor, J., Szaszi, I.: Reconfigurable control structure to prevent the rollover of heavy vehicles. Control Eng. Pract. **13**(6), 699–711 (2005)
10. Vu, V.T., Sename, O., Dugard, L., Gaspar, P.: $\mathcal{H}\infty/LP\ V$ controller design for an active anti-roll bar system of heavy vehicles using parameter-dependent weighting functions. Heliyon **5**, e01827 (2019)
11. Vu, V.T., Sename, O., Dugard, L., Gaspar, P.: Enhancing roll stability of heavy vehicle by LQR active anti-roll bar control using electronic servo-valve hydraulic actuators. Veh. Syst. Dyn. **55**(9), 1405–1429 (2017)
12. Ervin, R.D.: Influence of size and weight variables on the stability and control properties of heavy trucks. SAE Trans. **92**(831163), 629–654 (1983)

Studying Effect of Adding Buffer Gases to TRIES Gas on the Electron Transport Coefficients

Pham Xuan Hien[1], Phan Thi Tuoi[2], Tang Cam Nhung[3], and Do Anh Tuan[4(✉)]

[1] Department of Mechanical Engineering, University of Transport and Communications, Hanoi, Vietnam
hienpx@utc.edu.vn
[2] Faculty of Electronics and Electrical Engineering, Hung Yen University of Technology and Education, Hung Yen, Vietnam
[3] Thai Nguyen University of Technology, Thai Nguyen, Vietnam
[4] Department of Science and Technology Management and International Cooperation, Hung Yen University of Technology and Education, Hung Yen, Vietnam
tuandoanh@utehy.edu.vn

Abstract. The electron transport coefficients in TRIES-Kr, TRIES-He, TRIES-Xe and TRIES-Ne mixtures were firstly calculated by using two-term approximation of the Boltzmann equation for the E/N (ratio of the electric field E to the neutral number density N) range of 0.1–1000 Td (Townsend). The influences of adding different amounts of inert gases on the electron transport coefficient of TRIES-inert gases mixtures are investigated for industrial purposes, specifically in plasma polymerization and plasma enhancing chemical vapour deposition.

Keywords: Triethoxysilane · Boltzmann equation analysis · Electron transport coefficients

1 Introduction

Organosilicon compounds, which include hexamethyldisilazane [HMDS, $(CH_3)_3SiNHSi(CH_3)_3$], hexamethyldisiloxane [HMDSO, $(CH_3)_3SiOSi(CH_3)_3$], tetrae-thylorthosilicate [TEOS, $Si(OC_2H_5)_4$] and triethoxysilane [TRIES, $HSi(OC_2H_5)_3$], are organometallic compounds containing carbon–silicon bonds. As mentioned in [1] these compounds are widely used in chemical vapor deposition of silicon dioxide by thermal decomposition. TEOS gas is an organosilicon compound which is investigated by many researchers. TEOS, both in pure and mixture forms, has been extensively employed in abundant procedures, for example, deposition of silicon oxide, deposition of organosilicon films, silica nanoparticles using thermal decomposition, plasma polymerization [2–8]. Recently, Tuan and Jeon [9] have reported the most reliable set of electron collision cross sections for TEOS molecule and the electron transport coefficients in TEOS-Ar and TEOS-O_2 mixtures. Tuan [10] has also calculated and analyzed the electron transport

K.-U. Sattler et al. (Eds.): ICERA 2019, LNNS 104, pp. 693–703, 2020.
https://doi.org/10.1007/978-3-030-37497-6_80

coefficients in mixtures of TEOS gas with inert gases and suggested that these results were necessary for plasma assisted thin-film deposition. Dobkin *et al.* [1] studied mechanisms of deposition of SiO_2 from TEOS and TRIES and reported that their behaviors were generally similar.

Moreover, [11] suggested that TRIES can be used as an alternative for SiO_2 and TEOS for achieving better quality in building a SiO_2 insulating layer. In practical applications, the specific gas is often mixed with inert gases to obtain more effects. The electron transport coefficients, which include the electron drift velocity, the density-normalized longitudinal diffusion coefficient, the ratio of longitudinal diffusion coefficient to electron mobility and the Townsend first ionization coefficient are necessary for input data in discharge modeling. In the present work, these electron transport coefficients in mixtures of TRIES and inert gases were calculated by using a Boltzmann equation analysis.

2 Analysis

For computation of the electron transport coefficients in mixtures of TRIES gas and inert gases such as Kr, Xe, He and Ne, a two-term solution of Boltzmann's equation was applied. The electron drift velocity and the longitudinal diffusion coefficient (D_L) are calculated in the time-of-flight (TOF) parameters. The attachment coefficient and the ionization coefficient are calculated in the steady state Townsend (SST) parameters. These calculations based on the reliable sets of electron collision cross sections of TRIES [12], Kr [13], He [13], Ne [13] and Xe [14].

The present two-term solution of the Boltzmann's equation given by Tagashira *et al.* [15] has been also previously used for the pure gases and binary mixture gases [9, 11]. A description of the Boltzmann equation analysis in the previous publications [9, 11] is also briefly following represented.

The electron transport coefficients in gases can be achieved in a hypothetical manner once the Boltzmann equation for electrons in gases is resolved [16]. The general shape of the Boltzmann equation for electrons in gases is

$$\frac{\partial f}{\partial t} + \vec{v}\nabla_r f + \vec{a}\nabla_v f = \left(\frac{\partial f}{\partial t}\right)_{coll} \quad (1)$$

where $f = f(\vec{r}, \vec{v}, t)$ is the distribution function of the locative location r and time and with velocity of electrons v, $\vec{a}$ is the acceleration due to external forces, and $(\partial f/\partial t)_{coll}$ is the collision induced rate of change of the number of electrons per unit volume of phase space.

The connection between the electron drift velocity Wand electron energy distribution function is

$$W = -\frac{1}{3}\left(\frac{2}{m}\right)^{1/2}\frac{eE}{N}\int_0^{\infty}\frac{\varepsilon}{q_m(\varepsilon)}\frac{df(\varepsilon, E/N)}{d\varepsilon}d\varepsilon \quad (2)$$

where ε is the energy of the electron (in eV), m is the electron mass, e is the elementary charge and $q_m(\varepsilon)$ is the momentum transfer cross section (in cm^2).

The density-normalized longitudinal diffusion coefficient is formularized as

$$ND_L = \frac{V_1}{3N}\left(E\int_0^{\infty}\frac{\varepsilon}{q_T}\frac{\partial}{\partial\varepsilon}(F_1\varepsilon^{-1/2})d\varepsilon + \int_0^{\infty}\frac{\varepsilon^{1/2}}{q_T}F_0 d\varepsilon\right) - (\varpi_0 A_2 - \varpi_1 A_1 - \varpi_{02}) \quad (3)$$

In this formula, V_1 is electron speed and q_T is the total cross section. F_n is the energy distributions of numerous orders among electrons while ϖ_n (n = 0, 1, 2) are their eigen values. V_1, ϖ_n, ϖ_{0n}, and A_n are given by

$$V_1 = \left(\frac{2e}{m}\right)^{1/2} \quad (3.1)$$

$$\varpi_0 = V_1 N\int_0^{\infty}\varepsilon^{1/2}q_i F_0 d\varepsilon \quad (3.2)$$

$$\varpi_1 = -\frac{V_1 E}{3N}\int_0^{\infty}\frac{\varepsilon}{q_T}\frac{\partial}{\partial\varepsilon}(F_0\varepsilon^{-1/2})d\varepsilon + (\varpi_0 A_1 - \varpi_{01}) \quad (3.3)$$

$$\varpi_{0n} = V_1 N\int_0^{\infty}\varepsilon^{1/2}q_i F_n d\varepsilon \quad (3.4)$$

$$A_n = \int_0^{\infty}F_n d\varepsilon \quad (3.5)$$

where q_i is the ionization cross section.

The Townsend first ionization coefficient is defined as

$$\alpha/N = \frac{1}{W}\left(\frac{2}{m}\right)^{1/2}\int_I^{\infty} f(\varepsilon, E/N)\varepsilon^{1/2}q_i(\varepsilon)d\varepsilon \quad (3.6)$$

where I is the ionization onset energy and $q_i(\varepsilon)$ is the ionization cross section.

3 Results and Discussion

The variances of the electron transport coefficients in TRIES-Kr, TRIES-Xe, TRIES-He, TRIES-Ne mixtures for various concentration mixtures were shown in Figs. 1, 2, 3, 4, 5, 6, 7, 8, 9, 10, 11 and 12. The electron transport coefficients increase as the concentration of inert gases increase.

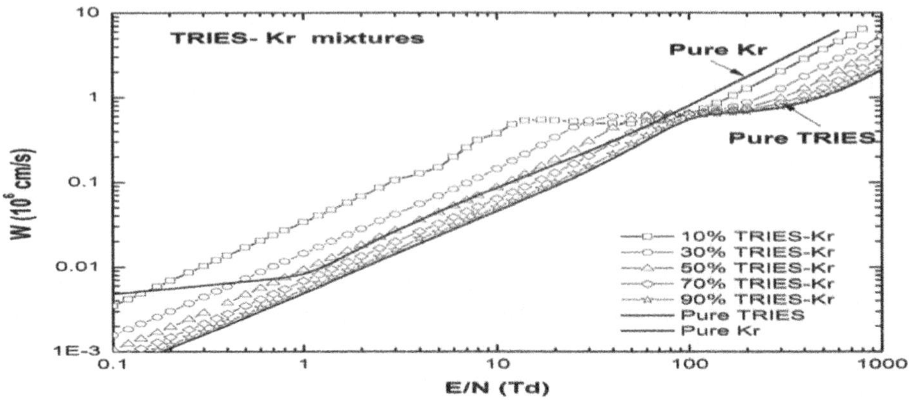

Fig. 1. The electron drift velocity, W, as a function of E/N in the TRIES-Kr mixtures.

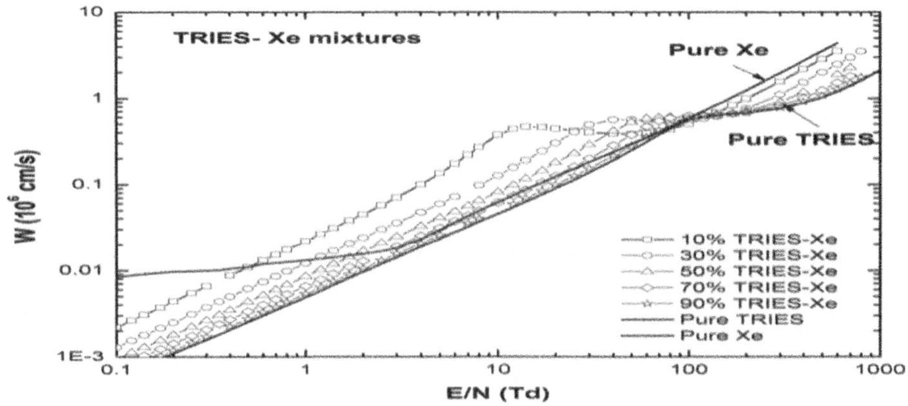

Fig. 2. The electron drift velocity, W, in TRIES-Xe mixtures.

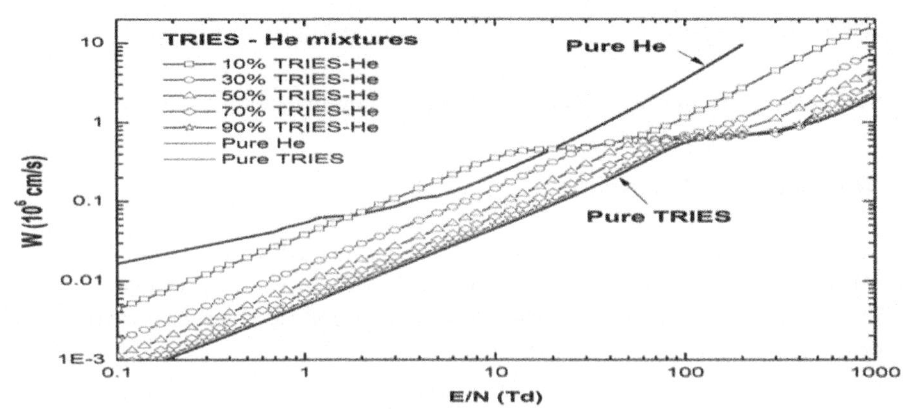

Fig. 3. The electron drift velocity, W, in the TRIES-He mixtures.

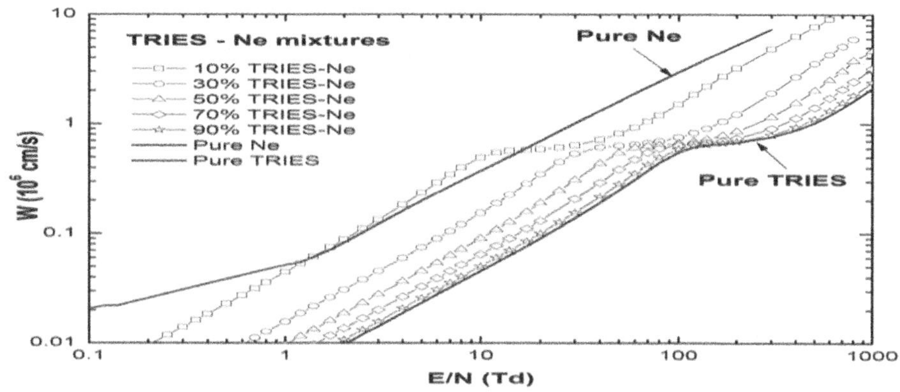

Fig. 4. The electron drift velocity, W, in the TRIES-Ne mixtures.

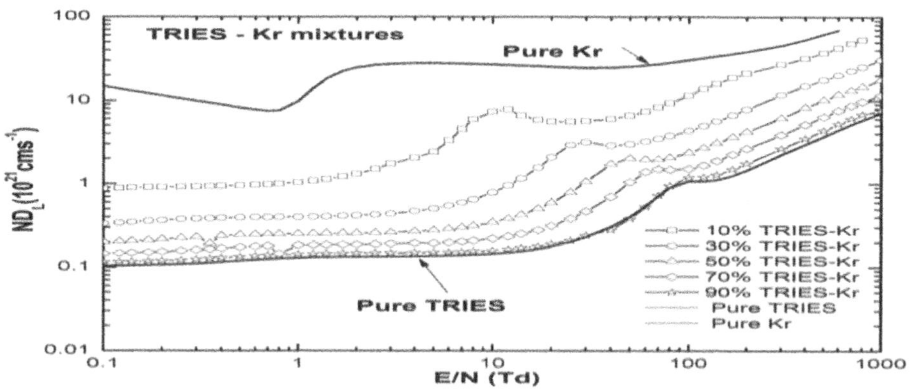

Fig. 5. The density-normalized longitudinal diffusion coefficient, ND_L, in the TRIES-Kr mixtures.

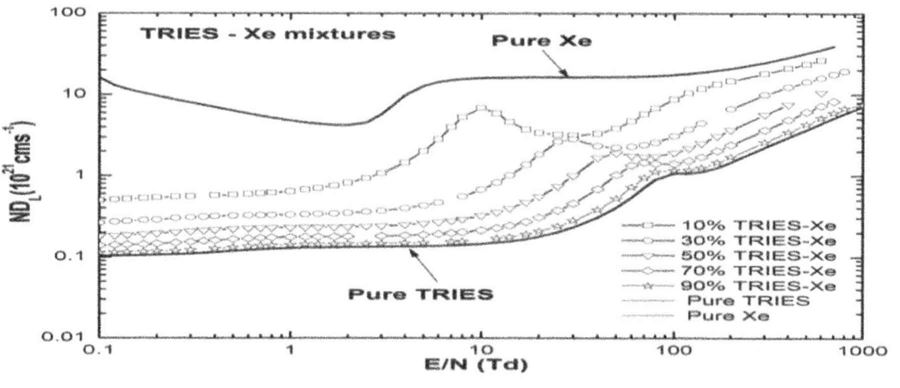

Fig. 6. The density-normalized longitudinal diffusion coefficient, ND_L, in the TRIES-Xe mixtures.

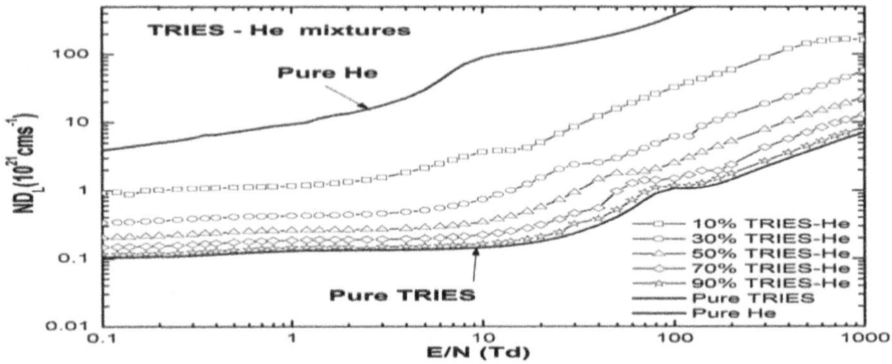

Fig. 7. The density-normalized longitudinal diffusion coefficient, ND_L, in the TRIES-He mixtures.

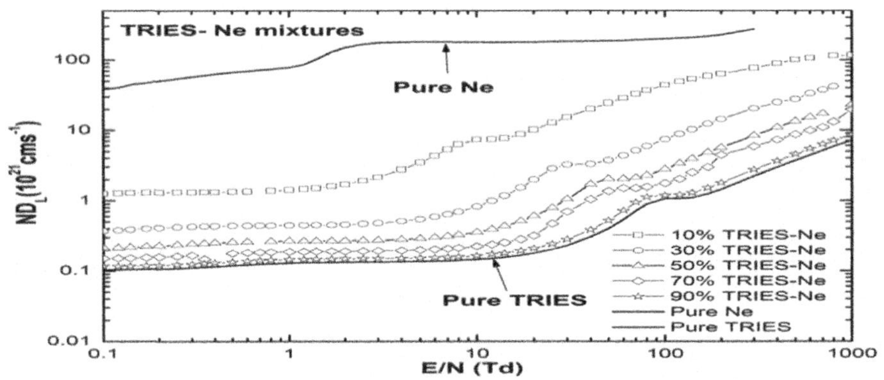

Fig. 8. The density-normalized longitudinal diffusion coefficient, ND_L, in the TRIES-Ne mixtures.

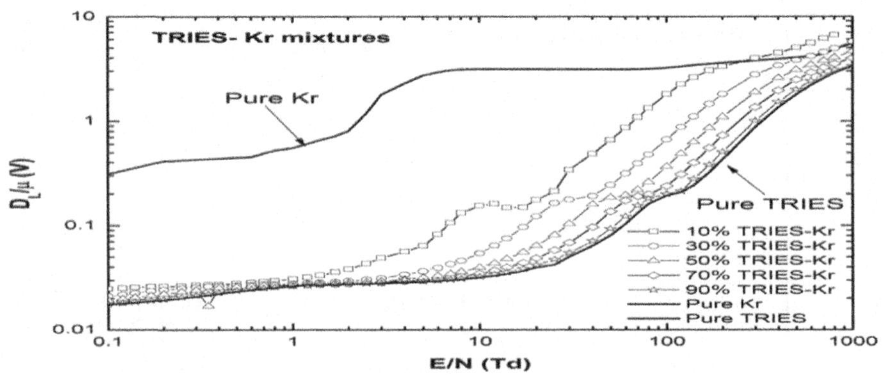

Fig. 9. Ratio of the longitudinal diffusion coefficient to the mobility, D_L/μ, as a function of E/N in the TRIES-Kr mixtures.

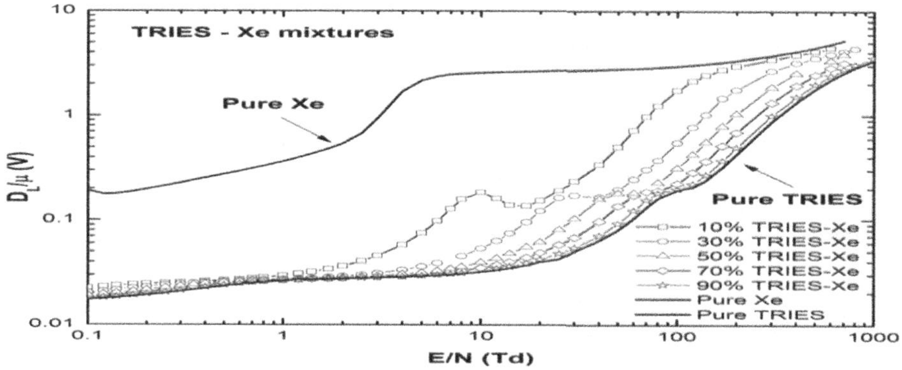

Fig. 10. Ratio of the longitudinal diffusion coefficient to the mobility, D_L/μ, in the TRIES-Xe mixtures.

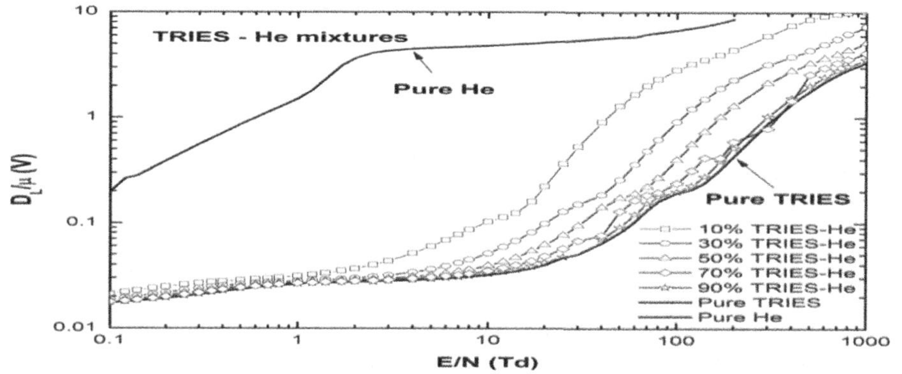

Fig. 11. Ratio of the longitudinal diffusion coefficient to the mobility, D_L/μ, in the TRIES-He mixtures.

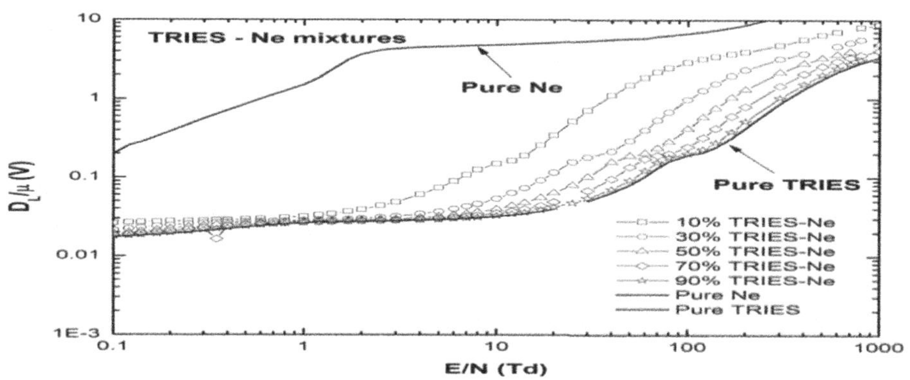

Fig. 12. Ratio of the longitudinal diffusion coefficient to the mobility, D_L/μ, in the TRIES-Ne mixtures.

The electron drift velocity in electric field is an important electron transport parameter which is characterized for the conductivity of the gases or gas mixtures. The electron drift velocities in TRIES-Kr, TRIES-Xe, TRIES-He and TRIES-Ne mixtures as functions of E/N were shown in Figs. 1, 2, 3 and 4. The pure TRIES has lower electron drift velocity whereas the inert gases has higher values for constant E/N. The mixtures electron drift velocities stand between them over the range of E/N > 85Td for TRIES-Kr mixtures, and over the range of E/N > 105 Td for TRIES-Xe mixtures. In the range of E/N < 85 Td for TRIES-Kr mixtures and in the range of E/N < 105 Td for TRIES-Xe mixtures, the values of W in these mixtures are higher than those in pure TRIES, Kr and Xe gases. As mentioned in [10], the electron drift velocity mainly depends on the momentum transfer and vibrational excitation cross sections. Within this range of electric field strength, the vibration excitation is occured, and the values of the momentum transfer cross section of TRIES molecule are higher than that of Kr and Xe atoms when the momentum transfer cross sections of these two atoms significantly go down to the minimum points. Therefore, the shapes of W in both pure TRIES and its mixtures have same tendency. As shown in Figs. 3 and 4, most values of W in TRIES-He and TRIES-Ne mixture gases are suggested to be between those of the pure gases over the all range of E/N, exception in 10% TRIES-He and 10% TRIES-Ne mixtures. The reasons could be suggested that in these E/N ranges, the values of the momentum transfer cross section of TRIES molecule are higher than that of He and Ne atoms while the momentum transfer cross sections of He and Ne gases are slightly increased.

The calculated density-normalized longitudinal diffusion coefficients and ratio of the longitudinal diffusion coefficients to the electron mobility in TRIES-Kr, TRIES-Xe, TRIES-He and TRIES-Ne mixtures are shown in Figs. 5, 6, 7, 8, 9, 10, 11 and 12. These coefficients in the mixtures are suggested to be between those of the pure gases over the all range of E/N. The calculated Townsend first ionization coefficients in TRIES-Kr, TRIES-Xe, TRIES-He, and TRIES-Ne mixture gases are shown in Figs. 13, 14, 15 and 16. In Figs. 13 and 14, the values of α/N in TEOS-Kr and TEOS-Xe mixture are suggested to be between those of the pure gases over entire range of

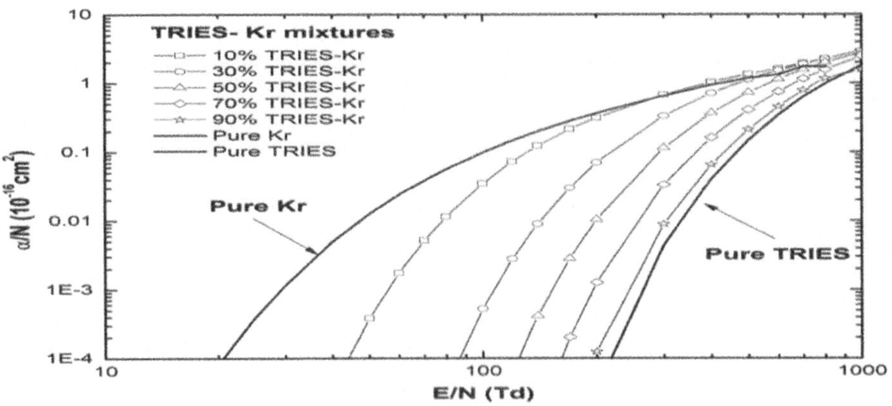

Fig. 13. The first Townsend ionization coefficient, α/N, as a function of E/N in the TRIES-Kr mixtures.

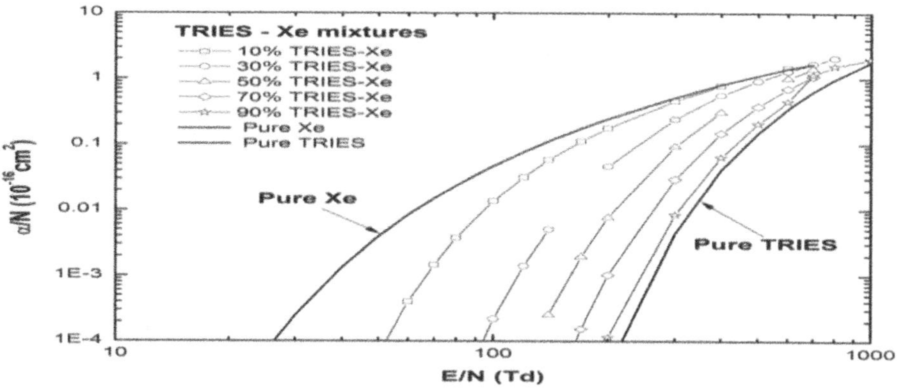

Fig. 14. The first Townsend ionization coefficient, α/N, in the TRIES-Xe mixtures

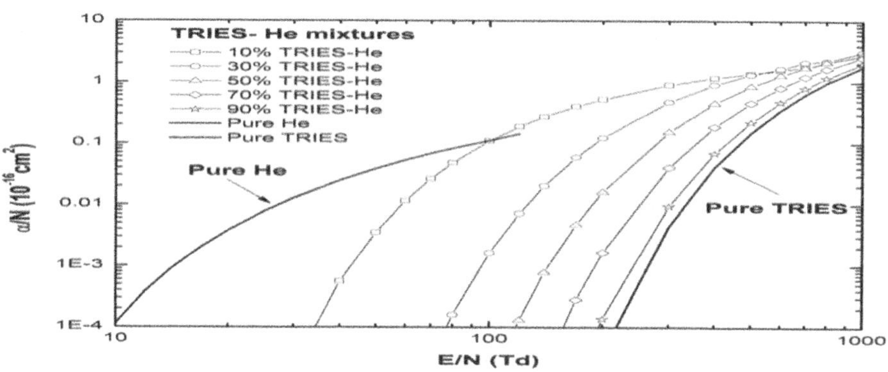

Fig. 15. The first Townsend ionization coefficient, α/N, in the TRIES-He mixtures.

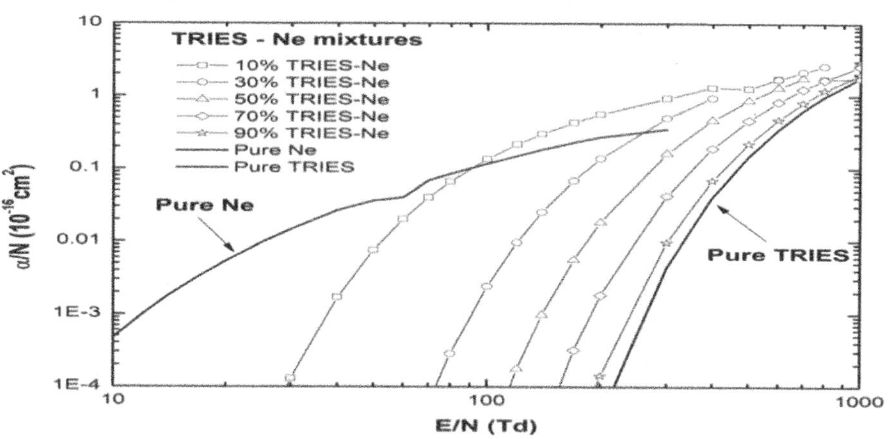

Fig. 16. The first Townsend ionization coefficient, α/N, in the TRIES-Ne mixtures.

E/N. For TRIES-He and TRIES-Ne mixture gases in Figs. 15 and 16, the α/N values are also suggested to be between those of the pure gases over the all range of E/N < 90 Td and E/N < 85 Td, respectively. In these cases, the reasons could be suggested that ionization cross sections of He and Ne gases are much greater than that of TRIES gas.

4 Conclusions

The electron transport coefficients of E/N range of 0.1–1000 Td for TRIES-Kr, TRIES-Xe, TRIES-He and TRIES-Ne mixtures were calculated and analyzed, using the two-term approximation of Boltzmann equation. The behavior of the electron transport coefficients were also discussed in more detail. These results were obtained based on accurate electron collision cross section sets for TRIES molecule, Kr, Xe, He, and Ne atoms. Accordingly, these findings are considered as being handy and essential for diverse industrial purposes, specifically in plasma polymerization and plasma enhanced chemical vapour deposition.

References

1. Dobkin, D.M., Mokhtari, S., Schmidt, M., Robinson, L., Sherman, A.: Mechanisms of deposition of SiO_2 from TEOS and related organosilicon compounds and ozone. J. Electrochem. Soc. **142**(7), 2332–2340 (1995)
2. Zajíčkova, L., Ohlidal, I., Janča, J.: Plasma-enhanced chemical vapour deposition of thin films from tetraethoxysilane and methanol: optical properties and XPS analyses. Thin Solid Films **280**, 26–36 (1996)
3. Lo, C.H., Hung, W.S., Huang, S.H., Guzman, M.D., Rouessac, V., Lee, K.R., Lai, J.Y.: Plasma deposition of tetraethoxysilane on polycarbonate membrane for pervaporation of tetrafluoropropanol aqueous solution. J. Membr. Sci. **329**(5), 138–145 (2009)
4. Nöthe, M., Bolt, H.: Plasma processes and film deposition using tetraethoxysilane. Surf. Coat. Technol. **131**(1–3), 102–108 (2000)
5. Sawada, Y., Ogawa, S., Kogoma, M.: Synthesis of plasma-polymerized tetraethoxysilane and hexamethyldisiloxane films prepared by atmospheric pressure glow discharge. J. Phys. D **28**, 1661–1669 (1995)
6. Holtgrave, J., Riehl, K., Abner, D., Haaland, P.D.: Ion chemistry in tetraethylorthosilicate $(C_2H_5O)_4Si$. Chem. Phys. Lett. **215**(6), 533–548 (1993)
7. Takizawa, K., Mori, Y., Miyatake, N., Murata, K.: Characteristics of monopole antenna plasmas for TEOS PECVD. Thin Solid Films **516**(11), 3605–3609 (2008)
8. Granier, A., Vervloet, M., Aumaille, K., Vallée, C.: Optical emission spectra of TEOS and HMDSO derived plasmas used for thin film deposition. Plasma Source Sci. Technol. **12**(1), 89–96 (2003)
9. Tuan, D.A., Jeon, B.H.: Electron collision cross sections for the tetraethoxysilanemolecule and electron transport coefficients in tetraethoxysilane-O_2 and tetraethoxysilane-Armixtures. J. Phys. Soc. Jpn. **81**(6), 064301-1-8 (2012)
10. Tuan, D.A.: Analysis of electron transport coefficients in binary mixtures of TEOS gas with Kr, Xe, He and Ne gases for using in plasma assisted thin-film deposition. J. Electr. Eng. Technol. **11**(2), 455–462 (2016)

11. Kudoh, Y., Homma, Y., Sakuma, N., Furusawa, T.: Directional plasma CVD technology for sub-quarter-micron feature size multilevel interconnections. Jpn. J. Appl. Phys. **37**, 1145–1149 (1998)
12. Tuoi, P.T., Hien, P.X., Tuan, D.A.: Electron collision cross sections for the TRIES molecule and electron transport coefficients in TRIES-Ar and TRIES-O_2 mixtures. J. Korean Phys. Soc. **73**(12), 1855–1862 (2018)
13. Hayashi, M.: Luminous layers in the prebreakdown region of low pressure noble gases. J. Phys. D **15**(8), 1411–1418 (1982)
14. Suzuki, M., Taniguchi, T., Yoshimura, N., Tagashira, H.: Momentum transfer cross section of Xenon deducted from electron drift velocity data. J. Phys. D **25**(1), 50–56 (1992)
15. Tagashira, H., Sakai, Y., Sakamoto, S.: The development of electron avalanches in Argon at high E/N values. II. Boltzmann equation analysis. J. Phys. D **10**(7), 1051–1064 (1977)
16. Tuan, D.A.: Determination of electron collision cross sections for F_2, Cl_2 molecules, and electron transport coefficients in mixture gases as prospective substitutes for the SF_6 gas in industrial applications. Ph.D. Dissertation, Dongguk Univ., Korea (2012)

Taguchi Based Process Parameters Optimization in Vibration Assisted Die Sinking Electrical Discharge Machining

Huu Phan Nguyen[1], Tien Long Banh[2], T. Muthuramalingam[3], Ngoc Pi Vu[5], Quang Dung Le[4], Le Xuan Hung[5(✉)], and Dinh Khai Nguyen[4]

[1] Hanoi University of Industry, Hanoi, Vietnam
[2] Hanoi University of Science and Technology, Hanoi, Vietnam
[3] SRM Institute of Science and Technology, Kattankulathur, India
[4] Hungyen University of Technology and Education, Hải Dương, Vietnam
[5] Thai Nguyen University of Technology, Thai Nguyen, Vietnam
lexuanhung@tnut.edu.vn

Abstract. The integration of low frequency vibration with the workpiece can enhance the machining efficiency in die sinking electrical discharge machining. Since the adaptation of vibration increases the complexity in such process, the optimization of the technological parameters is very important. In the present work, a research attempt has been made on Taguchi based monotonous optimization of vibration assisted EDM process while machining SKD61 die steel. The process parameters such as current (I), pulse on time (T_{on}), pulse off time (T_{of}), frequency of vibration (F) have been selected as input factors for investigation. Material removal rate has been chosen as quality indicator. From the experimental investigation, it has been found that material removal has been significantly improved by integrating vibration into the workpiece. The optimal combination of process parameters has been found with the accuracy of 6.90%.

Keywords: MRR · SKD61 · EDM · Vibration

1 Introduction

Many technological solutions have been introduced to overcome the limitations of DSEDM such as optimizing technological parameters, mixing powder into dielectric fluid and integrating vibration into machining process. Among them, the vibration assisted conventional EDM has been attracting the attention of researchers, especially the vibration integration with the workpiece holder in EDM. At present, the number of optimal research results in DSEDM with published vibration integration is very small. Therefore, in order to improve the effectiveness of this solution, it is necessary to implement the optimization of process parameters in the process.

The vibration integration in DSEDM can be done by two technological ways such as to electrode or the workpiece. The low frequency vibration has produced high MRR and lower surface related parameters while machining W9Mo2Cr4V material [1]. The vibration frequency has increased material removal rate (MRR) and electrode wear rate

K.-U. Sattler et al. (Eds.): ICERA 2019, LNNS 104, pp. 704–709, 2020.
https://doi.org/10.1007/978-3-030-37497-6_81

(EWR) along with lower overcut and taper angle in DSEDM process [2]. The vibration assisted DSEDM process can greatly improve the efficiency of process by pushing the chip out of the discharge gap. The better surface quality has been observed with oil as dielectric fluid than the water as dielectric medium in DSEDM process [3]. The amplitude and frequency of vibration have a strong influence on the productivity and quality of machined surfaces in DSEDM [4, 5]. The low frequency vibration could create the useful higher duration discharge pulse with lower duration of short circuit pulse to enhance the stability of the process [6]. It can improve the machining productivity with lower tool wear rate [7]. Although the effectiveness of vibration assignment in DSEDM process has been analyzed in many literatures, only few works are available about the vibrations to workpiece [8]. Among those literature, only very little attention has been given to optimization of process parameters in the process. Many solutions have been proposed to solve the optimal problem in DSEDM such as Taguchi, RSM, ANN, etc. Taguchi is the most suitable method for deriving optimal process parameters in DSEDM. Hence Taguchi method has been proposed in DSEDM process with integrated vibration.

In the present study, an attempt has been made to optimize the technological parameters in DSEDM with low frequency vibration assigned on machining SKD61 as workpiece using Taguchi method.

2 Experimental Design and Methodology

SKD61 steel has been used as workpiece specimen with the size of 25 × 25 × 30 mm. The copper electrode has been chosen with the diameter size of 25 mm. D323 oil has been selected as the dielectric solution. The technological parameters used in this study are shown in Table 1.

The experimental investigations were conducted on a CHEMER DSEDM machine type (CM 323C). The workpiece was attached to the vibration protection fixture of the vibration unit to facilitate stable and accurate transmission of vibrations to the workpiece as shown in Figs. 1 and 2. The vibration unit (Modal: Exciter 4824, Brüel & Kjær, Denmark) was used to investigate the influence of vibrations in EDM process. The amplitude of the vibrations for a chosen frequency value has been chosen as 0.75 μm. The precise weight of the workpiece was determined using a digital scale (Model: Vibra AJ-203 SHINKO, Japan).

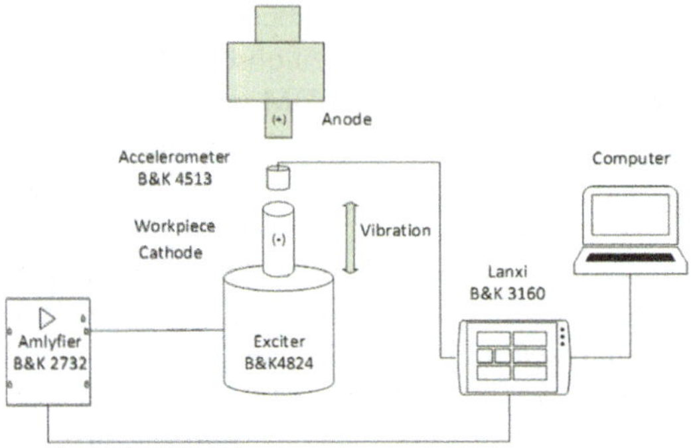

Fig. 1. The schematic experimental chart depicting the vibration setup and data logger

Fig. 2. Experimental diagram

Taguchi method is widely used to optimize the target unit in DSEDM process. The present study deals with four input parameters with three levels, L_9 has been chosen based on Taguchi's appropriate orthogonal matrix as shown in Table 1. The S/N of MRR has been evaluated according to "Larger is better".

Table 1. Matrix of experiment

Trial	I	T_{ON}	T_{OF}	F	MRR	S/N
1	3	12	5.5	128	2.333	−40.0000
2	3	25	12.5	256	2.827	−34.4249
3	3	50	25	512	3.564	−33.1515
4	6	12	12.5	512	4.460	−29.1186
5	6	25	25	128	3.154	−35.3910
6	6	50	5.5	256	5.470	−24.7314
7	8	12	25	256	5.855	−23.7417
8	8	25	5.5	512	9.564	−21.7237
9	8	50	12.5	128	5.205	−27.7443

3 Results and Discussion

The various machining experiments have been performed repeatedly three times to reduce the effect of noise on the experimental results. The average of the values have been considered for the analysis and tabulated in Table 1. The experimental results have been analyzed and processed using Minitab 18 software package.

Selection of Optimal Process Parameters: The level of the optimal process parameters is shown in Fig. 3. The results showed that the optimal process parameters in DSEDM with low frequency vibration were assigned SKD61 as I = 8 A, T_{ON} = 50 µs,

T_{OF} = 5.5 μs and F = 512 Hz. The value of the quality criteria at the optimum condition has been determined by (Eq. 1). The MRR has been calculated from the regression equation ($MRR_{Optimal}$) as 9.129 mm^3/min. The experimental verification with the optimal set of process parameters was obtained as $MRR_{OPTexp.}$ = 9.765 mm^3/min. The difference between the results of the calculation has been found as 6.90%.

$$MRR_{OPT} = I_3 + TON_3 + TOF_1 + F_3 - 3.T \tag{1}$$

Microstructure Analysis of Machined Surfaces at Optimal Process Parameter: The Profile of machined surface and surface roughness value in EDM at optimal conditions with and without low frequency vibration as shown in Fig. 4. The height of the rough peaks of the machining surface after EDM with low frequency vibration is more uniform. This may contribute on improving the wear resistance of the machining surface. The post-EDM and EDM machining surfaces with low frequency vibrations are a collection of many craters and their sizes are different due to random energy distribution as shown in Fig. 5. The generated spark energy during pulse cycles causes material to melt and evaporate on the surface of the workpiece. The machining surfaces also appear many small spherical particles over the machined surface as shown in Fig. 6. This has led to increase the surface roughness value. The electrode material are melted, evaporated and cooled quickly by dielectric fluid. Hence the removing of spherical particles from the machining surface is very difficult. Therefore the machining surface after EDM is required to be polished by grinding. The number of adhesion particles appearing on the machining surface after EDM with vibration is much less than as that of without vibration as shown in Fig. 6. The low frequency vibration has been integrated into the workpiece tends to have a positive impact on the process of pushing the chip out of the discharge gap. This helps the chip to be removed more easily and the adhesion conditions of the debris particles to the machining surface has been more difficult.

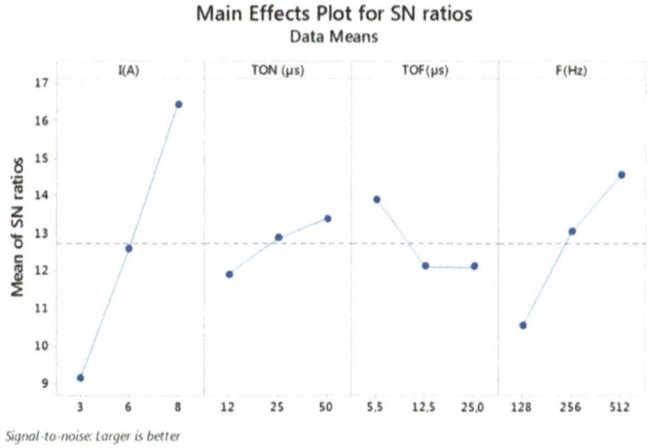

Fig. 3. Effect of process parameters to S/N of MRR

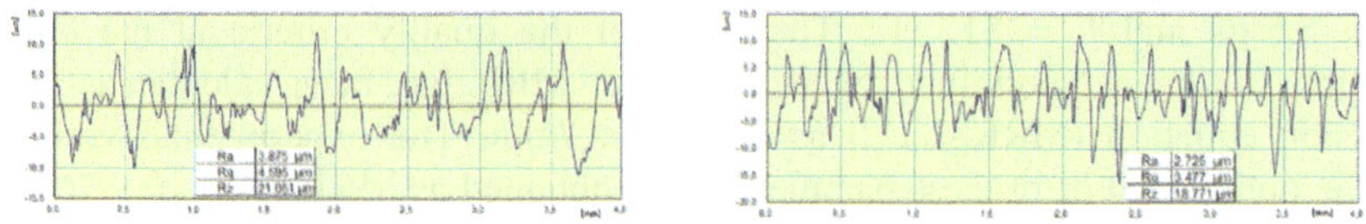

(a) F = 0Hz (b) F= 512 Hz

Fig. 4. Profile of machined surface

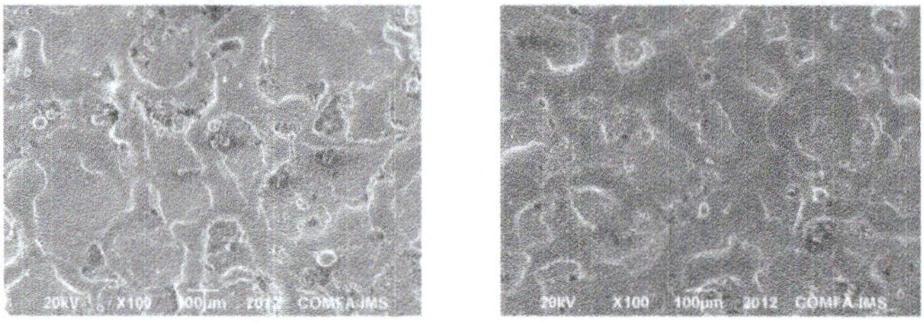

(a) F=0Hz (b) F=512Hz

Fig. 5. Topography of SKD61 machined after EDM

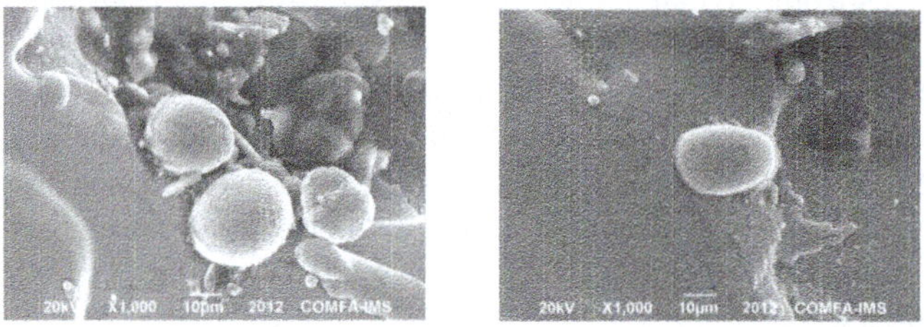

(a) F=0Hz (b) F=512Hz

Fig. 6. Particle adhesion of SKD61 machined after EDM

4 Conclusions

- Feasibility of low-frequency vibration on workpiece in EDM processes for SKD61 die steel by using cooper electrode has been proven.
- The optimal set of process parameters is I = 8 A, T_{ON} = 50 µs, T_{OF} = 5.5 µs and F = 512 Hz. MRR was found to be 9.129 mm^3/min with the accuracy of 6.90%.
- The low frequency vibration is integrated into the workpiece in EDM has contributed to improve machining surface quality and reducing finishing costs.

However, the new research results surveyed for the single-objective problem and optimal efficiency of this technology may be higher for the multi-objective problem. This will be the direction for further research.

Acknowledgements. The work described in this paper was supported by Thai Nguyen University of Technology.

References

1. Zhu, G., Zhang, M., Zhang, Q., Song, Z.C., Wang, K.: Machining behaviors of vibration-assisted electrical arc machining of $W_9Mo_3Cr_4V$. Int. J. Adv. Manuf. Technol. **96**(1–4), 1073–1080 (2018)
2. Unune, D.R., Mali, H.S.: Experimental investigations on low frequency workpiece vibration in micro electro discharge drilling of Inconel 718. In: Proceedings of 6th International & 27th All India Manufacturing Technology, Design and Research Conference 2016, pp. 1413–1417 (2016)
3. Mwangi, J., Ikua, B.W., Nyakoe, G.N., Zeidler, H., Kabini, S.K.: Effect of low frequency vibration in electrical discharge machining of AlSiC metal matrix composite. J. Sustain. Res. Eng. **1**(2), 45–50 (2014)
4. Prihandana, G.S., Mahardika, M., Hamdi, M., Mitsui, K.: Effect of low-frequency vibration on workpiece in EDM processes. J. Mech. Sci. Technol. **25**(5), 1231–1234 (2011)
5. Kumar, S., Grover, S.: Optimisation strategies in ultrasonic vibration assisted electrical discharge machining: a review. Int. J. Precis. Technol. **7**(1), 51–83 (2017)
6. Todkar, A.S., Sohani, M.S., Kamble, G.S., Nikam, R.B.: Effects of vibration on electro discharge machining processes. Int. J. Eng. Innov. Technol. **3**(1), 270–275 (2013)
7. Iwai, M., Ninomiya, S., Suzuki, K.: Improvement of EDM properties of PCD with electrode vibrated by ultrasonic transducer. Procedia CIRP **6**, 146–150 (2013)
8. Maity, K.P., Choubey, M.: A review on vibration-assisted EDM, micro-EDM and WEDM. Surf. Rev. Lett. **26**(5) (2018). https://doi.org/10.1142/s0218625x18300083

The Research and Calculation for the Selection of Motor and Battery for Five Seat Passenger Car When Replacing Engine by an Electric Motor

Nguyen Huy Truong(✉)

Institute of Military Mechanical Engineering, Cau Giay, Ha Noi, Vietnam
tnkvietnam.ltd@gmail.com

Abstract. The article presents a calculation methodology for the optimizing selection of electric motor and power of battery for five seat passenger car when replacing a gasoline engine by an electric motor.

Keywords: Electric motor · Battery · Electric vehicle (EV) · Hybrid electric vehicle (HEV)

1 Introduction

Today, the traditional automotive manufacturers all over the world have introduced a large number of hybrid cars and electric cars, which are becoming commercialized. Along with the new process of research and development, the utilization of old chassis which used fossil fuel and converted for using on electric vehicles is also another the direction of research. Therefore, the calculation for the selection of electric power unit when switching from a fuel engine to an electric motor is a potential direction in research as well as a suitable solution for developing countries with a small automotive industry like our country.

The paper is about a process of selection of the electrical power unit by replacing the gasoline engine of the vehicle; so basically, the structure of the vehicle systems are already available. In addition, the electrical engines are manufactured by the manufacturers and also supplied according to their type and capacity. Therefore, the selection of electrical power unit has been done in three steps:

Step 1 - Select a potential electric motor with suitable capacity;
Step 2 - Calculate the power of battery;
Step 3 - Calculate the distance reserves.

2 Select Electric Motor

The selected car model for application is the five seat passenger car, with the following basic performance parameters (Tables 1 and 2):

K.-U. Sattler et al. (Eds.): ICERA 2019, LNNS 104, pp. 710–719, 2020.
https://doi.org/10.1007/978-3-030-37497-6_82

Table 1. Basic performance parameters of the passenger car [2]

Parameter	Unit	Value
The equipped weight	kg	1150
Total weight	kg	1550
Maximum speed	km/h	132
Acceleration time from 0 to 100 km/h	s	25
Engine		
- *Maximum power at 5400 r/min*	kW (hp)	58,8 (80)
- *Maximum torque at 3000 r/min*	N.m	124
Gear box	4 speed	i_{h1} = 3,342; i_{h2} = 1,989 i_{h3} = 1,289; i_{h4} = 1,00
Distribution gearbox		i_{p1} = 1,2; i_{p2} = 2,135
Final ratio		i_0 = 5,16

Table 2. The specific parameters of BLDC motors

Parameter	Unit	Value	
		Motor I	Motor II
Voltage	V	72	72
Electric current (I_e)	A	850	450
Max power (N_e)	kW at r/min	82,6 at 5800 r/min	55,77 at 5618 r/min
Max torque (M_e)	N.m	235,1	117,55

The selection of electric motor for replacement are brushless DC electric motor (BLDC motors), with the following specific parameters (Figs. 1 and 2):

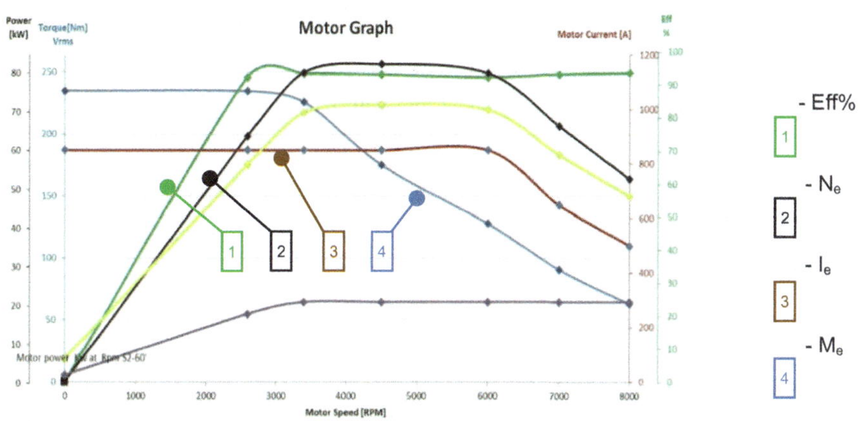

Fig. 1. Electric motor graph for motor I

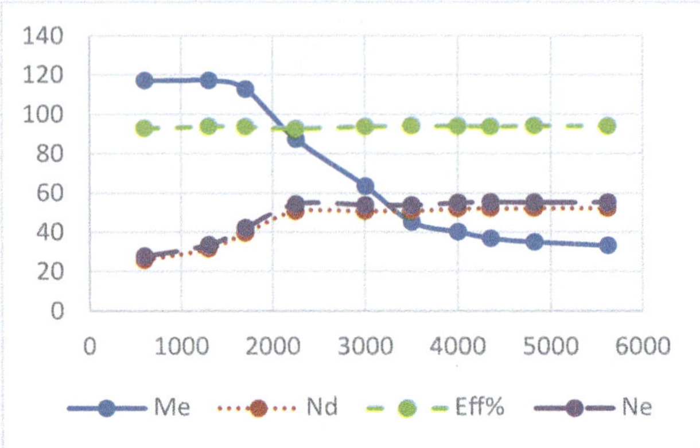

Fig. 2. Electric motor graph for motor II

We calculate on two models: In model I, we do not use the gearbox and distribution gearbox of the old car. Electric motors will be connected directly to the axles end wheels (electric-motor-powered with single-gear transmission). This model we use electric motor I. In model II, we do not use the gearbox of the old car. Electric motors will be connected directly to the distribution gearbox (electric-motor-powered with two-gear transmission). This model we use electric motor II.

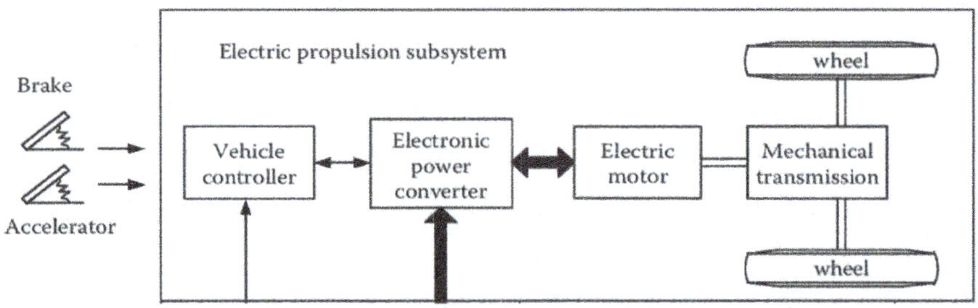

Fig. 3. Electric propulsion subsystem with mechanical transmission

Electric propulsion subsystem with mechanical transmission is shown in the Fig. 3.

3 Calculation

3.1 Vehicle Dynamic Factor

By references [4], when a vehicle drives on a road with a relatively small grade and at a constant speed, the tractive effort and resistance equilibrium and gradeability can be written (Fig. 4):

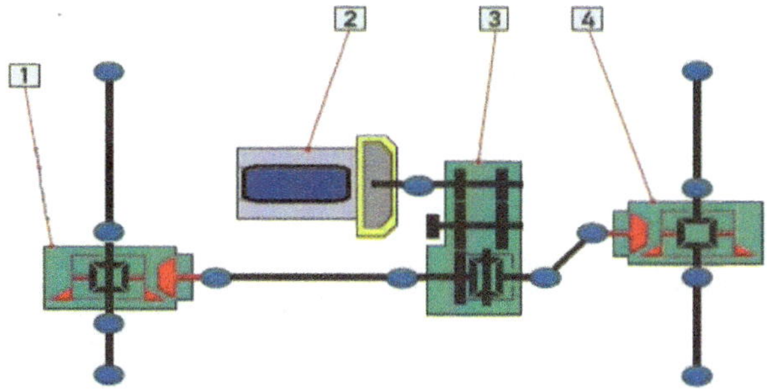

1-Front axle; 2-Electric motor; 3-Gearbox (two-gears); 4-Rear axle

Fig. 4. Electric-motor-powered passenger car with two-gears transmission

$$d = \frac{F_t - F_w}{Mg} = \frac{\frac{T_p i_0 i_h i_p \eta_t}{r_d} - \frac{1}{2}\rho_a C_D A_f V^2}{Mg} \quad (1)$$

- For electric-motor-powered with single-gear: $i_h = 1$, $i_p = 1$);
- For electric-motor-powered with two-gear: $i_h = 1$, $i_{p1} = 1{,}2$, $i_{p2} = 2{,}135$ (distribution gearbox of the old car).

3.2 Acceleration Performance

The acceleration performance of a vehicle is usually described by its acceleration time and distance covered from zero speed to a certain high speed. Using Newton's second law, the acceleration of the vehicle can be written:

$$a = \frac{dV}{dt} = \frac{F_t - F_f - F_w}{M\delta} = \frac{\frac{T_p i_0 i_h i_p \eta_t}{r_d} - Mgf_r - \frac{1}{2}\rho_a C_D A_f V^2}{M\delta} = \frac{g}{\delta}(d - f_r) \quad (2)$$

3.3 Acceleration Time and Distance

From Eq. (2), the acceleration time, ta, and distance, Sa, from low-speed V1 to high-speed V2 can be written, respectively, as

$$t_a = \int_{V_1}^{V_2} \frac{M\delta}{\frac{T_p i_0 i_h i_p \eta_t}{r_d} - Mgf_r - \frac{1}{2}\rho_a C_D A_f V^2} dV \quad (3)$$

and

$$S_a = \int_{V_1}^{V_2} \frac{M\delta V}{\frac{T_p i_0 i_h i_p \eta_t}{r_d} - Mgf_r - \frac{1}{2}\rho_a C_D A_f V^2} dV \quad (4)$$

Using MATLAB software applications, calculation by formula (1), (2), (3) and (4), we have results (Figs. 5, 6, 7 and 8):

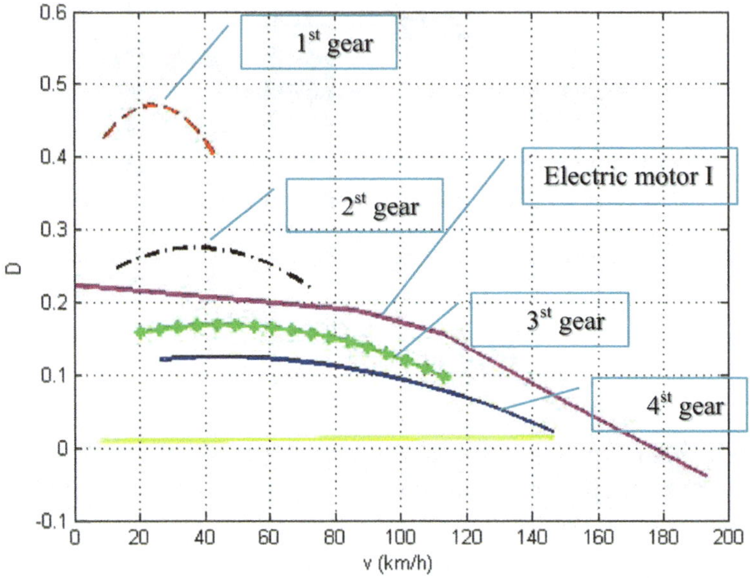

Fig. 5. Vehicle dynamic factor for Electric motor I.

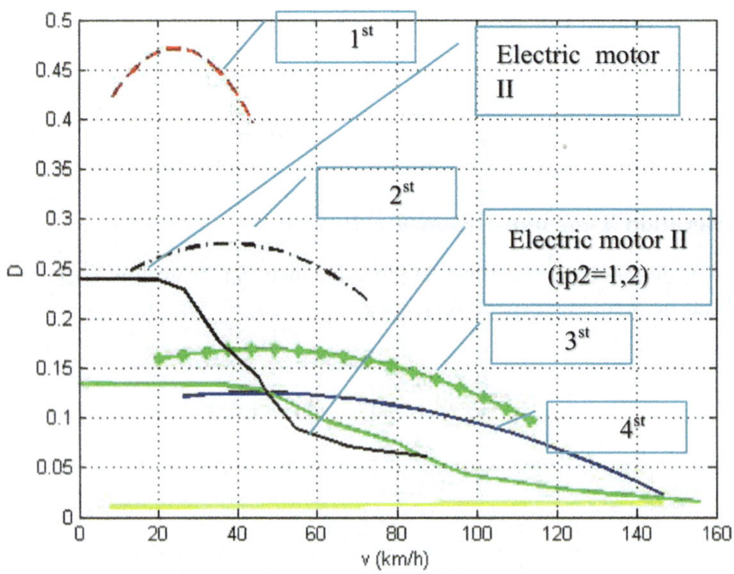

Fig. 6. Vehicle dynamic factor for Electric motor II.

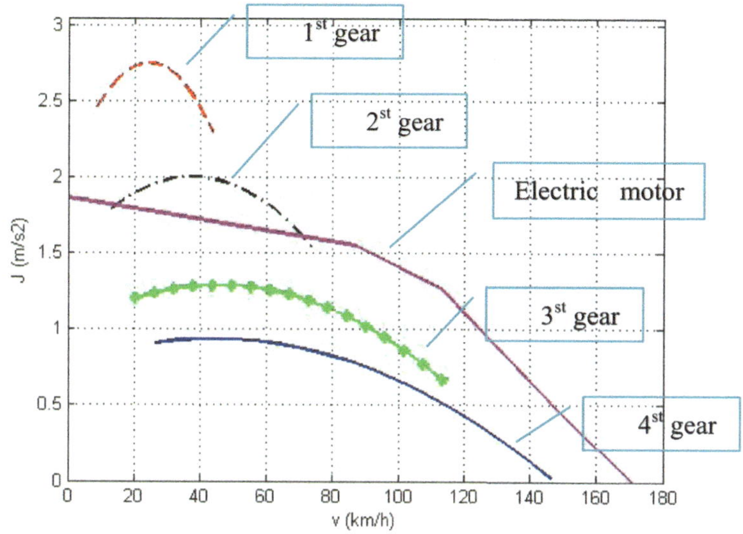

Fig. 7. Acceleration performance for Electric motor I.

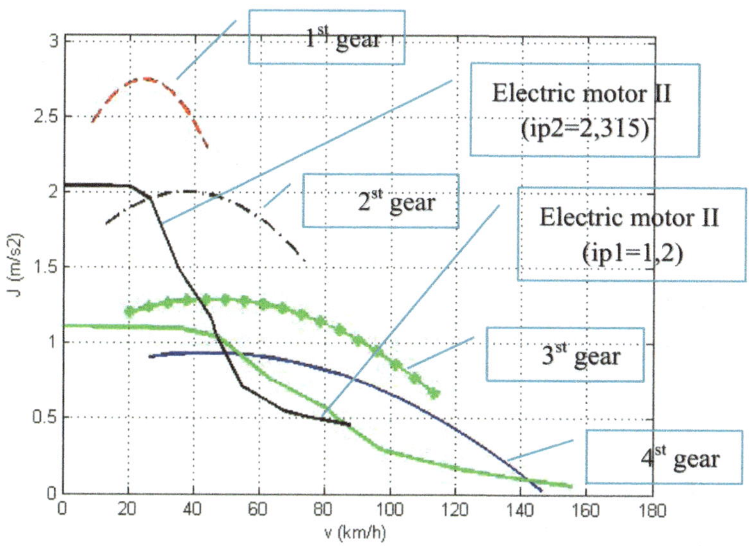

Fig. 8. Acceleration performance for Electric motor II.

The results of dynamics factors as well as the acceleration of electric car in the case of using a distribution gearbox (model II) at a lower transmission is better than not using a distribution gearbox (model I). Because the torque is increased through the gearbox in lower transmission. However, the dynamics factors values of electric car at high speeds in model II are smaller than model I, so the ability to accelerate of electric car in model I is much better (accelerating to 100 km/h less than 10 s, compared to the model II is nearly 40 s) (Figs. 9 and 10).

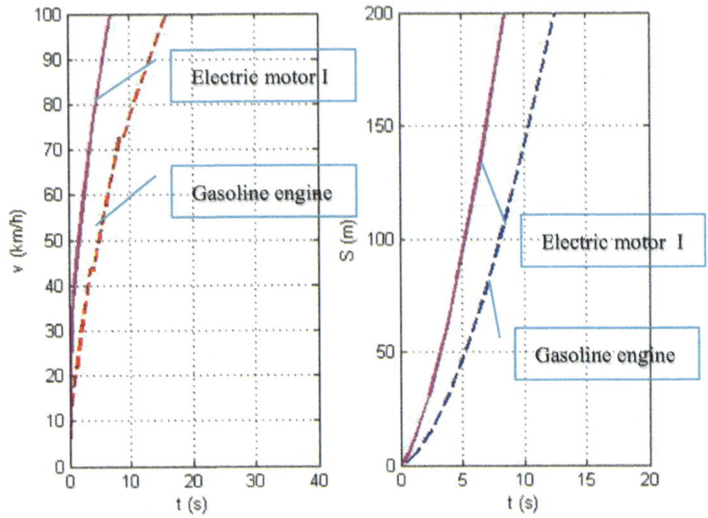

Fig. 9. Acceleration time and distance for Electric motor I.

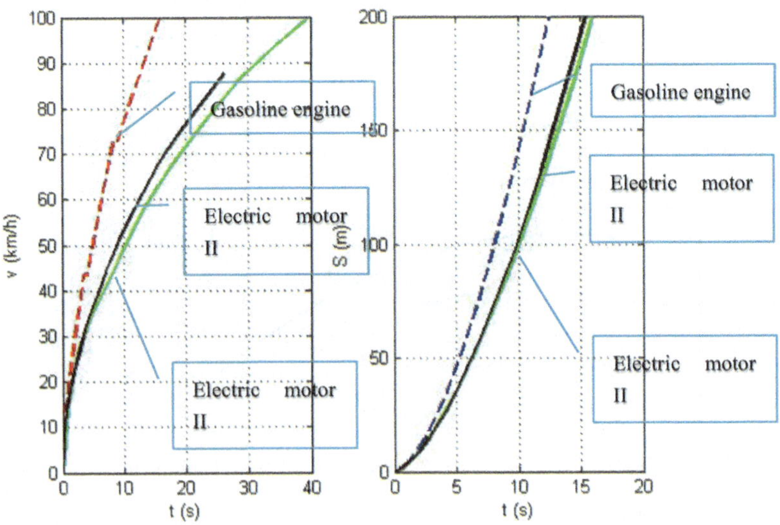

Fig. 10. Acceleration time and distance for Electric motor II.

4 Calculate the Power of Battery and Distance Reserves

4.1 Calculate the Power of Battery

The basic formula for determining distance is: (KWh of pack/wh/mi) = Distance.

Watt-Hour per Mile (Wh/mi) or Watt-Hour per km (Wh/km):

The basic rule of thumb for vehicle is [5]:

- Small Vehicle 250–300 Wh/mi (155–186 Wh/km)
- Small Pickup 350–400 Wh/mi (217,5–248,5 Wh/km)

The calculation is: Volts x (Amp Draw/km/h) = Wh/km

Battery Pack Size (KWh): Pack Voltage x Amp-Hour rating of battery = KWh

We can not use all of our battery Pack or we will kill our batteries fast. To extend the life of the battery pack, we do not want to discharge the batteries more than 80%.

In addition, because an EV will discharge the batteries faster than the manufacturer tested and rated, we get an effect called "Pt"-Peukerts effect. However if we use Lead-Acid batteries the Peukerts effect if considerable, where we only get to use about 55% of the power in the Battery. (Pt: Lead-Acid = 0,55; LiFePO4 = 1,0).

$$\text{Usable Pack size} = \text{KWh} \times 0,80 \times \text{Pt}\,(\text{KWh}) \tag{5}$$

We use electric motor BLDC 72 V, for should in use will be used Batteries: 12–12 V Lead-Acid, rated at 100 Ah or 6–12 V Lead-Acid, rated at 200 Ah.

$$\text{Usable Pack} = 14,4 \times 0,8 \times 0,55 = 6,336\,\text{KWh} = 6336\,\text{Wh}$$

$$\text{Pack Voltage} = 6 \text{ batteries} \times 12\,\text{V each} = 72\,\text{V} \tag{6}$$

$$\text{Pack Size} = 72\,\text{V} \times 200\,\text{Ah} = 14,400\,\text{Wh} = 14,4\,\text{KWh} \tag{7}$$

$$\text{Usable Pack} = 14,4 \times 0,8 \times 0,55 = 6,336\,\text{KWh} = 6336\,\text{Wh} \tag{8}$$

4.2 Distance

In model I:

From experience, we know that electric motor using a 72 V system will draw around 150 amps at 80 km/h [5]. Therefore, the

$$\text{Wh/m usage} = 72\,\text{V} \times 150\,\text{Amps}/80\,\text{km/h} = 135\,(\text{Wh/km}) \tag{9}$$

The distance our car will travel on this setup is: 6336 Wh/135 Wh/km ≅ 46,93 km. So, to increase the distance, we need to increase the number of batteries or switch to use LiFePO4 Batteries. To reduce the electric current, we need to use a high voltage motor. If we had a lithium pack of equal voltage and Ah, the range would be 85,33 km (because Peukerts effect does not play a role).

$$14,4 \times 0,8 = 11,52\,\text{KWh}\,(11520\ \text{Wh})\,\text{usable}/135\,\text{Wh/km}\ @85,33\,\text{km} \tag{10}$$

In model II:

From experience, we know that electric motor using a 72 V system will draw around 90 amps at 80 km/h. Therefore, the

$$\text{Wh/m usage} = 72\,\text{V} \times 90\,\text{Amps}/80\,\text{km/h} = 81\,(\text{Wh/km}) \tag{11}$$

The distance our car will travel on this setup is: 6336 Wh/81 Wh/km ≅ 78,22 km. If we had a lithium pack of equal voltage and Ah, the range would be

$$14,4 \times 0,8 = 11,52\,\text{KWh}\ (11520\ \text{Wh})\,\text{usable}/81\ \text{Wh/km} \cong 142,22\,\text{km} \tag{12}$$

4.3 To Calculate This in Reverse (Using LifePO4 Cells)

We will use the 81 Wh/km avg (at 80 km/h). Wh per km/pack voltage = 81/72 = 1,125 Ah per km, so we would need Ah per km x km per charge needed x 1,2 (so you still had 20% charge left after the drive) 1,125 × 142,22 × 1,2 ≅ 192 Ah batteries at 72 V needed to go the 142,22 km.

5 Conclusion

- The above results can be explained because in case we did not use the gearbox, we used electrical power unit with the capacity and torque nearly double the case of using gearboxes. And this is related to the increased capacity of the power supply, which in turn increases the vehicle weight and the charging time. Therefore, the selection of electrical power unit and proper power capacity is the decisive factor of the improvement.
- Reflecting the calculation results, we find that the reuse of the distribution gearbox has two transmission numbers on the vehicle, which would be more appropriate, meeting both travel distance and battery storage capacity.
- The selection of the electrical engine which has the suitable function to replace the original fossil-fuel-powered engine is perfectly feasible. By the calculation, the replacement of electrical power unit in old vehicles using fossil fuels has not only utilised the old generation vehicles, but also taking advantage of the benefit of electric motors and corresponding with the new researches on electric vehicles as well as their application in Vietnam's condition.

References

1. Hieu, N.P., Lap, V.D.: Theory of Military Vehicles. Institute of Military Technology, Zielonka (2002)
2. Bosch, R.: Automotive Handbook. Robert Bosch GmbH, Karlsruhe (2000)
3. Chan, C.C., Chau, K.T.: Modern Electric Vehicle Technology. Oxford University Press, New York (2001)
4. Ehsani, M., Gao, Y., Longo, S., Ebrahini, K.M.: Modern Electric, Hybrid Electric, and Fuel Cell Vehicles, 3rd edn. Taylo & Francis Group, London (2018)
5. Leitman, S., Brant, B.: Build Your Own Electric Vehicle, 3rd edn. Mc Graw Hill Education, New York (2013)
6. Rand, D.A.J., Woods, R., Dell, R.M.: Batteries for Electric Vehicles. Research Studies Press Ltd, Autin (1998)
7. Gao, Y., Maghbelli, H., Ehsani, M. et al.: Investigation of proper motor drive characteristics for military vehicle propulsion. In: SAE Journal, Warrendale, PA, No. 2003-01-2296 (2003)
8. Rahman, Z., Ehsani, M., Butler, K.L.: An investigation of electric motor drive characteristics for EV and HEV propulsion systems. In: SAE Journal, Warrendale, PA, No. 2000-01-3062 (2003)
9. Rahman, K.M., Ehsani, M.: Performance analysis of electric motor drives for electric and hybird electric vehicle application. IEEE Power Electron. Transp. 49–56 (1996)
10. https://www.go-ev.com/Products.html

Vibroacoustic Response of a Finite Simply Supported Double-Composite Plate Filled with an Air Cavity

Pham Ngoc Thanh[1(✉)] and Tran Ich Thinh[2]

[1] Viet Tri University of Industry, Viet Tri City, Phu Tho Province, Vietnam
[2] Hanoi University of Science and Technology,
1 Dai Co Viet Road, Ha Noi, Vietnam
thanhpham1986@gmail.com, thinh.tranich@hust.edu.vn

Abstract. Based on the classical laminated plate theory, the vibroacoustic behavior of a simply supported finite orthotropic double-composite plate with enclosed air cavity is investigated analytically. Using the method of modal decomposition, a double Fourier series solution is obtained to characterize the vibroacoustic performance of the structure. The sound transmission loss (STL) is calculated from the ratio of incident to transmitted acoustic powers. The insulation ability of finite double-composite plate is shown. Specifically, the focus is placed on the effects of several key system parameters on sound transmission loss including the composite materials, the laminate configurations, the faceplate thickness, as well as the air cavity thickness. All of them are systematically examined.

Keywords: Vibroacoustic analysis · Double-laminated composite plate · Sound transmission loss

1 Introduction

The double-plate is widely used in construction structures, ships, turbofan, aerospace, etc. Therefore, research on sound insulation capacity of double-plate has received much attention from researchers in the world to produce the most optimal structures with the best sound insulation ability and the best applicability.

For decades, the vibroacoustic behavior of an infinite or finite double-plates is an interesting research topic with the different approaches used. Traditionally, Maidanik [1] analyzed the vibration behavior of a complex structure under force or sound excitation by the statistical energy analysis method (SEA). However, the SEA method is less effective at relatively low frequencies on account of its pre-assumption that enough structural modes need to be excited. Ruzzene [2] investigated the acoustic properties of sandwich beams in terms of structural response and sound transmission reduction index, which is more efficient for low frequencies but required high calculation costs for high frequencies by a finite element method model (FEM). The first experimental test on the sound insulation properties of double-panel structures was carried out by London [3]. Later, Carneal and Fuller [4] presented an analytical and

K.-U. Sattler et al. (Eds.): ICERA 2019, LNNS 104, pp. 720–732, 2020.
https://doi.org/10.1007/978-3-030-37497-6_83

experimental study of active structural acoustic control of noise transmission through double panel systems. The experimental measurements in [4] were carried out on a clamped aluminum double-plate. Chazot and Guyader [5] studied the sound transmission loss through finite double-panels by using a patch-mobility method. This method and FEM give the same results when comparing transmission loss of double aluminum plates. In addition, to improve the sound isolation properties of double-panel partitions, various active control strategies have been proposed by Bao and Pan [6]. Sgard et al. [7] studied the sound transmission loss (STL) through the finite-sized, double-panel sound barrier with poroelastic linings by FEM. Villot et al. [8] have proposed a new way of calculating reflectance and transmitting power of finite-sized structures based on spatial widowing of plane waves.

The vibroacoustic behavior of the double-plates filled with absorbent materials has been extensively investigated. For example, Brouard et al. [9] proposed a general method to model sound propagation in layered systems such as fluid-saturated porous layers, while Lauriks et al. [10] developed a transfer matrix model to study the transmission loss through panels with solid porous layers, with the Biot model employed to describe sound propagation through the porous material. In [11] a finite element formulation for sound transmission through double wall sandwich panels with viscoelastic core is presented by Larbi et al. Panneton and Atalla [12] proposed a three-dimensional (3D) finite element model to calculate the loss of sound transmission through multilayer structures containing porous absorbent materials. Making use of two-dimensional (2D) elasticity theory, Chonan and Kugo [13] addressed a model to examine the sound transmission characteristics of a three-layer panel excited by plane waves. Kang et al. [14] used the method of Gauss distribution function for incident energy to investigate the STL of multilayered panels such as double-plate structures embedded with porous materials. Bolton et al. [15] calculated random incidence transmission loss through double-leaf panels lined with elastic porous materials.

More recently, the problem of vibroacoustic behavior of a double-plate structures of finite extent was solved by considering simple boundary conditions [16, 17] on the basis of modal superposition theory from different viewpoints. Lu and Xin [17] presented an analytical approach to investigate the sound transmission across rectangular metallic (isotropic) double-panel structure containing an air cavity under various boundary constraints by using the method of modal function and the weighted residual (Galerkin) method. Numerous analytical investigations on the STL through a finite isotropic (metallic) double-plate have been performed [8, 10, 13]. However, to the author's best knowledge, there exist a few studies on the sound transmission across finite-sized, simply supported double- composite plates. More recently, Thinh and Thanh calculated the loss of sound propagation through a finite simply supported [18] and clamped [19] composite plate and also considered the effect of some parameters on the sound insulation ability of the plate.

The objective of this study is to develop a model to describe accurately the sound transmission across a finite simply supported laminated double-composite plate filled with air. Thereby, assessing the influence of several key system parameters on sound transmission loss including the composite materials, the laminate configurations, the faceplate thickness, as well as the air cavity thickness are systematically examined. Since then, giving the most reasonable structure for sound insulation to be applied in practice.

2 Theoretical Formulation

2.1 Geometry and Assumptions

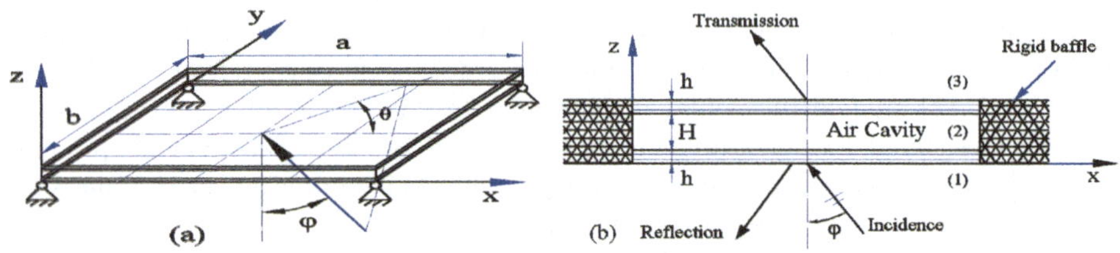

Fig. 1. Schematic of STL through a double-composite plate: (a) global view and (b) side view.

Supposedly, a finite rectangular double-composite plate with air cavity is in an infinite large acoustic rigid baffle. The two single-plates are orthotropic laminated composite and have similar geometric parameters and mechanical properties. The bottom and upper plates (Fig. 1) have thicknesses h and are separated by an air cavity of thickness H. The double-plate partition divides the spatial space into two fields, i.e., sound incidence field ($z < 0$) and sound transmitting field ($z > H + 2h$).

A plane sound wave varying harmonically in time is oblique (with the incident angle φ and azimuth angle θ) and stimulates the vibration of the bottom plate. This vibration changes the pressure in the air cavity and causes vibration of the upper plate then the sound wave is transmitted into the upper domain.

2.2 Laminated Composite Plate Dynamics

The vibroacoustic behavior of an orthotropic symmetric double- composite plate with air cavity (Fig. 1) induced by sound excitation is governed by [19–21]:

$$D_{11}\frac{\partial^4 w_1(x,y;t)}{\partial x^4}+2(D_{12}+2D_{66})\frac{\partial^4 w_1(x,y;t)}{\partial x^2\partial y^2}+D_{22}\frac{\partial^4 w_1(x,y;t)}{\partial y^4}+m^*\frac{\partial^2 w_1(x,y;t)}{\partial^2 t}-j\omega\rho_0[\Phi_1(x,y,z;t)-\Phi_2(x,y,z;t)]=0 \quad (1)$$

$$D_{11}\frac{\partial^4 w_2(x,y;t)}{\partial x^4}+2(D_{12}+2D_{66})\frac{\partial^4 w_2(x,y;t)}{\partial x^2\partial y^2}+D_{22}\frac{\partial^4 w_2(x,y;t)}{\partial y^4}+m^*\frac{\partial^2 w_2(x,y;t)}{\partial^2 t}-j\omega\rho_0[\Phi_2(x,y,z;t)-\Phi_3(x,y,z;t)]=0 \quad (2)$$

where $w_{1,2}$ are the transverse displacements of the two single-plates; D_{ij} *(ij = 11,12,22,66)* are the flexural rigidities; m^* is the surface density of the upper and bottom plates; ρ_0 is the air density, ω is the angular frequency of the incident sound and Φ_i *(i = 1,2,3)* denote the velocity potentials for the acoustic fields in the proximity of the two plates, corresponding to the sound incidence, the air cavity, and the transmitting field, respectively.

The flexural rigidity of the laminated composite plate is determined by (see any textbook of Mechanics of composite materials):

$$D_{ij} = \frac{1}{3}\sum_{k=1}^{n} Q_{ij}^{k}\left(z_{k+1}^{3} - z_{k}^{3}\right) \tag{3}$$

where the reduced stiffnesses of the kth layer are defined as:

$$Q_{11} = \frac{E_1}{1 - \upsilon_{12}\upsilon_{21}}; Q_{12} = \frac{\upsilon_{12}E_2}{1 - \upsilon_{12}\upsilon_{21}}; Q_{22} = \frac{E_2}{1 - \upsilon_{12}\upsilon_{21}}; Q_{66} = G_{12}; \frac{\upsilon_{12}}{E_1} = \frac{\upsilon_{21}}{E_2} \tag{4}$$

and E_1, E_2, G_{12}, ν_{12} are the kth layer elastic constants.

The displacements of the two composite faceplates induced by the incident sound can be expressed as:

$$w_1\left(x, y; t\right) = w_{01}.e^{-j(k_x x + k_y y - \omega t)}; \ w_2\left(x, y; t\right) = w_{02}.e^{-j(k_x x + k_y y - \omega t)} \tag{5}$$

The acoustic velocity potential in the incidence field (Fig. 1) is defined by [10, 16, 17]:

$$\Phi_1(x, y, z; t) = I.e^{-j(k_x x + k_y y + k_z z - \omega t)} + \beta.e^{-j(k_x x + k_y y - k_z z - \omega t)} \tag{6}$$

where the first term represents the velocity potential of the incident acoustic wave and the second term represents the velocity potential of the reflected acoustic waves, and I and β are the amplitudes of the incident (positive-going) and the reflected plus radiated (negative-going) waves, respectively. Similarly, the velocity potential in the air cavity can be written as:

$$\Phi_2(x, y, z; t) = \varepsilon.e^{-j(k_x x + k_y y + k_z z - \omega t)} + \psi.e^{-j(k_x x + k_y y - k_z z - \omega t)} \tag{7}$$

where ε is the amplitude of the positive-going wave and ψ is the amplitude of the negative-going wave. In the transmitting field adjacent to the transmitting upper plate, there exist no reflected waves, and therefore the velocity potential in the transmitted waves is given by:

$$\Phi_3(x, y, z; t) = \gamma.e^{-j(k_x x + k_y y + k_z z - \omega t)} \tag{8}$$

where ξ is the amplitude of the transmitted (positive-going) wave.

These wavenumbers are determined by the incident angle φ and azimuth angle θ of the incident sound wave as

$$k_x = k_0 \sin\varphi\cos\theta; k_y = k_0 \sin\varphi\sin\theta; k_z = k_0 \cos\varphi \tag{9}$$

where $k_O = \omega/c_O$ is the acoustic wave number in air and c_O is the acoustic speed in the air.

The boundary conditions for the simply supported double-composite plate can be expressed as:

$$\begin{aligned} &\text{At } x = 0,\ a, \quad w_1 = w_2 = 0,\ \frac{\partial^2 w_1}{\partial x^2} = \frac{\partial^2 w_2}{\partial x^2} = 0; \\ &\text{At } y = 0,\ b, \quad w_1 = w_2 = 0,\ \frac{\partial^2 w_1}{\partial y^2} = \frac{\partial^2 w_2}{\partial y^2} = 0 \end{aligned} \tag{10}$$

At the air-plate interface, the normal velocity is continuous, yielding the corresponding velocity compatibility condition equations:

$$\begin{aligned} &At\ z = 0,\ -\frac{\partial \Phi_1}{\partial z} = j\omega w_1;\ z = h,\ -\frac{\partial \Phi_2}{\partial z} = j\omega w_1; \\ &z = h + H,\ -\frac{\partial \Phi_2}{\partial z} = j\omega w_2;\ z = 2h + H,\ -\frac{\partial \Phi_3}{\partial z} = j\omega w_2 \end{aligned} \tag{11}$$

The flexural motions of the bottom and upper plates induced by a time-harmonic incident plate sound wave can be expressed in the form of double series as:

$$w_1(x, y; t) = \sum_{m=1}^{\infty}\sum_{n=1}^{\infty} \varphi_{mn}(x, y) u_{1,mn}(t);\ w_2(x, y; t) = \sum_{m=1}^{\infty}\sum_{n=1}^{\infty} \varphi_{mn}(x, y) u_{2,mn}(t) \tag{12}$$

where the modal function φ_{mn} and modal displacements $u_{i,mn}$ *(i = 1,2)* for simply supported boundary conditions are given by [22]:

$$\varphi_{mn}(x, y) = \sin\left(\frac{m\pi x}{a}\right) \sin\left(\frac{n\pi y}{b}\right) \tag{13}$$

$$u_{1,mn}(t) = \alpha_{1,mn} e^{j\omega t},\ u_{2,mn}(t) = \alpha_{2,mn} e^{j\omega t} \tag{14}$$

where $\alpha_{1,mn};\ \alpha_{2,mn}$ are the modal coefficients of the bottom and supper plates, respectively.

Thus, the velocity potentials for the sound incident region, air cavity, and sound transmitting region may be expressed as:

$$\Phi_1(x, y, z; t) = \sum_{m=1}^{\infty}\sum_{n=1}^{\infty} I_{mn} \varphi_{mn} e^{-j(k_z z - \omega t)} + \sum_{m=1}^{\infty}\sum_{n=1}^{\infty} \beta_{mn} \varphi_{mn} e^{-j(-k_z z - \omega t)} \tag{15}$$

$$\Phi_2(x, y, z; t) = \sum_{m=1}^{\infty}\sum_{n=1}^{\infty} \varepsilon_{mn} \varphi_{mn} e^{-j(k_z z - \omega t)} + \sum_{m=1}^{\infty}\sum_{n=1}^{\infty} \psi_{mn} \varphi_{mn} e^{-j(-k_z z - \omega t)} \tag{16}$$

$$\Phi_3(x, y, z; t) = \sum_{m=1}^{\infty}\sum_{n=1}^{\infty} \gamma_{mn} \varphi_{mn} e^{-j(k_z z - \omega t)} \tag{17}$$

where the coefficients I_{mn}, β_{mn}, ε_{mn}, ψ_{mn} and γ_{mn} for simply supported double-composite plate are determined by:

$$\chi_{mn} = \frac{4}{ab}\int_0^b\int_0^a \chi e^{-j(k_x x + k_y y)} \sin\frac{m\pi x}{a}\sin\frac{n\pi x}{b}dxdy \tag{18}$$

Here, the symbol χ can be referred to any of the coefficients I, β, ε, ψ and γ.

Using the displacement continuity condition at air-plate interfaces [23] and Bernoulli's equation [21], substituting Eqs. (15)–(17) into Eqs. (11) leads to:

$$\beta_{mn} = I_{mn} - \frac{\omega}{k_z}\alpha_{1,mn}; \varepsilon_{mn} = \frac{\omega\left(\alpha_{1,mn}e^{jk_zH} - \alpha_{2,mn}\right)}{k_z\left(e^{-jk_z(h-H)} - e^{-jk_z(H+h)}\right)} \tag{19}$$

$$\psi_{mn} = \frac{\omega\left(\alpha_{2,mn} - \alpha_{1,mn}e^{-jk_zH}\right)}{k_z\left(e^{jk_z(h-H)} - e^{jk_z(H+h)}\right)}; \gamma_{mn} = \frac{\omega\alpha_{2,mn}e^{jk_z(H+2h)}}{k_z} \tag{20}$$

Substituting Eqs. (12) and (19)–(20) into Eqs. (1)–(2) and applying the orthogonal properties of modal functions, one gets:

$$\ddot{u}_{1,mn}(t) + \omega_{1,mn}^2 u_{1,mn}(t) - \frac{j\omega\rho_0}{m^*}\left[(I_{mn} - \varepsilon_{mn})e^{-j(k_z z - \omega t)} + (\beta_{mn} - \psi_{mn})e^{-j(-k_z z - \omega t)}\right] = 0 \tag{21}$$

$$\ddot{u}_{2,mn}(t) + \omega_{2,mn}^2 u_{2,mn}(t) - \frac{j\omega\rho_0}{m^*}\left[(\varepsilon_{mn} - \gamma_{mn})e^{-j(k_z z - \omega t)} + \psi_{mn}e^{-j(-k_z z - \omega t)}\right] = 0 \tag{22}$$

The natural frequencies of the two single orthotropic laminated composite plates are determined by [24]:

$$\omega_{i,mn}^2 = \frac{\pi^4}{m^* b^4}\left[D_{11}m^4\left(\frac{b}{a}\right)^4 + 2(D_{12} + 2D_{66})m^2n^2\left(\frac{b}{a}\right)^2 + D_{22}n^4\right](i = 1, 2) \tag{23}$$

Using Eqs. (14), (21) and (22) can be rewritten in matrix form as:

$$\begin{bmatrix} X_{11} & X_{12} \\ X_{21} & X_{22} \end{bmatrix}\begin{Bmatrix} \alpha_{1,mn} \\ \alpha_{2,mn} \end{Bmatrix} = \begin{Bmatrix} Y \\ 0 \end{Bmatrix} \tag{24}$$

with:

$$X_{11} = \omega_{1,mn}^2 - \omega^2 - \frac{j\omega^2\rho_0}{k_z m^*}\left[\frac{e^{jk_zH}}{e^{-jk_z(h-H)} - e^{-jk_z(H+h)}} - \frac{e^{-jk_zH}}{e^{jk_z(h-H)} - e^{jk_z(H+h)}}\right]$$
$$X_{12} = \frac{j\omega^2\rho_0}{k_z m^*}\left[\frac{1}{e^{-jk_z(h-H)} - e^{-jk_z(H+h)}} - \frac{1}{e^{jk_z(h-H)} - e^{jk_z(H+h)}}\right]$$
$$X_{21} = \frac{j\omega^2\rho_0}{k_z m^*}\left[\frac{e^{jk_zH}}{e^{-jk_z(h-H)} - e^{-jk_z(H+h)}} - \frac{e^{-jk_zH}}{e^{jk_z(h-H)} - e^{jk_z(H+h)}}\right]$$
$$X_{22} = \omega_{2,mn}^2 - \omega^2 - \frac{j\omega^2\rho_0}{k_z m^*}\left[\frac{1}{e^{jk_z(h-H)} - e^{jk_z(H+h)}} - \frac{1}{e^{-jk_z(h-H)} - e^{-jk_z(H+h)}} + e^{jk_z(H+2h)}\right]$$
$$Y = \frac{2j\omega\rho_0 I_{mn}}{m^*} \tag{25}$$

After solving the system of Eq. (24), the unknown coefficients $\alpha_{1,mn}$ and $\alpha_{2,mn}$ are determined, all the other quantities such as w_1, w_2 and the coefficients (β_{mn}, ε_{mn}, ψ_{mn} and γ_{mn}) are also determined. Thence, the analysis of sound transmission across a simply supported double-composite plate is completely solved.

3 Definition of Sound Transmission Loss

The power of incident sound is defined by [12]:

$$\begin{aligned}\Pi_1 &= \frac{1}{2}\text{Re}\iint_A p_1 v_1^* dA \\ &= \frac{\rho_0\omega^2}{2c_0}\left| 4I^2\iint_A e^{-2j(k_x x + k_y y)}dA - 4I\frac{\omega}{k_z}\sum_{m,n=1}^{\infty}\alpha_{1,mn}\iint_A e^{-j(k_x x+k_y y)}\sin\left(\frac{m\pi x}{a}\right)\sin\left(\frac{n\pi y}{b}\right)dA \right. \\ &\quad \left. + \frac{\omega^2}{k_z^2}\sum_{m,n=1}^{\infty}\sum_{k.l=1}^{\infty}\alpha_{1,mn}\alpha_{1,kl}\iint_A \sin\left(\frac{m\pi x}{a}\right)\sin\left(\frac{n\pi y}{b}\right)\sin\left(\frac{k\pi x}{a}\right)\sin\left(\frac{l\pi y}{b}\right)dA \right|\end{aligned} \tag{26}$$

The transmitted sound power is defined by [12]:

$$\begin{aligned}\Pi_3 &= \frac{1}{2}\text{Re}\iint_A p_3 v_3^* dA \\ &= \frac{\rho_0\omega^4}{2c_0k_z^2}\left|\sum_{m,n=1}^{\infty}\sum_{k,l=1}^{\infty}\alpha_{2,mn}\alpha_{2,kl}\iint_A \sin\left(\frac{m\pi x}{a}\right)\sin\left(\frac{n\pi y}{b}\right)\sin\left(\frac{k\pi x}{a}\right)\sin\left(\frac{l\pi y}{b}\right)dA\right|\end{aligned} \tag{27}$$

where v_1^*, v_3^*, p_1, p_3 are the local acoustic velocity, the sound pressure in the incident field and the transmitted field, respectively.

The sound transmission loss across the laminated double-composite plate is defined by:

$$STL = 10\log_{10}\left(\frac{\Pi_1}{\Pi_3}\right) \tag{28}$$

4 Numerical Results and Discussion

4.1 Validation

For validation, the present analytical solutions are compared with the experimental results of Lu and Xin [17], as shown in Fig. 2. The double-plate considered consists of two identical aluminum (isotropic) faceplates. The dimensions of the plates are: length of the plate $a = 0.3$ m, width of the plate $b = 0.3$ m. The plate has thickness $h = 0.001$ m while the thickness of the air cavity is $H = 0.08$ m. The mechanical properties of aluminum materials are: $E = 70$ GPa; $\rho = 2700$ kg/m^3; $\nu = 0.33$. The airspeed of sound, c = 343 m/s; $\rho_0 = 1.21$ kg/m^3; the amplitude of the acoustic velocity potential for the incident sound is $I_0 = 1$ m^2/s.

Looking at Fig. 2, we can see that the current predictions are closely matched with the experimental measurements of [17]. The obvious difference between theory and experiment is attributed to a number of factors, for example, the incident wave has not satisfied the condition of a plane wave or the connection of the structure or due to interference between waves during the experiment.

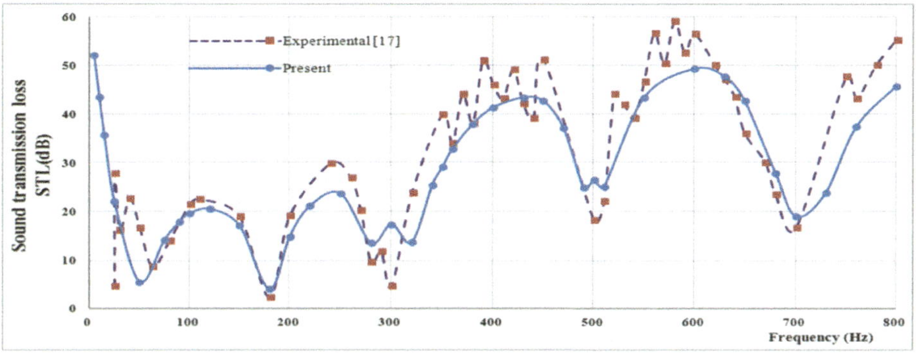

Fig. 2. Comparison between the present prediction with the experimental result of [17].

4.2 Influence of Composite Materials on STL

The influence of composite materials on STL through a finite simply supported double-composite plate is studied in this section by selecting four types of composite materials:

Boron/Epoxy, Glass/Epoxy, Graphite/Epoxy, and Kevlar/Epoxy. The double-plate consists of two identical orthotropic laminated composite faceplates. Laminate configuration of the bottom and the upper plate is *[0/90/0/90]*$_s$. The mechanical properties of composite material are shown in Table 1.

Table 1. Composite materials properties and the geometrical dimensions

Composite	E_1 *(GPa)*	E_2 *(GPa)*	G_{12} *(GPa)*	ν_{12}	ρ *(Kg/m³)*	$a \times b\ (m^2)$	$h\ (m)$	H *(m)*
Boron/Epoxy	204.000	18.500	5.590	0.23	2000	1 × 1	5.10^{-3}	0.08
Glass/Epoxy	40.851	10.097	3.788	0.27	1946			
Graphite/Epoxy	181.000	10.300	7.170	0.28	1600			
Kevlar/Epoxy	76.000	5.500	2.300	0.34	1460			

Figure 3 shows that the STL value of Boron/Epoxy materials is the largest compared to the remaining materials and the STL value of Kevlar/Epoxy materials is the smallest at frequencies lower than 100 Hz because in this region, surface density is the deciding factor (the stiffness-control zone). At frequencies greater than 100 Hz, the STL value of Glass/Epoxy material is larger than that of other materials when it passes the plate-air cavity-plate resonance and the STL curves of the four materials operate according to specific rules when the plate-air cavity-plate resonance is synchronized.

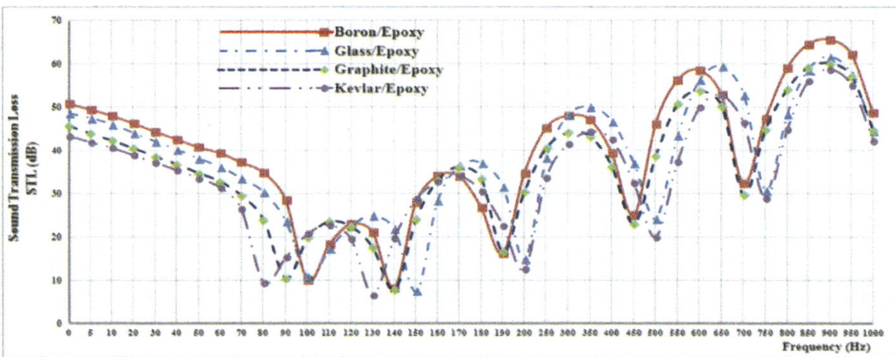

Fig. 3. Influence of composite materials on STL of a simply supported double-composite plate with air cavity.

4.3 Influence of Lamination Scheme on STL

In order to quantify the effects of lamination scheme on STL through the double-composite plate with an enclosed air cavity, four following configurations of the bottom and upper Glass/epoxy laminated composite plates are selected: *[0/90/0/90]*$_s$, *[0/0/0/0]*$_s$, *[90/90/90/90]*$_s$ and *[90/0/0/90]*$_s$. The geometrical and material parameters shown in the Table 1.

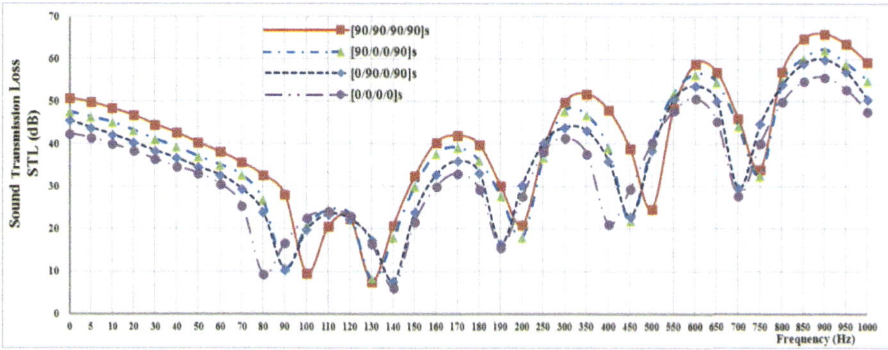

Fig. 4. Influence of laminate configuration on STL of a simply supported double-Glass/epoxy plate with air cavity.

As can be seen from Fig. 4, the lamination scheme [90/90/90/90]s has enhanced the STL better than the other patterns for all range of considered frequency.

4.4 Influence of Faceplate Thickness (H) on STL

To quantify the influence of faceplate thickness, the STL versus frequency curve is presented in Fig. 5 for a finite double-Glass/Epoxy plate. Three values of plate thickness are chosen: h = 2, 5 and 10 mm. The air cavity is fixed at H = 0.08 m. The geometrical and material parameters are shown in the Table 1.

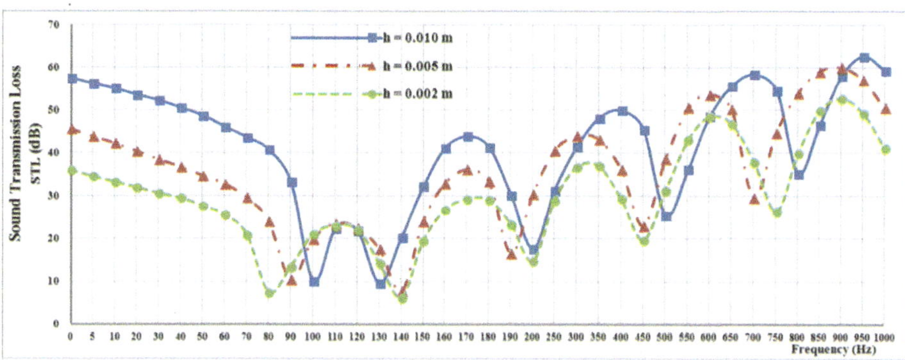

Fig. 5. Influence of laminate configuration on STL of a simply supported double-Glass/epoxy plate with air cavity.

According to Fig. 5, the STL value increases sharply when increasing the thickness of the bottom and upper plates. Effect of thickness of faceplate for STL is particularly strong at frequencies lower than 100 Hz. This is a very important region when designing finite dual sound insulation plates in practice. At higher frequencies, when peaks and poles appear in this mode, this is attributed to the strong interaction of individual plate behavior with the overall system performance for finite system. The position of double-plate resonances in Fig. 5 moves to higher frequencies when the thickness of the faceplate is increased.

4.5 Influence of Air Cavity Thickness (H) on STL

To demonstrate the influence of air cavity thickness on STL, the STLs are calculated for a finite double-Glass/Epoxy plate with selected values of air cavity thickness: $H = 0.04$, 0.06, and 0.08 m; $h_1 = h_2 = 0.01$ m, as shown in Fig. 6. The geometrical and material parameters are shown in the Table 1.

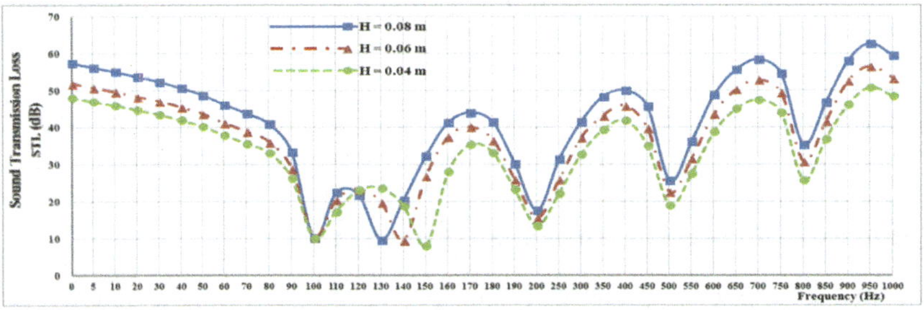

Fig. 6. Influence of laminate configuration on STL of a simply supported double-Glass/epoxy plate with air cavity.

As can be seen in Fig. 6, the first dip position does not depend on the air cavity thickness because it completely depends on the surface density of the plate. However, the second dip position changes drastically when the air cavity thickness increases (moving toward the lower frequency, Fig. 6) due to the fact that the plate-cavity-plate resonance plays a major role in this case. Their remaining positions are almost unchanged because the plate-cavity-plate is operating synchronously. Therefore, by tailoring the thickness of air cavity, it is possible to design finite double-plate partitions with better sound insulation properties over a wide frequency range.

5 Conclusions

In this investigation, an analytical model was developed to study the sound transmission loss through a finite double-laminated composite plate with the simply supported boundary conditions excited by a plane sound wave varying harmonically. The analytical model has been validated by comparing the present results of STL with previously published data on the double-plates. The influence of several key system parameters on STL including the density of composite materials, the laminate configurations, the faceplate thickness, as well as the air cavity thickness is systematically examined. From the results obtained, some conclusions can be drawn:

- The theoretical predictions are in good agreement with existing results.
- The surface density of composite materials influences considerably on STL of a finite simply supported double-composite plates.
- For the Glass/epoxy double-composite plate, the plies being arranged in a $[90/90/90/90]_s$ pattern of the bottom and upper plates appear to outperform other considered lamination schemes in terms of sound insulation.
- Increasing the plate thickness (equivalent to increasing the surface mass) enhances considerably the sound insulation property of the simply supported double-plates. The influence of plate thickness on STL is particularly strong for finite systems at low frequencies.
- As the thickness of the air cavity increases, the sound insulation capacity of the double plate also increases but is not as strong as increasing the thickness of bottom and upper plate.

Acknowledgments. This research is funded by the Vietnam National Foundation for Science and Technology Development (NAFOSTED) under grant number: **107.02-2018.07.**

References

1. Maidanik, G.: Response of ribbed panels to reverberant acoustic fields. J. Acoust. Soc. Am. **34**, 809–826 (1962)
2. Ruzzene, M.: Vibration and sound radiation of sandwich beams with honeycomb truss core. J. Sound Vib. **277**, 741–763 (2004)
3. London, A.: Transmission of reverberant sound through double walls. J. Acoust. Soc. Am. **22**, 270–279 (1950)
4. Carneal, J.P., Fuller, C.R.: An analytical and experimental investigation of active structural acoustic control of noise transmission through double panel systems. J. Sound Vib. **272**, 749–771 (2004)
5. Chazot, J.D., Guyader, J.L.: Prediction of transmission loss of double panels with a patch-mobility method. J. Acoust. Soc. Am. **121**, 267–278 (2007)
6. Bao, C., Pan, J.: Experimental study of different approaches for active control of sound transmission through double walls. J. Acoust. Soc. Am. **102**, 1664–1670 (1997)
7. Sgard, F.C., Atalla, N., Nicolas, J.: A numerical model for the low-frequency diffuse field sound transmission loss of double-wall sound barriers with elastic porous linings. J. Acoust. Soc. Am. **108**, 2865–2872 (2000)
8. Villot, M., Guigou, C., Gagliardini, L.: Predicting the acoustical radiation of a finite size multi-layered structures by applying spatial windowing on infinite structures. J. Sound Vib. **245**, 433–455 (2001)
9. Brouard, B., Lafarge, D., Allard, J.F.: A general method of modelling sound propagation in layered media. J. Sound Vib. **183**(1), 129–142 (1995)
10. Lauriks, W., Mees, P., Allard, J.F.: The acoustic transmission through layered systems. J. Sound Vib. **155**(1), 125–132 (1992)

11. Larbi, W., Deü, J.F., Ohayon, R.: Vibroacoustic analysis of double-wall sandwich panels with viscoelastic core. Comput. Struct. **174**, 92–103 (2016)
12. Panneton, R., Atalla, N.: Numerical prediction of sound transmission through finite multilayer systems with poroelastic materials. J. Acoust. Soc. Am. **100**, 346–354 (1996)
13. Chonan, S., Kugo, Y.: Acoustic design of a three layered plate with high sound interception. J. Sound Vib. **89**(2), 792–798 (1991)
14. Kang, H.J., Ih, J.G., Kim, J.S., et al.: Prediction of sound transmission loss through multilayered panels by using Gaussian distribution of directional incident energy. J. Acoust. Soc. Am. **107**(3), 1413–1420 (2000)
15. Bolton, J.S., Shiau, N.M., Kang, Y.J.: Sound transmission through multi-panel structures lined with elastic porous materials. J. Sound Vib. **191**(3), 317–347 (1996)
16. Leppington, F.G., Broadbent, E.G., Butler, G.F.: Transmission of sound through a pair of rectangular elastic plates. IMA J. Appl. Math. **71**(6), 940–955 (2006)
17. Lu, T.J., Xin, F.X.: Vibro-Acoustics of Lightweight Sandwich. Science Press Beijing and Springer, Heidenberg (2014)
18. Thinh, T.I., Thanh, P.N.: Vibro-acoustic response of an orthotropic composite laminated plate. In: Proceedings of National Conference on Mechanics, Ha Noi, pp. 1142–1150 (2017)
19. Thinh, T.I., Thanh, P.N.: Vibroacoustic behavior of a finite clamped orthotropic laminated composite double-plate with an enclosed air cavity. In: National Conference on Engineering Mechanics, Ha Noi, April 2019
20. Pellicier, A., Trompette, N.: A review of analytical methods, based on the wave approach, to compute partitions transmission loss. Appl. Acoust. **68**, 1192–1212 (2007)
21. Frampton, K.D.: The effect of flow-induced coupling on sound radiation from convected fluid loaded plates. J. Acoust. Soc. Am. **117**, 1129–1137 (2005)
22. Leissa, A.W.: Vibration of Plates. Acoustical Society of America, New York (1993)
23. Howe, M.S., Shah, P.L.: Influence of mean flow on boundary layer generated interior noise. J. Acoust. Soc. Am. **99**, 3401–3411 (1996)
24. Reddy, J.N.: Mechanics of Laminated Composite Plates and Shells, 2nd edn. Theory and Analysis. CRC Press, Boca Raton (2004)

Author Index

K.-U. Sattler et al. (Eds.): ICERA 2019, LNNS 104, pp. 733–735, 2020.
https://doi.org/10.1007/978-3-030-37497-6

The manufacturer's authorised representative in the EU is Springer Nature Customer Service Centre GmbH, Europaplatz 3, 69115 Heidelberg, Germany. If you have any concerns regarding our products, please contact ProductSafety@springernature.com

Printed and bound by CPI Group (UK) Ltd, Croydon, CR0 4YY
30/09/2026
02225230-0020